LABORATORY INVESTIGATIONS IN Anatomy & Physiology

STEPHEN N. SARIKAS, Ph.D.

with

William C. Ober, M.D.
Art Coordinator and Illustrator

Claire W. Garrison, R.N., and Anita Impagliazzo, M.A.
Illustrators

Ralph T. Hutchings
Biomedical Photographer

Shawn Miller
Organ and Animal Dissector

Mark Nielsen
Organ and Animal Dissection Photographer

San Francisco Boston New York
Cape Town Hong Kong London Madrid Mexico City
Montreal Munich Paris Singapore Sydney Tokyo Toronto

Executive Editor: Leslie Berriman
Project Editors: Mary Ann Murray, Michael Roney, Blythe Robbins
Development Manager: Claire Alexander
Development Editors: Lori K. Garrett, Megan Rundel
Editorial Assistant: Jon Duke
Managing Editor: Deborah Cogan
Production Supervisor: Caroline Ayres
Production Management and Composition: Carlisle Communications, LTD.
Production Editor: Becky Barnhart
Design Manager: Mark Ong
Interior Designer: Gary Hespenheide
Cover Designer: tani hasegawa
Photo Editor: Donna Kalal
Senior Manufacturing Buyer: Stacey Weinberger
Executive Marketing Manager: Lauren Harp

Cover Photo Credit: Anatomical Travelogue / Photo Researchers, Inc.

Safety Notification

The Author and Publisher believe that the lab experiments described in this publication, when conducted in conformity with the safety precautions described herein and according to the school's laboratory safety procedures, are reasonably safe for the student to whom this manual is directed. Nonetheless, many of the described experiments are accompanied by some degree of risk, including human error, the failure or misuses of laboratory or electrical equipment, mismeasurement, chemical spills, and exposure to sharp objects, heat, bodily fluids, blood, or other biologics. The Author and Publisher disclaim any liability arising from such risks in connection with any of the experiments contained in this manual. If students have any questions or problems with materials, procedures, or instructions on any experiment, they should always ask their instructor for help before proceeding.

Library of Congress Cataloging-in-Publication Data

Sarikas, Stephen N.
Laboratory investigations in anatomy & physiology / Stephen N. Sarikas with William C. Ober . . . [et al.].
p. cm.
Includes bibliographical references and index.
ISBN 0-8053-5321-6
1. Human physiology. 2. Human anatomy. I. Title: Laboratory investigations in anatomy and physiology. II. Ober, William C. III. Title.

QP34.5.S26 2007
612--dc22

2005053533

ISBN 0-8053-5321-6
10 9 8 7 6 5 4 3 2 1—QWD—11 10 09 08 07 06
www.aw-bc.com

Contents

The Muscular System

The Nervous System

The Endocrine System

The Cardiovascular System

The Reproductive Systems

Preface

The anatomy and physiology laboratory is a fundamental course for anyone pursuing a career in nursing, physical or occupational therapy, athletic training, exercise physiology, dental hygiene, or other allied health professions. The laboratory experience offers a unique and extraordinary vision of human structure and function. However, the considerable volume of material covered and the complexity of the scientific concepts introduced present significant obstacles for many students. Designing innovative strategies that make sophisticated ideas and information easy to understand is an ongoing challenge.

Connected learning combines traditional classroom learning with educational projects that provide students with relevant and hands-on investigations outside the classroom. The concept and practice of connected learning form the cornerstone of the academic mission at Lasell College in Newton, Massachusetts, where I teach anatomy and physiology, and it is the model that also defines my approach to teaching anatomy at the Boston School of Occupational Therapy at Tufts University. In this manual, my connected learning approach links basic scientific knowledge with applied science and clinical issues. Based on the premise that students are productive learners who can be motivated to apply their knowledge of anatomy and physiology to a lifelong commitment of intellectual exploration in science, this approach cultivates a broad range of knowledge and technical skills necessary for developing a dynamic professional career.

To be successful in learning and applying anatomical and physiological knowledge, students are challenged to develop critical, independent thinking skills and to become engaged in the subject in ways that go beyond classroom boundaries. This manual attempts to meet this objective, primarily through a connected learning pedagogy, which is realized in the following ways:

- The laboratory activities are written in a clear narrative style that encourages students to understand relevant concepts rather than memorize trivial facts. Understanding rather than memorizing is a critical first step to becoming a productive learner.
- The objective to be achieved in each laboratory activity is directly linked to information that students learn in lecture.
- The "Questions to Consider" at the end of each laboratory activity challenge students to critically analyze the work they have just completed.
- The "Clinical Correlations" boxes provide direct links between basic concepts in human biology and real issues in health and medicine.
- The "What's in a Word" feature succinctly connects common word roots with words and phrases used in anatomy and physiology to simplify learning.
- "Form a Hypothesis . . . Assess the Outcome" is an occasional feature that allows students to predict the outcome of an activity before starting, to collect and analyze data during the activity, and to test the validity of the prediction when the activity is completed.

This laboratory manual has been designed to serve the laboratory course that accompanies the two-semester anatomy and physiology lecture course. The manual is written to correspond to all current two-semester anatomy and physiology textbooks. Those students and instructors using Frederic H. Martini's *Fundamentals of Anatomy & Physiology*, Seventh Edition, will recognize here some of the exceptional art from that text by William C. Ober, M.D., and Claire W. Garrison, R.N., Dr. Martini's renowned biomedical illustrators.

The material in this laboratory manual is organized into 31 exercises that can be conveniently scheduled over two semesters. Although the number of exercises is far less than the number of exercises found in other currently available lab manuals, this manual includes all of the critical information students need to successfully complete a two-semester anatomy and physiology laboratory course. The text is logically organized and clearly written, and important concepts are described with the appropriate detail. The unique narrative style of the activities, the "learning by doing" approach, and the emphasis on the core exercises students have to complete in the lab to be successful in anatomy and physiology combine to make this a manual that addresses student and instructor needs.

Three Versions of the Laboratory Manual

The manual is available in three separate versions, allowing instructors to tailor their preference on including organ dissections only, cat dissections, or fetal pig dissections.

The Main Version covers the full two-semester A&P curriculum, including dissections of the cow eye and of the sheep heart, brain, and kidney. The Cat Version includes all of the same material as the Main Version plus an additional section of ten cat dissection exercises encompassing the major body systems. The Pig Version, similarly, includes all of the material from the Main Version with a separate section of ten fetal pig dissection exercises.

Organization

The laboratory manual contains 31 exercises, plus the ten additional dissection exercises at the end of each of the Cat and Pig Versions. Large systems, such as the skeletal, muscular, and nervous systems, appear across several exercises. Every exercise begins with a set of Laboratory Objectives and a list of Materials. A general introduction to the exercise gives students a preview of what they are about to learn. The exercises are each divided into a series of laboratory activities that focus on areas of specific study. The activities are self-contained, and instructors may easily assign only certain activities within an exercise.

The Visual Introduction on the following pages presents the key features of this laboratory manual.

Visual Introduction

Art Program A major feature of this laboratory manual is the outstanding full-color art program with illustrations by William C. Ober, M.D., and Claire W. Garrison, R.N., and biomedical photographs by Ralph T. Hutchings. Here, as with their renowned work on the various anatomy and physiology textbooks by Frederic H. Martini, these experts in visual representation have designed and developed the art to guide students from familiar to detailed structures.

Side-by-Side Pieces

Artist-rendered illustrations are paired with high-quality photos or micrographs, allowing students to see two complimentary views into a structure. The illustration calls out details and distinctions while the photo shows students the "real thing" as they would see it in a lab.

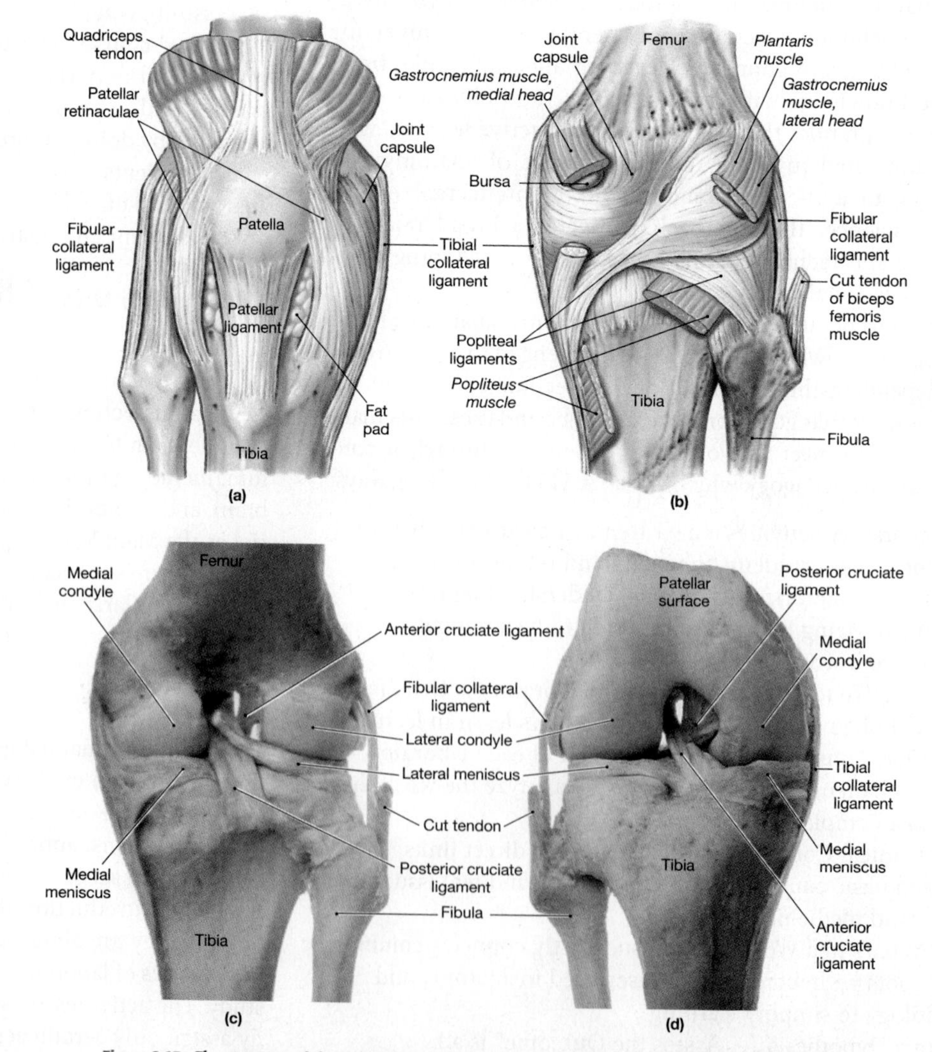

Figure 9.13 The structure of the right knee joint. a) Diagram of a superficial anterior view, in the extended position; **b)** diagram of a superficial posterior view, in the extended position; **c)** cadaver dissection of a deep posterior view, in the extended position; **d)** cadaver dissection of a deep anterior view in the flexed position.

Histology Images

Throughout the manual, an emphasis on high-quality histology micrographs helps guide students in their own microscopic observations.

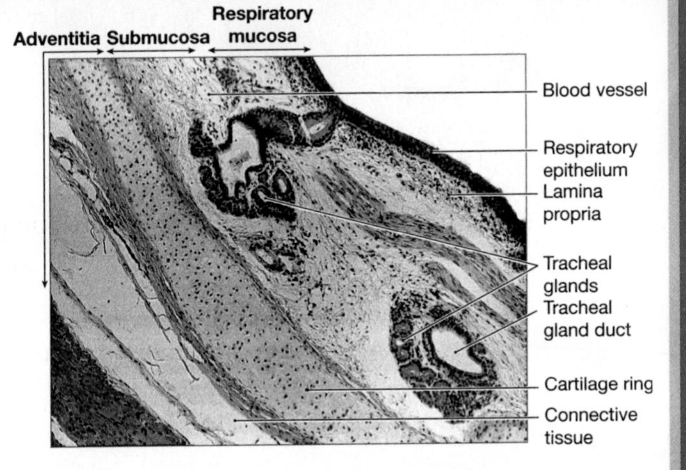

Figure 24.8 Microscopic structure of the trachea. The three tissue layers in the tracheal wall are the respiratory mucosa, submucosa, and adventitia. The C-shaped tracheal rings are embedded in the adventitia.

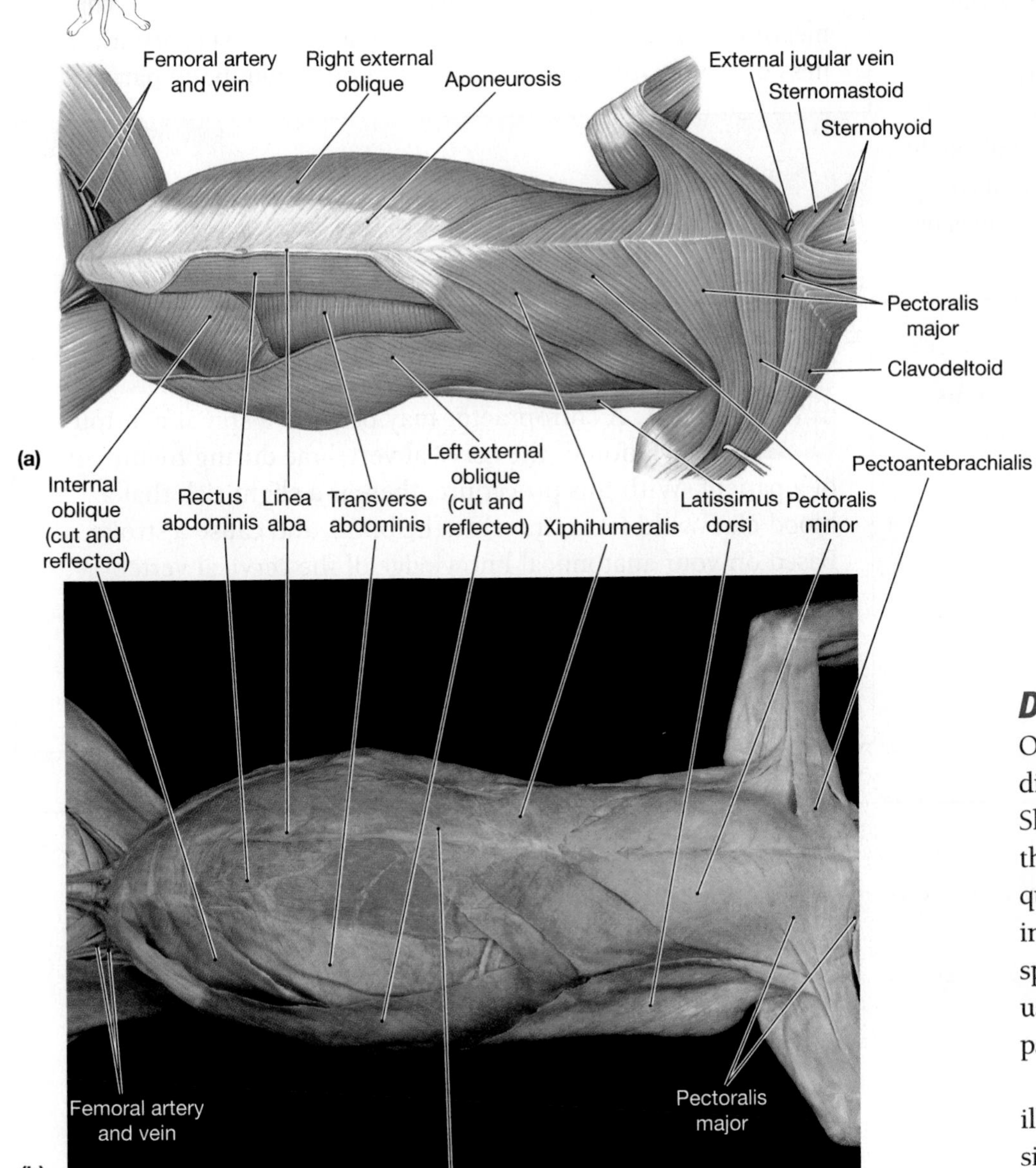

Dissection Art and Photos

One of the anatomy field's premier dissector/photographer duos, Shawn Miller and Mark Nielsen of the University of Utah, provided exquisite photos of cat, fetal pig, and individual organ dissections, taken specifically for this laboratory manual. The cat and fetal pig photos are paired with Ober and Garrison art.

The cat and fetal pig photos and illustrations appear in the Cat Version and the Pig Version, respectively, of this laboratory manual.

Features

In keeping with the goal of creating a concise laboratory manual, the features beyond the narrative and the art focus on the most important, relevant, and interesting information, just enough to provide students with core information that can support and add excitement to their study.

What's in a Word

What's in a Word boxes appear in all exercises. These boxes provide the language background of some of the more common or interesting anatomical and physiological terms. Not only do they provide a little color to the study of anatomy and physiology but the presentations of the roots of terms also shed light on their meanings or provide useful mnemonic devices.

WHAT'S IN A WORD The term *proprioception* is derived from two Latin words: *proprius*, meaning "one's own," and *capio*, meaning "to take." Proprioception is a general sensory awareness of the position and movement of one's own body parts.

Questions to Consider

Every activity is followed by a Questions to Consider section that prompts students to think more deeply about the concepts they have just explored.

QUESTIONS TO CONSIDER 1. A chiropractor may perform a spinal manipulation of the cervical vertebrae during treatment of a patient. With this procedure, there is a slight risk that a blood clot could form, travel to the brain, and cause a stroke. Based on your anatomical knowledge of the cervical vertebrae, explain how this could occur. ______________________________

____

____

Clinical Correlation

Clinical Correlation boxes appear in all exercises, offering students relevant, real-world applications of the knowledge they are gaining in the lab.

CLINICAL CORRELATION

Cholesterol can accumulate in our blood vessels, setting the stage for atherosclerosis, a disease in which the blood vessels become increasingly narrowed, reducing the blood flow through them and gradually starving the tissues that they supply. This is a leading cause of heart attacks and strokes. Plant cells, unlike animal cells, lack cholesterol in their membranes. This structural difference is one reason why people who are concerned about high cholesterol should add more fruits and vegetables to their diets.

5. When you are rested and comfortable, stand in an erect position, adjacent to the vertical reference line, but this time close your eyes. Remain standing for 2 minutes.
Form a Hypothesis Before you begin, make a prediction about the results. ______

6. Your laboratory partner should, once again, note any body movements that you make during the test. ______

7. **Assess the Outcome Make comparisons of the results from the two tests. In particular, contrast the number and type of movements that were made and note any evidence that a loss of balance had occurred when the eyes were open versus closed. Did your results support or refute your hypothesis?**

Form a Hypothesis, Assess the Outcome

Appearing as needed throughout the manual, Form a Hypothesis instructions provide opportunities for students to try their hand at predicting outcomes, an important clinical skill. These always precede a laboratory activity, with questions that guide students to think critically about the laboratory work they will carry out. Follow-up Assess the Outcome questions ask students to evaluate their hypotheses and check their accuracy.

ACTIVITY 14.6 Electroencephalography Using the Biopac Student Lab System

BIOPAC Systems, Inc.

Setup and Calibration

1. Position three electrodes on your lab partner's scalp as shown in Figure 14.10. Part your lab partner's hair tightly and firmly press the disposable electrode in the middle of each part. Attach electrode leads (SS2L) as illustrated. After the electrodes are attached, a swim cap or self-adhering wrap can be used to help keep them in place.
2. Have your lab partner lie on his or her back with the head tilted comfortably to one side. Instruct your partner to minimize movement throughout the experiment.
3. Attach the electrode leads (SS2L) to Channel 1 and turn on the Acquisition Unit.
4. Start the Biopac Student Lab System, choose Lesson 3 (L03-EEG-1), and click OK. When prompted, enter a unique file name and click OK.

BIOPAC Activities

To keep pace with the increasing use of technology in the lab, the laboratory manual includes four activities using the Biopac Student Lab System, with clear, concise directions for using the clinically oriented BIOPAC equipment. All BIOPAC activities incorporated into the exercises are called out with a special icon.

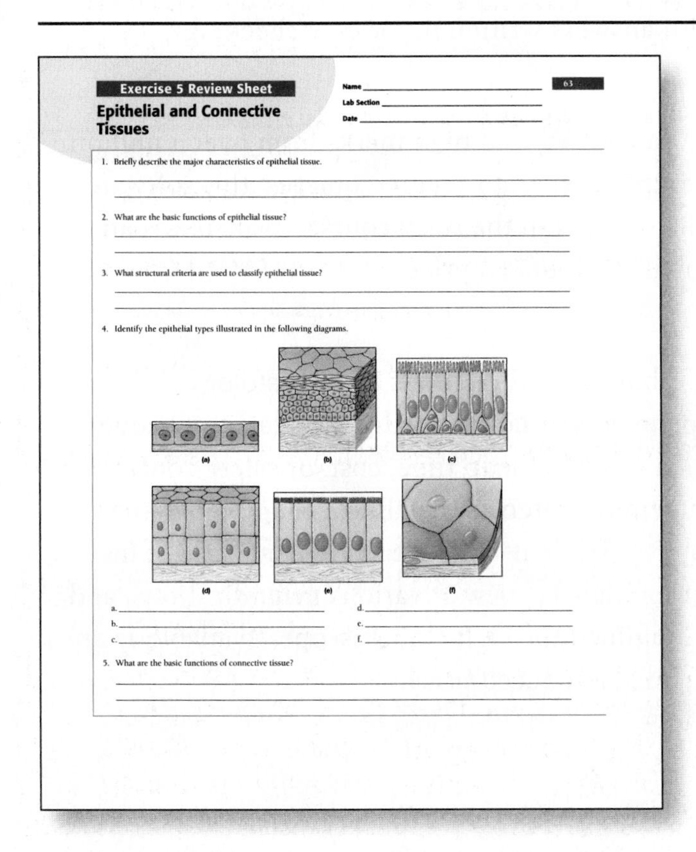
Exercise 5 Review Sheet

Epithelial and Connective Tissues

Name ______
Lab Section ______
Date ______

63

1. Briefly describe the major characteristics of epithelial tissue.
2. What are the basic functions of epithelial tissue?
3. What structural criteria are used to classify epithelial tissue?
4. Identify the epithelial types illustrated in the following diagrams.

(a) (b) (c) (d) (e) (f)

a. ______ d. ______
b. ______ e. ______
c. ______ f. ______

5. What are the basic functions of connective tissue?

Review Sheets

Review Sheets conclude each exercise with a series of matching, labeling, and short-answer questions, as well as some coloring activities, to help students evaluate their understanding of the material covered in the lab.

Supplements

Instructor's Manual

The Instructor's Manual was written by Lori Garrett, an experienced instructor in the A&P classroom and laboratory who was also this laboratory manual's Development Editor. With her unique perspective and thorough knowledge of the laboratory manual, she provides comprehensive support, with a particularly useful section introducing instructors to the laboratory manual and describing how best to take advantage of its unique approach to the material. The Instructor's Manual also contains an optional Nutrient Assessment take-home exercise that instructors can photocopy and assign, as well as teaching tips, common pitfalls, instructions for mixing solutions, answers to the Questions to Consider feature, and media references to two optional media programs, InterActive Physiology and PhysioEx. A final section reproduces the Review Sheets from the laboratory manual, with answers written in for easy checking.

InterActive Physiology®

This engaging physiology tutorial media program has received high marks from over a million students. Rich with detailed animations and complete with stimulating quizzes, this software program helps students learn the most difficult concepts in the A&P course. Available as an optional item to be packaged with the lab manual at a reduced price.

PhysioEx

This unique media program consists of twelve laboratory simulations and a histology slide tutorial. The lab simulations can be used to supplement wet labs, allowing students to conduct experiments that may be difficult to perform in a wet lab due to time, cost, or safety concerns. PhysioEx also allows students to repeat experiments as often as they like, changing parameters along the way. The histology slide tutorial gives students access to a rich database of histology slides, which they can study with or without labels, view at various magnifications, and move around as though the slides were being examined through a microscope. Available as an optional item to be packaged with the lab manual at a reduced price.

Acknowledgments

Many of my colleagues from across the United States and Canada donated their time, expertise, and thoughtfulness to review the manuscript. Their valuable comments and suggestions guided me through the writing and revision process. I am profoundly indebted to the following people for their support:

Julius Afolabi, *Savannah State University*
Mark Alston, *University of Tennessee, Knoxville*
Lynne Anderson, *Meridian Community College*
Theresa M. Arburn, *Palo Alto College*
Patricia Ashby, *Scottsdale Community College*
Eugene L. Bass, *University of Maryland, Eastern Shore*
Steven Basset, *Southeast Community College*
Theresa S. Bidle, *Hagerstown Community College*
Moges Bizuneh, *Ivy Tech Community College*
Richard M. Blaney, *Brevard Community College*
Jeff Blodig, *Johnson County Community College*
Joanna D. Borucinska, *University of Hartford*
Sara Brenizer, *Shelton State Community College*
Nishi Bryska, *University of North Carolina, Charlotte*
Pamela J. Carlton, *College of Staten Island/CUNY*
Christy Carmack, *Davidson Community College*
Ruth Conley, *Shepherd University*
Rosemary L. Davenport, *Gulf Coast Community College*
Bonnie Dean, *West Virginia State University*
Jason J. Dechant, *University of Pittsburgh*
Michael A. Dorset, *Cleveland State Community College*
Phillip Eichman, *University of Rio Grande*
Alan W. Erdahl, *Riverland Community College*
David L. Evans, *Pennsylvania College of Technology*
Gibril Fadika, *Hampton University*
Ann M. Findley, *University of Louisiana, Monroe*
Carl D. Frailey, *Johnson County Community College*
Lori K. Garrett, *Danville Area Community College*
Paul Gier, *Huntingdon College*
Chaya Gopalan, *St. Louis Community College*
Ewa Gorski, *Community College of Baltimore County, Catonsville*
Michael T. Griffin, *Angelo State University*
Linda Griffin Gingerich, *St. Petersburg College*
Clare Hays, *Metropolitan State College of Denver*
Kerry Henrickson, *Cochise College*
Jim Hershey, *Community College of Baltimore County, Essex*
Glenda Hill, *El Paso Community College*
Kerrie Hoar, *University of Wisconsin, La Crosse*
Yvette Huet-Hudson, *University of North Carolina, Charlotte*
Gary Johnson, *Madison Area Technical College*
Shelley Jones, *Florida Community College*
Susan J. Landesman, *Mt. Hood Community College*
Stephen Lebsack, *Linn-Benton Community College*
Terry J. Lee, *Garden City Community College*
Ellen J. Lehning, *Jamestown Community College*
Peggy LePage, *North Hennepin Community College*
Emlyn Louis, *Miami Dade Community College*
Elizabeth L. Lucyszyn, *Medaille College*
Jeannine M. Matz, *Mercy College of Health Sciences*
Danette McCandless, *Davidson Community College*
Karen McCort, *El Paso Community College*
Cherie McKeever, *Montana State University, Great Falls College of Technology*
Janice Meeking, *Mount Royal College*
Sarah L. Milton, *Florida Atlantic University*
Bill Montgomery, *College of Southern Maryland, La Plata*
Katie Morrison-Graham, *Lane Community College*
Thuy Nguyen, *Florida Community College, Jacksonville*
Betsy Peitz, *California State University, Los Angeles*
Diane Pelletier, *Green River Community College*
Julie Pilcher, *University of Southern Indiana*
Inga Pinnix, *Florida Community College, Jacksonville*
Tricia Reichert, *Colby Community College*
Kathleen Richardson, *Portland Community College*
Eugene Rutheny, *Westchester Community College*
Allen Sanborn, *Barry University*
Sharon Simpson, *Broward Community College*
John J. Stahura, *Manor College*
Phil Stephens, *Villanova University*
Barbara A. Stoos, *Mercy College of Northwest Ohio*
Dennis Strete, *McLennan Community College*
Jeffrey Taylor, *North Central State College*
Kent R. Thomas, *Wichita State University*
Sue Thornton, *Carolinas College of Health Sciences*
Kimberly Turk, *Mitchell Community College*
Patricia Turner, *Howard Community College*
Valerie Vander Vliet, *Lewis University*
Jane Wallace, *Chattanooga State Technical Community College*
Margaret A. Weck, *St. Louis College of Pharmacy*
Joe Wheeler, *North Lake College*
Heather Wilson-Ashworth, *Utah Valley State College*
Mark D. Womble, *Youngstown State University*
Jeanne M. Workman, *Duquesne University*
Ann W. Wright, *Canisius College*
Nina C. Zanetti, *Siena College*

One of the hallmarks of a Benjamin Cummings book is the collection of vivid photographic images and illustrations that bring life and vitality to the text. I am most grateful to a number of talented people for their magnificent artistic contributions to this laboratory manual. Some of the art in the manual comes from

Frederic H. Martini's *Fundamentals of Anatomy & Physiology,* Seventh Edition. The Martini art was masterfully produced by biomedical illustrators William Ober and Claire Garrison. Bill, Claire, and their new partner, Anita Impagliazzo, also worked closely with me to produce superb illustrations specifically for this lab manual. I am also grateful to the renowned anatomical photographer, Ralph Hutchings, who contributed a number of precise and beautiful photographs, and to Robert Tallitsch, who provided many stunning light micrographs. The talented dissector/photographer team of Shawn Miller and Mark Nielsen, from the University of Utah, contributed, arguably, the highest quality dissection photographs of the cat, the pig, and individual organs seen in any anatomy and physiology laboratory manual.

I appreciate the expertise of Doug Hirzel, of Cañada College in Redwood City, California, who contributed the BIOPAC activities that appear in the manual. I am grateful to BIOPAC Systems, and in particular, Frazer Findlay, who authorized the adaptation of several BIOPAC activities, and Jocelyn Kremer, who reviewed all the BIOPAC activities and offered useful suggestions.

The publication of this laboratory manual represents the work of many dedicated and creative individuals. I am especially grateful and fortunate to have had the opportunity to collaborate with a team of talented people from Benjamin Cummings who supported my work throughout every phase of this project. First and foremost, I thank Daryl Fox, Vice President and Publisher, who told me he recognized something special in my writing style and my approach to teaching anatomy and physiology and invited me to write this manual.

Special recognition goes to Leslie Berriman, Executive Editor, who managed all aspects of this project with great competence and skill. I thank her for trusting my instincts as a scientist and writer. Leslie's sensitivity to my ideas and confidence in my work were most gratifying.

I am especially indebted to Mary Ann Murray, my first Project Editor, whose patience, experience, and attention to detail were instrumental in putting this project on the right course during the early stages. A period of transition marked the arrival of Michael Roney as my second Project Editor. Mike demonstrated his keen editorial skills by keeping my project on a steady course without missing a beat as he joined the project team. His expertise in attending to all the fine details and coordinating a complex production schedule was vital to the successful completion of this book. I also express my gratitude to Blythe Robbins, Assistant Editor, who ably managed all the important behind-the-scenes tasks.

Megan Rundel and Lori Garrett served as Development Editors for this laboratory manual. I thank Megan for her important contributions in the development of the special features that appear in the manual. I am grateful for Lori's keen insight and tireless effort during the lengthy and often arduous task of writing revisions. Her expertise shines in every laboratory exercise. Lori also wrote the excellent *Instructor's Manual* that accompanies this laboratory manual.

Claire Alexander managed the development of this manual in the critically important early manuscript stages, and I am grateful to her for that.

I thank Caroline Ayres, Production Supervisor, and Deborah Cogan, Managing Editor, for keeping the multifaceted production process well coordinated and on schedule. I am also grateful to Becky Barnhart and her expert team at Carlisle Communications for their precise copyediting and talented page layout. I appreciate the unique talents of Gary Hespenheide for the attractive text and interior design and of tani hesegawa, who skillfully designed the cover of this manual. I am also thankful for the contributions of Mark Ong, who supervised the interior and cover design processes.

I thank Lauren Harp, Executive Marketing Manager, for planning and directing the presentation of this laboratory manual to my anatomy and physiology colleagues in colleges and universities. I also express my gratitude to all of the Addison Wesley/Benjamin Cummings sales reps for skillfully representing this new lab manual to instructors.

I am very fortunate to have in my life two special individuals who helped me to complete this project with their unconditional love and emotional support. My stepson, Anthony Atamanuik, guided me through some of the more difficult periods with his comic relief. Anthony often engaged me in lengthy discussions on particular topics or sections that I was writing. His unique perspective helped me to articulate my ideas more clearly, and, perhaps, I provided him with material for new comedy skits! It is gratifying to know that he is proud of my accomplishments, as I am proud of his. Marlena Yannetti, my wife and life partner for the past 25 years, was the glue that kept this project together over the years of its development. She offered guidance, keen insight, sound advice, and loving support. She helped me to manage my time and calmed me during periods of extreme pressure and stress. Her presence and support were so important to me, and without her, this laboratory manual would never have been written. Marlena, you're the best, and I love you.

It is difficult to produce a laboratory manual of this magnitude and complexity that is free of mistakes or omissions. Any errors or oversights are my responsibility and do not reflect the work of the editors, reviewers, artists, or production staff. I encourage faculty and students to send any relevant comments or suggestions about the content of this laboratory manual directly to me at the address given below. I will give all of your ideas serious consideration when I prepare for the second edition.

Stephen N. Sarikas
Professor of Science
Lasell College
1844 Commonwealth Ave.
Newton, MA 02466
ssarikas@Lasell.edu

About the Author

Stephen N. Sarikas received his Ph.D. in Anatomy from Boston University School of Medicine. His past research interests have included studies on the histochemistry of egg capsules in two salamander species (*Ambystoma* sp.), the development, maturation, and distribution of small-granule APUD cells in the mammalian lung, and membrane-intermediate filament interactions in transitional epithelium during the contraction–expansion cycle of the mammalian urinary bladder. His current research is a multiyear study on HIV/AIDS awareness among college students. Presently, Dr. Sarikas is Professor of Science at Lasell College in Newton, Massachusetts, where he teaches courses in anatomy and physiology, biology, environmental science, and history of science. He is also Lecturer of Occupational Therapy at the Boston School of Occupational Therapy, Tufts University, where he teaches a graduate level anatomy course. He is a member of the Human Anatomy and Physiology Society and the American Association of Anatomists. Dr. Sarikas lives in Chelsea, Massachusetts, where he serves as chairperson of the Chelsea Conservation Commission. He and his wife enjoy working in their garden, running, entertaining friends, and watching the Red Sox beat the Yankees. They regularly travel to New York City, where their son, Anthony, lives and performs comedy, and Montreal, where they have many close friends.

Exercise 1

Body Organization and Terminology

Laboratory Objectives

On completion of the activities in this exercise, you will be able to:

- Describe and demonstrate anatomical position.
- Use anatomical terminology to describe relative positions of structures in the human body.
- Describe and demonstrate the various anatomical planes and sections.
- List the organ systems and identify the organs in each.
- Name the anatomical regions of the human body.
- Identify the body cavities and the organs that are located in each.

Materials

- Human torso model, with dissectible parts
- Various anatomical models of organs and organ systems, with dissectible parts
- Fresh vegetables that are long and cylindrical in shape (e.g., cucumbers or eggplants)
- Small kitchen knife or scalpel
- Human skeleton or skull
- Large, clear plastic bags
- Coloring pencils

Your laboratory experience will reinforce the concepts and principles of anatomy and physiology with meaningful hands-on activities. It is during this time that you will get a true sense of the three-dimensional qualities, anatomical relationships, and physiological functions that define a living organism. You will acquire knowledge of a wide variety of topics in human biology ranging from the chemical basis of life and cell biology to the anatomy and physiology of the major organ systems. This knowledge will help you to understand complex biological processes, to connect structure with function at all organizational levels, and to collect and critically analyze data in a laboratory setting. To begin your exploration of the human body, you should become familiar with the basic ideas of anatomy and physiology.

Human **anatomy** is the study of the structure of the human body and the relative relationships between body parts. Anatomy can be divided into two main fields of study.

- **Gross (macroscopic) anatomy** deals with the study of structures that can be viewed with the unaided eye. Gross anatomy can be studied from a **regional approach** in which the structures of a specific region, such as the head and neck, or thorax, are examined. Alternatively, specific organ systems, such as the respiratory system or the digestive system, can be studied separately by using a **systemic approach**.
- **Microscopic anatomy** deals with the study of structures that can only be seen with magnification. The two main fields of microscopic anatomy include **cytology**, the study of cell structure, and **histology**, the study of tissue structure.

Human **physiology** is the study of normal function in the human body. The foundation for studying function is firmly rooted in the field of **cell physiology**, which deals with the metabolic processes inside cells and the special interactions between cells. However, physiological study also includes the function of specific organs and of organ systems **(systemic physiology)**.

WHAT'S IN A WORD The word *anatomy* is derived from the Greek words, *ana* (= "apart") and *tome* (= "a cutting"). The two words, together, mean "a cutting apart." The best way to study the structure of an organism is to dissect it or "cut it apart."

The word *physiology* is also derived from two Greek words: *physis* (= "nature") and *logos* (= "study"). The two words, together, mean "study nature." Physiology is the study of natural processes in the body.

To gain complete understanding of human biology, knowledge of both anatomy and physiology is essential. The two disciplines are closely linked by three important biological themes.

1. The human body can be studied from six increasingly complex **levels of organization**.
   - At the **chemical level**, the chemical bonds between atoms give rise to molecules. The bonding of small molecules results in the formation of large organic macromolecules (carbohydrates, proteins, lipids, and nucleic acids), which, in turn, are organized to form cellular structures called organelles.
   - At the **cellular level**, organelles are organized in a unique way to form cells. The cell represents the fundamental unit of life, characterized by the ability to reproduce, grow, and perform metabolic processes. The cellular level thus can be singled out as being distinctive and special.
   - At the **tissue level**, collections of cells are grouped to perform a similar function. It is at this level that we first see an organizational pattern that is multicellular.
   - At the **organ level**, two or more tissues are arranged into organs such as the heart, or stomach. Each organ has a well-defined, three-dimensional structure and a specific bodily function.
   - At the **organ system level**, a collection of organs functions as a unit to carry out a collection of related body activities.
   - The **organism level** is the highest level of organization and includes the structure and function of all the organ systems in the body.

2. **Structure and function** are complementary in nature. In almost all cases, a solid understanding of the anatomy of a body part will lead to a clearer understanding of its physiology. Consider the following example: Skeletal muscles contain the contractile proteins actin and myosin that are organized in a specific overlapping fashion. This unique arrangement allows muscle cells to contract and produce body movements.
3. Cellular metabolic activities are responsible for maintaining a relatively stable internal environment, called **homeostasis.** For organ systems to function properly, the body's concentrations of water, nutrients, oxygen, carbon dioxide, and ions, as well as body temperature and blood pressure, must remain at fairly constant levels. The body is always making adjustments to maintain a steady state. However, chronic stress, poor diet, disease, and injury can severely disrupt homeostasis, leading to serious health effects and possibly death.

WHAT'S IN A WORD The word *homeostasis* is derived from two Greek words: *homoios,* which means "alike," and *stasis,* which means "a standing." Thus, homeostasis refers to a similar position or a situation that does not change. For normal function, the body's internal environment must not change significantly. Dramatic fluctuations away from homeostatic conditions are often caused by disease and/or injury.

Anatomical Terms

To be successful in this course, you must become familiar with the general organization of the human body and learn the common anatomical language that is used to describe that organization. Most anatomical names are derived from Latin or Greek and have remained uniform throughout most of the over 2000 years of the study of anatomy.

Human anatomy is described with reference to the **anatomical position**, a universally accepted standard position for the body. In this basic position, the individual is standing erect with the head and eyes directed forward. The upper limbs are by the sides, with palms facing forward, and the lower limbs are together with the toes facing forward (Figure 1.1a).

Several anatomical terms are used to describe the location of one body part in relation to another. These terms are always used to illustrate the relative position of a structure when the body is in anatomical position. For example, to explain the relative position of the heart to the esophagus, you could state that the heart is **anterior** (closer to the front) to the esophagus. Alternatively, you could say that the esophagus is **posterior** (closer to the back) to the heart. Other terms that are used to express relative position are described in Table 1.1 and illustrated in Figure 1.2. Carefully review these terms and be sure that you understand their meanings before you proceed. Make it a habit to periodically review Table 1.1 as the course progresses.

ACTIVITY 1.1 Using Anatomical Terms to Describe Body Organization

1. Ask your lab partner to stand in the anatomical position.
2. With your lab partner's body in the anatomical position, use the directional terms listed in Table 1.1 to describe the relationships of the pairs of external structures listed below.
   a. Left eye to left ear ______________________
   b. Thumb to little finger ______________________
   c. Right ankle joint to right knee joint ______________
   d. Left elbow joint to right elbow joint ______________
3. Refer to a torso model. Use directional terms to describe the relationships of the following pairs of internal organs. You will have to "dissect" the torso model to identify some of these structures.
   a. Left kidney to spleen ______________________
   b. Right lung to right lobe of the liver ______________
   c. Pancreas to stomach ______________________
   d. Ascending colon to descending colon ____________

QUESTIONS TO CONSIDER 1. When the human body is in the anatomical position, the palms of the hands are facing forward. In more acceptable anatomical terms, you could state that the palms are the anterior surfaces of your hands, and accordingly, the thumbs are the most lateral digits. Suppose that the accepted anatomical position is to have the palms facing backward. How would this new position change your description of the palms and thumbs? Based on your answer to the previous question, why do you think it is important to have a universally accepted anatomical position?

____

____

____

2. Often, the terms in Table 1.1 are combined to provide a more specific description of location. For instance, one can say that the heart is **superomedial** to the ascending colon. What does the term *superomedial* say about the heart's anatomical relation to the ascending colon?

____

____

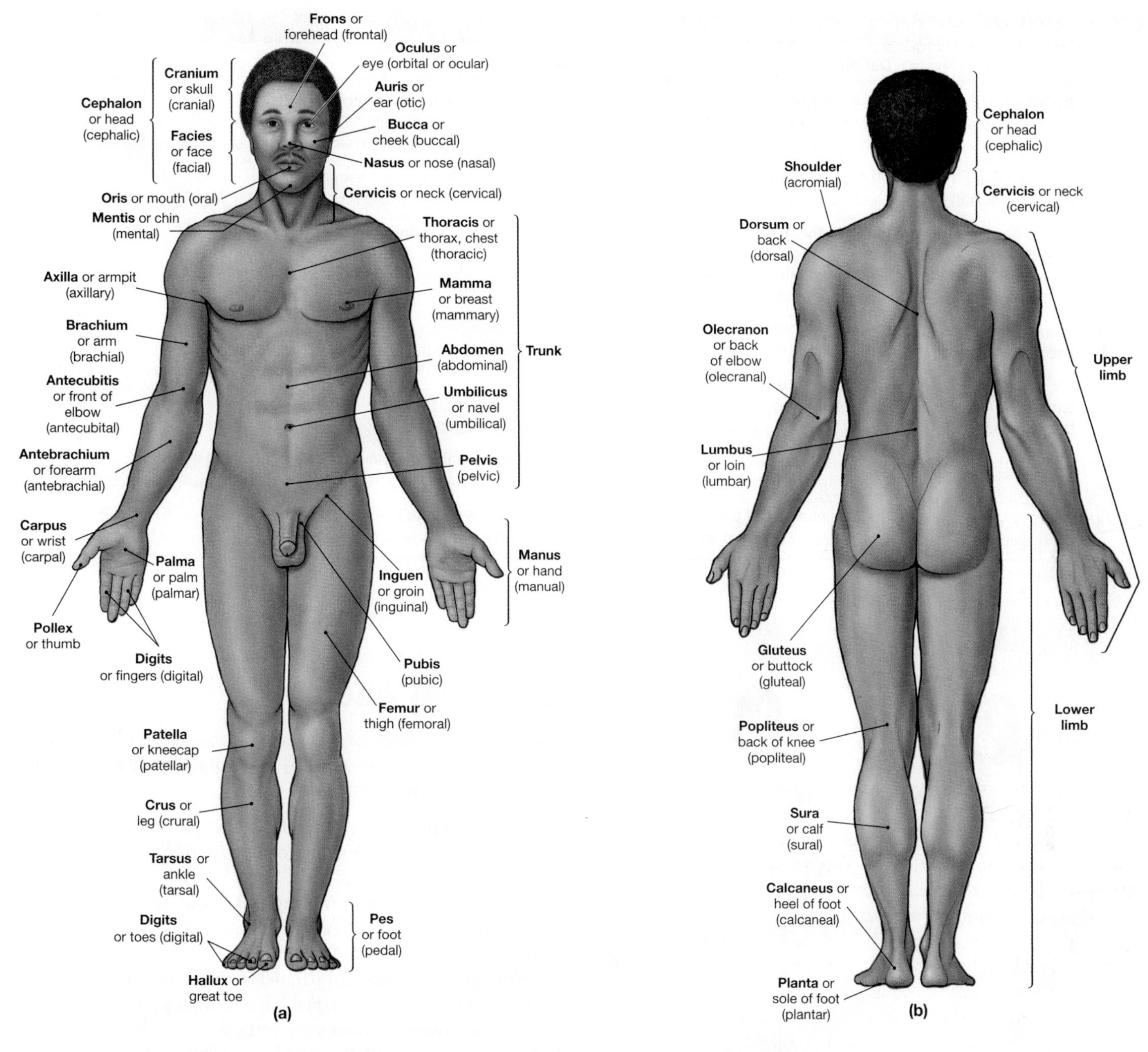

Figure 1.1 Anatomical body regions illustrated from the anatomical position. Anatomical terms are in boldface type, common names are in plain type, and anatomical adjectives are in parentheses. **a)** Anterior view; **b)** posterior view.

Table 1.1 Anatomical Terms of Relationship and Comparison

Term	Definition	Example
1. a. *Superior* (cranial) b. *Inferior* (caudal)	Closer to the head Closer to the feet	The lungs are superior to the stomach. The liver is inferior to the heart. (These terms are not typically used to describe structures on the extremities.)
2. a. *Anterior* (ventral) b. *Posterior* (dorsal)	Closer to the front Closer to the back	The trachea is anterior to the esophagus. The vertebral column is posterior to the heart.
3. a. *Medial* b. *Lateral* c. *Intermediate*	Closer to the midline Farther from the midline Between a more medial and more later6al structure	The nose is medial to the cheeks. The spleen is lateral to the pancreas. The clavicle is intermediate to the sternum and the shoulder.
4. a. *Proximal* b. *Distal*	Closer to the trunk Farther from the trunk	The shoulder is proximal to the elbow. The wrist is distal to the elbow. (These terms are most often used to describe structures on the extremities.)
5. a. *Superficial* (external) b. *Deep* (internal)	Closer to or on the surface Farther from the surface	The skin is superficial to the skeletal muscles. The bones are deep to the skin.
6. a. *Parietal* b. *Visceral*	Pertaining to the wall of a body cavity Pertaining to an organ	The membrane lining the pleural cavity is the parietal pleura. The membrane that covers the surface of the lungs is the visceral pleura.
7. a. *Ipsilateral* b. *Contralateral*	On the same side of the body On the opposite side of the body	The right lung is ipsilateral to the gallbladder. The left arm is contralateral to the right leg.

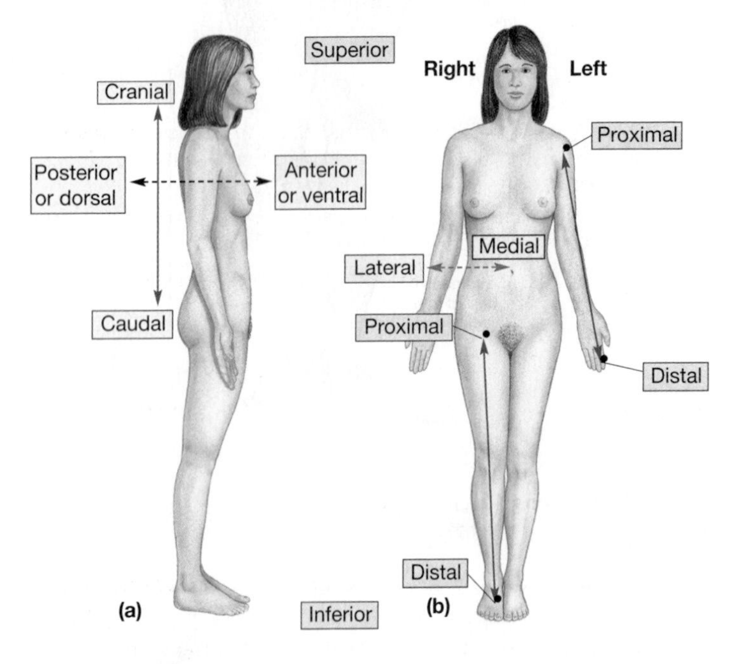

Figure 1.2 Terms used to describe anatomical relationships and comparisons. a) Anterior view of the body in anatomical position; **b)** lateral view of the body in anatomical position.

Sectional Anatomy

Three types of imaginary planes pass through the body in the anatomical position (Figure 1.3). Each of the three planes forms a right angle with the other two.

- **Sagittal planes** are vertical planes that pass longitudinally through the body from anterior to posterior and divide the body into left and right parts. There are two types of sagittal planes. The **median** or **midsagittal plane** passes through the midline and divides the body into equal left and right halves (Figure 1.3).
- **Parasagittal planes** run parallel to the median plane and divide the body into unequal left and right parts.
- **Coronal (frontal) planes** are vertical planes that pass longitudinally through the body from left to right. They divide the body into anterior and posterior portions.
- **Transverse (horizontal) planes** are planes that divide the body into superior and inferior portions.

The surfaces that are formed by cuts made in the various planes are called **sections**. There are three types of anatomical sections.

- **Longitudinal sections** run lengthwise in the direction of the long axis of the body. Longitudinal sections can include **sagittal sections**, which are produced by cuts made along sagittal planes, and **coronal sections**, from cuts made along coronal planes.

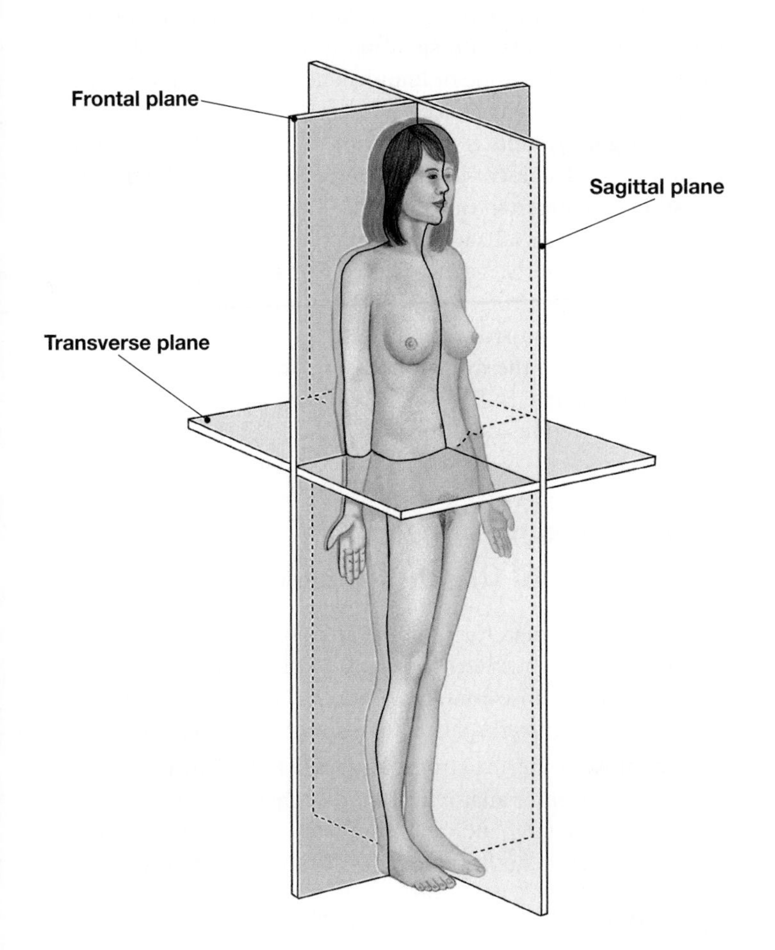

Figure 1.3 The three main anatomical planes. The planes are illustrated with the body in the anatomical position. Each plane intersects the other two at a right angle.

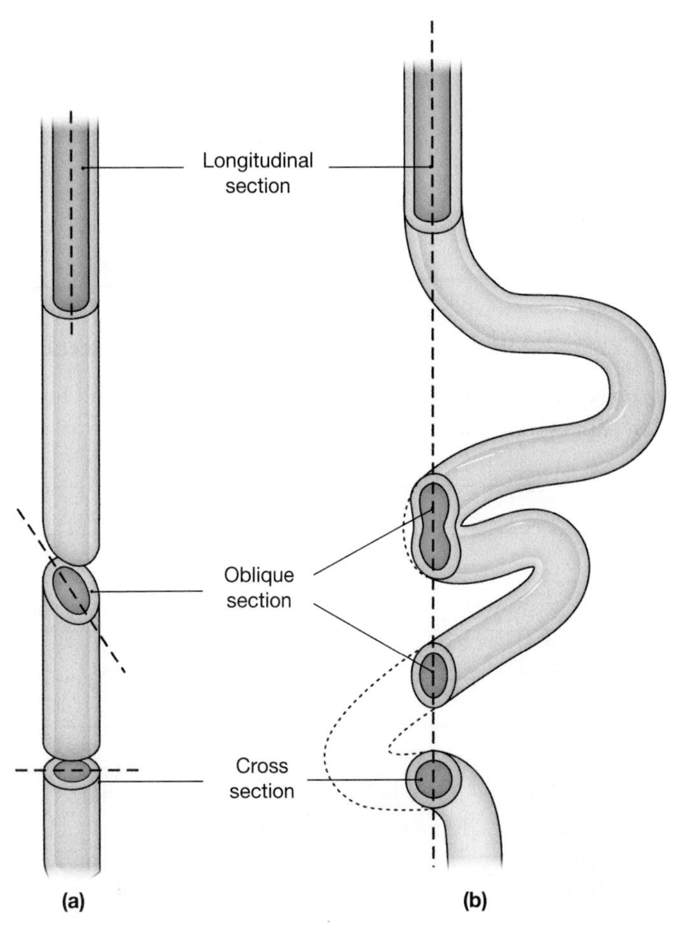

Figure 1.4 Types of sections in a tube structure such as the small intestine. If the tube is folded, as shown in **b)**, more than one type of section can be produced by a single cut in the same plane.

- **Transverse (cross) sections** run at right angles to the longitudinal axis of the body. They are formed by cuts made along transverse planes.
- **Oblique sections** are not formed by cuts made along any of the basic anatomical planes as described earlier. Instead, these sections slant or deviate from these planes and intersect them at angles less than 90°.

Sections are typically used to describe cuts made in specific structures, rather than the whole body. For example, if you are viewing a microscope slide of the trachea, you may be looking at a cross section (cs) or a longitudinal section (ls). Sometimes, the label on the slide will tell you what type of section you are observing (i.e., cs or ls). In a highly folded structure such as the small intestine, however, you often can find more than one type of section on the same microscope slide (Figure 1.4).

ACTIVITY 1.2 Defining Anatomical Planes and Sections

1. Identify the anatomical planes on torso models, and on your own or your lab partner's body.
2. Observe several anatomical models with removable parts. As you remove each part, determine what type of section has been made. Remember that you must place the structure in anatomical position to determine the answer.
3. Obtain a vegetable that has a long cylindrical shape (e.g., cucumber, zucchini, eggplant). Using a knife or scalpel, start near one end of the vegetable and cut, in order, a cross section, oblique section, and longitudinal section. Observe the surfaces that you have produced with these sectional cuts.

4. In the space below, make a drawing of each section that you have produced.

Cross section **Oblique section** **Longitudinal section**

5. For each section, describe its general appearance and how it differs from the others.

____

____

____

____

____

____

QUESTIONS TO CONSIDER 1. The thoracic cavity contains the heart and lungs. Explain, in a general way, how a view of the thoracic cavity along the midsagittal plane would differ from a view of the thoracic cavity along a transverse plane.

____

____

____

____

2. The epididymis is a highly convoluted (folded) muscular tube in the male reproductive system. When observing this structure with a microscope, cross sections, oblique sections, and longitudinal sections can be seen in the same field of view. How can you account for such an observation?

____

____

____

The Organ and Organ System Levels of Organization

An **organ** is a distinct structure that contains at least two, but often all four, types of tissues and carries out specific functions. Most organs are located within body cavities that are closed to the outside. For example, the small intestine contains all four types of tissues and performs the final steps for breaking down (digesting) nutrients into small molecules and absorbing these molecules into the blood or lymph. The small intestine is located within the abdominal cavity.

An **organ system** is a collection of organs that works as a team to complete a common objective. For example, the small intestine is an organ in the digestive system, which also includes structures in the oral cavity, pharynx, esophagus, and most organs in the abdominal cavity. These organs are responsible for ingesting food, digesting and absorbing nutrient molecules, and eliminating undigested wastes.

The organs in the human body are organized into 11 different organ systems. In Figure 1.5, each system is illustrated and its general functions described.

ACTIVITY 1.3 Identifying Organs and Organ Systems

1. Obtain a human torso model and observe the anatomical relations of the internal organs. Notice how adjacent organs are in close contact with one another, and that very little unoccupied space remains in the body cavities.
2. Identify all the structures that are listed below on a torso model or other anatomical models that are available in the laboratory. (Depending upon the type of models in your lab, you may not be able to locate all the structures.)

Aorta	Skull
Brain	Small intestine
Heart	Spinal cord
Kidneys	Spleen
Liver	Stomach
Lungs	Testes
Ovaries	Tonsils
Pancreas	Trachea
Skeletal muscles	Urinary bladder
Skin	Uterus

3. Place the structures that you just identified in the appropriate organ system listed below. Refer to Figure 1.5 for assistance in completing this activity.

Integumentary system ______________________

Skeletal system ______________________

Muscular system ______________________

Nervous system ______________________

Endocrine system ______________________

Cardiovascular system ______________________

THE INTEGUMENTARY SYSTEM

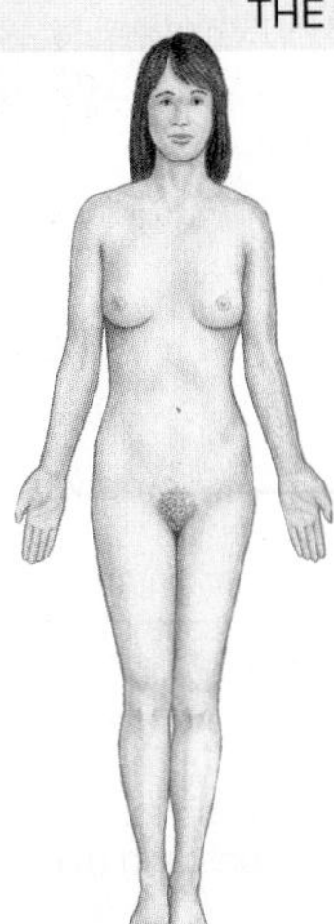

Major Organs:
- Skin
- Hair
- Sweat glands
- Nails

Functions:
- Protects against environmental hazards
- Helps regulate body temperature
- Provides sensory information

THE NERVOUS SYSTEM

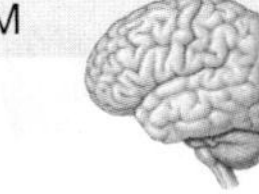

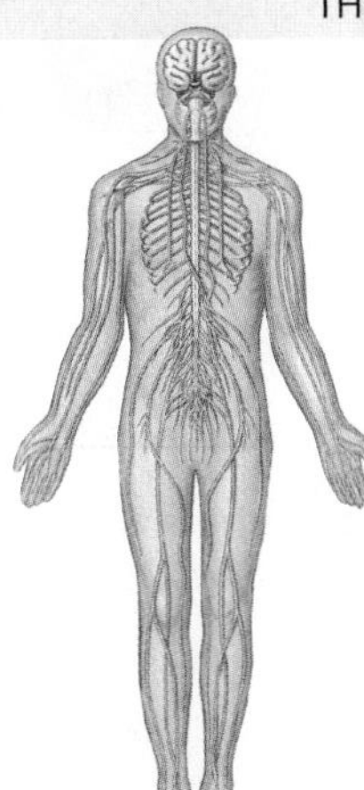

Major Organs:
- Brain
- Spinal cord
- Peripheral nerves
- Sense organs

Functions:
- Directs immediate responses to stimuli
- Coordinates or moderates activities of other organ systems
- Provides and interprets sensory information about external conditions

THE SKELETAL SYSTEM

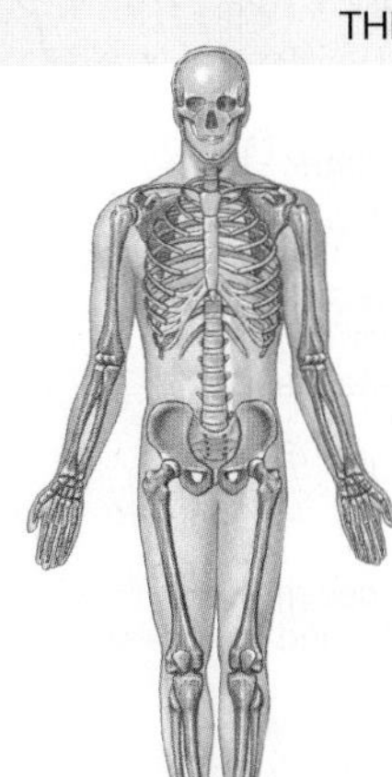

Major Organs:
- Bones
- Cartilage
- Associated ligaments
- Bone marrow

Functions:
- Provides support and protection for other tissues
- Stores calcium and other minerals
- Forms blood cells

THE ENDOCRINE SYSTEM

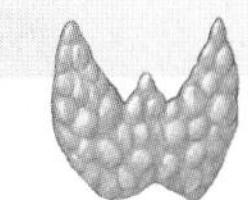

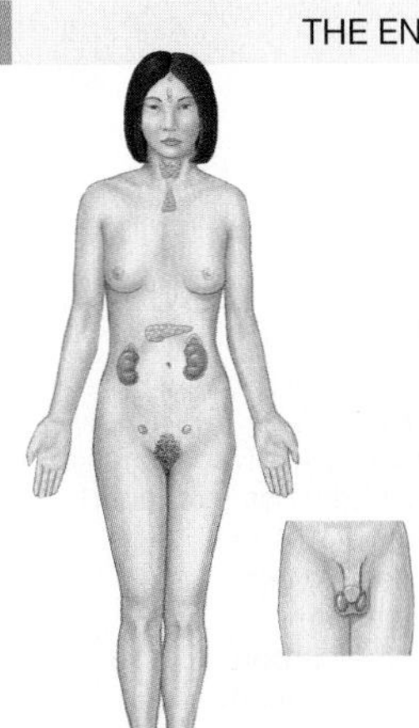

Major Organs:
- Pituitary gland
- Thyroid gland
- Pancreas
- Adrenal glands
- Gonads (testes and ovaries)
- Endocrine tissues in other systems

Functions:
- Directs long term changes in the activities of other organ systems
- Adjusts metabolic activity and energy use by the body
- Controls many structural and functional changes during development

THE MUSCULAR SYSTEM

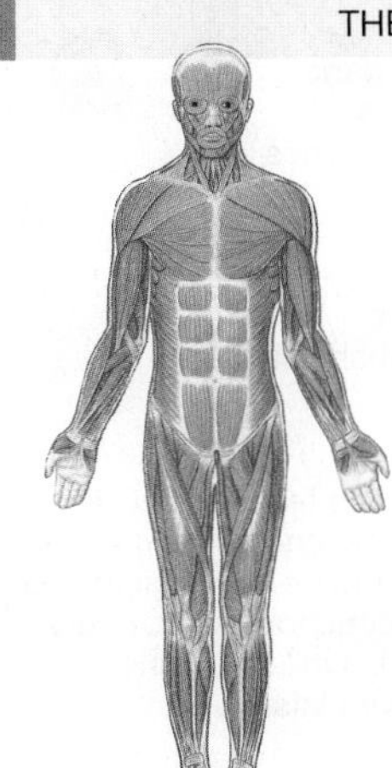

Major Organs:
- Skeletal muscles and associated ligaments

Functions:
- Enables movement
- Provides protection and support for other tissues
- Generates heat that maintains body temperature

THE CARDIOVASCULAR SYSTEM

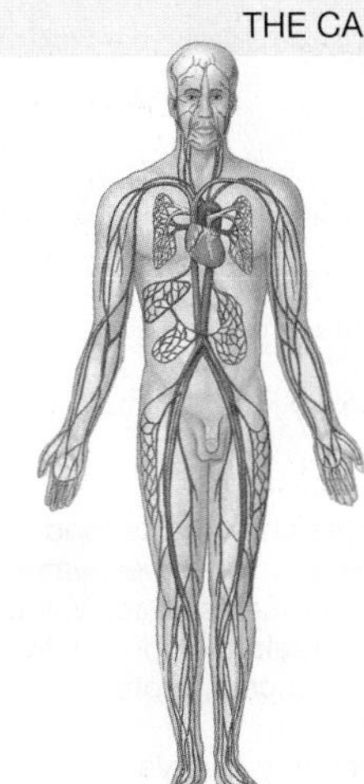

Major Organs:
- Heart
- Blood
- Blood vessels

Functions:
- Distributes blood cells, water, and dissolved materials, including nutrients, waste products, oxygen, and carbon dioxide
- Distributes heat and assists in control of body temperature

Figure 1.5 The 11 organ systems in the body. For each system, the organs are illustrated and the functions are summarized. The anatomy of the reproductive system is different in each sex. All other organ systems have the same structures. (*continues*)

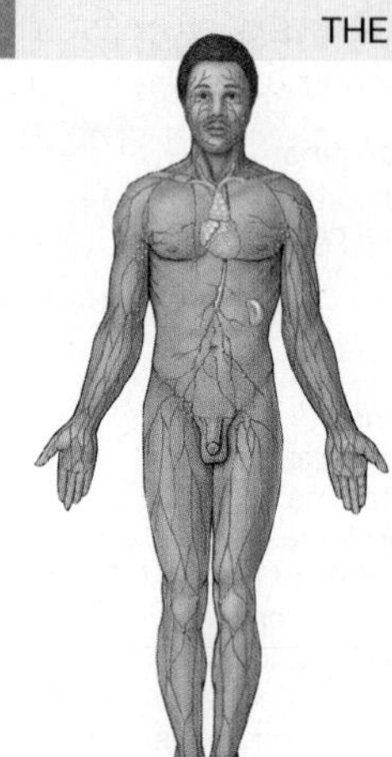

THE LYMPHATIC SYSTEM

Major Organs:
- Spleen
- Thymus
- Lymphatic vessels
- Lymph nodes
- Tonsils

Functions:
- Defends against infection and disease
- Returns tissue fluids to the bloodstream

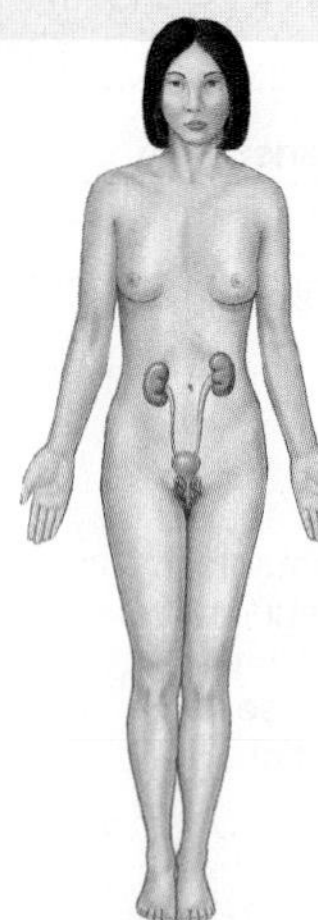

THE URINARY SYSTEM

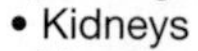

Major Organs:
- Kidneys
- Ureters
- Urinary bladder
- Urethra

Functions:
- Excretes waste products from the blood
- Controls water balance by regulating volume of urine produced
- Stores urine prior to voluntary elimination
- Regulates blood ion concentrations and pH

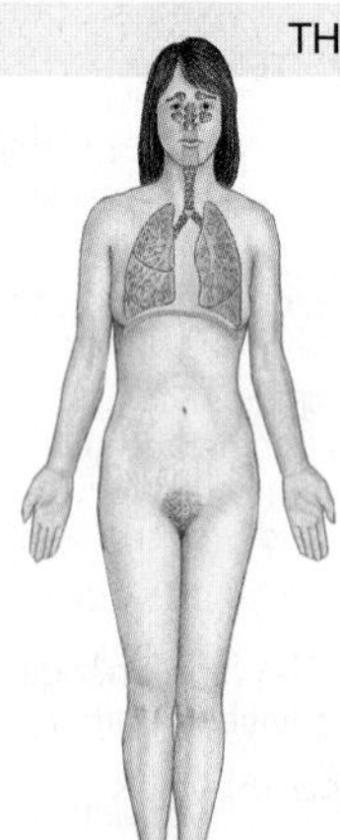

THE RESPIRATORY SYSTEM

Major Organs:
- Nasal cavities
- Sinuses
- Larynx
- Trachea
- Bronchi
- Lungs
- Alveoli

Functions:
- Delivers air to alveoli (sites in lungs where gas exchange occurs)
- Provides oxygen to bloodstream
- Removes carbon dioxide from bloodstream
- Produces sounds for communication

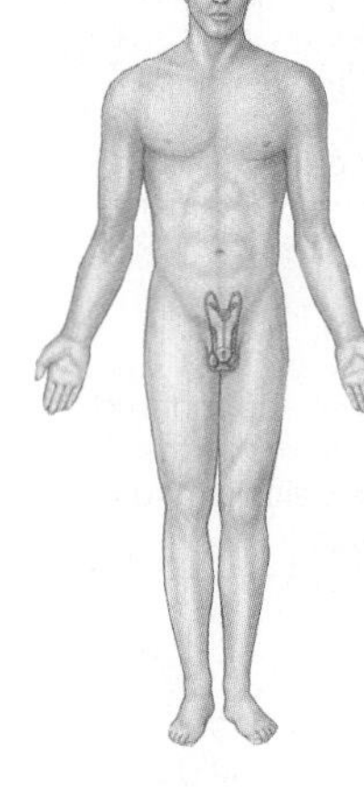

THE MALE REPRODUCTIVE SYSTEM

Major Organs:
- Testes
- Epididymis
- Ductus deferens
- Seminal vesicles
- Prostate gland
- Penis
- Scrotum

Functions:
- Produces male sex cells (sperm) and hormones

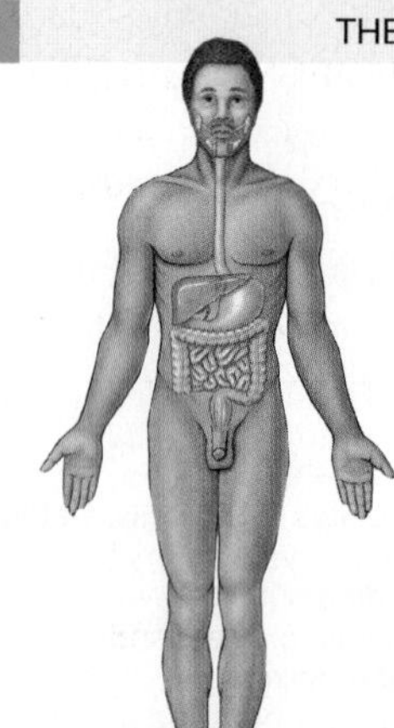

THE DIGESTIVE SYSTEM

Major Organs:
- Teeth
- Tongue
- Pharynx
- Esophagus
- Stomach
- Small Intestine
- Large intestine
- Liver
- Gallbladder
- Pancreas

Functions:
- Processes and digests food
- Absorbs and conserves water
- Absorbs nutrients (ions, water, and the breakdown products of dietary sugars, proteins, and fats)
- Stores energy reserves

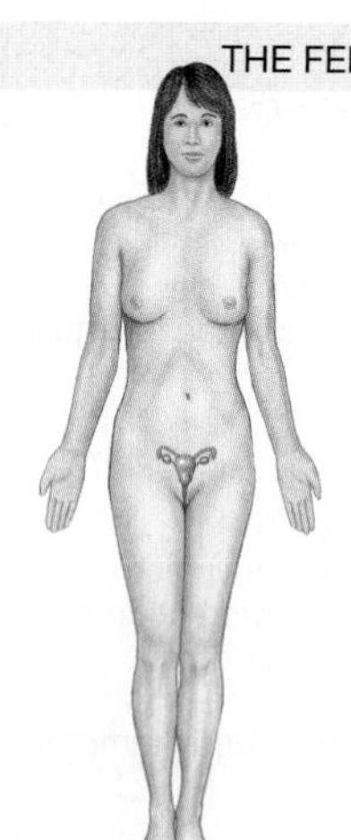

THE FEMALE REPRODUCTIVE SYSTEM

Major Organs:
- Ovaries
- Uterine tubes
- Uterus
- Vagina
- Labia
- Clitoris
- Mammary glands

Functions:
- Produces female sex cells (oocytes) and hormones
- Supports developing embryo from conception to delivery
- Provides milk to nourish newborn infant

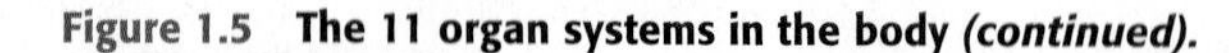

Figure 1.5 The 11 organ systems in the body *(continued)*.

Lymphatic system ____________________

Respiratory system ____________________

Digestive system ____________________

Urinary system ____________________

Male reproductive system ____________________

Female reproductive system ____________________

CLINICAL CORRELATION

It is convenient to study the organ systems as discrete entities, but from a functional perspective, each organ system is closely integrated with other systems. Consider the following examples.

- The lymphatic system defends the organs in other systems against infection and plays a pivotal role in tissue repair after an injury.
- The digestive system provides nutrients for cells in all organ systems. These nutrients are transported by the cardiovascular system.

Because the organ systems are so closely connected in function, many diseases present symptoms with a wide range of systemic effects. For example, diabetes mellitus, a disease that is characterized by the inability of cells to take up glucose, forces the body to break down vital proteins and lipids to produce enough energy for metabolism. As a result, many degenerative changes occur throughout the body, leading to a myriad of medical problems including blindness, kidney failure, reduced blood flow to the limbs, and heart disease.

As you can see from these examples, during normal function and during periods of disease, the activities of each organ system are influenced and sometimes controlled by the activities of the others. Begin to understand and learn to appreciate this close integration of function.

QUESTION TO CONSIDER In the previous exercise, you grouped various organs by organ system. Review your groupings and identify any organs that appear in more than one organ system. Comment on the functional significance of organs having a role in more than one system.

____

____

____

Regional Anatomy

The body can be divided into two major divisions: the **axial division** and the **appendicular division**. The axial division is the central part of the body and includes the head, neck, and trunk. The appendicular division includes the upper and lower extremities. Both the axial and appendicular divisions can be subdivided into numerous smaller regions, each with a specific anatomical and common name. Familiarity with these terms will help you later to locate and learn the names of other structures. For example, the *axilla* is the region of the body that is commonly referred to as the armpit. The *axillary* artery and vein travel through this region of the body, and the *axillary* lymph nodes are located in this area as well. The major anatomical regions are illustrated in Figure 1.1.

The abdomen and pelvis, together often referred to as the **abdominopelvic** region, can be divided into even smaller segments. Clinicians divide this region into quadrants that are formed by two imaginary, perpendicular lines intersecting at the umbilicus. The four quadrants are the **right upper quadrant**, **right lower quadrant**, **left upper quadrant**, and **left lower quadrant** (Figure 1.6a). Anatomists usually describe the abdominopelvic area in a more specific manner by dividing it into nine smaller regions. These regions are illustrated in Figure 1.6b.

CLINICAL CORRELATION

The quadrant system is clinically important because it can be used to identify the general location of underlying organs (Figure 1.6c). For example, the appendix is a wormlike extension attached to the cecum at the origin of the large intestine. The appendix is located in the right lower quadrant. If a patient complains of a persistent pain in this region, it could indicate an inflammation of the appendix, or **appendicitis.**

ACTIVITY 1.4 Identifying Anatomical Regions

1. Using your own body, or that of a lab partner, identify the anatomical regions in which you will likely find the following structures. Refer to Figure 1.1 for assistance in completing this exercise.
   a. Inguinal canal ____________________
   b. Brachial artery ____________________
   c. Femoral vein ____________________
   d. Facial nerve ____________________
   e. Thoracic vertebrae ____________________
   f. Carpal bones ____________________
   g. Cranial bones ____________________
2. Using the torso models in the laboratory, identify the abdominopelvic quadrants. Identify all the organs, or parts of organs, found within each quadrant and list them in the appropriate section in Table 1.2. Use Figure 1.6 as a reference.
3. Using the torso models, identify the more specific nine abdominopelvic regions. Identify all the organs, or parts of organs, found within each region and list them in the appropriate section in Table 1.2. Use Figure 1.6 as a reference.

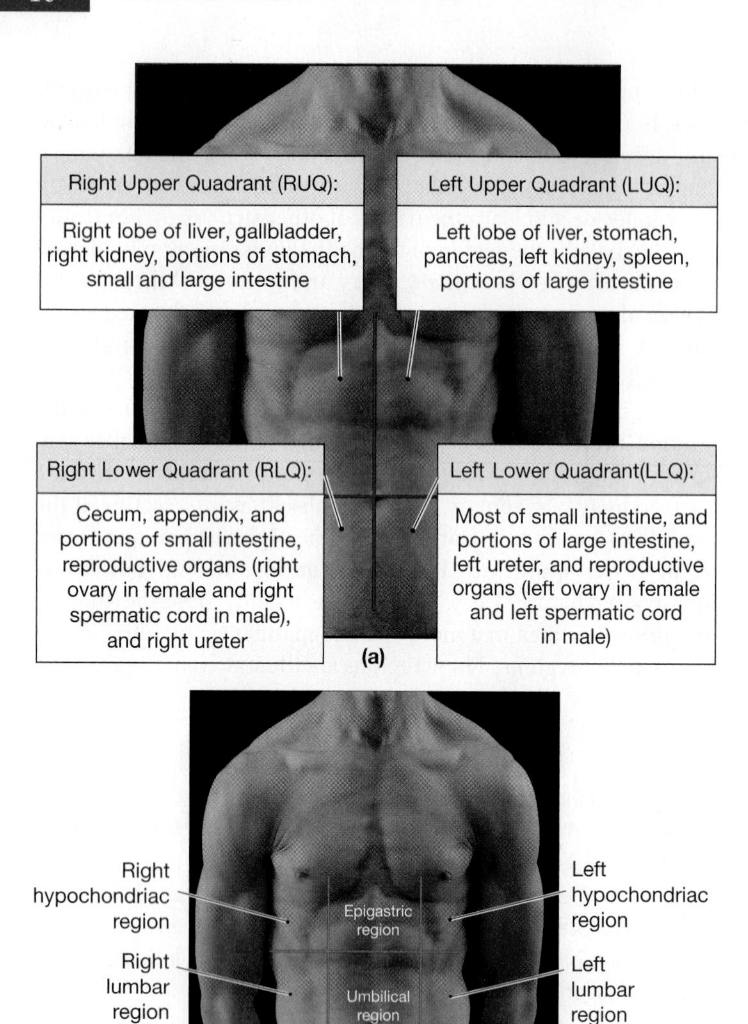

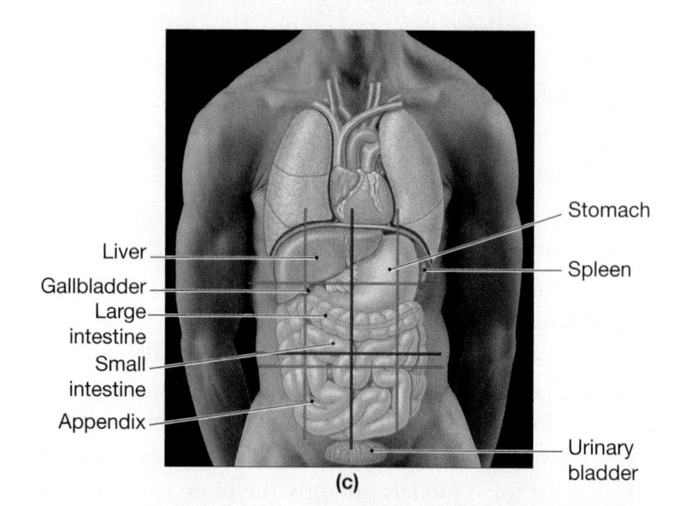

Figure 1.6 Two methods used to divide the abdominopelvic region. **a)** The abdominopelvic quadrants used by clinicians; **b)** the nine more specific regions used by anatomists; **c)** the abdominopelvic organs with the boundaries of the four quadrants (blue) and nine regions (red) superimposed over them.

Table 1.2 The Abdominopelvic Regions and Underlying Organs

Region	Organs found in each region
Four Regions (Quadrants)	
1. Right upper quadrant	
2. Left upper quadrant	
3. Right lower quadrant	
4. Left lower quadrant	
Nine Regions	
1. Right hypochondriac	
2. Right lumbar	
3. Right inguinal	
4. Epigastric	
5. Umbilical	
6. Hypogastric (pubic)	
7. Left hypochondriac	
8. Left lumbar	
9. Left inguinal	

QUESTION TO CONSIDER Which method of dividing the abdominopelvic region do you find to be more useful from an anatomical perspective? From a clinical perspective? Explain.

____

____

____

Body Cavities

The axial division of the body contains two major body cavities: the **dorsal (posterior) cavity** and the **ventral (anterior) cavity**. They are closed to the outside and contain the body's vital organs.

The dorsal cavity contains the central nervous system. It is subdivided into the **cranial cavity** that is formed by the cranial bones in the skull, and the **vertebral (spinal) cavity** that is a bony canal formed by consecutive vertebrae in the vertebral column (Figures 1.7a and c).

The **diaphragm** divides the ventral cavity into a superior compartment, known as the **thoracic cavity**, and an inferior compartment, known as the **abdominopelvic (peritoneal) cavity** (Figure 1.7a). The thoracic cavity is further subdivided into two lateral **pleural cavities**, which surround the lungs, and a centrally located region known as the **mediastinum** (Figure 1.7d). The mediastinum contains the **pericardial cavity**, which surrounds the heart, as well as the thoracic portions of the aorta, trachea, and esophagus. The abdominopelvic cavity contains organs involved

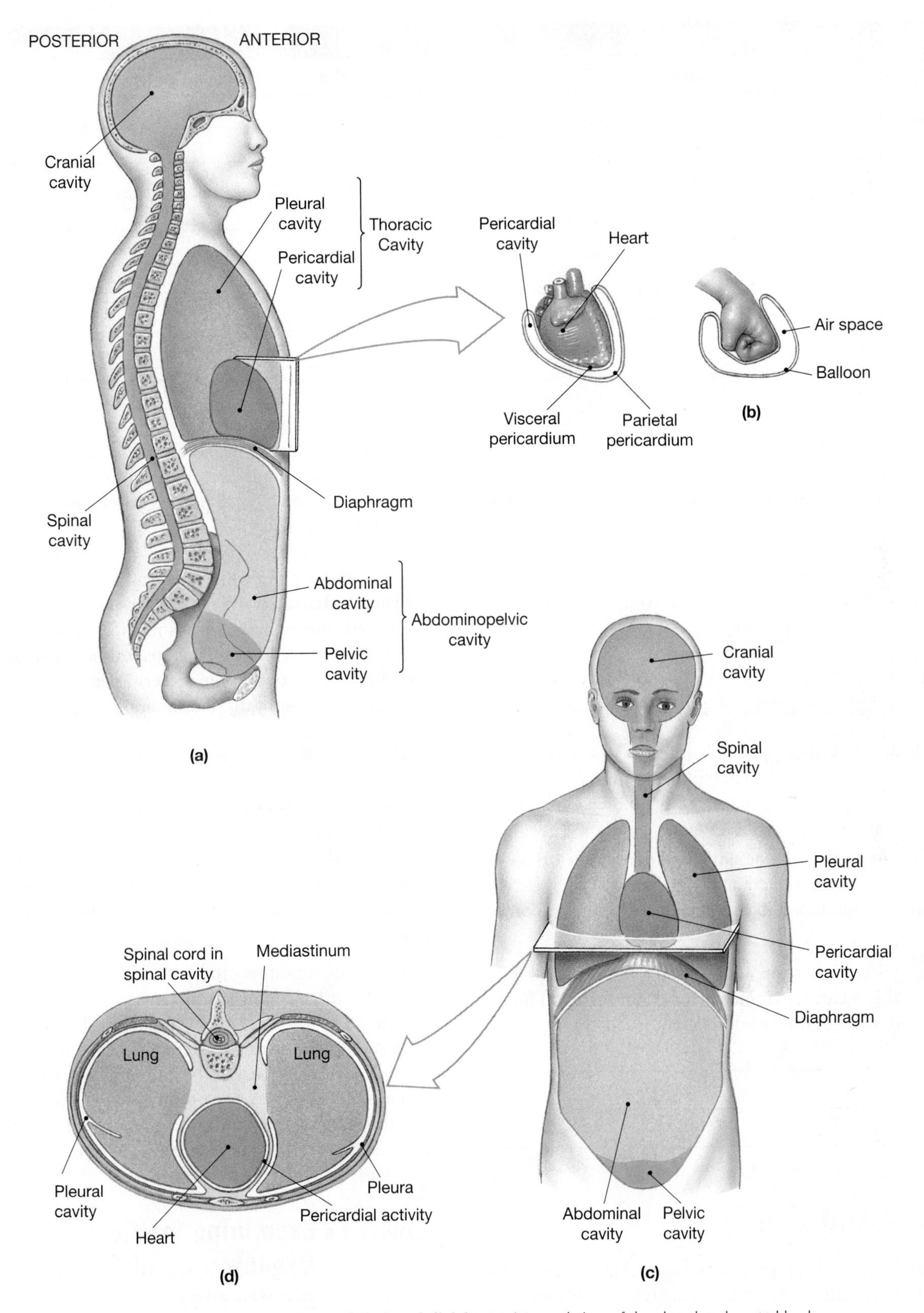

Figure 1.7 Major body cavities and their subdivisions. a) Lateral view of the dorsal and ventral body cavities; **b)** the relationship between the heart and pericardial cavity is similar to a fist that is pushed into a balloon. The fist represents the heart. The wrist and forearm represent the large blood vessels that are connected to the heart; **c)** anterior view of the dorsal and ventral body cavities; **d)** transverse section through the thoracic cavity showing the pericardial cavity within the mediastinum and the two lateral pleural cavities.

Table 1.3 The Major Body Cavities

Major cavity	Subdivisions	Organs found in each cavity
1. Dorsal (posterior) cavity	a. Cranial cavity	______________
	b. Vertebral (spinal) cavity	______________
2. Ventral (anterior) cavity	a. Thoracic cavity	______________
	1. Pleural cavities	______________
	2. Pericardial cavity	______________
	b. Abdominopelvic (peritoneal) cavity	______________
	1. Abdominal cavity	______________
	2. Pelvic cavity	______________

with digestion, excretion, and reproduction. Although there is no physical division, for descriptive purposes, it is artificially subdivided into the superior **abdominal cavity** and the inferior **pelvic cavity** (Figures 1.7a and c).

In addition to the major body cavities, a number of smaller functional cavities are located in the head. These include the **orbital cavities** for the eyeballs, the **nasal cavity** and **paranasal sinuses**, and the **oral cavity** or mouth.

ACTIVITY 1.5 Exploring the Body Cavities

1. Using the torso models in the laboratory, identify the major body cavities in the human body (Figure 1.7). Determine which organs are located in each subdivision and list them in Table 1.3.
2. Obtain a human skeleton or skull and identify the orbital, nasal, and oral cavities. On a torso model, identify these same cavities, and if possible, the paranasal sinuses as well.

QUESTION TO CONSIDER Using a skeleton, demonstrate and explain how the skeletal and muscular systems are vital for protecting the organs found within the major body cavities.

____

____

____

Serous Membranes

Serous membranes are thin, protective layers that line the walls of the pleural, pericardial, and abdominopelvic cavities and the outside walls of the organs within them. The portion of a serous membrane that lines a cavity wall is the **parietal layer**; the portion that covers an internal organ is the **visceral layer**. The serous membranes are assigned specific names according to the body cavity in which they are found.

- The pericardium is the serous membrane in the pericardial cavity. The **parietal pericardium** lines the wall of the cavity and the **visceral pericardium** covers the heart wall (Figure 1.7b).
- The pleura is the serous membrane in the pleural cavities. The **parietal pleura** lines the wall of the cavity, and the **visceral pleura** covers the surfaces of the lungs.
- The peritoneum is the serous membrane in the abdominopelvic cavity. The **parietal peritoneum** lines the wall of the cavity, and the **visceral peritoneum** covers the surfaces of most of the abdominopelvic organs.

WHAT'S IN A WORD The term, *parietal* is derived from the Latin word *paries*, which means "wall." The term *visceral* is derived from the Latin word *viscus*, which means "internal organ." These terms are used often in anatomy to describe structures associated with internal organs or body cavity walls. Thus, the "visceral" layer of a serous membrane covers an internal organ and the "parietal" layer of a serous membrane lines a body cavity wall.

Serous membranes produce and secrete a watery fluid that keeps the surfaces moist at all times. This serous fluid reduces friction when the opposing parietal and visceral layers slide along each other during movements of the internal organs. For example, when the heart is beating, the heart wall is constantly in motion. For the heart to pump blood efficiently, it is essential that the parietal pericardium and visceral pericardium slide along each other with as little friction as possible.

ACTIVITY 1.6 Examining the General Organization of Serous Membranes

1. Obtain a large, clear plastic bag and close off the open end.
2. With one hand, make a tight fist and slowly push it into the wall of the closed plastic bag (Figure 1.7b).

3. Your fist represents an organ within a body cavity. Notice that when you push your fist into the closed bag, you form two layers of plastic, separated by a space (Figure 1.7b).
4. The plastic bag represents a serous membrane as follows:
   - The plastic layer in contact with your fist (the internal organ) represents the **visceral layer**.
   - The outside layer of plastic represents the **parietal layer**.
   - The space between the two plastic layers is the **serous cavity**, and would contain the serous fluid.
5. With your fist pushed into the bag, notice that the two layers of plastic are continuous with each other at the wrist. In other words, the inner plastic layer reflects onto the outer plastic layer.
6. Consider a real example that demonstrates the fist and plastic bag model—**the heart within the pericardial cavity** (Figure 1.7b).
   - The **visceral pericardium** (inner plastic layer) covers the **heart** (your fist).
   - The **parietal pericardium** (outer plastic layer) lines the wall of the pericardial cavity.
   - The **pericardial cavity** is the space between the visceral pericardium and parietal pericardium (space between plastic layers).
   - The visceral pericardium is continuous with the parietal pericardium where the great vessels (the aorta, for example) are connected to the heart (the two plastic layers are continuous at the wrist).

QUESTION TO CONSIDER What potential problems do you think could develop if the watery secretions produced by the visceral pericardium and parietal pericardium were less than normal?

___

___

___

Exercise 1 Review Sheet

Body Organization and Terminology

Name ____________________

Lab Section ____________________

Date ____________________

1. Describe what is meant by the anatomical position. Why is it important to examine the body in this position?

2. Compare the three basic types of anatomical planes.

3. What is the difference between a median (midsagittal) plane and a parasagittal plane?

4. Draw dashed lines in the diagram below to demonstrate how longitudinal, cross, and oblique sections would be made.

Questions 5–10: Based on the usage of the underlined anatomical terms of relationship, describe the meaning of the following sentences.

5. In the arm, the biceps brachii muscle lies <u>anterior</u> to the brachialis muscle.
6. In the forearm, the radial artery is <u>lateral</u> to the ulnar artery.
7. In the skin, the epidermis is <u>superficial</u> to the dermis.
8. In females, the body of the uterus is just <u>superior</u> to the urinary bladder; in males, the prostate gland is just <u>inferior</u> to the bladder.
9. In the lower extremity, the knee joint is <u>proximal</u> to the ankle joint, but <u>distal</u> to the hip joint.
10. In the brain, the cerebellum is <u>inferior</u> to the occipital lobes of the cerebrum and <u>posterior</u> to the brainstem.

Questions 11–15: Match the organ in column A with the appropriate organ system in column B.

A	B
11. Trachea ____________	a. Urinary system
12. Brain ____________	b. Respiratory system
13. Small intestine ____________	c. Nervous system
14. Kidney ____________	d. Lymphatic system
15. Aorta ____________	e. Cardiovascular system
	f. Digestive system

Questions 16–20: Match the organ in column A with the appropriate body cavity in column B.

A	B
16. Lung ________	a. Cranial cavity
17. Stomach ________	b. Pelvic cavity
18. Spinal cord ________	c. Pleural cavity
19. Urinary bladder ________	d. Abdominal cavity
20. Heart ________	e. Pericardial cavity
	f. Vertebral cavity

Exercise 2

Care and Use of the Compound Light Microscope

Laboratory Objectives

On completion of the activities in this exercise, you will be able to:

- Correctly carry, clean, and store a compound light microscope.
- Identify the parts of a compound light microscope and explain their functions.
- Use the proper method for placing a microscope slide on the stage, prior to viewing.
- Adjust the magnification and light source of a compound light microscope.
- Correctly focus the image that you are viewing.
- Calculate the total magnification of the lenses that you are using to view an object.
- Describe the principles of inversion of image and depth of field.
- Measure the diameter of the field of view and estimate the size of structures in a tissue section.

Materials

- Compound light microscopes
- Prepared microscope slides of various tissues
- Prepared microscope slides of the letter *e*
- Prepared microscope slides of intersecting colored threads
- Immersion oil
- Clear millimeter ruler
- Lens paper

The microscope is one of the most important tools used in the study of anatomy and physiology. It is used to view tissue and cellular structures that are too small for examination with the naked eye. Clinically, microscopes have become essential diagnostic tools and are used regularly during surgical procedures.

Light microscopes use optical lenses to bend or refract the light waves that pass through an observed object (a section of tissue, for instance). As a result, the image of the object is magnified, typically 40 to 1000 times. The two basic types of light microscopes are the **simple microscope**, which has a magnifying system that utilizes only one lens, and the **compound microscope**, which uses a series of lenses in combination (Figure 2.1). The microscopes you will be using in the laboratory are compound microscopes.

In the typical compound light microscope, the **light source (substage light)** is located in the **base** of the instrument. A beam of light from this source passes through a **condenser lens** located just below the **stage**, upon which a microscope slide is placed for viewing. The condenser lens concentrates the beam of light before it passes through the specimen. Next, the light beam passes through the **objective** and **ocular lenses**, which magnify the image for the observer.

Objects that are magnified by a light microscope will have more clarity of detail only if **resolution** is also increased. Resolution, also referred to as **resolving power**, is the ability to distinguish close objects as separate and distinct. The unaided human eye has a resolving power of about 0.1 millimeter (mm). This means that a person can distinguish two objects that are 0.1 mm apart as distinct entities. If they are less than 0.1 mm apart, they are perceived as being a single object. A good compound microscope can increase resolution to 0.001 mm, so the resolving power is 100 times greater than the unaided eye.

Care and Use of the Compound Light Microscope

The microscopes that you will be using to study microscopic anatomy are expensive precision instruments. They can be damaged if they are not handled with prudence and care. Strict adherence to the following general rules of maintenance will extend the life of your microscope.

- When carrying a microscope, hold it firmly in front of you, with one hand on the **arm** and the other hand supporting it under the base. Always set it down gently, without sliding, onto the table.
- Clean microscope lenses regularly so blurred images or dirt are not confused with cell structures. Always use lens paper for cleaning microscope lenses. Other types of tissues, such as Kimwipes or towels, can scratch the lens. To remove an oily smudge, moisten the lens paper with lens cleaner before cleaning.
- Avoid touching the lenses with your fingers because oils from the skin can damage the lens. If you accidentally touch the lens, clean it immediately with lens paper moistened with lens cleaner.
- When you complete your microscopic observations, remove the last slide you viewed and return it to the appropriate slide tray or storage box. Make sure the lowest power objective lens is in position, neatly wrap the electric cord, and cover the microscope with a dust cover. Use the two-hand method described earlier to return the microscope to its storage area.

Before you begin to use the light microscope, first review the location and function of its parts. Refer to a microscope in the laboratory and to Figure 2.1 as you identify and study the various microscope components.

Figure 2.1 A typical compound light microscope. The major operating components are illustrated.

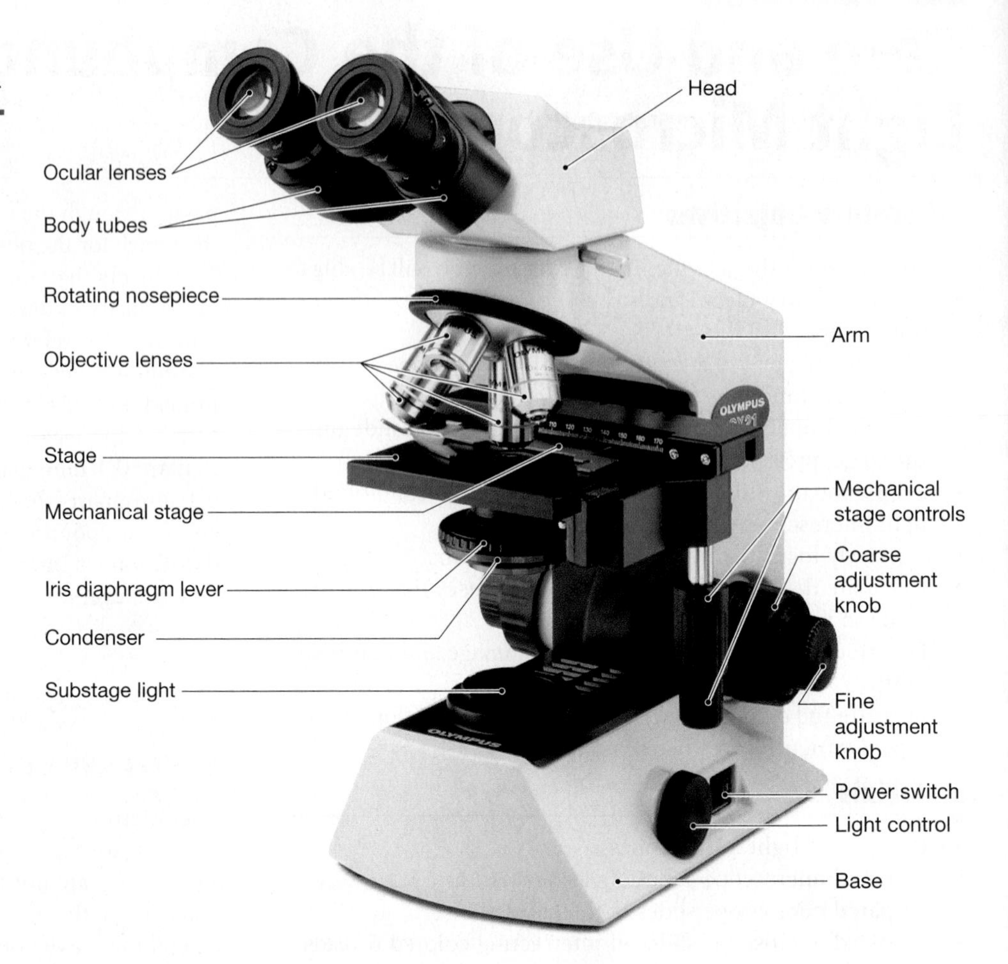

ACTIVITY 2.1 Learning the Parts of a Light Microscope

1. Plug in the electric cord and switch on the substage light source built into the base. There may be a rolling on-off light switch that will allow a range of brightness. If so, always start with a high light intensity and adjust the brightness with the **iris diaphragm**.
2. The lever of the **iris diaphragm** can be seen under the microscope **stage.** Look through the ocular (eyepiece) lens and observe what happens when you move the lever to open and close the diaphragm. Alternatively, some microscopes have an iris diaphragm composed of a wheel that can be rotated between different-sized openings.
3. The microscope stage is the platform upon which a microscope slide is placed. The stage has a central opening or aperture through which light passes to reach the specimen.
4. Typically, the stage is equipped with a clamping device, called a **mechanical stage**, which holds the slide securely and is used to position the object to be viewed directly over the stage aperture. Below the stage, on the side of the microscope, identify the two **mechanical stage control knobs.** Turn each of these knobs and observe how the mechanical stage moves. One knob moves the mechanical stage forward and backward; the other knob moves it to the left and right.
5. Locate the condenser lens, between the stage and the iris diaphragm. After light passes through the aperture (opening) of the iris diaphragm, it travels through the condenser lens. The condenser lens will concentrate the light before it passes through the tissue specimen.
6. Identify the **nosepiece**, located at the base of the **head**. The nosepiece is a revolving structure that holds two to four objective lenses. After traveling through the condenser lens and the tissue specimen, light passes through an objective lens and magnifies the image that you see. The magnification of each objective lens, often called the "power," is stamped on its rim. The typical laboratory microscope will be equipped with at least three objective lenses: a low-power (10X) lens, a high-power (40X to 45X) lens, and an oil immersion (100X) lens (X stands for "times"). Some microscopes may also have a scanning lens with a magnification of 4X. To quickly identify the different objective lenses, notice that, typically, the longer the lens, the higher the magnification. On many microscopes, the objective lenses are also marked with thin rings of varying colors.
7. Identify the ocular or eyepiece lenses. As light passes through these lenses, the image is magnified again before it reaches

your eyes. For most laboratory microscopes, the ocular lenses are 5X, 10X, or 15X. For the microscopes that you will be using, the magnification is probably 10X. Similar to the objective lenses, you should find the magnification stamped on the rims of the lenses. Some microscopes in your laboratory may be equipped with **pointers.** If present, the pointers are usually attached to the inside of the casing of one ocular lens. Some pointers can be moved by rotating the eyepiece.

8. The **total magnification** of the microscope with a particular objective lens in place can be calculated when the ocular lens magnification is multiplied by the objective lens magnification, as follows:

   Total magnification = magnification of ocular lens × magnification of objective lens.

9. On each side of the base of the microscope are two knobs, the larger **coarse adjustment knob** and the smaller **fine adjustment knob**. These knobs are used to bring the specimen into focus.
10. With the lowest power objective lens in place, turn the coarse adjustment knob one complete revolution and observe what occurs. Depending on your microscope, either the nosepiece or the stage moves up and down. In either case, notice that the distance between the nosepiece and stage changes.
11. Turn the fine adjustment knob one complete revolution and observe what occurs. If you see no change, do not be concerned. The fine adjustment knob also changes the distance between the nosepiece and stage, but the adjustments are much more subtle than with the coarse adjustment knob.

QUESTIONS TO CONSIDER

1. Based on your observation, describe the function of the iris diaphragm.

____

____

2. Suppose you are viewing a section of lung with a 10X ocular lens and a 40X objective lens. Calculate the total magnification of the section that you are viewing.

____

CLINICAL CORRELATION

Do people who normally wear eyeglasses need to wear them while using a microscope? If you are near- or farsighted, you do not need your eyeglasses because focusing the microscope will correct your vision. However, if you have an astigmatism, an image distortion caused by an irregular curvature in the cornea, you should wear your glasses because the microscope cannot make the proper correction.

ACTIVITY 2.2 Viewing a Specimen with the Compound Microscope

1. Position the microscope with the ocular lenses pointing toward you. Switch on the light source and immediately check the brightness. If the light appears to be too bright, make an adjustment with the on-off switch and/or the iris diaphragm.
2. Rotate the lowest power objective lens into position.
3. Using the coarse adjustment knob, move the nosepiece up (or the stage down) to make room for placing a slide in position. The distance between the objective lens and the microscope stage is called the **working distance.** When you are inserting or removing a slide, the working distance should be at its maximum.
4. Obtain a prepared microscope slide of any type of sectioned tissue and place it on the microscope stage. Make sure the slide is secured in position by the mechanical stage.
5. Adjust the position of the slide with the two mechanical stage control knobs so that the tissue you are about to view is centered over the stage aperture.
6. Turn the coarse adjustment knob to bring the objective lens and stage as close as possible. Notice that the coarse adjustment knob will stop turning before the objective lens and slide come into contact. This will only occur when a scanning (4X) or low-power (10X) objective lens is in position. Therefore, to guard against damaging a lens or breaking a slide, always begin with the lowest power objective lens that is on your microscope. With a scanning or low-power objective lens in position and the stage as close as possible, you can now begin to focus the tissue section.
7. Look into the microscope through the ocular lenses. If you are using a binocular microscope, which has two separate eyepieces, you may have to adjust the ocular lenses by moving them closer together or farther apart. This can be done by turning the adjustment knob, located just below or between the ocular lenses, or by manually moving the lenses closer or farther apart. The adjustment is correct when the images from both eyes fuse into one circular image.
8. The illuminated area that you are viewing is called the **field of view.** Adjust the iris diaphragm so that an adequate amount of light passes through the tissue section.
9. If your condenser lens is adjustable, keep it just below the stage, as far up as it will go. For most work in the laboratory, the condenser lens should be kept in this position. Experiment with this lens by moving it to various positions under the stage. What changes do you see in the field of view when the condenser lens is moved?

____

10. While looking into the ocular lenses, focus the image by slowly turning the coarse adjustment knob. When the image is approximately in focus, use the fine adjustment knob to bring it to an exact focus.
11. Select an area on the tissue section that you would like to view at a higher magnification. Move the slide with the mechanical stage control knobs, so that the selected area is in the center of the field of view.
12. Rotate the nosepiece so that a higher magnification (40X to 45X) objective lens is in position. Make sure that the lens will not hit the slide before moving it into position. **(Do not use the 100X oil immersion lens at this time.)**
13. Look into the ocular lenses to see if any focusing adjustments must be made. If your microscope is **parfocal**, this means that once the initial focus is made with a low-power objective lens, the image will remain in focus when switching between objective lenses. If your microscope is not parfocal, make any focusing adjustments with the fine adjustment knob. **Do not attempt to focus with the coarse adjustment knob under high power.** When you move to a higher magnification, what change occurs in the overall size of the field of view?

____

14. As you move to a higher magnification, you may have to adjust the iris diaphragm to allow more light to reach the specimen.

WHAT'S IN A WORD The prefix *par* in the word *parfocal* is derived from Latin and means "equal." Thus, a parfocal microscope will have "equal focus" or will remain in focus at all magnifications.

QUESTIONS TO CONSIDER 1. The microscopes that you are using in the laboratory are either **monocular** or **binocular.** A monocular microscope has only one ocular lens, and thus, only one eye can be used for viewing. On the other hand, a binocular microscope has two ocular lenses, one for each eye. Many students using binocular microscopes claim that it is easier to view a specimen by using only one ocular lens and one eye. However, your laboratory instructor will advocate the use of both ocular lenses while viewing a tissue section. Microscope work will become easier and more enjoyable once you are accustomed to binocular observation. Do you agree or disagree with this assertion? Explain why.

____

____

____

2. Why must the iris diaphragm be adjusted when you increase (or decrease) magnification?

____

____

3. During the previous activity, you were asked to describe any changes in the field of view when switching to a different objective lens. How does a change of objective lens affect the light that passes through the tissue specimen? (Hint: Recall the function of the condenser lens.)

____

____

Inversion of Image

When viewing a tissue section with a microscope, the image that you observe is not only **inverted** (turned upside down) but also **reversed** (turned from side to side). The following activity demonstrates this principle.

ACTIVITY 2.3 Viewing the Letter *e*

1. With the course adjustment knob, maximize the working distance on your microscope.
2. Place the low-power (10X) objective lens in position.
3. Place a prepared microscope slide with the letter *e* on the stage in the *right side up position* and centered over the aperture. **Do not view the slide through the ocular lens yet.** In the space below, make an illustration of the letter *e* as viewed with the unaided eye.
4. Position the condenser lens so that it is as far up as it will go.
5. Adjust the iris diaphragm to produce a brightly illuminated field of view.
6. Focus the letter *e* with the coarse and fine adjustment knobs.
7. Observe the letter *e* in the field of view. In the space below, make an illustration of the *e* as it actually appears under low power.
8. Verify the following changes in the orientation of the letter *e* when viewed under a microscope.
   - The image has been inverted (top to bottom).
   - The image has been reversed (left to right).
9. Use the mechanical stage control knob to move the slide to the right. In which direction does the *e* move when you view the letter through the microscope?

____

10. Use the mechanical stage control knob to move the slide away from you. In which direction does the *e* move when you view the letter through the microscope?

___

11. Maximize the working distance by turning the coarse adjustment knob and remove the slide.

QUESTION TO CONSIDER Review the observations that you made in Activity 2.3 and discuss how they demonstrate the concept of "inversion of image."

___

___

___

Depth of Field

Tissue sections on prepared microscope slides are sliced very thinly so that light may pass through them. Despite the translucence of the section, most tissues will still have several layers of cells. By careful use of the fine adjustment knob, you can sometimes observe the different layers of a tissue specimen. Thus, when one layer moves out of focus and becomes blurry, another layer, above or below the first layer, moves into focus and becomes sharp and clear. The **depth of field** refers to the thickness (depth) of the tissue layer that is currently in focus.

When you examine prepared slides of various tissues, be aware of the three-dimensional qualities of the structures you are viewing. By using the fine adjustment knob, you can focus on objects located at different levels within the section.

ACTIVITY 2.4 Perceiving Depth of Field

1. Obtain a slide with three different colored threads, all intersecting at a common point.
2. Position the slide so that the intersection point is in the center of the stage aperture.
3. Under low power (4X or 10X objective lens), use the coarse adjustment knob to focus the threads at the point where they all intersect. You should be able to easily identify the different thread colors.
4. Switch to the high power objective lens (40X to 45X).
5. With the fine adjustment knob, focus on each colored thread separately. Notice that when you are finely focused on one thread, the other two are blurred.
6. After focusing on all three threads, determine the order of colors from top to bottom.
   - Color of top thread ___
   - Color of middle thread ___
   - Color of bottom thread ___
7. Your observations illustrate the depth of field of a specimen.

QUESTION TO CONSIDER Speculate on how the depth of field changes (increases or decreases) as the total magnification increases. Explain why.

___

___

___

___

The Oil Immersion Lens

The oil immersion objective lens (100X) provides the highest magnification on your microscope and is useful for studying very small cells, such as bacteria, and structures within cells. A drop of oil is placed on the microscope slide and the lens is immersed into the drop. The oil concentrates the light, resulting in increased resolution at high magnification. Without the oil, you will be unable to focus a tissue specimen with the oil immersion lens.

Work cautiously when using immersion oil. If you get oil on your hands, wash them immediately with soap and water. If you get oil on the other lenses, clean them immediately with lens paper moistened with lens cleaner.

ACTIVITY 2.5 Using the Oil Immersion Lens

1. Obtain a prepared slide of any sectioned tissue.
2. After focusing the slide with the low power, select any structure that you wish to view with higher magnification and center it in the field of view.
3. Switch to the high power lens and view the structure to make sure it remains in the center of the field. If your microscope is not parfocal, you will have to refocus the image when you change the objective lens.
4. Raise the nosepiece (or lower the stage) slightly with the coarse adjustment knob and carefully place a drop of immersion oil on the center of the coverslip over the specimen.
5. Carefully rotate the nosepiece to move the oil immersion lens into position.
6. Carefully lower the nosepiece (or elevate the stage) with the coarse adjustment knob so that the oil immersion lens just touches the drop of oil. **It is important that you carry out this step very carefully. If you lower the nosepiece too quickly, you might shatter the slide and damage the lens.**
7. When the lens has made contact with the oil, look into the ocular lens and focus with the fine adjustment knob only. **Do not focus with the coarse adjustment once the lens has been immersed in the oil drop.**

8. At this magnification, more light will be needed to illuminate the field. You can increase the illumination of the field by adjusting the iris diaphragm. Although the field of view is reduced, the resolving power is increased, allowing the investigator to identify very small cellular structures as distinct and separate.
9. When you are finished viewing, raise the nosepiece (or lower the stage) with the coarse adjustment knob and remove the slide. Clean the oil immersion lens with lens paper and place the low power lens into position. Remove excess oil from the slide with a Kimwipe.

QUESTIONS TO CONSIDER 1. What advantage do you gain by viewing a tissue section with the oil immersion lens?

____

____

2. What are the possible limitations of using the oil immersion lens?

____

____

____

Total Magnification and the Diameter of the Field of View

When you switch to a higher power objective lens to increase the total magnification, the diameter of the field of view (field diameter) will decrease proportionately. The relationship between total magnification and field diameter can be expressed mathematically with the following equation:

$$M_1D_1 = M_2D_2$$

M_1 is the initial total magnification; D_1 is the field diameter at M_1; M_2 is the second total magnification when the objective lens is changed; and D_2 is the field diameter at M_2.

The above relationship is useful for predicting the field diameter when the total magnification is increased (or decreased). For example, suppose that at a total magnification of 100X, the diameter of the field of view is 4.0 mm. If you increase the total magnification to 200X, the field diameter at the greater magnification can be estimated, as follows:

$$M_1 = 100X; D_1 = 4.0 \text{ mm}; M_2 = 200X; D_2 \text{ is unknown}$$

$$(100X)(4.0 \text{ mm}) = (200X)(D_2)$$

$$D_2 = (100X)(4.0 \text{ mm}) \div 200X = 2.0 \text{ mm}$$

If the diameter of the field of view is known, the size of structures in tissue sections can be estimated. For example, consider a structure (e.g., a sweat gland in skin) that extends across one tenth the diameter of the field of view. If the field diameter is known to be 2.0 mm, then the diameter of the object will be one tenth of 2.0 mm, or 0.2 mm.

ACTIVITY 2.6 Determining the Diameter of the Field of View

1. Place a clear millimeter ruler across the aperture on the microscope stage.
2. With the scanning (4X) objective lens in position, focus the ruler lines and make sure that they are crossing the widest portion of the field of view.
3. Align the beginning of a millimeter interval with the left edge of the field of view (Figure 2.2).
4. Estimate the diameter in millimeters (mm). Record the magnification and field diameter in Table 2.1.
5. Repeat steps 1 through 4 with the low power (10X), and high power (40X to 45X), objective lenses in position. Record the magnification and field diameter for each lens in Table 2.1. How does the diameter change as you increase the magnification of the objective lens?

____

6. Place a prepared slide with any type of tissue on the microscope stage and view it with the scanning (4X) objective lens (Figure 2.3a).
7. Select a specific structure in the tissue that you are viewing.
8. Estimate the proportion of the field diameter across which the structure extends.

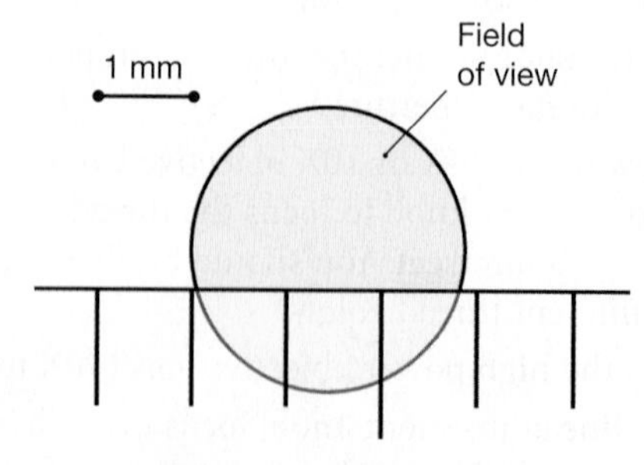

Figure 2.2 Measuring the microscope field of view diameter using a transparent millimeter ruler. The ruler lines cross the widest portion of the field of view. The beginning of a millimeter interval is aligned with the left edge of the field of view.

9. Use your earlier measurements of the field diameter (Table 2.1) to estimate the structure's size (i.e., diameter of the structure).

   Size of structure (mm)

   ____

10. Move the slide so that the structure is centered.
11. Switch to the low power (10X) objective lens and view the structure at a higher magnification (Figure 2.3b).
12. Once again, estimate the proportion of the field diameter across which the structure extends.
13. Predict the size of the structure by using the data recorded in Table 2.1. How does this calculation compare with your earlier estimation of structure size at the lower magnification?

    ____

    ____

Table 2.1 Measuring the Diameter of the Field of View

Type of objective lens	Magnification of objective lens	Magnification of ocular lens	Total magnification	Diameter of the field of view (mm)
Scanning	______	______	______	______
Low power	______	______	______	______
High power	______	______	______	______

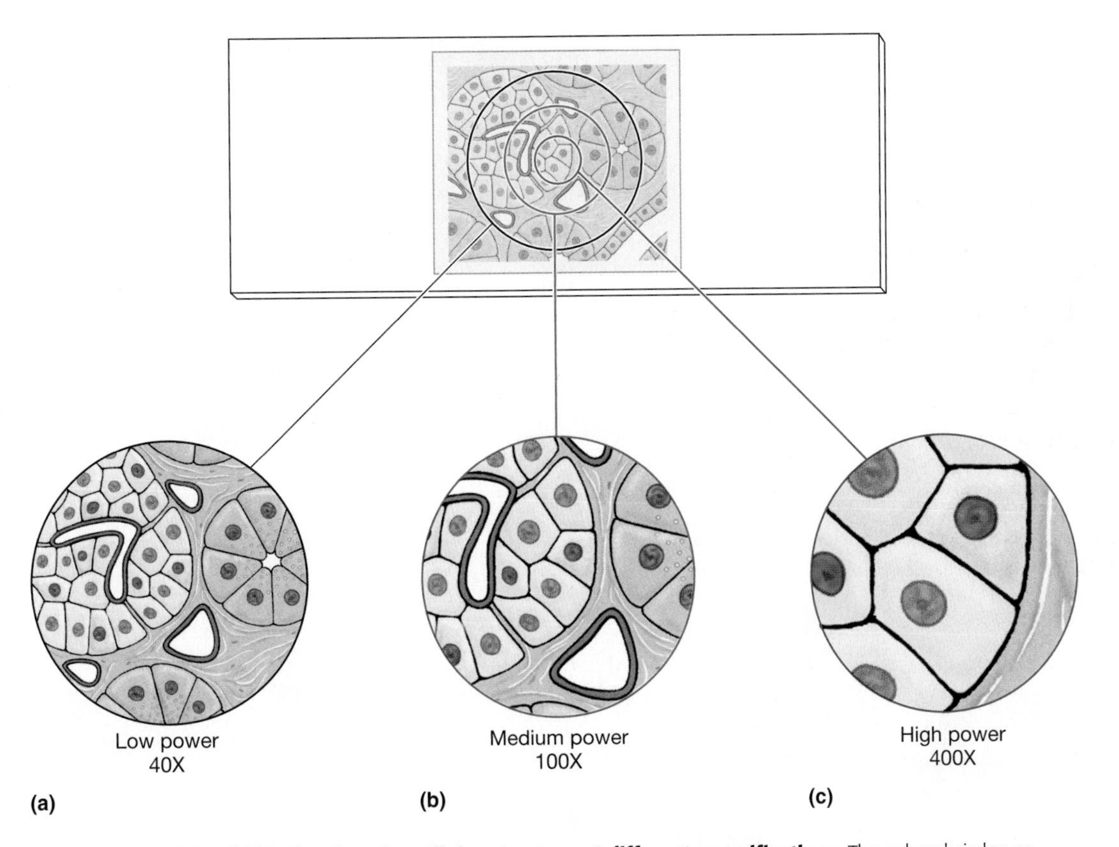

Figure 2.3 Measuring the size of a cellular structure at different magnifications. The colored circles on the slide represent the fields of view at different magnifications: **a)** the field of view represented by the black circle on the slide; **b)** the field of view represented by the red circle on the slide; **c)** the field of view represented by the blue circle on the slide. As the magnification increases, the field diameter decreases. Notice that the selected cells enclosed by the circle in **a** take up an increasingly larger proportion of the field of view as magnification increases.

14. Make sure that the structure you are viewing remains in the center of the field of view.
15. Switch to the high power objective lens and view the structure at this higher magnification (Figure 2.3c).
16. Once again, estimate the proportion of the field diameter across which the structure extends.
17. Predict the size of the structure by using the data recorded in Table 2.1. How does this calculation compare with your first two estimations of structure size with the scanning and low-power objective lenses?

____

____

QUESTION TO CONSIDER Review your results in Table 2.1. Test the relationship between total magnification and field diameter by inserting your data into the equation, $M_1D_1 = M_2D_2$. Do your results support the relationship? Explain.

____

____

____

Exercise 2 Review Sheet

Care and Use of the Compound Light Microscope

Name ______________________

Lab Section ______________________

Date ______________________

1. What is meant by resolving power?

Questions 2–6: Describe the function of the following structures.

2. Condenser lens

3. Objective lenses

4. Iris diaphragm

5. Coarse and fine adjustment knobs

6. Mechanical stage

7. You are viewing a microscope slide. The microscope you are using is equipped with an ocular lens that has a 10X magnification. The objective lens you are using has a 65X magnification. What is the total magnification at which you are viewing the slide?

8. Before placing a slide on the microscope stage, you observe the tissue with the naked eye and notice that it has the following shape.

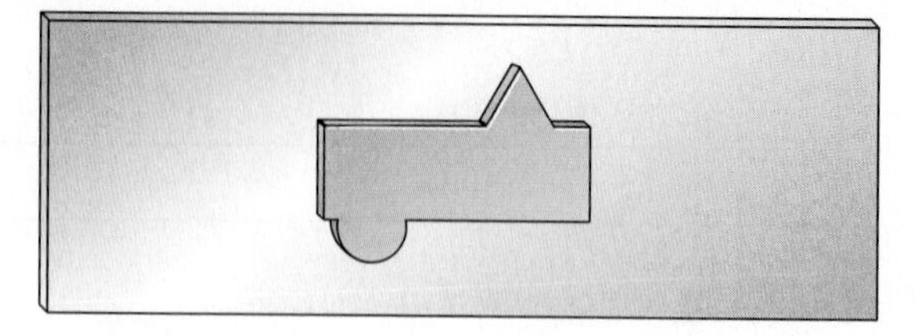

Next, you view the slide under low magnification so that the entire tissue section can be seen in the field of view. In the space below, illustrate how the tissue section will appear under the microscope.

9. If you are viewing a tissue section and decide to move from a lower to a higher magnification, how will the field of view change?

____

____

10. Why is it important to avoid touching the microscope lenses with your fingers?

____

____

11. Describe the correct procedure for cleaning a microscope lens (e.g., cleaning immersion oil from the oil immersion objective lens).

____

____

12. Explain what is meant by the depth of field.

____

____

Exercise 3

Cell Structure and Cell Division

Laboratory Objectives

On completion of the activities in this exercise, you will be able to:

- Provide details on the structure and organization of the cell membrane.
- Describe the structure and function of the cell nucleus.
- Identify the major cell organelles and explain their functions.
- Explain the events that define the life cycle of a cell.
- Identify and describe the stages of mitosis.

Materials

- Compound light microscopes
- Microscope slides
- Coverslips
- Toothpicks
- 10% methylene blue stain
- Prepared microscope slides of cheek cells
- Prepared microscope slides of various structures
- Animal cell model
- Electron micrographs of various cell structures
- Coloring pencils

The term *cell* was first used in 1665 when Robert Hooke was examining a thin slice of cork for a demonstration at the Royal Society of London. He observed evenly spaced rows of boxes that reminded him of the "cells," or living quarters, for monks in a monastery. Hooke did not actually observe living cells because the cork tissue was dead. All that remained was the cell wall, a structure found in plant cells and bacteria. With the advent of the first microscope in 1673, Anton van Leeuwenhoek was the first to identify living cells. However, significant advances in cell biology did not occur until the first part of the 19th century, when microscopes with stronger magnification and resolving power were developed.

At the beginning of the 19th century, medical studies on the human body paved the way for the development of what is known as **cell theory**. It was during this time that a fundamental understanding of cell structure and function began to emerge. Gradually, the work of various scientists contributed to the modern cell theory, which includes the following concepts.

- Cells are the structural building blocks of all living organisms.
- All cells arise from preexisting cells.
- The cell is the basic unit of life.
- In a multicellular organism, each cell maintains its own metabolism, independent of other cells, yet individual cells depend on other cells for survival.
- The activities of all cells in an organism are essential and highly coordinated.

Cell Structure

In your study of anatomy and physiology, you will focus on the structure and function of **eukaryotic cells**, complex cells with well-defined nuclei and specialized cell organelles, as these are the cells found in the organs and tissues of the human body.

WHAT'S IN A WORD The Greek term *karyon* refers to a nucleus and the prefix *eu-* means "good." A *eukaryotic* cell literally means a cell with "a good nucleus."

Human cells must be observed with the aid of a microscope. For example, even though their fibers can extend the entire length of the spinal cord, many nerve cells are too small to be seen without a microscope. Among the smallest cells in the body are red blood cells, which average about 7 micrometers (μm) in diameter (1 micrometer = 1/1000 of a millimeter).

Although they are quite variable in size, shape, and function, most cells share many common structures. Figure 3.1 illustrates and Table 3.1 describes the major components of eukaryotic cells.

The Cell Membrane

The **cell membrane** is composed of phospholipids, proteins, and carbohydrates and acts as a **selectively permeable** barrier between the cell and its external environment. At normal body temperature, the membrane has a fluid nature and, therefore, is very flexible. Figure 3.2 illustrates the structure of the cell membrane.

The **phospholipid bilayer**, composed of two layers of phospholipid molecules, is the primary component of the cell membrane. Layers are arranged so that the hydrophilic ("water-loving") head regions (phosphate groups), which readily interact with water, are on the outer and inner surfaces and in direct contact with the watery environments of the extracellular spaces and cytosol. The hydrophobic ("water-hating") tail regions, composed of two fatty acid molecules, do not interact with water and, therefore, are directed internally.

Figure 3.1 Basic structure of a eukaryotic cell. The nucleus contains DNA, the cell's genetic material. The cytoplasm contains various organelles, each with specialized functions.

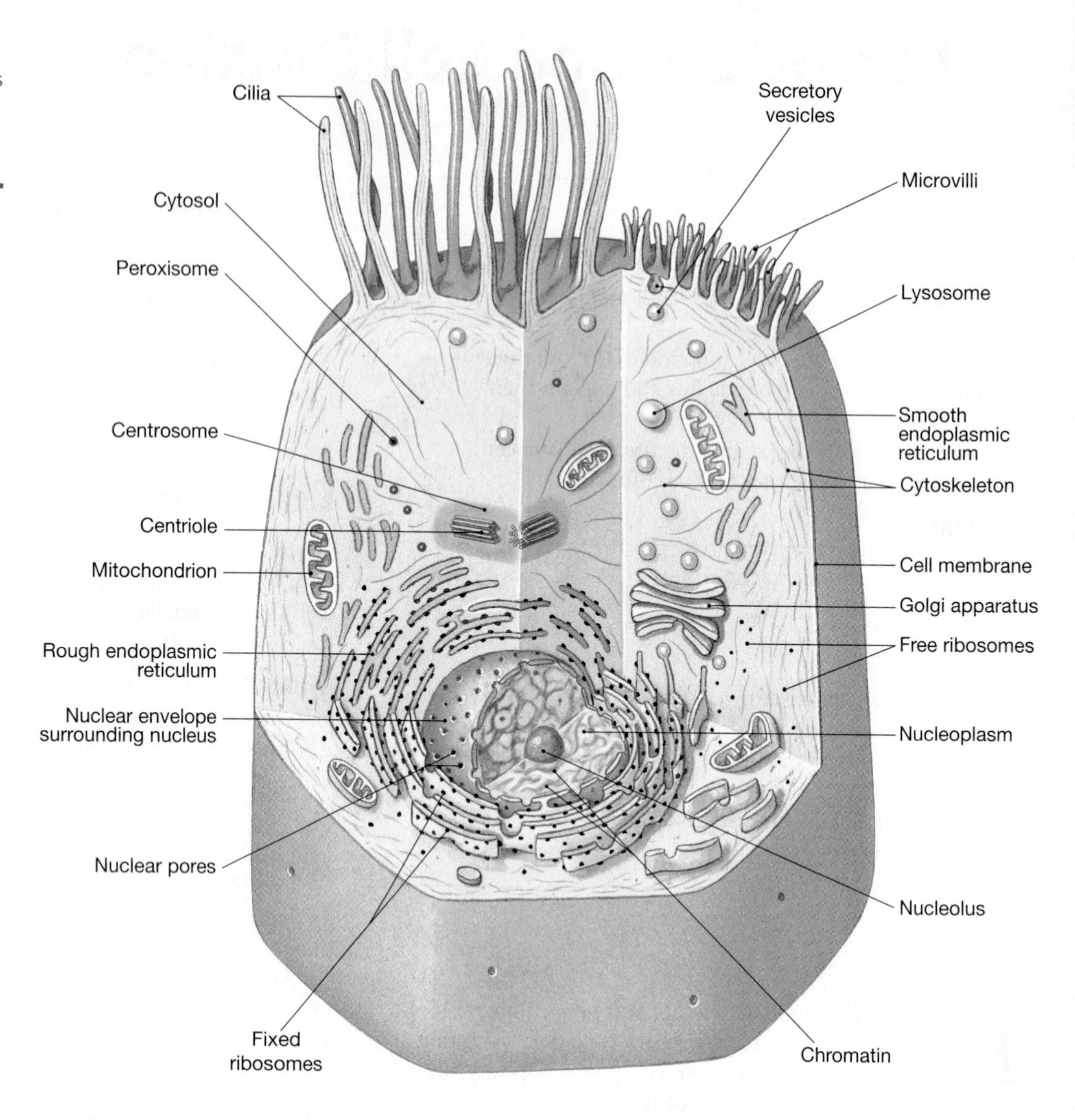

Cholesterol, a second lipid component, is interspersed between phospholipids in both layers of the membrane. Cholesterol provides some degree of stability to the phospholipid bilayer structure.

CLINICAL CORRELATION

Cholesterol can accumulate in our blood vessels, setting the stage for atherosclerosis, a disease in which the blood vessels become increasingly narrowed, reducing the blood flow through them and gradually starving the tissues that they supply. This is a leading cause of heart attacks and strokes. Plant cells, unlike animal cells, lack cholesterol in their membranes. This structural difference is one reason why people who are concerned about high cholesterol should add more fruits and vegetables to their diets.

In addition to lipids, the membrane also contains various types of proteins. **Integral proteins** are firmly inserted into the lipid bilayer. The majority of these proteins pass through the entire width of the membrane and protrude from each surface (**transmembrane proteins.**) Other integral proteins are only partially embedded into the outer or inner surfaces. Many integral proteins serve as membrane transporters that allow water, ions, and other larger molecules to pass from one side of the membrane to the other. Others are receptors that allow specific molecules, such as hormones, to attach to the cell membrane.

In contrast to integral proteins, **peripheral proteins** do not have a strong attachment to the membrane, but instead rest loosely on the inner surface. An important group of peripheral proteins are the filaments located along the inside surface of the membrane. These structures form a supporting meshwork of fibers that prevent the membrane from breaking apart.

Carbohydrate molecules are found only along the outside surface of the cell membrane. They are attached either to integral proteins, forming **glycoproteins**, or to lipids, forming **glycolipids**. Carbohydrate molecules arranged in this way form a sticky coat along the outer surface known as the **glycocalyx**. Some carbohydrate molecules are receptors while others form recognition sites that allow cells to identify each other. For example, this al-

Table 3.1 Functional Summary of Major Cellular Structures

Structure	Function(s)
I. Nucleus	
1. Nucleolus	Production of ribosomes.
2. Chromatin	Contains DNA, which is the genetic material for the cell.
3. Nuclear envelope	a. Separates nucleoplasm from cytoplasm. b. Regulates passage of material into and out of the nucleus.
II. Cell membrane	a. Separates cellular structures from the tissues and fluids outside the cell. b. Regulates the transport of substances into and out of the cell.
III. Cytoplasm	
1. Membranous organelles	
a. Rough endoplasmic reticulum (rER)	Produces proteins that are secreted by the cell, incorporated into the cell membrane, or used by lysosomes.
b. Smooth endoplasmic reticulum (sER)	a. Synthesis of lipids and steroids. b. Detoxification of poisons and drugs.
c. Golgi apparatus	Modifies and packages proteins that are produced by the rER.
d. Transport vesicles	Transport proteins from the rER to the Golgi apparatus.
e. Secretory vesicles	Transport proteins from the Golgi apparatus to the cell membrane for cellular secretion.
f. Lysosomes	a. Digest old, worn-out organelles. b. Destroy harmful bacteria, viruses, and toxins.
g. Peroxisomes	a. Neutralize toxins produced by cellular metabolism or taken in from the outside. b. Break down fatty acids.
h. Mitochondria	Produce most of the cell's energy in the form of ATP.
2. Nonmembranous organelles	
a. Free ribosomes	Produce proteins used for cellular metabolism.
b. Cytoskeleton	Provides strength and flexibility to the cell and support for the various other organelles.
c. Centrioles	a. Produce microtubules for the cytoskeleton. b. Form the bases from which cilia and flagella are produced. c. Form the mitotic spindle for mitosis.
d. Cilia	Moves substances over cell surfaces.
e. Flagella	Used for cell locomotion; in humans, found only on sperm cells.
f. Microvilli	Increases surface area along cell membrane of cells that absorb substances.

lows cells in the immune system to recognize and not attack your body's own cells.

WHAT'S IN A WORD The prefix *glyco-* is derived from the Greek word *glykys*, which means "sweet." Any term with this prefix refers to a substance that contains sugar (carbohydrate) molecules.

The Nucleus

The **nucleus** is typically the largest structure inside the cell (Figure 3.1). It contains the cell's DNA and directs all cellular activities. The nucleus is surrounded by a **nuclear envelope** of a double phospholipid bilayer. This double membrane is dotted with numerous openings called **nuclear pores**, which allow the passage of various substances such as ions, proteins, and RNA. The gel-like matrix in the nucleus is called the **nucleoplasm**. Suspended in the nucleoplasm is the **nucleolus**, where **ribosomes** are made. The nucleus also contains **chromatin**, made of complex molecules of DNA and proteins. In nondividing cells, the chromatin is stretched out and appears as fine threads throughout the nucleoplasm. In a dividing cell, the chromatin condenses into discrete structures known as **chromosomes**.

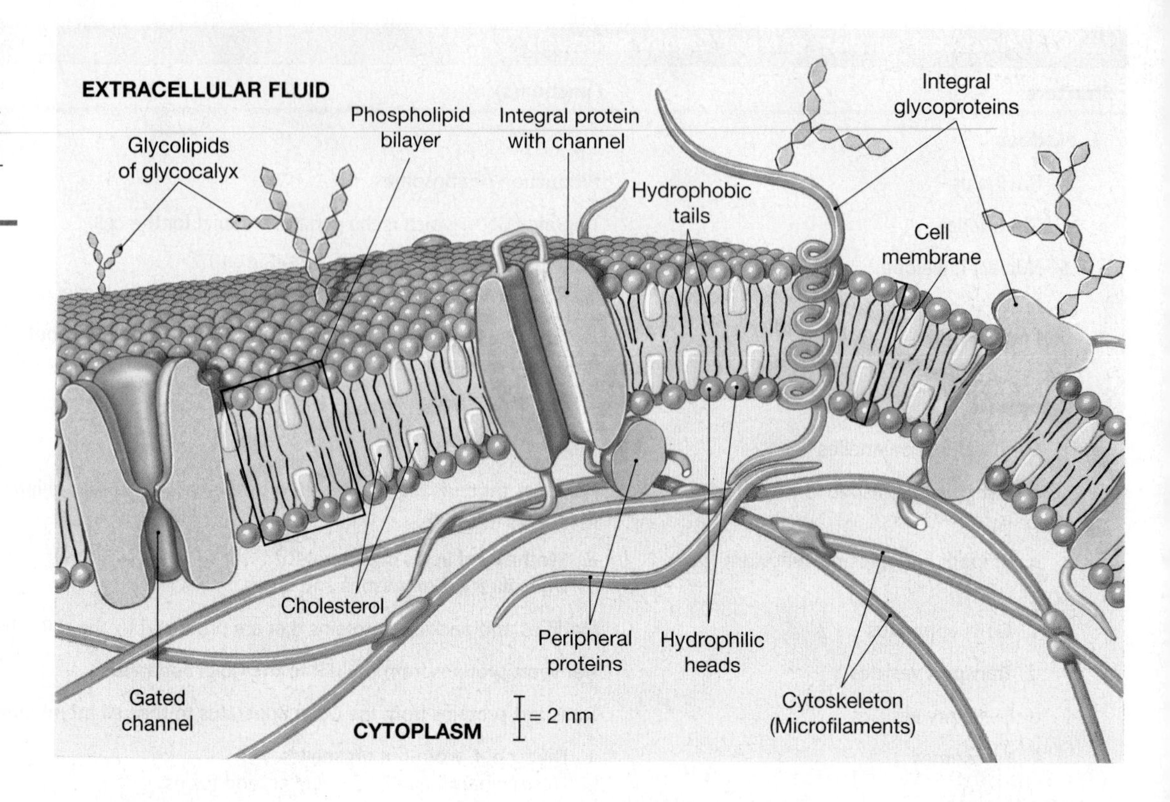

Figure 3.2 Molecular components of the cell membrane. The basic structural feature of the membrane is the phospholipid bilayer. Other components include proteins, cholesterol, and carbohydrates.

The Cytoplasm

The **cytoplasm** is the gel-like cell matrix located outside the nucleus. It consists of a fluid portion, the **cytosol**, and a particulate portion that includes the various **organelles** (Figure 3.1). The cell organelles are divided into two main categories.

1. **Membranous organelles** are surrounded by a phospholipid membrane, similar in structure to the cell membrane. These membrane-bound structures have internal compartments with chemical environments that may be different than the surrounding cytosol.
2. **Nonmembranous organelles** are not surrounded by a membrane and, thus, are in direct contact with the cytosol.

Most eukaryotic cells contain the following organelles, which are illustrated in Figure 3.1 and summarized in Table 3.1.

Membranous Organelles

1. The **endoplasmic reticulum (ER)** is a highly folded membrane structure that encloses a network of fluid-filled cavities called **cisternae. Rough endoplasmic reticulum (rER)** is continuous with the nuclear membrane and is studded with numerous **ribosomes**, the sites of protein synthesis. The rER manufactures proteins that will be secreted by the cell, added to the cell membrane, or used as digestive enzymes within **lysosomes** (see below). Smooth **endoplasmic reticulum (sER)** is a continuation of the rER. Lacking ribosomes, the sER has no role in protein synthesis. Rather, the sER synthesizes lipids and detoxifies poisons and various drugs.
2. The **Golgi apparatus (Golgi complex)** is a series of flattened membranous sacs, resembling a stack of pancakes. Proteins, produced in the rER, are packaged in **transport vesicles** and transferred to the Golgi apparatus. Inside the Golgi apparatus, the proteins are modified into final protein products and repackaged in **secretory vesicles.** The secretory vesicles transport the proteins to the cell membrane where they are secreted or incorporated into the membrane itself (Figure 3.3). Other vesicles that pinch off from the Golgi apparatus transport digestive enzymes to the lysosomes.
3. **Lysosomes** are filled with digestive (hydrolytic) enzymes. These organelles are responsible for digesting away old, worn-out organelles, and destroying potentially harmful bacteria, viruses, and toxins.
4. **Peroxisomes** contain enzymes that neutralize toxins produced by cellular metabolism or taken in from the outside (e.g., alcohol). They are also involved in the breakdown of fatty acids. A by-product of their activity is hydrogen peroxide, which is toxic to cells. Peroxisomes convert hydrogen peroxide to oxygen and water by using the enzyme **catalase.**

CLINICAL CORRELATION

We often use hydrogen peroxide to cleanse small wounds. The product we buy at the store is a very weak solution, usually only 3% to 5% (meaning it is 95% to 97% water). The bubbling action that you see is caused when the peroxide contacts

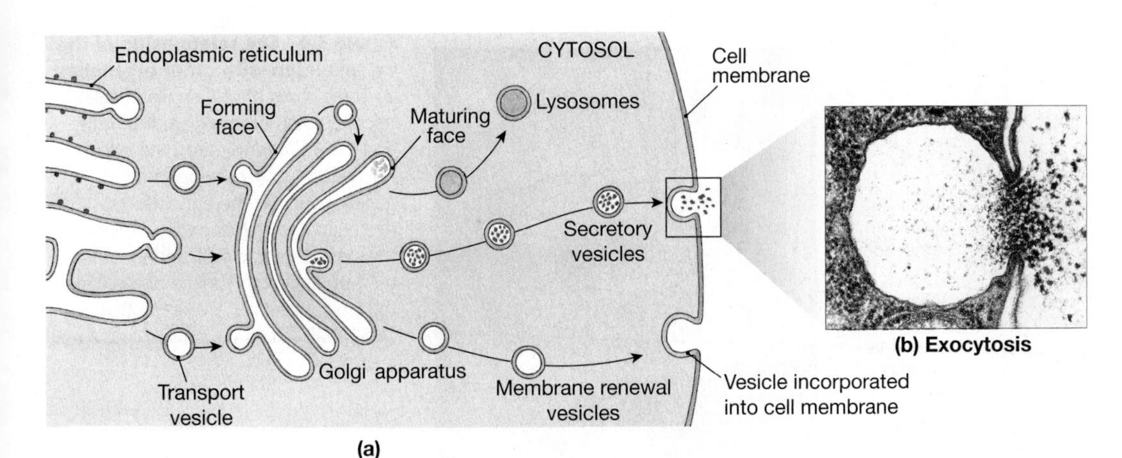

Figure 3.3 The cellular secretory process. a) Cell organelles that have a role in cell secretion; **b)** secretory vesicles release proteins at the cell membrane by exocytosis.

catalase that is released from damaged cells. The enzyme causes the formation of oxygen gas bubbles. This bubbling action helps dislodge debris that enters a wound site and could lead to an infection.

5. **Mitochondria** are rodlike structures surrounded by a double membrane, similar to the nuclear membrane. The inner membrane is unique because of its numerous inward folds known as **cristae.** The gel-like material enclosed by the inner membrane is called the **matrix.** The matrix contains DNA and ribosomes, so mitochondria can synthesize some of their own proteins.

   Mitochondria produce the majority of the cell's energy through the breakdown of organic molecules and subsequent release of chemical bond energy. This extracted energy is used to make **adenosine triphosphate (ATP)** molecules, the fuel for cellular metabolism.

Nonmembranous Organelles

1. In addition to the ribosomes attached to the rER, cells also have vast amounts of **free ribosomes.** These structures synthesize proteins for metabolic processes within the cytosol.
2. The internal framework of the cell is called the **cytoskeleton.** It consists of three types of protein filaments that provide strength and flexibility to the cell and support for the organelles. The three filament types are **microtubules, intermediate filaments,** and **microfilaments** (Figure 3.4).
3. **Centrioles** are paired cylindrical structures, arranged at right angles to each other, and located at one end of the nucleus in an area of cytoplasm known as the **centrosome.** Each centriole is composed of nine triplets of microtubules that are arranged to form a tube or cylinder (Figure 3.5a). Centrioles produce the microtubules of the cytoskeleton and the microtubule spindle used during mitosis and meiosis. They also form the bases (**basal bodies**) from which cilia and flagella are produced.
4. **Cilia** and **flagella** are surface projections composed of nine pairs of microtubules surrounded by a central pair (Figure 3.5b). Cilia are short projections that propel substances over the cell surface. Flagella are long projections used for cell locomotion. Sperm cells are the only human cells that possess flagella.
5. **Microvilli** are tiny, fingerlike projections that greatly increase the surface area of the cell membrane. They are prominent in cells that absorb substances, such as the absorptive cells lining the small intestine. Microvilli contain bundles of microfilaments which are anchored to the **terminal web,** a filamentous band that runs just below the cell surface (Figure 3.4).

   Table 3.1 summarizes functions of the major structures found in a eukaryotic cell.

ACTIVITY 3.1 Preparing a Wet Mount of Cheek Epithelial Cells

1. Gently scrape the inner lining of your cheek with the broad end of a flat toothpick.
2. Stir the toothpick vigorously in a drop of distilled water on a clean microscope slide.
3. Gently scrape your cheek two or three more times and stir into the water drop. Be sure to use a new toothpick each time.
4. Cover the drop with a coverslip lowered onto the slide at an angle to avoid forming air bubbles.
5. Observe with the light microscope, first on low and then on high power, and sketch what you observe in the space below.

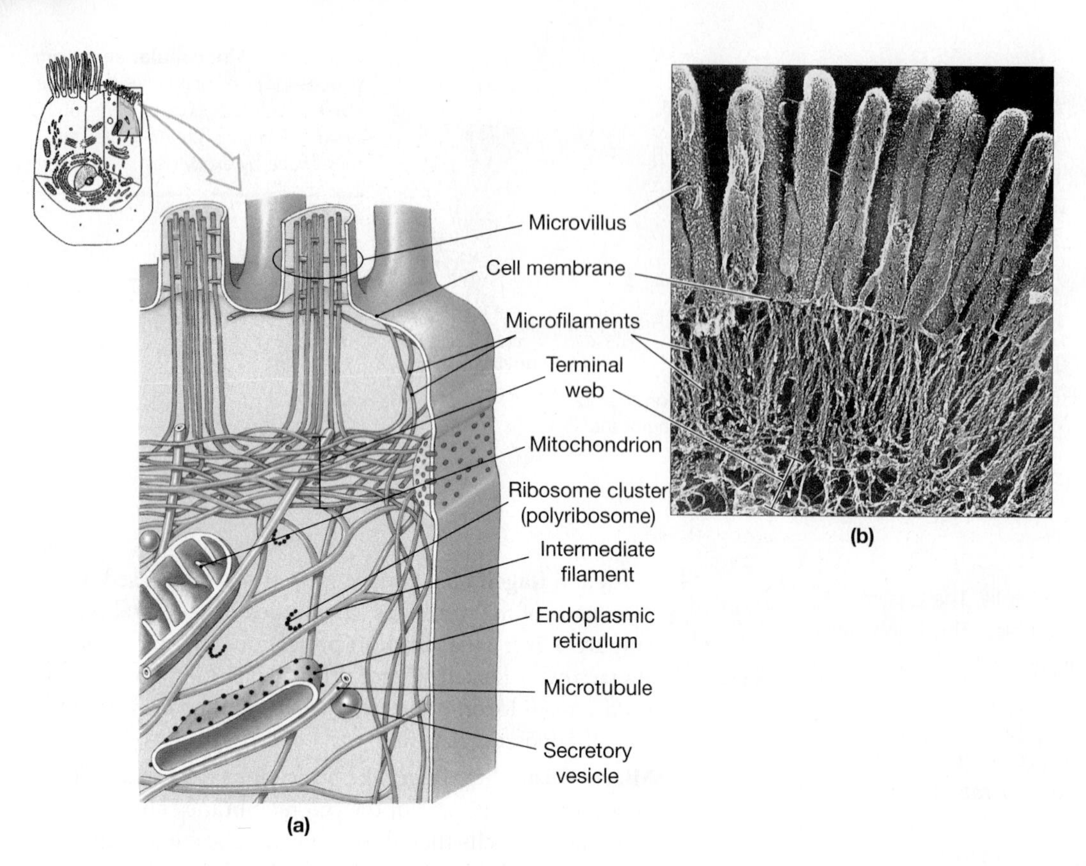

Figure 3.4 The relationship of the cytoskeleton with other organelles. a) Three types of protein filaments—microtubules, intermediate filaments, and microfilaments—form the cytoskeleton, which provides support for organelles such as the endoplasmic reticulum and mitochondria. **b)** Electron micrograph of microvilli and the terminal web along the surface of a cell in the small intestine (LM × 80000).

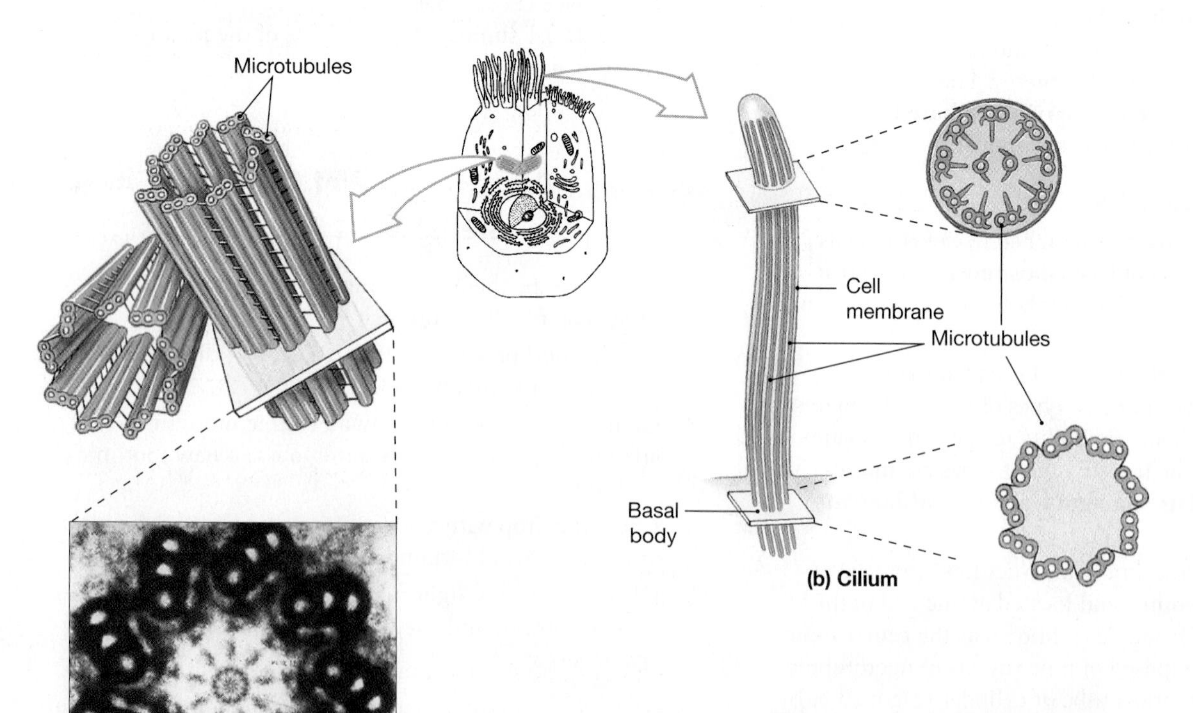

Figure 3.5 The structure of centrioles, basal bodies, and cilia. a) A pair of centrioles arranged at right angles to each other. In cross section, nine triplets of microtubules can be identified in each centriole (LM × 150000). **b)** A cilium arising from a basal body. Each cilium consists of nine pairs of microtubules that surround a central pair. The basal body has a microtubular arrangement that is similar to a centriole.

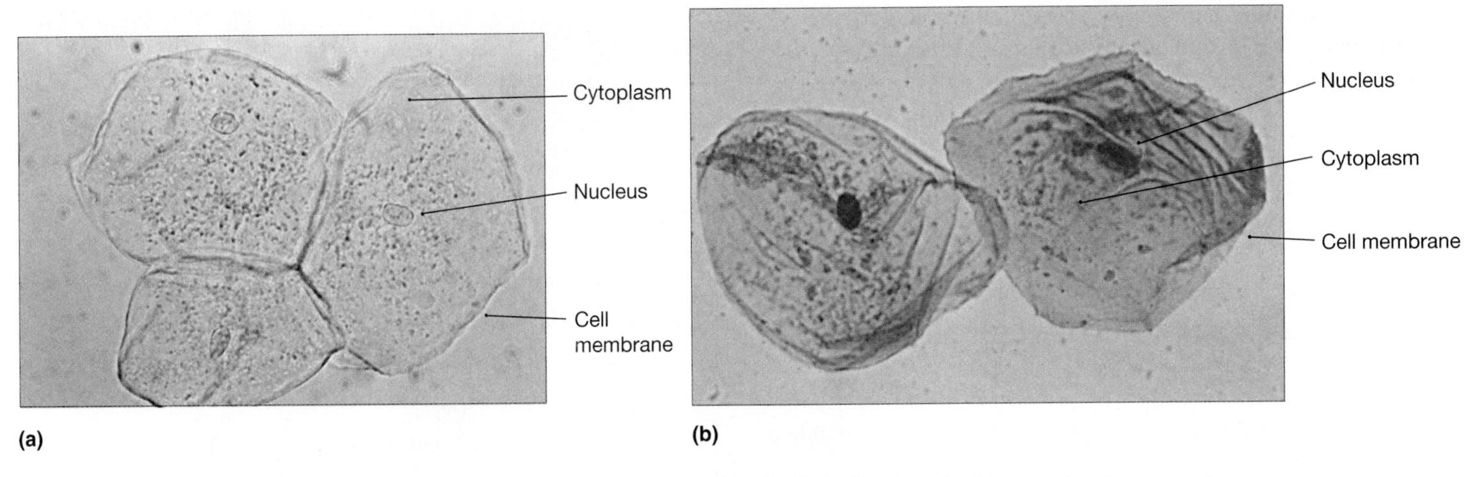

Figure 3.6 Light microscopic views of human cheek epithelial cells. These flat (squamous) cells are part of the mucous membrane that covers the oral cavity. **a)** Unstained cells (LM × 500); **b)** cells stained with methylene blue (LM × 500).

6. Repeat steps 1 through 3, this time adding a drop of 10% methylene blue stain to your preparation.
7. Cover the preparation with a coverslip and observe the cells with the microscope. If there is too much liquid on the slide, place a paper towel at the edge of the coverslip to wick some of the excess away.
8. Add to your sketch any additional structures that you see in the stained cells.
9. Under high power, identify the following structures.
   a. Nucleus
   b. Nucleolus
   c. Chromatin
   d. Nuclear membrane
   e. Cytoplasm
   f. Cell membrane
10. Compare your observations of your own cheek cells with a prepared slide of the same cell type, or with the photographs in Figure 3.6.

QUESTIONS TO CONSIDER 1. What effect did methylene blue have on the cells in your preparation?

____

____

2. What advantage may be gained by staining cells?

____

____

ACTIVITY 3.2 Light Microscopic Observations of Various Cell Types

1. Observe other cell types by studying several prepared slides provided in the laboratory.
2. In the space below, make sketches of the cells and identify any structures that you recognize. For each drawing, identify the tissue or organ that you observed.

QUESTION TO CONSIDER Make a list of the different cell types that you observed in the previous activity and identify one unique characteristic in each.

Cell Type	Unique Characteristic

ACTIVITY 3.3 Examining a Model of a Typical Eukaryotic Cell

1. Study the model of a typical eukaryotic animal cell. If a model is not available, refer to Figure 3.1 or other illustrations that are available in the laboratory. The model (or figure) illustrates most of the organelles described earlier.
2. Try to correlate what you see on the model (or figure) with your previous observations with the light microscope.

QUESTION TO CONSIDER Examining a model of the cell allows you to study the spatial arrangements of the nucleus and organelles in three dimensions. How does this compare with the two-dimensional view that you observed with the light microscope? Are there any advantages to seeing structures in three dimensions? Explain.

____

____

____

ACTIVITY 3.4 Electron Microscopic Observations of Cells

1. Examine electron micrographs of cells that are available in the laboratory.
2. Note the difference in magnification between these micrographs and the observations you made during the microscope activities that you completed earlier.

   Maximum magnification with your microscope:

   ___________×

   Minimum magnification of the electron micrographs you view: ___________×

   Maximum magnification of the electron micrographs you view: ___________×

3. Attempt to identify the structures that are discussed in the figure legends that accompany the micrographs.

QUESTION TO CONSIDER Make a comparison of the cell structural detail that you observed at the light microscopic and electron microscopic levels. What structures are you able to examine in greater depth with the electron microscope than with the light microscope?

____

____

____

The Cell Cycle and Cell Division

The regularly recurring series of life processes that are performed by any cell is called the **cell cycle**. Each cycle begins when the cell is formed and is completed when the cell divides. The cell cycle is divided into two time periods (Figure 3.7).

1. **Interphase** is the period when the cell is not dividing.
2. **Cell division** is the period during which identical copies of the original cell are produced. Cell division includes two processes.
   - **Mitosis**—the division of a cell's nuclear material
   - **Cytokinesis**—the division of cytoplasmic components

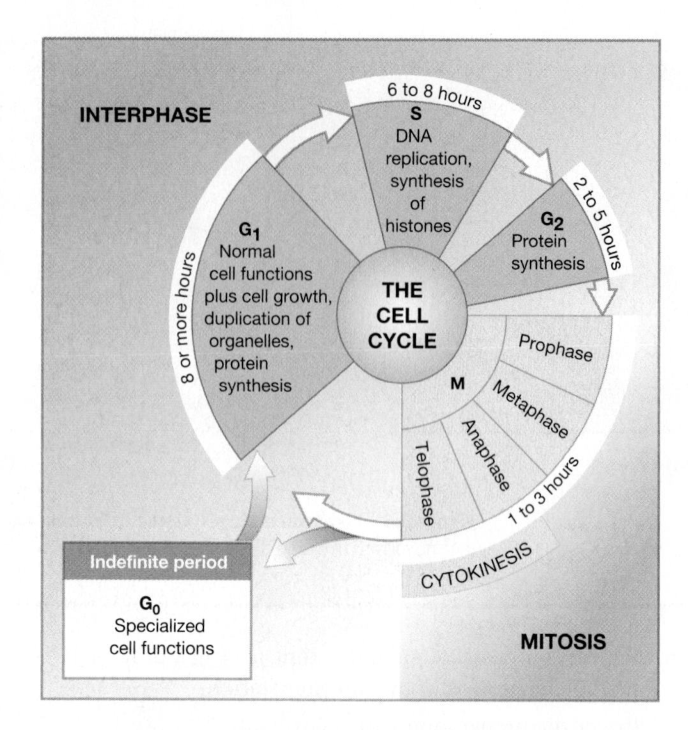

Figure 3.7 The generalized life cycle of a cell. Interphase includes the G_1, S, and G_2 phases. Mitosis is divided into four stages: prophase, metaphase, anaphase, and telophase. Highly specialized cells that do not divide (such as most nerve cells and skeletal muscle cells) remain in the G_0 stage indefinitely.

During interphase, a cell conducts its normal metabolic activities and prepares itself for cell division. This portion of the cycle consists of three major divisions.

1. During the **G_1 phase** (G = growth), the cell conducts its regular cellular activities. The cell prepares for DNA replication and cell division by beginning to replicate its centrioles and cell organelles. The duration of G_1 is quite variable and its length determines the timing of the entire cell cycle. For example, in rapidly dividing cells, such as skin cells, G_1 lasts only a few hours; in slowly dividing cells, such as liver or kidney cells, it can last weeks or months. Highly specialized cells, such as most nerve cells, never divide and are said to be in **G_0 phase.**
2. During the **S phase** (S = synthesis), the DNA replicates itself. This process must occur if the two new cells are to receive identical copies of the genetic material. The S phase typically lasts from 6 to 8 hours.
3. During the **G_2 phase**, enzymes and other proteins that are needed for cell division are produced. Centriole replication, which began during G_1, is completed. This phase usually lasts 2 to 5 hours.

At the conclusion of interphase, the cell enters the mitotic (M) phase. Mitosis is analogous to photocopying, in that cells divide to produce genetically identical copies of themselves. For one-celled organisms, mitosis is a form of reproduction, allowing a population to replace its members and increase its numbers. In multicellular organisms, such as humans, mitosis provides growth, wound

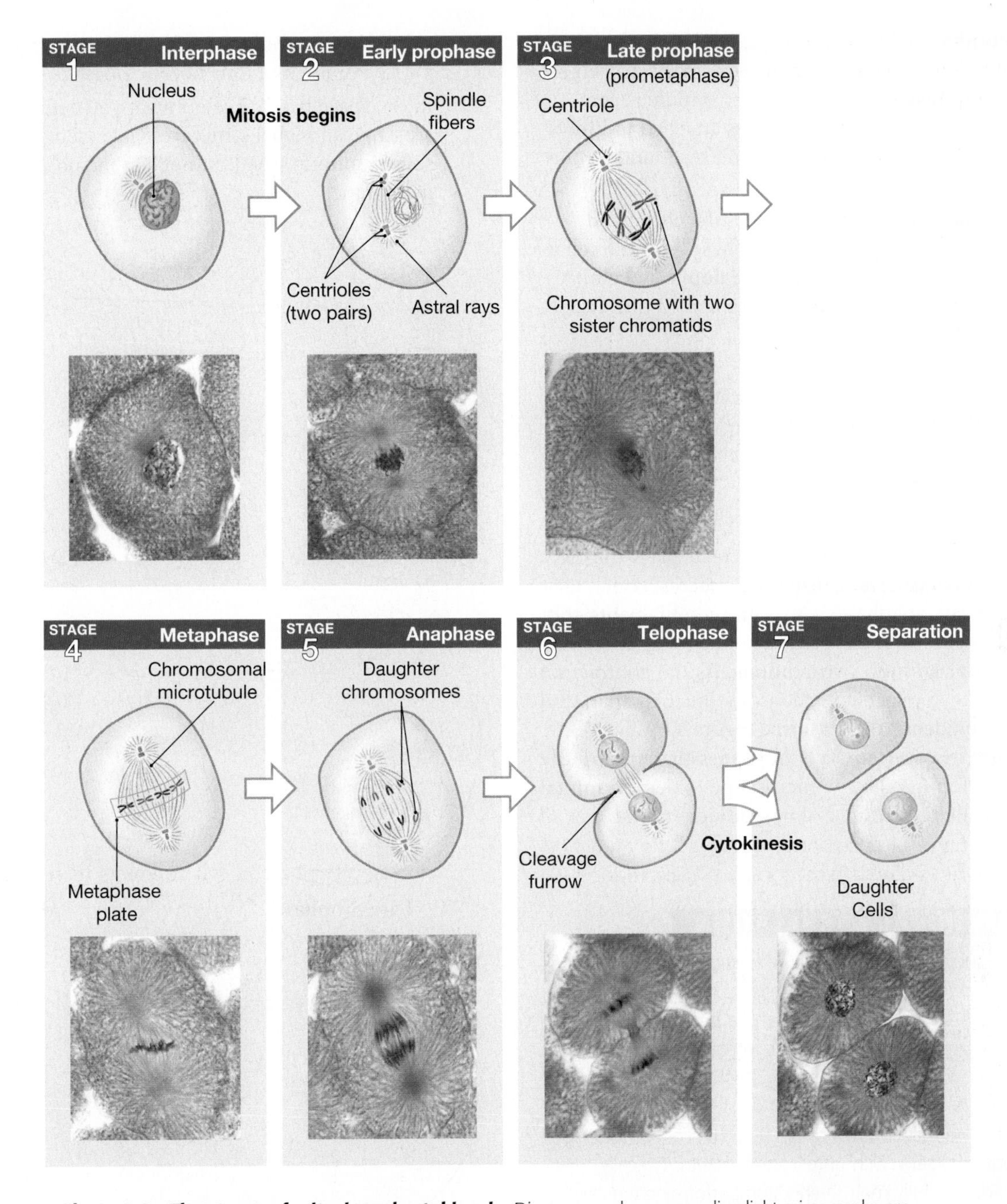

Figure 3.8 The stages of mitosis and cytokinesis. Diagrams and corresponding light micrographs are whitefish blastula cells. For simplicity, only two pairs of chromosomes are shown in the illustrations. **Stage 1:** interphase (LM × 500); **stages 2** and **3:** Prophase (LM × 500); **stage 4:** metaphase (LM × 500); **stage 5:** anaphase (LM × 500); **stage 6:** telophase and the beginning of cytokinesis (LM × 500); **stage 7:** the end of cytokinesis and the separation of the two new cells (LM × 500).

healing, and remodeling. In the embryo, mitosis is the primary mechanism for developing new structures and growth of the individual. Mitosis takes 1 to 3 hours to complete and is subdivided into four major phases.

1. **Prophase** is the longest phase in mitosis. For descriptive purposes, it is often divided into two smaller periods: **early prophase** (Figure 3.8, stage 2) and **late prophase** (Figure 3.8, stage 3). During early prophase, the chromatin in the nucleus becomes highly coiled and condensed, forming **chromosomes**. Because the DNA has replicated, each chromosome consists of two identical **chromatids**, held together at a region known as the **centromere**. Other events that occur include the disappearance of the nucleolus, separation of the centrioles, and development of microtubules for the formation of the **mitotic spindle (spindle fibers)**. The mitotic spindle is used to separate the two chromatids on each chromosome. During late prophase, the centrioles continue to separate by migrating to opposite ends of the cell. The nuclear membrane breaks apart and the mitotic spindle is completed.

2. During **metaphase** (Figure 3.8, stage 4), the chromosomes line up, end to end, along the equator of the cell to form the **metaphase plate.** The mitotic spindle is completely formed with some microtubules attached to **kinetochores**, which are DNA-protein complexes found on the centromere of each chromosome.
3. During **anaphase** (Figure 3.8, stage 5), the chromatid pairs separate and are pulled to opposite ends of the cell as the mitotic spindles shorten. Thus, the duplicated chromosomes of the original cell become separated into two identical sets of chromosomes.
4. During **telophase** (Figure 3.8, stage 6), a nuclear membrane forms around each new set of chromosomes and a nucleolus develops in each new nucleus. The chromosomes unravel to form fine threads of chromatin and the microtubular spindle disappears.

The formation of two new cells is completed when the cytoplasm is divided into two approximately equal parts during a process known as **cytokinesis.** During this process, which usually begins during late anaphase, a ring of microfilaments contracts along the cell's equator (cleavage furrow), effectively squeezing the cytoplasm into two portions. As the contraction continues, the two cytoplasmic portions are pinched apart and two new genetically identical cells form (Figure 3.8, stage 7).

Mitosis occurs most extensively in tissues that are rapidly growing and developing. Thus, microscope slides of animal embryos are excellent resources for studying the stages of mitosis.

CLINICAL CORRELATION

Cells usually are subjected to local and genetic controls on their rates of growth and reproduction. If these rates increase, the tissue where the cells are located enlarges and a **benign tumor** may develop. The growth of these tumors remains restricted within a connective-tissue capsule, and thus usually cause no harm. **Cancer** refers to a large number of disorders in which mutations disrupt the normal controls of cell reproduction, producing increasingly abnormal cells, with unrestricted growth. Typically, cancerous cells do not fully mature and stop performing their normal functions. These cells produce **malignant tumors,** and over time replace healthy cells and prevent organs from functioning properly. Some cancerous cells may break away from the primary tumor and spread, often through the bloodstream, to other locations in a process called **metastasis.**

ACTIVITY 3.5 Examining the Stages of Mitosis

1. Examine a prepared slide that illustrates mitosis in the whitefish **blastula.** A blastula is an early embryological stage of the vertebrate embryo. It is a rapidly growing structure, so mitotic cells will be abundant. You should be able to identify cells in interphase and all phases of mitosis. As you examine the slides, use Figure 3.8 as a guide for identifying the four mitotic phases.
2. In the space below, sketch what you actually see in the microscope and label structures (if present) such as chromosomes, mitotic spindles, nuclear membrane, and cell membrane.

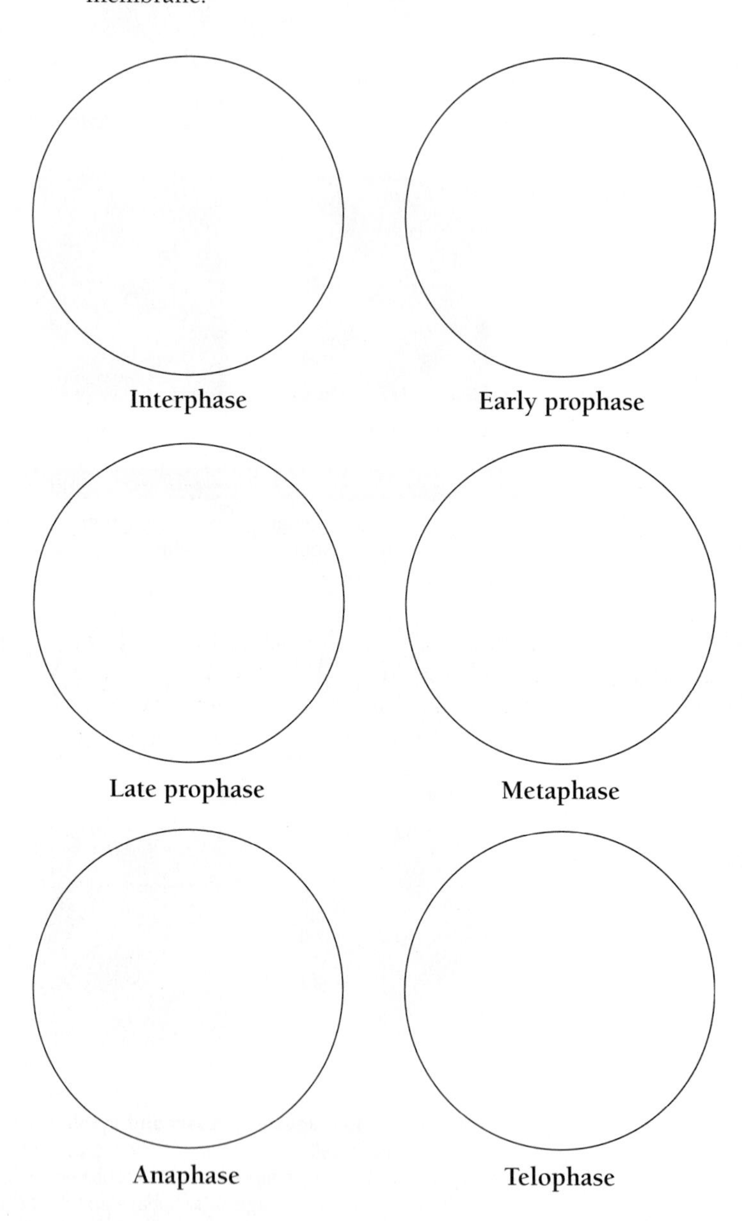

QUESTION TO CONSIDER Explain why the cells in the epidermis (superficial layer) of your skin must be constantly dividing by mitosis. (Hint: What happens when you scratch your skin?)

____

____

____

Exercise 3 Review Sheet

Cell Structure and Cell Division

Name ______________________

Lab Section ______________________

Date ______________________

1. What are the main points of the cell theory?

____

____

____

2. Briefly describe the structure of the cell membrane. Why is it important for normal cell function?

____

____

3. Discuss the difference between a membranous organelle and a nonmembranous organelle. Provide examples of each type.

____

____

____

Questions 4–10: Identify the labeled structures in the diagram by writing the name next to the appropriate number in column A of the table on page 38. In column B, describe the function of the structure. Highlight the structure with the color indicated in column C.

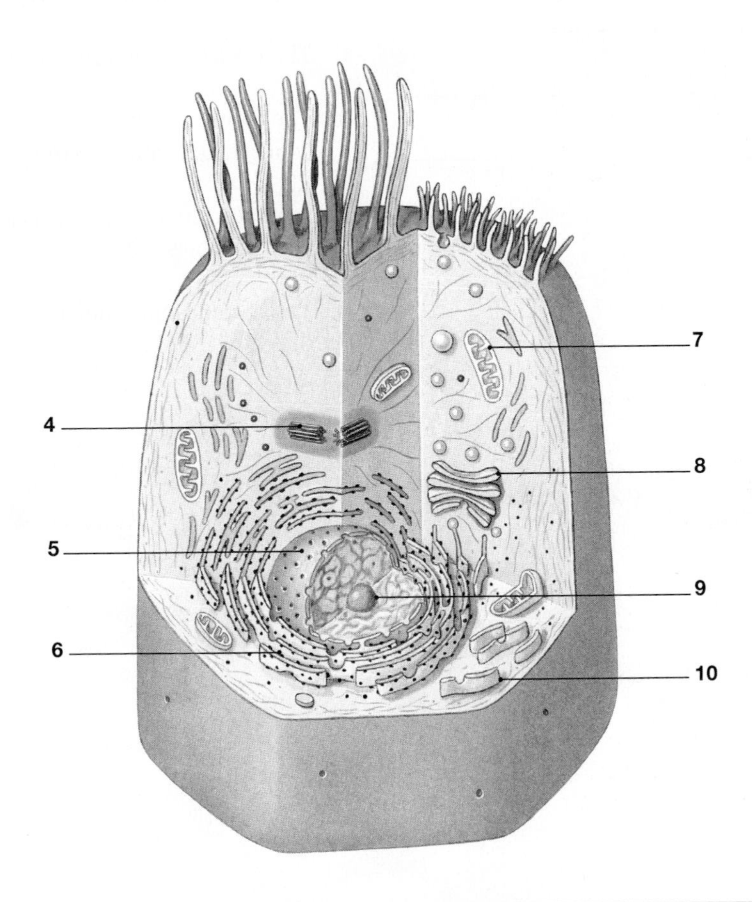

A. Structure	B. Function	C. Color
4.		green
5.		yellow
6.		red
7.		blue
8.		brown
9.		purple
10.		orange

11. Describe the various phases of interphase. What types of cells enter the G_0 phase?

___

___

___

Questions 12–16: Match the phase of mitosis in column A with the appropriate event in column B.

A

12. Early prophase ___
13. Late prophase ___
14. Metaphase ___
15. Anaphase ___
16. Telophase ___

B

a. The chromatid pairs separate and are pulled to opposite ends of the cell.
b. The DNA in the nucleus replicates.
c. The nuclear membrane breaks down.
d. The chromatin molecules in the nucleus become highly condensed, forming chromosomes.
e. A nuclear membrane forms around each new set of chromosomes.
f. The chromosomes line up, end to end, along the equator of the cell.

Exercise 4

Membrane Transport

Laboratory Objectives

On completion of the activities in this exercise, you will be able to:

- Differentiate between the various transport processes that occur across cell membranes.
- Predict which type of molecules and materials will be moved by each movement process.
- Predict the direction and rate of movement when provided information about membrane permeability, concentration gradient, and temperature.
- Discuss osmosis in terms of hypotonic, hypertonic, and isotonic solutions.
- Complete a written laboratory report.

Materials

- Clean white paper
- 3 petri dishes
- Pencil
- Distilled water near freezing (0° to 10° C)
- Distilled water at room temperature (about 23° C)
- Distilled water near boiling (90° to 100° C)
- Methylene blue or toluidine blue dye
- Millimeter ruler
- Pipettes
- Dialysis tubing
- Water
- 10% starch solution
- Iodine potassium iodide (IKI) solution
- Elastic bands
- 500 ml beaker
- Ring stand with clamp
- Thistle tube
- 25% and 50% molasses solutions
- Marking pen

The cell membrane is a selectively permeable barrier, which means that some substances are allowed to pass across the membrane while others are not. For example, the cell membrane will allow various nutrients, such as glucose and oxygen, to pass into the cell, but will block various toxins. Metabolic cellular wastes, such as lactic acid and carbon dioxide, can freely exit the cell, but vital enzymes are prevented from leaving. Thus, the cell membrane plays the vital role of regulating the substances that enter and exit the cell.

The movement of substances across a cell membrane can occur by either **passive** or **active processes.** Passive processes do not require input of cellular energy (ATP). Rather, they are driven by the constant motion of molecules (kinetic energy) and the tendency of substances to move along a **concentration gradient**, from an area of high concentration to an area of low concentration. In cells, for example, if the concentration of a substance in the cytoplasm (intracellular fluid) is greater than the concentration in the extracellular fluid, that substance will tend to move from the cytoplasm to the extracellular fluid. However, this will only occur if the cell membrane is permeable to that substance.

These processes will proceed until equilibrium is reached, a condition at which there is no longer a concentration gradient. When that occurs, the molecules and atoms continue to move spontaneously, but equally in all directions. Without the concentration gradient, there is no longer a net directional movement.

Passive processes include the following mechanisms (Table 4.1).

- **Simple diffusion** (Figure 4.1)
- **Osmosis** (Figure 4.2)
- **Facilitated diffusion** (Figure 4.3)
- **Filtration**

Active processes are transport mechanisms that require cellular energy (ATP). Active processes are used to transport substances that are too large to pass through membrane pores, are not soluble in lipids, or are being transported against a concentration gradient, from an area of low concentration to an area of high concentration. Active processes include the following mechanisms (Table 4.1).

- **Active transport** (Figure 4.4).
- **Vesicular transport:** Endocytosis (Figure 4.5) and exocytosis (Figure 4.6).

WHAT'S IN A WORD The term *endocytosis* is derived from three Greek words. *Endo* means "within" (from the word *endon*), *cyto* refers to a cell (from the word, *kytos*), and *osis* means "condition." If you put the three words together, endocytosis literally means "within the cell's condition," and refers to movement into the cell. In the word *exocytosis,* the prefix *exo* is the Greek word for "outside." Exocytosis means "outside the cell's condition," and refers to movement into the cell.

Because active transport requires cellular energy, and vesicular transport requires living cells and vesicle formation, we will not examine them in this laboratory exercise. Rather, we will focus on two of the passive processes: diffusion and osmosis. The following procedures are designed to demonstrate the principles of movement through membranes. Although you will work with a partner to collect data, your instructor may require each of you

Table 4.1 Membrane Transport Mechanisms

Mechanism	Description	Examples
Passive processes (ATP not required)		
1. Simple diffusion (Figure 4.1)	Net movement of a substance from a region of high concentration to a region of low concentration.	Transport of oxygen and carbon dioxide across a cell membrane.
2. Osmosis (Figure 4.2)	Simple diffusion of water across a selectively permeable membrane.	Movement of water into and out of cells to maintain osmotic balance.
3. Facilitated diffusion (Figure 4.3)	Same as simple diffusion, but the transported substance must attach to a transport protein in the cell membrane.	Transport of glucose into cells.
4. Filtration	Movement of an aqueous (water) solution along a pressure gradient, from an area of high hydrostatic pressure to an area of low hydrostatic pressure.	Movement of fluids out of capillary beds; the filtering of blood by the kidney.
Active processes (ATP required)		
1. Active transport (Figure 4.4)	Movement of substances against a concentration gradient (from a region of low concentration to a region of high concentration). A transport protein molecule in the membrane is required.	Sodium-potassium pump and other ion pumps
2. Vesicular (bulk) transport	Movement across a membrane of large particles and macromolecules. The transported substance is enclosed by a vesicle.	
a. Endocytosis (Figure 4.5)	Large particles and macromolecules are transported into a cell.	
i. Phagocytosis	Endocytosis of undissolved particles such as bacteria, cell debris, or large proteins.	Ingestion of bacteria and viruses by macrophages and white blood cells.
ii. Pinocytosis	Endocytosis of liquid droplets with dissolved solutes.	Absorption of some nutrients by cells in the small intestine.
iii. Receptor-mediated endocytosis	Endocytosis in which the transport vesicle must bind to a specific membrane receptor.	Transport of cholesterol into cells.
b. Exocytosis (Figure 4.6)	Secretion of a substance from a cell.	Release of hormones by endocrine cells; release of neurotransmitter by nerve cells.

to write your own individual laboratory report that critically analyzes the results of your experiments. If your instructor does not provide specific instructions for this, some guidelines for writing a laboratory report appear in the Appendix.

Diffusion

As described earlier, diffusion across a cell membrane is a passive process. The required energy is derived from the kinetic energy (energy of motion) of the molecules or atoms that are being transported, so cellular energy from ATP is not needed. The rate at which diffusion proceeds depends on the following factors.

- *Size of the particles.* Small particles diffuse at a faster rate than large ones because they are less resistant to moving and, once in motion, continue to move more easily than large particles.
- *Temperature of the system.* As temperature increases, molecular motion and kinetic energy also increase. Consequently, the rate of diffusion will also accelerate. Conversely, when the temperature decreases, the rate of diffusion decreases.
- *Polarity of the substance being transported.* The interior of the cell membrane consists of nonpolar fatty acids and cholesterol. Substances that are lipid soluble, such as oxygen, carbon dioxide, and fats, will readily cross the membrane by simple diffusion (Figure 4.1b). Substances that

(a)

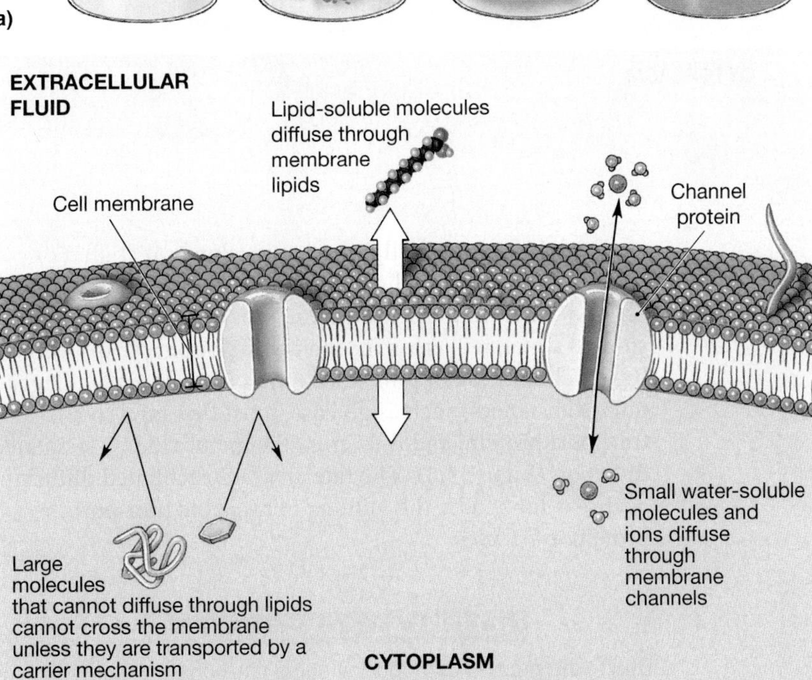

(b)

Figure 4.1 Simple diffusion in an open space and across a membrane. Simple diffusion occurs because molecules are constantly moving and colliding with other molecules. Although molecular motion is random, a substance tends to move along a concentration gradient, from a region of high concentration to a region of low concentration. **a)** Simple diffusion in an open container. Placing a colored sugar cube in a beaker of water establishes a concentration gradient. As the cube dissolves, the sugar molecules tend to move along the concentration gradient. Equilibrium is reached when the sugar molecules are uniformly distributed throughout the beaker. **b)** Simple diffusion across a cell membrane. Substances that are soluble in lipids will travel freely across the phospholipid bilayer of the membrane. Ions and small polar molecules will pass through protein channels. Large molecules that are soluble in lipids cannot cross the membrane by simple diffusion.

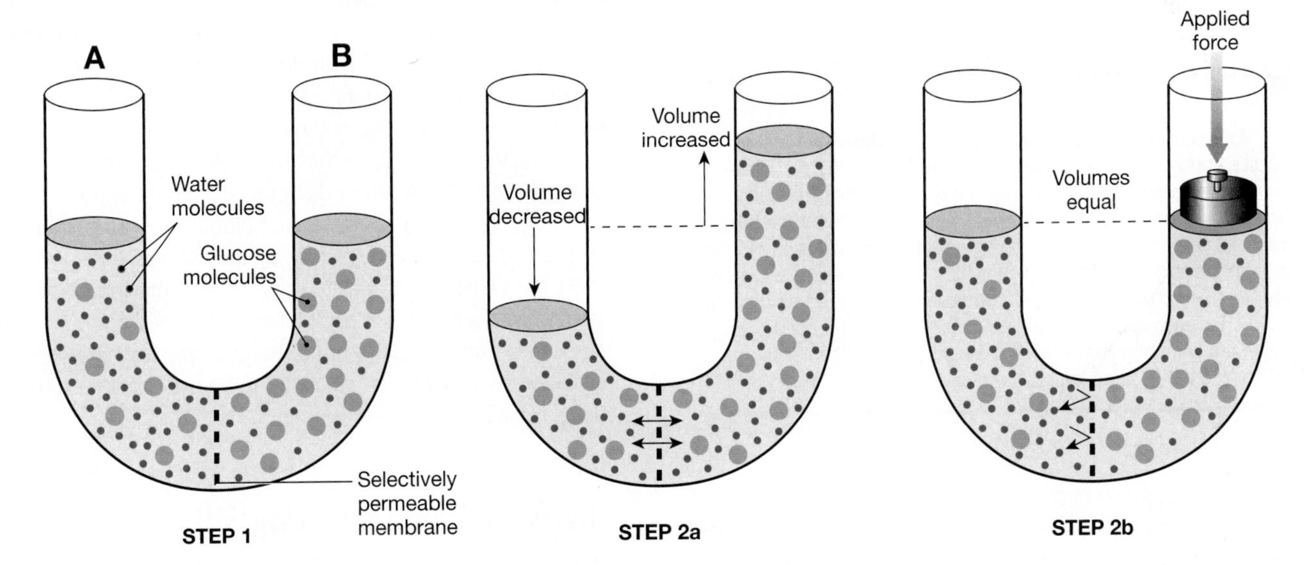

Figure 4.2 Diffusion of water: osmosis. Osmosis is a special case of simple diffusion in which water moves across a selectively permeable membrane, along its own concentration gradient. When two solutions are separated by a membrane, water tends to move from the solution with the lower solute concentration (the hypotonic solution) to the solution with the higher solute concentration (the hypertonic solution). **Step 1:** Two glucose solutions are separated by a selectively permeable membrane. Solution A is less concentrated and is hypotonic compared with solution B, which is more concentrated and hypertonic. **Step 2a:** Water tends to move from solution A to solution B until the solute concentrations on both sides are equal. When equilibrium is reached, the solutions become isotonic and water passes uniformly in both directions. Note that the fluid volumes on both sides have changed. **Step 2b:** A visual demonstration of the osmotic pressure, which is the hydrostatic pressure in a solution that opposes osmosis.

Figure 4.3 Diffusion of large polar molecules: facilitated diffusion. Facilitated diffusion is a process that allows large, lipid-insoluble substances to cross a cell membrane. It operates under the same principles as simple diffusion. However, the substance to be transported (i.e., glucose) must first bind to a transport protein. The protein carrier then undergoes a conformational change to allow the substance to cross the membrane.

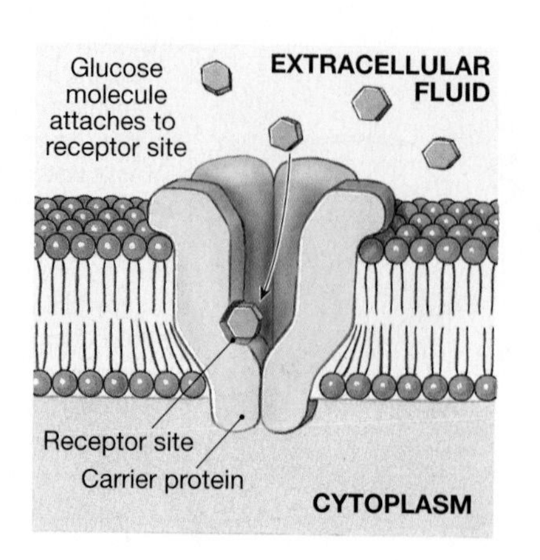

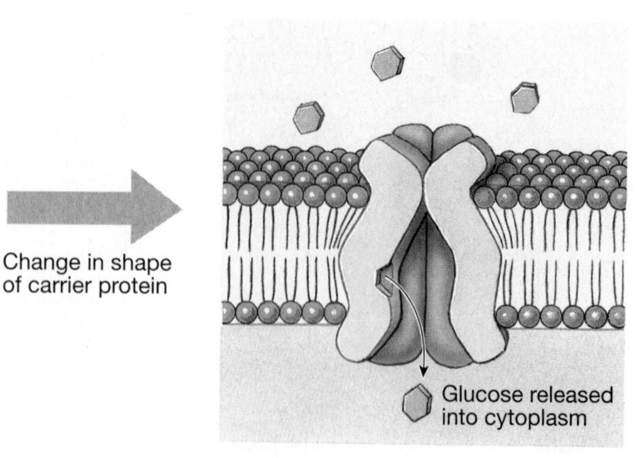

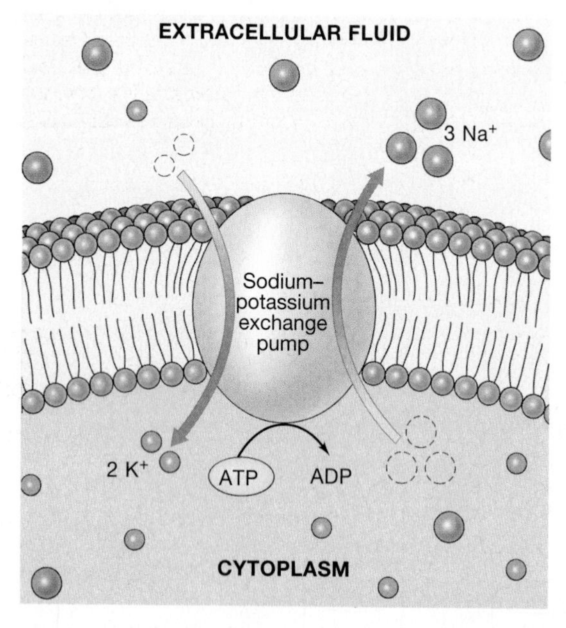

Figure 4.4 An example of active transport: the sodium-potassium pump. In active transport, substances are transported across the cell membrane with the aid of cellular energy (ATP) and a carrier protein. This transport mechanism is not dependent on a concentration gradient, so it can be used to move substances against their concentration gradient, from the region of lower concentration to the region of higher concentration. In the process, cellular energy derived from ATP is used. The sodium-potassium pump is an example of active transport. For each cycle, three sodium ions (Na^+) are pumped out of the cytoplasm and two potassium ions (K^+) are pumped in. The transport of each ion is coupled to the same transport protein.

are not lipid soluble will diffuse quite slowly, if at all. For example, ions and small polar molecules (molecules with a slight difference in electric charge from one part to another), such as water, will cross the membrane by simple diffusion by traveling through protein channels (Figure 4.1b). Larger polar substances, such as glucose, must first bind to a transport protein, and thus cross the membrane by facilitated diffusion (Figure 4.3). The rate at which facilitated diffusion occurs is limited by the number of available transport protein molecules.

CLINICAL CORRELATION

During the process of producing urine, the kidneys are able to remove metabolic waste products from the blood. In the kidney, many substances, including glucose, are filtered out of the blood into the renal tubules. Some of these substances remain in the tubules and are expelled in the urine. Because glucose is so important to our cellular metabolism, it is reabsorbed into the blood via facilitated diffusion. In a healthy individual, it is usually completely reabsorbed, so that no glucose is present in the urine. Individuals who have diabetes mellitus, however, have excessive amounts of glucose in their blood. As a result, there are far more glucose molecules moving through the kidneys than there are transport proteins to move it back into the blood. Consequently, glucose is not completely reabsorbed and a notable amount is excreted in the urine. Thus, simply testing a person's urine for the presence of glucose can be used to screen for diabetes mellitus.

ACTIVITY 4.1 Demonstrating Simple Diffusion

1. Place a piece of clean white paper on the lab bench. With a pencil, mark three dots on the paper, spaced at least 15 cm (6 in) apart. (You might need two sheets of paper to space the dots appropriately.)
2. Obtain three glass petri dishes and number them 1, 2, and 3.

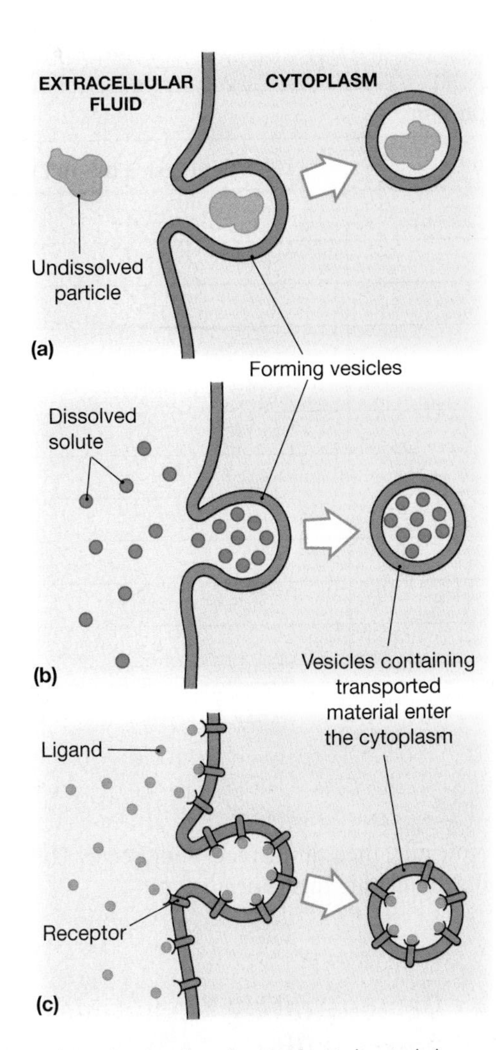

Figure 4.5 Three forms of endocytosis. Endocytosis is a vesicular transport process that allows cells to take in particles that are too large to pass into the cell by other means. Cellular energy derived from ATP molecules is required. Substances are transported within vesicles that are created by invaginations of the cell membrane. Thus, the substance does not actually travel across the cell membrane to enter the cytoplasm. Three forms of endocytosis are illustrated. **a)** During phagocytosis, the cell ingests large insoluble particles. **b)** During pinocytosis, the cell takes in liquid droplets containing large dissolved molecules such as proteins. **c)** During receptor-mediated endocytosis, the transported substance (ligand) must first bind to a specific protein receptor embedded in the cell membrane.

3. Place the dishes on the piece of white paper so that each dish is centered over one of the three dots.
4. Add distilled water to each petri dish until the bottom surface is completely covered according to the following instructions.
   - Petri dish 1: Add water that is at 23°C (room temperature).
   - Petri dish 2: Add water that is at 90° to 100°C (just below the boiling point for water).
   - Petri dish 3: Add water that is at 0° to 10°C (just above the freezing point for water).

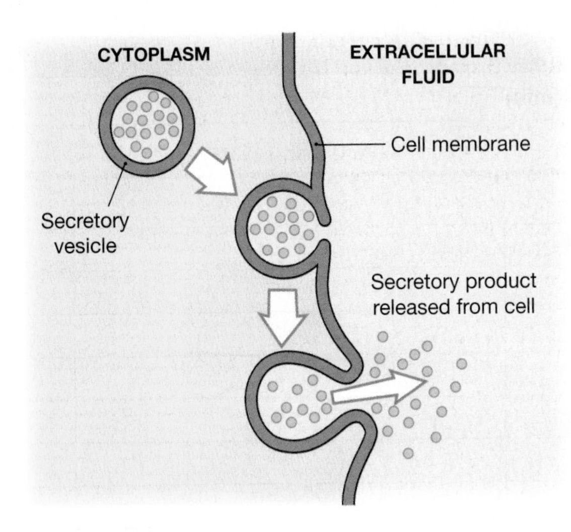

Figure 4.6 Release of a substance by exocytosis. Exocytosis is a vesicular transport process in which cells transport a secretory product, usually a protein, from the cytoplasm to the extracellular fluid. The transported substance is packaged inside a secretory vesicle. After the vesicle fuses to the cell membrane, it opens to the outside and the substance is released.

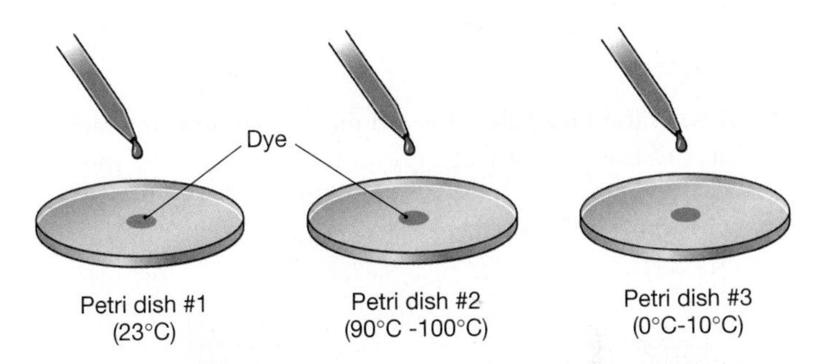

Figure 4.7 Placement of dye in water-filled petri dishes to demonstrate simple diffusion. A drop of dye is carefully placed in the center of each petri dish. The dishes are placed on a sheet of white paper so diffusion of the dye molecules can be clearly observed.

5. **Form a Hypothesis** Before you add a drop of dye to each petri dish, make a prediction about the diffusion rate at each water temperature.
   - Predict the water temperature at which the diffusion rate will be the greatest. ______________
   - Predict the water temperature at which the diffusion rate will be the least. ______________
6. Add a drop of blue dye (methylene blue or toluidine blue) to each petri dish, carefully placing each drop as close to the center of the dish (over the pencil dot) as possible (Figure 4.7).
7. For each petri dish, observe the diffusion of the blue dye in the water. At 3-minute intervals, compare the relative rates of diffusion for each temperature by measuring the distance (millimeters) between the origin of the dye and the diffusion front.
8. Record your data in Table 4.2.

Table 4.2 Effect of Temperature on Diffusion Rate

Time (min)	Distance from dye origin (mm)		
	Petri Dish 1 (23°C)	Petri Dish 2 (90°–100°C)	Petri Dish 3 (0°–10°C)
0	0	0	0
3			
6			
9			
12			
15			
18			
21			
24			
27			
30			

9. **Assess the Outcome** Record the temperature at which the diffusion rate was the greatest __________ and the least ____________. Do your experimental results agree with your earlier prediction? _________

____

QUESTIONS TO CONSIDER

1. What effect does temperature have on the diffusion rate? Explain why this effect occurs.

____

____

2. In the activity that you have just completed, what must happen for equilibrium to be reached? Was equilibrium reached at any of the water temperatures that were used?

____

____

Diffusion and Membrane Permeability

Diffusion through a membrane has the following two requirements.

1. Concentration gradient
2. Membrane that is permeable to the substance being moved

Even with a big difference between the concentrations on each side of a membrane, if the membrane is not permeable to that particular molecule, then diffusion cannot occur. The following activity will demonstrate this concept.

ACTIVITY 4.2 Demonstrating Membrane Permeability

1. Obtain a 10-cm strip of dialysis tubing. Soak it in water for a few minutes so it will be easier to open.
2. Secure one end by using a clip or wrapping it with multiple turns of an elastic band to form a tight seal.
3. Open the other end of the bag and add enough 10% starch solution to fill it about one third of its volume. Carefully secure the remaining end with a clip or several turns of an elastic band.
4. Rinse the outside of the bag to remove any starch that might have spilled.
5. Put your bag of starch into a beaker and fill the beaker with warm water until the bag is fully submerged.
6. Add iodine potassium iodide (IKI) solution into the beaker until the water is a medium amber color. The IKI solution is used to test for the presence of starch. The solution will turn dark blue or black if starch is present.
7. Observe the bag after 10 minutes and record the color of the solutions in the beaker and in the dialysis tubing bag.
   - What color is the solution in the beaker?____________
   - What color is the solution in the bag? ____________

QUESTION TO CONSIDER Based on your results (solution colors) from the previous activity, identify the substances (if any) that diffused across the dialysis tubing. Provide an explanation for the results.

____

____

____

Osmosis

Osmosis is a special case of simple diffusion. Specifically, it is the term for the diffusion of water through a selectively permeable membrane. The net movement of water across a membrane will be from an area of higher water concentration to an area of lower water concentration (Figure 4.2). Another way of saying this is that the net movement of water will be from a solution with a lower total solute concentration (a **hypotonic solution**) to a solution with a higher total solute concentration (a **hypertonic solution**). A **solute** is a substance that dissolves when added to another substance. The substance that dissolves the solute is known as a **solvent**. Water is considered the **universal solvent** because it dissolves all other polar substances. For example, our body fluids (e.g., blood, lymph, cytoplasm, and extracellular fluid), which are mostly water, will dissolve various polar compounds such as glucose, proteins, and electrolytes.

Under normal conditions, the cells in our body are exposed to **isotonic solutions**. That is, the solute concentration of the cytoplasm is about the same as the solute concentration of extracellular fluids. Under these conditions, equilibrium is established, meaning the amount of water that enters a cell is about equal to the amount that leaves. Maintaining osmotic balance is an important homeostatic mechanism for maintaining cell shape and ensuring proper cell function.

CLINICAL CORRELATION

We all know that alcohol consumption can lead to intoxication, but most of us do not realize that similar symptoms can occur from excessive water consumption, a condition called **water intoxication.** As essential as water is to life, overconsumption of water can be fatal. Normally, our kidneys eliminate excess water from the body. Under various conditions, though, too much water could be consumed at once. For example, severe dehydration, such as from prolonged vomiting and diarrhea or from excessive sweating, causes the loss not only of water but also of electrolytes (solutes). If a very large amount of water is consumed in a short time, without replacing electrolytes, our kidneys may not be able to clear the water at the desired rate and the blood will become hypotonic. As a result, cells in the brain, which are hypertonic to our now diluted blood, will take in too much water by osmosis. If this happens, the brain cells will swell, similar to when the brain swells from serious head trauma, and death can occur.

ACTIVITY 4.3 Observing Osmosis

1. Cut an 8 cm (3 inch) piece of dialysis tubing and separate the two layers by soaking it in tap water.
2. Slide one blade of the scissors between the two layers and cut along the folded margin to form a single-layer membrane.
3. If you are using a thistle tube that can be separated, place the dialysis membrane tightly over the mouth of a thistle tube. Keeping the membrane taut, secure it to the thistle tube with several wrappings of an elastic band. If your thistle tube does not come apart, see alternative instructions below.
4. Separate the mouth of the thistle tube from the cylindrical portion. Using a pipette, carefully fill the mouth with a molasses solution until the solution is about to overflow. Half the class should use a 50% molasses solution, the other half a 25% solution.
5. Reconnect the cylindrical portion of the thistle tube to the mouth. Make sure the connection is tightly secured. The meniscus of the molasses solution should rise slightly up the cylinder. If an air bubble is trapped in the thistle tube, the molasses will recede back into the mouth. If this occurs, detach the cylinder and add more molasses to the mouth. Proceed to step 6.
6. Mark the initial level of the meniscus of the molasses solution with a marking pen.
7. Lower the thistle tube into a beaker of distilled water so that the dialysis membrane is submerged. Secure the thistle tube in this position with a ring stand and clamp, being careful not to break the delicate tube (Figure 4.8). If the water solution becomes discolored, it indicates a leak in the dialysis tubing. If a leak does occur, you must take apart the assembly and start over.

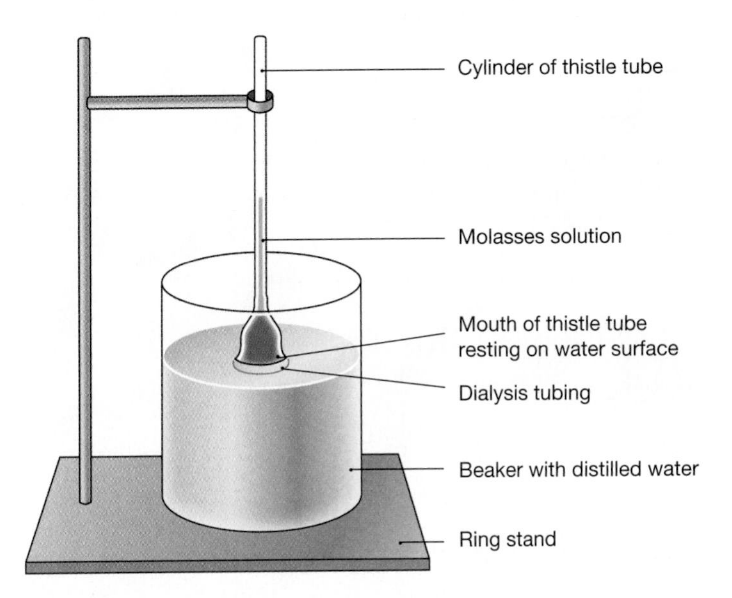

Figure 4.8 A thistle tube assembly for the demonstration of osmosis. The distilled water in the beaker and the molasses in the thistle tube are separated by the dialysis tube membrane. The thistle tube is held securely in place with a clamp attached to the ring stand.

8. Using your initial mark as your zero point, note the change in the level of the meniscus (in centimeters) every 10 minutes for a 1-hour period (longer, if time permits). Record your results in Table 4.3.
9. Obtain results for both the 25% and 50% molasses solutions by sharing your data with other students in the laboratory.
10. For each concentration, what was the total distance that the molasses solution traveled up the thistle tube?
    - 25% molasses ____________
    - 50% molasses ____________

Table 4.3 Osmosis Demonstration

	Meniscus level (cm)	
Time (min)	**25% Molasses**	**50% Molasses**
0	0	0
10		
20		
30		
40		
50		
60		

QUESTIONS TO CONSIDER

1. Explain why the molasses solutions stopped rising.

____

____

2. Were there any differences between the 25% and 50% molasses in the rate at which the solutions rose?

____

____

3. If there were any differences, can you explain the reason for them?

____

____

4. As described earlier, cells are normally exposed to isotonic body fluids. What do you think would happen if cells were exposed to the following?
    - Hypotonic solution such as distilled water

____

____

    - Hypertonic solution such as 10% glucose

____

____

Exercise 4 Review Sheet

Membrane Transport

Name ______________________

Lab Section ______________________

Date ______________________

1. Explain the fundamental difference between passive and active processes of membrane transport.

____

____

____

2. Why is membrane transport vital for normal cell function?

____

____

____

3. Explain the difference between simple diffusion and facilitated diffusion.

____

____

____

4. Facilitated diffusion and active transport are both classified as carrier-mediated transport processes. Explain why.

____

____

____

Questions 5–7: Consider the following scenario.

After measuring the solute concentrations of two fluid regions, A and B, separated by a selectively permeable membrane, it is determined that region A contains the hypertonic solution and region B contains the hypotonic solution.

5. What can you say about the relative solute concentrations of regions A and B?

____

____

____

6. In which direction will osmosis occur?

____

____

____

7. Assume fluid regions A and B are isotonic.

   a. What can you say about the solute concentrations in each region?

   ___

   ___

   ___

   b. How can you describe the passage of water across the membrane?

   ___

   ___

   ___

8. Explain the difference between endocytosis and exocytosis.

   ___

   ___

   ___

9. Identify the differences and similarities between the three forms of endocytosis.

   ___

   ___

   ___

Exercise 5

Epithelial and Connective Tissues

Laboratory Objectives

On completion of the activities in this exercise, you will be able to:

- Describe the general characteristics of epithelial and connective tissues.
- Recognize examples of epithelial and connective tissues.
- Describe specific characteristics of each type of epithelial tissue, including its structure, functions, and locations.
- Describe specific characteristics of each type of connective tissue, including its structure, functions, and locations.

Materials

- Compound light microscope
- Colored markers or pencils
- Prepared microscope slides:
  - Artery, vein, nerve
  - Kidney, human
  - Small intestine
  - Esophagus
  - Human skin
  - Trachea
  - Urinary bladder
  - Mesenchyme
  - Areolar (loose) connective tissue
  - Adipose tissue
  - Reticular tissue
  - Dense regular (fibrous) connective tissue
  - Elastic tissue
  - Elastic cartilage
  - Fibrocartilage
  - Bone
  - Human blood smear

A **tissue** is a group of cells that has a similar embryological origin and works as a unit to carry out a specialized function. **Histology** is the biological discipline devoted to the microscopic study of tissues. The body has four major categories of tissues.

- Epithelial tissue
- Connective tissue
- Muscular tissue
- Nervous tissue

In this exercise, the structure and function of epithelial and connective tissues will be examined. You will study the histology of muscular tissue in Exercise 10, and nervous tissue in Exercise 13.

Epithelial Tissue (Epithelium)

Epithelial tissue covers all body surfaces, inside and out. It covers the outside surface of organs and forms the inner lining of hollow organs. It lines all body cavities and the outside surface of the body (the epidermis of the skin). In addition, epithelium is the primary tissue of glands.

The underside of epithelium is anchored to connective tissue by a layer of noncellular material called the **basement membrane**, or **basement lamina**. In most cases, epithelia lack blood vessels, but are nourished by capillaries in the underlying connective tissue. Epithelial cells are constantly exposed to conditions of abrasion and injury. For example, when you scratch your skin, epithelial cells are sloughed off the surface. Earlier, when you prepared a wet mount of your cheek cells (Exercise 3), you scraped epithelial cells from the surface of the oral cavity. Thus, epithelial cells are constantly being lost and subsequently replaced by regular mitotic divisions. Epithelial cells are usually tightly packed and joined by cell junctions such as desmosomes. Because of this feature, epithelial tissues are effective in providing protection for the organ they cover. In addition to their ability to protect structures, epithelial tissues have other important functions, as listed in Table 5.1.

Epithelia are classified by using two structural criteria.

1. *Number of cell layers.* A **simple epithelium** consists of a single layer of cells, whereas a **stratified epithelium** has several layers.
2. *Shape of the cells.* A **squamous epithelium** contains flattened, irregular-shaped cells. A **cuboidal epithelium** contains cube-shaped cells in which the height and width are about equal. A **columnar epithelium** contains elongated cells (column shaped) in which the height is much greater than the width.

Several different types of epithelia exist in the body and can be categorized by considering the two structural criteria together

Table 5.1 Functions of Epithelial Tissue

Function	Example
1. Protection	Epidermis of the skin
2. Secretion	Release of glandular products by glands
3. Absorption	Uptake of nutrients by absorptive cells in the small intestine
4. Sensory reception	Taste buds on the surface of the tongue
5. Filtration	Filtering of blood by Bowman's capsules in the kidney

(Table 5.2). For example, an epithelium with a single layer of cube-shaped cells is classified as **simple cuboidal**; an epithelium with several layers of mostly flattened cells is classified as **stratified squamous**.

ACTIVITY 5.1 Microscopic Observations of Epithelial Tissue

A. Simple Squamous Epithelium

1. Obtain a slide of arteries, veins, and nerves in cross section.
2. Scan the slide under low magnification until you locate the circular profile of an artery (Figure 5.1). Unlike the nerve, the artery and vein will both be hollow structures. The artery will have the thicker wall of the two blood vessels.
3. Examine the inner wall of the artery. The internal space (the **lumen** of the artery) may contain some brownish-red or pink-staining red blood cells. The arterial wall (surrounding the lumen) is composed of three distinct layers. The innermost layer, appearing as a wavy dark-staining line, is the tunica intima. Notice that the tunica intima contains a single layer of flattened cells. It is a **simple squamous epithelium** (Figure 5.1). All epithelia that line

Table 5.2 Types of Epithelial Tissues

Type	Example
1. Simple squamous	a. Inner lining of blood vessels (endothelium) b. Lining of body cavities (mesothelium)
2. Simple cuboidal	a. Renal tubules (kidney) b. Secretory portion of glands c. Liver cells
3. Simple columnar	Inner lining of much of the digestive tract (stomach, small and large intestines)
4. Pseudostratified columnar	Inner lining of trachea and large airways in lung
5. Stratified squamous	a. Epidermis of the skin b. Inner lining of esophagus c. Inner lining of vaginal canal
6. Stratified cuboidal	Ducts of sweat, mammary, and salivary glands
7. Stratified columnar	a. Small portion of male urethra b. Some large ducts of glands
8. Transitional	Inner lining of urinary tract (urinary bladder, ureter, and part of urethra.)

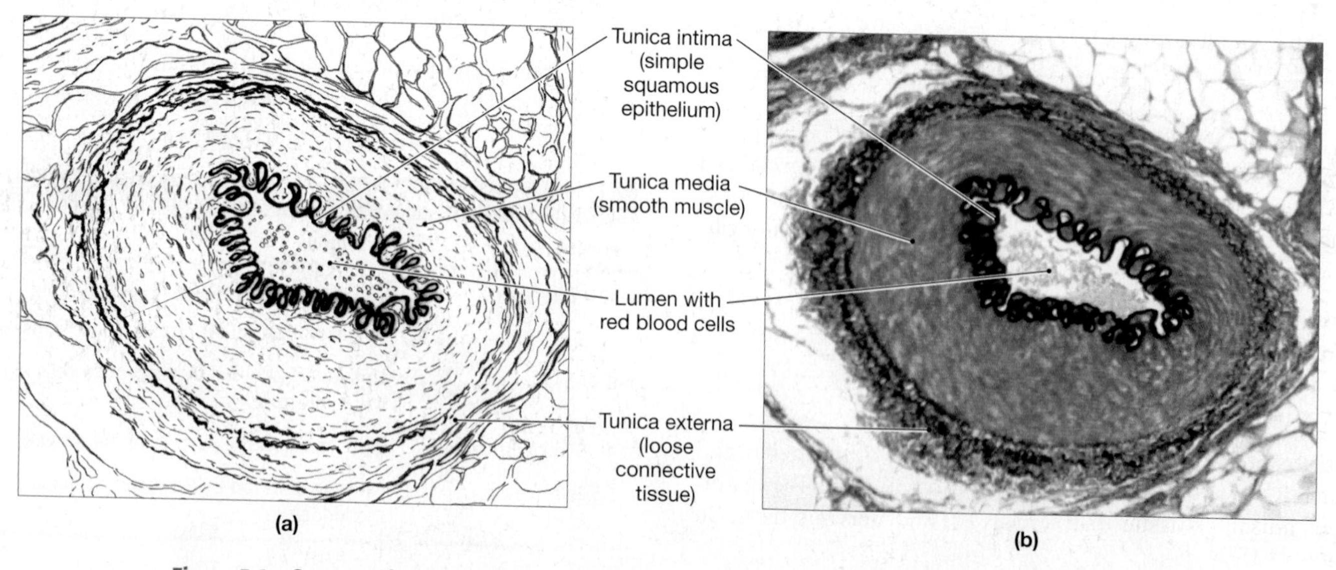

Figure 5.1 Cross section of an artery. Simple squamous epithelium in the tunica intima lines the lumen of the blood vessel. **a)** Illustration of an artery; **b)** corresponding light micrograph (LM × 80).

blood vessels are simple squamous, and are given the special name **endothelium**.

4. Identify other tissue layers of the arterial wall. The middle layer, the tunica media, consists of muscle tissue (smooth muscle), which you will study in a later exercise. The outer layer, the tunica externa (adventitia), contains loose connective tissue.

B. Simple Squamous and Simple Cuboidal Epithelium

1. Obtain a slide of the kidney.
2. Scan the slide under low magnification until you locate a field of view that contains numerous **renal corpuscles** (Figure 5.2a). Each corpuscle consists of a tuft of capillaries, called a **glomerulus**, that is surrounded by a **Bowman's capsule**. Blood flowing through the glomerulus is filtered by Bowman's capsule. Blood filtering is the first step in the production of urine by the kidney. Under the microscope, each glomerulus will appear as a spherical mass of pink- to red-staining tissue surrounded by a clear space. Most of what you see in this field of view are cross sections of other kidney tubules.
3. Switch to high magnification and examine a glomerulus and Bowman's capsule more carefully. Like all blood vessels, the walls of the glomerular capillaries are lined by simple squamous epithelium (the endothelium). You likely cannot clearly see the arrangement of this epithelium with this preparation; however, many of the nuclei visible in the

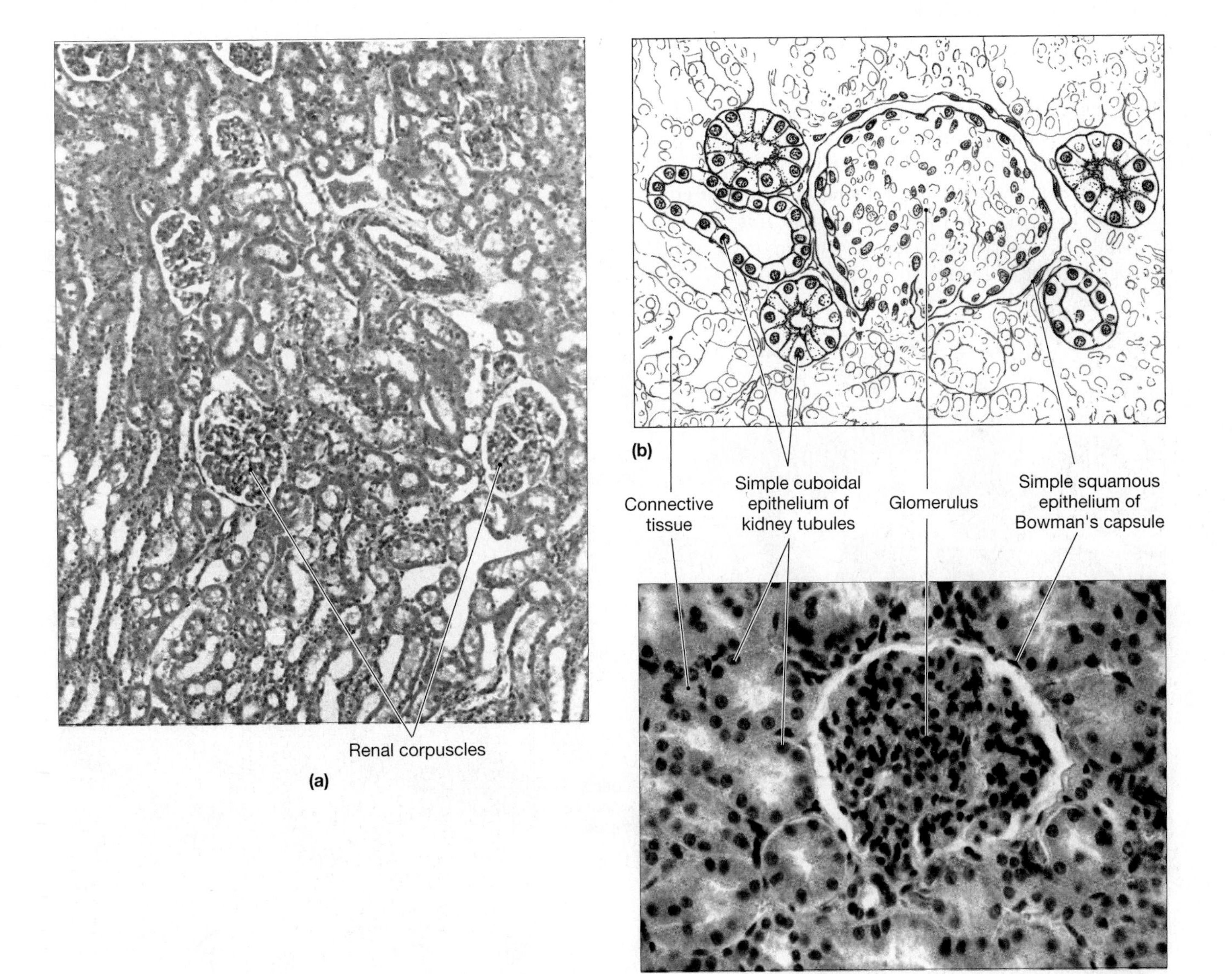

Figure 5.2 Renal corpuscles and tubules in the kidney. In a corpuscle, the epithelium lining a Bowman's capsule is simple squamous. The epithelium lining the wall of a kidney tubule is simple cuboidal. **a)** Low-power light micrograph of renal corpuscles and kidney tubules (LM × 150); **b)** illustration of a renal corpuscle; **c)** corresponding light micrograph (LM × 400).

glomerulus belong to the endothelial cells. The lining of the Bowman's capsule is also a simple squamous epithelium. If you look carefully across the open space from the glomerulus, you can identify the single layer of squamous cells that line the outer wall of the capsule (Figures 5.2b and c).

4. Stay at high magnification and locate the many cross sections of renal tubules that surround Bowman's capsule. You may have to move the slide slightly to get a better view of these structures. These tubules have an internal space (the lumen) that is lined by a single layer of cubed-shaped cells. This is an example of a **simple cuboidal epithelium** (Figures 5.2b and c). Note that the renal tubules are separated by thin layers of connective tissue.

C. Simple Columnar Epithelium

1. Obtain a slide of the small intestine in cross or longitudinal section.
2. Scan the slide under low magnification and locate the fingerlike projections that extend into the lumen. These projections are lined by a single layer of column-shaped cells (Figure 5.3). This is an example a **simple columnar epithelium**. In addition to the small intestine, this type of epithelium is found in other digestive organs such as the stomach and the large intestine.
3. Observe the location of the nuclei in the epithelial cells. Note that they are all situated near the base of the cells. This characteristic is an indication that the epithelium contains only one cell layer. Note the distinctive, oval-shaped **goblet (mucous) cells** (Figure 5.3). Depending on the preparation, these cells may appear clear or stained blue or red. Goblet cells are common to this type of epithelium. The mucus they secrete enhances the digestive process and protects the epithelium from harsh digestive chemicals.
4. Note the other tissue types in the wall of the small intestine. Immediately below the epithelium is a layer of loose connective tissue. Next, there are two outer layers of smooth muscle, which most likely appear pinkish or reddish. The outer layer of the epithelial wall contains a simple squamous epithelium (microscopic sections may not show all of these layers).
5. Switch to high power and observe the free margin of the epithelial cells. Observe the darker staining fringe along this margin. This structure, known as the **brush border**, is a special feature of the epithelium in the small intestine (Figure 5.3). Actually, the brush border is a series of fingerlike projections known as **microvilli**. They function to increase the surface area of small intestinal cells to maximize absorption of nutrients.

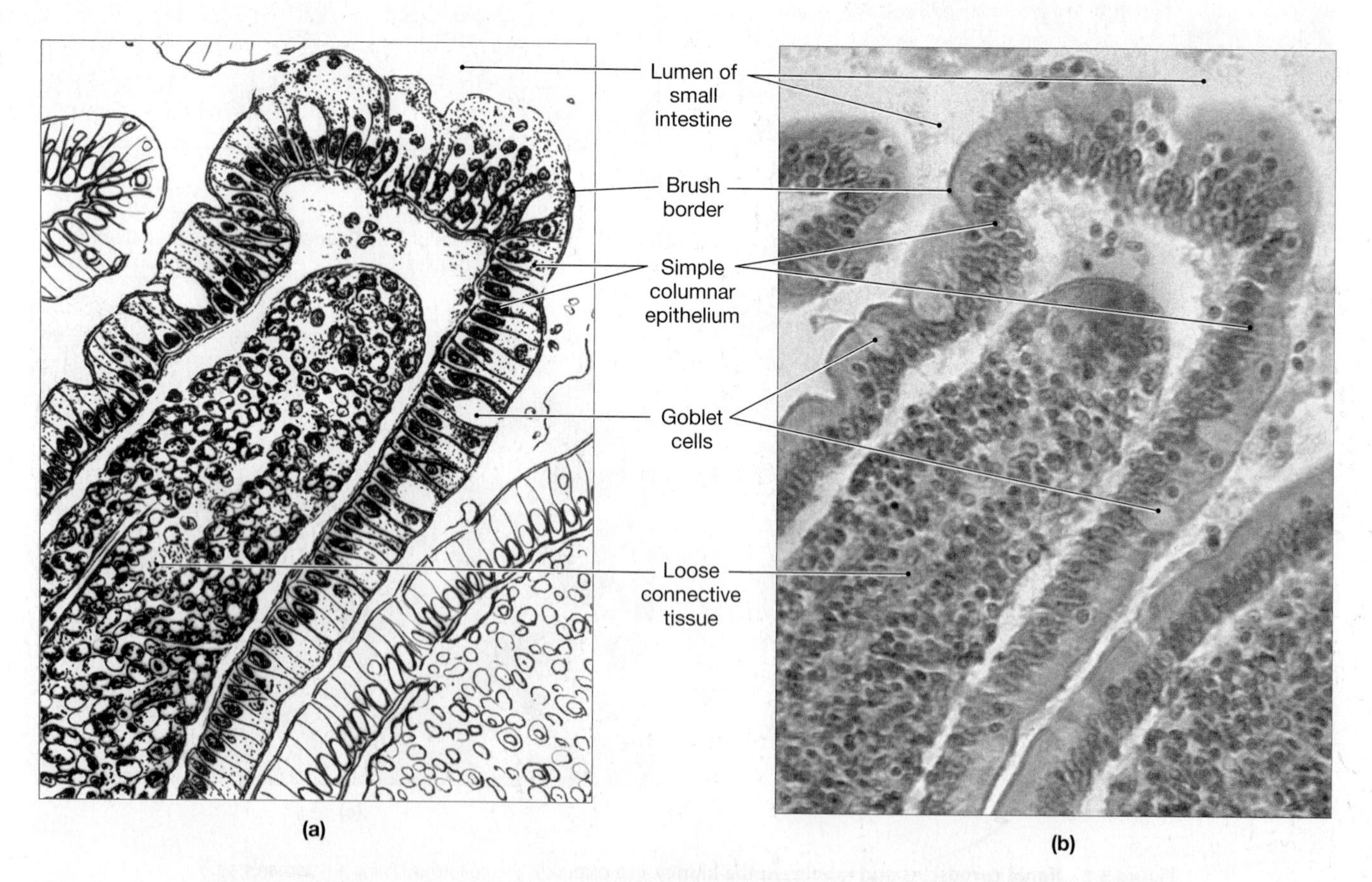

Figure 5.3 Simple columnar epithelium in the small intestine. a) Illustration of the epithelium with goblet cells; **b)** corresponding light micrograph (LM × 400).

D. Stratified Squamous Epithelium

1. Obtain a slide of the esophagus.
2. View the section under low power. The upper margin (assuming you placed the slide on the stage with the label right side up) should be the epithelium lining the lumen of the esophagus. Note that this epithelium has many layers of cells. The deeper layers contain cube-shaped cells but the superficial layers consist of flattened cells. (You may have to switch to a higher magnification to observe the transition in cell shape.) This is an example of a **stratified squamous epithelium** (Figure 5.4).
3. Scan the section and search for other tissue layers. Directly below the epithelium is a layer of connective tissue, and beneath the connective tissue are two layers of smooth muscle. Depending on the preparation, your slide may not have all these layers.

E. Stratified Squamous, Keratinized Epithelium

1. Obtain a slide of human skin.
2. View the section under low power. In the field of view, the upper margin (assuming you placed the slide on the stage with the label right side up) is the epithelium of the skin. This epithelium, known as the **epidermis** of the skin, is a **stratified squamous** type, with a thick layer of dead cells filled with keratin fibers at the surface (Figure 5.5). This layer of dead cells is known as the **keratinized layer**. The thickness of this layer varies in different regions of the body. It is very thick in the skin covering the palms of the hands and the soles of the feet, but quite thin on the anterior abdominal wall and anterior forearm. As always, directly below the epithelium is a layer of connective tissue.

F. Pseudostratified Columnar Epithelium

1. Obtain a slide of the trachea.
2. View the section under low power. Move the slide so that a section of the epithelium that lines the lumen of the trachea is in the center of the field of view.
3. Switch to high power to view the epithelium more closely. Observe that the nuclei in the epithelial cells are located at different levels, giving the impression that there are several layers of cells (Figure 5.6). Actually, this epithelium contains only one cell layer because all the cells touch the basement membrane, although some do not reach the surface. The nuclei are at different levels because some cells are wider near the base while others are wider near the surface. This epithelium is called **pseudostratified columnar** (*pseudostratified* means "false stratification"). It is also referred to as the **respiratory epithelium** because it lines the nasal cavity, trachea, and large airways in the lung.
4. Like the simple columnar epithelium in the digestive tract, pseudostratified columnar epithelium contains numerous goblet cells that secrete mucus. Identify these cells in the epithelium of the trachea (Figure 5.6).
5. Switch to high power to view the epithelium more closely. Identify cilia projecting from the surface of most of the epithelial cells. The cilia will appear as thin, hairlike structures emerging from the surface of the cells (Figure 5.6).

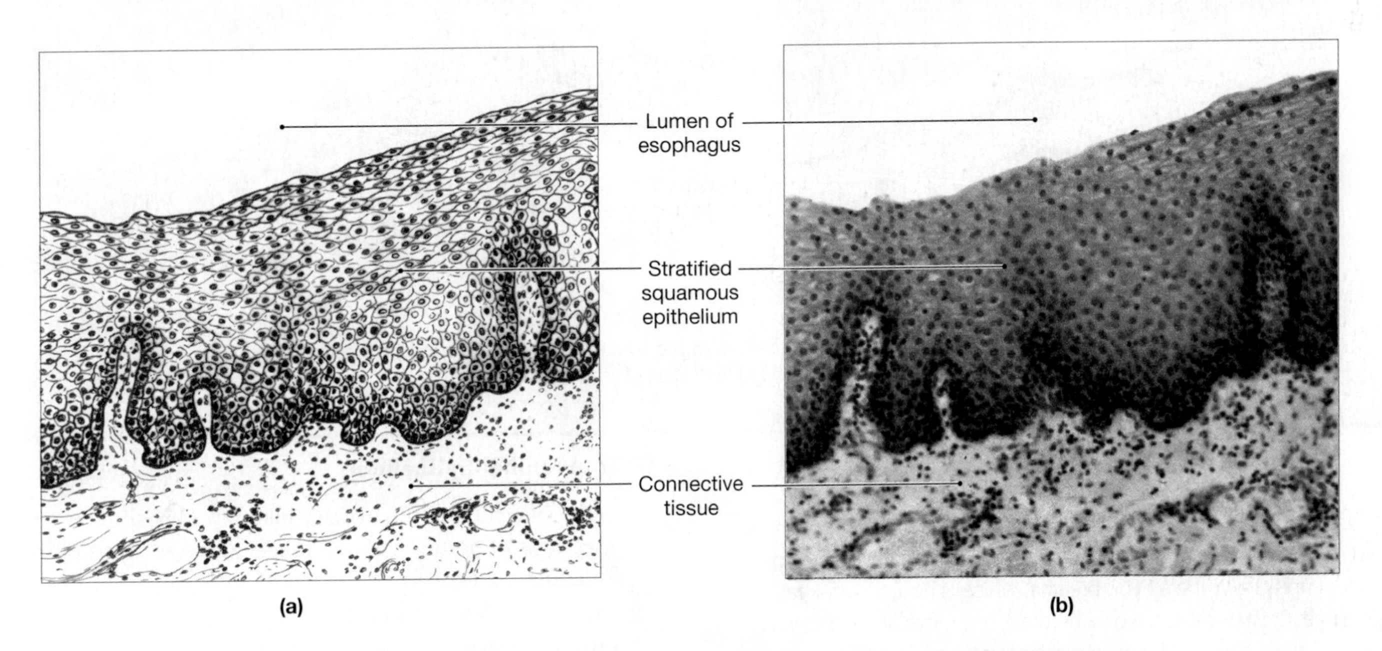

Figure 5.4 Stratified squamous epithelium in the esophagus. From the base of the epithelium to the surface, there is a gradual transition in the shape of the cells. At the base, the cells are cuboidal, but at the surface they are squamous. **a)** Illustration of the epithelium and underlying connective tissue; **b)** corresponding light micrograph (LM × 100).

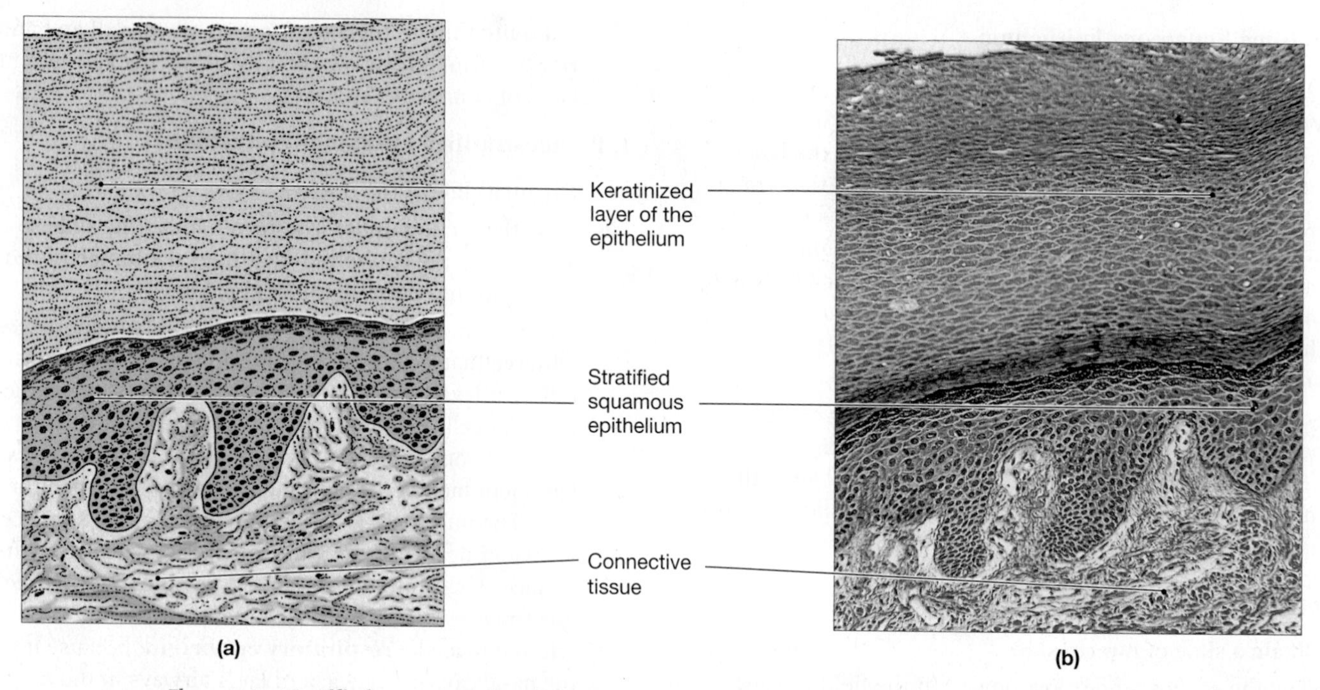

Figure 5.5 Stratified squamous, keratinized epithelium. This type of epithelium is found in the epidermis of the skin. **a)** Illustration of stratified squamous epithelium with a keratinized layer; **b)** corresponding light micrograph (LM × 100).

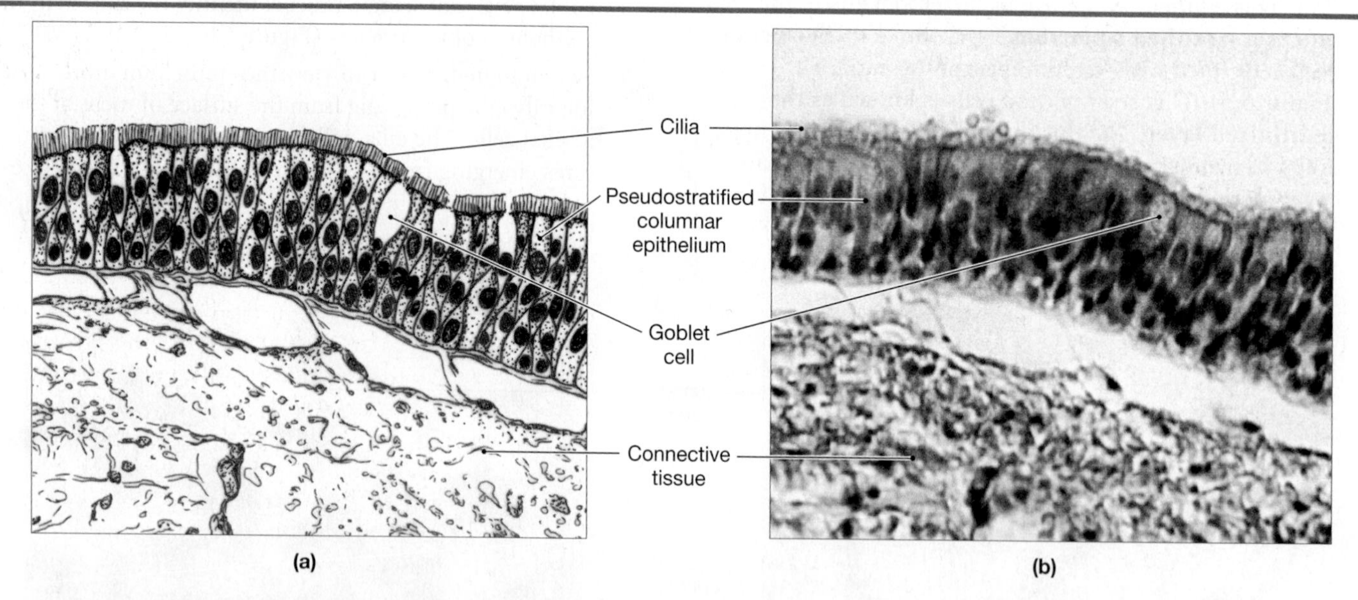

Figure 5.6 Pseudostratified columnar epithelium in the trachea. The cell shape and the location of nuclei give the appearance of several cell layers (stratification.) **a)** Illustration of the epithelium; **b)** corresponding light micrograph (LM × 400).

CLINICAL CORRELATION

The mucus in the respiratory tract has multiple functions. It moisturizes the airways so the cells lining them do not dry out. It also adds moisture to the air you breathe, which is necessary for gas exchange. Oxygen must dissolve before it can move across the wall of the air sacs in your lungs to reach your blood. In addition, the mucus, which is somewhat sticky, traps particulate material that is breathed in. This debris is then removed, by the cilia, to be sure it does not reach and obstruct the delicate air sacs within your lungs.

G. Transitional Epithelium

1. Obtain a slide of the urinary bladder. This slide has two sections of tissue. Both are longitudinal sections of the bladder wall.
2. View both sections under low power and identify the epithelium that lines the lumen at the top of each section. Note that the epithelium is strikingly different in both sections. In one section, the bladder wall is thrown into numerous folds and the epithelium is relatively thick. In the other section, the folding pattern in the bladder wall

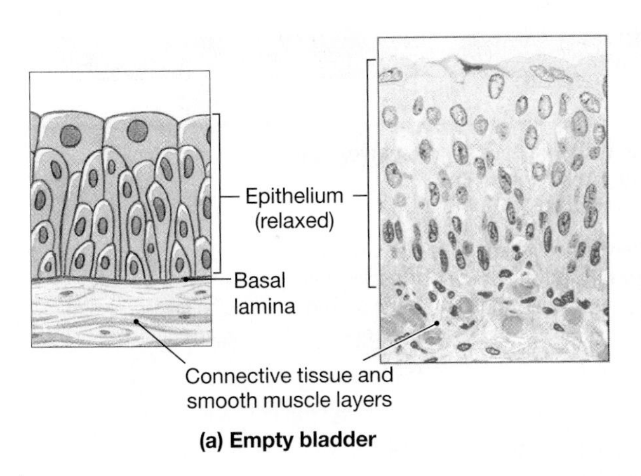

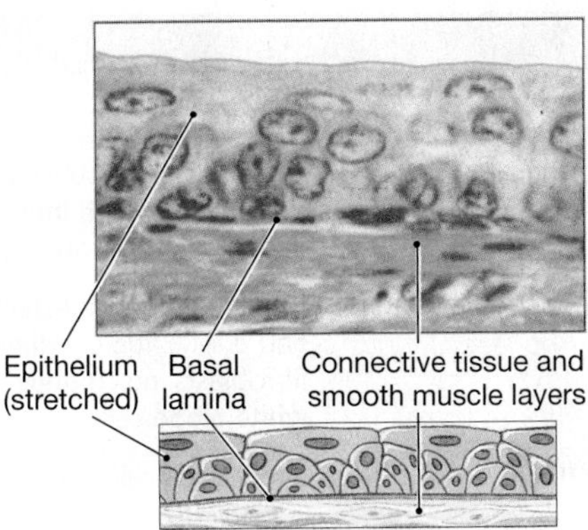

Figure 5.7 Transitional epithelium in the urinary bladder. a) Contracted state (empty bladder); **b)** distended state (full bladder).

has disappeared and the epithelium appears very thin (Figure 5.7). You are observing a special type of epithelium that is unique to the urinary tract (urinary bladder and ureters), known as **transitional epithelium**. It is also called **urothelium**, because this epithelium is only found in the urinary system. The term *transitional* implies some kind of change. The cells in this epithelium are able to alter their shapes with changing physical conditions. For instance, when the bladder is empty, the bladder wall, including the epithelium, is contracted. In this condition, the epithelium is thick and appears to have several layers of stacked cuboidal cells (Figure 5.7a). When the bladder is filled with urine, the wall expands. In this condition, the epithelium stretches out so that only two to three layers of squamous cells are visible (Figure 5.7b).

QUESTION TO CONSIDER During the previous activity, you observed that Bowman's capsule, where blood is filtered in the kidney, is lined by a simple squamous epithelium, which is a very thin cell layer. You also observed that the lumen of the esophagus, a passageway for food to the stomach, is lined by a stratified squamous epithelium, which is a much thicker layer. Why do you think these two structures have epithelia that are so strikingly different?

____

____

____

Connective Tissue

Connective tissues are found in all parts of the body and are quite diverse in both structure and function (Table 5.3.) Connective tissues connect structures and provide support and protection for vital organs. In addition, they fill spaces between structures, store fat, defend the body from infections, and repair tissue damage.

The cells in connective tissue are not packed tightly together as in epithelial tissue. Instead, they are separated from each other by an intercellular material known as the **matrix**. The matrix consists of a **ground substance** and various protein fibers. The consistency of the ground substance varies from a liquid to a solid and contains various types of adhesion proteins and complex protein-carbohydrate molecules known as proteoglycans. Adhesion proteins bind connective tissue cells to other components in the matrix, and proteoglycans trap water molecules in the matrix. The matrix also contains three types of fibers: **collagen**, **elastic**, and **reticular.** Both collagen and reticular fibers contain the fibrous protein collagen, whereas elastic fibers contain the fibrous protein elastin. Together, these fibers provide support and resiliency to the tissue.

The many different types of cells found in connective tissues reflect the structural and functional diversity of these tissues. Undifferentiated cells known as **blast cells** (*blast* means "forming") are actively dividing cells that produce the fibers and ground substance. They include the **fibroblasts**, in loose and dense connective tissues, **chondroblasts** in cartilage, **osteoblasts** in bone, and **hemocytoblasts** in blood. The hemocytoblasts always remain in the blast cell stage, actively producing new blood cells. The other blast cells can differentiate into their respective mature forms, **fibrocytes**, **chondrocytes**, and **osteocytes**, and change back to the blast forms during periods of tissue repair and injury. Other important connective tissue cells are as follows:

1. *Fat cells (adipocytes).* Fat cells are specialized cells that store nutrient fat molecules and are located in adipose tissue.
2. *Macrophages.* Macrophages are specialized for phagocytosis. They act as scavengers by clearing out foreign particles, thus providing an important defense against infection. Macrophages are located in loose connective tissues and a variety of organs, including the lungs, liver, spleen, and brain.
3. *Mast cells.* Mast cells are located in loose connective tissue, near blood vessels. They produce and secrete heparin, an anticlotting substance, and histamine, a chemical that promotes inflammation due to allergies or damage to blood vessels.

Table 5.3 Types of Connective Tissues

Category	Type	Description	Location
A. Embryonic connective tissue	1. Mesenchyme	Gel-like ground substance with fine protein fibers and immature cells; gives rise to all other connective tissues.	Embryo/fetus.
B. Connective tissue proper	2. Areolar connective tissue	Gel-like ground substance with collagen and elastic fibers; cell types include fibroblasts, macrophages, mast cells, and white blood cells.	Between skin and underlying muscles; between muscles; directly beneath most epithelial layers; between adjacent organs.
	3. Adipose tissue	Specialized cells for fat storage.	Deep skin layers; walls of organs; spaces around joints; bone cavities.
	4. Reticular connective tissue	Gel-like ground substance; network of reticular fibers with intervening reticular cells.	Walls of lymphatic organs.
	5. Dense regular connective tissue	Fibroblasts arranged in parallel rows between densely packed bundles of collagen fibers.	Tendons, ligaments, and aponeuroses.
	6. Dense irregular connective tissue	Irregularly arranged collagen and elastic fibers with intervening fibroblasts.	Dermis of skin; capsules around organs; coverings around brain, spinal cord, and nerves.
	7. Elastic tissue	Parallel bundles of elastic fibers with fibroblasts interspersed between them.	Elastic ligaments between vertebrae.
C. Solid connective tissue	8. Cartilage	Chondrocytes within cavities, called lacunae; lacunae separated by a solid matrix with varying amounts of collagen and elastic fibers. Three types: hyaline cartilage, fibrocartilage, and elastic cartilage.	Ends of long bones; costal cartilages of ribs; cartilages of nose, trachea, larynx, epiglottis, and ear; intervertebral discs; discs in joint cavities.
	9. Bone	Osteocytes within lacunae; lacunae separated by a solid matrix containing collagen fibers and calcium salts.	Bones of the skeleton.
D. Fluid connective tissue	10. Blood	Blood cells in a fluid matrix (blood plasma).	Within blood vessels.

4. *White blood cells (leukocytes)*. The various types of white blood cells are transported by the blood to all regions of the body. In response to injury or infection, they can migrate into other connective tissues, where they have important roles in both the immune and inflammatory responses.
5. *Plasma cells*. When the body is exposed to an infectious agent, B-lymphocytes (a type of white blood cell) differentiate into plasma cells. Plasma cells produce antibodies.
6. *Reticular cells*. Reticular cells are specialized fibroblasts found in reticular connective tissue.

ACTIVITY 5.2 Microscopic Observations of Connective Tissue

A. Mesenchyme

1. Obtain a slide of **mesenchyme**, which is the first connective tissue to form in the developing embryo. Mesenchyme gives rise to all the other types of connective tissues.
2. View the slide under low power and identify the **mesenchymal cells**. The clear space separating the cells is the fluid ground substance of the matrix. The matrix also contains many protein filaments (Figure 5.8).
3. Switch to the high power and view the mesenchymal cells more closely. For each cell, identify the nucleus and the nucleolus within it. Depending on the quality of your slide, you may also be able to identify the lightly stained cytoplasm of some cells. Mesenchymal cells are star shaped, with several cell processes emerging from a central cell body. If your slide is a high-quality preparation, you should be able to identify cell processes on some cells.

B. Areolar Connective Tissue

1. Obtain a slide of **areolar connective tissue**.
2. View the slide under low power and notice the loose arrangement of connective tissue fibers and cells. The clear spaces between structures represent the gel-like matrix in which the cells and fibers are suspended (Figure 5.9).

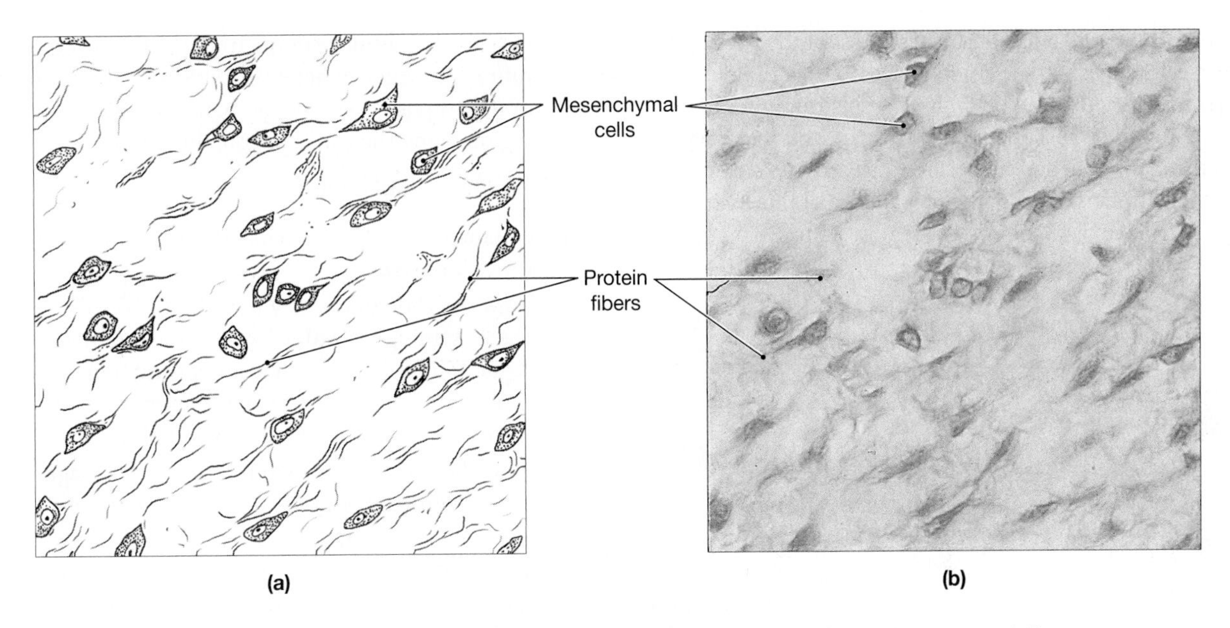

Figure 5.8 Mesenchyme. This embryonic tissue gives rise to all other types of connective tissue. **a)** Illustration of mesenchyme; **b)** corresponding light micrograph (LM × 400).

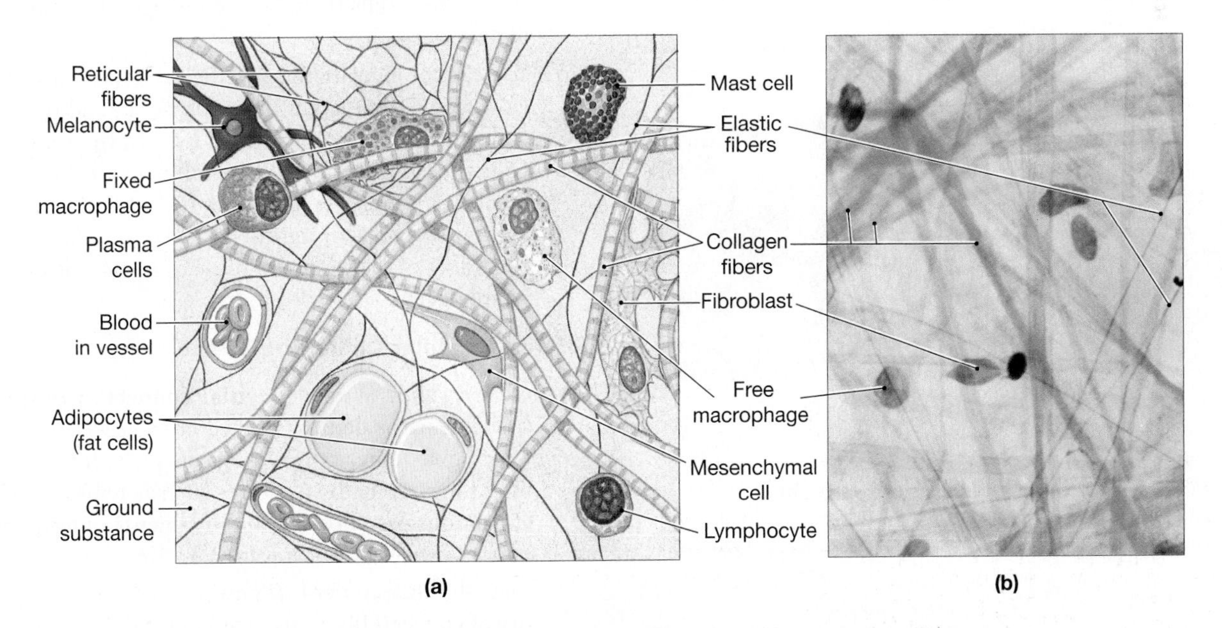

Figure 5.9 Areolar (loose) connective tissue. a) Illustration showing the various types of cells and protein fibers in this tissue; **b)** corresponding light micrograph (LM × 480).

3. Examine the visible fibers. The thick, pink-staining fibers are the **collagen fibers**. The very fine, dark-staining (blue-purple) fibers are **elastic fibers**.
4. Although there are many different cell types in this connective tissue, most cells in this preparation are **fibroblasts.** They can be identified by the purple-staining nuclei scattered throughout the specimen. The cytoplasm of the cells is stained very weakly, and therefore, it is difficult to identify.
5. Gently pinch the skin on your arm. Notice that when the skin is lifted up, the underlying muscle is not affected, because a layer of areolar connective tissue is located between the skin and muscle and allows for independent movement between the two. Areolar connective tissue also forms a thin layer directly below epithelial cells and around capillaries.

C. Adipose (Fat) Tissue

1. Obtain a slide of **adipose (fat) tissue.**
2. View the slide under low power. Adipose tissue contains very little intercellular matrix. You are viewing a region of closely

packed **adipose cells**, or **adipocytes**. Most of the volume of adipose cells appears to be empty spaces, with a thin, darkly stained rim of cytoplasm along the margins (Figure 5.10). In living tissue, these spaces are filled with fat, and the remainder of the cytoplasm and nuclei are squeezed into a thin rim along the periphery. During the preparation of this slide, the fats were extracted, so the spaces are now empty.

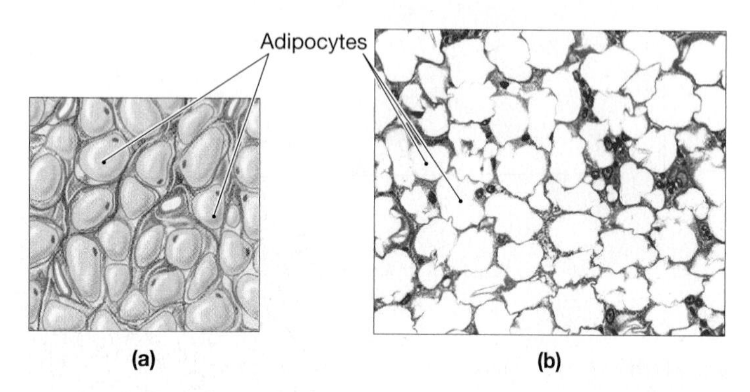

Figure 5.10 Adipose (fat) tissue. Adipocytes are specialized cells that are used for fat storage. **a)** Illustration of adipose tissue; **b)** corresponding light micrograph (LM × 100).

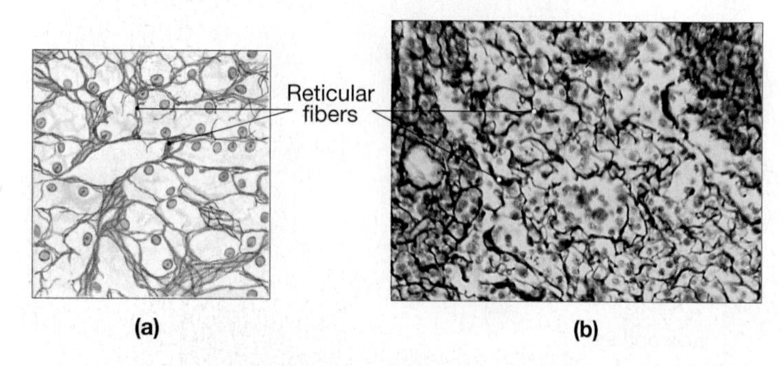

Figure 5.11 Reticular tissue. This tissue provides a supporting framework for many organs in the body. **a)** Illustration of reticular tissue; **b)** corresponding light micrograph (LM × 100).

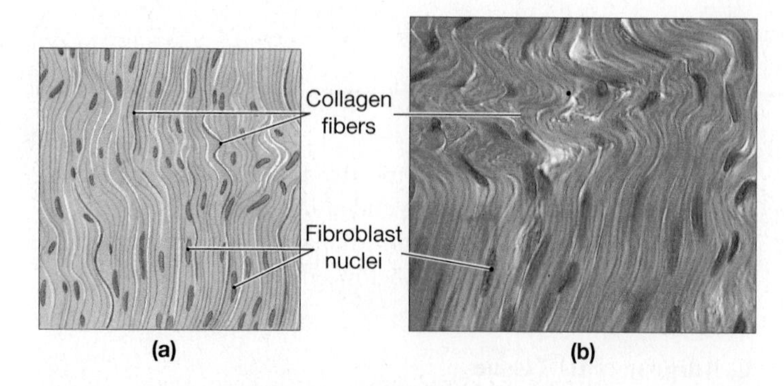

Figure 5.12 Dense regular connective tissue. This type of tissue is found in tendons and ligaments. **a)** Illustration of a tendon; **b)** corresponding light micrograph (LM × 400).

3. Observe the adipose tissue under high power. At this magnification, you can get a better view of the cytoplasmic rim of the adipose cells. As you scan the slide, try to find a region along the rim that contains a bulging nucleus.

D. Reticular Connective Tissue

1. Obtain a slide that is prepared specifically to demonstrate **reticular tissue**.
2. View the slide under low power. Observe the network of black-staining **reticular fibers** coursing throughout the tissue. These fibers form a supporting framework for organs in the lymphatic system (lymph nodes, thymus, and spleen) as well as in the liver and kidney.
3. View the slide under high power. At this magnification, you can more clearly see the arrangement of the reticular fiber network (Figure 5.11).
4. Various types of cells are found in reticular tissue. On the slides that you are viewing, the nuclei of cells can be easily identified, but the surrounding cytoplasm stains very faintly and is often difficult to see. Many of the cells directly associated with the reticular fibers are **reticular cells**. Other cells include macrophages, fibroblasts, and lymphocytes.
5. You cannot identify the specific cell types on your slides. However, lymphocytes in lymph nodes and the spleen are concentrated in regions called **lymphatic nodules**. If your slide is a section of a lymph node or the spleen, identify a lymphatic nodule. A nodule can be found by looking for areas of tightly packed, dark-staining cell nuclei. Most of the nuclei in these aggregations belong to lymphocytes.

E. Dense Regular Connective Tissue

1. Obtain a slide of **dense regular connective tissue**. This tissue contains densely packed bundles of collagen fibers, which give it great strength. That is why it is ideally suited for strong, resilient structures such as **tendons**, which attach muscles to bones, and **ligaments**, which connect bones to other bones at joints.
2. View the slide under low power and note the large quantity of collagen fibers arranged in parallel bundles.
3. Center the specimen in the field of view, then switch to high power. At this magnification, you can see parallel bundles of collagen fibers more clearly. Fibroblasts (dark staining cells) are arranged in parallel rows between the fibrous bundles (Figure 5.12).

F. Dense Irregular Connective Tissue

1. Obtain a slide of human skin. You examined this slide earlier when you studied the structure of stratified squamous, keratinized epithelium in the epidermis.
2. View the slide under low power. Identify the epidermis once again. Directly below the epidermis is the dermis of the skin. The dermis is composed of **dense irregular connective tissue** (Figure 5.13). This type of connective

tissue consists mostly of collagen fibers although elastic and reticular fibers are also present. Unlike the arrangement in tendons and ligaments (dense regular connective tissue), the fibers in this tissue form a branching and interlacing meshwork. Compare the appearance of this tissue with the parallel bundles of fibers in dense regular connective tissue (Figure 5.12).

G. Elastic Tissue

1. Obtain a slide of **elastic tissue**, which is a type of dense regular connective tissue that contains a large amount of elastic fibers. This tissue is found in areas that need flexibility, such as the ligaments that support the vertebral column and large arteries such as the aorta.
2. View the slide first under low power, then under high power. Note the bundles of elastic fibers, usually dark staining, that predominate in this tissue (Figure 5.14).

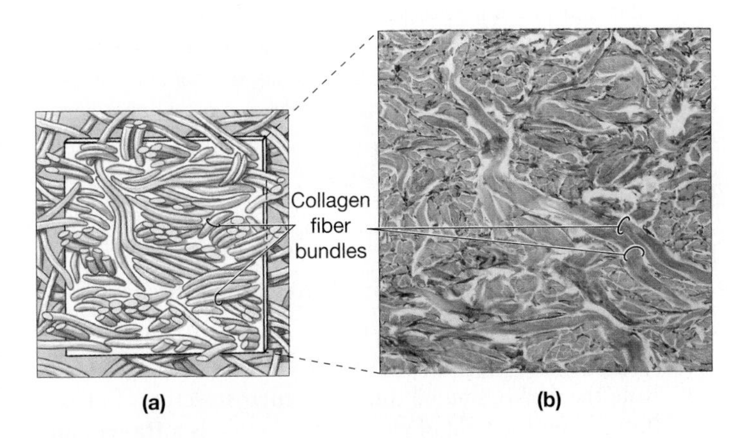

Figure 5.13 Dense irregular connective tissue. Collagen fibers form a highly branched meshwork. **a)** Illustration of dense irregular connective tissue in the dermis of the skin; **b)** corresponding light micrograph (LM × 150).

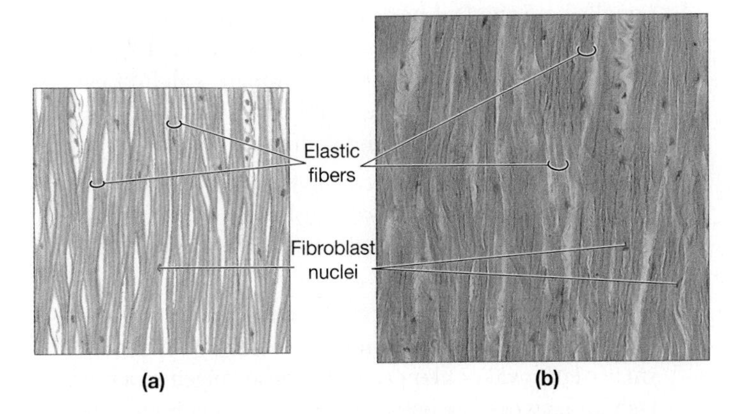

Figure 5.14 Elastic tissue. This tissue is abundant in structures that need to be highly flexible, such as large arteries and ligaments that connect vertebrae. **a)** Illustration of elastic tissue; **b)** corresponding light micrograph (LM × 250).

H. Cartilage

1. Obtain a slide of the trachea.
2. View the slide under low power. Locate the pseudostratified columnar epithelium that you examined before, which lines the lumen of the trachea. Underlying the epithelium is a thick layer of dense irregular connective tissue. The thinner epithelial layer appears slightly darker than the underlying connective tissue.
3. Next you will find a light purple-staining region of **hyaline cartilage** (Figure 5.15a). The trachea is distinguished by a series of C-shaped cartilage rings in its wall. You are viewing a portion of one cartilage ring.

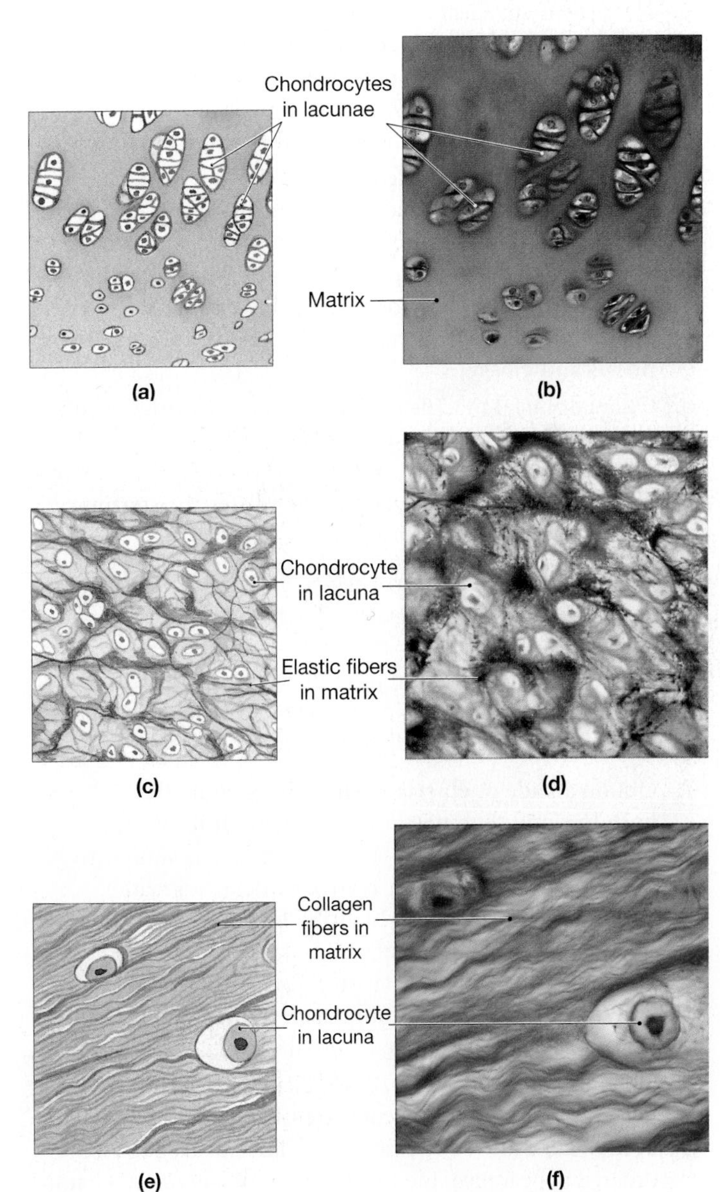

Figure 5.15 Three types of cartilage. In cartilage, chondrocytes are located in lacunae, which are separated by a solid matrix material. **a and b)** Hyaline cartilage (LM × 600); **c and d)** elastic cartilage (LM × 600); **e and f)** fibrocartilage (LM × 600).

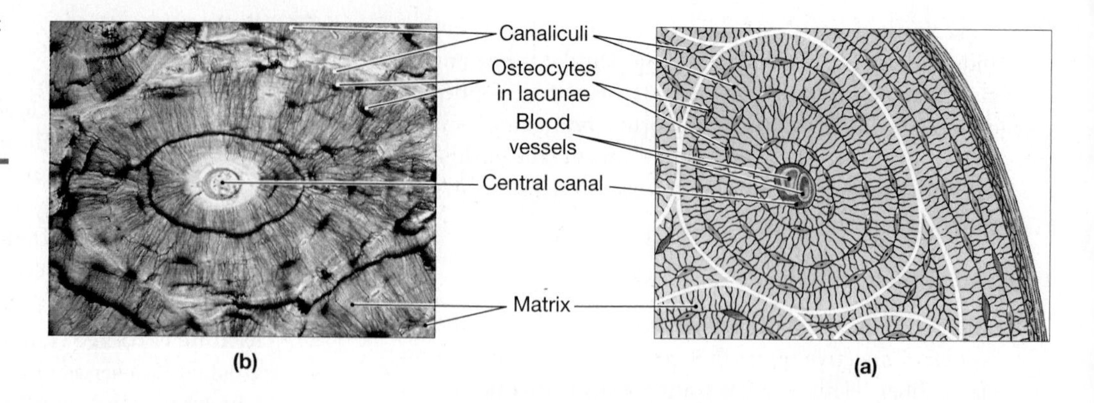

Figure 5.16 Compact bone. The basic structural unit of compact bone is the Haversian system. **a)** Illustration of compact bone; **b)** corresponding light micrograph (LM × 300).

4. Note the many clear spaces in the cartilage. These spaces are cavities known as **lacunae** (singular = **lacuna**). Within each lacuna is a **cartilage cell**, or **chondrocyte** (Figures 5.15a and b). Note that some lacunae occur singly, but many are clumped together in small groups.
5. The lacunae are separated from each other by a solid matrix material. The matrix contains large amounts of water, adhesion proteins, proteoglycans, and varying amounts of collagen and elastic fibers. The matrix in hyaline cartilage contains many collagen fibers, but they are not clearly visible. Instead, the matrix appears amorphous and glossy.
6. On anatomical models or yourself, identify other structures that contain hyaline cartilage:
   - cartilage of the nose
   - costal cartilages, connecting the ribs to the sternum
   - cartilage of the larynx

WHAT'S IN A WORD The term *hyaline* is derived from the Greek word *hyalos,* which means "glass." The name reflects the glassy, homogenous quality of the matrix in hyaline cartilage.

The term *lacuna* is derived from the Latin word, *lacus,* which refers to a hollow or a lake. In cartilage and bone, a lacuna is a space or cavity in which a cell is located.

7. Obtain a slide of **elastic cartilage.** Examine the slide on both low and high power and compare it to the appearance of hyaline cartilage. Elastic cartilage is similar to hyaline cartilage, but the matrix contains a greater amount of elastic fibers. Identify the dark-staining elastic fibers, which give the matrix a fibrous appearance (Figures 5.15c and d). Elastic cartilage is located in areas where elasticity is important, such as the epiglottis and the auricle (pinna) of the external ear.
8. Obtain a slide of **fibrocartilage.** Examine the slide on both low and high power and compare its appearance to that of both hyaline and elastic cartilage. Fibrocartilage has a more ordered appearance. Identify the rows of lacunae, containing chondrocytes, alternating with rows of thick collagen fibers (Figures 5.15e and f). Fibrocartilage is a good shock absorber, and is located in areas where protection from heavy pressure is needed, such as the intervertebral discs between the vertebrae, and the menisci on the knee joint.

CLINICAL CORRELATION

The matrix in cartilage contains a lot of water. Because cartilage generally does not have a good blood supply, the water is vital for helping nutrients to diffuse through the matrix. The lack of blood supply in cartilage is starkly contrasted to the rich blood supply found in bone. In fact, damaged cartilage can be far more difficult to heal than a fractured bone. In addition, most bones in the body form from hyaline cartilage. If the cartilage is damaged before the bone has fully grown, growth of the damaged bone may be abruptly halted.

I. Bone

1. Obtain a slide with a section of ground bone tissue.
2. View the slide under low power (Figure 5.16). You are observing a cross section of **compact bone.**
3. Note the darkly stained (black) circular structures found throughout the field of view. Each of these is a **Haversian (central) canal**, located in the center of a **Haversian system**. These canals allow the passage of numerous blood vessels and nerves throughout the compact bone.
4. Note several concentric layers of **bone lamellae** (singular = **lamella**) surrounding each Haversian canal. The lamellae are rings of bone matrix organized around the Haversian canal. The bone matrix contains collagen fibers, calcium salts (calcium phosphate, calcium carbonate, calcium hydroxide), and a variety of ions such as fluoride, magnesium, and sodium.
5. Identify additional bone lamellae that do not appear to be surrounding Haversian canals. Instead, they are located between adjacent Haversian systems. These areas are called **interstitial lamellae.**
6. Similar to cartilage, small cavities or **lacunae** are present in bone. Identify lacunae on the slide. They appear as small, dark oval or elongated spaces, arranged in a circular fashion between the bone lamellae. In living bone tissue, **osteocytes** (mature bone cells) are located within the lacunae. However, in a ground bone preparation, such as the slides you are viewing, all organic material is removed, so the osteocytes are absent.

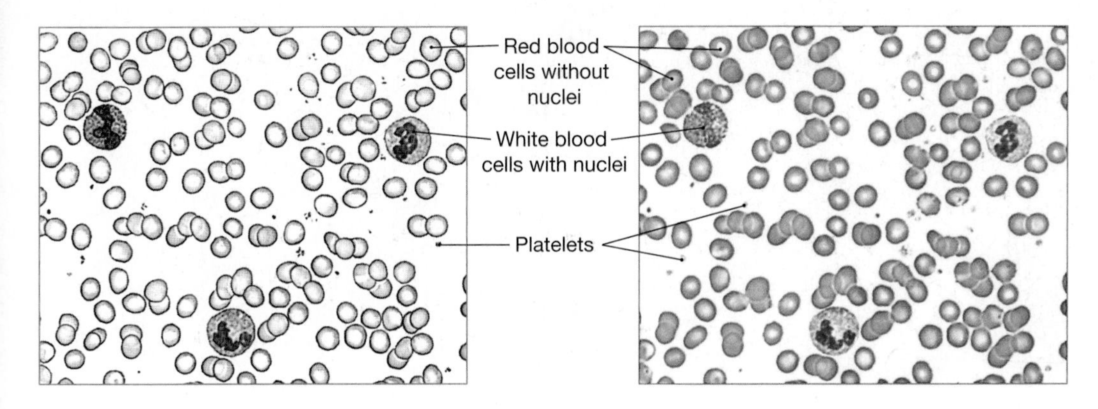

Figure 5.17 Human blood smear. Blood is a connective tissue with a liquid matrix material (blood plasma). **a)** Illustration of a human blood smear; **b)** corresponding light micrograph (LM × 300).

7. Switch to high power and focus on several lacunae within one Haversian system. At this magnification, identify the thin black threads that appear to originate from the lacunae and course through the bone matrix. These structures are the **canaliculi** (singular = **canaliculus**), which are microscopic canals passing through the lamellae. The canaliculi transmit the cell processes of osteocytes (the cell processes were removed during the slide preparation). Notice how canaliculi from one lacuna tend to merge with the canaliculi from other nearby lacunae, thus forming a vast network of connecting channels. By traveling through the canaliculi, cell processes of neighboring osteocytes can communicate with each other and with the blood supply in the Haversian canal.

WHAT'S IN A WORD The word *lamella* is a form of the Latin word *lamina,* which means "plate." A bone lamella is a layer or plate of bone tissue. The word *canaliculus* is derived from the Latin word *canalis,* which means "canal." A canaliculus is a microscopic canal in bone tissue.

J. Blood

1. Obtain a slide of a human **blood** smear. Blood is classified as a fluid connective tissue. The matrix is the blood plasma, which consists of water and numerous dissolved solutes (e.g., proteins, ions, gases, hormones, nutrients).
2. View the slide under low power and notice that the entire field is filled with blood cells.
3. The vast majority of the blood cells are the pinkish-staining **red blood cells**, or **erythrocytes.** As you scan the slide, you should be able to identify the larger and more darkly stained **white blood cells**, or **leukocytes.**
4. Move the slide so that some white blood cells are in the center of the field of view. Switch to high power to examine the blood cells more closely (Figure 5.17).
5. Note that the red blood cells lack a nucleus. They lose their nuclei as they mature so that more space is available to store hemoglobin, the molecule needed to transport oxygen. White blood cells retain their nuclei, which are stained a deep blue or purple.
6. There are five basic white blood cell types. You will be studying the blood cells in more detail in Exercise 19. For now, scan the slide under high power and see if you notice any structural differences between the various white blood cells that you see.
7. You might also see small cell fragments in the blood smear. These fragments are platelets, or thrombocytes, and are essential for blood clotting.

QUESTION TO CONSIDER During the previous activity, you observed connective tissues with diverse structures and functions. Consider the structure and function of the following three connective tissues: areolar connective tissue, bone, and blood. Identify their similarities and differences.

Similarities:

___

___

Differences:

___

___

Exercise 5 Review Sheet

Epithelial and Connective Tissues

Name ______________________________

Lab Section ______________________________

Date ______________________________

1. Briefly describe the major characteristics of epithelial tissue.

____

____

2. What are the basic functions of epithelial tissue?

____

____

3. What structural criteria are used to classify epithelial tissue?

____

____

4. Identify the epithelial types illustrated in the following diagrams.

(a) (b) (c)

(d) (e) (f)

a. ______________________________

b. ______________________________

c. ______________________________

d. ______________________________

e. ______________________________

f. ______________________________

5. What are the basic functions of connective tissue?

____

____

6. What is the connective tissue matrix?

____

____

7. Compare the structure of the three types of cartilage and list two locations where each is found.

____

____

____

Questions 8–13: Complete the following table.

Connective Tissue Type	Location
8. Mesenchyme	
9. Adipose tissue	
10.	Tendons and ligaments
11.	Intervertebral discs
12.	Skeleton
13. Areolar connective tissue	

Questions 14–21: Match the connective tissue cells in column A with the appropriate description in column B.

A	B
14. Macrophages ____________	a. Transported in blood; help to fight off infections
15. White blood cells ____________	b. Specialized fibroblasts; produce reticular fibers
16. Adipocytes ____________	c. Cells found in bone tissue
17. Plasma cells ____________	d. Similar to some white blood cells; attacks foreign particles by phagocytosis
18. Mast cells ____________	e. Cells found in cartilage
19. Reticular cells ____________	f. Promote inflammation by secreting heparin and histamine
20. Osteocytes ____________	g. Cells that produce antibodies
21. Chondrocytes ____________	h. Storage sites for fat molecules

Questions 22–26: Coloring exercises. For each diagram, identify the structures by using the color that is indicated.

22.

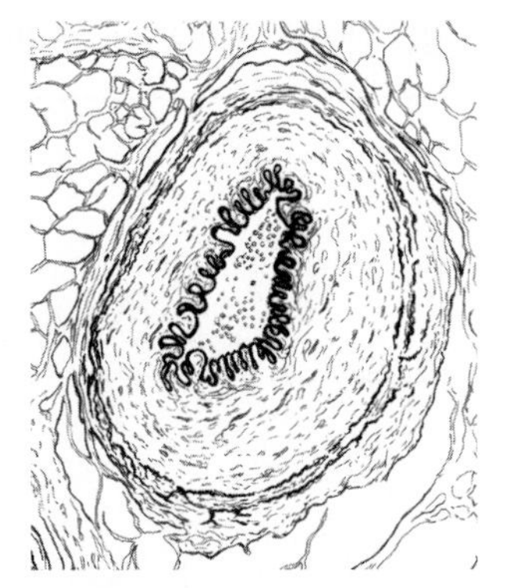

Cross section of an artery.

- Simple squamous epithelium in the tunica intima = **green**
- Smooth muscle in the tunica media = **red**
- Loose connective tissue in the tunica adventitia = **yellow**

23.

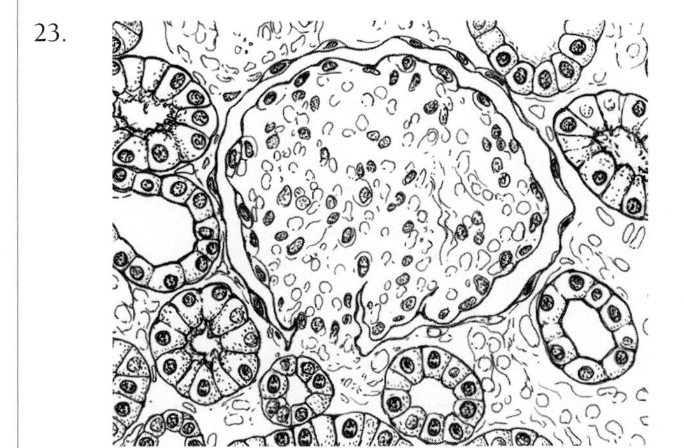

Renal tubules and a renal corpuscle in the kidney.

- Simple squamous epithelium lining Bowman's capsule = **green**
- Simple cuboidal epithelium lining the kidney tubules = **red**
- Connective tissue = **yellow**

24.

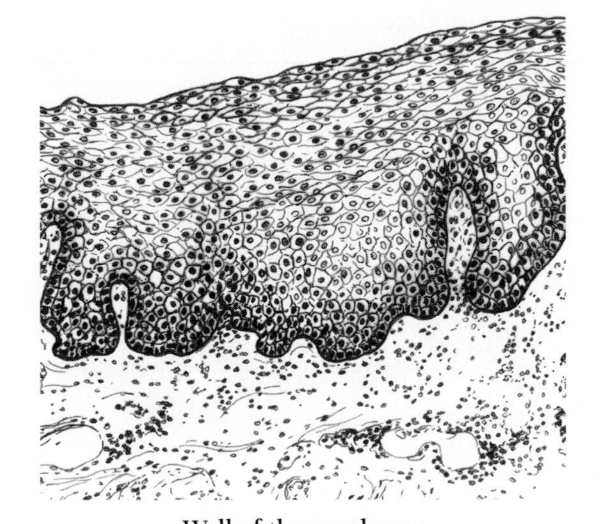

Wall of the esophagus.

- Epithelium = **red**. What type? ______________________
- Connective tissue = **green**

25.

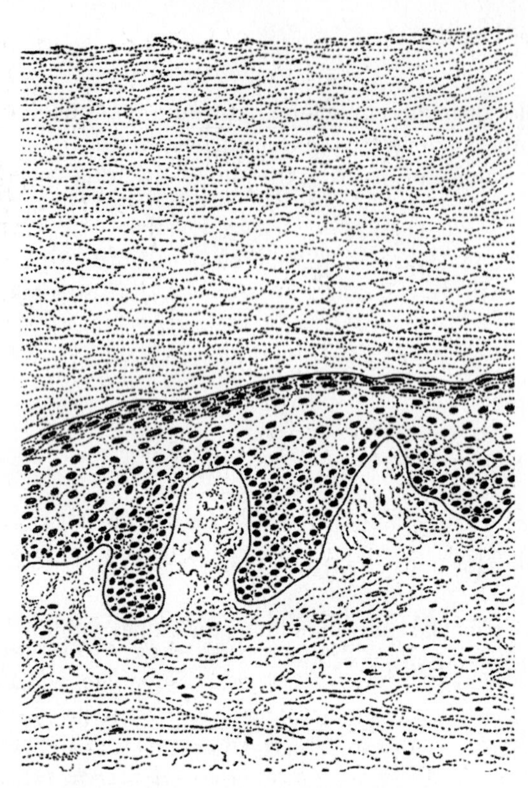

The skin.

- Epidermis of skin = **red.** What type of epithelium?

  ____

- Dermis of skin = **green.** What type of connective tissue?

  ____

26.

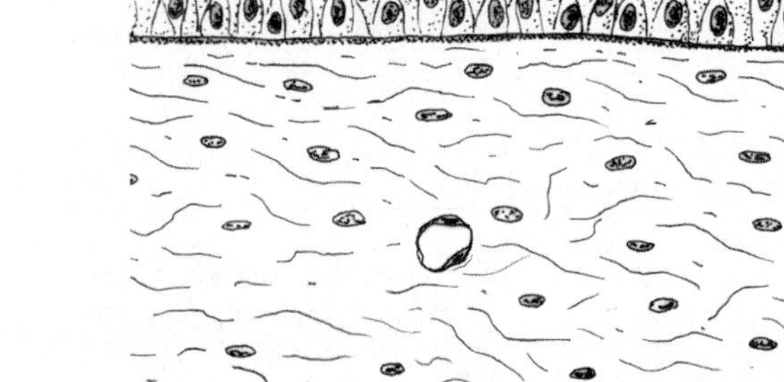

Wall of the trachea.

- Epithelium = **green.** What type of epithelium?

  ____

- Dense irregular connective tissue = **red**
- Cartilage:
  - a. Matrix = **yellow**
  - b. Chondrocytes within lacunae = **blue**
- What type of cartilage?

  ____

Exercise 6

The Integumentary System

Laboratory Objectives

On completion of the activities in this exercise, you will be able to:

- Describe the organization of the epidermis, dermis, and hypodermis.
- Identify the accessory structures of the skin.
- Describe the organization and distribution of, and relationships between, accessory structures in the skin.

Materials

- Handheld magnifying glass
- Compound light microscope
- Prepared microscope slides:
  - Scalp
  - Scalp, pigmented
  - Skin, foot, human
- Anatomical model of a skin section
- Colored pencils

As you study the structure and function of the integumentary system, you enter new levels of organization of the human body—the **organ level** and the **organ system level**. An **organ** is a structure that contains two or more of the primary tissue types, and often all four. Each tissue type works collaboratively with the others to perform specific functions. An organ system consists of two or more organs that work together in a coordinated manner to perform specific functions. The **cutaneous membrane**, or **skin**, is an organ. Embedded within it are various **accessory structures**, some of which are also considered organs (Figure 6.1). The cutaneous membrane and accessory structures form the organ system known as the **integumentary system.**

The Cutaneous Membrane

In terms of total surface area, the skin is the largest organ of the body. It consists of two tissue regions: the outer **epidermis** and the inner **dermis** (Figure 6.1).

The surface layer of skin, the epidermis, is a stratified, squamous keratinized epithelium. In **thick skin**, which covers the palms of the hand and soles of the feet, the epidermis contains five cell layers—the **stratum germinativum**, **stratum spinosum**, **stratum granulosum**, **stratum lucidum**, and **stratum corneum** (Figure 6.2). In **thin skin**, which covers other parts of the body, the stratum lucidum and often the stratum granulosum are absent.

The stratum germinativum contains a population of actively dividing cells. As new cells form, they displace the cells that are already present, pushing the older cells toward the surface, where they gradually accumulate keratin and eventually die. Thus, the outermost layer of skin, the stratum corneum (keratinized layer), consists of several layers of dead, keratin-filled

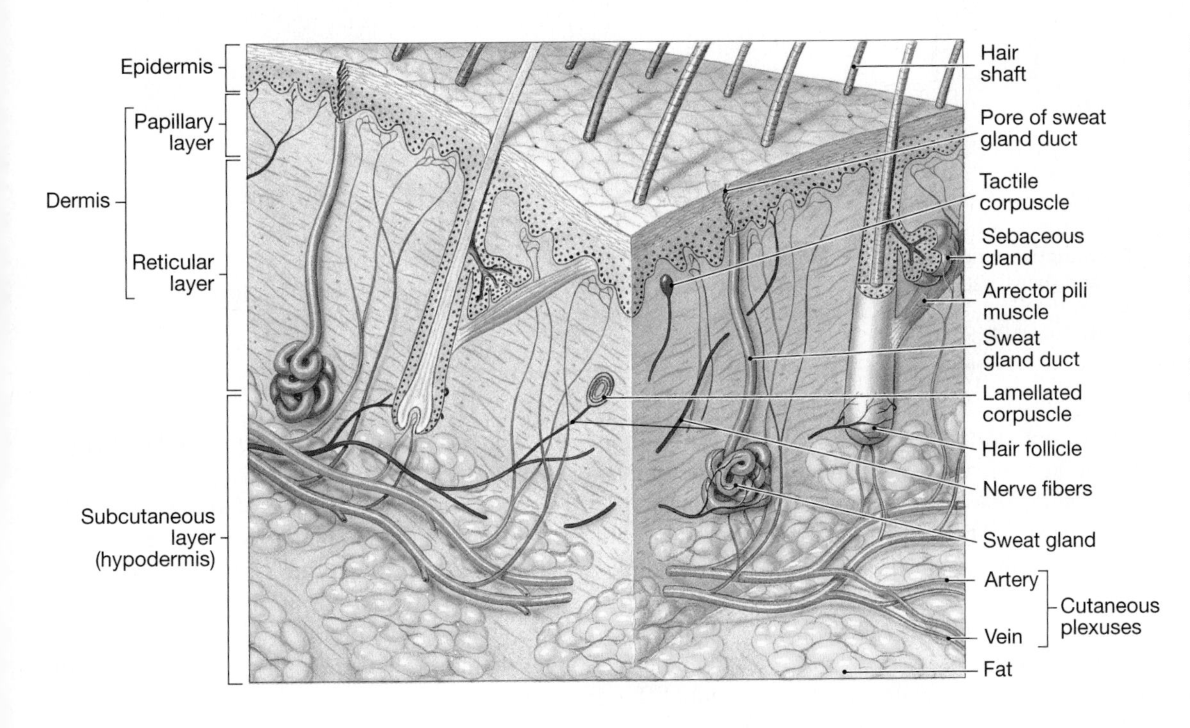

Figure 6.1 General organization of the integumentary system. The epidermis is a stratified squamous, keratinized epithelium. The dermis contains areolar and irregular dense connective tissue. Various accessory structures are embedded in the dermis. The hypodermis is deep to the dermis and contains areolar connective tissue and adipose tissue.

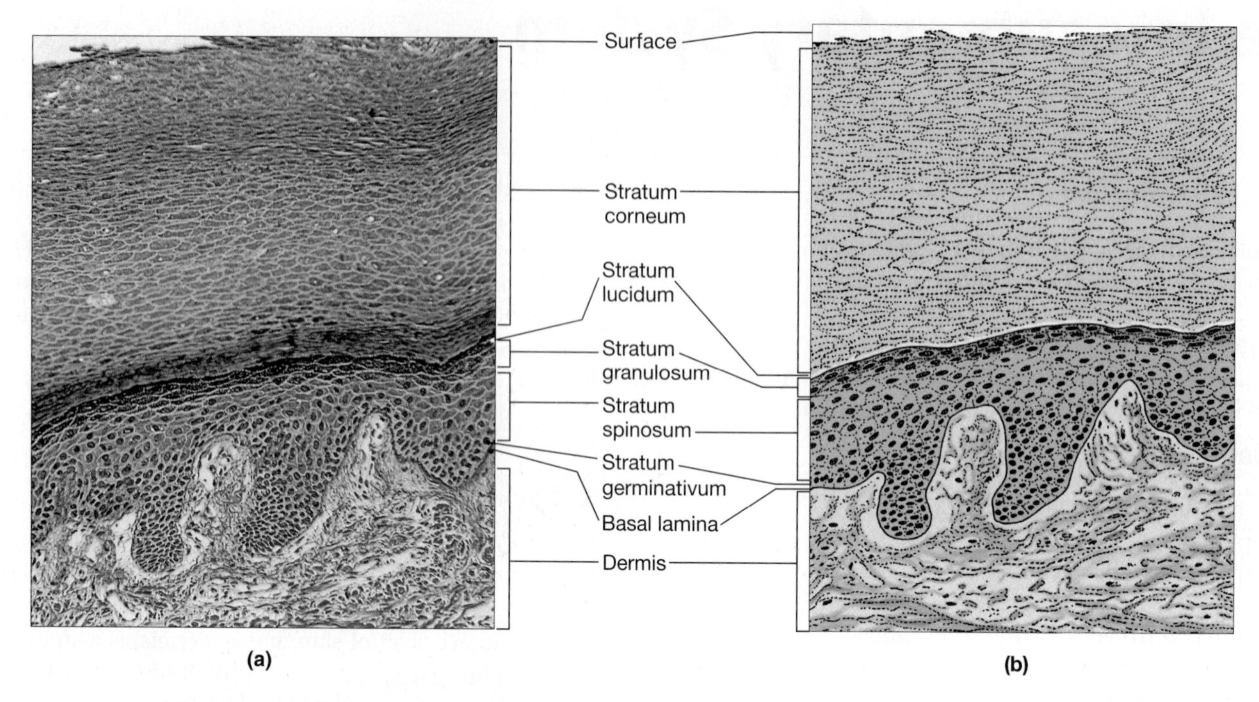

Figure 6.2 Cell layers in the epidermis of thick skin. a) Light micrograph of thick skin (LM × 100); **b)** corresponding illustration.

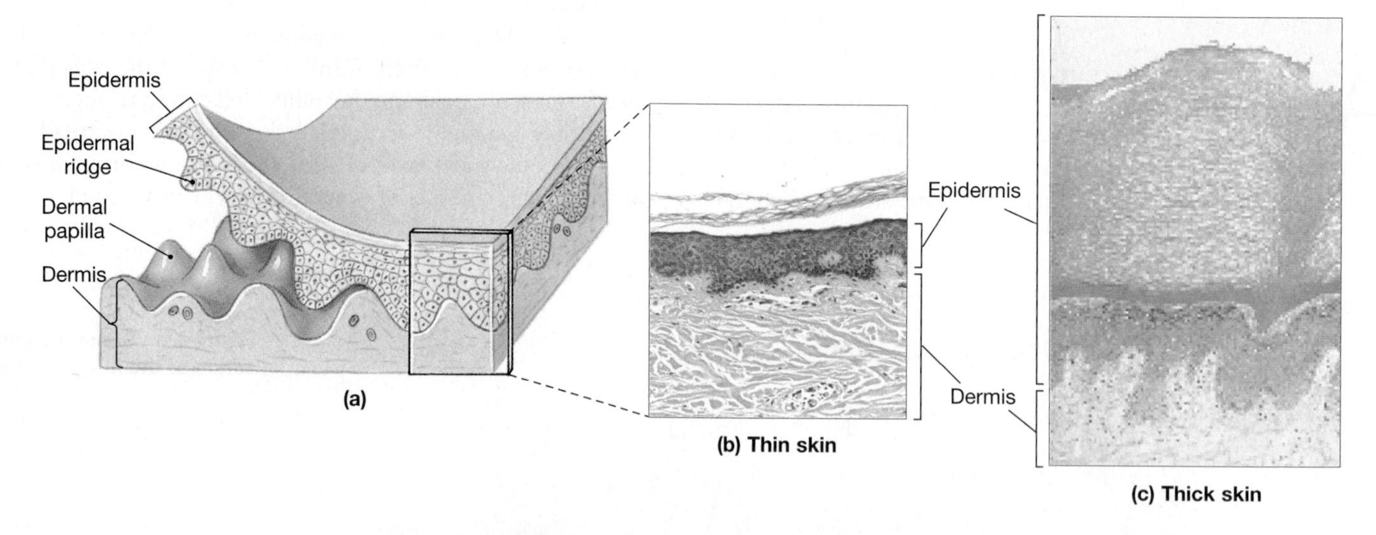

Figure 6.3 Comparative structure of thin skin and thick skin. a) Illustration demonstrating the pattern of epidermal ridges and dermal papillae along the border between the epidermis and dermis; **b)** light micrograph of thin skin (LM × 50); **c)** light micrograph of thick skin. Notice the difference in thickness between the stratum corneum in thin skin and thick skin (LM × 50).

cells (Figure 6.2). Keratin, a waterproof protein, protects the body from excessive water loss. The keratinized layer is much thicker in the soles and palms (thick skin) than in other parts of the body (thin skin; Figure 6.3). The cells in the keratinized layer continuously slough off as more cells are pushed to the surface.

Deep to the epidermis is a connective tissue layer called the dermis. This region is divided into two layers. The relatively thin **papillary layer**, composed largely of areolar connective tissue, directly borders the epidermis. Deep to the papillary layer is the **reticular layer**, which is composed of dense irregular connective tissue and contains the **accessory structures of skin** (Figure 6.1).

The epidermis forms an undulating border with the papillary layer, along which **dermal papillae** project between **epidermal ridges** (Figure 6.3). In the fingertips, the border between the papillary layer and the epidermis produces a distinctive pattern of epidermal ridges that form the unique fingerprints of an individual (Figure 6.4).

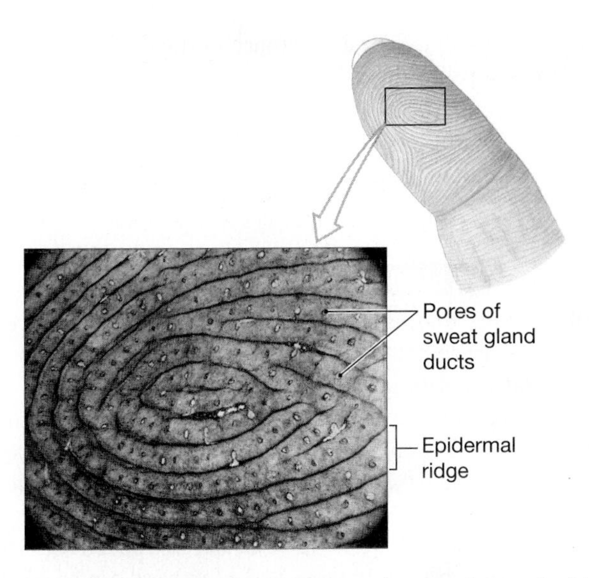

Figure 6.4 Formation of a fingerprint. An electron micrograph of the skin covering the fingertip. The unique pattern of the epidermal ridges produce the fingerprint (LM × 50).

Underlying the dermis is the **hypodermis** (Figure 6.1). This layer is composed of areolar connective tissue and adipose tissue. The hypodermis is usually not considered to be a part of the integumentary system; however, its fibers are continuous with those in the dermis, so it is usually discussed with the skin. It also contains blood vessels and nerves that supply the skin. The hypodermis is sometimes referred to as the **superficial fascia** or **subcutaneous tissue.** It connects the skin to underlying muscles.

WHAT'S IN A WORD The term *dermis* is derived from the Greek word *derma,* which means "skin." The words *epidermis* and *hypodermis* are derived by adding various roots to *-dermis.* The prefix *epi-* means "upon" (derived from Greek), and, indeed, the epidermis is located upon the dermis. *Hypo-* means "under" (also originating from Greek); the hypodermis is deep to the dermis.

CLINICAL CORRELATION

Some medications are administered by hypodermic injection, meaning they are injected deep to the dermis. Others, such as nicotine, nitroglycerine, and some hormone replacements, are administered transdermally (meaning "across the dermis") by applying a patch that contains the medication to the skin. The medication is then absorbed "across" the skin.

ACTIVITY 6.1 Examining Fingerprint Patterns

1. Observe the ridges on the epidermis that covers your fingertips.
2. Use a magnifying glass to examine the patterns on two of your fingers more closely (Figure 6.4). How do the patterns on your two fingers compare with each other?

____

____

3. Examine the ridges on the same fingers on your lab partner's hand. How do the patterns on your two fingers compare to those on your lab partner's hand?

____

____

QUESTION TO CONSIDER What factors might alter a person's fingerprint patterns during his or her life-time?

____

____

ACTIVITY 6.2 Examining the Microscopic Structure of Skin

1. Obtain a slide of the scalp.
2. View the section under low power. You are observing a section of thin skin.
3. Locate the darkly stained epidermis at one edge of the section. It consists of a **stratified squamous, keratinized epithelium** (Figures 6.1 and 6.3b). Most of the cells in the epidermis are called **keratinocytes**.
4. Switch to high power and observe the epidermis more closely. On conventional microscope slide preparations, it is difficult to identify all of the epidermal cell layers mentioned earlier (Figure 6.2). However, two of these layers can be located.
   - The stratum corneum (keratinized layer) is the outermost layer of the epidermis. It contains several layers of dead keratinocytes whose cytoplasm is filled with keratin. The stratum corneum forms a protective, water-resistant covering, but the cells in this layer are eventually shed from the surface. In the scalp, the keratinized layer is very thin (Figure 6.3b).
   - The stratum germinativum, or stratum basale, is a single layer of cuboidal cells located along the base on the epidermis. These cells regularly divide to replace the old keratinocytes that are sloughed off the stratum corneum.
5. Locate the dermis just below the epidermis (Figures 6.1 and 6.3). Observe that the narrow papillary layer, bordering the epidermis, contains areolar connective tissue. Also note that the reticular layer comprises the majority of the dermis and consists of dense irregular connective tissue. In the dermal layer, you will find the accessory structures of the skin (Figure 6.1). The accessory structures will be studied in Activity 6.3.
6. Deep to the dermis, locate the hypodermis (Figure 6.1). On some sections, this layer might be missing. If it is present, observe that it is composed largely of loose connective tissue and adipose tissue.
7. Obtain a slide of the pigmented scalp.
8. View the slide under low power and locate the epidermis.

9. Some cells in the epidermis contain **melanin**, which is a protein pigment that appears dark brown on your slide (Figure 6.5). Melanin is produced by special cells called **melanocytes**, found deep in the stratum germinativum. The pigment is secreted within vesicles, called **melanosomes**, and transferred to the keratinocytes in the more superficial layers of the epidermis. On your slide, describe the distribution of the pigment within the epidermis.

____

____

10. Obtain a slide of skin from the sole of the foot.
11. View the section under low power. You are observing a section of thick skin.
12. Locate the epidermis at one edge of the section. Notice that the keratinized layer is very thick in the sole of the foot (Figure 6.3c).
13. Scan the slide under low power and identify the three tissue layers that you observed on the slide of the scalp: epidermis, dermis, and hypodermis.

QUESTIONS TO CONSIDER 1. How does the thickness of the epidermis, in general, and the keratinized layer, specifically, in the thin skin of the scalp compare with that in the thick skin of the sole?

____

____

2. How does the thickness of the dermis of the scalp compare to that of the sole of the foot?

____

____

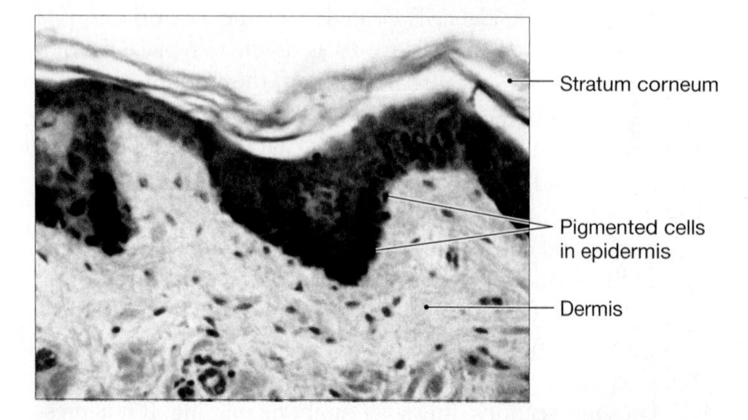

Figure 6.5 The distribution of melanin in the epidermis of thin skin. Melanin is produced in melanocytes and transferred to keratinocytes. The two cell types are indistinguishable in this light micrograph (LM × 200).

3. Why would you expect a difference in the thickness of the skin of the scalp and the sole?

____

____

CLINICAL CORRELATION

Malignant melanoma is a type of cancer that causes unregulated reproduction of the melanocytes. Because these cells produce melanin, a dark brown pigment, melanomas are often easily spotted because of their dark coloration. Although these malignancies may grow slowly at first, they become aggressive and, if not caught early, have a high mortality rate. Melanin protects us from genetic damage caused by ultraviolet radiation, which can lead to cancer. People who have skin with a high level of melanin have maximum protection against this damage. The risk of melanoma is about 20 times higher for people of European descent than for people of African descent. However, when it does occur in people of African descent, it is more often deadly. Early detection is the key to a cure, and that requires frequent self-examination of all body areas, not just those exposed to sunlight.

Accessory Structures of the Skin

Accessory structures of the skin are well established in the dermis of the skin. Portions of some structures, such as hair follicles and the ducts of sweat glands, extend deeply into the hypodermis or superficially into the epidermis. The accessory structures include hair follicles, sebaceous glands, arrector pili muscles, sweat glands, sensory receptors, and nails.

ACTIVITY 6.3 Examining Accessory Structures of the Skin

1. Once again, examine a microscope slide of thin skin from the scalp.
2. Scan the slide under low power and identify the various accessory structures in the dermis (Figure 6.6).
3. Identify numerous **hair follicles** in the dermis (Figure 6.6). Hair follicles are tubular compartments, each one containing a **hair**. The follicle walls are continuous with the epidermis and extend deeply into the dermis and hypodermis. On your slide, identify some follicles that are directly connected to the epidermis. Because of the plane of the slide section, other follicles may not appear to have this connection. The outer walls of hair follicles are connective tissue sheaths derived from the dermis. The deepest portion of a hair follicle is a knoblike structure called the **hair bulb** (Figure 6.6). Identify these structures on your slide. The hair bulb contains the **hair matrix**, a region of actively dividing cells that give rise to the **hair root**. As the hair root approaches the

surface of the skin, it becomes the **hair shaft**. A portion of the hair shaft is that part of the hair that we can see. Like cells in the keratinized layer of the epidermis, cells in the hair root and shaft become keratinized and die. The **hard keratin** that covers the hair is stronger and longer lasting than the **soft keratin** in the epidermis.

4. Locate some **sebaceous glands** and verify that they are connected to the walls of hair follicles. The glandular cells are arranged like clusters of grapes; they are lightly stained or clear with dark-staining nuclei (Figure 6.6).

CLINICAL CORRELATION

Sebaceous glands are found in all regions of the body except the palms of the hands and the soles of the feet. The glandular product is a lipid mixture called **sebum.** Secretion of sebum occurs when the glandular cells rupture, a process known as **holocrine secretion.** The sebum empties into a short straight duct, which drains into a hair follicle. In hairless regions of the body, sebaceous glands, called **sebaceous follicles,** secrete sebum into ducts that run directly to the skin's surface. Sebum acts as a softening agent that lubricates the skin and the hair. It also may serve as a protective agent against some bacterial and fungal infections. At puberty, under the influence of increasing levels of sex hormones, sebaceous glands enlarge and the secretion of sebum increases. As a result, pimples or acne lesions may form on the skin.

5. Attempt to identify the smooth muscle fibers of an **arrector pili muscle**. These structures can be seen attached to the connective tissue sheaths of hair follicles. They pass obliquely toward the papillary layer of the dermis, but because of the plane of section, the entire lengths of these muscles are rarely seen. Sebaceous glands are normally found in the triangular regions formed by hair follicles, arrector pili muscles, and the epidermis (Figure 6.6).

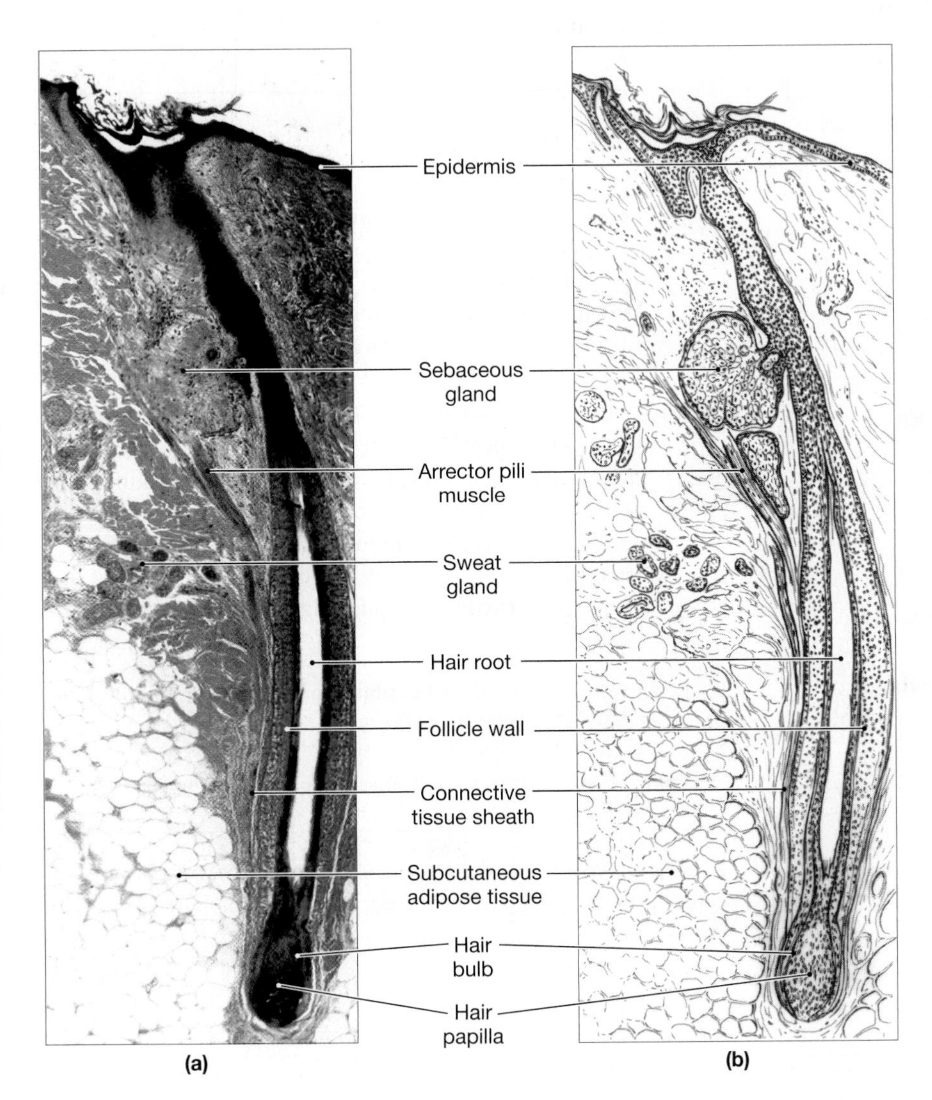

Figure 6.6 Accessory structures of skin. a) Light micrograph (LM × 100) and **b)** corresponding illustration of thin skin, showing hair follicles, sebaceous glands, sweat glands, and arrector pili muscles in the dermis. Notice the typical position of a sebaceous gland within a triangular region formed by a hair follicle, arrector pili muscle, and the epidermis.

CLINICAL CORRELATION

In response to cold temperatures or certain emotions such as fear or anger, arrector pili muscles contract. This action moves hair follicles to an upright position and dimples the skin to form "goose bumps." In addition, the muscle contractions promote the secretion of sebum from sebaceous glands.

6. Locate numerous **sweat glands** scattered throughout the dermis between hair follicles (Figure 6.7). The lighter stained regions of the sweat glands are the glandular cells that produce and secrete sweat. The darker staining portions are the sweat gland ducts. The ducts travel through the dermis and epidermis and empty onto the skin surface. Notice that the glandular regions consist of pseudostratified or cuboidal epithelial cells. The ducts are lined by a stratified cuboidal epithelium. The sweat glands that you are observing are **merocrine (eccrine) sweat glands**, which function throughout life and are important for regulating body temperature. **Apocrine sweat glands** are another type of sweat gland, but their locations are restricted to regions around the nipples, the axilla (armpit), and the anogenital regions of the body. They will not be present on your slide.

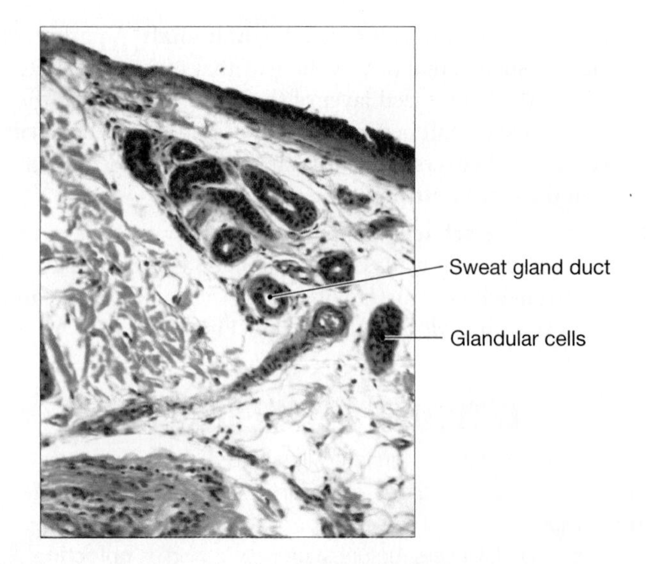

Figure 6.7 Merocrine (eccrine) sweat gland. Glandular cells secrete sweat into ducts that travel to the surface of the skin.

CLINICAL CORRELATION

Merocrine sweat glands are vital components of the body's temperature regulating mechanism. Sweat is released through tiny pores directly onto the surface of the skin. When the water in sweat evaporates, excess heat is released and the body is cooled.

Apocrine sweat glands are less numerous than merocrine sweat glands and they do not become functional until puberty. Their activity increases during periods of stress, pain, or sexual arousal and their secretions, which are released into hair follicles, have a distinct odor. The precise function of apocrine glands in humans is unknown, although some researchers believe that they are comparable to the sexual scent glands of other animals.

7. Obtain a slide of thick skin from the sole of the foot.
8. View the section first under low power, then switch to high power for a closer examination.
9. How does the presence or absence of the following accessory structures in the scalp compare to those in the sole?
   a. Hair follicles

   ____

   b. Sebaceous glands

   ____

   c. Arrector pili muscles

   ____

   d. Sweat glands

   ____

10. On the dermis of thick skin, attempt to identify two types of sensory receptors.
    - **Tactile (Meissner's) corpuscles** are nerve endings wrapped in a connective tissue sheath. They are sensitive to fine or light touch. Tactile corpuscles are abundant in various hairless regions of the skin (lips, nipples, fingertips, palms and soles, external genitalia). Attempt to identify these structures in the papillary layer of the dermis, immediately deep to the epidermis (Figures 6.1 and 6.8a and b)
    - **Lamellated (pacinian) corpuscles** are nerve endings enclosed in a relatively large, onion-shaped connective tissue sheath. They respond to deep pressure sensations. Attempt to identify a lamellated corpuscle deep in the dermis or in the hypodermis (Figures 6.1 and 6.8a and c).

QUESTIONS TO CONSIDER

1. Provide an explanation why some people have oily hair and others have dry hair.

____

____

2. Why do you think that tactile (Meissner's) corpuscles are located more superficially in the dermis than lamellated (pacinian) corpuscles?

____

____

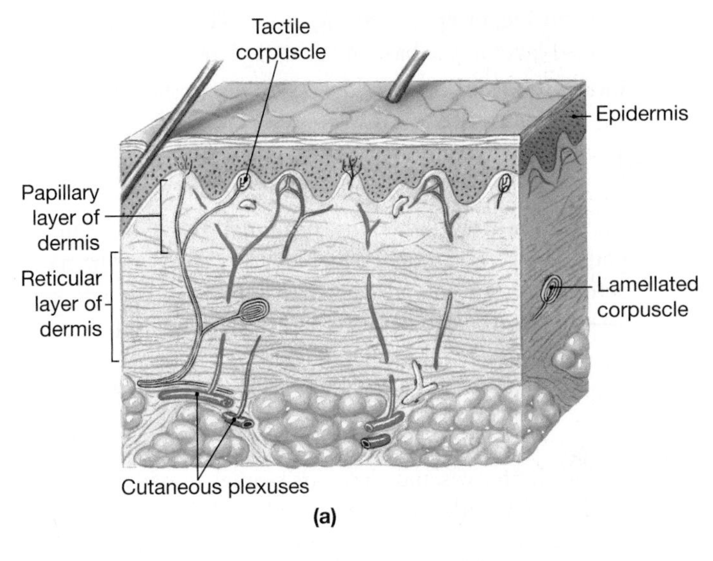

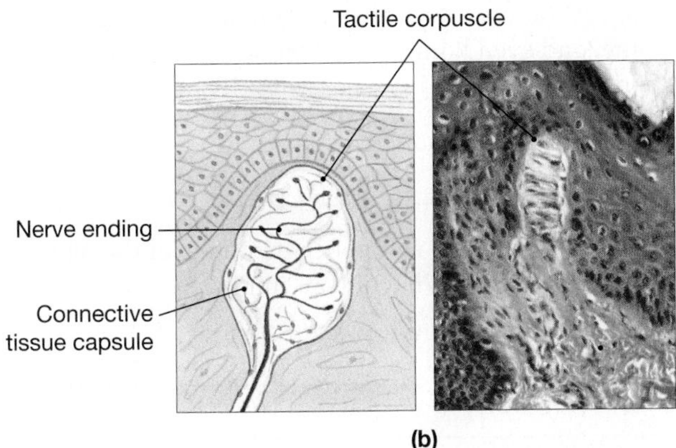

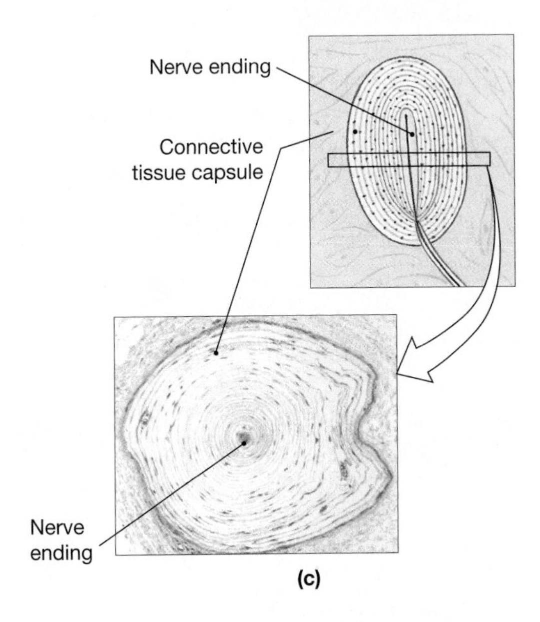

Figure 6.8 Sensory receptors in the dermis of the skin. a) Illustration of thin skin, showing the positions of tactile (Meissner's) corpuscles in the papillary layer of the dermis and lamellated (pacinian) corpuscles in the reticular layer of the dermis; **b)** illustration and light micrograph of a tactile corpuscle (LM × 100); **c)** illustration and light micrograph of a lamellated corpuscle (LM × 100).

ACTIVITY 6.4 Examining the Anatomical Model of Skin

1. Study the skin model on display in the laboratory and compare your observations with the microscopic examination of skin.
2. Verify that the structures listed here can be identified both on the model and under the microscope.
   - Epidermis (stratified squamous, keratinized epithelium)
   - Dermis
   - Hypodermis
   - Accessory structures
     - Hair follicles
     - Sebacous glands
     - Sweat glands
     - Arrector pili muscles

QUESTIONS TO CONSIDER 1. For people who are overweight or obese, one surgical option is the removal of excess fat by a procedure called liposuction. From what tissue layer is excess fat removed during this procedure? Explain.

____

____

2. Fat that is removed by liposuction can redevelop over time. Thus, unless there is a change in lifestyle, the procedure provides only a temporary solution for obesity. What lifestyle changes should be made so that the benefits of liposuction are more enduring?

____

____

ACTIVITY 6.5 Examining the Structure of Nails

1. Inspect the nails on your fingers and identify the following structures (Figure 6.9).
   - The **nail plate** is the visible portion of the nail. Similar to hair, it consists of dead cells filled with hard keratin. It rests on a region of the epidermis known as the **nail bed**.
   - The **free edge** of the nail plate is the part that you periodically clip. It extends over the **hyponychium**, which is a region of skin with a thickened keratinized layer.
   - The **lunula** is the white crescent-shaped region at the base of the nail plate. It may be difficult to see on smaller nails but is prominent on the large nail plate of the thumb.

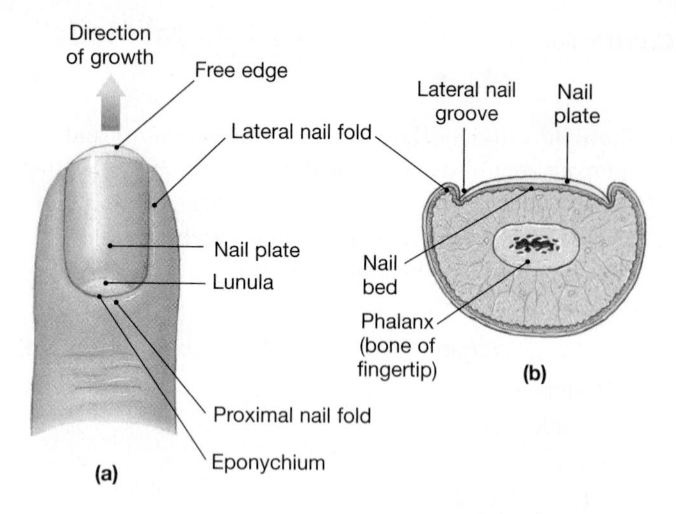

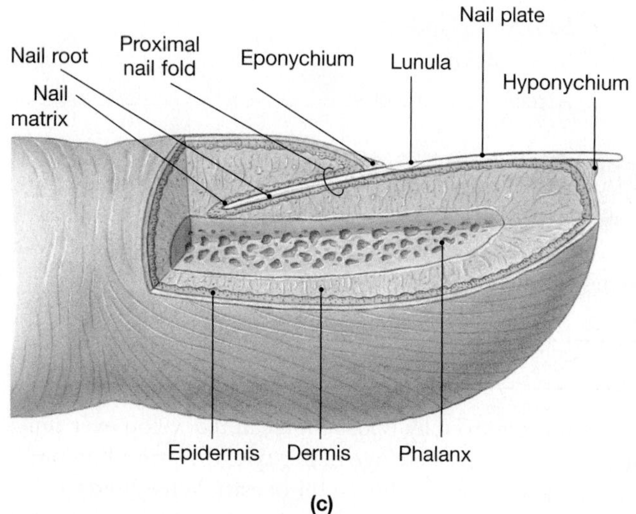

Figure 6.9 Structure of a nail. a) Dorsal view; **b)** cross section; **c)** longitudinal section.

- The **cuticle**, or **eponychium**, is a fold of the skin's keratinized layer at the base of the nail plate.
- **Lateral nail folds** are folds of skin along the left and right margins of the nail plate.

2. Refer to Figure 6.9 to identify the following structures that are not visible on your fingers.
   - The **nail matrix** is the most proximal part of the nail plate. It is a region of actively dividing cells. As the nail cells are produced and pushed distally, they become keratinized (with hard keratin) and form the nail plate.
   - The **nail root** is positioned deep to the eponychium. It anchors the nail plate to the underlying nail bed.

QUESTIONS TO CONSIDER

1. Hooves and claws are similar to nails in that they both are composed of epidermal cells filled with hard keratin. Hooves are used for locomotion and claws are used for several functions such as climbing, digging, hanging, grasping, and even killing prey. Identify functions that nails provide for humans.

____

____

2. Observe your fingernails and notice that most of the nail plate has a pinkish color. What do you think accounts for this appearance?

____

____

Exercise 6 Review Sheet

The Integumentary System

Name ______________________________

Lab Section ______________________________

Date ______________________________

1. Explain why the cutaneous membrane (skin) is classified as an organ.

2. Explain why the cutaneous membrane and the embedded accessory organs are classified as an organ system.

3. Briefly describe the structural features of the epidermis, dermis, and hypodermis.

4. What is the function of sebaceous glands?

5. Describe the structural and functional relationships between sebaceous glands, hair follicles, and arrector pili muscles.

6. Compare the functions of merocrine (eccrine) sweat glands and apocrine sweat glands.

7. Describe the structure of a nail. How is its structure similar to the structure of a hair follicle?

Questions 8–9: Coloring exercises. In the following diagrams, identify the structures by using the color that is indicated.

8.

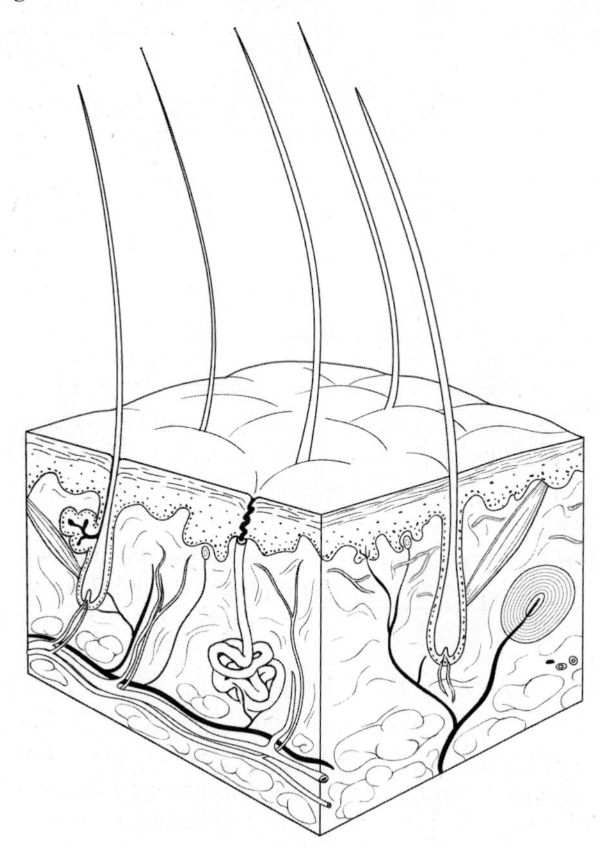

Skin from the scalp.

- Epidermis = **blue**
- Dermis = **purple**
- Hypodermis = **gray**
- Hair follicle = **green**
- Sebaceous gland = **red**
- Sweat gland = **brown**
- Arrector pili muscle = **orange**
- Lamellated corpuscle = **yellow**
- Tactile corpuscle = **black**
- Hair root = **pink**
- Hair shaft = **tan (light brown)**

9.

Epidermis and dermis of skin.

- Keratinized layer of the epidermis = **green**
- Basal layer of the epidermis = **red**
- Papillary layer of dermis = **blue**
- Reticular layer of dermis = **yellow**

Exercise 7

Introduction to the Skeletal System and the Axial Skeleton

Laboratory Objectives

On completion of the activities in this exercise, you will be able to:

- Classify bones according to their shapes.
- Recognize and describe the microscopic structure of compact and spongy bone.
- Differentiate between the axial and appendicular skeleton.
- Identify the bones of the axial skeleton and their bony markings.
- Describe the arrangement of the bones of the axial skeleton, relative to each other.

Materials

- Anatomical model of compact bone structure
- Compound light microscopes (optional)
- Microscope slides of ground human bone (optional)
- Humerus, longitudinal section
- Cranial bones, cross section
- Human skeleton
- Disarticulated bones of the human skeleton
- Human skull
- Model of the brain
- Slender probe
- Fetal skull
- Assorted vertebrae including C1, C2, and at least one typical vertebra from the cervical, thoracic, and lumbar regions
- Sacrum
- Typical rib
- Colored pencils

The skeletal system includes all the bones of the body, along with the cartilage, tendons, and ligaments associated with the articulations between the bones (the joints). Although it appears to be inactive, bone is actually a dynamic tissue with many important functions. For example, bones provide a rigid scaffolding that supports the body and protects vital organs. Skeletal muscles use bones as levers to make body movements possible. Bones play a vital role as storage sites for various minerals, such as calcium and phosphate. Furthermore, many bones contain red bone marrow, where most blood cell formation (**hematopoiesis**) is carried out.

Classification of Bones

Bones can be classified according to their varied sizes and shapes as well as by their location on the skeleton.

ACTIVITY 7.1 Classifying Bones According to Shape

1. Obtain a whole skeleton, or a complete collection of disarticulated bones. Quickly inspect the bones and note the wide range of sizes and shapes.
2. Categorize the bones according to the following shapes.
   - **Long bones** have long longitudinal axes so that the length of the bone is much greater than the width (Figure 7.1a). A typical long bone contains an elongated shaft, known as the **diaphysis**, with two expanded, knoblike ends, known as the **epiphyses** (Figure 7.1a). Examples of this type include the bones in the arm (**humerus**), forearm (**radius** and **ulna**), thigh (**femur**), and leg (**tibia** and **fibula**).
   - **Short bones** are those in which the length and width are about equal,so they appear cube shaped. They include the bones of the wrist (**carpal bones**; Figure 7.1e) and ankle (**tarsal bones**).
   - **Flat bones** are thin, platelike structures. This group includes the **cranial bones** (bones that protect the brain; Figure 7.1b), the **sternum**, the **ribs**, and the **scapulae**.
   - **Irregular bones** have a variety of shapes and include the **facial bones** of the skull, the **vertebrae** (Figure 7.1d), and the **pelvic bones**.
   - **Sesamoid bones** are embedded in tendons at articulations. The **patella** (kneecap, Figure 7.1f) is a sesamoid bone found in all individuals. The number of other sesamoid bones varies between individuals.
   - **Sutural bones** develop between the joints (sutures) of cranial bones (Figure 7.1c). Their number also varies between individuals.

WHAT'S IN A WORD The word *sesamoid* is derived from the Greek word *sesamoeides,* which means "like a sesame." Most sesamoid bones are small and roughly resemble the shape of a sesame seed. The patella resembles a small plate or shallow disk, which is what the term *patella* means in Latin.

QUESTIONS TO CONSIDER 1. The length of a rib is much longer than its width, like a long bone, and yet it is classified as a flat bone. Why do you think that ribs are classified as flat bones?

____

____

____

2. The metacarpal bones (in the palms), metatarsal bones (in the soles), and phalanges (in the fingers and toes) are much

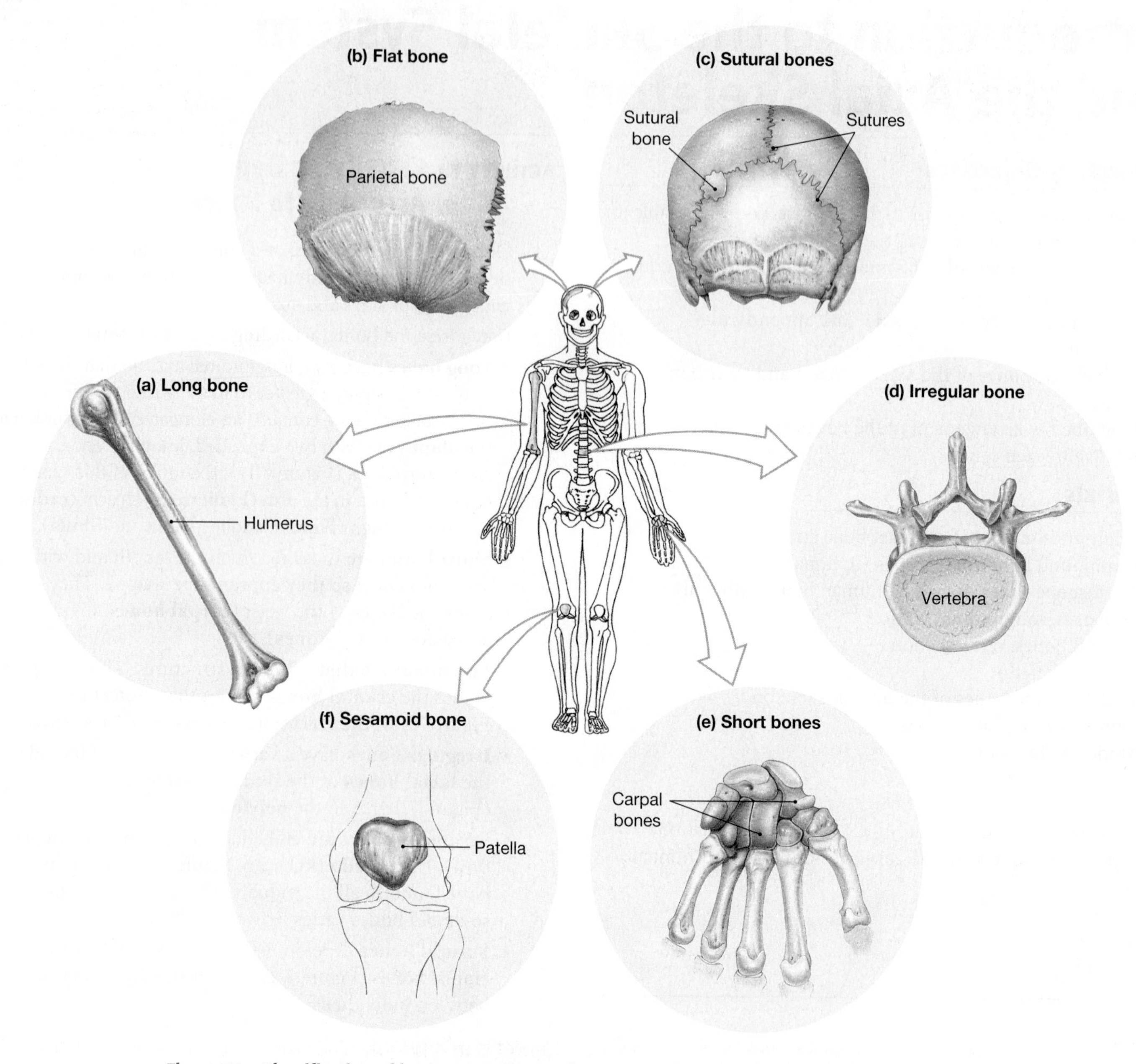

Figure 7.1 Classification of bones according to shape. Types of bones include **a)** long bones such as the humerus in the arm; **b)** flat bones such as a cranial bone in the skull; **c)** sutural bones such as those that can form between the sutures in the skull; **d)** irregular bones such as a vertebra; **e)** short bones such as the carpal bones in the wrist; and **f)** sesamoid bones such as the patella at the knee joint.

shorter than most ribs, and yet they are classified as long bones. Why do you think these bones are considered to be long bones?

____

____

ACTIVITY 7.2 Classifying Bones According to Location

1. Obtain a whole skeleton, or a complete collection of disarticulated bones.

2. Identify the bones that comprise the following two categories.
   - The **axial skeleton** consists of the bones that form the vertical axis of the body. These bones include the **facial** and **cranial bones** in the skull, the **vertebral column**, the **sternum**, and the **ribs** (Figure 7.2a). The bones shaded in yellow in Figure 7.2b illustrate the axial skeletal bones in relation to the entire skeleton.
   - The **appendicular skeleton** includes the bones of the upper and lower extremities (the appendages). Each upper extremity includes the scapula and **clavicle** (pectoral girdle), the humerus in the arm, the radius

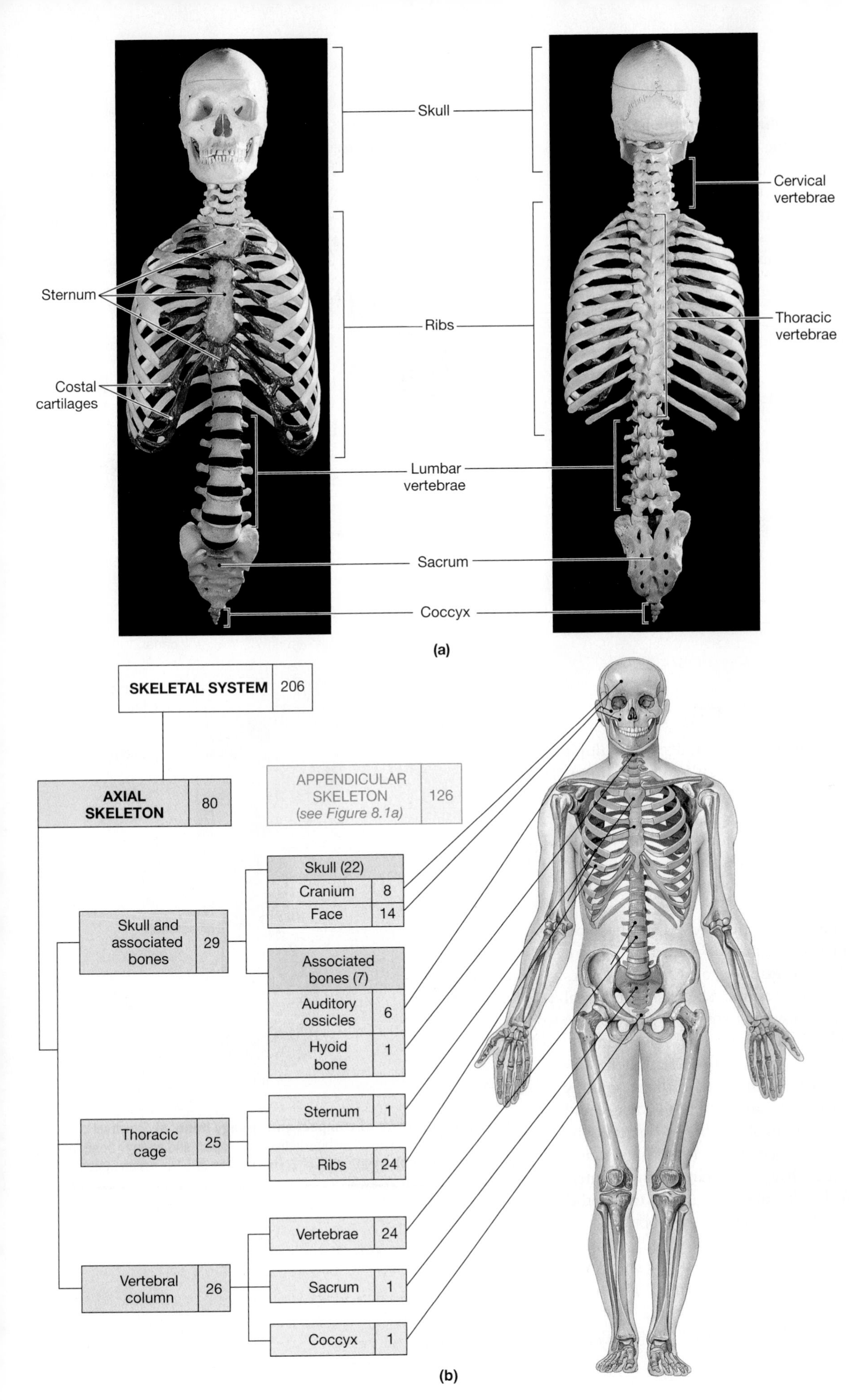

Figure 7.2 Bones of the axial skeleton. a) Anterior and posterior views of the axial skeleton, which includes bones of the skull, vertebral column, and thoracic cage; **b)** anterior view of the axial skeleton, shaded in yellow, in relation to the appendicular skeleton.

and ulna in the forearm, the carpal bones in the wrist and the **metacarpals** and **phalanges** in the hand. Each lower extremity includes the pelvic bones (**ilium**, **ischium**, and **pubis**), the femur in the thigh, the patella at the knee joint, the tibia and fibula in the leg, and the tarsal bones, **metatarsals**, and phalanges in the foot. The unshaded bones in Figure 7.2b illustrate the appendicular skeletal bones in relation to the entire skeleton.

QUESTIONS TO CONSIDER On the skeleton, identify the locations of the flat bones.

1. According to location on the skeleton, how would most flat bones be classified?

____

2. What bones are exceptions to your answer to question 1?

____

3. Based on the location of most flat bones, what important function do you think they serve?

____

Structure of Bone Tissue

The **intercellular matrix** of bone tissue is arranged in multiple layers called **lamellae**. The matrix contains **collagen fibers**, which form 25% of the dry weight, and **calcium salts**, primarily calcium phosphate and calcium carbonate. Collagen fibers give bone its flexibility and calcium salts contribute hardness and rigidity. Pound for pound, these materials act together to make bone as strong as, but much more flexible than, steel or reinforced concrete.

Two types of bone tissue exist in the body. The first type, **compact bone**, is an extremely dense material with a texture that resembles ivory. It forms the hard exterior covering of all bones and most of the diaphyses of long bones. The basic functional unit of compact bone is a microscopic structure known as a **Haversian system**, or **osteon**. In a typical bone, Haversian systems appear as tall columns of tissue that run parallel to the longitudinal axis (Figure 7.3a). The second type of bone, **spongy (cancellous) bone**, fills the interior regions of most bones and forms a thin layer along the diaphyses of long bones. Spongy bone does not possess the regular arrangement of Haversian systems. Instead, bone lamellae form an irregular arrangement of interconnecting bony struts called **trabeculae** (singular = **trabecula**) with spaces surrounding the latticework of bony tissue (Figure 7.3a). The spaces between trabeculae are filled with bone marrow. In the fetal skeleton, most of these spaces contain **red bone marrow**, where new blood cells are formed. Following birth, some red marrow is invaded by fat tissue and is converted to **yellow bone marrow**, a site for fat storage. Spongy bone is not

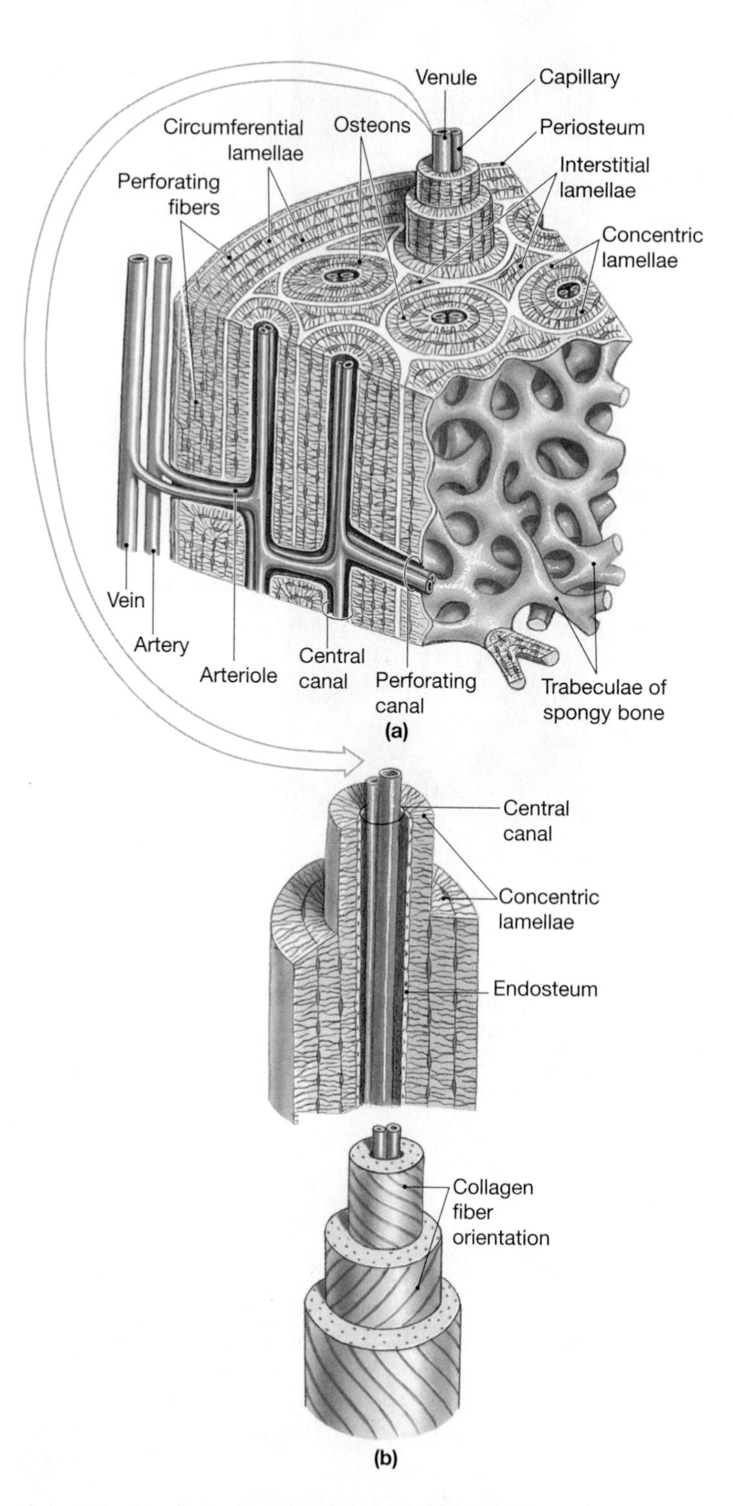

Figure 7.3 Structure of compact and spongy bone. a) Structural differences and anatomical relationships between compact and spongy bone in a typical long bone; **b)** organization of bone lamellae in a Haversian system.

as strong as compact bone, but its porous structure makes it suitable to cushion the forces of impact generated while walking or running.

WHAT'S IN A WORD In Latin, the word *trabecula* means "beam." Spongy bone is composed of interconnecting trabeculae (or beams) of bone tissue.

ACTIVITY 7.3 Examining the Microscopic Structure of Bone

You examined the microscopic structure of bone tissue in Exercise 5. During this activity you will review this structure.

1. Obtain an anatomical model of compact bone, which is an enlarged replica of a microscopic section of bone tissue. If a model is not available, return to Activity 5.2 and complete the microscopic observations for bone tissue. If possible, you may wish to compare the structures that you can identify on the model with structures that you can observe under the microscope.
2. On the model, identify the following structures that are found in compact bone (Figure 7.3a).
   - Numerous Haversian systems (osteons) are circular columns of bone tissue that run parallel to each other.
   - A **central (Haversian) canal** travels through the center of each Haversian system.
   - The central canals are connected by cross channels known as **perforating (Volkmann's) canals**.
   - Small **arteries** travel through the central and perforating canals. Veins, nerves, and lymphatic vessels also travel through these passageways, but they may not be shown on the model.
   - Each Haversian system is composed of several bone layers (bone lamellae). The lamellae form concentric rings of bone tissue around the central canal (Figure 7.3b).
   - **Osteocytes** (bone cells) are located between the bone lamellae and within small cavities called **lacunae.**
   - The osteocytes give rise to cell processes that travel through narrow passageways called **canaliculi**. Identify the cell processes within the canaliculi. Observe that the processes of nearby cells come together and form cell junctions with each other.

Along the inner free margin of the model, identify the region of spongy bone (Figure 7.3a). Note the different appearance of spongy bone compared to compact bone. Identify the complex meshwork of bony struts, known as trabeculae, and the spaces between them.

QUESTIONS TO CONSIDER 1. Central (Haversian) and perforating (Volkmann's) canals form a network of passageways within compact bone. Why are these canals important for the normal functioning of bone tissue?

____

____

____

2. During the previous activity you observed the cell processes of osteocytes traveling within canaliculi. Within the canaliculi, the processes of neighboring osteocytes can link together by forming cell junctions with each other. What do you think is the significance of these cell junctions?

____

____

____

CLINICAL CORRELATION

Part of the aging process involves a decrease in the activity of osteoblasts, which are the cells that deposit new bone matrix. As a consequence, we start to lose some bone mass. If the condition progresses, eventually enough bone mass is lost so that the bone's normal functioning is impaired. This clinical condition is called *osteoporosis,* which literally means "porous bone" and, in fact, the bones become visibly more spongy.

ACTIVITY 7.4 Examining the Gross Anatomy and Arrangement of Compact and Spongy Bone

1. Obtain a long bone, such as a humerus or femur, that has been cut along its longitudinal axis in a coronal plane (Figure 7.4a). Note the external layer of solid compact bone. This layer is relatively thick where it surrounds the diaphysis but becomes much thinner around each epiphysis.
2. The more porous spongy bone predominates at the two epiphyses. Note the complex network of bony trabeculae. Remember that, in living bone, the spaces between the trabeculae are filled with bone marrow.
3. The hollow space inside the diaphysis is the **medullary cavity**. During bone development, this cavity is filled with spongy bone and red bone marrow. As the bone matures, the bone is resorbed and the red bone marrow is converted to yellow bone marrow. Thus, in the adult, the medullary cavity of a typical long bone becomes a storage site for fat.
4. Obtain a cranial bone that has been cut so it can be viewed in cross section (Figure 7.4b). Flat bones, such as these, consist of an internal layer of spongy bone called the **diploë** sandwiched between two thin layers of compact bone. Examine this relationship, and remember that the cavities in the diploë are filled with red bone marrow.

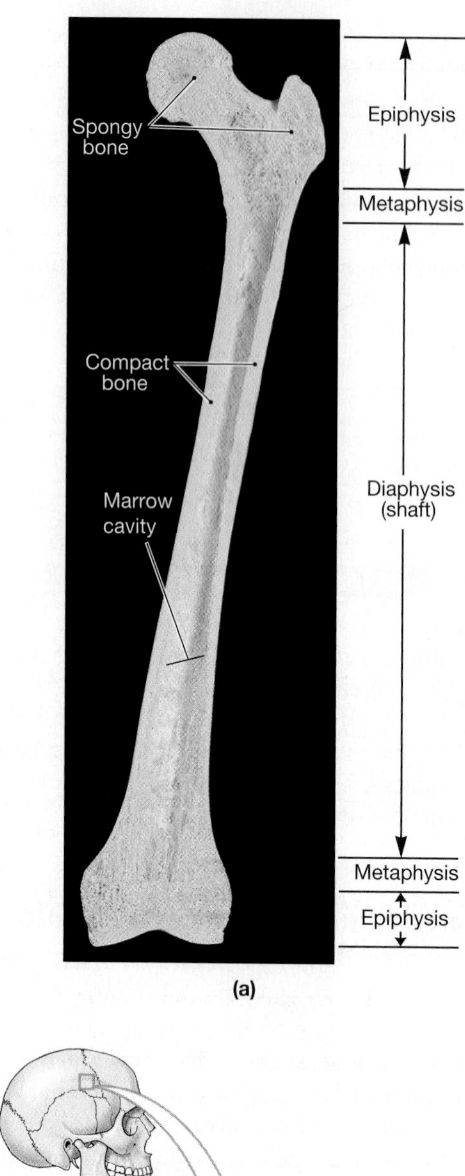

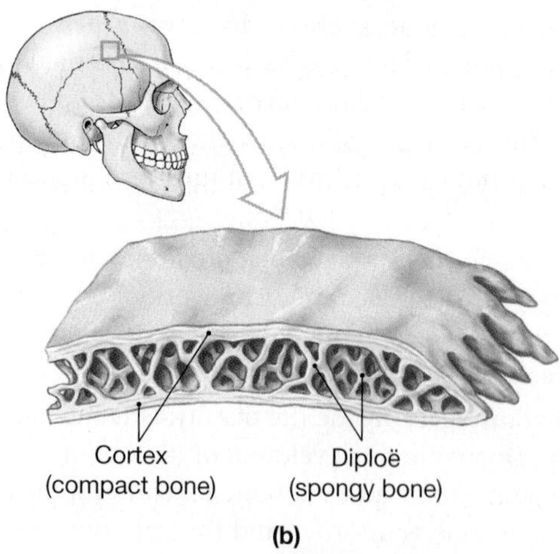

Figure 7.4 Arrangement of compact and spongy bone in the human skeleton. a) Longitudinal section of a femur, an example of a long bone; **b)** cross section of a cranial bone, an example of a flat bone.

WHAT'S IN A WORD The word *diploë* is derived from the Greek word *diplous*, which means "double." In a flat bone, diploë is the name given to spongy bone because it is located between two layers of compact bone.

QUESTIONS TO CONSIDER 1. In the previous activity, you observed the varying thickness of the compact bone layer that surrounds a long bone. Explain why the compact bone layer is much thicker around the diaphysis compared to the epiphyses.

___

___

___

2. If bones were composed entirely of compact bone, muscles would have to contract with greater force to produce movements. Explain why.

___

___

___

CLINICAL CORRELATION

In the fetal skeleton, most of the spaces in spongy bone contain red bone marrow, where new blood cells are formed. Following birth, some red marrow is invaded by adipose tissue and is converted to yellow bone marrow, which is a site for fat storage. In the adult skeleton, red marrow is abundant in the spongy bone of the ribs, sternum, cranial bones, bodies of vertebrae, clavicles, scapulae, pelvic bones, and proximal epiphyses of the long bones, such as the femur and humerus. Most other spongy bone is filled with yellow marrow, but in times of blood cell deficiency, yellow marrow can be converted to red marrow to increase blood cell production.

Bones of the Axial Skeleton

For the remainder of this laboratory exercise, you will explore the bones of the axial skeleton, and in the next exercise, you will study the appendicular skeleton. In addition to identifying individual bones, you will also be required to locate unique structural features on the bones. All bones possess distinctive landmarks that are designed for specific functions. For example, certain landmarks are used as articulating surfaces to form a joint (an articulation) with another bone on the skeleton. Examples of these include **condyles**, **heads**, and **facets**. Other bony features, such as **trochanters**, **tubercles**, and **processes**, provide areas of attachment for tendons and ligaments. Various types of depressions and openings provide passageways for nerves and blood vessels. These include **foramina** (singular = **foramen**), **fissures**, and **grooves**. A more complete list of bony landmarks is shown in Table 7.1.

Table 7.1 Types of Bony Landmarks

Bony landmarks	Description
I. Articulating surfaces	
A. Head	Rounded expansion, connected to a narrow neck
B. Facet	Smooth, flat surface
C. Condyle	Rounded projection
D. Ramus	Armlike projection
E. Fossa	Shallow depression
II. Attachment sites for tendons and ligaments	
A. Tuberosity	Elevated projection, with roughened surface
B. Crest	Prominent, narrow ridge
C. Line	Narrow ridge, less prominent than a crest
D. Trochanter	Large, irregularly shaped process (on femur only)
E. Tubercle	Rounded projection or process
F. Epicondyle	Elevated area, above a condyle
G. Spine	Slender, pointed projection
III. Depressions and openings	
A. Meatus	Bony canal
B. Sinus	Cavity within a bone
C. Groove	Narrow channel
D. Fissure	Slitlike opening
E. Foramen	Round or oval hole or opening

The Skull

The skull is composed of two distinct groups of bones. The eight **cranial bones (neurocranium)** form the walls and floor of the **cranial cavity**, where the brain and associated structures are located. Most of these bones are flat or slightly curved. The 14 facial bones have irregular shapes. These bones form the face and the walls of orbital and nasal cavities and provide bony sockets for the teeth.

Learning the bones of the skull and their various parts can be quite challenging, so go slowly and refer regularly to Figures 7.5 and 7.6. Most of the bones can be identified from more than one view. Consequently, to appreciate the three-dimensional quality of the skull, you must examine it from several positions.

ACTIVITY 7.5 Identifying the Bones of the Skull

Anterior View of the Skull

1. Obtain a human skull and identify the bones visible from an anterior view (Figure 7.5a).
   - The **frontal bone** forms the forehead.
   - The **zygomatic bones** form the prominences of the cheeks.
   - The **maxillary bones** form the upper jaw.
   - The **mandible** forms the lower jaw.
   - The **nasal bones** form the bony portion of the nose.
2. Identify the **nasal conchae**, which form bony projections on the lateral walls of the nasal cavity. The **superior** and **middle nasal conchae** are parts of the **ethmoid bone**; the **inferior nasal concha** is a separate bone.
3. Locate the **nasal septum** that divides the nasal cavity into left and right parts. It is composed of the **perpendicular plate of the ethmoid** and the **vomer**.
4. The **bony orbit** is formed by portions of seven bones: frontal, ethmoid, lacrimal, sphenoid, maxillary, zygomatic, and palatine.

Lateral View of the Skull

1. From a lateral view of the skull (Figure 7.5b), observe that the lateral wall of the calvaria (skullcap) is formed by portions of four bones: frontal, parietal, sphenoid, and temporal.
2. Examine the **zygomatic arch** (cheekbone) and note that it is formed by processes of the zygomatic and temporal bones.
3. Locate the **temporal fossa**, which is the squamous (flat) part of the temporal bone, superior to the zygomatic arch. The **infratemporal fossa** is deep to the zygomatic arch and posterior to the maxilla.
4. Identify the sutures that are seen in this view. The **coronal suture** joins the frontal bone and parietal bones. The **squamous suture** joins the parietal bone and the temporal bone.
5. Follow these sutures to locate the **pterion** (**temple**). This H-shaped formation of sutures unites the frontal, parietal, sphenoid, and temporal bones. Note that the bones in this area are relatively thin and can fracture easily.

Superior View of the Skull

1. From a superior view of the skull (Figure 7.5c), observe that the roof of the calvaria is formed by portions of the frontal, parietal, and occipital bones.
2. Identify the sutures seen in this view. The coronal suture is the joint between the two parietal bones and the frontal bone. The **sagittal suture** is the joint between the two parietal bones. The **lambdoid suture** is the joint between the two parietal bones and the occipital bone.

Posterior View of the Skull

1. From the posterior view (Figure 7.5d), observe that the posterior wall of the calvaria is formed by the occipital bone (**occiput** or **back of head**) and portions of the parietal bones.
2. Locate the lambdoid suture that joins the parietal bones and the occipital bone.
3. Identify the **superior nuchal line**, which marks the superior limit of the neck. It extends laterally from each side of the **external occipital protuberance**.

Inferior View of the Skull

1. From an inferior view of the skull (Figure 7.5e), observe that the hard palate is formed by the palatine processes of the maxillary bones and the horizontal plates of the palatine bones.

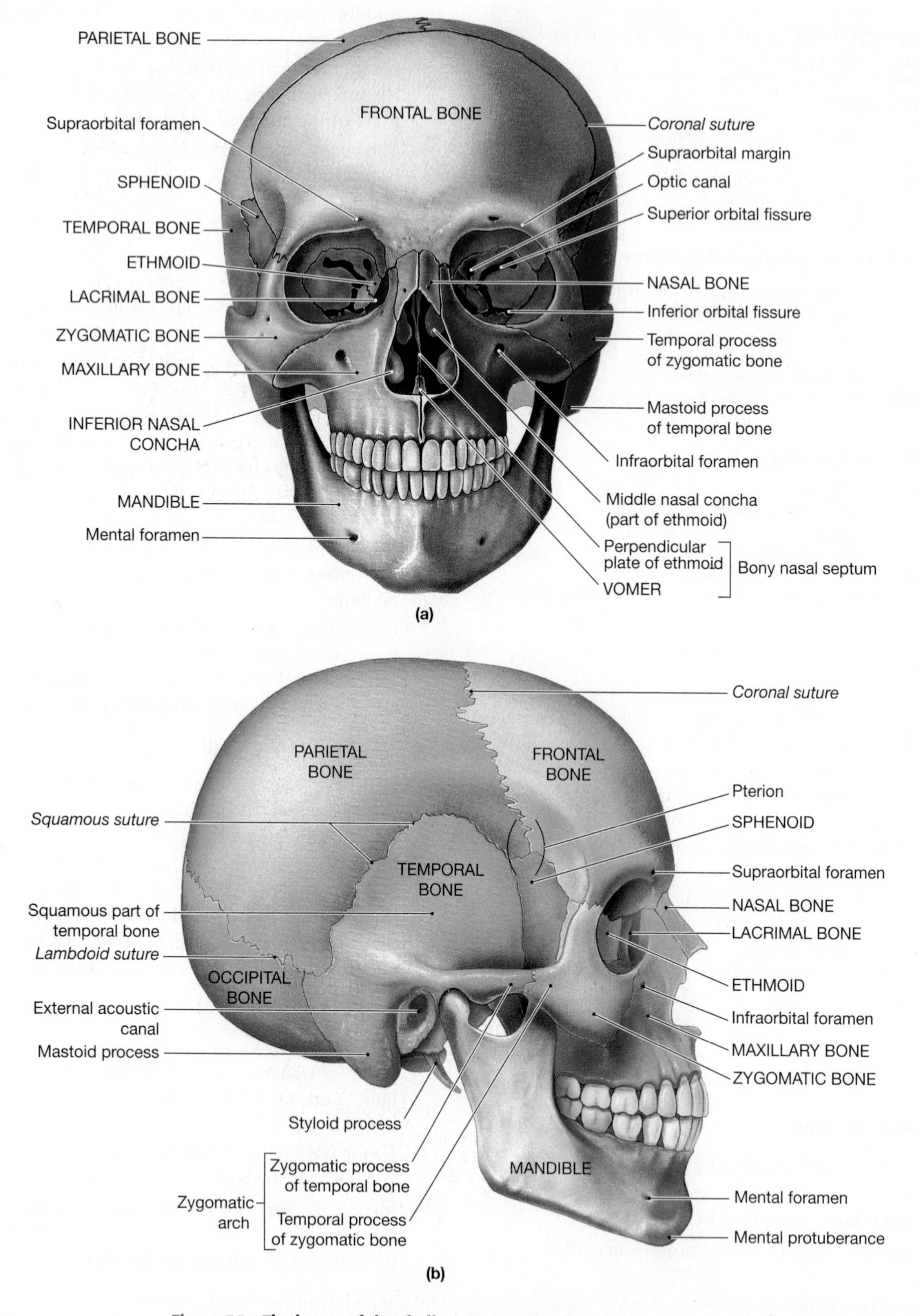
PARIETAL BONE
FRONTAL BONE
Supraorbital foramen
Coronal suture
Supraorbital margin
SPHENOID
Optic canal
TEMPORAL BONE
Superior orbital fissure
ETHMOID
NASAL BONE
LACRIMAL BONE
Inferior orbital fissure
ZYGOMATIC BONE
Temporal process of zygomatic bone
MAXILLARY BONE
Mastoid process of temporal bone
INFERIOR NASAL CONCHA
Infraorbital foramen
Middle nasal concha (part of ethmoid)
MANDIBLE
Mental foramen
Perpendicular plate of ethmoid
VOMER
Bony nasal septum
(a)
Coronal suture
PARIETAL BONE
FRONTAL BONE
Pterion
Squamous suture
SPHENOID
TEMPORAL BONE
Supraorbital foramen
Squamous part of temporal bone
NASAL BONE
LACRIMAL BONE
Lambdoid suture
OCCIPITAL BONE
ETHMOID
External acoustic canal
Infraorbital foramen
Mastoid process
MAXILLARY BONE
ZYGOMATIC BONE
Styloid process
Zygomatic process of temporal bone
MANDIBLE
Zygomatic arch
Mental foramen
Temporal process of zygomatic bone
Mental protuberance
(b)

Figure 7.5 The bones of the skull. a) Anterior view; **b)** lateral view. *(continues)*

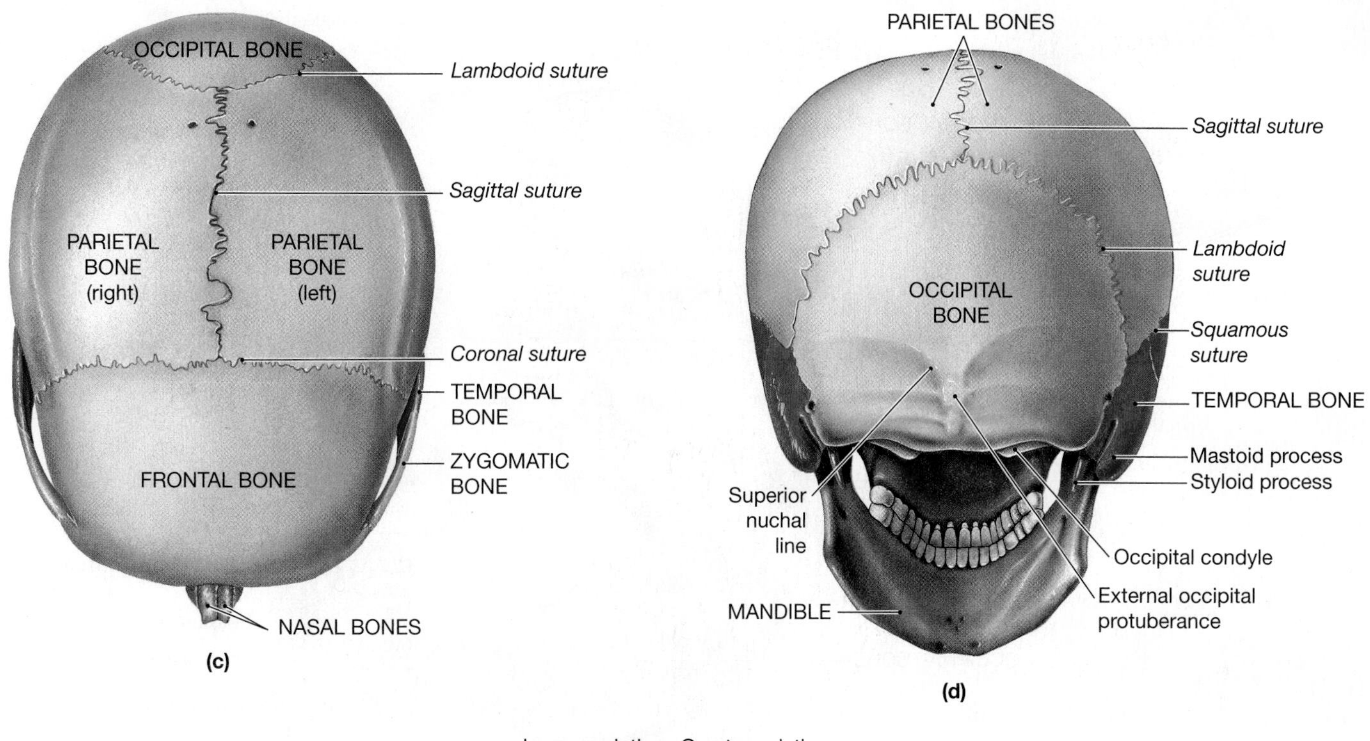

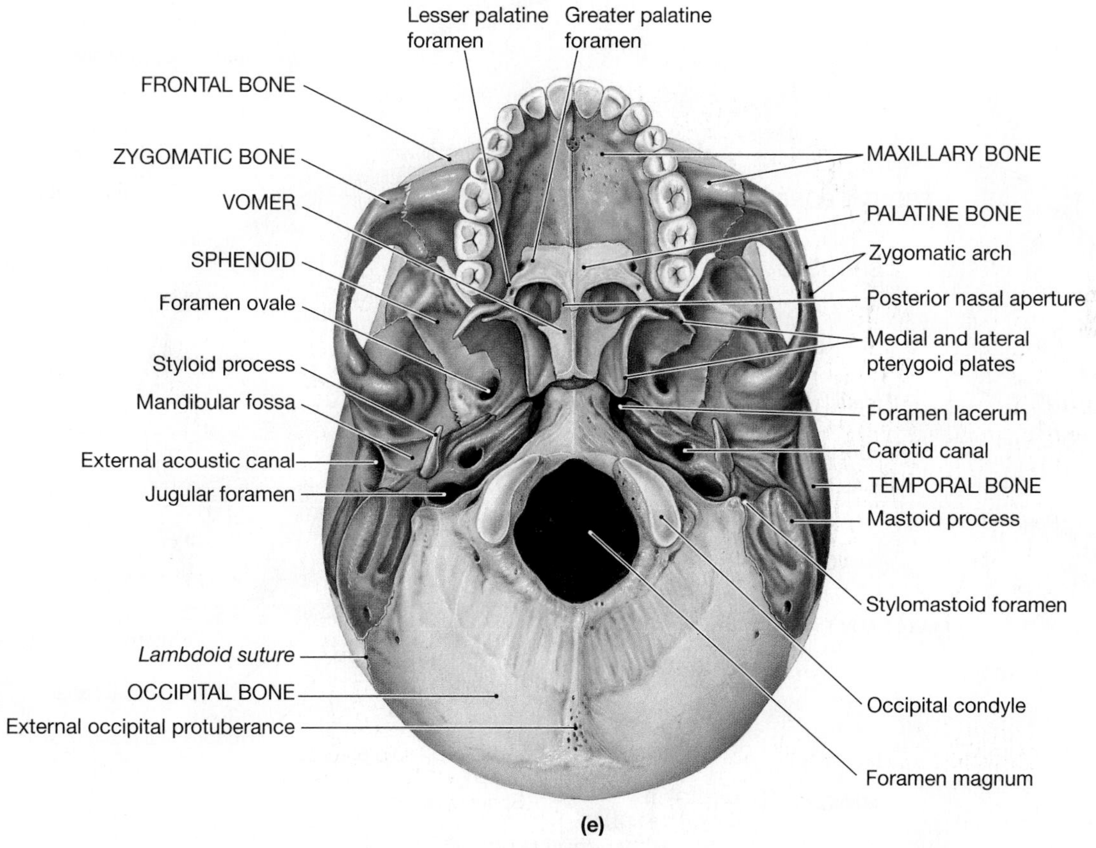

Figure 7.5 The bones of the skull *(continued)*. **c)** Superior view; **d)** posterior view; **e)** inferior view.

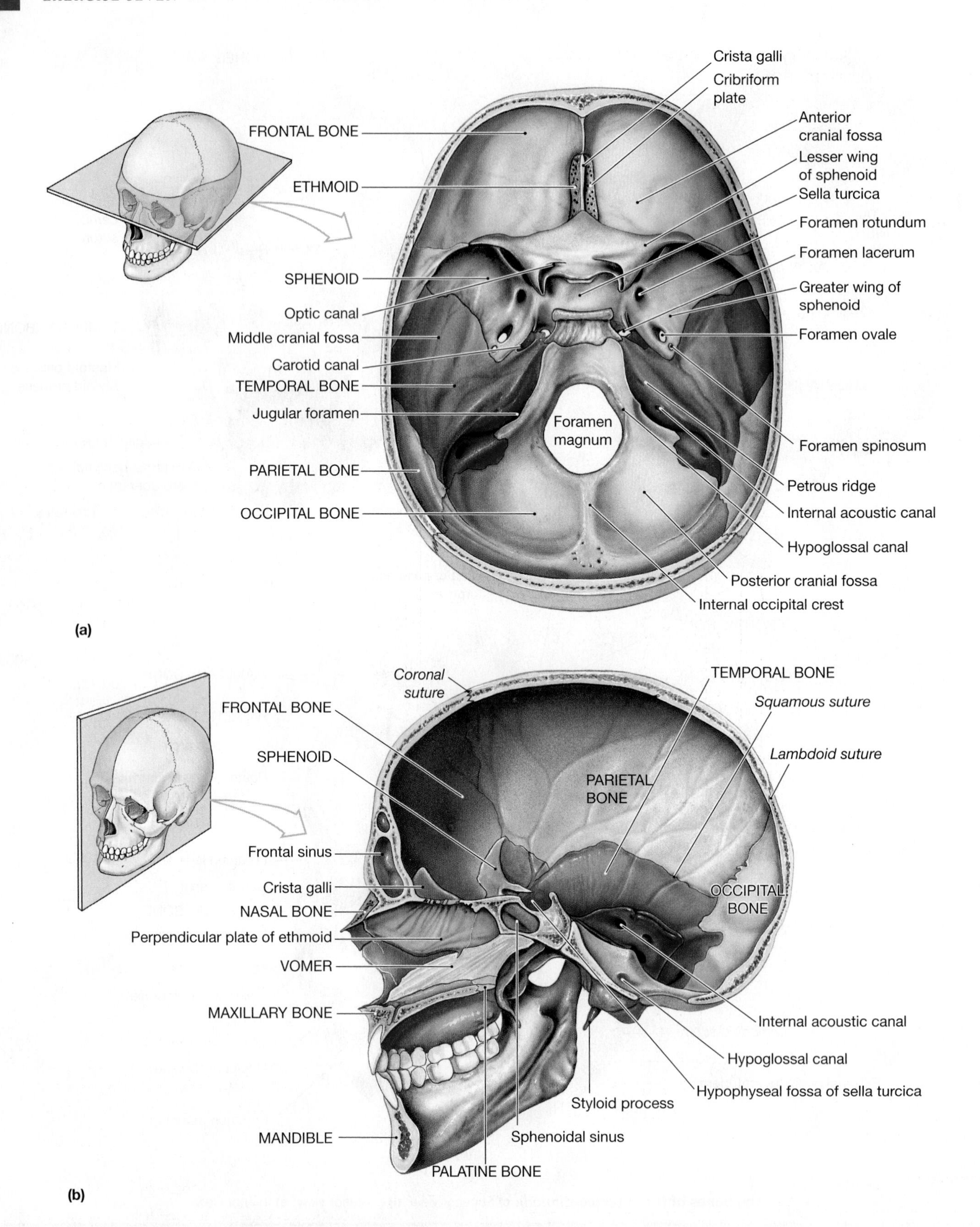

Figure 7.6 Sectional views of the skull. a) Horizontal section revealing the floor of the cranial cavity; **b)** midsagittal section revealing the internal surfaces of the cranial bones, the mandible, and bones of the nasal septum.

2. Observe that the **posterior nasal apertures (choanae)** are separated by the **vomer.**
3. Note that the sphenoid, temporal, and occipital bones form a portion of the basicranium (base of the cranial cavity).

Internal View of the Skull

1. Remove the top of the calvaria to expose the **floor of the cranial cavity** (Figure 7.6a). Note that the floor of the cranial cavity is formed by portions of the frontal, ethmoid, sphenoid, temporal, and occipital bones.
2. Locate the three bony depressions in the floor: the **anterior**, **middle**, and **posterior cranial fossae** (singular = **fossa**).
3. Obtain a model of the brain and place it within the cranial cavity. Verify the following:
   - The anterior cranial fossa is occupied by the frontal lobes of the brain.
   - The middle cranial fossa is occupied by the temporal lobes of the brain.
   - The posterior cranial fossa is occupied by the cerebellum and brainstem.
4. Obtain a skull that has been cut in half to form a midsagittal section. If this model is not available, refer to Figure 7.6b. Identify the following features.
   - The bony portion of the nasal septum is formed by the **perpendicular plate of the ethmoid** and the vomer (Figures 7.5a and 7.6b).
   - **Paranasal sinuses** are internal cavities that are found in some bones of the skull. They open directly to the nasal cavity. Identify the **frontal sinus** and the **sphenoidal sinus.** Other paranasal sinuses are located in the maxillary and ethmoid bones.
   - The **hard palate** is formed by portions of the maxillary and palatine bones (Figures 7.5e and 7.6b). Notice that the hard palate separates the nasal cavity from the oral cavity (mouth).

QUESTIONS TO CONSIDER 1. Based on shape, how are most cranial bones classified? Are there any exceptions to this rule?

____

____

____

2. The walls of the orbits are formed by portions of several bones. Observe this region of the skull. Are there any structural weaknesses in this area that you can identify?

____

____

____

3. The failure of the two maxillary bones to completely fuse along the midline of the hard palate is known as a **cleft palate.** Surgery to correct this condition is usually performed between 6 and 12 months after birth. Until that time, what problems do you predict could result from a cleft palate?

____

____

____

ACTIVITY 7.6 Identifying the Bony Landmarks and Openings of the Skull

1. On the skull, identify the various bony landmarks that are listed for each bone in Table 7.2. In the table, notice that the view in which you can best identify a bony landmark along with the appropriate figure is given in parentheses.
2. Observe that the skull has numerous holes (foramina), canals, and fissures. These apertures serve as passageways for cranial nerves and blood vessels. Identify the major foramina, canals, and fissures listed in Table 7.3. Use the recommended figure, listed in parentheses in the table, to assist you in identifying the openings. Follow the course of each opening by inserting a slender probe from one end and noticing where the other end emerges. Be very cautious with the skull so you do not damage it. Do not use any writing utensils as pointers because they will mark the skull.

QUESTION TO CONSIDER In the previous two activities, you identified the bones and the bony landmarks on the skull. Why is it essential that you study the skull by observing it from several views?

____

____

____

ACTIVITY 7.7 Examining the Fetal Skull

1. Obtain a fetal skull and locate the cranial and facial bones that you identified in the adult skull (Figure 7.7).
2. Identify the major sutures and notice that they are not fully developed in the fetal skull. The ossification process (bone formation) is incomplete and adjacent bones are connected by regions of fibrous connective tissue, known as **fontanels.** Identify the following fontanels in the fetal skull (Figure 7.7).
   - **Anterior (frontal) fontanel**
   - **Posterior (occipital) fontanel**
   - **Anterolateral (sphenoidal) fontanel**
   - **Posterolateral (mastoid) fontanel**

WHAT'S IN A WORD In French, the word *fontanel* refers to a fountain or spring. The "soft spots" on a baby's skull are called fontanels because a pulse can be felt in these areas, especially in the anterior fontanel.

Table 7.2 Bones of the Skull

	Bone/structure	Bony landmarks / Special features (best view)
1. Cranial bones	Frontal bone	Supraorbital margin (anterior view; Figure 7.5a)
	Parietal bones (2)	External occipital protuberance (posterior view; Figure 7.5d)
	Temporal bones (2)	Styloid process (inferior view; Figure 7.5e) Mastoid process (inferior view; Figure 7.5e) Mandibular fossa (inferior view; Figure 7.5e) Petrous ridge (internal view; Figure 7.6a) Zygomatic process (lateral view; Figure 7.5b)
	Occipital bone	Occipital condyles (inferior view; Figure 7.5e)
	Sphenoid bone	Sella turcica (internal view; Figure 7.6a) Greater wing (internal view; Figure 7.6a) Lesser wing (internal view; Figure 7.6a) Medial pterygoid plate (inferior view; Figure 7.5e) Lateral pterygoid plate (inferior view; Figure 7.5e)
	Ethmoid bone	Crista galli (internal view; Figure 7.6a) Cribriform plate (internal view; Figure 7.6a) Perpendicular plate (nasal cavity; Figures 7.5a, 7.6b) Superior concha (nasal cavity; not illustrated) Middle concha (nasal cavity; Figure 7.5a)
2. Facial bones	Nasal bones	
	Maxillae	
	Zygomatic bones	Temporal process (lateral view; Figure 7.5b)
	Mandible	Condylar process (lateral view; Figure 7.5b) Coronoid process (lateral view; Figure 7.5b) Body (lateral view; Figure 7.5b) Ramus (lateral view; Figure 7.5b)
	Lacrimal bones	
	Palatine bones	
	Inferior nasal conchae	
	Vomer	

Table 7.3 Foramina, Canals, and Fissures in the Skull

Opening	Bone location	Best view
Foramina in cribriform plate	Ethmoid	Anterior cranial fossa (Figure 7.6a)
Optic canal	Sphenoid	Middle cranial fossa (Figure 7.6a)
Foramen rotundum	Sphenoid	
Foramen ovale	Sphenoid	
Foramen spinosum	Sphenoid	
Foramen lacerum	Between occipital and temporal	
Carotid canal	Temporal	
Internal acoustic canal (meatus)	Temporal	Posterior cranial fossa (Figure 7.6a)
Jugular foramen	Between occipital and temporal	
Foramen magnum	Occipital	
Hypoglossal canal	Occipital	
Superior orbital fissure	Sphenoid	Anterior view of skull (Figure 7.5a)
Inferior orbital fissure	Between maxilla and sphenoid	
Mental foramen	Mandible	
Supraorbital foramen (or notch)	Frontal	
Infraorbital foramen	Maxilla	
External acoustic canal (meatus)	Temporal	Lateral view of skull (Figure 7.5b)
Greater palatine foramen	Palatine	Inferior view of skull (Figure 7.5e)
Lesser palatine foramen	Palatine	
Mandibular foramen	Mandible	Medial view of mandible (not illustrated)

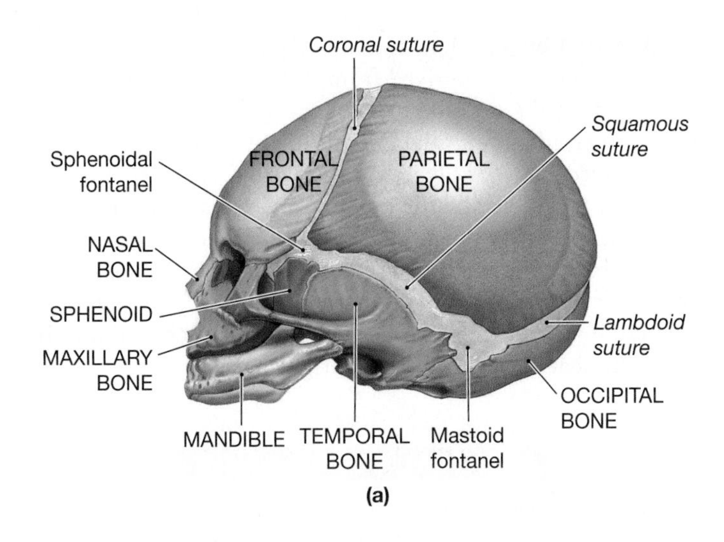

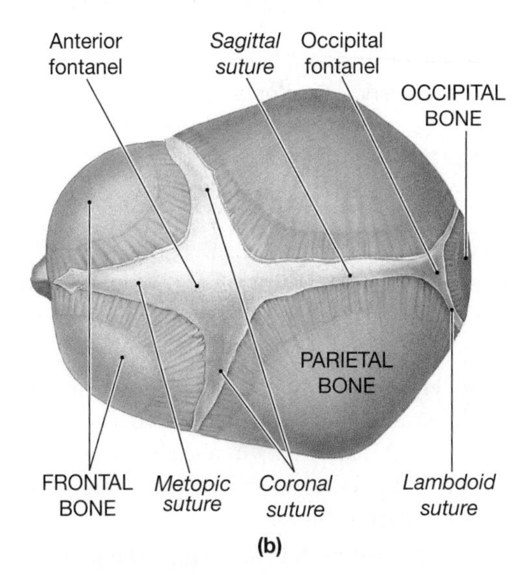

Figure 7.7 The fetal skull. a) Lateral view; **b)** superior view. Locations of the major fontanels are indicated.

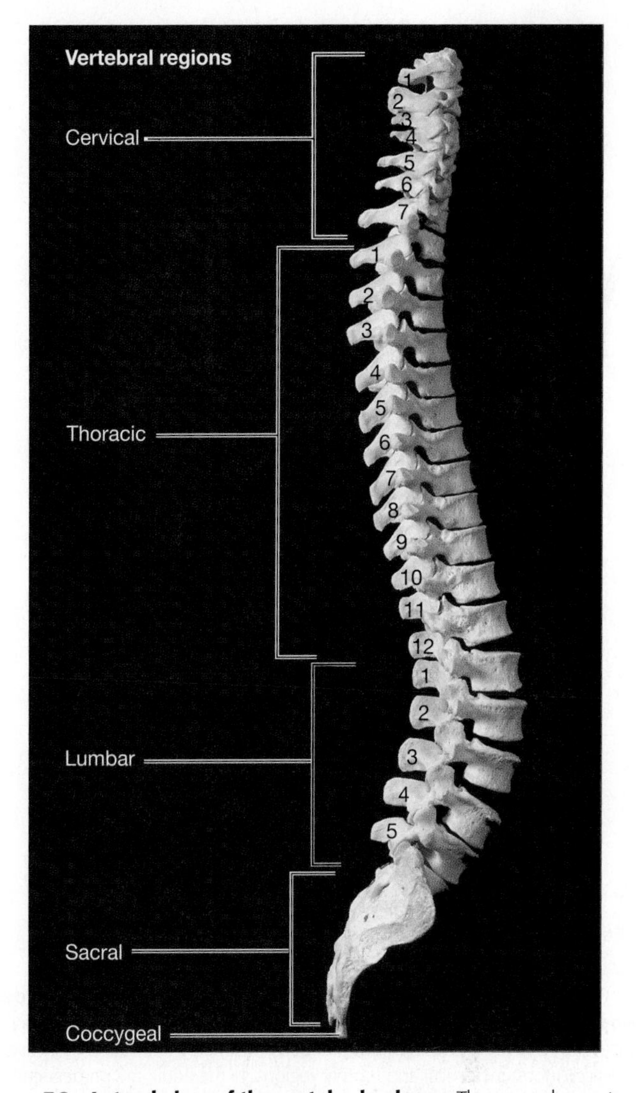

Figure 7.8 Lateral view of the vertebral column. The normal curvatures of the column are illustrated.

CLINICAL CORRELATION

The fontanels provide some flexibility in the head during the birth process. The bones are allowed to slightly slide across each other to reshape the skull as it passes through the birth canal. In addition, the fontanels also accommodate brain growth in the fetus and infant.

QUESTION TO CONSIDER What potential problems can you predict if the sutures between cranial bones formed prematurely after birth?

The Vertebral Column

The **vertebral column (spine)** provides support for the head, neck, and trunk and is instrumental in maintaining erect posture while standing and sitting. In addition, it provides protection for the spinal cord. During development, the vertebral column is formed by 33 individual **vertebrae** (singular = **vertebra**). However, as a result of bone fusion, the adult spine has only 26 bones and is divided into five regions (Figure 7.8), as follows:

- Cervical vertebrae (7)
- Thoracic vertebrae (12)
- Lumbar vertebrae (5)
- Fused vertebrae of the **sacrum** (5)
- Fused vertebrae of the **coccyx** (4)

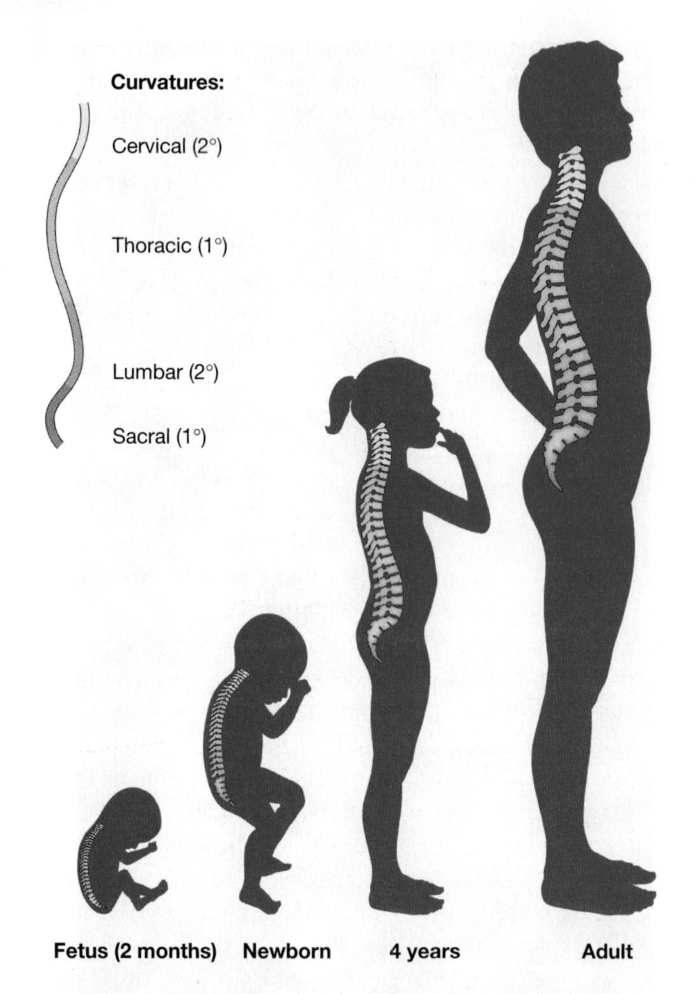

Figure 7.9 Curvatures of the vertebral column at various times of development. In the 2-month fetus, the primary curvature of the entire column is concave anteriorly. As development proceeds, secondary curvatures (concave posteriorly) form in cervical and lumbar regions. In the adult, the primary curvature remains in thoracic and sacral portions only.

ACTIVITY 7.8 Examining the General Features of the Vertebral Column

1. Obtain a complete human skeleton and notice the general position of the vertebral column. Identify the different regions of the vertebral column (Figure 7.8). The cervical vertebrae are in the neck. The thoracic vertebrae, which articulate (form joints) with the ribs, are located in the thorax. The lumbar vertebrae are in the lower back. The fifth lumbar (L5) vertebra articulates with the sacrum; and inferior to the sacrum is the small coccyx.
2. Observe the four normal curvatures of the adult vertebral column (Figures 7.8 and 7.9, adult). Note that in the thoracic and sacral regions, the concave surface of the curvature faces anteriorly (the curve opens to the front). These are the **primary curvatures** of the vertebral column. The cervical and lumbar regions curve in the opposite direction (the concave surface faces posteriorly). These are the **secondary curvatures** of the vertebral column.

WHAT'S IN A WORD The terms *primary curvature* and *secondary curvature* are derived from the structural changes that occur during the development of the vertebrae (Figure 7.9). During fetal life, the initial curvature of the vertebral column is C shaped and concave anteriorly. This so-called primary curvature is retained by the thoracic and sacral portions of the vertebral column. During infancy and childhood, the cervical and lumbar portions of the vertebral column become progressively concave posteriorly. These secondary curvatures are adaptations to support the head (cervical vertebrae) and the torso (lumbar vertebrae) in an upright position.

CLINICAL CORRELATION

The vertebral column has four normal curvatures, but it can also become abnormally curved. **Kyphosis** is an exaggerated thoracic curvature that often gives an individual a hunchback appearance. **Lordosis** is an exaggerated lumbar curvature and results in a swayback appearance. **Scoliosis** is an abnormal lateral deviation of the vertebral column, which is normally straight from side to side. Any of these problems with alignment can lead to chronic and severe pain.

3. Obtain a typical vertebra and identify the following structural features, which are listed in Table 7.4 and illustrated in Figure 7.10.
   - The **vertebral body** is the main anterior bony mass of each vertebra and is the part through which body weight is transmitted. On a skeleton, notice that the vertebral bodies become progressively larger from a superior to inferior direction.
   - The **vertebral arch** is formed by the **pedicles** and **laminae.** The pedicles project posteriorly from the body and form the lateral walls of the arch. The laminae travel medially from the pedicles to form the roof.
   - The **vertebral foramen** is the hole that is formed by the posterior surface of the body and the vertebral arch. On a skeleton, notice that the vertebral foramina of all the vertebrae form the **vertebral canal**, which transmits the spinal cord.
4. Identify the following bony processes on a typical vertebra (Figure 7.10).
   - One **spinous process** projects posteriorly from the union of the laminae.
   - Two **transverse processes**, one on each side, project laterally from the union of a pedicle and lamina.
   - Two **inferior articular processes**, one on each side, project inferiorly from a pedicle.
   - Two **superior articular processes**, one on each side, project superiorly from a pedicle.
5. On the skeleton, identify the **intervertebral foramina** between adjacent vertebrae (Figure 7.10e). These openings allow the spinal nerves to exit the vertebral canal.

Table 7.4 Bones of the Vertebral Column

Bone	Bony landmarks or special features
I. General structure	
A. Typical vertebra	Body Lamina Pedicle Vertebral foramen Spinous process Transverse process Superior articular process Inferior articular process
B. Sacrum	Transverse lines Sacral foramina Median sacral crest Lateral sacral crest Sacral canal Auricular surface Sacral promontory
C. Coccyx	
II. Special features	
A. Cervical vertebra	Transverse foramina Bifid (forked) spinous process (except C1 and C7) Small body
B. C1 (atlas)	No body No spinous process
C. C2 (axis)	Dens (odontoid process)
D. Thoracic vertebra	Spinous process is long and directed inferiorly Articulating surfaces (facets) for ribs on body and transverse processes
E. Lumbar vertebrae	Large, robust body and spinous process Very slender transverse processes
F. Sacrum	Composed of five fused vertebrae
G. Coccyx (tailbone)	Composed of three to five fused vertebrae

6. On a skeleton, locate the following articulations (joints) between adjacent vertebrae:
   - **Intervertebral discs** separate the bodies of adjacent vertebrae (Figures 7.10d and e). The discs are composed primarily of fibrocartilage. They are absent between the first and second cervical vertebrae and between the fused vertebrae in the sacrum and coccyx.
   - **Synovial joints** (joints with a fluid-filled joint capsule) are found between inferior articular processes of the superior vertebra and the superior articular processes of the inferior vertebra (Figures 7.10d and e). Examine the position of these joints on an intact skeleton. Obtain two vertebrae from the same region and stack them so you can see how the articular processes from adjacent vertebrae form a joint.
7. Although they are not shown on the skeleton or other models in the laboratory, realize that the vertebrae are also connected to each other by numerous ligaments. The most important ligaments connect the vertebral bodies, laminae, and spinous processes. These connections are instrumental in allowing the vertebrae in the column to act as a single unit.
8. Obtain a **sacrum** (Figure 7.11). This bone develops as five separate vertebrae. Over time, the vertebrae fuse together to form one bone. Identify the following features.
   - On the anterior surface (Figure 7.11c), the four **transverse lines** indicate the areas of fusion between the original bodies of the sacral vertebrae.
   - On the dorsal surface (Figure 7.11a) are two prominent ridges. The **median sacral crest** is the central bony ridge that is formed by the fusion of the spinous processes. Along each lateral margin, the **lateral sacral crests** mark the fusion of the transverse processes. The lateral sacral crests are enlarged and thickened (Figure 7.11b) so they can transmit the weight of the upper parts of the body to the ilium and on to the lower extremity.
   - The **sacral promontory** (Figures 7.11b and c), a bulge on the anterior margin of the base, is an important landmark for obstetric pelvic measurements.
   - The **sacral canal** is the continuation of the vertebral canal. The roof of the sacral canal is formed by the fusion of the laminae of the sacral vertebrae.
   - The ventral and dorsal **sacral foramina** (Figures 7.11a and c) are openings that represent the original intervertebral foramina. They transmit the sacral spinal nerves.
9. Locate the **coccyx** (Figure 7.11) on a skeleton (it is often broken). This bone consists of three to five (usually four) fused vertebrae. It represents a vestige of an embryonic tail that usually disappears by the eighth week of development.
10. On the whole skeleton, observe the relative position of the sacrum and coccyx. Locate the following articulations between the sacrum and other structures.
    - An intervertebral disc is located between the body of the fifth lumbar vertebra and the superior surface (the base) of the sacrum.
    - An intervertebral disc is located between the inferior surface (the apex) of the sacrum and the coccyx.
    - On each side, the **auricular surface** of the sacrum (Figure 7.11b) articulates with the pelvic girdle to form the **sacroiliac joint.**

1. Suggest a reason for the increasing size of the vertebral bodies from a superior to inferior direction.

____

____

2. As a result of fusion of the five sacral vertebrae, the sacrum provides relatively large surface areas for muscle attachments.

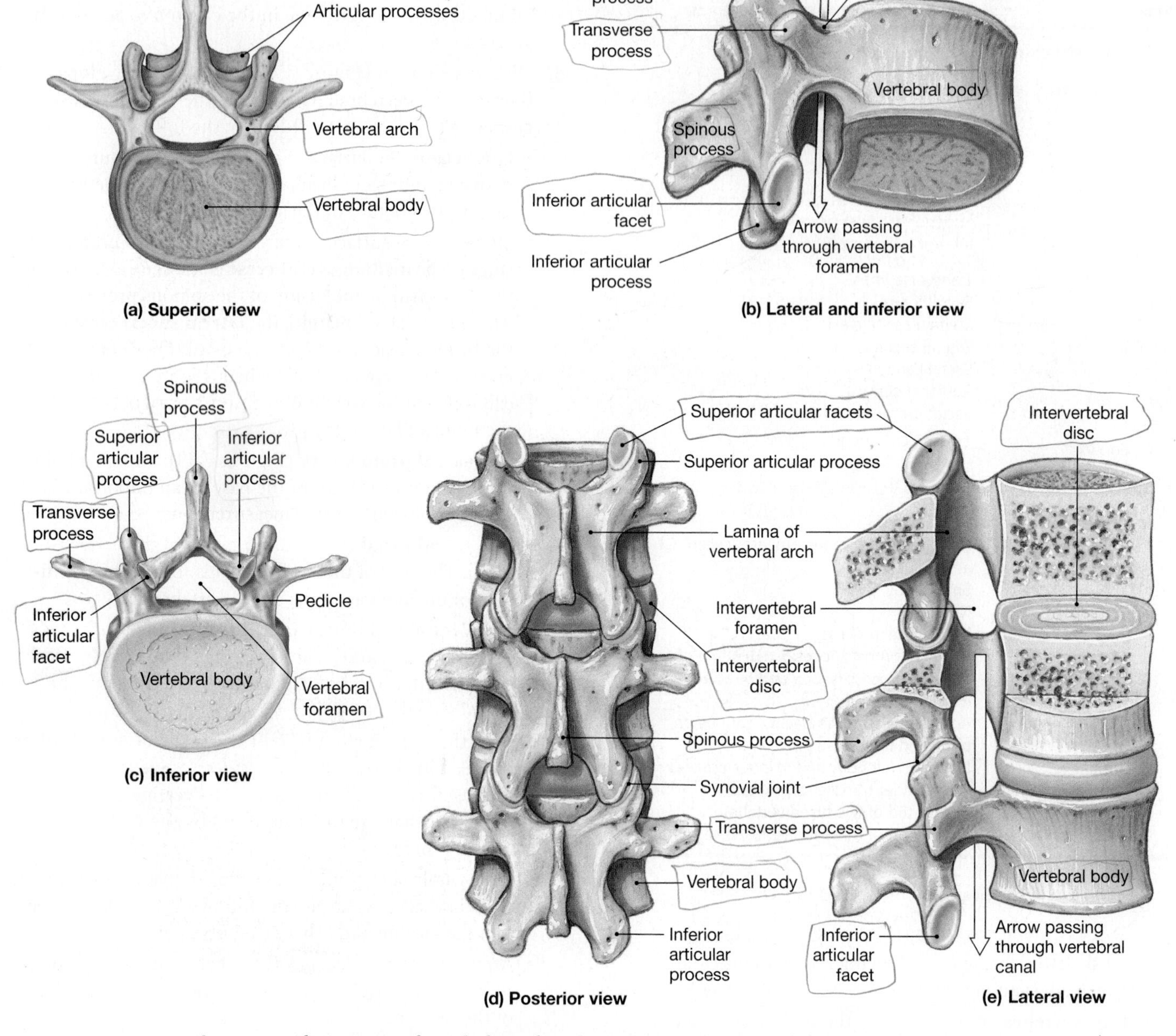

Figure 7.10 The structure of a typical vertebra. a) Superior view of a vertebra; **b)** inferiolateral view of a vertebra; **c)** inferior view of a vertebra; **d)** posterior view of three vertebrae showing the articulations between them; **e)** three articulating vertebrae showing a midsagittal section superiorly, and a lateral view inferiorly.

Explain why this feature is significant. (Hint: Look ahead to the muscle chapter in your text or Exercise 11 in this manual to determine the general function of the muscles that have sacral attachments.)

____

____

____

ACTIVITY 7.9 Examining the Unique Features of Specific Vertebrae

Cervical Vertebrae

1. On the whole skeleton or a model of the vertebral column, identify the seven **cervical vertebrae** (Figure 7.12a). They are designed to support the head and allow for its movement.

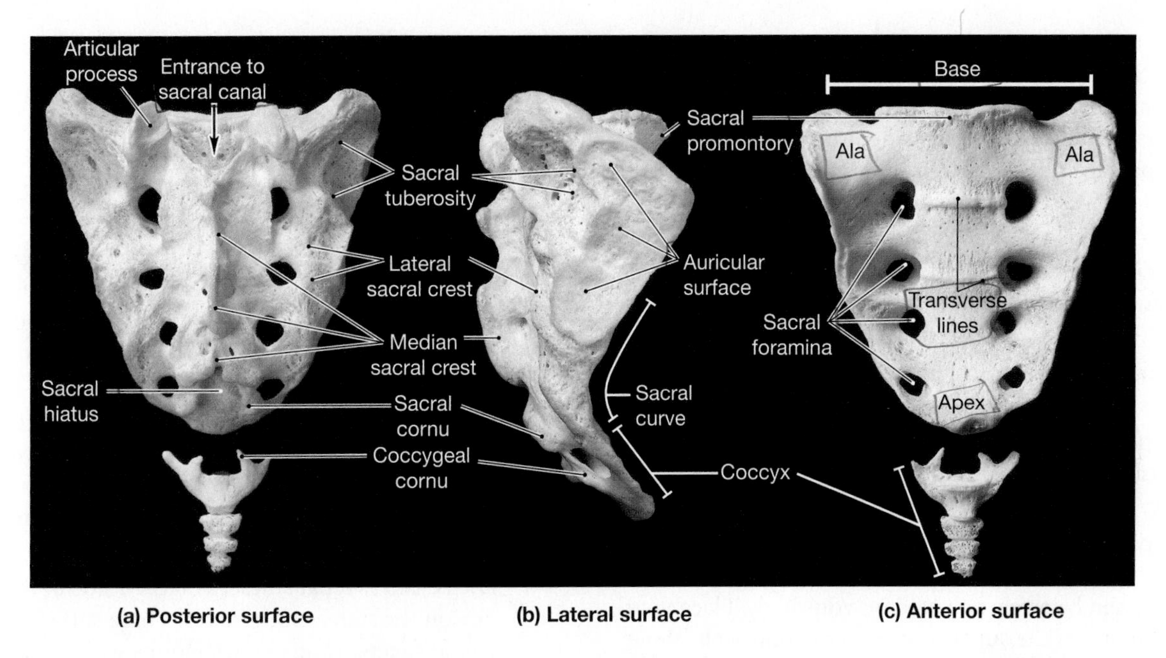

Figure 7.11 The sacrum and coccyx. a) Posterior view; **b)** lateral view; **c)** anterior view.

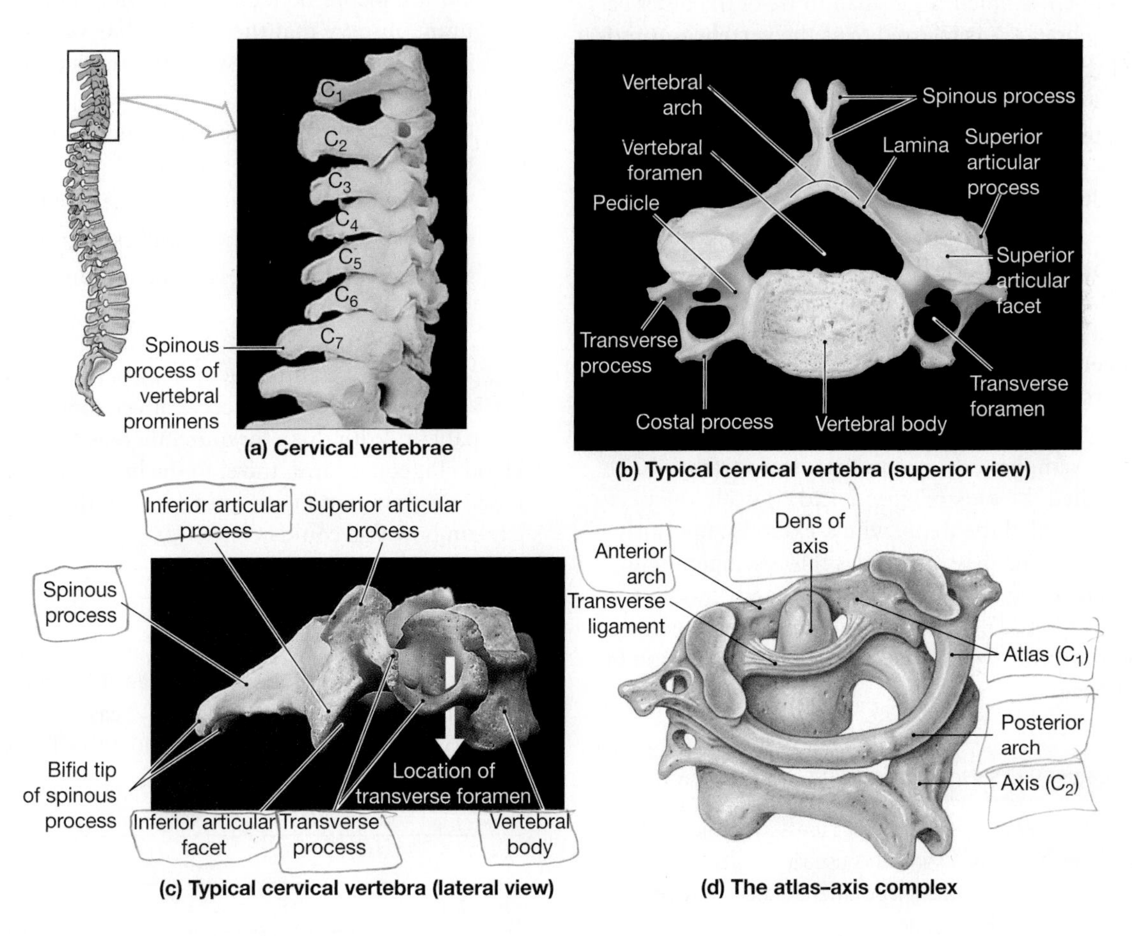

Figure 7.12 The cervical vertebrae. a) Lateral view of all the cervical vertebrae; **b)** superior view of a typical cervical vertebra; **c)** lateral view of a typical cervical vertebra; **d)** the relationship between the atlas (C1) and axis (C2).

2. Identify the **transverse foramina**, which are openings on the transverse processes of all cervical vertebrae. On the model of the vertebral column, note that the transverse foramina, on each side, form a bony canal for the passage of the **vertebral arteries** and **veins**. These blood vessels supply and drain blood from the brain.
3. Obtain examples of the third through seventh cervical vertebrae and observe the following special features (Table 7.4, Figures 7.12b and c).
   - The transverse foramina, as described, are unique to the cervical vertebrae.
   - The spinous processes on the third through sixth cervical vertebrae are relatively short and bifurcate at the tip.
   - The spinous process of the last cervical vertebrae (C7) is relatively long and does not bifurcate. The tip of this process can be easily palpated on your body. Place your index finger at the superior margin of your neck, along the midline, and slowly move your finger inferiorly. The first prominent bony protuberance that you feel is the tip of the spinous process of C7. Because the spinous process of C7 is much larger than those of the other cervical vertebrae, C7 is referred to as the **vertebra prominens** (Figure 7.12a).
4. Obtain an example of the first cervical vertebra (C1), known as the **atlas** (Figure 7.12d). The atlas is a ringlike structure with a short anterior arch and long posterior arch. It lacks a body and a spinous process.
5. Obtain a skull and locate the occipital condyles at the base (Figure 7.5e). Position the atlas so that it articulates with the occipital condyles. The articulation between the atlas and the occipital condyles of the skull is called the **atlantooccipital joint**. The "yes" nodding motion of the head occurs at this joint. Identify the atlantooccipital joint on the skeleton.
6. Obtain an example of the second cervical vertebra (C2), which is called the **axis** (Figure 7.12d). Locate the peglike process, called the **dens**, which extends superiorly from the body. The dens protrudes into the opening of the axis and is held against the anterior arch by a ligament (Figure 7.12d). This arrangement forms the **atlantoaxial joint**, where the "no" rotational motion of the head occurs.

CLINICAL CORRELATION

Whiplash injuries occur as a result of small tears in the ligament that reinforces the atlantoaxial joint. This usually results from abrupt and excessive front-to-back movements, such as when you are hit from the rear in a car accident.

Thoracic Vertebrae

1. On the whole skeleton or a model of the vertebral column, identify the 12 **thoracic vertebrae** (Figure 7.13a). These vertebrae are easily identified by their articulations with the 12 pairs of ribs. Notice that the vertebrae increase in size as they proceed inferiorly.
2. Observe that, in terms of general appearance, T1 through T4 are similar to cervical vertebrae, and T9 through T12 are similar to lumbar vertebrae. The middle four thoracic vertebrae (T5 through T8) are considered "typical."
3. Obtain examples of the typical thoracic vertebrae. Notice that the spinous process of a thoracic vertebra projects inferiorly (Figure 7.13c), the body has a distinctive heart shape, and the vertebral foramen is circular (Figure 7.13b).
4. Locate the two flat surfaces on each side of the body and one on the transverse process. These surfaces, called **costal facets**, form synovial joints with the ribs.

Lumbar Vertebrae

1. On the whole skeleton or a model of the vertebral column, observe that the five **lumbar vertebrae** are located in the lower back, between the thorax and pelvis (Figure 7.14a).
2. Obtain samples of these bones and observe the following (Figures 7.14b and c):
   - The oval or kidney-shaped bodies are relatively large.
   - The short, blunt spinous processes project dorsally.
   - The transverse processes are long and slender.

QUESTIONS TO CONSIDER 1. A chiropractor may perform a spinal manipulation of the cervical vertebrae during treatment of a patient. With this procedure, there is a slight risk that a blood clot could form, travel to the brain, and cause a stroke. Based on your anatomical knowledge of the cervical vertebrae, explain how this could occur. ______________________

____

____

2. The inferior two thoracic vertebrae, T11 and T12, look similar to the lumbar vertebrae and could easily be mistaken as such. Explain how you can readily identify T11 and T12 as thoracic vertebrae.

____

____

____

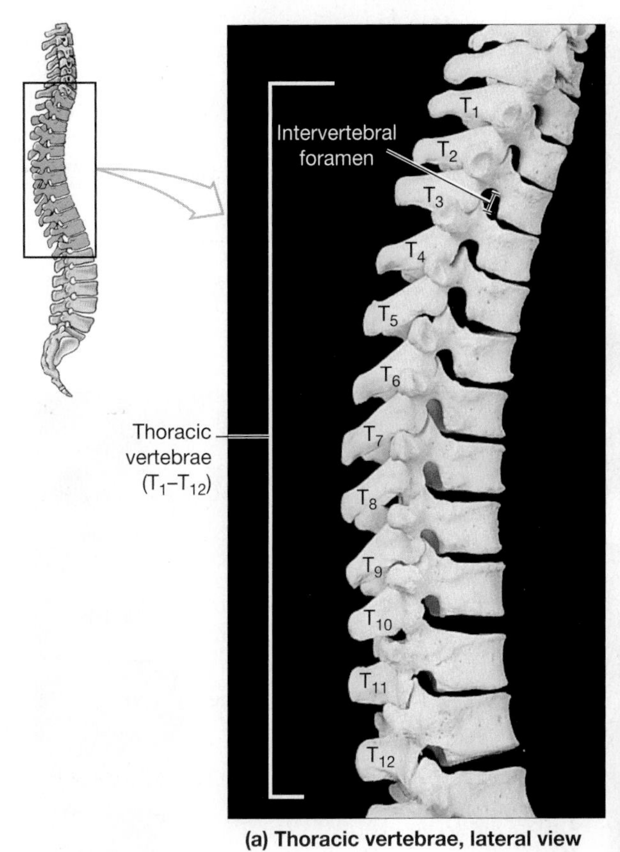

(a) Thoracic vertebrae, lateral view

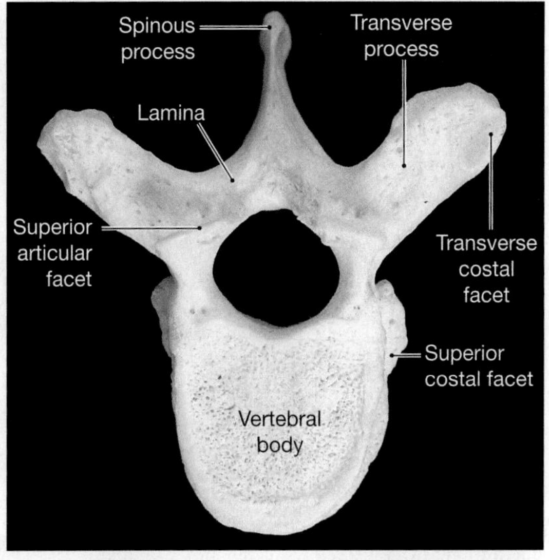

(b) Thoracic vertebra, superior view

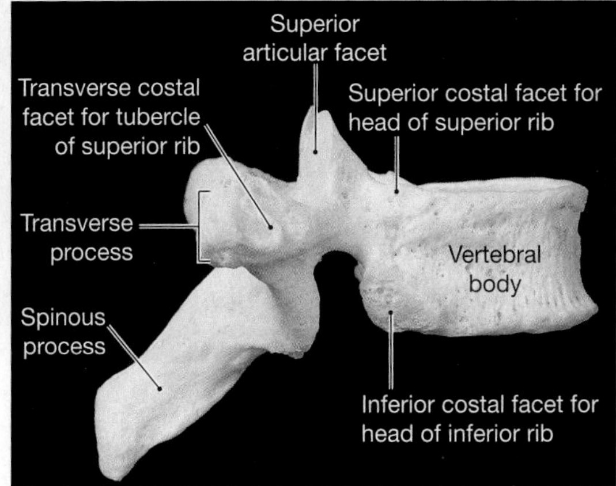

(c) Thoracic vertebra, lateral view

Figure 7.13 The thoracic vertebrae. a) Lateral view of all the thoracic vertebrae; **b)** superior view of a typical thoracic vertebra; **c)** lateral view of a typical thoracic vertebra.

The Thoracic Cage

The thoracic cage includes the 12 pairs of ribs and the sternum. These structures surround and protect the lungs, the heart, and other structures. In addition, movements of the ribs and sternum are essential for the breathing process.

ACTIVITY 7.10 Examining the Structure of the Thoracic Cage

1. On the whole skeleton, locate the **sternum** on the anterior surface of the thoracic cage and identify its three parts: the **manubrium**, the **body**, and the **xiphoid process** (Figure 7.15).
2. The manubrium is the broad superior segment. Notice that it articulates with the clavicles and the first pair of ribs. The superior margin of the manubrium is called the **jugular notch**.
3. The elongated middle portion is the body. Notice that it is the largest portion of the sternum and articulates with the second through seventh pairs of ribs.
4. The xiphoid process, the smallest portion of the sternum, is attached to the inferior end of the body. It remains cartilaginous well into adulthood.
5. Identify the joint between the manubrium and body. This articulation acts as a hinge that allows the body to move anteriorly during inhalation. The angle formed at the joint is the **sternal angle.** It remains cartilaginous well into adulthood.

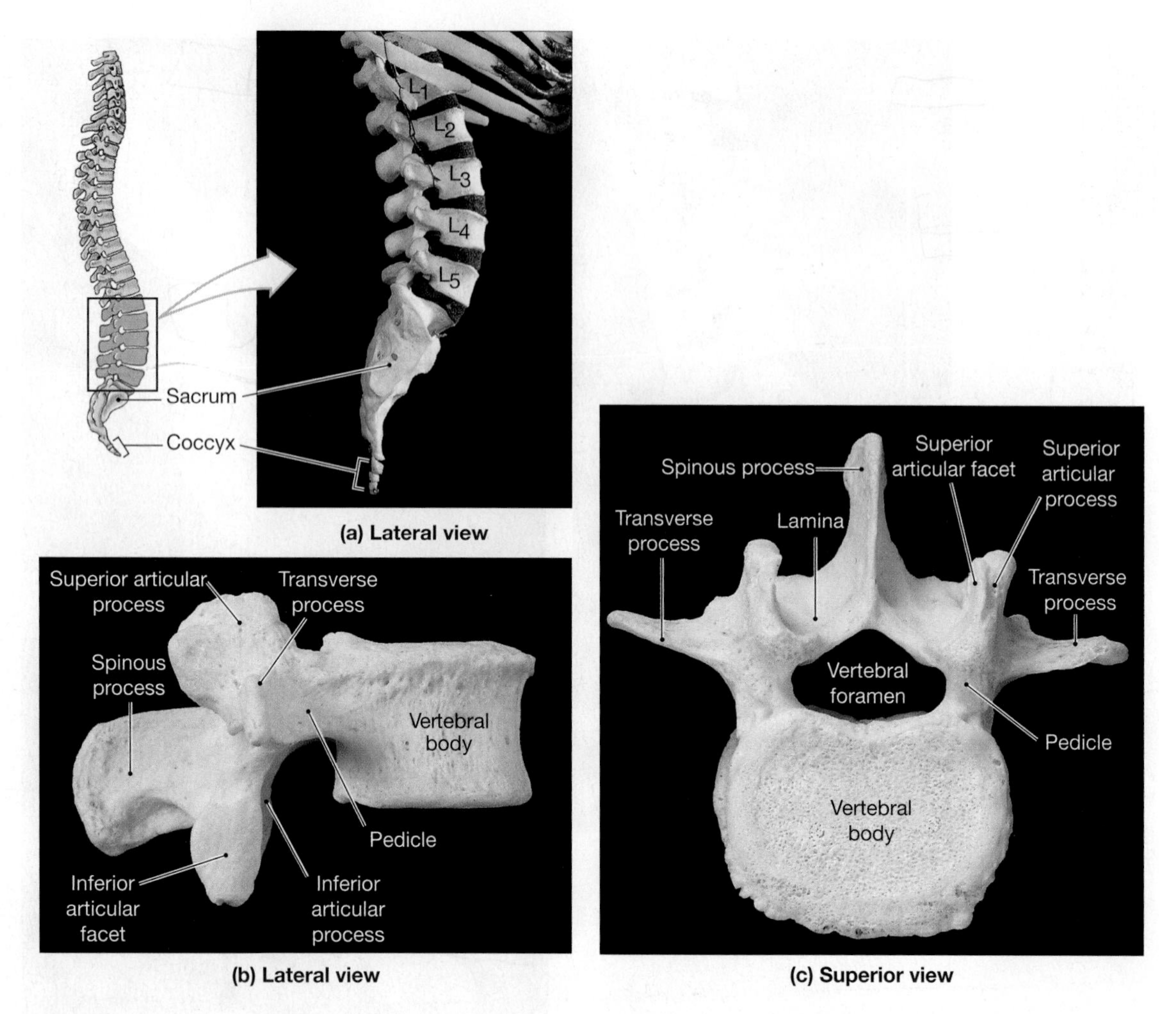

Figure 7.14 The lumbar vertebrae. a) Lateral view of all the lumbar vertebrae; **b)** lateral view of a typical lumbar vertebra; **c)** superior view of a typical lumbar vertebra.

6. Examine the arrangement of the 12 pairs of ribs. Observe that the length of the ribs gradually increases from ribs 1 through 7, then decreases from ribs 8 through 12.
7. Notice that all the ribs articulate with the thoracic vertebrae. Specifically, each rib articulates with its numerically corresponding vertebra, the intervertebral disc, and the vertebra superior to it. For example, rib 5 articulates with T5, T4, and the intervening disc (Figure 7.16).
8. Anteriorly, note that the ribs are attached to pieces of cartilage, referred to as **costal cartilages** (Figure 7.15). The ribs are categorized according to how the costal cartilages are attached to the sternum (Figure 7.15).
   - Ribs 1 through 7 are called **true (vertebrosternal) ribs** because their costal cartilages are *directly* attached to the sternum.
   - Ribs 8 through 12 are called **false ribs** because they do not have a direct attachment to the sternum. There are two types of false ribs.
     - Notice that the costal cartilages for ribs 8 through 10 are connected to each other and attach to the sternum via the costal cartilage of rib 7. These ribs have an indirect attachment to the sternum and are called **vertebrochondral ribs.**
     - Ribs 11 and 12 are called **floating ribs** because they have no connection to the sternum.

WHAT'S IN A WORD The 11th and 12th pairs of ribs are called *floating ribs,* but that term can be misleading. It implies that these ribs are floating unattached along the side. They are, in fact, very much anchored in place, attached posteriorly to the 10th through 12th thoracic vertebrae, and anchored anteriorly

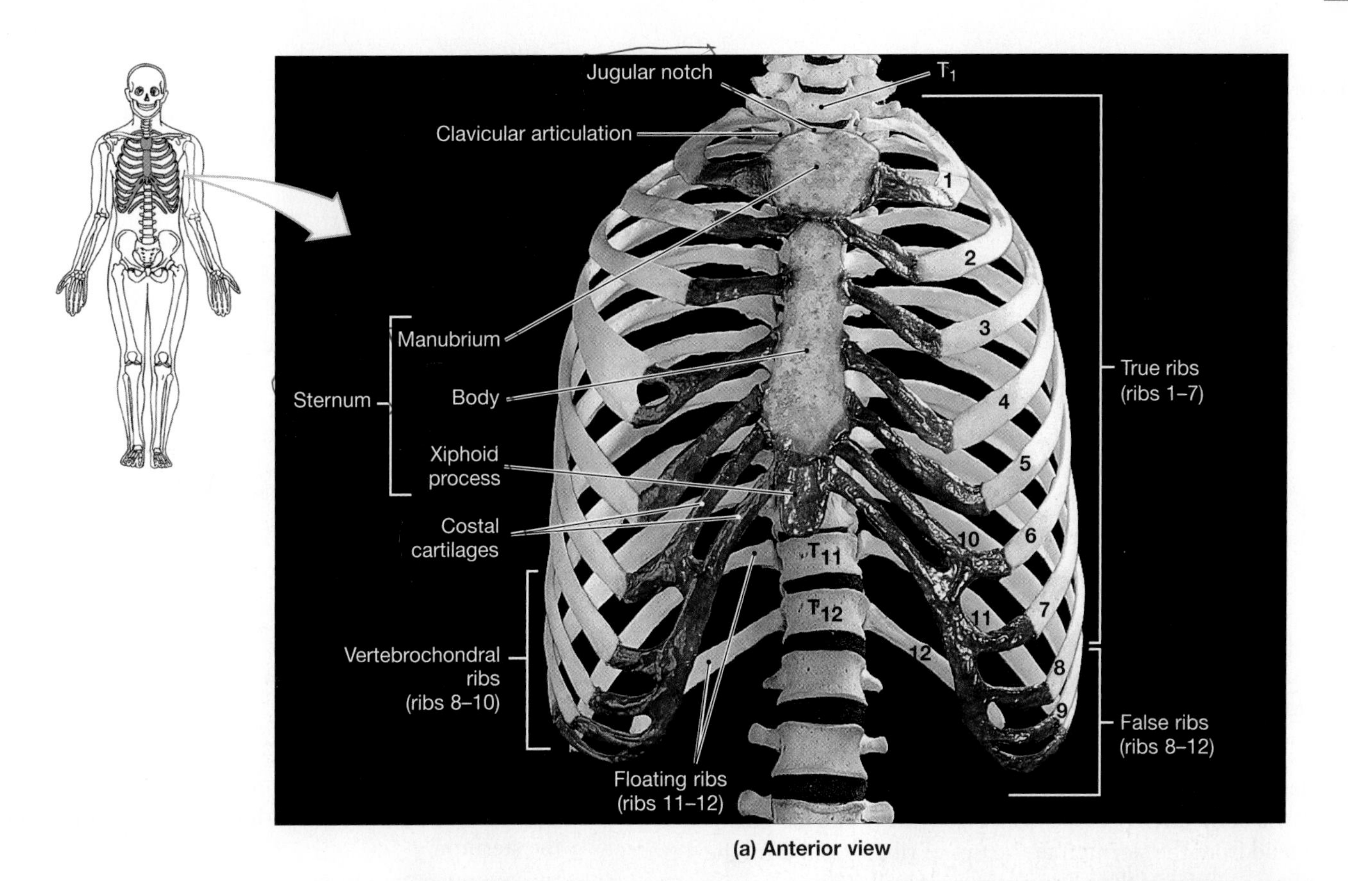

(a) Anterior view

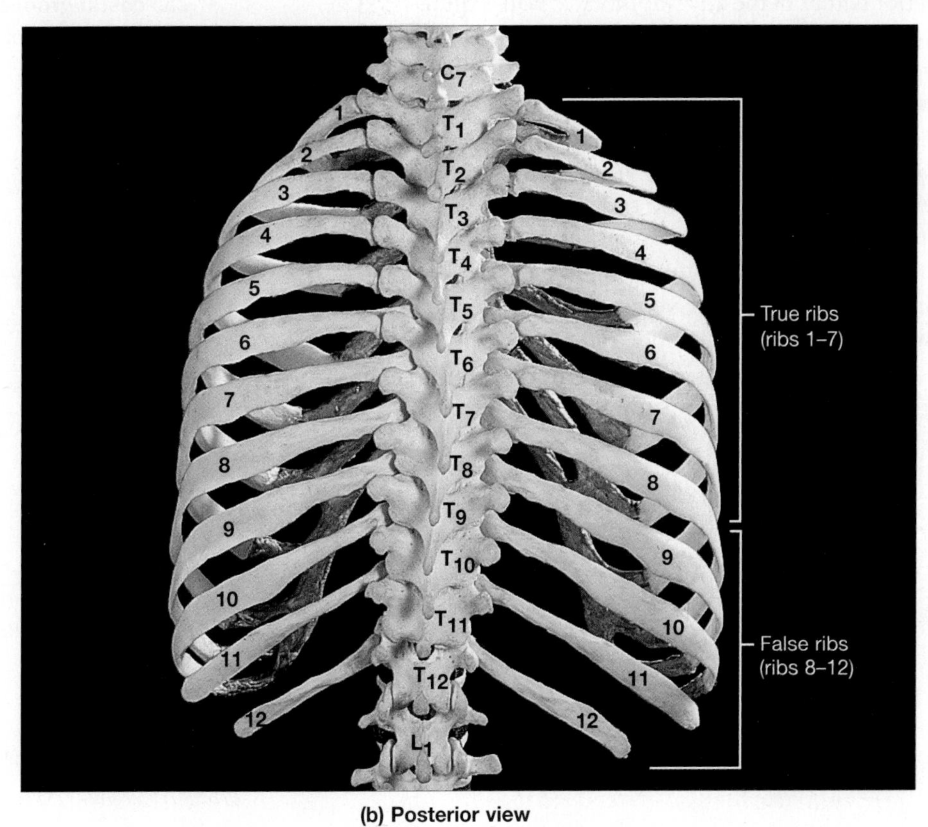

(b) Posterior view

Figure 7.15 The thoracic cage. a) Anterior view; **b)** posterior view.

Figure 7.16 Articulations of a true rib with the vertebral column and the sternum. Most ribs articulate with two thoracic vertebrae. True ribs have a direct connection with the sternum by way of a costal cartilage.

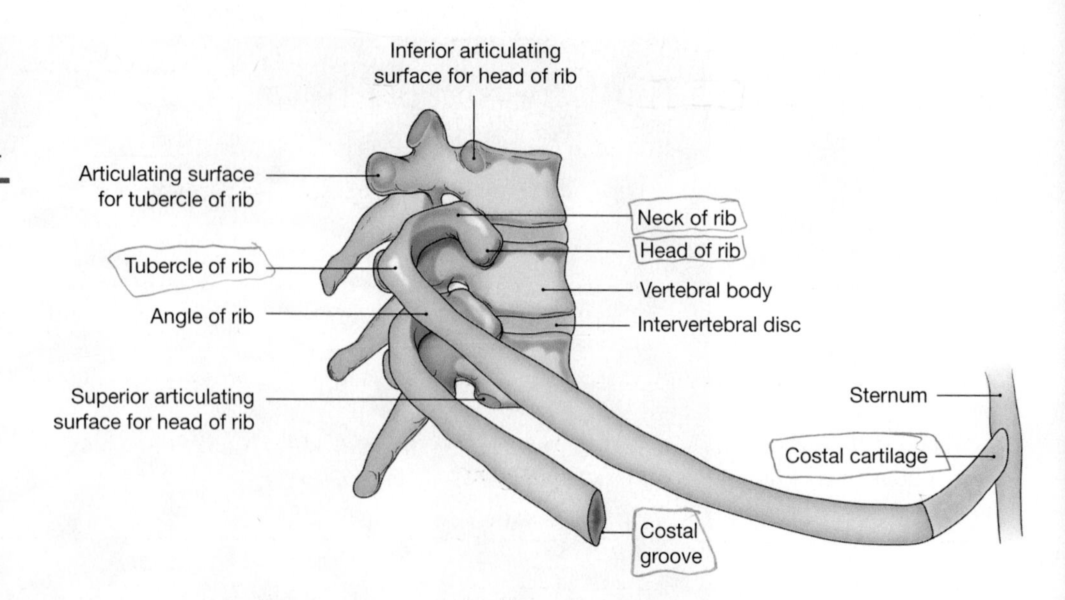

by the muscles of the abdominal wall. Because the articulations with thoracic vertebrae are their only connections to bone, floating ribs are also called *vertebral ribs.*

9. Identify the left and right **costal margins** that are formed by the costal cartilages of ribs 7 through 10. The margins extend from the sternum to the tips of ribs 10 and form the inferior border of the anterior thoracic wall (Figure 7.15).
10. The spaces between the ribs are called **intercostal spaces** (Figure 7.15). Each space contains three layers of **intercostal muscles**, which play an important role in respiration.
11. Obtain an example of a "typical" rib (ribs 3 through 9) and differentiate between the **vertebral end** and the **sternal end**. At the vertebral end, identify the bony landmarks listed and illustrated in Figure 7.16.
    - The **head** articulates with the body of the numerically corresponding thoracic vertebra, the intervertebral disc, and the body of the superior vertebra.
    - The **tubercle** articulates with the transverse process of the numerically corresponding thoracic vertebra.
    - The **neck** is the short length of bone between the head and tubercle.
    - The **shaft**, or **body**, is the main part of the rib.
    - Beginning at the vertebral end, the **angle of the rib** is where the shaft curves anteriorly and inferiorly toward the sternum.
    - The **costal groove** runs along the inferior border of a rib. The groove provides a pathway for intercostal arteries, veins, and nerves.

QUESTION TO CONSIDER In the activity that you just completed, you observed that the ribs articulate with two different vertebrae. Can you find any exceptions to this rule?

___

___

___

Exercise 7 Review Sheet

Introduction to the Skeletal System and the Axial Skeleton

Name ____________________

Lab Section ____________________

Date ____________________

1. Discuss how bones are classified according to shape. Give at least one example of a bone in each category.

____

____

____

2. Describe the microscopic structure of compact bone. How does the structure of compact bone differ from the structure of spongy bone?

____

____

____

3. Identify the bones that comprise the axial skeleton.

____

____

____

4. What are the primary functions of red bone marrow and yellow bone marrow? In which bones is red bone marrow most abundant?

____

____

____

Questions 5–18: Identify the labeled bones in the diagram.

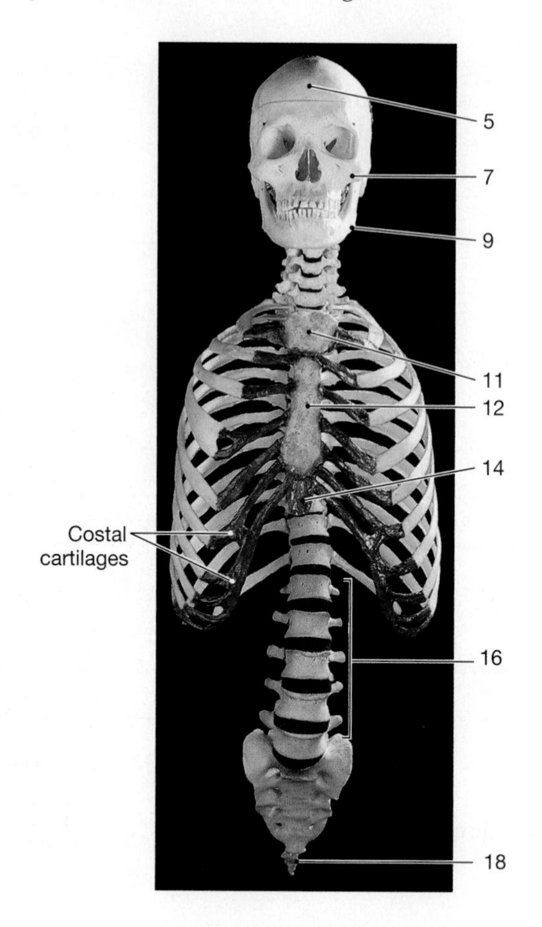

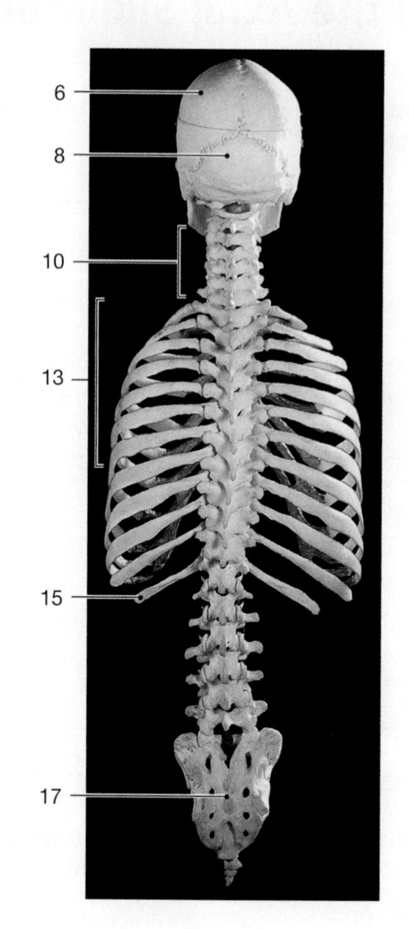

5. ______________________

6. ______________________

7. ______________________

8. ______________________

9. ______________________

10. ______________________

11. ______________________

12. ______________________

13. ______________________

14. ______________________

15. ______________________

16. ______________________

17. ______________________

18. ______________________

Questions 19–25: Identify the labeled bones in the diagram.

19. ______________________________

20. ______________________________

21. ______________________________

22. ______________________________

23. ______________________________

24. ______________________________

25. ______________________________

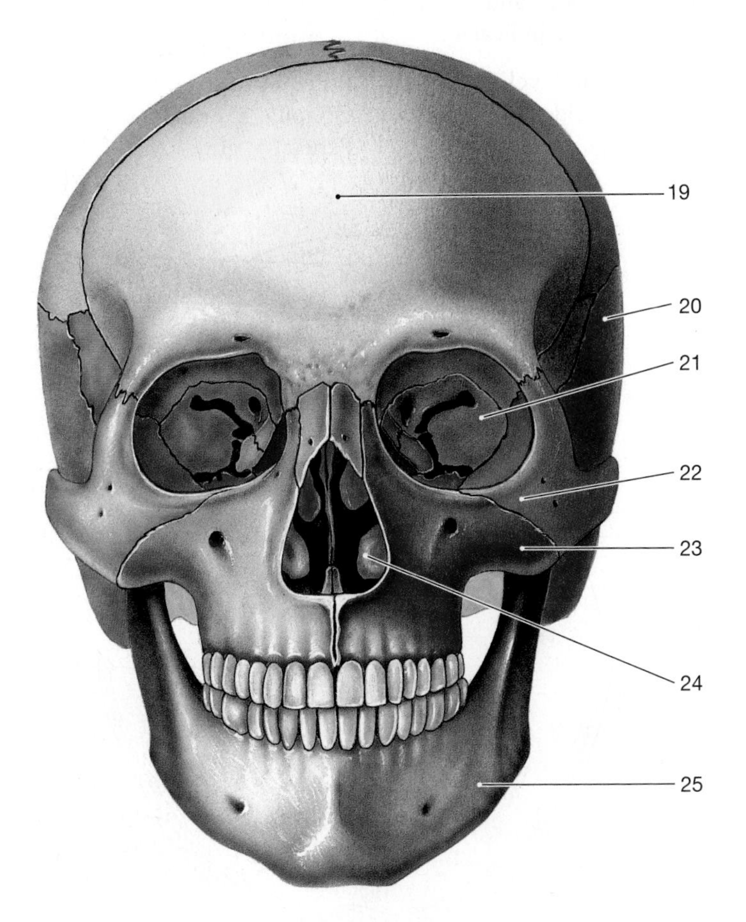

26. In the diagram at right, identify the bones by using the colors that are indicated.

Frontal = **yellow**
Sphenoid = **purple**
Zygomatic = **blue**
Maxillary = **orange**
Lacrimal = **red**
Mandible = **gray**
Parietal = **pink**
Occipital = **orange**
Temporal = **red**
Nasal = **purple**
Ethmoid = **green**

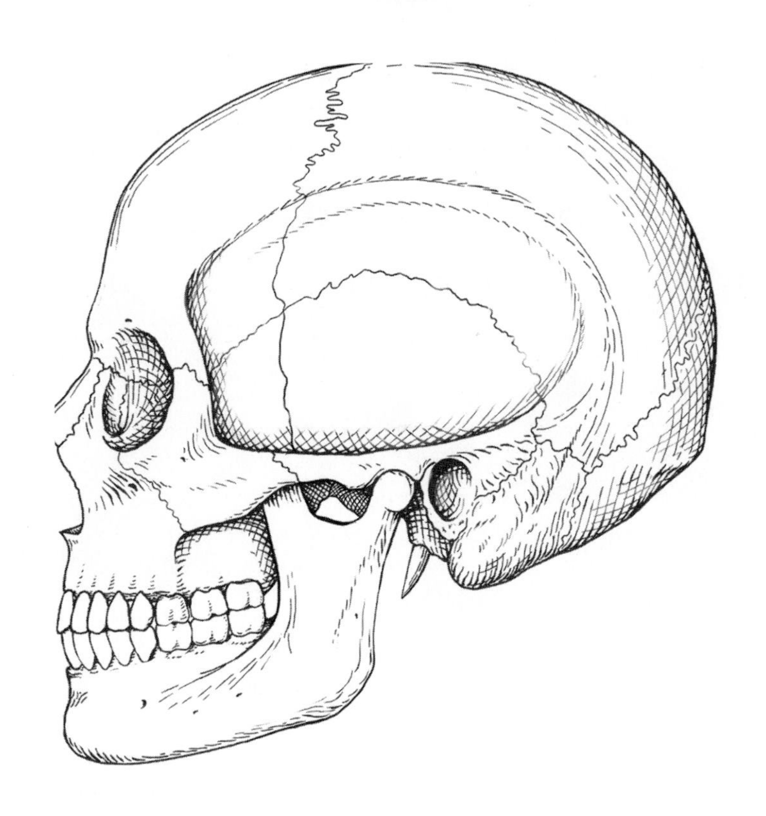

Questions 27–38: In the following diagram, identify the labeled bony landmarks and openings. Identify the bones by using the colors that are indicated.

Frontal = **yellow**	Ethmoid = **green**
Sphenoid = **purple**	Occipital = **orange**
Temporal = **red**	Parietal = **gray**

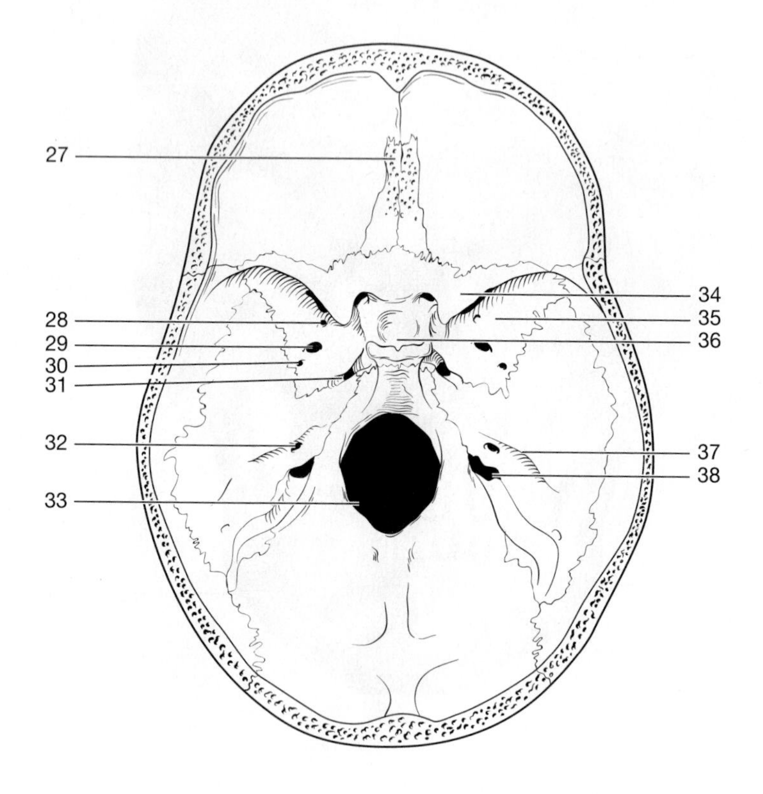

27. ______________________

28. ______________________

29. ______________________

30. ______________________

31. ______________________

32. ______________________

33. ______________________

34. ______________________

35. ______________________

36. ______________________

37. ______________________

38. ______________________

Questions 39–48: In the following diagram, identify the labeled bony landmarks and openings. Identify the bones by using the colors that are indicated.

Palatine = **yellow**
Sphenoid = **purple**
Temporal = **red**
Maxillary = **orange**
Zygomatic = **blue**
Occipital = **yellow**
Vomer = **brown**

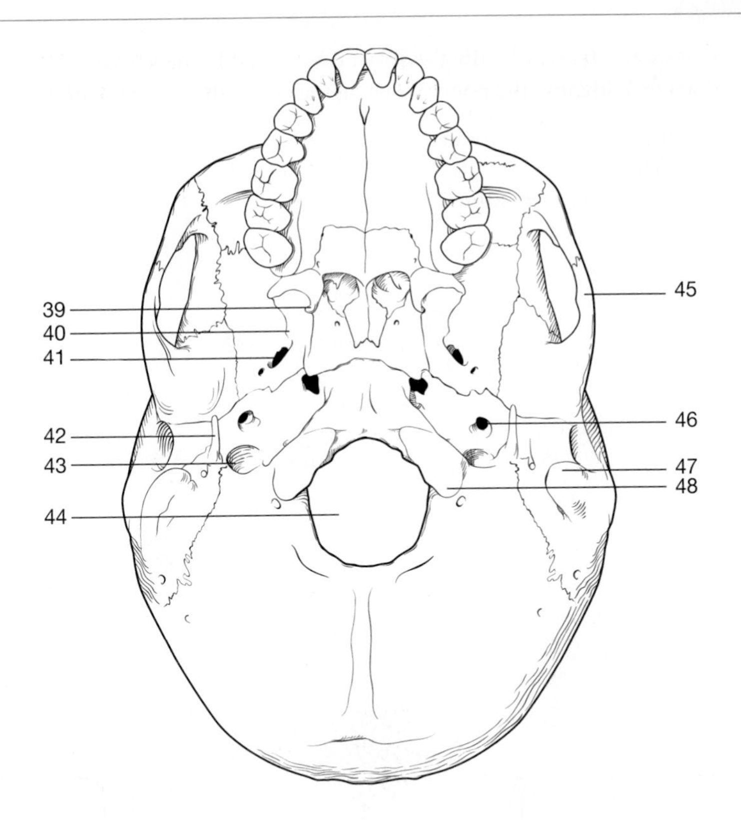

39. ______________________________

40. ______________________________

41. ______________________________

42. ______________________________

43. ______________________________

44. ______________________________

45. ______________________________

46. ______________________________

47. ______________________________

48. ______________________________

Questions 49–57: Identify the structures labeled in the following diagrams. Identify the bones by using the colors that are indicated.

Cervical vertebra = **green**
Thoracic vertebra = **red**
Lumbar vertebra = **blue**

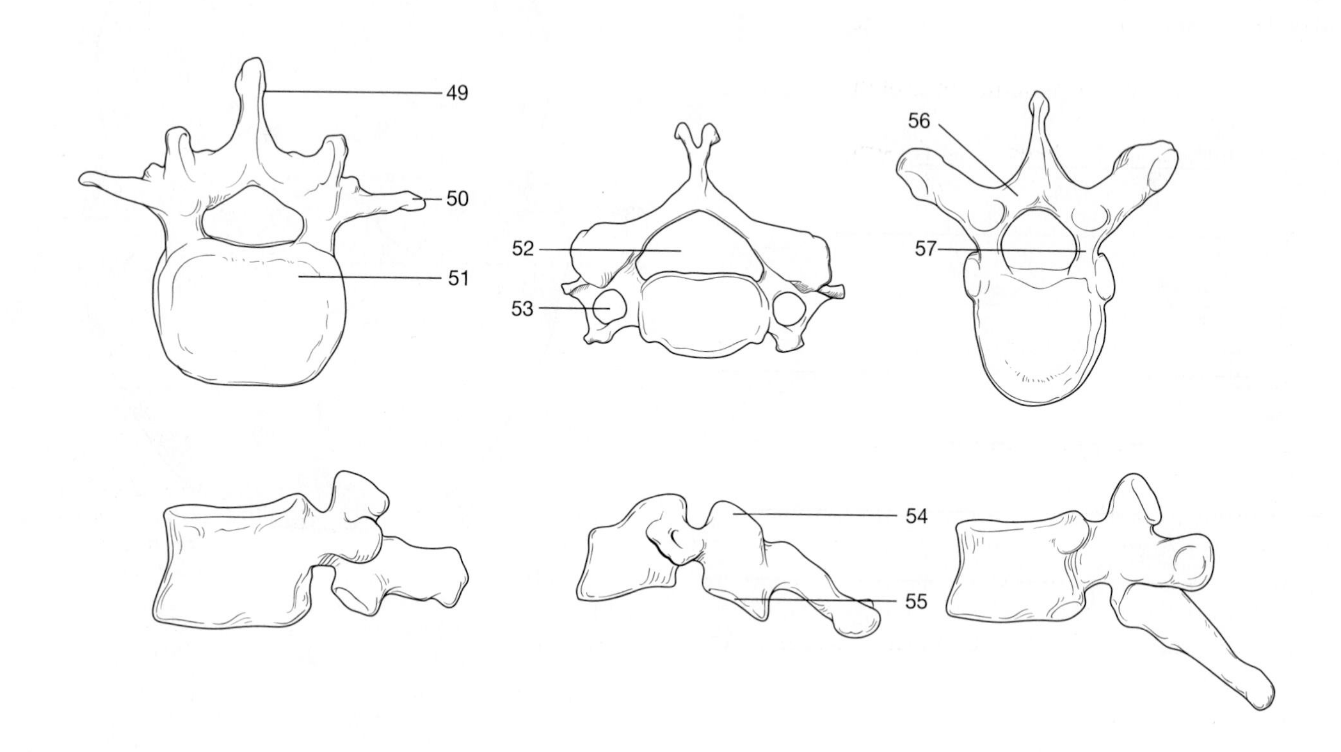

49. ____________________

50. ____________________

51. ____________________

52. ____________________

53. ____________________

54. ____________________

55. ____________________

56. ____________________

57. ____________________

Exercise 8

The Appendicular Skeleton

Laboratory Objectives

On completion of the activities in this exercise, you will be able to:

- Identify the bones of the axial skeleton and their bony landmarks.
- Understand the arrangement of the bones of the axial skeleton, relative to each other.
- Describe similarities and differences in the structure and function of the upper and lower extremities.

Materials

- Human skeleton
- Model of the knee
- Disarticulated bones of the upper and lower extremities
- Colored pencils

The **appendicular skeleton** comprises the bones of the upper and lower extremities (the appendages). Each upper extremity includes the pectoral girdle and the bones of the arm, forearm, wrist, and hand. Each lower extremity includes the pelvic girdle and the bones of the thigh, leg, and foot (Table 8.1; Figure 8.1).

Table 8.1 Bones of the Upper Extremity

	Bone	Bony landmarks
Pectoral girdle	Scapula	Acromion process Coracoid process Glenoid fossa Spine of scapula Supraspinous fossa Infraspinous fossa Subscapular fossa
	Clavicle	Sternal end Acromial end
Arm	Humerus	Head of humerus Anatomical neck Greater tubercle Lesser tubercle Intertubercular groove Deltoid tuberosity Medial epicondyle Lateral epicondyle Coronoid fossa Radial fossa Olecranon fossa Capitulum Trochlea
Forearm	Radius	Head of radius Radial tuberosity Ulnar notch Styloid process of radius
	Ulna	Olecranon process Coronoid process Trochlear notch Radial notch Head of ulna Styloid process of ulna
Wrist and hand	Carpal bones (from lateral to medial) proximal row: scaphoid–lunate–triquetrum–pisiform distal row: trapezium–trapezoid–capitate–hamate	
	Metacarpal bones	
	Phalanges	

The Upper Extremity

The upper extremity contains four segments. These segments and the bones found in each are as follows:

- The **pectoral girdle** consists of a ring of bone formed by the two **clavicles** (collar bones) and **scapulae** (shoulder blades). The ring is completed anteriorly by the manubrium of the sternum, but is incomplete posteriorly. The clavicle and scapula on each side of the body represent one half of the pectoral girdle (Figure 8.1a).
- The arm is the region between the shoulder and elbow joints. It contains one large long bone known as the **humerus.**
- The forearm is the region between the elbow and wrist joints. It contains two smaller long bones, the medial **ulna** and the lateral **radius.**
- The hand is the part of the upper extremity that is distal to the forearm. It contains 8 **carpal bones** at the wrist, 5 **metacarpal bones** in the palm of the hand, and 14 **phalanges** in the digits (thumb and fingers).

The pectoral girdle forms a bony strut that suspends the upper extremities lateral to the trunk. It is stabilized almost entirely by muscles that connect it to the sternum, ribs, and vertebrae. The only bony articulations between the pectoral girdle and the axial skeleton occur between the clavicles and the manubrium of the sternum. This arrangement provides the upper extremity with a wide range of motion, particularly at the shoulder joint, but in return, sacrifices some stability.

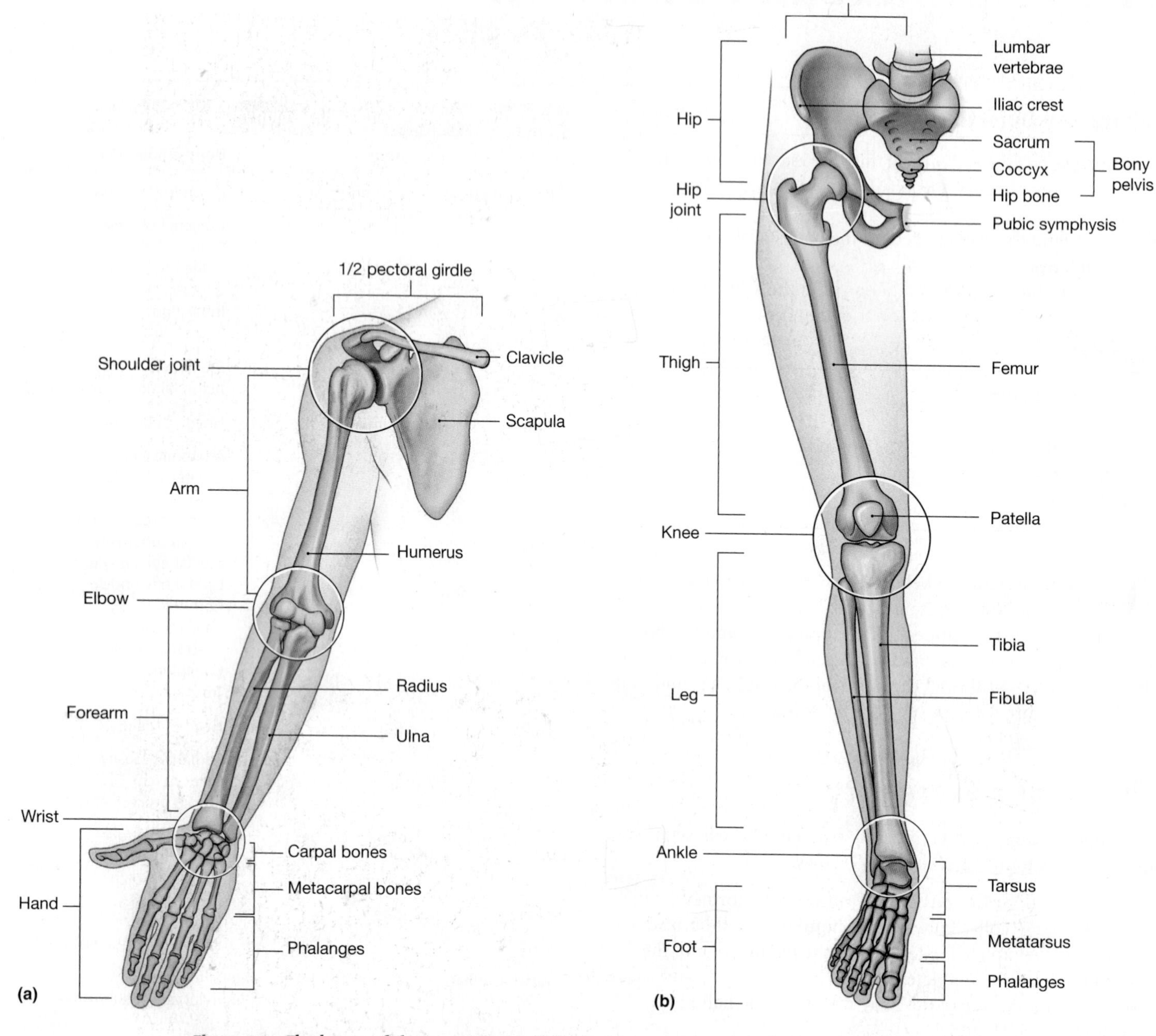

Figure 8.1 The bones of the appendicular skeleton. a) anterior view of the bones of the upper extremity; **b)** anterior view of the bones of the lower extremity.

ACTIVITY 8.1 Examination of the Upper Extremity

1. On a whole skeleton, identify all the bones of the upper extremity (Figure 8.1a; Table 8.1).
2. Note that the clavicle has two distinct curves (Figure 8.2). The medial or sternal end is concave posteriorly and has a pyramidal-shaped end. The lateral or acromial end is concave anteriorly and has a flattened end.
3. The articulation between the medial end of the clavicle and the manubrium of the sternum is called the **sternoclavicular joint** (Figure 8.2a). Realize that this joint is the only bony attachment between the upper extremity and the axial skeleton. All other attachments between bones of the axial skeleton (vertebral column, ribs, and sternum) and the upper appendicular skeleton (scapula, clavicle, and humerus) are musculotendinous. The articulation between the lateral end of the clavicle and the acromion process of the scapula is called the **acromioclavicular joint** (Figure 8.2a).
4. Identify the **scapula** (Figure 8.3) and note that it is generally a flat, triangular bone. It rests on the posterior thoracic wall between ribs 2 and 7. Realize that the

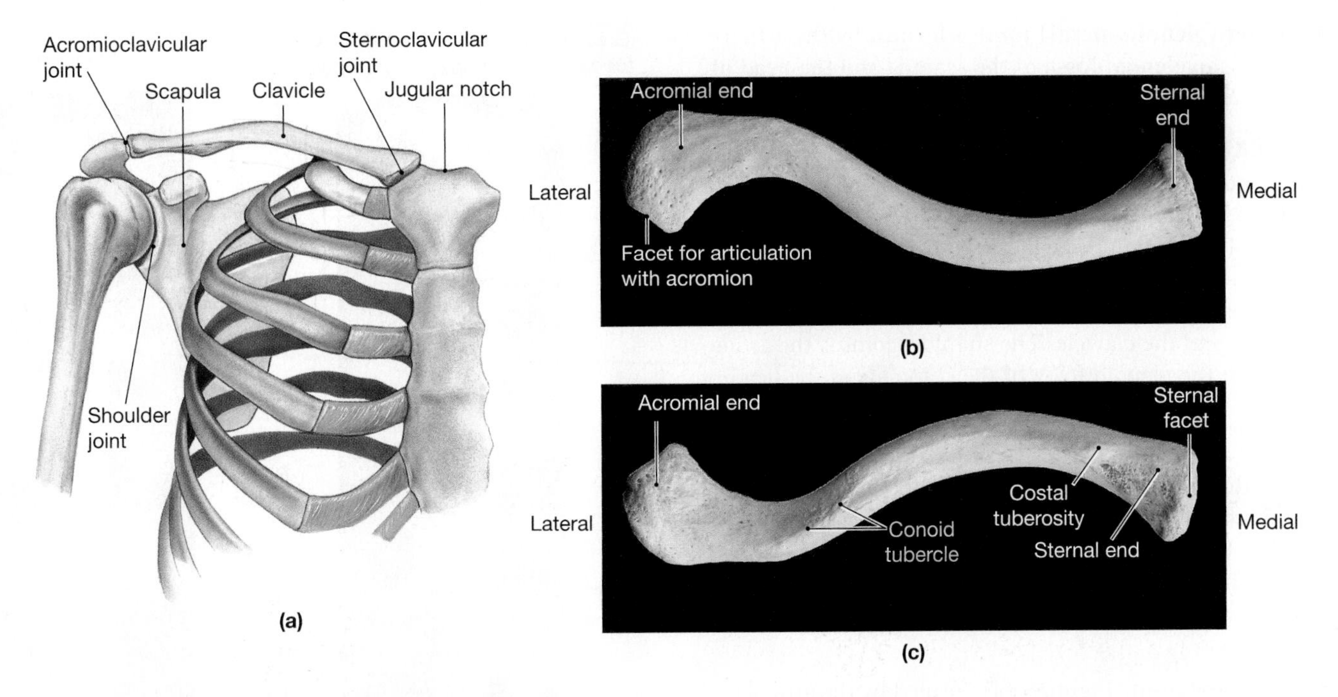

Figure 8.2 The right clavicle. a) The anatomical position of the clavicle, showing its articulations with the manubrium of the sternum and the acromion process of the scapula; **b)** superior view of the clavicle; **c)** inferior view of the clavicle.

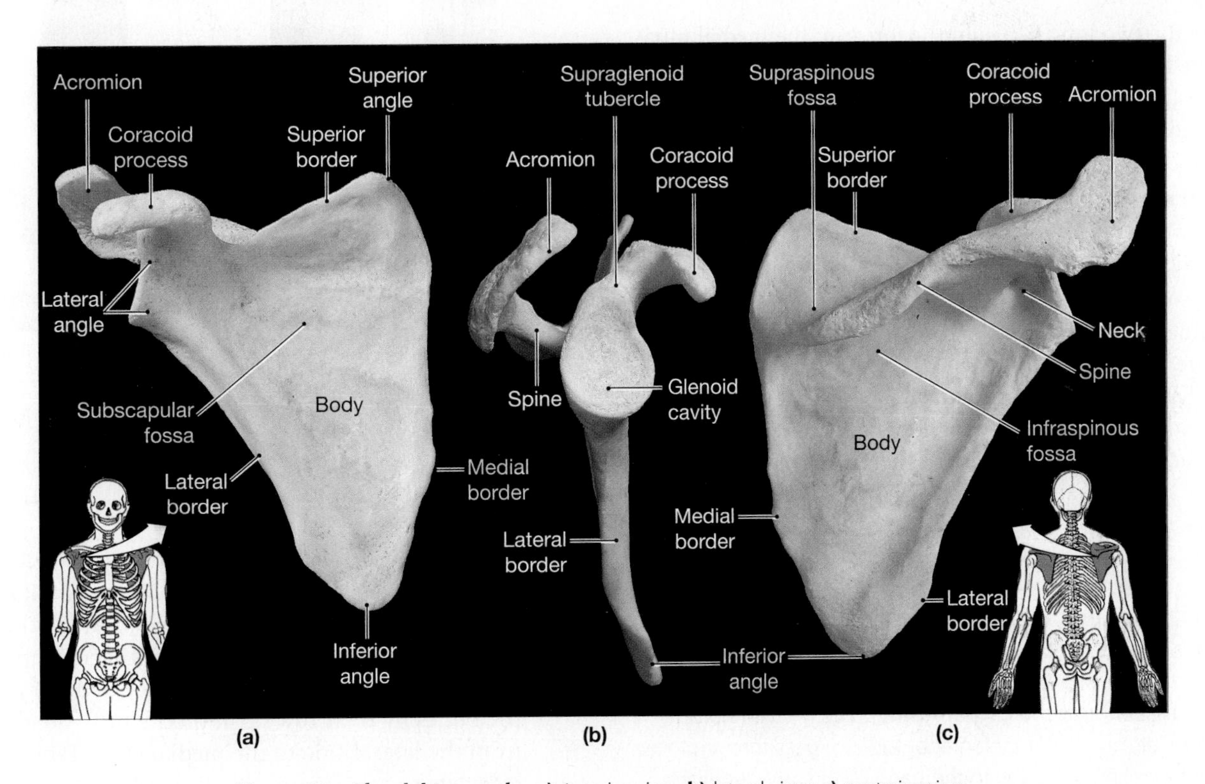

Figure 8.3 The right scapula. a) Anterior view; **b)** lateral view; **c)** posterior view.

shoulder (glenohumeral) joint is formed by the articulation of the glenoid fossa of the scapula and the head of the humerus (Figures 8.1a and 8.2a).

WHAT'S IN A WORD The names of the joints associated with the clavicle provide clues to help you determine how the articulation is formed. The word *sternoclavicular* refers to the sternum and clavicle. This is a joint in which the manubrium of the sternum articulates with the clavicle. Similarly, the term *acromioclavicular* refers to the articulation between the acromion process of the scapula and the clavicle. The shoulder joint is the articulation between the glenoid fossa of the scapula and the head of the humerus. Thus, another name for the shoulder is the *glenohumeral* joint.

5. Locate the humerus, which is the only bone found in the arm. This is the longest and largest bone of the upper extremity (Figure 8.4). It is involved in the formation of two major joints.
   - The shoulder joint, as described previously (Figures 8.1a and 8.2a)
   - The **elbow joint** (Figure 8.1a), formed by the articulations between the trochlea on the humerus (Figure 8.4a) and the trochlear notch on the ulna (Figure 8.5b), and between the capitulum on the humerus (Figure 8.4a) and the radial head on the radius (Figure 8.5b)
6. Locate the two long bones in the forearm: the medial ulna and the lateral radius (Figure 8.5). (Recall correct anatomical position.) Both bones take part in the formation of the elbow joint, as described previously.
7. At the **wrist joint** (Figures 8.1a and 8.6), note that the inferior surface of the radius articulates with the proximal row of carpal bones. Although it might appear otherwise, the ulna does not contribute an articulating surface at the wrist.
8. Examine the eight carpal bones at the wrist and note that they are arranged in two rows of four (Figure 8.6; Table 8.1). These bones are held together by ligaments that restrict their mobility to gliding movements. As a unit, carpal bones are shaped so that the dorsal surface is convex and the palmar (anterior) surface is concave.

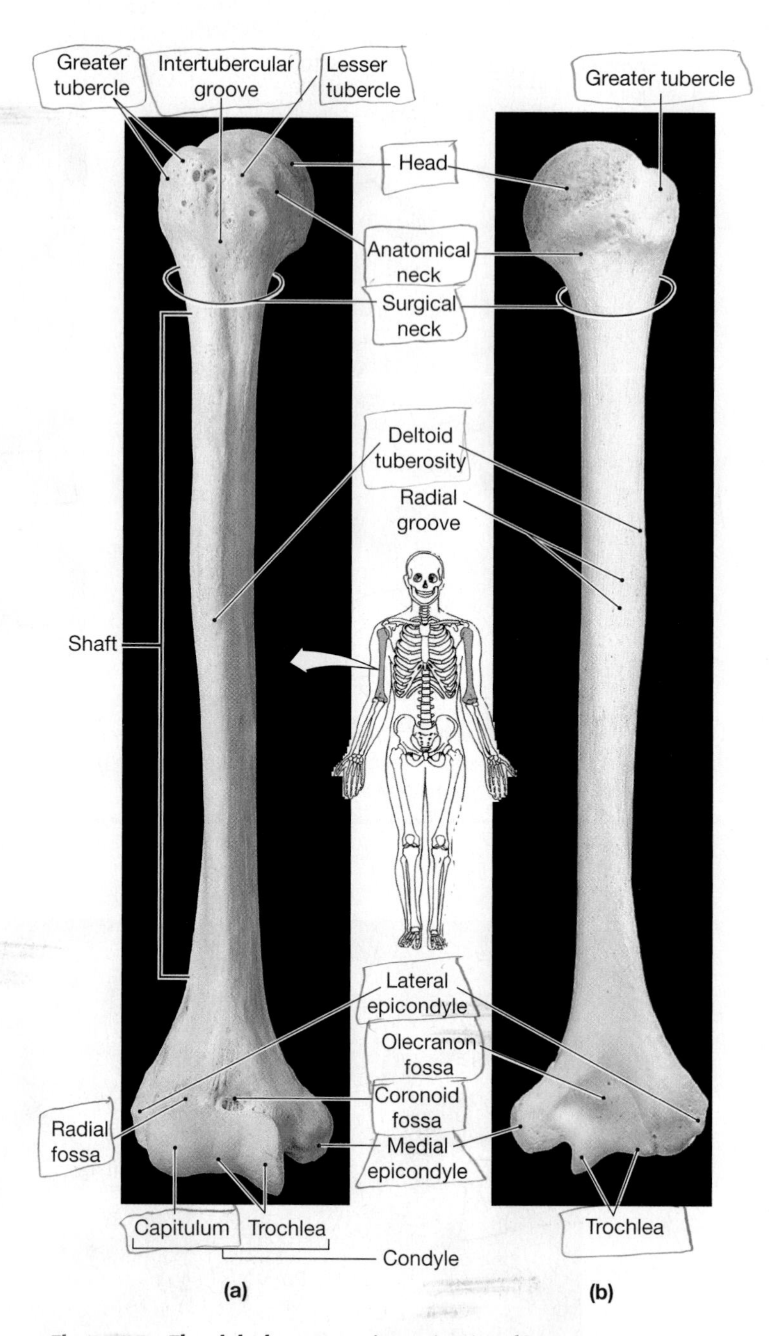

Figure 8.4 The right humerus. a) Anterior view; **b)** posterior view.

CLINICAL CORRELATION

At the wrist, a connective tissue bridge, called the **flexor retinaculum,** stretches over the concavity created by the anterior surfaces of the carpal bones. The combination of bones and connective tissue forms the **carpal tunnel.** The median nerve and tendons of muscles in the anterior forearm pass through the carpal tunnel on their way to the hand. With some repetitive motions, the tendons can become inflamed and swell, compressing the median nerve. This is called **carpal tunnel syndrome** and can cause numbness, severe pain, and impaired muscle function.

9. The best way to learn the names of the carpal bones is to be aware of their positions relative to each other. For example, learn to recognize the bones according to what row (proximal or distal) they are located. As you learn the bones by their row position, identify them, in order, from the lateral side to the medial side (Table 8.1), or vice versa.

WHAT'S IN A WORD It can be very challenging to recall the names of the carpal bones. Noting special appearances of the bones can be beneficial to help you get your bearings. In the proximal row, the *lunate* is a comma-shaped bone, somewhat

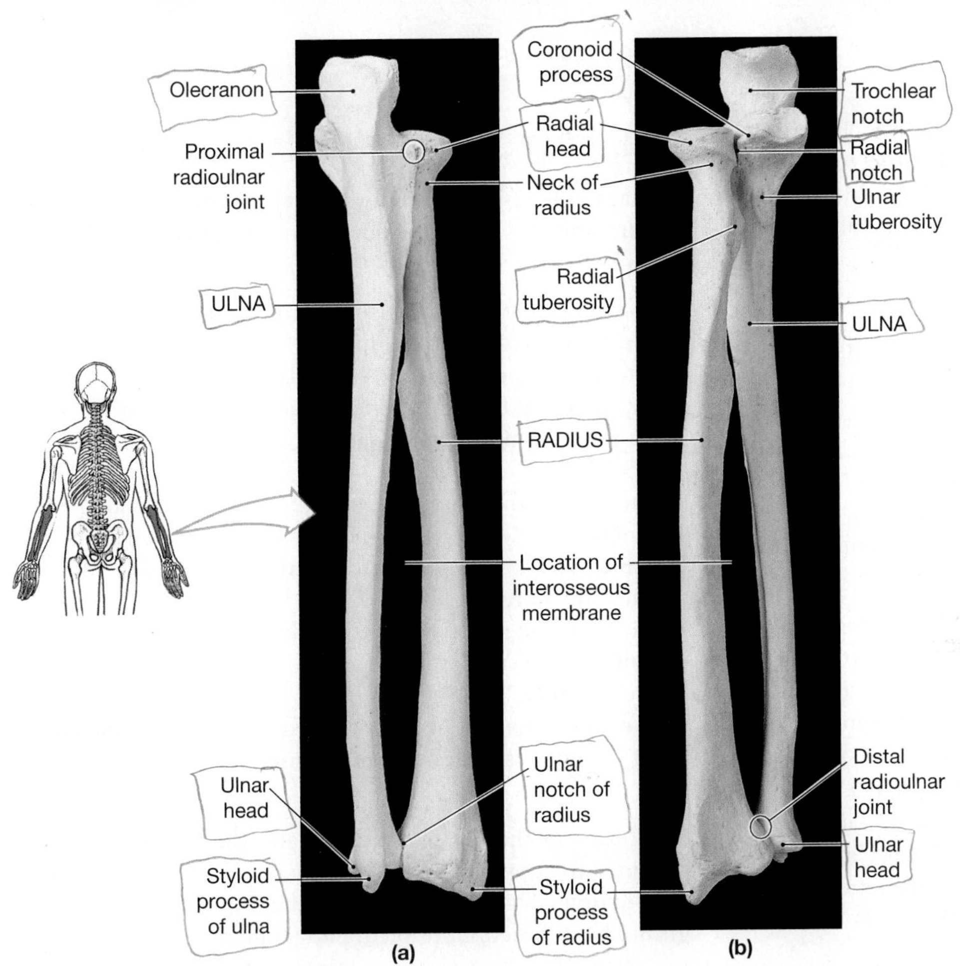

Figure 8.5 The bones of the right forearm. The radius is the lateral bone of the forearm. The ulna is the medial bone. **a)** Posterior view; **b)** anterior view.

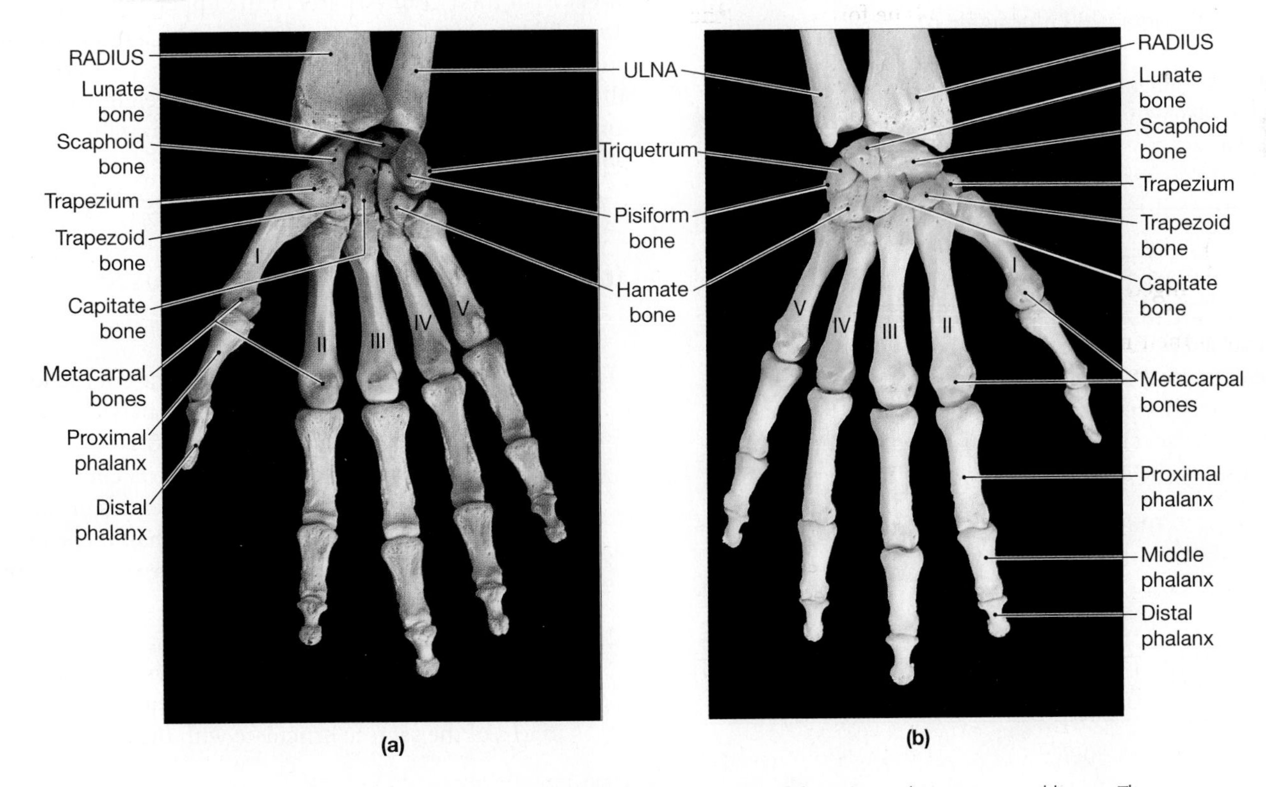

Figure 8.6 The bones of the right wrist and hand. The bones of the wrist are known as carpal bones. The bones of the hand include the metacarpals and phalanges. **a)** Anterior view; **b)** posterior view.

resembling the moon. In Latin, *luna* means "moon". The largest and most central of the carpals, located in the distal row, is the *capitate*. If you think of its position and size as making it the "head" bone, then its name makes sense, because in Latin, *caput* means "head." Finally, next to the capitate is the obviously hooked *hamate*, whose name is derived from the Latin word *hamatus*, which means "hooked."

Some students find it easier to recall these bones if they use a mnemonic device, such as "Sam likes to push the toy car hard." With this particular version, the first four words stand for the carpals in the proximal row, lateral to medial (scaphoid, lunate, triquetrum, pisiform). The next four words represent the distal row, again from lateral to medial (trapezium, trapezoid, capitate, hamate). Observe the carpals while you practice this mnemonic device, or write one of your own. ■

10. Locate the five metacarpal bones (Figure 8.6). The proximal ends (bases) of these miniature long bones articulate with the carpal bones. Distally, the heads articulate with the proximal phalanges of the digits to form the **metacarpophalangeal (MP) joints** (the knuckles; Figure 8.6).
11. Notice that, similar to the metacarpal bones, the phalanges are long bones with a proximal base and a distal head. The thumb contains only two phalanges; all the other digits contain three. The articulations between phalanges are called **interphalangeal (IP) joints** (Figure 8.6).
12. Obtain individual examples of all the bones in the upper extremity and identify the bony landmarks that are listed in Table 8.1 (Figures 8.2 through 8.6).

QUESTIONS TO CONSIDER 1. Speculate on the principal function of the clavicle. (Hint: Consider the consequences if the clavicle was absent.)

___

___

___

2. As you learned in Exercise 7, a long bone consists of a diaphysis (the elongated shaft) and two epiphyses (the knoblike ends). Examine a humerus, a radius, and an ulna. For each bone, identify the specific bony landmarks that form the proximal epiphysis and the distal epiphysis.

	Proximal epiphysis	Distal epiphysis
a. Humerus		
b. Radius		
c. Ulna		

The Lower Extremity

Similar to the upper extremity, the lower extremity also consists of four segments. These segments and the bones found in each are as follows (Figure 8.1b):

- The hip extends from the superior margin (iliac crest) of the **coxal bone** (os coxae or hip bone) to the hip joint. It contains the **pelvic girdle**, which is a ring of bone formed by the two coxal bones. Anteriorly, the coxal bones articulate at the **pubic symphysis**. Posteriorly, each articulates with the sacrum.
- The thigh is the region between the hip and knee joints. It contains one long bone, the body's largest, known as the **femur**. The knee is protected anteriorly by a sesamoid bone called the **patella**.
- The leg is the region between the knee and ankle joints. It contains two smaller long bones, the medial **tibia** and the lateral **fibula**.
- The foot is the part of the lower extremity that is distal to the leg. It contains 7 **tarsal bones**, 5 **metatarsal bones**, and 14 **phalanges**.

The bones of the lower extremity and their articulations are designed to support and transmit body weight and to provide locomotion. The pelvic girdle, in contrast to the pectoral girdle, has a much more secure attachment to the axial skeleton (compare the sacroiliac joint with the sternoclavicular joint). All of the bones and their articulations are generally larger than their counterparts in the upper extremity. In addition, the tarsal and metatarsal bones in the foot provide a supportive but flexible platform for the body. This arrangement allows for the even distribution of body weight through both lower limbs and makes it easier to maintain an upright position.

ACTIVITY 8.2 Examination of the Lower Extremity

1. On the whole skeleton, identify all bones of the lower extremity. They include the bones of the pelvic girdle, thigh, leg, and foot (Table 8.2, Figure 8.1b). Compare each bone of the lower extremity with its counterpart in the upper extremity. Observe that the bones of the lower extremity are larger and more robust than those in the upper extremity. For example, compare the pelvic and pectoral girdles.
2. Locate the sacrum, which is part of the axial skeleton. The part of the coxal bone that articulates with the sacrum is called the **ilium** (plural= **ilia**). Note that the auricular surfaces of the sacrum articulate with the auricular sur-

Table 8.2 Bones of the Lower Extremity

	Bone	Bony landmarks
Coxal bone	Ilium	Iliac crest Anterior superior iliac spine Anterior inferior iliac spine Posterior superior iliac spine Posterior inferior iliac spine Greater sciatic notch Iliac fossa Auricular surface
	Ischium	Ischial spine Ischial tuberosity Lesser sciatic notch Ischial ramus
	Pubis	Pubic crest Pubic tubercle Superior ramus of pubis Inferior ramus of pubis Pubic arch (angle) Symphysis pubis Acetabulum Obturator foramen
Thigh	Femur	Head of femur Neck of femur Greater trochanter Lesser trochanter Linea aspera Medial epicondyle Lateral epicondyle Medial condyle Lateral condyle Patellar surface Intercondylar fossa
Knee	Patella	
Leg	Tibia	Medial condyle Lateral condyle Tibial tuberosity Medial malleolus
	Fibula	Head of fibula Lateral malleolus
Foot	Tarsal bones: Talus Calcaneus Cuboid Navicular Medial cuneiform Intermediate cuneiform Lateral cuneiform	
	Metatarsals	
	Phalanges	

faces of the two ilia (Figure 8.7b) to form the **sacroiliac joints** (Figure 8.8). The resulting structure, which includes the pelvic girdle, sacrum, and coccyx, is called the **bony pelvis.** Notice that the bony pelvis contains elements of both the axial and appendicular skeletons.

3. Observe that the pelvic girdle consists of the right and left coxal bones (hip bones). Each coxal bone is formed by the fusion of three bones (Figure 8.7).
   - The ilium is the largest and most superior bone. The expansive and flat medial and lateral surfaces on this bone serve as attachment sites for muscles.
   - The **ischium** (plural = **ischia**) forms the posteroinferior portion of the coxal bone. It contains the **ischial tuberosity**, which helps to support the body while sitting.
   - The **pubis** forms the anterior portion of the coxal bone. Anteriorly, the two pubic bones are joined by a disc of fibrocartilage. This articulation is called the **pubic symphysis** (Figure 8.8).

WHAT'S IN A WORD The three bones that fuse to form the coxal bone are derived from Greek or Latin words for adjacent structures or body regions. Thus, the word *ilium* is derived from the Latin word for "groin." The term *ischium* is a modern Latin word, derived from *ischion,* the Greek word for "hip." The word *pubis* is derived from the Latin word *pubes,* which is the hair on the genitals.

4. Observe that the ilium, ischium, and pubic bone join together on the lateral surface of the coxal bone to form a deep bony socket known as the **acetabulum** (Figure 8.7a). The acetabulum articulates with the head of the femur to form the **hip joint** (Figure 8.1b).
5. On each ilium, note the **arcuate line**, which is a bony ridge that runs along the inferior border of the **iliac fossa** (Figure 8.7b).
6. Identify the **pelvic brim** (Figures 8.9a and b). It is defined by a circular margin that extends from either side of the sacral promontory, along the arcuate lines, and to the superior margin of the pubic symphysis. The pelvic brim forms the boundary between the two subdivisions of the bony pelvis (Figure 8.9b).
   - The **greater (false) pelvis** is the region superior to the pelvic brim and contains the organs in the inferior portion of the abdominal cavity.
   - The **lesser (true) pelvis** is the region inferior to the pelvic brim and contains the organs of the pelvic cavity.
7. Observe the **pelvic inlet** (Figure 8.9). This is the opening leading into the pelvic cavity. It is formed by the pelvic brim.
8. Next observe the **pelvic outlet** (Figure 8.9). This is the opening leading out of the pelvic cavity, formed by the inferior margins of the bony pelvis.

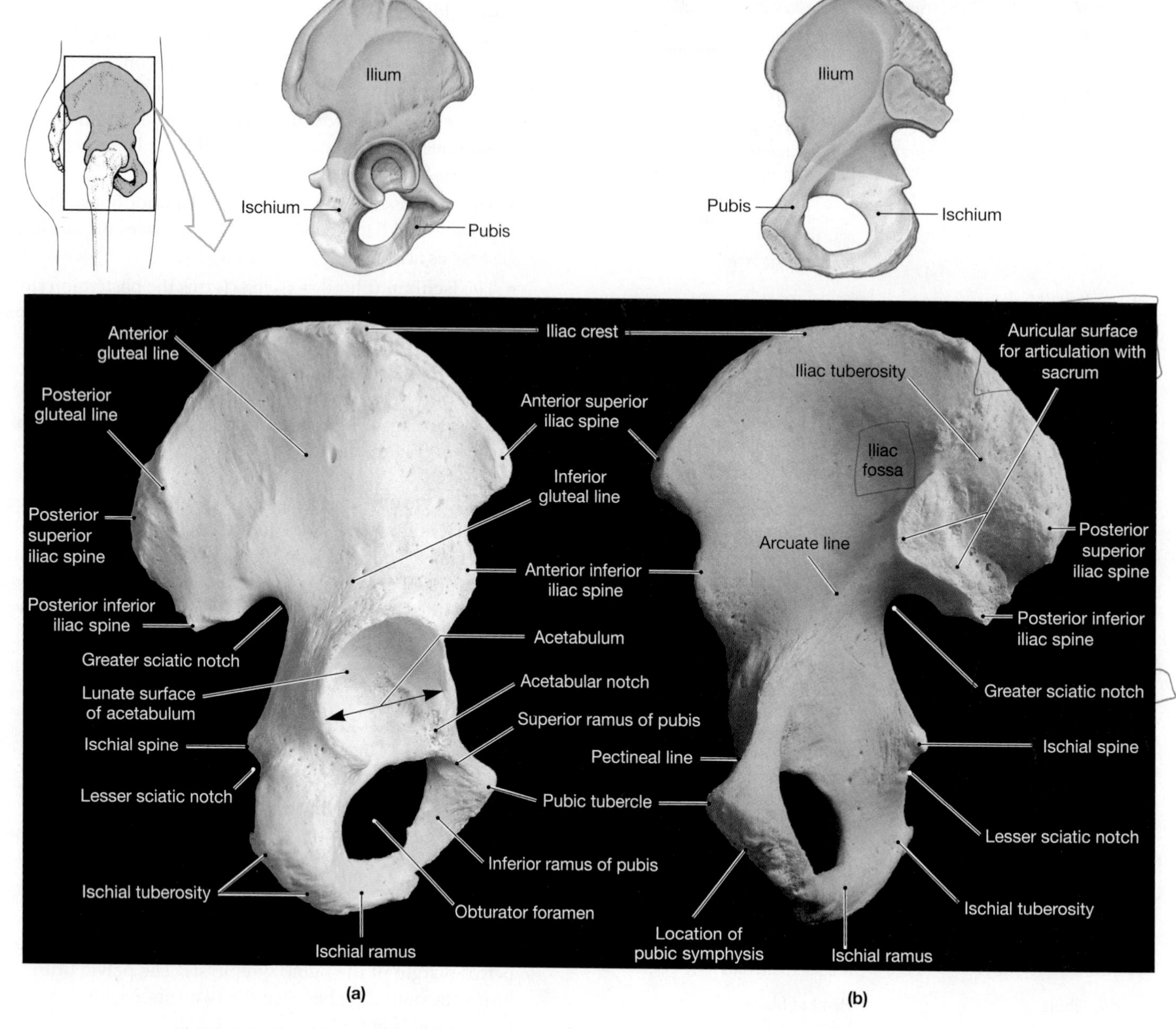

Figure 8.7 The right coxal bone. The coxal bone is also called the hip bone, the os coxae, or the innominate bone. **a)** Lateral view; **b)** medial view.

CLINICAL CORRELATION

If you compare the bony pelvis of a male with that of a female, some striking anatomical differences will be evident (Figure 8.8). Many of the structural differences in females facilitate childbirth. The anatomical features of the typical female pelvis include:

- Wide pelvic inlet and outlet
- Small sacral curvature
- Wide pubic arch (the angle formed by the two pubic bones)
- More laterally projected ilia

These features allow more space within the lesser pelvis to accommodate the developing baby, provide a broader base of support for the weight of the body and the displaced abdominal organs as the fetus and uterus enlarge, and form a wider exit through which the baby can pass at delivery.

9. Locate the **femur** (Figures 8.1c and 8.10), which is the thigh bone. The femur is the heaviest, largest, and longest bone of the body. A person's height can be estimated by measuring the length of the femur and multiplying that value by four.

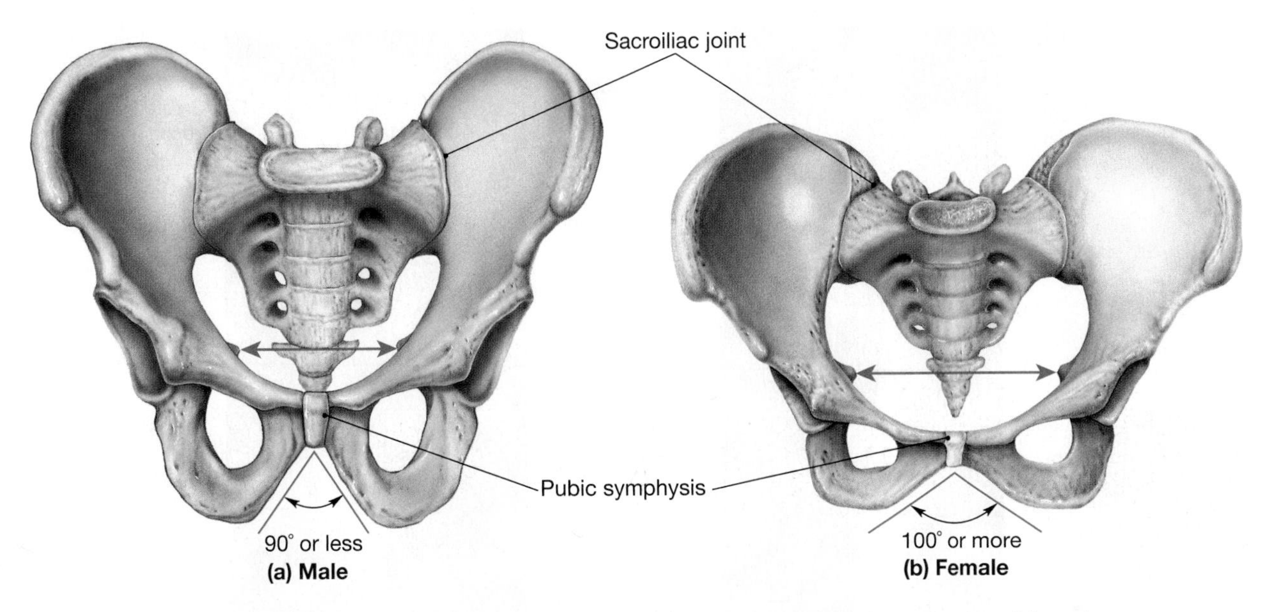

Figure 8.8 Anatomical comparison between the male and female pelvis. a) Anterior view of the male pelvis; **b)** anterior view of the female pelvis.

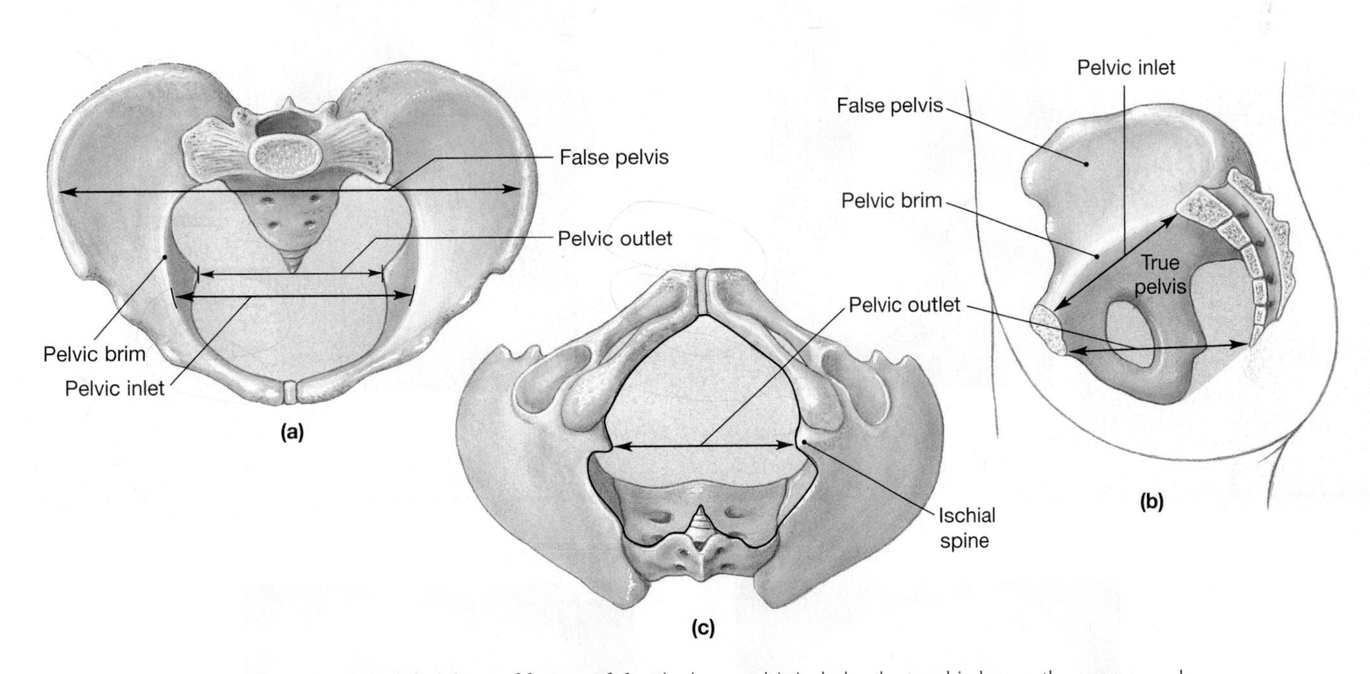

Figure 8.9 Regional divisions of bony pelvis. The bony pelvis includes the two hip bones, the sacrum, and the coccyx. The pelvic inlet is defined by a bony margin called the pelvic brim that begins posteriorly at the sacral promontory and travels along the arcuate lines on each side to the symphysis pubis, anteriorly. The pelvic brim forms the boundary between the false pelvis and true pelvis. **a)** Superior view of the bony pelvis; **b)** medial view of the bony pelvis; **c)** inferior view of the bony pelvis.

10. Observe how the **femoral head** at the proximal end of the bone articulates with the acetabulum to form the hip joint. Distally, the **medial** and **lateral condyles** of the femur articulate with the medial and lateral condyles of the tibia to form the **knee joint** (Figure 8.1b).

11. Observe the patella (Figure 8.11), which is a sesamoid bone that is actually located within the tendon of the quadriceps femoris muscles, a group of four muscles in the anterior thigh. The deep surface of the patella fits into the **intercondylar fossa**, a depression between the two

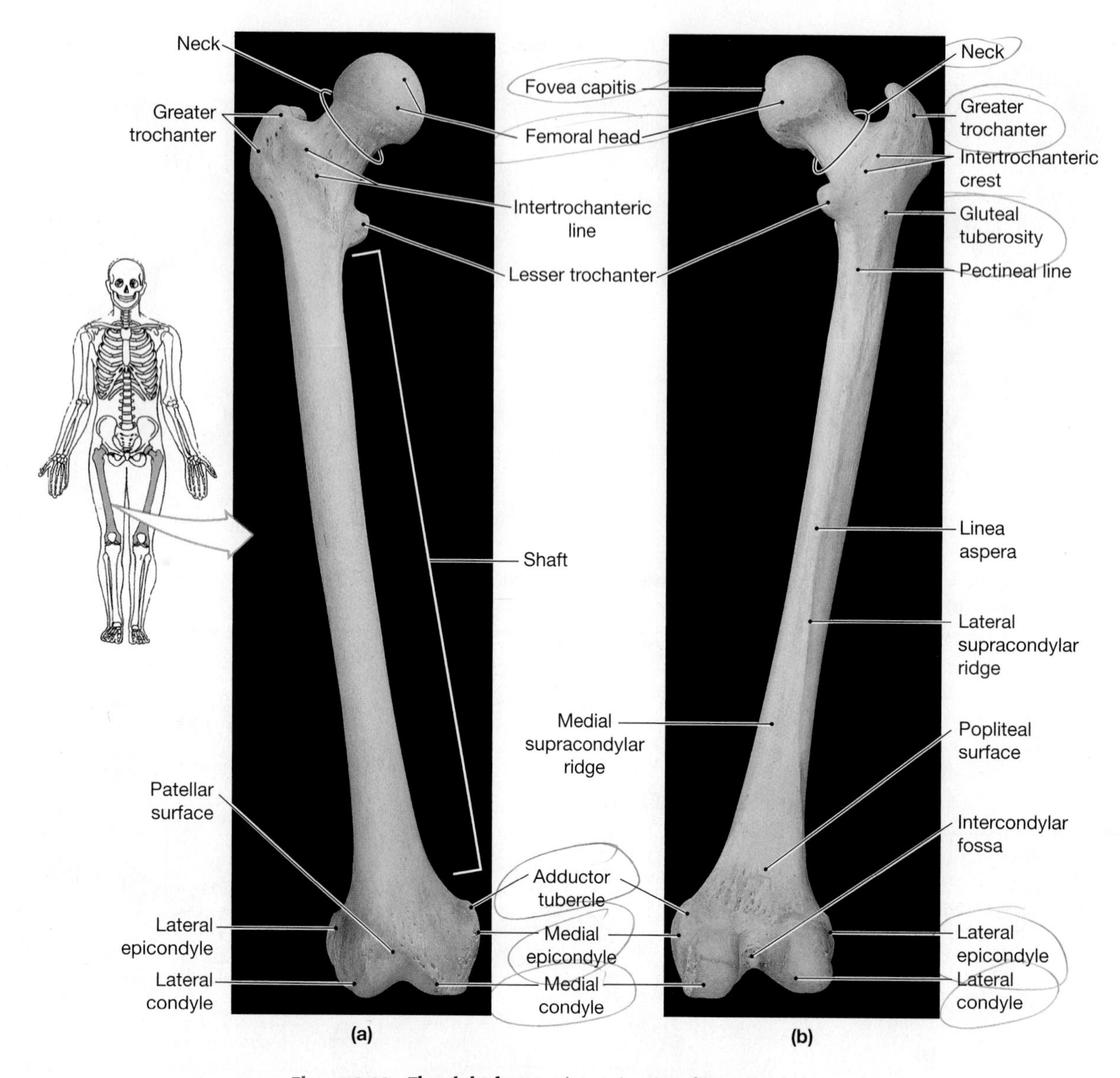

Figure 8.10 The right femur. a) Anterior view; **b)** posterior view.

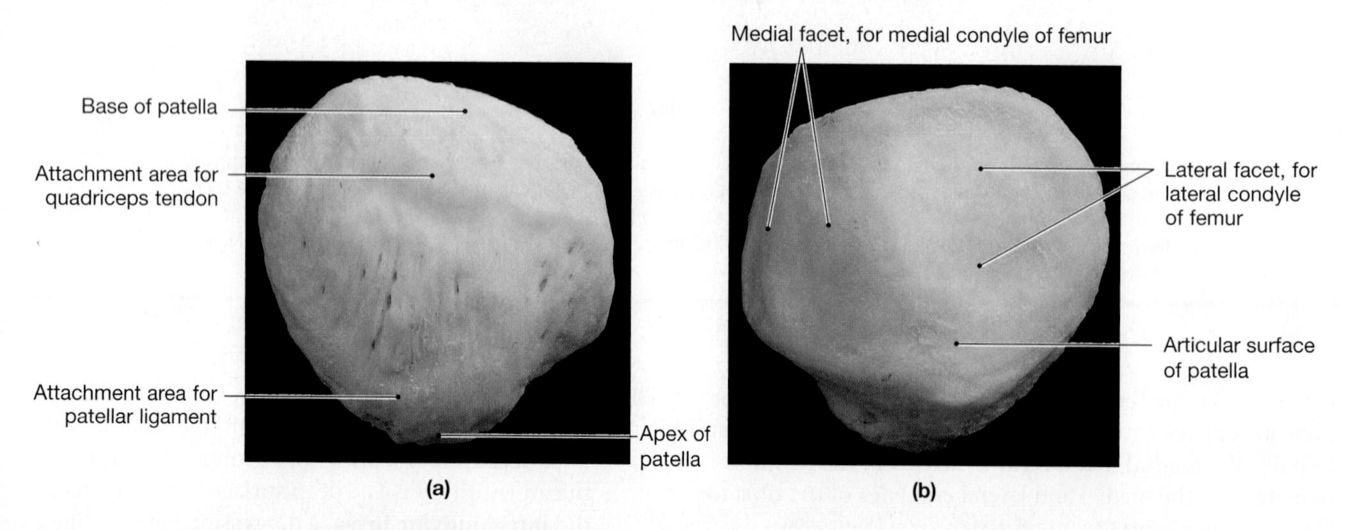

Figure 8.11 The right patella. The patella serves as an attachment for the quadriceps tendon and the patellar ligament, and articulates with the condyles on the femur. **a)** Anterior view; **b)** posterior view.

condyles on the femur. The patella forms a shield of protection across the front of the knee.

12. On the skeleton, locate the tibia and fibula, which are the two long bones in the leg (Figure 8.12). The more massive tibia (the shin bone) is located anteriomedially. Palpate along the anterior aspect of your leg and you will be able to feel the subcutaneous portion of the tibia. Feel the prominent bump on the anterior surface of the tibia slightly inferior to the knee. This **tibial tuberosity** marks the point of attachment for the patellar ligament (the continuation of the quadriceps tendon).
13. The fibula is lateral and roughly parallel to the tibia (Figure 8.12). Notice that the head of the fibula articulates with the lateral condyle of the tibia, but the fibula plays no role in the formation of the knee joint (Figure 8.1c). The fibula serves as an attachment site for muscles, but supports little or no body weight.

WHAT'S IN A WORD The fibula is a slender, long bone that resembles the hinged pin that is used to secure a brooch onto clothing. In Latin, the word *fibula* refers to a clasp or buckle.

14. The **ankle joint** (Figure 8.1b) is formed by articulations of the tibia and fibula with the **talus**, which is the most superior tarsal bone in the foot. Specifically, the **medial malleolus** and inferior surface of the tibia, and the **lateral malleolus** of the fibula, form a socket into which the head of the talus fits (Figure 8.12). The two malleoli can be easily palpated on your body. They are the two prominent bumps on either side of the ankle and are commonly referred to as the ankle bones.
15. The bones of the foot include 7 tarsal bones, 5 metatarsal bones, and 14 phalanges (Figure 8.13). Note the differences and general positions of these bones, with comparable bones in the hand.

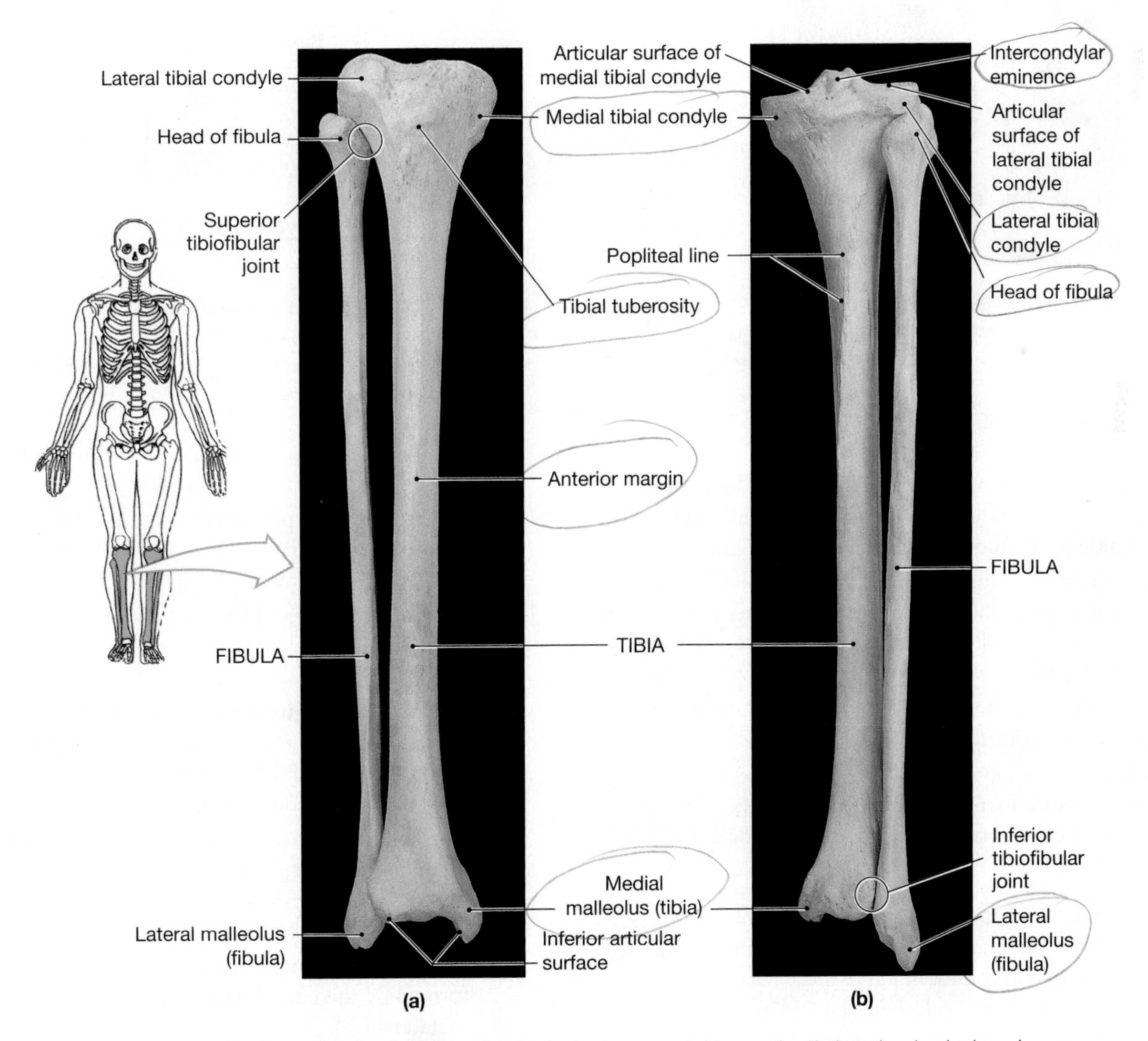

Figure 8.12 The bones of the right leg. The tibia is the larger medial bone. The fibula is the slender lateral bone. **a)** Anterior view; **b)** posterior view.

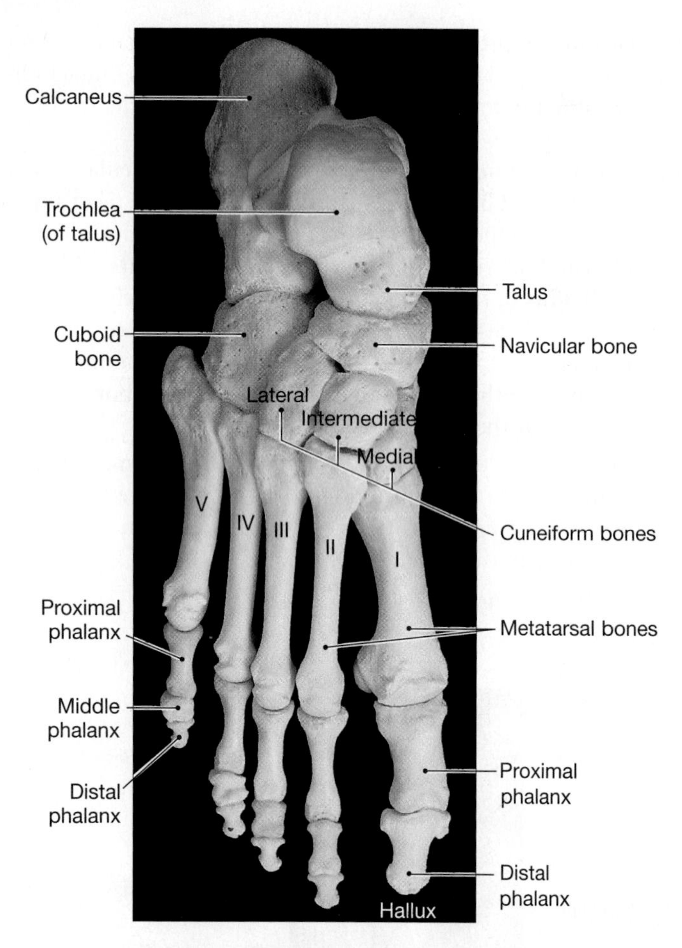

Figure 8.13 The bones of the right foot, dorsal view. The bones of the foot include the tarsal bones, metatarsal bones, and phalanges.

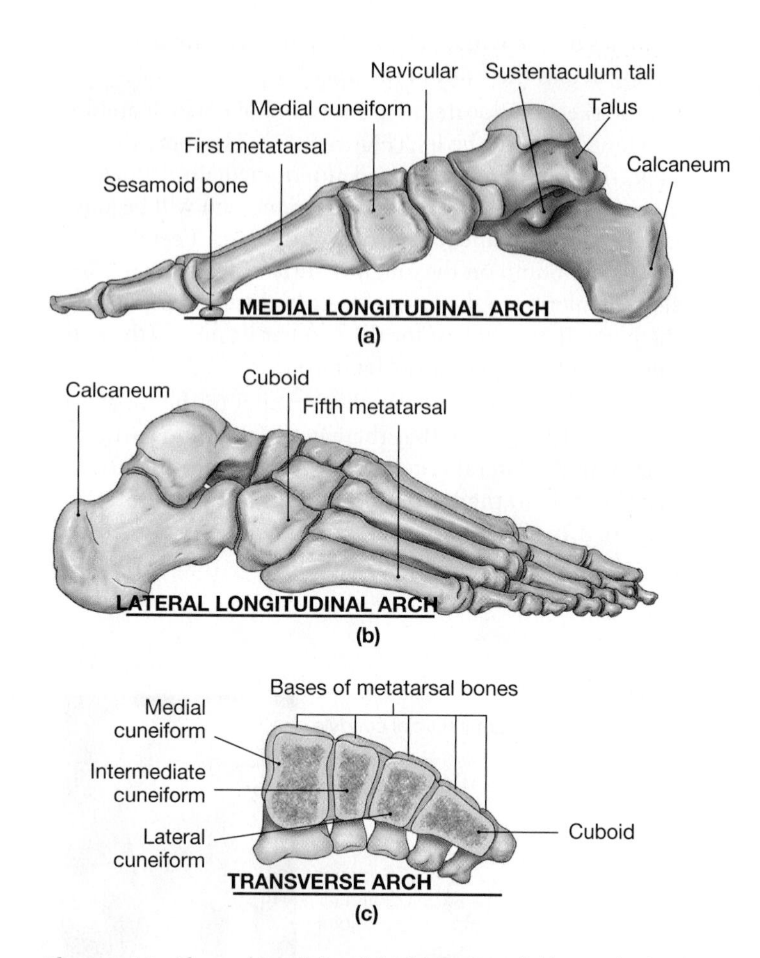

Figure 8.14 The arches of the right foot. a) Medial longitudinal arch; **b)** lateral longitudinal arch; **c)** transverse arch.

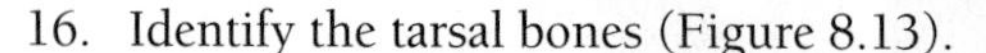

16. Identify the tarsal bones (Figure 8.13).
    - The talus is the most superior bone of the foot. As described, it articulates with the tibia and fibula to form the ankle joint, and receives the body's weight.
    - The **calcaneus** is the largest of the tarsal bones. It is positioned inferior to the talus and extends posteriorly to form the heel. Because of its location, the calcaneus transmits most of the body's weight from the talus to the ground.
    - Anterior to the calcaneus and talus, from lateral to medial, are the **cuboid** and **navicular bones**.
    - Anterior to the navicular and medial to the cuboid are the three **cuneiforms**. From lateral to medial, they are the lateral cuneiform, intermediate cuneiform, and medial cuneiform.

WHAT'S IN A WORD You may want to use a mnemonic device to learn the names of the tarsal bones, as you did the carpals. You might try "Tom can control not much in life." This stands for the tarsal bones as follows: talus, calcaneus, cuboid, navicular, medial cuneiform, intermediate cuneiform, and lateral cuneiform.

17. Identify the five long metatarsals (Figure 8.13) and note their position in your own foot.
18. Examine the short phalanges (Figure 8.13). Note that the great toe, like the thumb, has only two phalanges, while the other toes all have three.
19. Note that the bones of the foot are arranged to form two longitudinal arches, and one transverse arch (Figure 8.14).
    - The **medial longitudinal arch** is the highest arch. It travels along the medial aspect of the foot and is formed by the calcaneus, talus, navicular, medial cuneiform, and the most medial (first) metatarsal.
    - The **lateral longitudinal arch** is much lower than the medial arch. It runs along the lateral aspect of the foot and is formed by the calcaneus, cuboid, and the most lateral (fifth) metatarsal.
    - The **transverse arch** runs from side to side and is formed by the cuneiforms, cuboid, and bases of the metatarsals.

During walking and running, the arches are instrumental in absorbing shock and reducing the amount of work required by muscles. They also distribute the body weight through a greater area to reduce the load on the feet.

20. Obtain individual examples of all the bones in the lower extremity and identify the bony landmarks that are listed in Table 8.2 (Figures 8.7 through 8.13).

QUESTIONS TO CONSIDER

1. Can you provide a reason why the bones of the lower extremity are larger than those in the upper extremity?

___

___

___

2. Speculate on the mechanical function of the patella.

___

___

3. Although the bones of the foot are homologous to their counterparts in the wrist and hand, they serve a much different function. What is the fundamental difference in function between the hand and the foot?

___

___

___

___

Exercise 8 Review Sheet

The Appendicular Skeleton

Name ______________________

Lab Section ______________________

Date ______________________

1. a. List the bones that comprise the appendicular skeleton and categorize them by shape.

Bone	Type of Bone

b. In terms of shape, what type of bone is predominant in the appendicular skeleton?

____

Questions 2–12: Identify the labeled bones and joints in the following diagram.

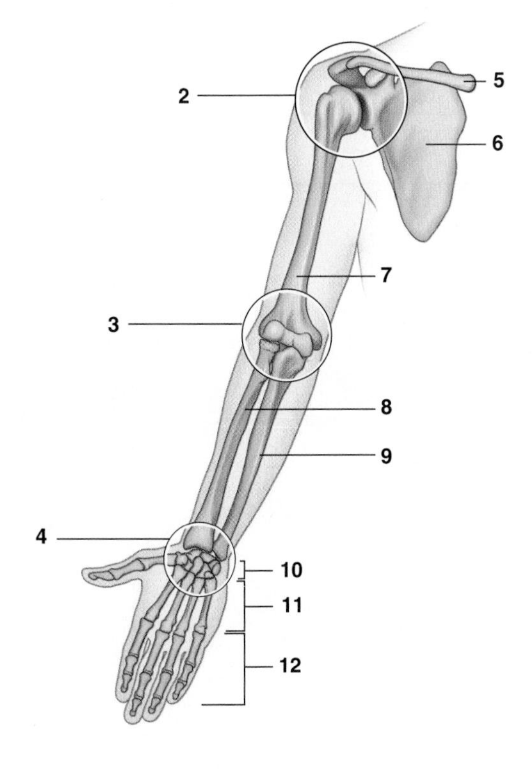

2. ______________________
3. ______________________
4. ______________________
5. ______________________
6. ______________________
7. ______________________
8. ______________________
9. ______________________
10. ______________________
11. ______________________
12. ______________________

Questions 13–25: Identify the labeled bones and joints in the following diagram.

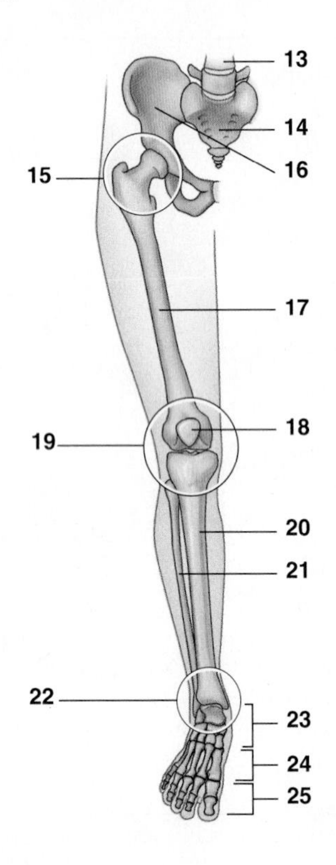

13. ______________________________

14. ______________________________

15. ______________________________

16. ______________________________

17. ______________________________

18. ______________________________

19. ______________________________

20. ______________________________

21. ______________________________

22. ______________________________

23. ______________________________

24. ______________________________

25. ______________________________

Questions 26–36: Identify the bony landmarks labeled in the following diagrams. Identify the bones by using the colors that are indicated.

a. Clavicle = **green**
b. Scapula = **red**
c. Humerus = **blue**

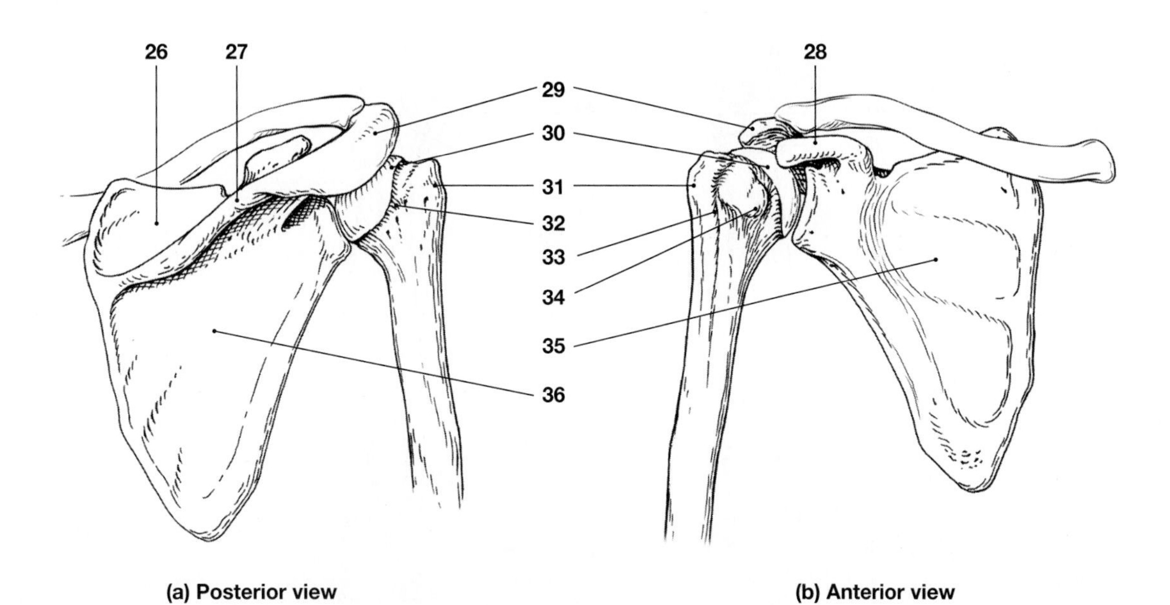

(a) Posterior view

(b) Anterior view

26. ______________________

27. ______________________

28. ______________________

29. ______________________

30. ______________________

31. ______________________

32. ______________________

33. ______________________

34. ______________________

35. ______________________

36. ______________________

Questions 37–45: Identify the bony landmarks labeled in the following diagrams. Identify the bones by using the colors that are indicated.

a. Humerus = **blue**
b. Radius = **red**
c. Ulna = **yellow**
d. Carpals = **green**
d. Metacarpals = **brown**
e. Phalanges = **orange**

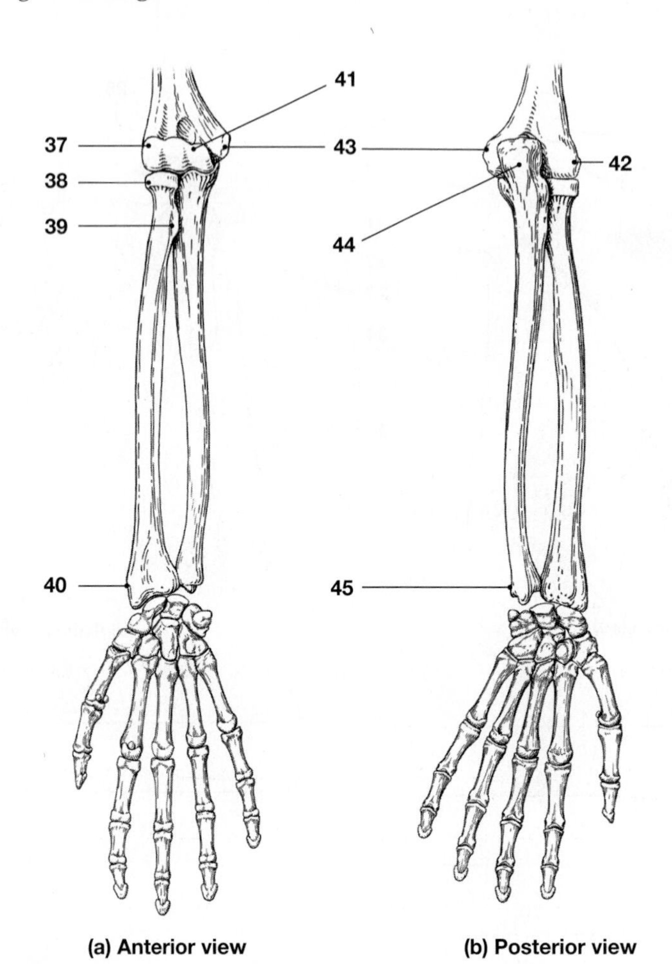

(a) Anterior view

(b) Posterior view

37. ______________________

38. ______________________

39. ______________________

40. ______________________

41. ______________________

42. ______________________

43. ______________________

44. ______________________

45. ______________________

Questions 46–58: Identify the bony landmarks labeled in the following diagrams. Identify the bones by using the colors that are indicated.

a. Ilium = **green**
b. Ischium = **red**
c. Pubis = **yellow**

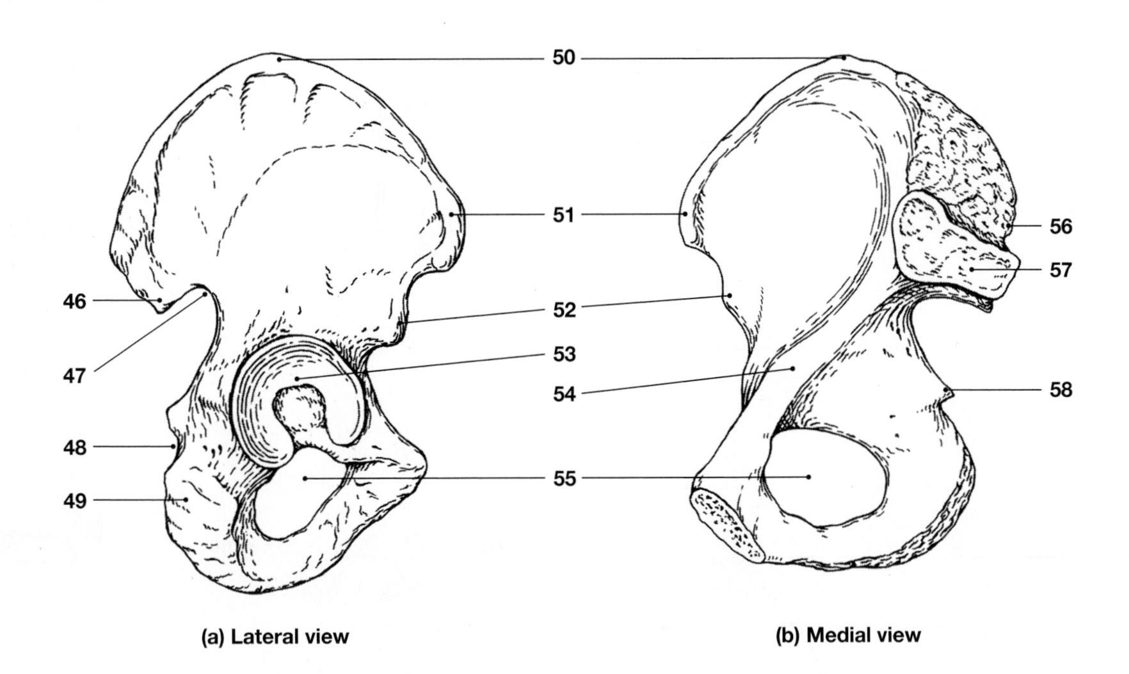

46. ______________________________

47. ______________________________

48. ______________________________

49. ______________________________

50. ______________________________

51. ______________________________

52. ______________________________

53. ______________________________

54. ______________________________

55. ______________________________

56. ______________________________

57. ______________________________

58. ______________________________

Questions 59–71: Identify the bony landmarks labeled in the following diagrams. Identify the bones by using the colors that are indicated.

a. Femur = **green**
b. Tibia = **red**
c. Fibula = **yellow**

59. ______________________________

60. ______________________________

61. ______________________________

62. ______________________________

63. ______________________________

64. ______________________________

65. ______________________________

66. ______________________________

67. ______________________________

68. ______________________________

69. ______________________________

70. ______________________________

71. ______________________________

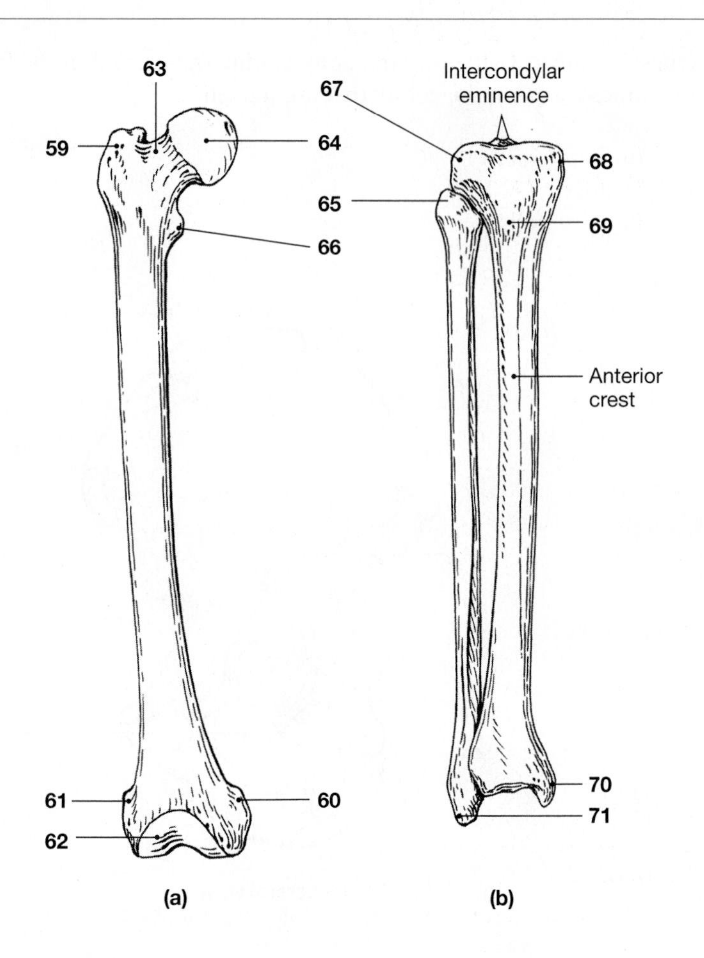

72. Identify the bones by using the colors that are indicated.

a. Calcaneus = **red**

b. Talus = **green**

c. Navicular = **yellow**

d. Cuboid = **blue**

e. Cuneiforms = **orange**

f. Metatarsals = **purple**

g. Phalanges = **brown**

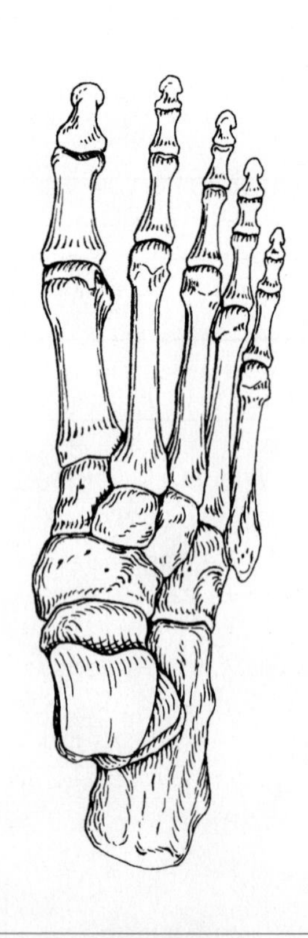

Exercise 9

Articulations

Laboratory Objectives

On completion of the activities in this exercise, you will be able to:

- Categorize joints according to their structure and movement.
- List, describe, and provide examples of the different types of fibrous and cartilaginous joints.
- Explain the location and function of the fetal fontanels.
- Describe the general structure of a typical synovial joint.
- Explain the six basic types of synovial joints and the movements possible at each type.
- Describe the specialized structures and functions of the shoulder, elbow, radioulnar, hip, knee, and ankle joints.
- Identify the various synovial joints on a skeleton and on models.

Materials

- Human skull
- Disarticulated human skull
- Human fetal skull
- Human skeleton
- Model of the lumbar spine
- Model of intervertebral disc
- Herniated disc model or illustration
- Adult humerus, longitudinal section, coronal plane
- Humerus, intact
- Shoulder joint model
- Elbow joint model with proximal radioulnar joint
- Hip joint model
- Knee joint model
- Ankle joint model

Articulations, or **joints**, are the functional junctions between bones. They bind various parts of the skeletal system together, allow bone growth and development, permit parts of the skeleton to change shape, and, with the skeletal muscles, provide body movement.

Articulations can be classified in two ways. First, they can be divided according to the degree of movement that is allowed. Some joints, called **synarthroses** (singular—**synarthosis**), are immovable. Other joints, classified as **amphiarthroses** (singular—**amphiarthrosis**), are slightly movable. **Diarthroses** (singular—**diarthrosis**) are freely movable joints.

Alternatively, articulations can be classified according to their structure (Table 9.1). **Fibrous joints** are held together by fibrous connective tissue. **Cartilaginous joints** are held together by cartilage. **Synovial joints** are the most complex and most common type of joint. They are best characterized by the presence of a fluid-filled cavity between the articulating bones. We will discuss the structure of synovial joints in detail later in this exercise. Both fibrous and cartilaginous joints can be either immovable (synarthroses) or slightly movable (amphiarthroses). All synovial joints are freely movable (diarthroses).

Fibrous Joints

In fibrous joints, the bones are held firmly together by fibrous connective tissue in which collagen fibers predominate. A joint cavity is not present, and little or no movement occurs between the bones. The three types of fibrous joints in the human body are as follows:

- Suture
- Syndesmosis
- Gomphosis

Sutures are very tight articulations between adjacent bones. This type of joint is only found in the skull. In the adult, the connective tissue fibers that connect the bones become completely ossified. Thus, the bones are fused together and no movement occurs between them. In the fetus and infant, ossification is not complete and the articulating bones are held together by areas of connective tissue fibers called **fontanels**. The fontanels, or "soft spots," allow for some movement between the bones. This flexibility is particularly important during the birth process, and while the brain is growing.

In a **syndesmosis** (plural = **syndesmoses**), the bones are held together by strong, fibrous connective tissue. The articulating surfaces of the bones may be either relatively small and held together by cordlike **ligaments** or broad and held together by fibrous sheets called **interosseous membranes**. The movement between bones can vary but usually is quite limited.

A **gomphosis** is a unique peg and socket joint at which no movement occurs. The only joints of this type are the articulations between the permanent teeth and the maxilla (upper teeth) and mandible (lower teeth). Each tooth is anchored to a bony socket (**dental alveolus**) by **periodontal ligaments**.

WHAT'S IN A WORD The term *gomphosis* is derived from the Greek word *gomphos*, which means "a bolt or nail." Think of the teeth as bolts that are inserted into the bones.

CLINICAL CORRELATION

The fibers of the periodontal ligaments are continuous with the periosteum that surrounds the bones of the jaw. Only the adult, or permanent, teeth have gomphoses holding them in place. Baby teeth lack this anchoring structure and, for that reason, become loose and fall out as the socket grows.

Table 9.1 Types of Joints in the Human Body

Type of joint		Examples
Structure	**Degree of Movement**	
A. Fibrous joints		
1. Suture	Immovable	Articulations between skull bones
2. Syndesmosis	Slightly movable	Coracoclavicular joint Distal tibiofibular joint Interosseous membranes between radius and ulna, and between tibia and fibula
3. Gomphosis	Immovable	Teeth anchored to bony sockets
B. Cartilaginous joints		
1. Synchondrosis	Immovable	Growth plates in long bones Articulation between first rib and manubrium of the sternum
2. Symphysis	Slightly movable	Intervertebral disks Symphysis pubis
C. Synovial joints	All freely movable	
1. Gliding (plane)		Intercarpal Intertarsal Acromioclavicular Sternoclavicular Sacroiliac Articular processes between vertebrae
2. Hinge		Elbow Knee Ankle Interphalangeal
3. Pivot		Proximal radioulnar Distal radioulnar Atlantoaxial
4. Condyloid		Atlanto-occipital Wrist (radiocarpal) Metacarpophalangeal (knuckles)
5. Saddle		Carpometacarpal of the thumb
6. Ball and socket		Shoulder Hip

ACTIVITY 9.1 Examining the Structure of Fibrous Joints

The Suture

1. Obtain an adult skull and inspect the sutures between the cranial bones (Figure 9.1). Notice the complex interdigitations that define the junctions between these bones.
2. Examine the disarticulated bones of an adult skull. Attempt to reconstruct the sutures of the cranial bones by correctly matching the interdigitations of adjacent bones.
3. Identify the major sutures between cranial bones in the adult skull (Figure 9.1).
    - The **coronal suture** is the joint between the frontal bone, anteriorly, and the two parietal bones, posteriorly.
    - The **lambdoid suture** is the joint between the occipital bone, posteriorly, and the two parietal bones, anteriorly.
    - The **sagittal suture** is the joint on the superior surface of the skull, between the two parietal bones.
    - The two **squamous sutures** are found on each side of the skull. Each forms a joint between the parietal bone, superiorly, and the temporal bone, inferiorly.

WHAT'S IN A WORD The "lambdoid" suture acquired its name because, in combination with the sagittal suture, it resembles the Greek letter, lambda (λ).

The "squamous" sutures obtained their names because the *squamous* (Latin for "scalelike" or flat") portions of the temporal bones articulate with the parietal bones.

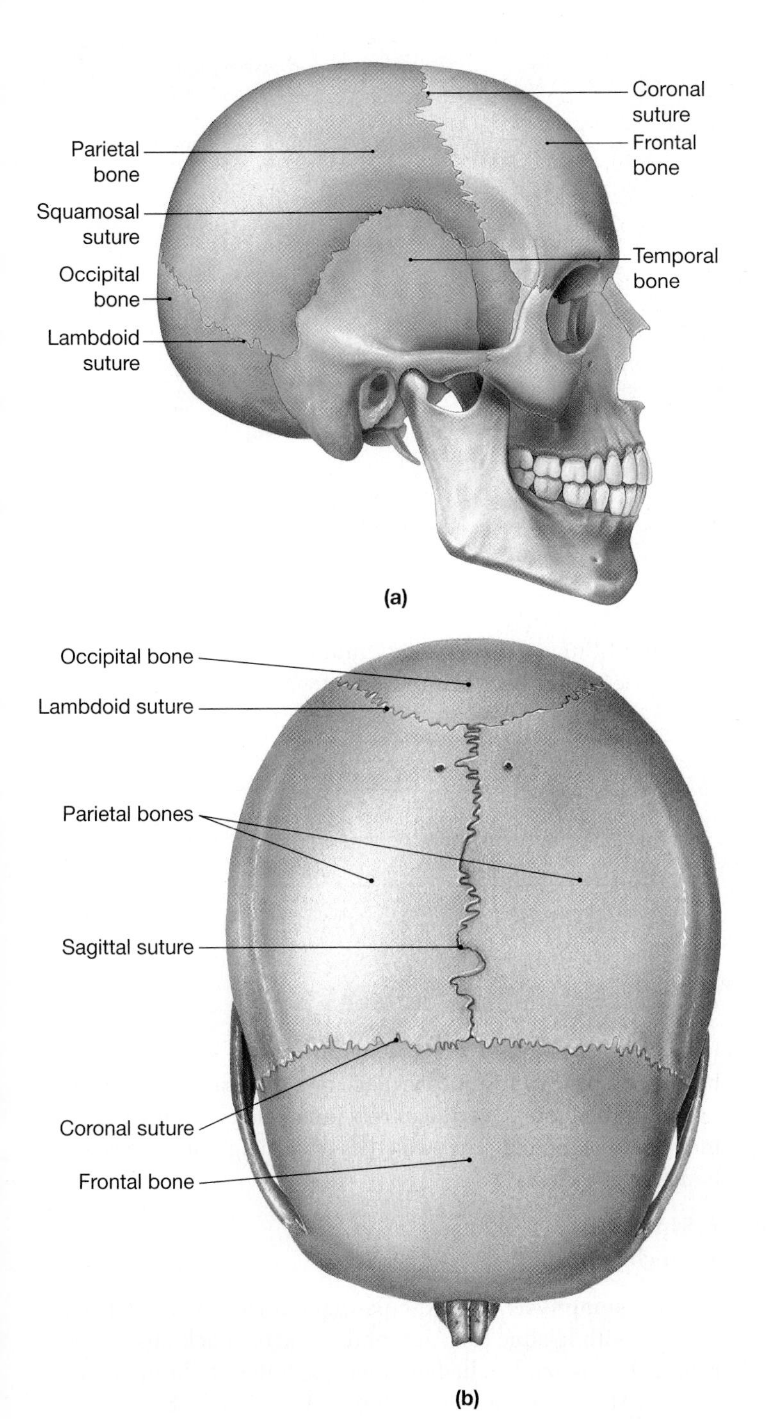

Figure 9.1 The major sutures on the skull. a) Lateral view; **b)** superior view.

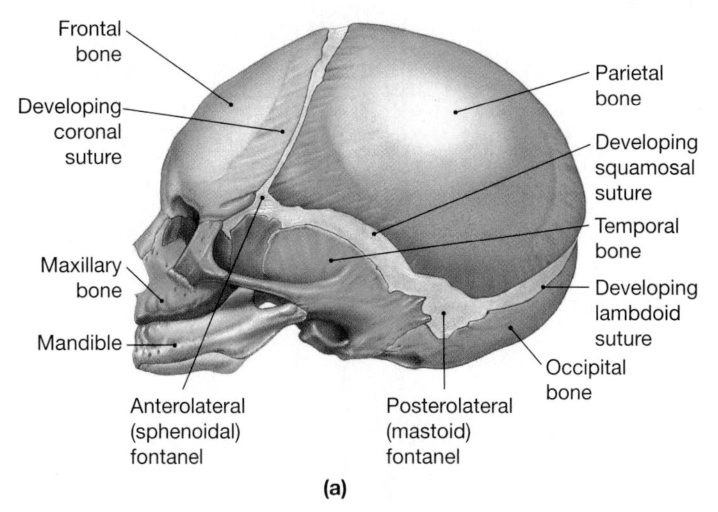

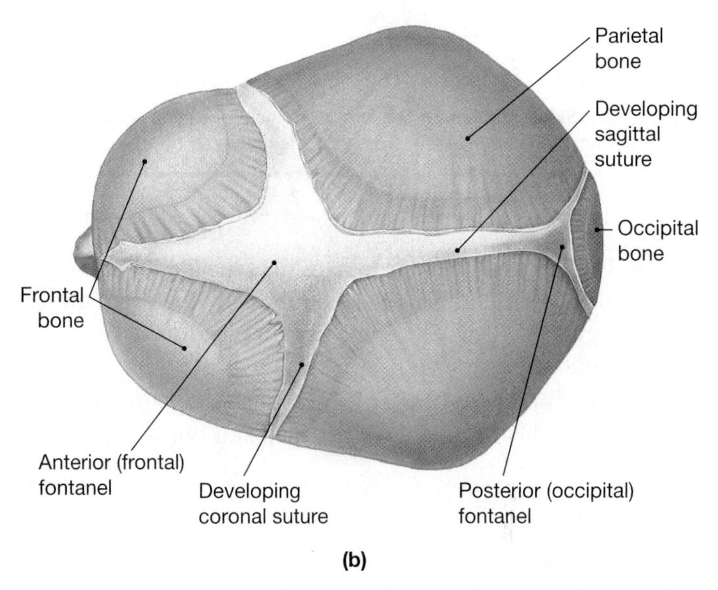

Figure 9.2 The fontanels and developing sutures on the fetal skull. a) Lateral view; **b)** superior view.

4. On the fetal skull, identify the same sutures as above, as they are developing (Figure 9.2).
5. Identify the major fontanels in the fetal skull (Figure 9.2).
   - The **anterior (frontal) fontanel** is located at the junction of the coronal and sagittal sutures.
   - The **posterior (occipital) fontanel** is located at the junction of the lambdoid and sagittal sutures.
   - The two **anterolateral (sphenoidal) fontanels** are each located at the junction of the coronal and squamosal sutures.
   - The two **posterolateral (mastoid) fontanels** are each located at the junction of the lambdoid and squamosal sutures.

The Syndesmosis

1. On the skeleton, examine the relationship between the radius and ulna, and between the tibia and fibula, in the anatomical position. In a living individual, the bones of the forearm and leg are connected by ligamentous sheets, known as **interosseous membranes**, to form syndesmoses (Figure 9.3).
2. On the whole skeleton, identify two other examples of syndesmoses.

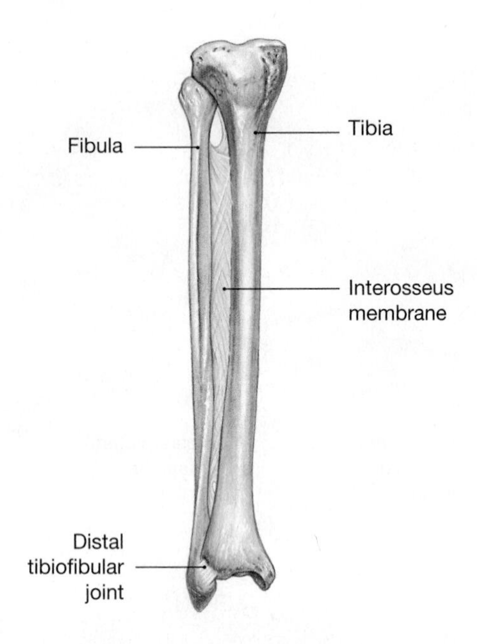

Figure 9.3 Syndesmoses in the leg. In this type of joint, the articulating bones are held together by an interosseous membrane or by ligaments.

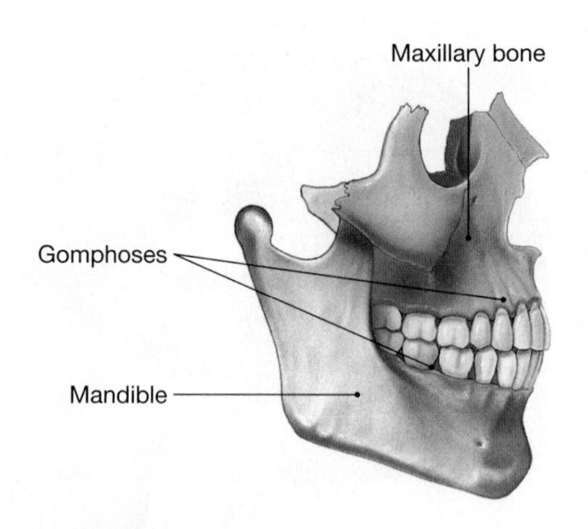

Figure 9.4 Gomphoses in the skull. Sockets in the maxillary bones and mandible articulate with the roots of permanent teeth.

- The **coracoclavicular joint** is the articulation between the clavicle and the coracoid process of the scapula (See Figure 9.11 later in the exercise to identify the coracoclavicular ligaments).
- The distal **tibiofibular joint** is the articulation between the distal ends of the tibia and fibula (Figure 9.3).

The Gomphosis

1. On the skull, identify the maxillary bones and the mandible.
2. Locate the bony sockets (alveoli) in these bones. The roots of the teeth articulate with these sockets to form gomphosis joints (Figure 9.4). Periodontal ligaments connect the teeth to the bones.

QUESTIONS TO CONSIDER 1. How do the developing sutures on the fetal skull differ from the adult sutures?

____

____

____

____

2. As you learned earlier in this activity, the interosseous membranes that connect the long bones in the forearm and those in the leg are syndesmoses. Describe the general relationship between the radius and ulna in the forearm and the tibia and fibula in the leg. How much movement occurs at the syndesmoses between these bones?

____

____

____

____

Cartilaginous Joints

In cartilaginous joints, the articulating bones are held together by either a plate of hyaline cartilage or a fibrocartilage disc. Similar to fibrous joints, cartilaginous joints lack a joint cavity and movement is limited. The two types of cartilaginous joints in the body are as follows:

- Symphysis
- Synchondrosis

At a **symphysis**, the articulating surfaces of the bones are covered with hyaline cartilage, and a disc of shock-absorbing fibrocartilage is sandwiched between the bones, holding them together. Movement at these joints is limited. Examples of symphyses include the **intervertebral discs** between adjacent vertebrae and the **symphysis pubis** between the two pubic bones in the pelvic girdle.

In a **synchondrosis**, a plate of hyaline cartilage unites the bones. It is a temporary joint because, in most cases, the cartilage is gradually replaced by bone. The primary function of a synchondrosis is to allow bone growth; movement does not occur at these articulations. The best example of a synchondrosis is the epiphyseal (growth) plate located between the diaphysis and epiphysis of a long bone.

CLINICAL CORRELATION

Most bones in the body form through a process called **endochondral ossification** in which a hyaline cartilage template is gradually replaced by osseous (bony) tissue. In long bones, the cartilage at the epiphyseal plate is where bone elongation occurs. As long as this synchondrosis is in place and healthy, the bone can lengthen as more cartilage is formed in that region. Eventually, usually by the early 20s, the cartilage is entirely replaced by bone and that bone has reached its final adult length. If, however, the cartilage is seriously injured during bone development, growth may end prematurely and the bone will remain shorter than its normal counterpart on the other side of the body.

ACTIVITY 9.2 Examining the Structure of Cartilaginous Joints

The Symphysis

1. Inspect the vertebral column on the skeleton. The bodies of each vertebra are separated by intervertebral discs (Figure 9.5).
2. Observe the placement of the intervertebral disc on a model of the lumbar vertebrae. Remove the superior vertebra from the model. Notice that the intervertebral disc has two parts (Figure 9.6a), as follows:
   - The outer region, composed of fibrocartilage, is called the **annulus fibrosus**.
   - The central core of the disc, a gelatinous mass, is called the **nucleus pulposus**.
3. Examine a model that demonstrates a herniated intervertebral disc (Figure 9.6b). (An illustration may be substituted if no model is available.) This condition occurs more frequently in older individuals. Most herniated discs are caused by deterioration of the annulus fibrosus posteriorly and laterally. As a result, the nucleus pulposus is displaced posterolaterally from its normal position and compresses the emerging spinal nerve. In most cases, herniated discs occur between lumbar vertebrae and may cause severe low back and leg pain.
4. On the whole skeleton, identify the articulation between the two pubic bones. This joint, the symphysis pubis

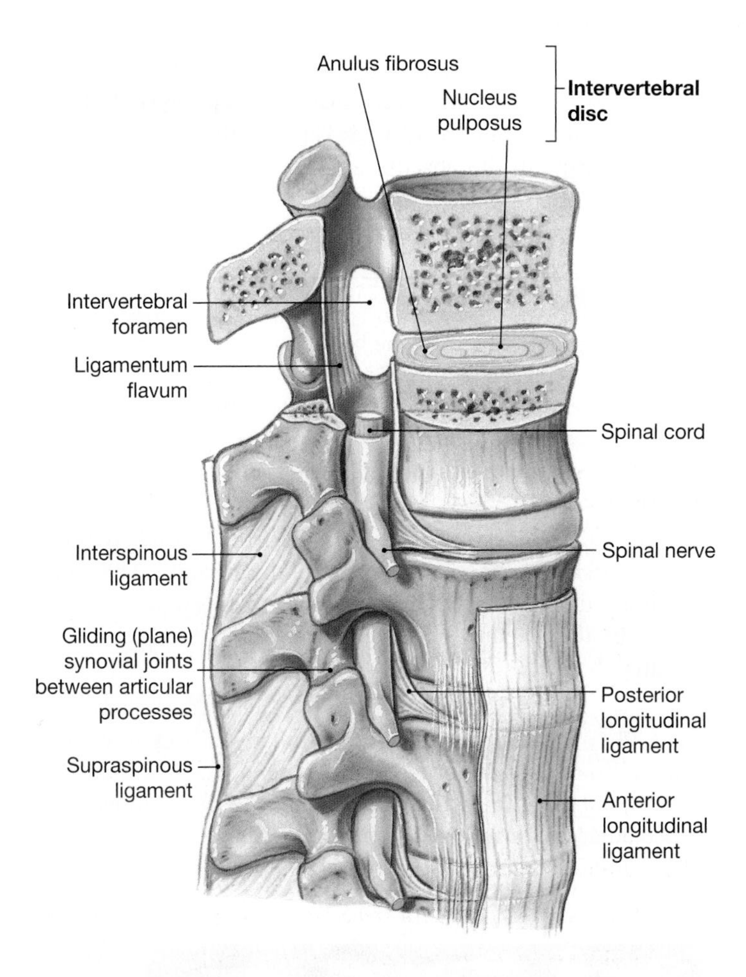

Figure 9.5 Articulations between adjacent lumbar vertebrae. Intervertebral discs, located between the bodies of adjacent vertebrae, form symphysis joints. Gliding synovial joints are formed between articular processes.

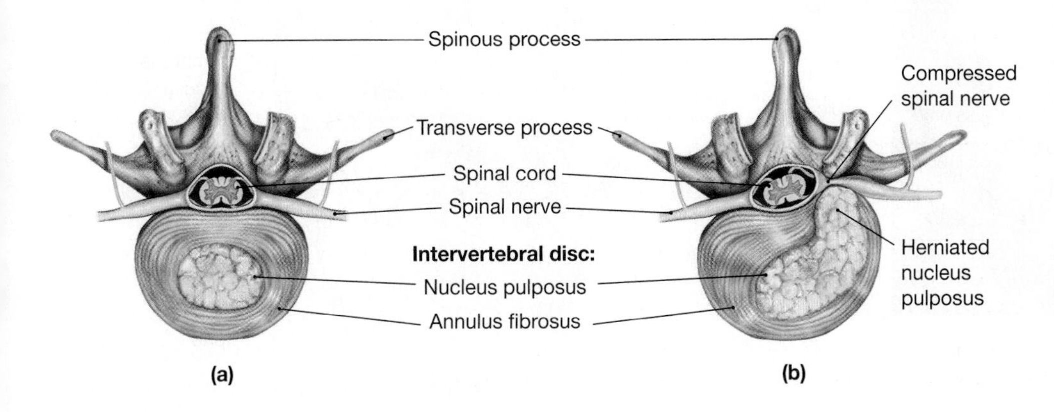

Figure 9.6 Cross section of an intervertebral disc. Each disc is composed of an outer ring of fibrocartilage (the annulus fibrosus) and a central gelatinous mass (the nucleus pulposus). **a)** The normal condition; **b)** a herniated disc in which a portion of the nucleus pulposus is displaced posteriorly.

(pubic symphysis), is another example of a cartilaginous joint (Figure 9.7). What type of cartilage is found at this joint?

The Synchondrosis

1. Inspect an adult humerus that has been cut along its longitudinal axis in the coronal plane. The **epiphyseal plate**, a type of synchondrosis, is absent from this bone. Explain the reason for this absence.

2. At the proximal end of the bone, carefully inspect the border between the diaphysis and epiphysis. In this region, a fine line of ossified tissue travels horizontally across the bone. This **epiphyseal line** marks the area where the epiphyseal plate was located during active bone growth (Figure 9.8).
3. On the whole skeleton, identify the articulation between the first rib and the manubrium of the sternum. This joint is another example of a synchondrosis. It is unique because it is the only permanent synchondrosis in the human body.

QUESTION TO CONSIDER 1. Place your hand on the lower back of your lab partner to examine the movement at an intervertebral disc. To do this you will have to detect movement at two adjacent vertebrae. When you are ready, have your lab partner move his or her vertebral column in various directions by performing the following movements.

- Bend forward at the waist and try to touch the toes.
- Return to the upright position, and bend backward at the waist.
- Return to the upright position and bend laterally at the waist to the right and to the left.
- Return to the upright position and rotate the trunk by slowly twisting from side to side.

a. How much movement occurred at the one intervertebral disc that you were testing?

b. Can all the movement you were observing be attributed to only one joint? Explain.

Synovial Joints

Most articulations in the body are synovial joints. These joints are capable of a variety of movements in several planes and are defined by the presence of a fluid-filled cavity between the articulating surfaces. Several basic structures characterize all synovial joints (Figure 9.8). The articulating bony surfaces are covered by a smooth layer of hyaline cartilage, known as the **articular cartilage**. This layer is important for its ability to absorb shock and reduce friction during movement. The bones are held together by a double-layered **joint capsule** that encloses a space between

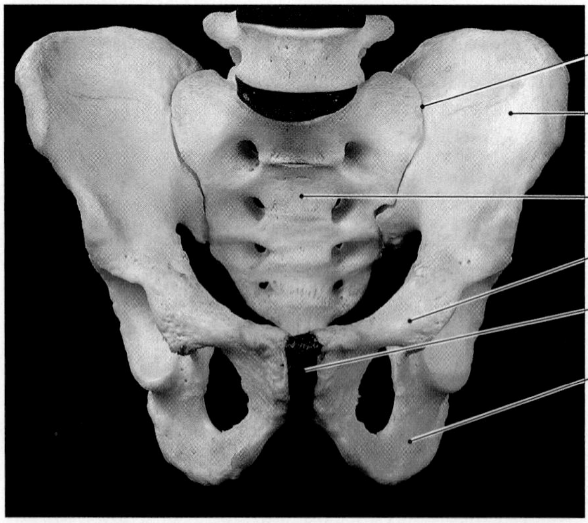

Sacroiliac joint
Ilium
Sacrum
Pubis
Pubic symphysis
Ischium

Figure 9.7 Articulations between bones in the pelvis. The sacroiliac joint is a gliding synovial joint between the sacrum and ilium. The symphysis pubis, between the pubic bones, and the intervertebral disc, between the 5th lumbar vertebra and the sacrum, are symphysis cartilaginous joints.

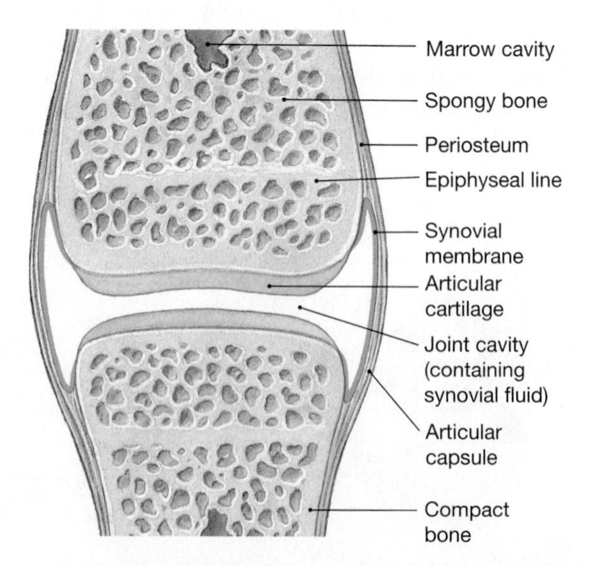

Figure 9.8 Structure of a typical synovial joint. Synovial joints have a more complex structure than fibrous and cartilaginous joints. The principal feature is the presence of a fluid-filled joint cavity.

the two articulating bones, known as the **joint cavity**. The outer layer of the joint capsule, known as the **articular (fibrous) capsule**, is composed of strong fibrous connective tissue and is attached to each bone in the joint. The inner layer is a thin lining of loose connective tissue called the **synovial membrane**. The synovial membrane lines all surfaces of the joint cavity except those covered by the articular cartilage. It produces **synovial fluid** that is secreted into the joint cavity. Synovial fluid is a viscous, lubricating fluid that reduces friction between the articular cartilages during movement. Synovial joints are strengthened by a variety of **ligaments**. Most ligaments are localized thickenings of the articular capsule but others, such as the cruciate ligaments in the knee, are independent structures.

In addition to these structures, many accessory structures can be found at synovial joints. At some joints, such as the hip or knee, shock-absorbing **fat pads** fill spaces between the fibrous capsule and the synovial membrane (see Figures 9.13a and 9.14a later in the exercise). Within a number of joints, **fibrocartilage pads** are present. These structures act as shock absorbers, help to evenly distribute the forces at the joint, and provide a better fit between articulating bones. Examples include the **glenoid labrum** in the shoulder (see Figure 9.10b), the **acetabular labrum** in the hip (see Figure 9.12a), and the **lateral** and **medial menisci** in the knee (see Figures 9.13c and d). **Bursae** are flattened sacs lined with a synovial membrane and filled with synovial fluid. They are located at strategic locations around synovial joints where tendons, ligaments, muscles, or skin rub against bone or neighboring soft tissue structures (refer to Figures 9.10 and 9.13b later in the exercise). Bursae provide additional cushioning at the joint and reduce the friction when adjacent structures rub against each other during movement. Some bursae are directly connected to the joint cavity, while others are independent structures. **Tendon sheaths** are modified bursae that form fluid-filled tubular sheaths around long tendons that are subjected to constant friction. They are found around the long tendons that cross the wrist and ankle to enter the hand and foot, respectively.

ACTIVITY 9.3 Understanding the Types of Movements at Synovial Joints

Specific terms of movement are used to describe the various actions that occur at synovial joints. It is essential that you become familiar with these terms before you begin Activity 9.4, during which you will analyze and compare movements at specific joints. Study the terms of movement listed in Table 9.2. Work with your lab partner by quizzing each other on the performance of each movement.

QUESTIONS TO CONSIDER Describe the following actions and identify the joint at which they occur.

a. Abduction of the thigh:

____

____

b. Flexion of the leg:

____

____

c. Circumduction of the arm:

____

____

d. Supination of the forearm:

____

____

e. Dorsiflexion of the foot:

____

____

ACTIVITY 9.4 Examining the Structure and Range of Motion at Synovial Joints

In this activity, you will learn how to categorize synovial joints by the shape of their articulating surfaces and the range of motion at the joint. You will compare the structure and function of the six synovial joint categories described herein (Figure 9.9, Table 9.1), identifying specific examples on the skeleton and observing the possible range of motion.

WHAT'S IN A WORD The names of many synovial joints are derived from the bones that form the articulation. Consider the following examples.

- The *intercarpal* joints are articulations between carpal bones in the wrist.
- The *metacarpophalangeal* joints are articulations between the metacarpal bones and the proximal phalanges in the hand. They are also called the knuckle joints.
- The proximal and distal *radioulnar* joints are articulations in the forearm between the radius and ulna.

There are exceptions, but this general rule should make it easier for you to learn the names and location of many joints.

Gliding (Planar) Joints

1. On a skeleton, identify the **intercarpal joints** at the wrist and the **intertarsal joints** in the foot. These articulations are examples of **gliding joints**.
2. Notice that the articulating surfaces of the bones are relatively small, flat surfaces. The limited actions at these joints include side-to-side and back-and-forth **sliding movements**.
3. Rub your palms together, first side to side and then back and forth. This resembles the movements at your

Table 9.2 Anatomical Terms of Movement

Term	Definition	Comments
A. Gliding movements		
1. Gliding	Articulating surfaces of two bones move back and forth or side to side	Limited gliding movements between carpal and tarsal bones.
B. Angular movements		
2. Flexion/extension		
a. Flexion	A bending action that decreases the angle between body parts	Compare flexion and extension of the elbow and knee.
b. Extension	A straightening action that increases the angle between body parts	
3. Abduction/adduction		
a. Abduction	Moving away from the median plane, in a coronal plane	Compare abduction and adduction of the arm and thigh.
b. Adduction	Moving toward the median plane, in the coronal plane	
4. Circumduction	A circular motion that combines flexion, extension, abduction, and adduction; the distal end of the part being moved describes a circle	Compare circumduction of the arm at the shoulder and circumduction of the thigh at the hip.
C. Rotational movements	Moving around the long axis of a bone	
5. Medial/lateral rotation		These movements usually pertain to the long bones of the upper and lower extremities; compare medial and lateral rotation of the humerus (arm) and femur (thigh).
a. Medial (internal) rotation	Anterior surface of moving part is brought toward the median plane	
b. Lateral (external) rotation	Anterior surface of moving part is brought away from the median plane	
6. Right/left rotation	Rotation of a body part to the right or left around the midline of the body	Rotation of the head, to signify "no" at the atlantoaxial joint.

(continues)

intercarpal and intertarsal joints (movement at the joints will be much more restrictive, however). Since these movements do not occur along a specific body plane or around an axis, gliding joints are referred to as **nonaxial joints**.

4. On the skeleton, identify the following gliding joints.
   - Acromioclavicular joint (see Figure 9.10 later in the exercise; identify the acromioclavicular ligament)
   - Sternoclavicular joint (Figure 9.9a)
   - Sacroiliac joint (Figure 9.7)
   - Joints between articular processes of adjacent vertebrae (Figure 9.5)

Hinge Joints

1. On a skeleton, identify the elbow, knee, and ankle joints. These articulations are examples of **hinge joints**.
2. Notice that one articulating surface is convex and the other is concave.
3. **Flexion** and **extension** are the two principal movements that occur at hinge joints. At the knee, however, slight medial and lateral rotation of the leg also occurs. Perform flexion and extension of your forearm and notice that these movements occur along a single plane or axis. Because movement is restricted to one plane, hinge joints are a type of **uniaxial (monaxial) joint**.
4. Like other hinge joints, movements at the ankle occur in one plane. However, the terms *flexion* and *extension* are not typically used. Rather, the movements at the ankle are called **plantar flexion** and **dorsiflexion**.
   - Plantar flexion occurs when you bend the foot at the ankle in the direction of the plantar (inferior) surface. When you stand on your tiptoes or point your toes downward, your feet are plantar flexed.
   - Dorsiflexion occurs when you bend the foot at the ankle in the direction of the dorsum (superior surface). When you stand on your heels and lift your toes upward, your feet are dorsiflexed.

Table 9.2 (continued)

Term	Definition	Comments
D. Special movements		
7. Supination/pronation		
a. Supination	Rotating forearm and hand laterally so that palm faces anteriorly distal radioulnar joints.[1]	These movements specifically apply the rotation of the radius around the ulna at the proximal and
b. Pronation	Rotating forearm and hand medially so that palm faces posteriorly	
8. Opposition/reposition		
a. Opposition	The thumb or pinky finger is brought over to touch another digit	These movements are specific for the hand; opposition and reposition of the thumb are critical for performing fine motor skills such as holding a pen to write.
b. Reposition	The thumb or pinky finger is brought back to anatomical position	
9. Eversion/inversion		
a. Eversion	Moving sole of foot away from median plane	These movements specifically apply to activity at intertarsal joints in the feet.
b. Inversion	Moving sole of foot toward median plane	
10. Protraction/retraction		
a. Protraction	Moving anteriorly	Compare protraction and retraction of the mandible.
b. Retraction	Moving posteriorly	
11. Elevation/depression		
a. Elevation	Raising a part superiorly	Compare elevation and depression of the scapula and the mandible.
b. Depression	Lowering a part inferiorly	
12. Dorsiflexion/plantar flexion		
a. Dorsiflexion	A bending action that elevates the soles, as when you stand on your heels	These movements occur at the ankle joint.
b. Plantar flexion	A bending action that elevates the heels, as when you stand on your toes	

[1] In a clinical setting, pronation and supination are often used to describe actions of the foot. However, anatomists do not officially recognize these terms to describe foot actions. When they are used, pronation refers to a combination of eversion and abduction of the foot, and supination to a combination of inversion and adduction of the foot.

5. On a skeleton or bony model of the lower extremity, plantar flex and dorsiflex the foot and observe the hingelike motion at the ankle joint. Next, **invert** and **evert** the foot, and notice that these movements occur at **intertarsal joints**, inferior to the ankle. It is common to mistakenly associate inversion and eversion with the ankle joint. However, the ankle is a uniaxial joint, where movement can occur in only one plane, and these movements occur in a different plane than plantar flexion and dorsiflexion.
6. On the skeleton, identify the **interphalangeal joints**, which are between the phalanges of the hands and feet. Verify the following characteristics of these joints.
   - Similar to other hinge joints, the convex surface of one bone articulates with the concave surface of another bone.
   - Flex and extend your fingers at the interphalangeal joints. These are the only two movements at these articulations, and both occur in the same plane.

Pivot Joints

1. On a skeleton, identify the **proximal** and **distal radioulnar joints**. These are **pivot joints**, in which a rounded or conical surface of one bone fits into a shallow depression of another bone. The bones are held in position by a strong ligamentous collar. On the skeleton, observe the articulations for the radioulnar joints.
   - At the proximal joint, the **head of the radius** articulates with the **radial notch** on the ulna.
   - At the the distal joint, the **head of the ulna** articulates with the **ulnar notch** on the radius.
2. Similar to hinge joints, pivot joints are uniaxial. The actions are restricted to **rotational movements** around a central axis. At the radioulnar joints, the rotational movements are **pronation** and **supination** of the forearm. Place yourself in anatomical position (palms facing anteriorly) and rotate your right forearm so that the palm

TYPES OF SYNOVIAL JOINTS	MOVEMENT	EXAMPLES
(a) Gliding joint 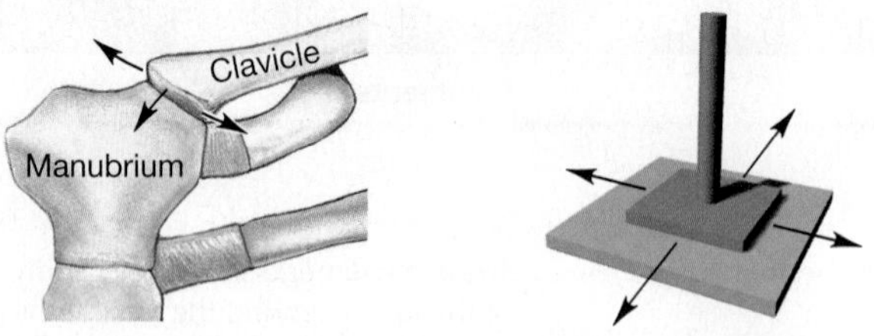	Slight nonaxial or multiaxial	• Acromioclavicular and sternoclavicular joints • Intercarpal and intertarsal joints • Vertebrocostal joints • Sacroiliac joints
(b) Hinge joint 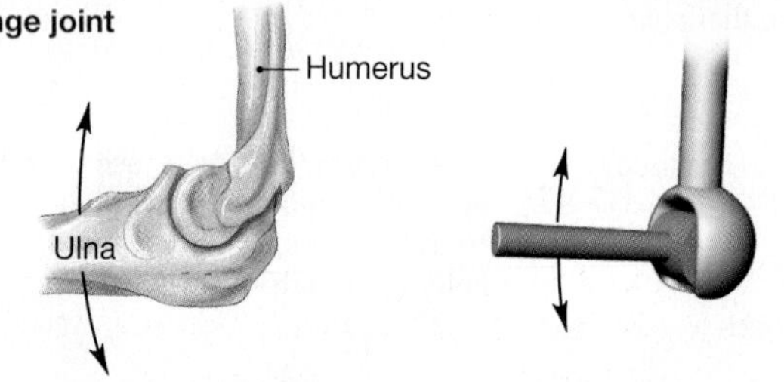	Monaxial	• Elbow joint • Knee joint • Ankle joint • Interphalangeal joint
(c) Pivot joint 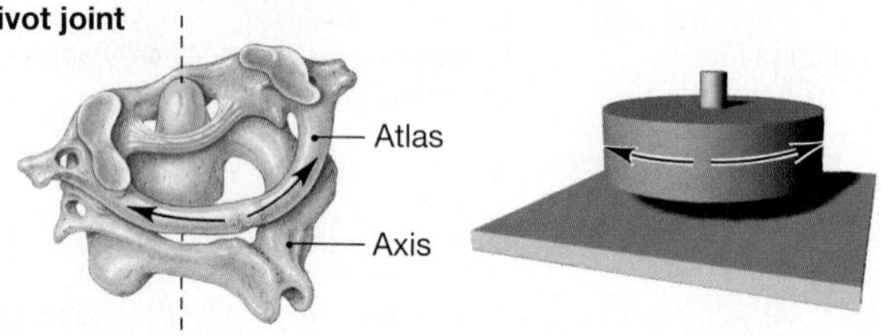	Monaxial (rotation)	• Atlas/axis joint • Proximal radioulnar joint
(d) Ellipsoidal joint 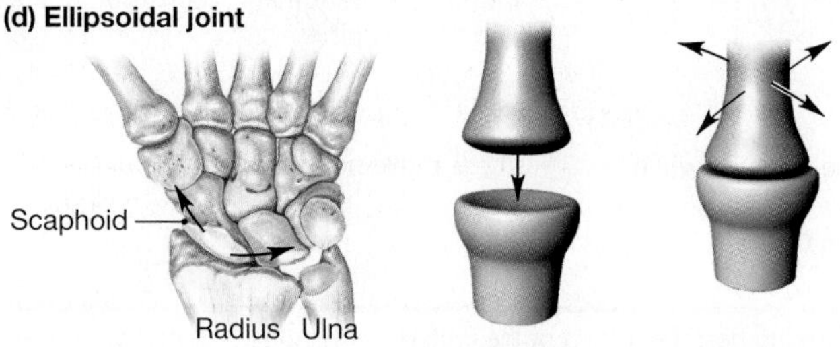	Biaxial	• Radiocarpal joint • Metacarpophalangeal joints 2–5 • Metatarsophalangeal joints
(e) Saddle joint 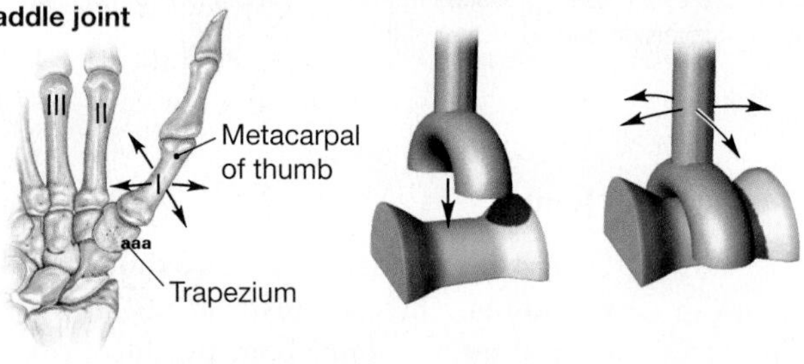	Biaxial	• First carpometacarpal joint
(f) Ball-and-socket joint 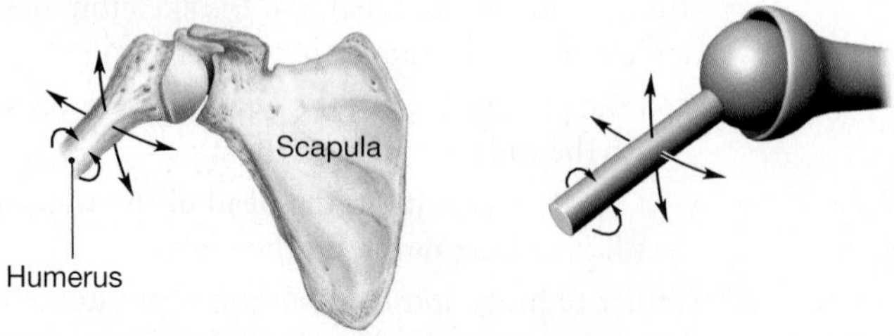	Triaxial	• Shoulder joint • Hip joint

Figure 9.9 Types of synovial joints. Synovial joint classification is determined by the shape of the articulating surfaces and the range of motion. **a)** Gliding joint; **b)** hinge joint; **c)** pivot joint; **d)** ellipsoidal joint; **e)** saddle joint; **f)** ball and socket joint.

faces posteriorly. This action is called pronation of the forearm. To supinate your forearm, rotate your forearm back to the anatomical position so that the palm faces anteriorly once again.

On a skeleton, repeat the pronation and supination movements on the right forearm. As you perform these actions, notice that the movement occurs at the two radioulnar joints, and as a result, the entire radius rotates around the ulna. Describe the relationship between the radius and ulna when the forearm is pronated, and when it is supinated.

____

____

3. On the skeleton, identify another example of a pivot joint, the **atlantoaxial joint** (Figure 9.9c). This joint is an articulation between the first two cervical vertebrae, the atlas (C1) and the axis (C2). Notice how the **dens of the axis** articulates with a shallow depression on the **anterior arch** of the atlas. At this joint, the head can be rotated to the left and right to indicate the "no" response.

Ellipsoidal (Condyloid) Joints

1. On a skeleton, identify a **radiocarpal (wrist) joint**. This articulation is an **ellipsoidal joint**.
2. At ellipsoidal joints, an oval-shaped condyle articulates with an elliptical cavity. At the wrist, a condyle is formed by the proximal row of carpal bones. The elliptical cavity is formed by the distal end of the radius. Notice that the ulna does not contribute an articulating surface to the wrist joint (Figure 9.9d).
3. At the wrist and other ellipsoidal joints, **flexion** and **extension** occur in one plane, and **abduction** and **adduction** in a second plane. Because movements pass through two different planes, these joints are classified as **biaxial**.
4. With your body in anatomical position, bend your right wrist so that your hand moves anteriorly. This movement is called *flexion* of the hand. To *extend* your hand, bend your wrist to move your hand posteriorly. Notice that these two movements occur in the same plane.
5. With your body in anatomical position, bend your right wrist so that your hand moves laterally. This movement is called *abduction* of the hand. To *adduct* your hand, bend your wrist to move your hand medially. Notice that these two movements occur in a plane that is perpendicular to the plane in which flexion and extension occur.
6. Very slowly perform the following sequence of movements at the wrist joint: flexion, abduction, extension, adduction. Repeat the sequence several times, increasing the speed after each turn until the movements become fluid. Notice that the tips of your fingers are moving in a circle. This combination of movements is called **circumduction**, an action in which the distal end of the part being moved describes a circle.
7. On the skeleton, identify other examples of ellipsoidal joints.
   - **Atlanto-occipital joint** between the atlas (first cervical vertebra) and the occipital bone of the skull
   - **Metacarpophalangeal (knuckle) joints**, between the metacarpal bones and the proximal phalanges in the hand

Saddle Joints

1. On a skeleton, identify a **carpometacarpal joint of the thumb**. This articulation is an example of a **saddle joint**. On each hand, this joint is formed by the articulation between the trapezium and the metacarpal bone of the thumb (Figure 9.9e).
2. Observe that the articulating surfaces of both the trapezium and the metacarpal bone have convex and concave regions, and each surface resembles a saddle. This special feature is common to all saddle joints.
3. The primary movements that occur at these biaxial joints are flexion and extension, and abduction and adduction. To demonstrate these movements, position your right hand with the palm directly in front of your face. Perform the following movements at the carpometacarpal joint.
   - *Flexion of the thumb.* Bend the thumb so that it passes directly across the palmar surface of your hand.
   - *Extension of the thumb.* Bend the thumb so that it moves away from the hand, but in the same plane as flexion (like a hitchhiker).
   - *Abduction of the thumb.* With the thumb positioned tightly against the palm and parallel to the index finger, bend it so that it moves toward your face.
   - *Adduction of the thumb.* With the thumb in the abducted position, bend it so that it moves toward the palm.

   Notice that flexion and extension occur in a plane that is perpendicular to the plane in which abduction and adduction occur.
4. The shape of the articular surfaces of this joint allows for much greater freedom of movement compared to other biaxial joints. At this joint, flexion and abduction are combined with slight medial rotation so that the thumb can be brought into contact with the palmar surface of the pinky finger. This action is called **opposition**. Bringing the thumb back to the anatomical position is called **reposition**. Perform these movements with your own thumb. Notice that you can also circumduct your thumb at the carpometacarpal joint.

Ball and Socket Joints

1. On a skeleton, identify the **shoulder** and **hip joints**. These articulations are examples of **ball and socket joints**, where a rounded head articulates with a cuplike concavity. Ball and socket joints are triaxial, allowing movements in

three planes: flexion and extension in the first plane, abduction and adduction in the second, and medial and lateral rotation in the third. Circumduction can also be performed at these joints.

2. From the anatomical position, perform the following movements at the shoulder joint.
   - *Flexion of the arm.* Move the arm (not the forearm) anteriorly in a sagittal plane.
   - *Extension of the arm.* Move the arm posteriorly in a sagittal plane.
   - *Abduction of the arm.* Move the arm away from the body (laterally) in a coronal plane.
   - *Adduction of the arm.* From the abducted position, move the arm toward the body (medially) in a coronal plane.
   - *Circumduction of the arm.* Combine the above four movements to circumduct the arm. Your palm should face anteriorly at all times. Notice that while you perform this action, your hand moves in a circular path.
   - *Medial rotation of the arm.* From the anatomical position, rotate the arm so that the anterior surface moves medially. Be sure you are moving your arm and not rotating the forearm.
   - *Lateral rotation of the arm.* From the anatomical position, rotate the arm so that the anterior surface moves laterally.

   Notice that flexion and extension occur in a plane that is perpendicular to the plane in which abduction and adduction occur. Medial and lateral rotation occur around a vertical axis that runs through the shaft of the humerus.

3. From the anatomical position, perform flexion/extension, abduction/adduction, circumduction, and medial/lateral rotation of the thigh at the hip joint. For the pairs of actions, compare the range of motion between the shoulder and hip joints.

   Flexion/extension:

   ____

   ____

   Abduction/adduction:

   ____

   ____

   Medial/lateral rotation:

   ____

   ____

QUESTIONS TO CONSIDER 1. Flexion and extension of the fingers can occur at the metacarpophalangeal (knuckle) joints and the interphalangeal joints. Abduction and adduction of the fingers (moving the fingers apart and bringing them back together) occurs at the metacarpophalangeal joints only. Based on the structure of these two types of joints, provide an explanation for the difference in function.

____

____

____

____

2. In the previous activity, you compared the range of motion between the shoulder and hip joints. Did your observations reveal any differences in the freedom of movement between the two joints? If so, provide a reason why these differences are significant. (Hint: Consider the general functions of the upper and lower extremities.)

____

____

____

____

ACTIVITY 9.5 Examining the Structure of Specific Synovial Joints

The Shoulder Joint

1. On the whole skeleton, examine the **shoulder (glenohumeral) joint** (Figure 9.10). Note that the joint is formed by the articulation between the **glenoid fossa** of the scapula and the **head of the humerus.** The glenoid fossa contains a rim of fibrocartilage, the **glenoid labrum**, that extends slightly beyond the edge of the genoid fossa to deepen the socket. The shoulder joint is stabilized mostly by a group of muscles surrounding the joint, collectively referred to as the **rotator cuff.**
2. Obtain an anatomical model of the shoulder joint. Along the superior aspect of the joint, identify the **coracohumeral ligament**. It connects the coracoid process of the scapula to the anatomical neck of the humerus
3. Find the **glenohumeral ligament** along the anterior aspect of the joint. It travels from the superior margin of the glenoid fossa on the scapula to the lesser tubercle and anatomical neck of the humerus.
4. Verify that the **coracoacromial ligament** connects the coracoid and acromion processes on the scapula. The combination of this ligament and the two bony processes forms the **coracoacromial arch**. Passing under the arch is the tendon of a rotator cuff muscle known as the **supraspinatus**.

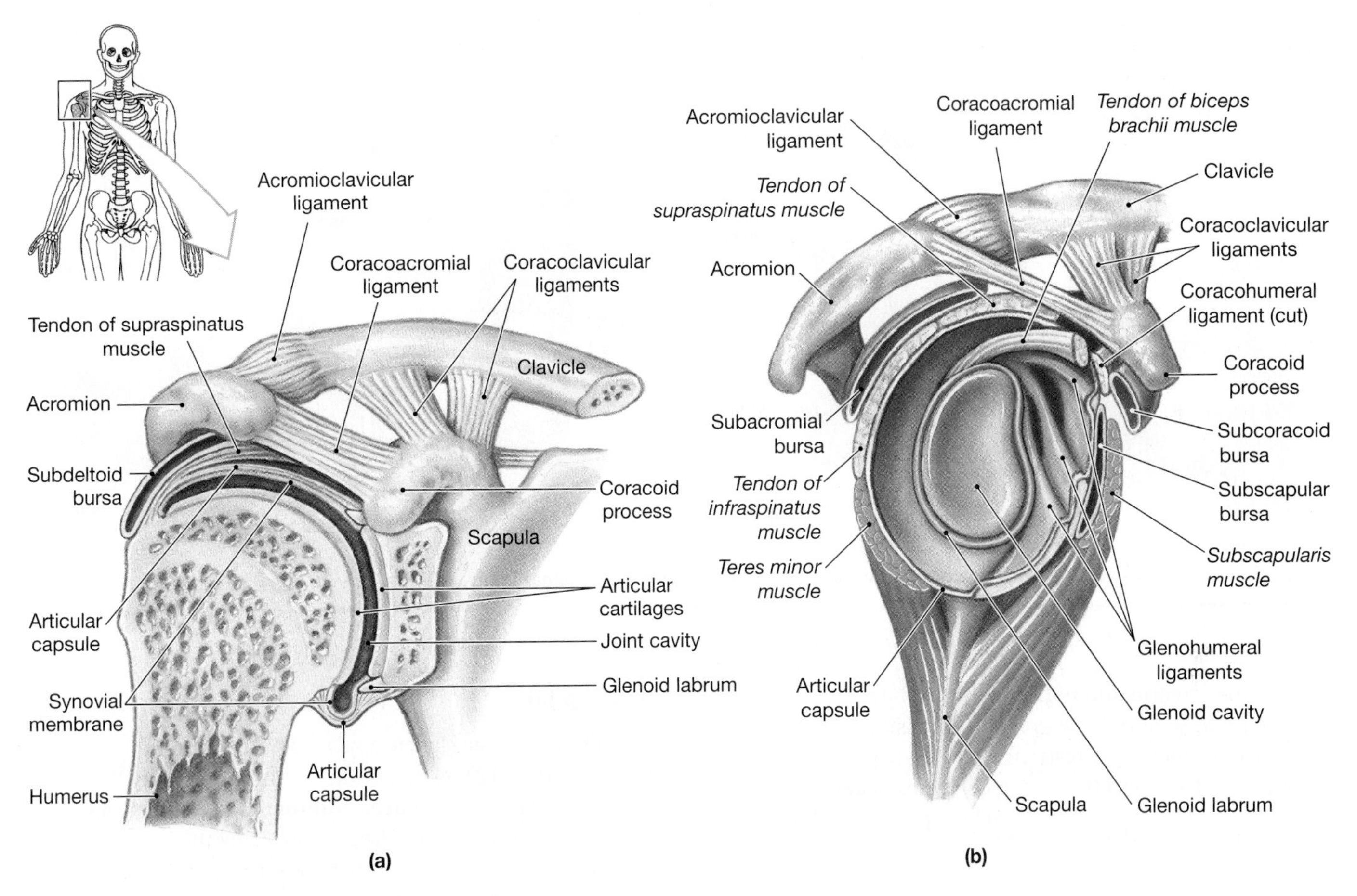

Figure 9.10 The structure of the right shoulder joint. a) Anterior view, showing a frontal section; **b)** lateral view with the humerus removed.

5. Identify the **acromioclavicular ligament**, which connects the lateral end of the clavicle to the acromion process. The acromioclavicular ligament provides support for the superior aspect of the shoulder. Technically, it is not a structural component of the shoulder. Rather, it is a part of the **acromioclavicular joint.** How would you classify the latter joint? ________________
6. Obtain a humerus and identify the **lesser** and **greater tubercles**, and the groove between them (the **intertubercular** or **bicipital groove**). The tendon of the long head of the biceps brachii muscle travels in this groove.
7. On the shoulder model, identify the **transverse humeral ligament**. Observe that this ligament forms a tunnel that covers the intertubercular groove. It keeps the biceps brachii tendon in its proper position.
8. On the model, identify the **coracoclavicular ligaments**. As you learned earlier, the ligamentous connections between the coracoid process and the clavicle is a syndesmosis. It provides support for the shoulder, even though it is not a structural component of the joint.

The Elbow Joint

1. On a skeleton, examine the bony components of the **elbow joint** (Figure 9.11). Confirm that the elbow joint involves two articulations.
   - Between the **trochlea** of the humerus and the **trochlear notch** of the ulna
   - Between the **capitulum** of the humerus and the **head of the radius**
2. Identify the two collateral ligaments of the elbow joint, each with a triangular shape. Locate the **radial (lateral) collateral ligament**. The apex of the triangle is attached to the lateral epicondyle of the humerus; the base blends and becomes continuous with the annular ligament. Locate the **ulnar (medial) collateral ligament**. It attaches proximally to the medial epicondyle of the humerus, and distally to the coronoid and olecranon processes on the ulna.

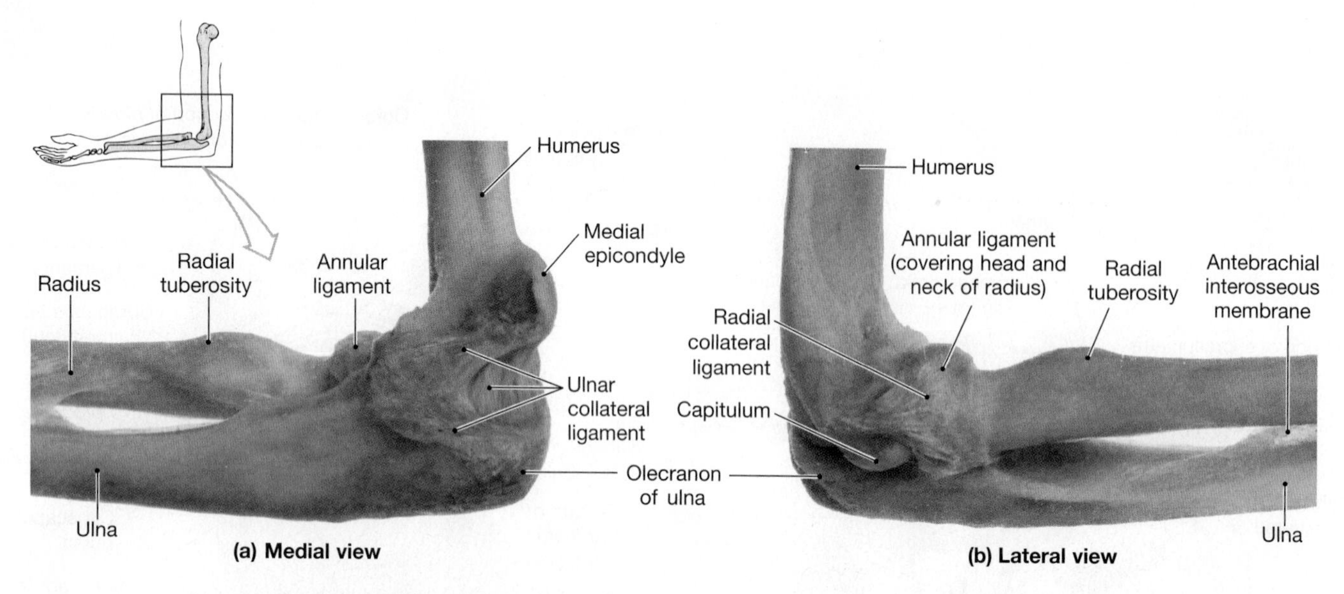

Figure 9.11 The structure of the right elbow and proximal radioulnar joints. a) Medial view; **b)** lateral view.

3. Move the forearm and notice that the elbow functions as a hinge joint, with only flexion and extension possible. Other movements are restricted by the shapes of the articulating surfaces and the strong collateral ligaments running on each side of the joint, which prevent abduction and adduction.

WHAT'S IN A WORD The **ulnar nerve** passes posterior to the medial epicondyle and is close to the ulnar collateral ligament. Injury to the medial epicondyle may compress the ulnar nerve and cause temporary numbness or a tingling sensation along the nerve's area of distribution—the medial aspect of the forearm and hand. For this reason, the medial epicondyle of the humerus is called the "funny bone."

The Radioulnar Joints

1. On a skeleton, identify the **head of the radius.** Observe that it articulates with the **radial notch** on the ulna to form the **proximal radioulnar joint**.
2. On a model of the elbow and proximal radioulnar joints, identify the **annular ligament** (Figure 9.11). This ligament attaches to the anterior and posterior margins of the radial notch and wraps around the head of the radius, keeping the radial head in its correct position. This joint is a pivot joint, with the head of the radius rotating (pivoting) within the annular ligament.
3. On the skeleton, examine the **distal radioulnar joint**, also a pivot joint. Identify the articular surfaces of this joint. They are the **head of the ulna** and the **ulnar notch** of the radius.
4. Pronation and supination of the forearm, discussed earlier, do not occur at the elbow joint proper, but at the two radioulnar joints.

The Hip Joint

1. On the whole skeleton, identify the **hip joint** (Figure 9.12). Note that the joint is formed by the articulation between the **acetabulum** of the hip bone and the **head of the femur.** The hip joint, like the shoulder joint, is a ball and socket joint; however, this joint is designed for stability, because it supports and transmits all of the weight of the upper body. Because of this, the socket is much deeper than that found at the shoulder, and strong ligaments firmly anchor the bones together. This limits the range of motion but provides a strong joint. As you examine the hip joint, compare it to the shoulder joint to note similarities and differences in the designs.
2. On the model of the hip joint, identify the **iliofemoral ligament.** It is a Y-shaped ligament that strengthens the anterior aspect of the hip joint. Proximally, it is attached to the ilium at the anterior inferior iliac spine and the rim of the acetabulum. Distally, it is attached to the femur at the intertrochanteric line.
3. Identify another anterior ligament, the **pubofemoral ligament.** Proximally, this structure is attached to the pubic part of the acetabular rim. It travels distally and blends in with the medial portion of the iliofemoral ligament.
4. The **ischiofemoral ligament** reinforces the posterior aspect of the joint. Its attachments include the ischial portion of the acetabular rim and the neck of the femur, medial to the base of the greater trochanter.
5. If you can view the inside of the hip joint, notice the acetabular labrum, which is a thickened rim of fibrocartilage that deepens the socket. This is much larger than the glenoid labrum in the shoulder. The ball and socket

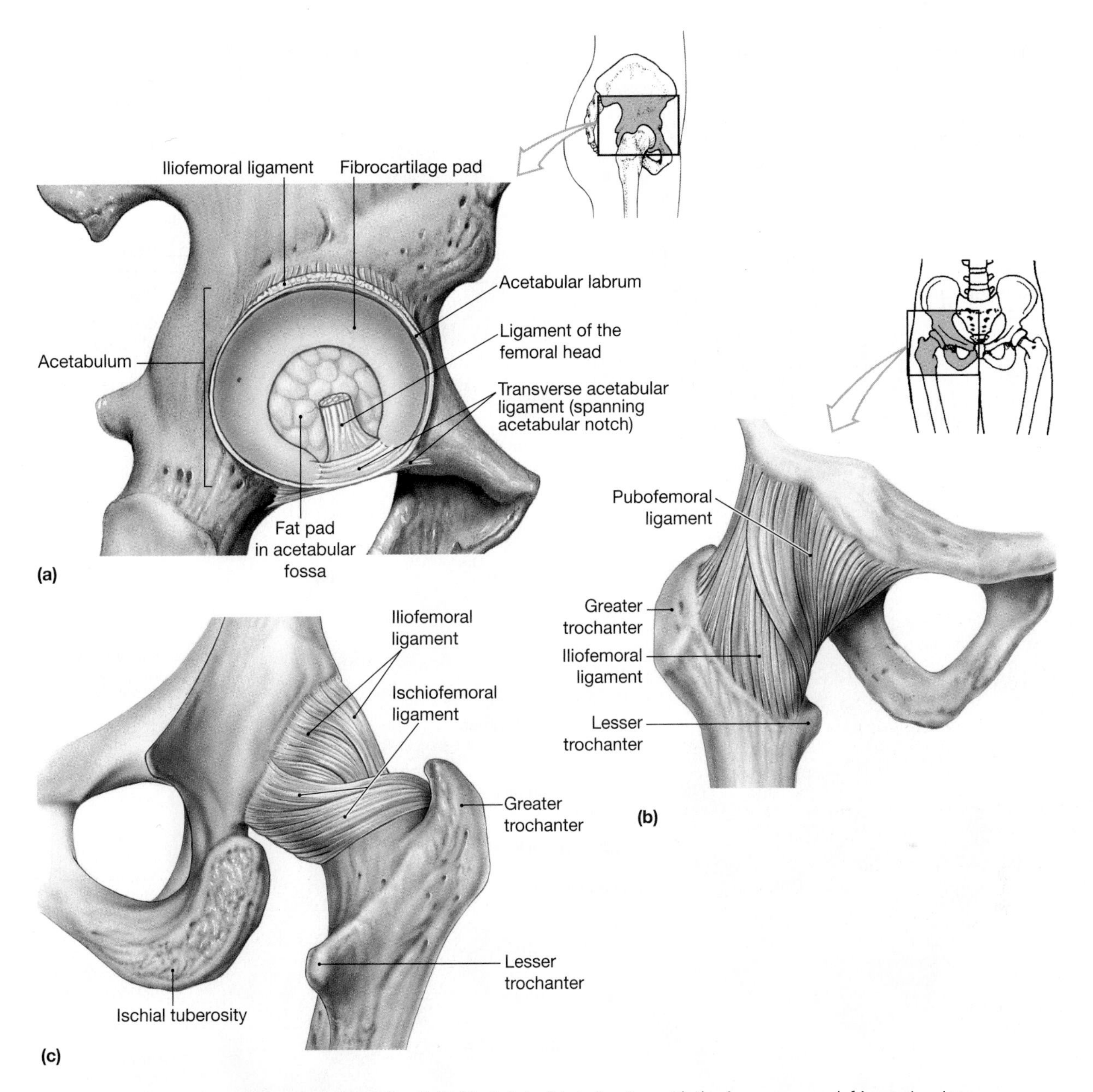

Figure 9.12 The structure of the right hip joint. a) Laterior view with the femur removed; **b)** anterior view; **c)** posterior view.

parts fit much more tightly together here than in the shoulder, and the head of the femur is tethered into the socket by the **ligament of the femoral head (ligamentum teres).**

The Knee Joint

1. On a skeleton, identify the bones that form the **knee joint** (Figure 9.13). Confirm that the knee joint involves the following three articulations.
   - Between the **lateral condyle** of the femur and the **lateral condyle** of the tibia
   - Between the **medial condyle** of the femur and the **medial condyle** of the tibia
   - Between the posterior surface of the **patella** and the **patellar surface** of the femur
2. The knee has a complex structure and actually involves two joints. There is a gliding joint between the **patella** (kneecap) and the femur. The main knee joint, though, is an ellipsoid joint between the femur and the tibia. Normally, an ellipsoid joint would allow flexion and extension, abduction and adduction, and circumduction. For stability, however, the knee joint has many specialized

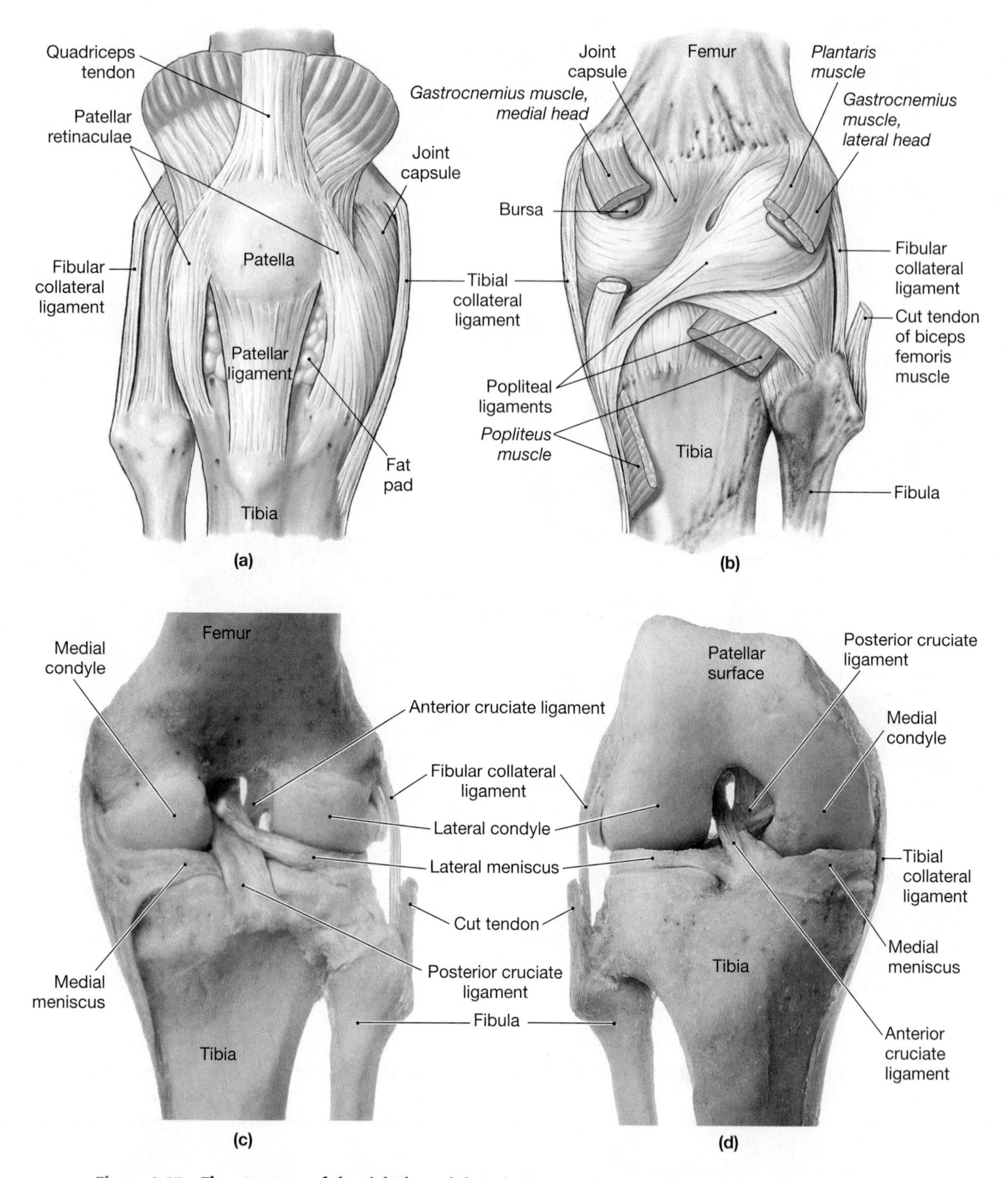

Figure 9.13 The structure of the right knee joint. a) Diagram of a superficial anterior view, in the extended position; **b)** diagram of a superficial posterior view, in the extended position; **c)** cadaver dissection of a deep posterior view, in the extended position; **d)** cadaver dissection of a deep anterior view in the flexed position.

ligaments and connections that restrict its movement, primarily to only flexion and extension, with slight rotation that allows the articulating surfaces to align more exactly (locking the knee). Because the movement is restricted, the knee functions more like a hinge joint and is often referred to as a modified hinge joint.

3. On a knee model, identify the **quadriceps tendon** that attaches the quadriceps femoris muscle to the patella. Note that the **patellar ligament** is the continuation of the quadriceps tendon. It extends from the patella to the tibial tuberosity.

   What is the difference between a tendon and a ligament?

   ____

4. Along the lateral side of the knee, identify the **lateral (fibular) collateral ligament**. Verify that the ligament extends from the lateral epicondyle of the femur to the head to the fibula.
5. Along the medial side of the knee, identify the **medial (tibial) collateral ligament**, which connects the medial epicondyle of the femur to the medial condyle of the tibia.
6. Move the patella and flex the knee to expose the two **cruciate ligaments**. Note that these ligaments are found within the joint capsule and cross each other. Both ligaments connect the tibia to the femur, but they are named according to their site of attachment on the tibia.
   - Identify the **anterior cruciate ligament**. It passes from the *anterior* aspect of the intercondylar area (*intercondylar* means "between the condyles") of the tibia to the lateral condyle of the femur.
   - Locate the **posterior cruciate ligament**. It travels from the *posterior* aspect of the intercondylar area of the tibia to the medial condyle of the femur.

   What do you think might be the function of these ligaments?

   ____

WHAT'S IN A WORD The term *cruciate* is derived from the Latin word *cruciatus*, which means "cross." Inside the knee joint, the two cruciate ligaments form a cross.

7. With the knee in the flexed position, identify the **lateral meniscus** and the **medial meniscus**. The two **menisci** are C-shaped fibrocartilage plates that rest on the articular surface of the tibia. Observe how the menisci form a deeper articulating surface on the tibia for the condyles of the femur. This arrangement increases the stability of the knee. In addition, the menisci have an important shock-absorbing function.

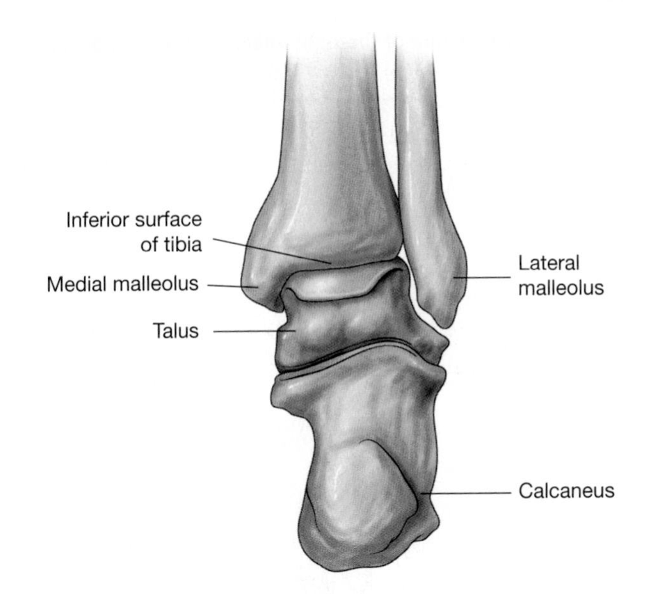

Figure 9.14 Skeletal components of the ankle joint. The talus articulates with a bony socket formed by the lateral malleolus, the inferior surface of the tibia, and the medial malleolus.

CLINICAL CORRELATION

The knee is a very complex joint that supports and transmits all of your body weight. It is under a lot of stress and, in spite of its stabilizing specializations, the knee can be injured. An abrupt movement that causes the knee to buckle medially can tear the medial collateral ligament and that injury, in turn, often tears the medial meniscus, which is attached to it. A knee injury generally known as "torn cartilage" occurs when cartilage fragments are dislodged from one or both menisci. In addition, the cruciate ligaments can be injured if the knee is moved too forcefully in an anterior or posterior direction.

The Ankle Joint

1. On the skeleton or a bony model of the lower extremity, observe the articulating surfaces that form the **ankle joint**. Notice that the **medial malleolus**, the **inferior surface of the tibia**, and the **lateral malleolus** of the fibula form a deep socket into which the **talus** fits (Figure 9.14).
2. On a bony model of the foot, observe the superior articulating surface of the talus. Notice that this surface is much wider anteriorly. The ankle is more stable when the foot is dorsiflexed because the tibia and fibula are exposed to a wider articulating surface on the talus. When a downhill skier is in the crouch position, the ankles are in a relatively stable position. On the other hand, the ankle is less stable when the foot is

Figure 9.15 The structure of the right ankle joint. a) Medial view; **b)** lateral view; **c)** posterior view.

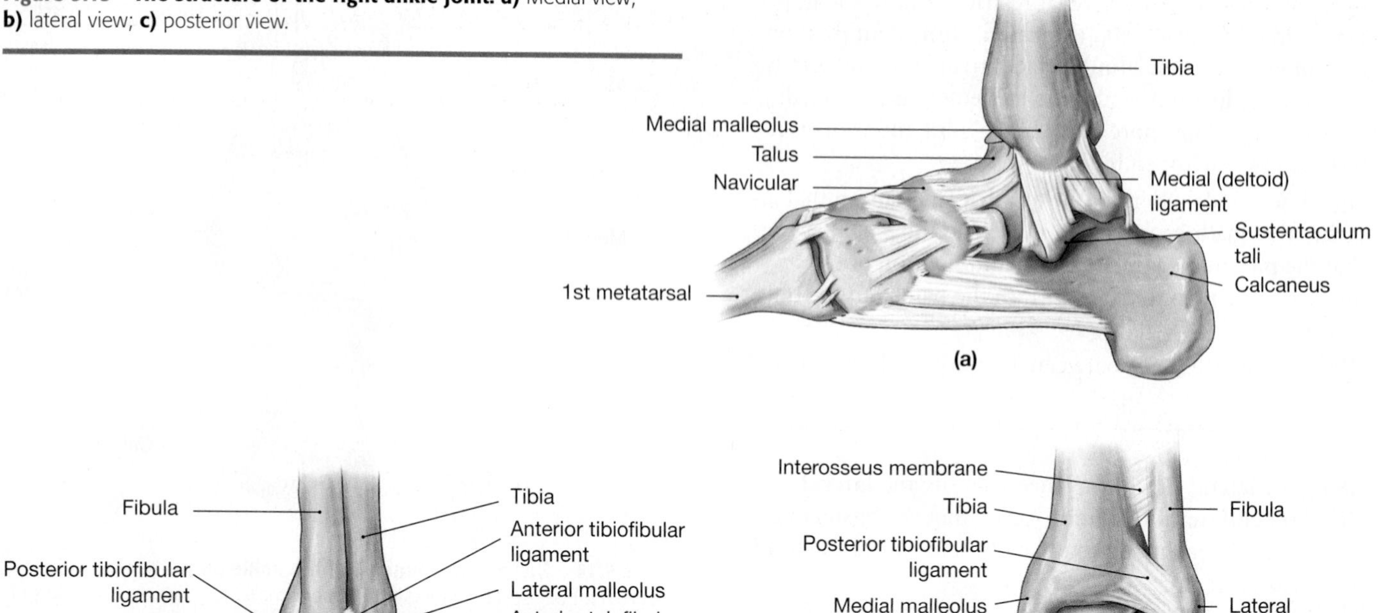

plantarflexed, because the articulating surface on the talus is narrower. When a ballerina dances en pointe, her feet are plantarflexed and the ankles are in a relatively unstable position. You can see why very stable ankles are a prerequisite for ballet!

3. On a model of the ankle joint, identify the collateral ligaments of the ankle. The **medial collateral (deltoid) ligament** is triangular shaped. The apex is attached to the medial malleolus on the tibia. From this region, the ligament fans out and attaches to the talus, navicular, and calcaneus on the foot (Figures 9.15a and c).
4. On the lateral side of the ankle, notice the three distinct collateral ligaments. They connect the lateral malleolus on the fibula with the talus and calcaneus on the foot. The lateral collateral ligaments are named according to the bones to which they attach. They include the **anterior talofibular ligament**, the **posterior talofibular ligament**, and the **calcaneofibular ligament** (Figures 9.15b and c).

CLINICAL CORRELATION

An **ankle sprain** refers to the stretching or tearing of ligaments that support the ankle. This type of injury is quite common among athletes who are involved in running or jumping sports (e.g., basketball, football, track and field) and typically occurs when excessive force is applied to the joint. Lateral ankle sprains occur most frequently. They are usually caused by excessive inversion of the ankle ("turning" the ankle) and result in damage to the lateral collateral ligaments, particularly the anterior talofibular ligament (Figure 9.15b).

Excessive eversion of the ankle can cause damage to the medial (deltoid) ligament (Figure 9.15a). However, this ligament is relatively strong and usually withstands the extreme force. A more likely scenario is the occurrence of a **Pott's fracture,** in which the tip of the medial malleolus is torn off the tibia, causing the talus to move laterally and the distal end of the fibula to be fractured. In general, therefore, eversion injuries are usually more serious than inversion injuries.

Table 9.3 Articulations and Actions of Synovial Joints

Joint	Articulating surfaces	Type of joint: uniaxial, biaxial, or triaxial	Actions
1. Shoulder (Figure 9.11)	Glenoid fossa articulates with the head of the humerus	Triaxial	Flexion/extension, abduction/adduction, and medial/lateral rotation of the arm
2. Elbow (Figure 9.12)			
3. Proximal radioulnar joint (Figure 9.12)			
4. Wrist (Figure 9.10d)			
5. Hip (Figure 9.13)			
6. Knee (Figure 9.14)			
7. Ankle (Figures 9.15 and 9.16)			

QUESTIONS TO CONSIDER

1. Table 9.3 provides the articulating surfaces, type of joint, and actions for the shoulder joint. Use the same format to provide similar information for the other synovial joints listed in the table.

2. At the knee joint, the patella acts as a pulley for the quadriceps tendon and patellar ligament. Given this information, what advantage does the patella provide for the knee? How would removal of the patella affect the knee's function? (Hint: To answer this question, first determine the function of a pulley.)

____

____

____

____

3. During the previous activity, you examined the structure of the shoulder joint and compared it to the hip joint. Based on this comparison, which joint do you believe is more vulnerable to a dislocation injury? Explain.

____

____

____

____

Exercise 9 Review Sheet

Articulations

Name ____________________

Lab Section ____________________

Date ____________________

1. Compare the structural features and the relative degree of movement of the three types of fibrous joints.

2. What is the difference between a synchondrosis and a symphysis?

3. What is the functional difference between uniaxial, biaxial, and triaxial joints?

Questions 4–9: Define the following terms.

4. Articular cartilage:

5. Synovial membrane:

6. Bursa:

7. Joint cavity:

8. Ligament versus tendon:

9. Meniscus:

10. Using the following table, provide two examples of a uniaxial joint, a biaxial joint, and a triaxial joint. List the type of movement(s) (flexion, extension, abduction, adduction, etc.) that occur at each type of joint.

Type of joint movement(s)	**Examples**
1. Uniaxial joint	a.
	b.
2. Biaxial joint	a.
	b.
3. Triaxial joint	a.
	b.

Exercise 10

Histology of Muscle Tissue

Laboratory Objectives

On completion of the activities in this exercise, you will be able to:

- Compare the functions of the three types of muscle tissue.
- Identify microscopic features of the three types of muscle tissue.
- Identify the components of a neuromuscular junction.
- Interpret the sequence of events that occurs at a neuromuscular junction.
- Explain the three-dimensional relationships between important structures in a skeletal muscle cell.
- Describe how skeletal muscle fibers contract, according to the sliding filament theory.

Materials

- Prepared microscope slides of skeletal, cardiac, and smooth muscle
- Compound light microscopes
- Dissecting microscopes
- Colored pencils
- Three-dimensional model of a neuromuscular junction
- Prepared microscope slide of a neuromuscular junction
- Three-dimensional model of a skeletal muscle fiber
- Glycerinated skeletal muscle preparation
- 0.25% ATP in distilled water
- Ion solution: 0.05M potassium chloride (KCl) and 0.001M magnesium chloride ($MgCl_2$) in distilled water
- 50% glycerol in distilled water
- Microscope slides and coverslips
- Forceps and dissecting needles

Muscle tissue performs a variety of important functions. **Skeletal muscles** are attached to bone by tendons and usually cross at least one joint. They are responsible for voluntary movements and have a significant role in maintaining posture and stabilizing joints. **Cardiac muscle** is located in the heart wall. It is responsible for the involuntary contractions of the heart chambers (the heartbeat) that pump blood out of the heart. **Smooth muscle** is located in the walls of internal organs, blood vessels, and airways. It is responsible for the involuntary movements associated with the function of various organs. For example, smooth muscle causes the constriction and dilation of blood vessels and airways and the muscular actions that move food along the alimentary canal (digestive tract). Contracting muscles, particularly skeletal muscles, generate a significant amount of heat, which is vital for maintaining a constant body temperature.

Four special characteristics of muscle cells (also referred to as muscle fibers) make their normal activities possible.

- **Excitability**. Muscle cells possess cell membranes that are excitable. This allows muscle tissue to receive and respond to stimuli from nerves or hormones.
- **Contractility**. Muscle cells contain contractile proteins that allow them to change in length, or contract, and thereby generate a force to do work.
- **Stretchability**. Muscle cells can be stretched, allowing some muscles to oppose the movements generated by others that are contracting.
- **Elasticity**. Muscles are able to return to their original shape after contracting or stretching.

Skeletal Muscle

Skeletal muscle fibers have two distinguishing anatomical characteristics.

- They possess alternating light and dark striations, which are formed by the unique arrangement of actin (thin) and myosin (thick) filaments.
- They are multinucleated cells. Several peripherally located nuclei are present in each muscle fiber.

ACTIVITY 10.1 Light Microscopic Structure of Skeletal Muscle

1. Obtain a prepared microscope slide of skeletal muscle.
2. View the slide with the low-power objective lens of a compound light microscope. You are observing different views of **skeletal muscle fibers**. Your slide most likely contains both a cross section and a longitudinal section of the fibers (Figure 10.1).
3. Switch to high power and observe both cross and longitudinal sections of skeletal muscle fibers more closely.
4. In the cross-sectional view, identify the darkly stained nuclei arranged around the periphery of each muscle fiber. Identify the connective tissue layer, the **endomysium**, which surrounds each muscle fiber. Note that a connective tissue layer known as the **perimysium** surrounds each bundle of muscle fibers (a muscle fascicle).
5. In the longitudinal view, identify the distinctive **striations**. They will appear as alternating light and dark bands across the width of each muscle fiber. The striations are formed by the arrangement of the thick and thin contractile protein filaments, within long

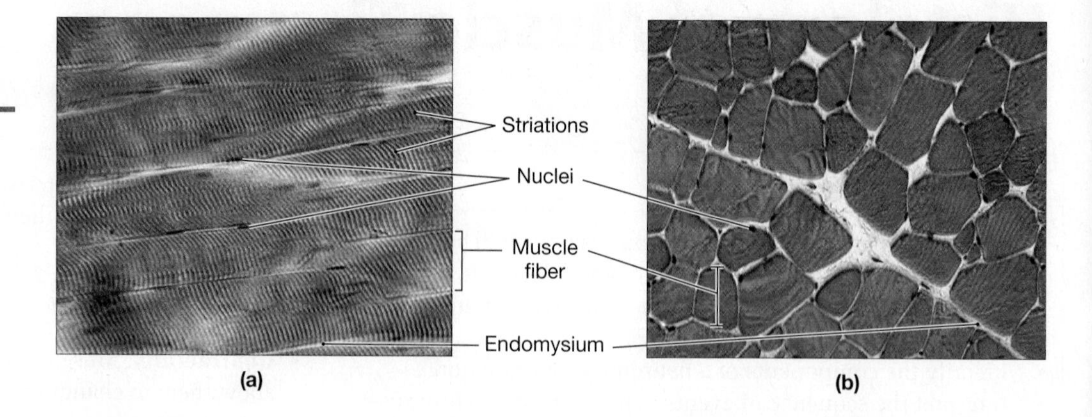

Figure 10.1 Microscopic structure of skeletal muscle. a) Longitudinal section (LM × 200); **b)** cross section (LM × 200).

cylindrical structures called **myofibrils**. The lighter **I bands** contain only thin filaments. The darker **A bands** contain overlapping thin and thick filaments. In the longitudinal section, note also the peripheral location of the nuclei.

WHAT'S IN A WORD The prefix *myo-* has its origins from the Greek word *mys*, for "muscle." Thus, myofibrils are the fibrils (bundles of filaments) of a muscle cell. Actin and myosin molecules are found in the thin and thick myofilaments, or muscle filaments. Any word with the prefix *myo-* is related to muscle.

QUESTION TO CONSIDER In the space provided, draw skeletal muscle cells as you view them in both cross and longitudinal sections. Label the following structures: (a) muscle fibers, (b) cell nuclei, (c) striations.

Cardiac Muscle

Unlike skeletal muscle, which typically consists of parallel fibers, cardiac muscle forms a network of branching fibers. In longitudinal sections, striations are visible but not as prominent as in skeletal muscle fibers. Cardiac muscle cells are linked together by **intercalated discs**. These highly specialized cell junctions are not found in other types of muscle tissue. At the intercalated discs, ions can move directly from one cell to the next, making muscle cells electrically coupled. As a result, electric impulses can spread rapidly, allowing cardiac muscle cells to contract almost simultaneously.

ACTIVITY 10.2 Light Microscopic Structure of Cardiac Muscle

1. Obtain a prepared slide of cardiac muscle.
2. View the slide with the low-power objective lens. You are observing **cardiac muscle fibers** in the heart wall. As you scan the slide, you may find some muscle fibers cut in cross section and others cut in longitudinal section (Figure 10.2).
3. Move to an area of the field of view that contains longitudinally sectioned muscle fibers and switch to high power. Notice that the muscle fibers form branches and that striations are present. Compared to skeletal muscle, the striations are not as well organized and therefore are more difficult to identify. You will have to make minor adjustments with the fine focus knob and look carefully to see them.
4. Identify the darker stained nuclei. Unlike skeletal muscle, the nuclei in cardiac muscle cells are centrally located. In most cases, there is only one nucleus in each cell, but you might find a few multinucleated fibers.
5. Using the high-power objective lens, locate the intercalated discs, which appear as lines traveling across the width of the muscle fibers. They are stained slightly darker than the cytoplasm of cells. By carefully adjusting the fine focus knob, you can see intercalated discs throughout the field of view.

QUESTION TO CONSIDER In the space provided, draw cardiac muscle cells as you view them in a longitudinal section. Label the following structures: (a) muscle fibers, (b) cell nuclei, (c) striations, (d) intercalated discs.

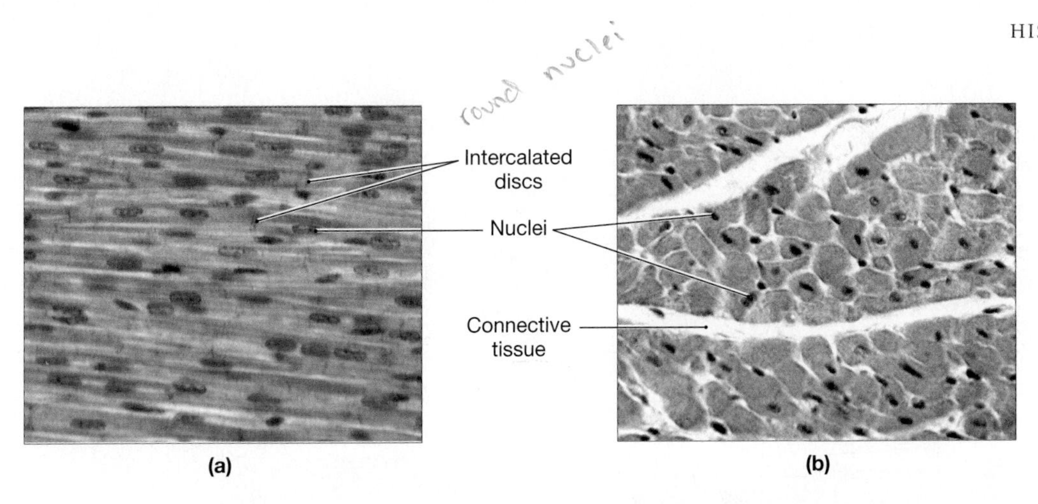

Figure 10.2 Microscopic structure of cardiac muscle. a) Longitudinal section (LM × 400); **b)** cross section (LM × 300).

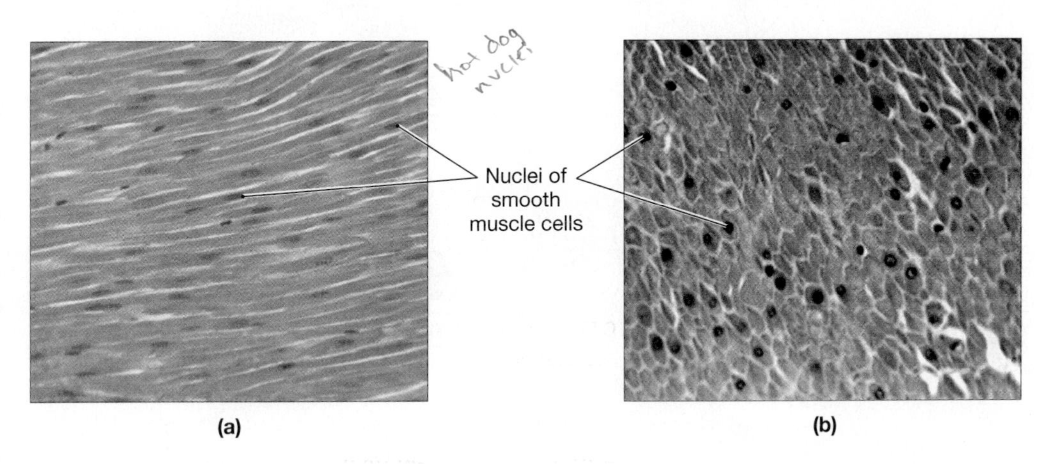

Figure 10.3 Microscopic structure of smooth muscle. a) Longitudinal section (LM × 200); **b)** cross section (LM × 200).

Smooth Muscle

Smooth muscle fibers are widest in the middle and tapered toward each end. They contain one centrally located nucleus and lack striations.

ACTIVITY 10.3 Light Microscopic Structure of Smooth Muscle

1. Obtain a prepared slide of smooth muscle.
2. Scan the slide with the low-power objective lens in position. In the field of view, you will be able to identify bundles of smooth muscle fibers cut in both longitudinal and cross-sectional views (Figure 10.3). Depending on how your slide is prepared, you might also see regions of connective tissue separating areas of muscle fibers.
3. View a longitudinal section of smooth muscle with the high-power objective lens.
4. Note that smooth muscle cells lack striations and contain one centrally located nucleus.
5. Examine an area that contains several layers of cells. The cells are relatively broad in the middle and taper toward the ends. Also observe that they are layered in an overlapping way similar to the arrangement of bricks in a brick wall.

QUESTION TO CONSIDER In the space provided, draw smooth muscle cells as you view them in a longitudinal section. Label the following structures: (a) muscle fibers, (b) cell nuclei.

Skeletal Muscle Cell Structure: A Closer Look

Skeletal muscle cells contain two principal contractile proteins.

- **Actin** is the main component of **thin filaments**.
- **Myosin** is the main component of **thick filaments**.

Together, the thin and thick filaments make up the **myofilaments** of the muscle cell. The myofilaments are packaged into parallel cylindrical bundles, called **myofibrils**, which extend the entire length of the muscle fiber and account for most of the cell's volume (Figure 10.4a). Since the myofilaments fill most of the sarcoplasm (muscle cell cytoplasm), the nuclei are pushed to the periphery of the cell.

Figure 10.4 Structure of a skeletal muscle fiber. a) Section of a muscle fiber illustrating the arrangement of the myofibrils; **b)** relationship of T-tubules and sarcoplasmic reticulum to the myofibrils; **c)** organization of thin and thick filaments in a sarcomere; **d)** electron micrograph of a sarcomere (LM × 20,000).

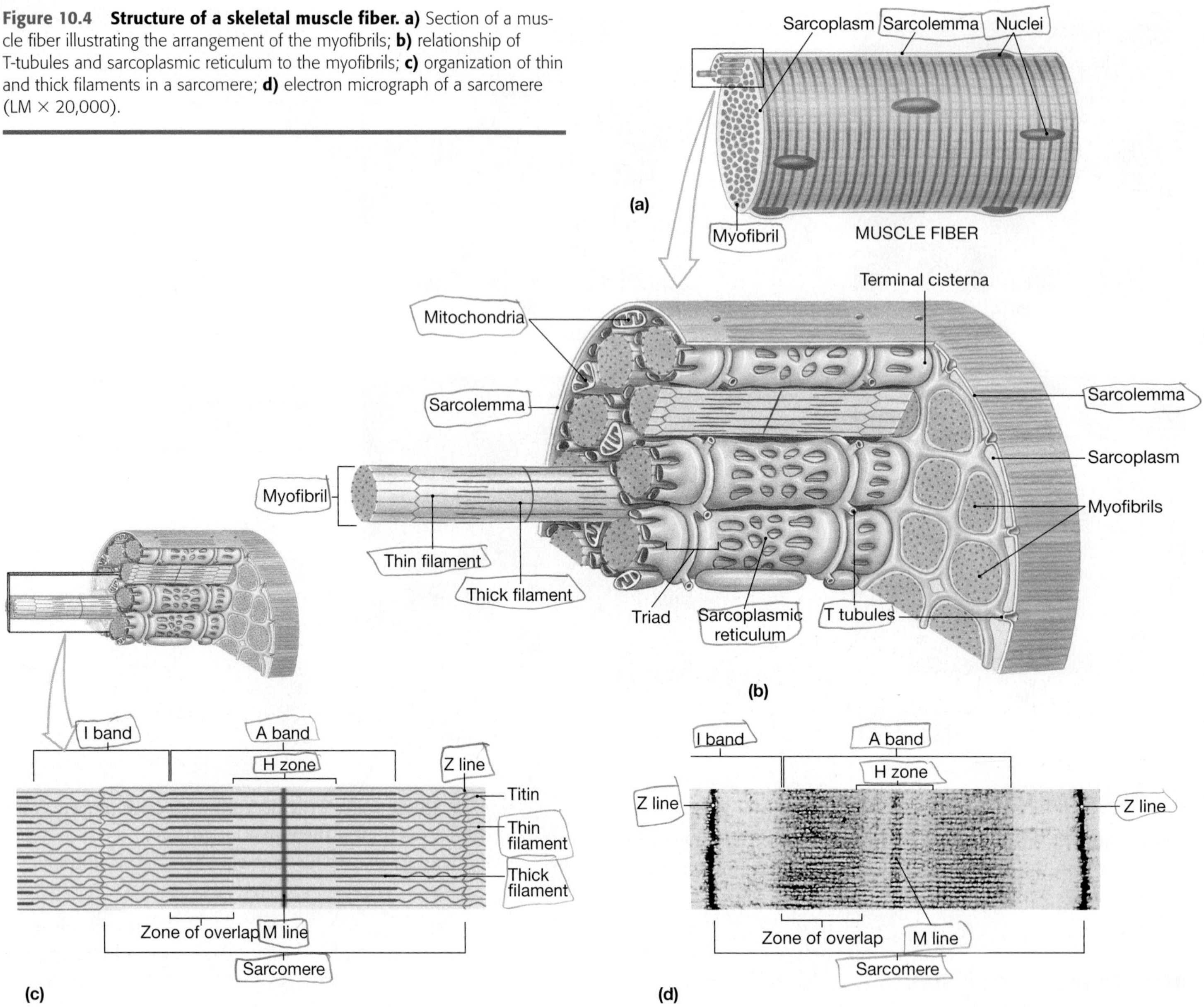

The membrane structures of a skeletal muscle fiber—T-tubules, sarcoplasmic reticulum, and cisterns—surround the myofibrils (Figure 10.4b). When electrical impulses called **action potentials** travel along the membranes of T-tubules, they promote the release of calcium ions from the sarcoplasmic reticulum and cisterns into the sarcoplasm. These actions are necessary to trigger the contraction of the muscle fiber.

ACTIVITY 10.4 Examining Skeletal Muscle Cells in Three Dimensions

1. Obtain a three-dimensional model of a skeletal muscle fiber (Figure 10.4b).
2. Identify the myofibrils, and note the special arrangement of the two types of myofilaments: thin filaments (containing actin) and thick filaments (containing myosin).
3. Identify the A bands, where actin and myosin overlap, and the I bands, where only actin is present. Also identify the Z lines and notice that they bisect the I bands (Figure 10.4c).
4. Locate the positions of two successive Z lines in a myofibril. Note that the interval between these two lines represents one **sarcomere**. Also observe that each myofibril in the muscle fiber is composed of a series of sarcomeres.
5. Identify the following membranous structures that surround the myofibrils (Figure 10.4b).
   - **Sarcolemma**, or cell membrane of the muscle cell
   - **T-tubules**, which are tubelike structures that are continuous with the sarcolemma and travel into the cytoplasm of the cell
   - **Sarcoplasmic reticulum**, which is similar to the smooth endoplasmic reticulum in other cells

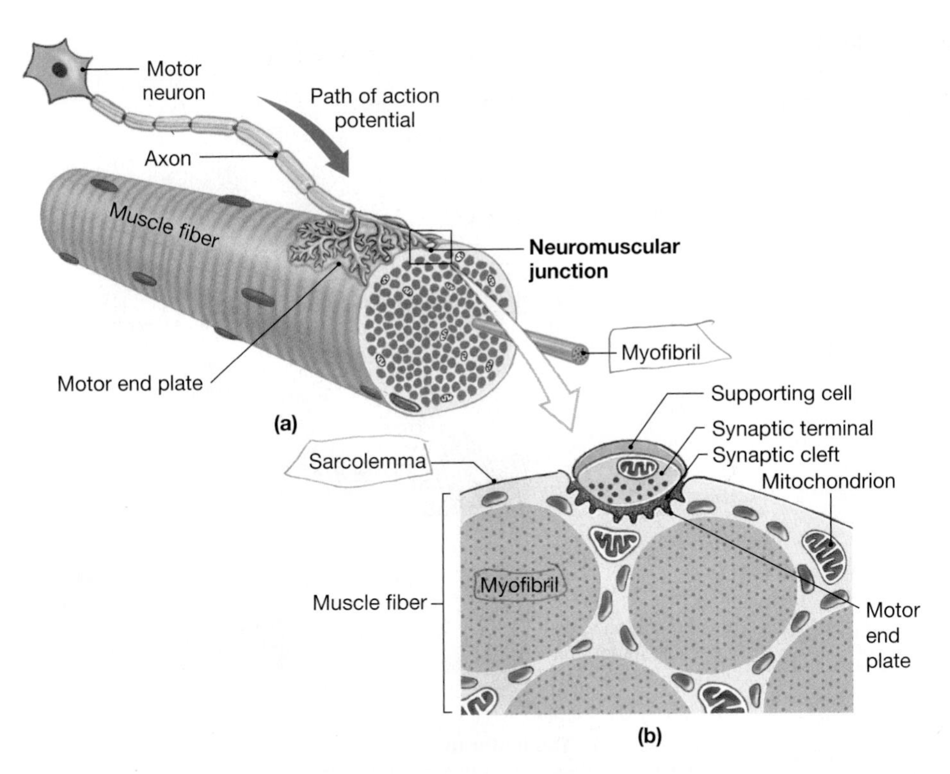

Figure 10.5 The neuromuscular junction.
a) Relationship between a motor neuron and a muscle fiber at a neuromuscular junction; **b)** closer view of a neuromuscular junction.

- **Cisternae** (singular = cistern), which are saclike extensions of the sarcoplasmic reticulum, and are located on either side of a T-tubule. The combination of two cisternae and one T-tubule is called a **triad**.

6. Verify that each muscle fiber contains numerous **mitochondria**.

CLINICAL CORRELATION

Duchenne's muscular dystrophy (DMD) is a genetic disease that is found almost exclusively in males. The disease, which appears during early childhood, results in progressive deterioration and weakness of skeletal muscles. The muscle fibers of individuals with DMD lack or have low amounts of the protein dystrophin that typically is attached to the inside surface of the sarcolemma, adjacent to the triads. Lack of this protein will cause muscle cell membranes to rupture easily and triggers a rapid influx of calcium ions into the sarcoplasm. In children with DMD, the prolonged period of high calcium levels disrupts the normal contraction and relaxation of muscle fibers and causes several muscle proteins to become damaged and nonfunctional. Most individuals afflicted with this disorder die by their late teens due to paralysis of the diaphragm and other respiratory muscles.

Recently, the gene responsible for the production of dystrophin was identified on the X chromosome. Since DMD is a sex-linked disease, a woman can have one abnormal dystrophin gene and not be affected because the defective gene is "masked" by the normal second X chromosome. She can, however, pass the disease on to her sons. Prenatal testing is now available to determine whether a woman possesses a defective dystrophin gene.

Why is it important for muscle cells to have large numbers of mitochondria?

____

____

The Neuromuscular Junction

Motor nerve cells (motor neurons), located in the central nervous system (brain or spinal cord), possess long fibers (axons) that send impulses along spinal nerves and some cranial nerves to reach skeletal muscles. The end of a motor nerve fiber (synaptic terminal) interacts with a muscle fiber at a **neuromuscular junction** (Figure 10.5). At the junction, muscle fibers are stimulated when a **neurotransmitter** substance called **acetylcholine (ACh)** is released from the synaptic terminal on the nerve fiber and binds to receptors on the motor end plate of the muscle cell. This process initiates a muscle contraction by passing the electrical impulse (action potential) from the motor neuron to the muscle fiber. Acetylcholine is then removed from the neuromuscular junction by the enzyme **acetylcholinesterase (AChE)**.

ACTIVITY 10.5 Examining the Neuromuscular Junction at a Skeletal Muscle Cell

1. Obtain a three-dimensional model of a neuromuscular junction (Figure 10.5).
2. Identify the sarcolemma (cell membrane) surrounding the muscle fiber and the myofibrils in the sarcoplasm

(cytoplasm). Identify the I bands, A bands, and Z lines in the myofibrils.

3. Identify the **motor nerve fiber (axon)** as it approaches the neuromuscular junction.
4. At the neuromuscular junction, identify the following structures (Figure 10.5).
   - The **synaptic terminal**, the expanded knob at the end of the nerve fiber
   - The **motor end plate**, the region of the sarcolemma associated with the neuromuscular junction (Notice the complex folding of the motor end plate membrane.)
   - The **synaptic cleft**, the narrow space between the cell membrane of the synaptic terminal and the motor end plate

WHAT'S IN A WORD The prefix *sarco-* is derived from the Greek word *sarx*, meaning "flesh." Any word that begins with *sarco-* refers to "muscular substance" or "resemblance to flesh." Thus, the *sarcolemma* is the cell membrane, the *sarcoplasm* is the cytoplasm, and the *sarcoplasmic reticulum* is the endoplasmic reticulum of a muscle cell.

5. Obtain a prepared microscope slide of a neuromuscular junction and view it with the low-power objective lens in position (Figure 10.6). In longitudinal view, identify the following:
   - The striations in several skeletal muscle fibers
   - A motor nerve fiber and its many branches (Notice that these branches innervate several muscle fibers at neuromuscular junctions.)
   - A **motor unit**, which includes a motor nerve fiber and all the muscle fibers it innervates

QUESTION TO CONSIDER Botulism poisoning is caused by the consumption of canned or smoked food contaminated by the bacteria *Clostridium botulinum*. These bacteria produce a potent toxin that prevents the release of acetylcholine at synaptic terminals. Comment on how botulism will affect the normal activities at neuromuscular junctions. Speculate on why botulism poisoning could be fatal.

____

____

____

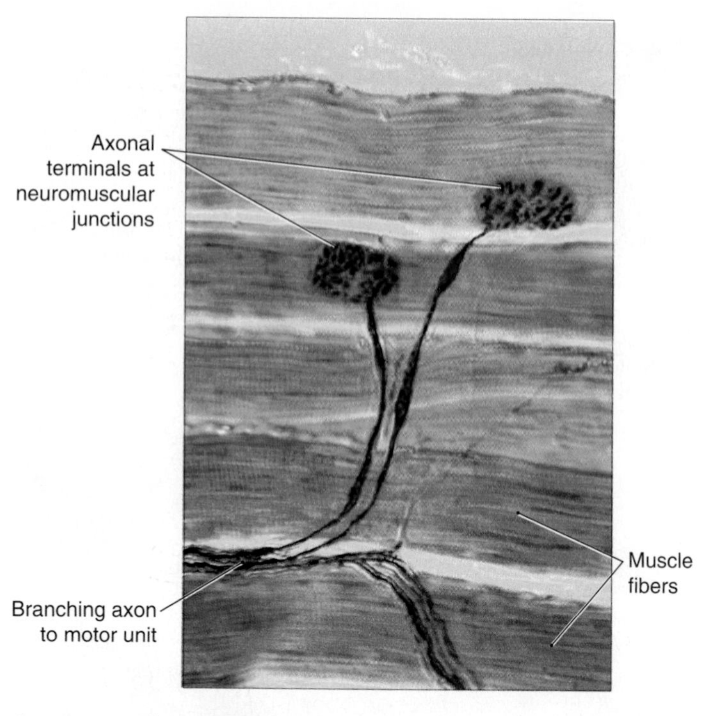

Figure 10.6 The motor unit. An axon from a motor neuron divides into several branches, each with an axon terminal. Each axon terminal forms a neuromuscular junction with a muscle fiber.

Skeletal Muscle Contraction

When a muscle contracts, its filaments do not change in length. Instead, the myofilaments slide along each other and the length of the sarcomeres changes (Figure 10.7). For a muscle fiber to contract, cross bridges of myosin molecules must first attach to binding sites on actin molecules. At rest, however, the binding sites are blocked by **tropomyosin**, an elongated protein that spirals around the actin filaments. Another protein, **troponin**, is attached to tropomyosin to form the **troponin-tropomyosin complex**. When a muscle cell is stimulated, calcium ions are released from the sarcoplasmic reticulum and cisternae into the surrounding sarcoplasm. When they enter the sarcoplasm, calcium ions bind to the troponin molecules. This interaction causes the troponin-tropomyosin complex to shift so that actin-binding sites are exposed and cross bridges on myosin filaments can attach to them. Using cellular energy (ATP), the cross bridges bend and pull on the actin filaments. As a result, the thin filaments slide along the thick filaments. The cross bridges then detach from the actin molecules and return to their original positions to begin a new cycle. This process is referred to as the **sliding filament theory**.

CLINICAL CORRELATION

For a muscle to relax, the calcium ions, which were released into the sarcoplasm prior to contraction, must return to the sarcoplasmic reticulum. This process requires the expenditure of ATP because the calcium ions reenter the sarcoplasmic reticulum by active transport. If ATP is not available, muscles enter a temporary period of continuous contraction or **muscle cramping.** The calcium ion concentration in the sarcoplasm remains high and myosin cross bridges are unable to detach from actin-binding sites.

In this activity, you will be comparing the effects on skeletal muscle contraction when muscle fibers are exposed to (1) ATP alone, (2) potassium and magnesium ions alone, and (3) a com-

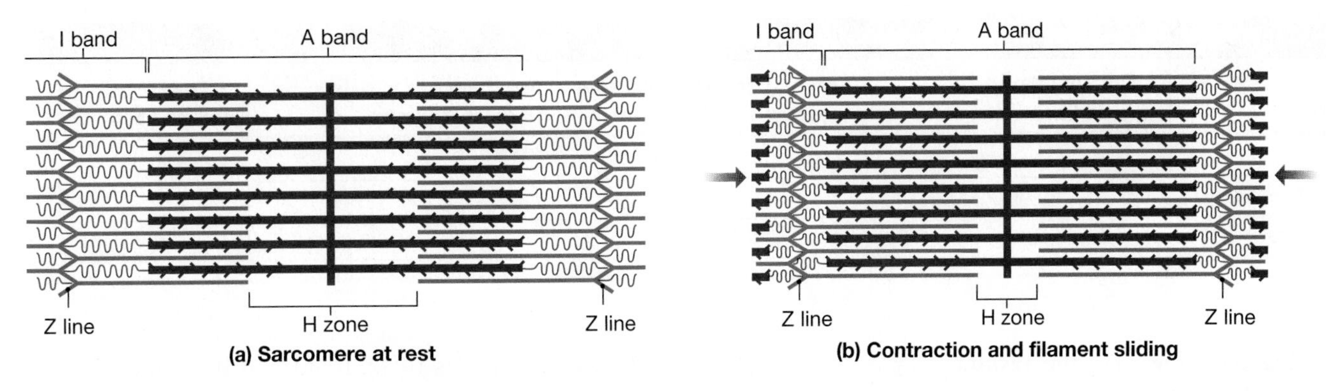

Figure 10.7 Changes that occur in a sarcomere during contraction of a skeletal muscle. a) A sarcomere in a muscle fiber at rest; **b)** a sarcomere in a contracted muscle fiber.

bination of ATP and ions. You will also be able to observe the microscopic structure of fresh (not preserved) skeletal muscle fibers.

You will be using glycerinated skeletal muscle fibers. During the glycerination process, ions and ATP are extracted from the muscle cells and the structure of troponin and tropomyosin is disrupted. The molecular change in the troponin-tropomysin complex keeps the actin binding sites open, so calcium ions are not required. However, potassium and magnesium ions are needed for the normal function of ATPase, the enzyme that is used to extract energy from ATP.

Form a Hypothesis Before you begin this activity, make a prediction about which of the three conditions listed above will result in the greatest contraction of the muscle fibers. Briefly describe the rationale that led to your prediction.

____

____

____

ACTIVITY 10.6 Observing the Contraction of Skeletal Muscle Fibers

1. Obtain a sample of glycerinated skeletal muscle fibers and place it under a dissecting microscope. With clean forceps or dissecting needles, gently tease the muscle fibers apart. Attempt to separate out four groups, each with two or three muscle fibers or, if possible, single muscle fibers. Place each group on a microscope slide and add a small drop of 50% glycerol to each to prevent dehydration. You should have a total of four slides.
2. Perform the following procedures.
   - *Slide 1:* Place a coverslip over the sample and observe the fibers with a compound light microscope. Begin your observations at low power and then move to high power. Identify the striations and the multiple nuclei in each fiber. Record your observations with a drawing in the space provided.
   - *Slide 2:* Without a coverslip, observe the fibers on slide 2 with a dissecting microscope. Place a millimeter ruler under the microscope and measure the length of the relaxed fibers. Record the result in Table 10.1. With the fibers still under the dissecting microscope, add a drop of 0.25% ATP solution and observe any changes that occur. After a period of 45 seconds, measure the length of the fibers and record the result in Table 10.1.
   - *Slide 3:* Repeat the procedure from slide 2, but substitute for the ATP solution an ion solution (a mixture of 0.05M potassium chloride [KCl] and 0.001M magnesium chloride [$MgCl_2$]). Record the lengths of the fibers before and after exposure to the ion solution in Table 10.1.
   - *Slide 4:* Once again, repeat the procedure from slide 2, but this time use a solution containing both ATP and ions. Record the lengths of the fibers before and after exposure to the ATP-ion solution in Table 10.1.
3. Use the following formulas to calculate the net change and the percentage change in the length of the muscle fibers exposed to each solution. Record these results in Table 10.1.
   - Net change = (length of relaxed muscle fibers) − (length of muscle fibers after exposure to solution)
   - Percentage change = (net change/length of relaxed muscle fibers) × 100

Table 10.1 Effect of ATP and Ions on Skeletal Muscle Contraction

Solution	Length of relaxed muscle fibers (mm)	Length of muscle fibers after exposure to solution (mm)	Net change in muscle fiber length (mm)	% Change in muscle fiber
ATP				
Ions ($KCl + MgCl_2$)				
ATP + ions				

Assess the Outcome **Do the results support your earlier prediction? Explain.**

____

____

____

QUESTIONS TO CONSIDER

1. Examine the data that you have collected in Table 10.1. Based on your results, what can you say about the chemical requirements for muscle contraction?

____

____

____

2. Examine Figures 10.4c, 10.4d, and 10.7, and imagine the lengths of the sarcomeres changing during muscle contraction. Which bands or zones will change in length and which will stay the same when contraction occurs?

____

____

____

Exercise 10 Review Sheet

Histology of Muscle Tissue

Name ______________________________

Lab Section ______________________________

Date ______________________________

Questions 1–3: Compare the three types of muscle tissue by completing the following table.

Muscle Type	Cell Structure	Location	Type of Contractions (voluntary/involuntary)	General Function
1. Skeletal				
2. Cardiac				
3. Smooth				

4. Describe how the microscopic striations in muscle fibers are formed by the arrangement of actin and myosin filaments. Draw a diagram to illustrate your written answer.

5. Describe the structure of the neuromuscular junction. Draw a diagram to accompany your written response.

6. Describe the events that occur at a neuromuscular junction when an action potential is transferred from a motor nerve fiber to a muscle cell.

7. Explain the role of the following structures in muscle cell contraction.

   a. Sarcolemma: ______________________________

   b. T-tubules: ______________________________

   c. Cisterns: ______________________________

   d. Sarcoplasmic reticulum: ______________________________

   e. Calcium ions: ______________________________

   f. ATP: ______________________________

Exercise 11

Gross Anatomy of the Muscular System

Laboratory Objectives

On completion of the activities in this exercise, you will be able to:

- Identify the bony attachments of skeletal muscles.
- Explain how skeletal muscles are organized according to body region or muscular compartments.
- Identify individual skeletal muscles and skeletal muscle groups of the human body.
- Understand the functions of individual muscles and muscle groups.

Materials

- Skulls
- Hand mirrors
- Anatomical models, charts, and/or atlases of the human body
- Human skeletons, or disarticulated human bones
- Pieces of string of various lengths ranging between 5 cm and 60 cm
- Masking tape
- Metric ruler

Although there are three types of muscle tissue (skeletal, cardiac, and smooth), the **muscular system** refers only to the skeletal muscles. The basic functional unit of skeletal muscle is the **muscle fiber** (a muscle fiber is the same as a muscle cell). Each fiber is enclosed by a layer of connective tissue called the **endomysium**. Fibers are organized into bundles called **muscle fascicles**, and each fascicle is enclosed by a layer of connective tissue called the **perimysium**. Muscle fascicles are grouped together to form a whole muscle. Each **whole muscle**, such as the biceps brachii in the arm, is enclosed by a connective tissue layer called the **epimysium** (Figure 11.1).

Most skeletal muscles are attached to bones by dense regular connective tissue in the form of cordlike **tendons** or membranous sheets called **aponeuroses**. With few exceptions, muscles cross one or more joints along their paths. When a muscle contracts, it causes an action at the joint(s) that it crosses. When an action occurs, one bony attachment, the **origin**, remains fixed or stationary while the other attachment, the **insertion**, will move. In the upper and lower extremities, the origin is usually proximal to the insertion.

Muscles can be classified according to the functions that they serve. A **prime mover**, or **agonist**, is a muscle that directly brings about a specific action; an **antagonist** muscle directly opposes that action. For example, consider flexion and extension of the elbow. For flexion, brachialis and biceps brachii are prime movers, while triceps brachii is the antagonist. However, for extension, the roles are reversed: Triceps brachii is the prime mover and brachialis and biceps brachii are the antagonists.

Muscles can also act as **synergists** (which means "working together") by promoting or assisting in a specific action, or by reducing unnecessary movements while the action is performed. For example, when you flex your fingers to make a fist, the wrist extensor muscles act as synergists by preventing wrist flexion.

A special class of synergist muscles, known as **fixators**, stabilizes joints or muscle origins so prime movers can act more efficiently. For example, muscles that attach the scapula to the axial skeleton, such as pectoralis minor, serratus anterior, and the rhomboids, stabilize the scapula and shoulder joint while other muscles move the arm.

Any given muscle can be involved in or influence several actions, and the functional role of a muscle depends on the specific movement under consideration. Thus, a muscle that is a synergist for one action can be a prime mover or antagonist for another. Keep this concept in mind as you identify the muscles and consider their actions.

The following laboratory activities will help you study the gross anatomy of the muscular system. Most of these follow a similar pattern. First, you will locate the muscles' bony attachments (origins and insertions) and trace the muscles' paths by running a piece of string between the connections.

To do this, you must locate important bony landmarks. If you do not recall the location of any of these, refer to an appropriate figure in either Exercise 7 or 8 of this lab manual, or look it up in your textbook or an anatomy atlas.

Next, you will identify the muscles by examining anatomical models, reviewing diagrams from an anatomy atlas or chart, performing a dissection, or exploring anatomy and physiology websites and CD-ROMs, depending on your instructor's directions.

Finally, you will analyze the muscles' actions either by performing them yourself or by observing as your lab partner performs them.

Although this lab is quite lengthy, if you do these activities carefully and with patience, you will learn how individual muscles are positioned and will appreciate how muscle groups are arranged and how they function.

Muscles of the Head

The muscles of the head are organized into different groups. Muscles of facial expression allow us to move our facial features. Muscles of mastication allow us to chew food. Extrinsic eye muscles control the movements of our eyes, and extrinsic tongue muscles control voluntary movements of the tongue, allowing us to eat and speak. Finally, muscles of the pharynx are involved in swallowing.

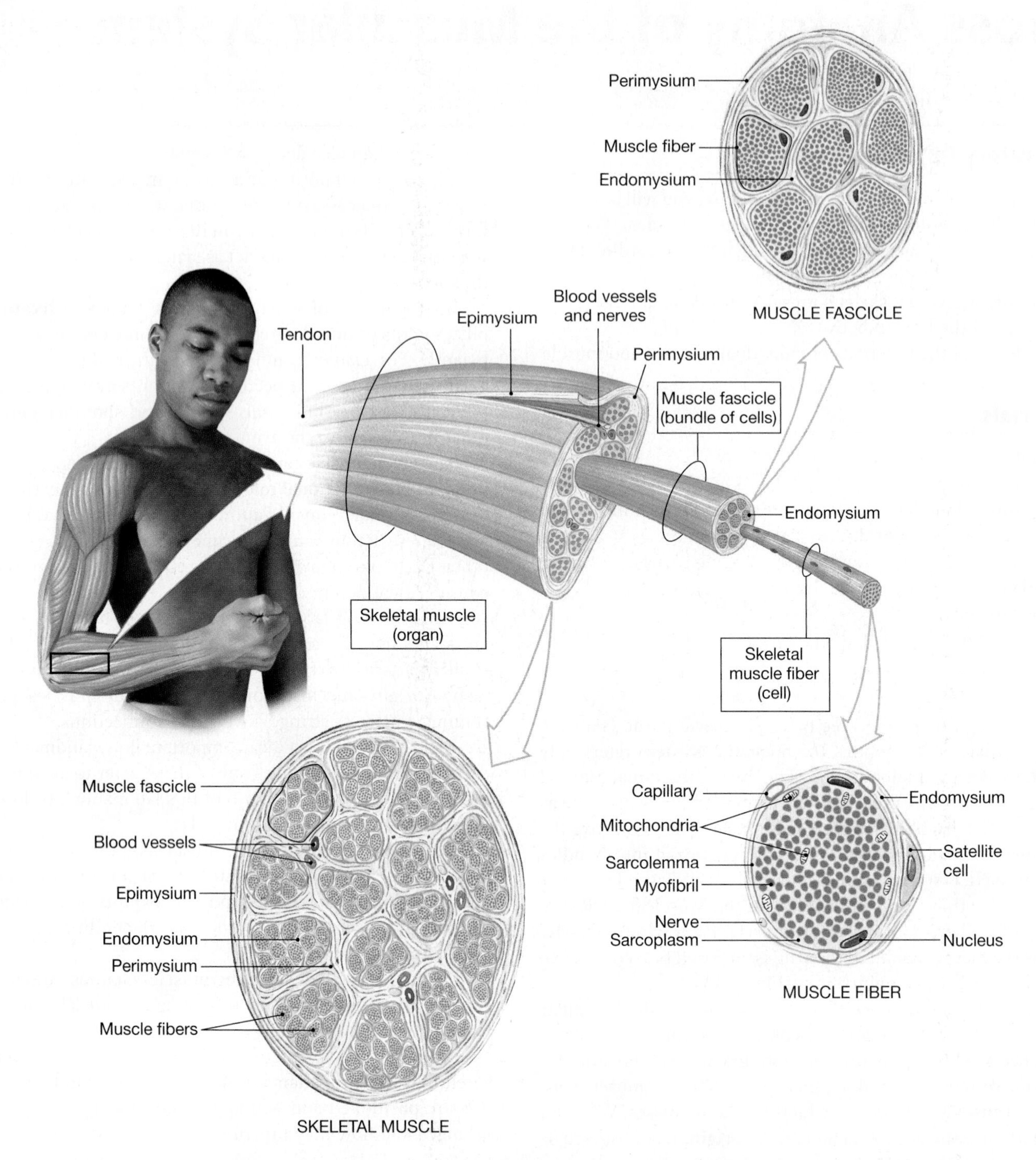

Figure 11.1 Levels of organization of a whole skeletal muscle. A whole muscle is enclosed by the epimysium and contains numerous bundles of muscle fibers, called muscle fascicles. Each fascicle is enclosed by the perimysium. Each muscle fiber in a fascicle is enclosed by the endomysium.

ACTIVITY 11.1 Identifying Muscles of the Head

Muscles of Facial Expression

The muscles of facial expression are located deep to the skin of the face, neck, and scalp. Most originate on bone and insert into the skin. All muscles of facial expression are innervated by the facial nerve (cranial nerve VII). The actions of these muscles are described in Table 11.1A, and illustrated in Figure 11.2.

1. Obtain a skull and review the arrangement of the facial bones. Recall that the bones articulate at sutures, which are immovable joints. Thus, rather than initiating movement at

Table 11.1 Muscles of the Head

Muscle Group/Muscle	Origin	Insertion	Action	Innervation
A. Muscles of facial expression				
1. Occipitofrontalis				
a. Frontal belly	Epicranial aponeurosis (aponeurosis covering parietal bone)	Skin of eyebrow and bridge of nose	Raises eyebrows; wrinkles forehead	Facial nerve (cranial nerve VII)
b. Occipital belly	Superior nuchal line on occipital bone	Epicranial aponeurosis	Pulls scalp posteriorly; wrinkles skin on posterior surface of neck	As above
2. Procerus	Nasal bones; lateral nasal cartilages	Bridge of nose; skin on forehead	Depresses eyebrow; wrinkles skin over bridge of nose	As above
3. Nasalis	Maxillary bone; alar cartilage of nose	Bridge of nose	Compresses nostrils; dilates nostrils	As above
4. Orbicularis oculi	Medial aspect of bony orbit	Skin around eyelids	Closes eyelid; acts during squinting and blinking	As above
5. Orbicularis oris	Maxillary bone; mandible	Lips	Closes and protrudes lips as in speaking, kissing, and whistling	As above
6. Buccinator	Maxillary bone; mandible	Blends with fibers of orbicularis oris	Compresses cheek while whistling or blowing	As above
7. Zygomaticus major	Zygomatic bone	Angle of mouth	Raises lateral corner of mouth for smiling	As above
8. Zygomaticus minor	Zygomatic bone	Upper lip	Raises upper lip	As above
9. Levator labii superioris	Inferior aspect of bony orbit	Blends with fibers of orbicularis oris	Elevates upper lip; dilates nostrils	As above
10. Depressor anguli oris	Anterolateral surface of mandible	Skin at angle of mouth	Depresses angle of mouth	As above
11. Depressor labii inferioris	Anterolateral surface of mandible	Skin of lower lip	Depresses lower lip	As above
12. Mentalis	Anterior surface of mandible	Skin of chin	Protrudes lower lip; wrinkles skin on chin	As above
13. Platysma	Superior thoracic wall	Mandible and skin of cheek	Tenses skin of the neck	As above
B. Muscles of mastication				
1. Masseter	Zygomatic arch	Lateral surface of mandibular ramus	Elevates mandible to close jaw; protrudes mandible	Mandibular branch of trigeminal nerve (cranial nerve V)
2. Temporalis	Temporal bone	Coronoid process of mandible	Elevates mandible to close jaw; retrudes mandible	As above
3. Medial pterygoid	Lateral pterygoid plate	Medial surface of mandibular ramus	Elevates mandible to close jaw; protrudes mandible; promotes side-to-side grinding movements	As above
4. Lateral pterygoid	As above	Neck of mandible	Protrudes mandible; depresses mandible to open jaw; promotes side-to-side grinding movements	As above

(continues)

Table 11.1 (continued)

Muscle Group/Muscle	Origin	Insertion	Action	Innervation
C. Extrinsic eye muscles				
1. Inferior rectus	Tendinous ring attached to posterior wall of bony orbit	Inferior surface of eyeball	Rotates eyeball Inferiorly	Oculomotor nerve (cranial nerve III)
2. Medial rectus	As above	Medial surface of eyeball	Rotates eyeball medially	As above
3. Superior rectus	As above	Superior surface of eyeball	Rotates eyeball superiorly	As above
4. Lateral rectus	As above	Lateral surface of eyeball	Rotates eyeball laterally	Abducens nerve (cranial nerve VI)
5. Superior oblique	As above	Superolateral surface of eyeball	Rotates eyeball Inferiorly and laterally	Trochlear nerve (cranial nerve IV)
6. Inferior oblique	As above	Inferolateral surface of eyeball	Rotates eyeball superiorly and laterally	Oculomotor nerve (cranial nerve III)
D. Extrinsic tongue				
1. Genioglossus	Medial surface of mandible	Tongue	Depresses and protracts tongue muscles	Hypoglossal nerve (cranial nerve XII)
2. Hyoglossus	Hyoid bone	As above	Depresses tongue	As above
3. Styloglossus	Styloid process	As above	Elevates and retracts tongue	As above
4. Palatoglossus	Soft palate	As above	Elevates tongue; depresses soft palate	Cranial branch of accessory nerve (cranial nerve XI)
E. Muscles of the Pharynx				
1. Muscles of the soft palate				
a. Levator veli palatini	Temporal bone on base of skull	Soft palate	Elevates soft palate	Accessory nerve (cranial nerve XI)
b. Tensor veli palatini	Sphenoid bone on base of skull	As above	Tenses the soft palate	Trigeminal nerve (cranial nerve V)
2. Pharyngeal constrictors				
a. Superior constrictor	Sphenoid bone; mandible	Median raphe (connective tissue band along posterior wall of pharynx)	Constricts the pharynx to move ingested food into the esophagus	Vagus nerve (cranial nerve X)
b. Middle constrictor	Hyoid bone	As above	As above	As above
c. Inferior constrictor	Thyroid and cricoid cartilages of larynx	As above	As above	As above
3. Laryngeal elevators	Soft palate; auditory tube; styloid process	Thyroid cartilage	Elevates the larynx	Glossopharyngeal nerve (cranial nerve IX); vagus nerve (cranial nerve X)

a joint, the muscles of facial expression move the skin on the head and neck. These actions change facial expressions as moods change.

2. Use an anatomical model, chart, or atlas to identify the following muscles of facial expression (Figure 11.2; Table 11.1A): frontalis, occipitalis, procerus, nasalis, orbicularis oculi, orbicularis oris, zygomaticus major and minor, levator labii superioris, depressor anguli oris, depressor labii inferioris, mentalis, and platysma.

3. Examine the following muscle actions by observing yourself in a mirror or by watching your lab partner.
   - Raise your eyebrows and notice that the skin of your forehead wrinkles. This action is produced by the frontal belly of the **occipitofrontalis muscle**, which covers the forehead.
   - Close one eye by blinking or squinting. This movement is initiated by the **orbicularis oculi**, a circular muscle that surrounds the eye.

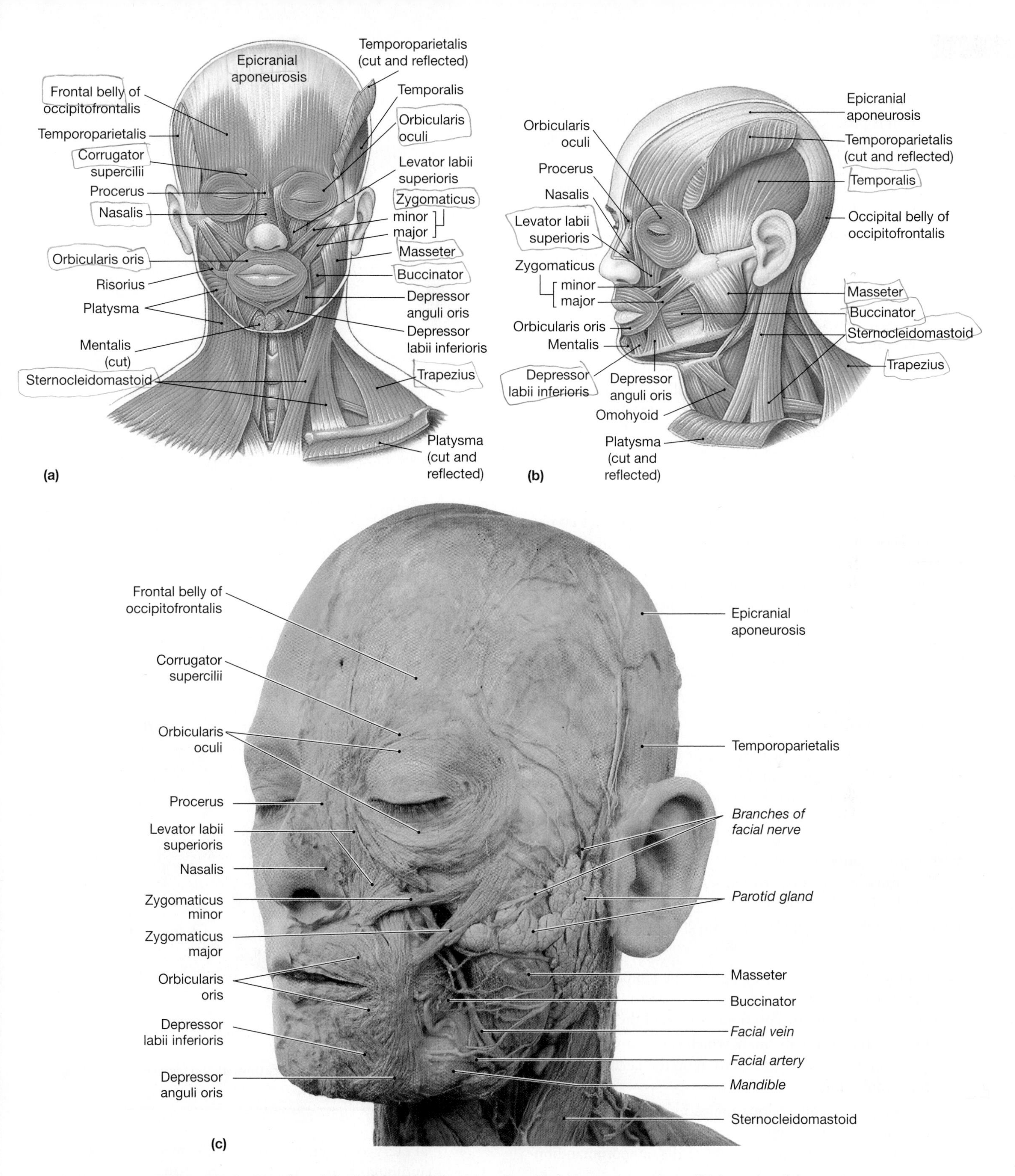

Figure 11.2 Muscles of the head and neck. a) Muscles of facial expression and superficial muscles of the neck, anterior view; **b)** muscles of the head and neck, anterolateral view; **c)** cadaver prosection of the muscles of facial expression, anterolateral view.

- Compress your cheeks by whistling or blowing. This action is produced by the **buccinator** muscle, which is the relatively large muscle of facial expression that can be palpated in your cheek wall.
- When you smile, **zygomaticus major** is the primary muscle that elevates the corners of your mouth.
- When you frown, **depressor anguli oris** is the primary muscle that depresses the corners of your mouth.
- While you speak with your lab partner, your lips are closed and protruded by the **orbicularis oris**, a circular muscle that surrounds your mouth. The orbicularis oris is also active during kissing and whistling.
- As you grit your teeth, palpate the skin along the anterior surface of your neck. Notice that the skin is very tense. This is caused by the contraction of the **platysma**, which extends from the superficial fascia of your pectoral region to your mandible, the skin on your cheek, and the angle of your mouth.

QUESTION TO CONSIDER Bell's palsy is a viral infection that causes an inflammation of the facial nerve (cranial nerve VII), and temporary paralysis of facial muscles. Speculate on what might occur if Bell's palsy affected the left facial nerve of an individual, but the right facial nerve functioned normally.

____

____

Muscles of Mastication

The four muscles of mastication are involved in chewing food. They act on the temporomandibular joint (TMJ) to elevate, depress, protract, and retract the mandible. The two large superficial muscles in this group include the **temporalis** and **masseter** (Figure 11.3a); the two deeper muscles are the **lateral** and **medial pterygoids** (Figure 11.3b).

The muscles of mastication are all innervated by the mandibular branch of the trigeminal nerve (cranial nerve V). Their actions and bony attachments are described in Table 11.1B.

1. On the skull, identify the origin of the temporalis along the surface of the temporal fossa. Next identify the muscle's insertion on the coronoid process of the mandible.
2. Identify the zygomatic arch, which is the origin of the masseter, then locate its insertion on the lateral part of the mandibular ramus.
3. Place a pen or pencil along the paths of these two muscles by connecting the origins and insertions for each. Notice that both muscles cross anterior to the temporomandibular joint (TMJ).
4. Use a piece of string to trace the path of the two pterygoid muscles. Both originate from the lateral pterygoid plate on the sphenoid. The lateral pterygoid runs almost horizontally to attach to the neck of the mandible, while the medial pterygoid runs inferiorly to the medial part of the mandibular ramus. Verify that the string crosses the TMJ.
5. Use an anatomical model, chart, or atlas to identify the muscles of mastication and verify the following:
   - The temporalis covers the temporal fossa. It has a broad origin, but tapers to a narrow insertion (Figure 11.3a).

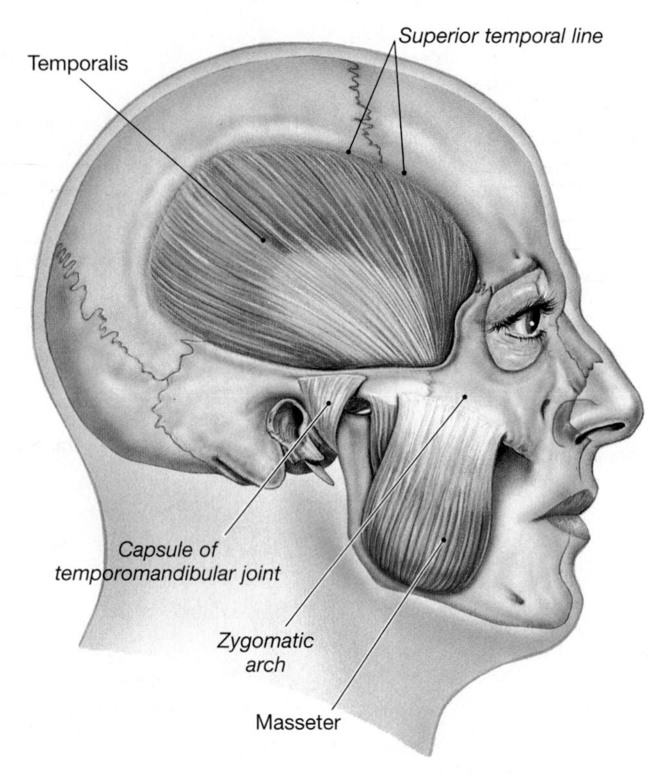

(a) Lateral view

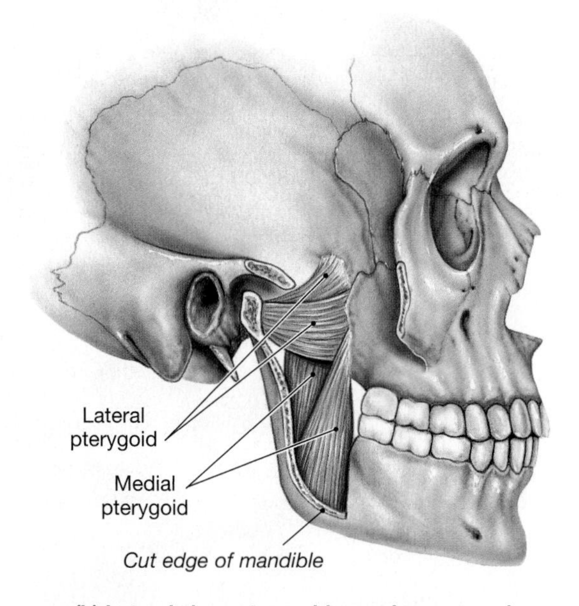

(b) Lateral view, pterygoid muscles exposed

Figure 11.3 Muscles of mastication. a) Superficial muscles, lateral view; **b)** deep muscles, lateral view.

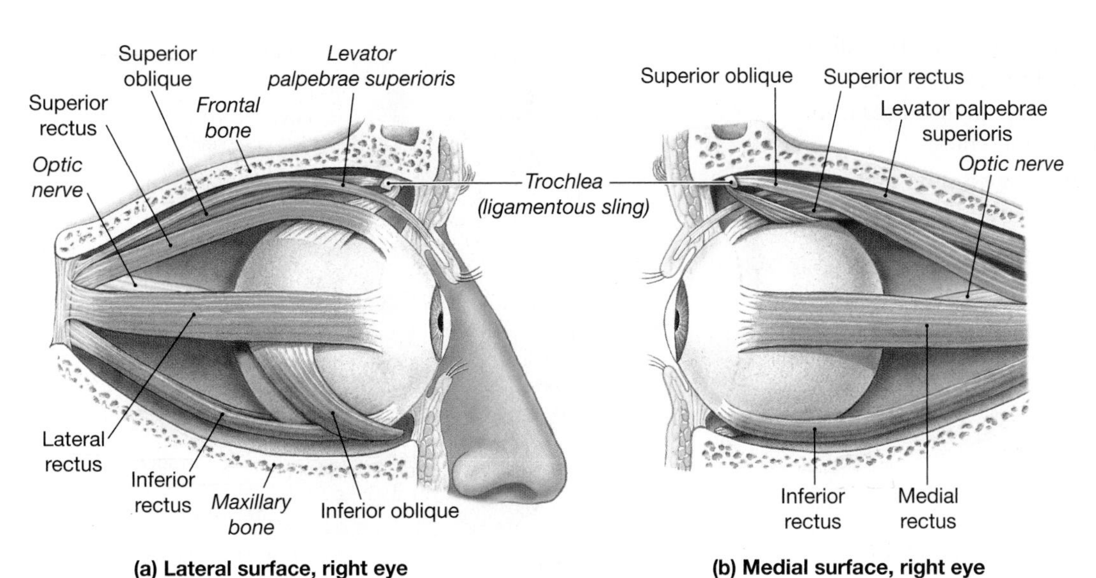

Figure 11.4 Extrinsic muscles of the right eye. a) Lateral view; **b)** medial view.

- The masseter is a relatively short but bulky muscle that connects the zygomatic arch to the mandible (Figure 11.3a).
- The pterygoids are small muscles that lie deep to the temporalis and masseter (Figure 11.3b).

6. Starting with your mouth opened wide, place two fingers of one hand on your temporal fossa. Elevate your mandible so that your teeth come together. As you perform this action, you should feel the temporalis muscle contract. Repeat this action, but put your fingers on your cheek, just anterior to your ear and inferior to the zygomatic arch. As you elevate your mandible, palpate the masseter muscle as it contracts.

WHAT'S IN A WORD The masseter is a very strong chewing muscle. The name of this muscle is derived from the Greek word for "chewer."

Pterygoid is derived from Greek and means "wing shaped." Observe the medial and lateral pterygoid muscles and note that they form winglike muscular suspensions between the sphenoid bone and mandible.

QUESTION TO CONSIDER Explain why the force of gravity will assist in depressing the mandible but will be antagonistic to elevating the mandible.

____

____

____

Based on your answer to the previous question, explain why three of the four muscles of mastication—the temporalis, masseter, and medial pterygoid—are involved in elevation of the mandible, but only one muscle, the lateral pterygoid, is involved in depression of the mandible.

____

____

____

Extrinsic Eye Muscles

The six extrinsic eye muscles (Figure 11.4; Table 11.1C) are responsible for voluntary eye movements. Innervation to these muscles is supplied by three cranial nerves: oculomotor (III), trochlear (IV), and abducens (VI).

1. On a skull, identify the two bony orbits. Observe the opening of the optic canal along the posterior wall. All the extrinsic eye muscles originate from a tendinous ring that surrounds this opening. From their origins, these small muscles insert onto the outer layer of the eyeball (Figure 11.4).
2. Use an anatomical model, chart, or atlas to identify the extrinsic eye muscles.
   - The **medial rectus** (plural = **recti**) attaches to the medial surface of the eyeball and moves the eye medially (inward).
   - The **lateral rectus** inserts on the lateral surface of the eyeball and moves the eye laterally (outward).
   - The **superior rectus** runs across the superior surface of the eyeball but at a slight angle, so it not only moves the eyeball superiorly but also rotates it medially.
   - The **inferior rectus** runs along the inferior surface of the eyeball, also at a slight angle, and moves the eye inferiorly and medially.
   - The **superior oblique** runs along the superior surface of the eyeball, angling in toward the nose. At this position, the muscle's tendon travels through a pulleylike

structure called the trochlea before curving laterally to insert on the superior surface of the eye. Be sure to look at its path on both eyes. Its unusual path causes it to rotate the eye both inferiorly and laterally.

- The **inferior oblique** runs rather like a sling under the eyeball, and its contraction rotates the eye both superiorly and laterally.

QUESTION TO CONSIDER The abducens nerve (cranial nerve VI) innervates the lateral rectus muscle. If the left abducens nerve is paralyzed, speculate on the primary effects to the individual.

___

___

___

Extrinsic Tongue Muscles

The four extrinsic tongue muscles (Table 11.1D; Figure 11.5) are involved in voluntary tongue movements during chewing, swallowing, and speaking. All are innervated by the hypoglossal nerve (cranial nerve XII).

1. On the skeleton, identify the bony origins of the following muscles.
   - **Genioglossus** originates from the medial surface of the mental (chin) area of the mandible.
   - **Styloglossus** originates from the styloid process of the temporal bone.
   - **Hyoglossus** originates from the hyoid bone.
   - **Palatoglossus** originates from the soft palate, which is a region of soft tissue extending posteriorly from the hard palate. It is not present on the skeleton but you can locate it on a model or illustration. To approximate the position of the soft palate, locate the posterior margin of the hard palate on the skull.
2. Hold a piece of string to each muscle origin and extend it to where the tongue would be. The various lengths of string approximate the paths of the four extrinsic tongue muscles.

WHAT'S IN A WORD The names of the extrinsic tongue muscles provide clues for identifying their attachment sites. The term *glossus* refers to the tongue, where all four muscles insert.

The terms *genio, stylo, palato,* and *hyo* refer to the mandible, styloid process, soft palate, and hyoid bone, respectively, which are the origins for these muscles. If you consider genioglossus, for example, you can predict from the name that it originates on the mandible (genio) and inserts on the tongue (glossus).

3. Use an anatomical model, chart, or atlas to identify the four extrinsic tongue muscles listed in step 1 (Figure 11.5), and verify the following:
   - Palatoglossus has a soft tissue origin and insertion.
   - Genioglossus is a fan-shaped muscle that comprises the bulk of the tongue.

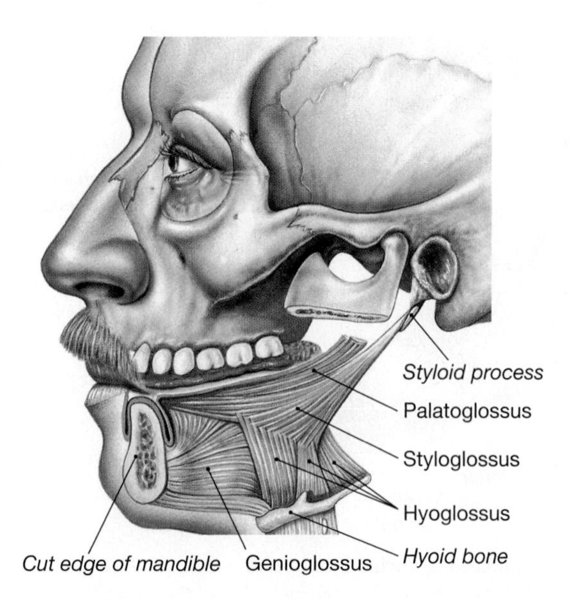

Figure 11.5 Extrinsic muscles of the tongue, lateral view. These muscles are involved in voluntary tongue movements during chewing, swallowing, and speaking.

- The styloglossus connects the tongue to the base of the skull (styloid process).
- The hyoid bone, where hyoglossus originates, is inferior to the mandible

QUESTION TO CONSIDER Identify some daily activities that require normal functioning of the extrinsic tongue muscles.

___

___

Muscles of the Pharynx

The muscles associated with the pharynx (Figure 11.6) are necessary for swallowing. They are organized into three groups.

- *Muscles of the soft palate* include **levator veli palatini** and **tensor veli palatini** and elevate and tighten the soft palate during swallowing.
- *Pharyngeal constrictors* include the superior, middle, and inferior pharyngeal constrictors and form a circular layer of muscle around the wall of the pharynx.
- *Laryngeal elevators* elevate the larynx during swallowing. The actions, bony attachments, and innervations of the muscles of the pharynx are summarized in Table 11.2E.

1. Use an anatomical model, chart, or atlas to identify the muscles of the pharynx. Note that many of these muscles (Figure 11.6) support the larynx by suspending it from the base of the skull.
2. Observe your lab partner while he or she is swallowing. Notice that during the swallowing process, the larynx is elevated toward the pharynx. You can observe this action by

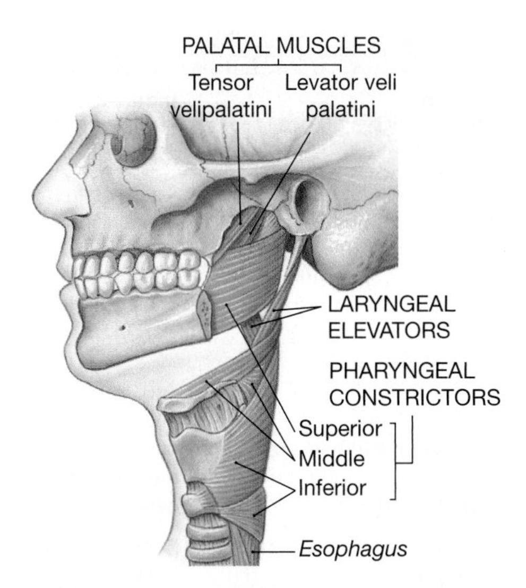

Figure 11.6 Muscles of the palate, larynx, and pharynx, lateral view. These muscles are active during swallowing.

identifying the thyroid cartilage along the anterior aspect of the neck and watching it move superiorly during swallowing. The act of swallowing also involves the actions of muscles that elevate the soft palate and constrict the pharynx.

QUESTION TO CONSIDER Elevation of the soft palate during swallowing closes the passageway that connects the pharynx to the nasal cavity. Why is it essential for this action to occur while eating or drinking?

____

____

Muscles of the Neck

The most prominent muscle in the neck is the **sternocleidomastoid**, which is a relatively long and superficial straplike muscle (Figure 11.2). It is an important anatomical landmark because it separates the neck into two triangular-shaped regions known as the **anterior** and **posterior triangles**.

CLINICAL CORRELATION

The sternocleidomastoid is innervated by the spinal root of the accessory nerve (cranial nerve XI). The branches that supply the sternocleidomastoid travel through the posterior triangle of the neck and are vulnerable to injury during surgical procedures in this area (e.g., a biopsy of lymph nodes). Damage to the accessory nerve can lead to weakness or loss of muscle function. Symptoms include difficulty in rotating the head to the opposite side or laterally flexing the neck to the same side.

ACTIVITY 11.2 Identifying Muscles of the Neck

Sternocleidomastoid

The bony attachments, innervation, and function of the sternocleidomastoid are described in Table 11.2A.

1. On the skeleton, identify the two origins of the sternocleidomastoid: the manubrium of the sternum and the medial end of the clavicle.
2. Identify the muscle's insertion on the mastoid process of the temporal bone.
3. Connect the origins and insertion with two pieces of string. The position of the strings marks the boundary between the anterior and posterior triangles of the neck.
4. Use an anatomical model, chart, or atlas to identify the sternocleidomastoid and note the following (Figure 11.2):
   - The sternocleidomastoid lies just deep to the platysma.
   - The sternocleidomastoid separates the anterior and posterior triangles of the neck.
   - Traveling just deep to the sternocleidomastoid are the **common carotid artery** and **internal jugular vein.**

Muscles of the Anterior Triangle

Muscles of the anterior triangle include a superficial and a deep group. The superficial group has two subdivisions.

- The **suprahyoid muscles** (Figure 11.7a) are superior to the hyoid bone and connect it to the skull. They are the **mylohyoid**, **geniohyoid**, **stylohyoid**, and **digastric.**
- The **infrahyoid muscles** (Figure 11.7a) are inferior to the hyoid bone and connect it to the sternum, clavicle, and scapula. They are the **sternohyoid**, **sternothyroid**, **thyrohyoid**, and **omohyoid.**

The deep muscles of the anterior triangle connect to the base of the skull and vertebrae C1 through T3. The two primary muscles in this group are the **longus capitis** and the **longus colli** (Figure 11.7b). The bony attachments, innervations, and functions of the anterior triangle muscles are outlined in Table 11.2B.

1. On the skeleton, locate the hyoid bone, the common insertion site for most of the superficial muscles. Identify the origins of the suprahyoid muscles on the mandible and temporal bone.
   - Mylohyoid originates from the body of the mandible, at the mylohyoid line.
   - Geniohyoid originates from the mandible at the chin.
   - Stylohyoid originates at the styloid process of the temporal bone.
   - Digastric has two heads of origin. The posterior belly arises from the mastoid process of the temporal bone; the anterior belly arises from the mandible at the chin.

Table 11.2 Muscles of the Neck

Muscle Group/Muscle	Origin	Insertion	Action	Innervation
A. Superficial muscle				
Sternocleidomastoid	Manubrium; medial one third of clavicle	Mastoid process; occipital bone near mastoid process	Flexion and lateral flexion of neck; rotates head so that face is turned superiorly and to the opposite side	Accessory nerve (cranial nerve XI); spinal nerves C2 and C3
B. Muscles of the anterior triangle				
Superficial muscles				
Suprahyoid muscles				
1. Mylohyoid	Inferior margin of mandible at the chin	Hyoid bone	Elevates hyoid bone, tongue, and floor of mouth during swallowing and speaking	Mandibular branch of trigeminal nerve (cranial nerve V)
2. Geniohyoid	As above	As above	Elevates the hyoid bone and pulls it anteriorly during swallowing and speaking	Cervical spinal nerves (C1)
3. Stylohyoid	Styloid process	As above	Elevates and retracts hyoid bone during swallowing and speaking	Facial nerve (cranial nerve VII)
4. Digastric	*posterior belly* from mastoid process; *anterior belly* from inferior margin of mandible	As above	Depresses mandible; elevates hyoid bone during swallowing and speaking	Trigeminal nerve (cranial nerve V) to anterior belly; facial nerve (cranial nerve VII) to posterior belly
Infrahyoid muscles				
1. Sternohyoid	Manubrium	As above	Depresses larynx and hyoid bone during swallowing	Cervical spinal nerves (C1–C3)
2. Sternothyroid	Manubrium	Thyroid cartilage of larynx	Depresses larynx during swallowing	As above
3. Thyrohyoid	Thyroid cartilage of larynx	Hyoid bone	Depresses hyoid bone; elevates larynx	Cervical spinal nerves (C1–C2)
4. Omohyoid (superior belly)	Clavicle and first rib	As above	Depresses and retracts hyoid bone	Cervical spinal nerves (C2–C3)
Deep muscles				
1. Longus capitis	Transverse processes of cervical vertebrae	Occipital bone at the base of skull	Flexes the head	Cervical spinal nerves (C1–C3)
2. Longus colli	Bodies of cervical and superior thoracic vertebrae	Tranverse processes of superior cervical vertebrae	Flexes the neck	Cervical spinal nerves (C2–C6)
C. Muscles of the posterior triangle				
1. Splenius	Spinous processes of inferior cervical and superior thoracic vertebrae	Mastoid process, occipital bone and transverse processes of superior cervical vertebrae	Laterally flexes, rotates, and extends head and neck	Cervical spinal nerves (C3–C5)
2. Scalenes (posterior, middle, and anterior)	Transverse processes of cervical vertebrae	First and second ribs	Lateral flexion of neck; elevates ribs 1 and 2 during forced inspiration	Cervical spinal nerves (C4–C8)
3. Levator scapulae	Transverse processes of superior four cervical vertebrae	Medial border of scapula near superior angle	Elevates scapula; Inferior rotation of scapula	Dorsal scapular nerves (C4–C5)
4. Omohyoid (inferior belly)	Clavicle and first rib	Superior margin of scapula	Depresses and retracts hyoid bone	Cervical spinal nerves (C2–C3)

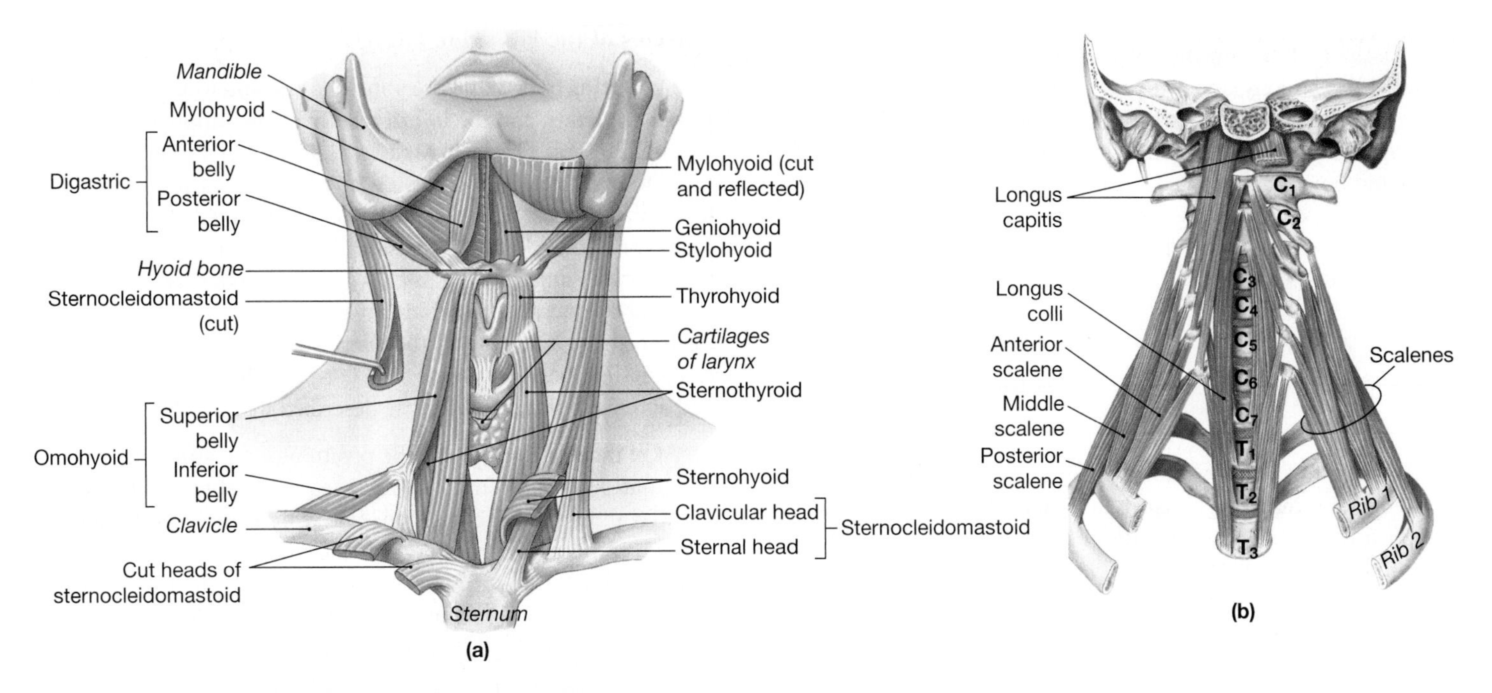

Figure 11.7 Muscles of the anterior triangle of the neck. a) Suprahyoid and infrahyoid muscles, anterior view; **b)** deep muscles of the anterior triangle of the neck and the scalene muscles in the posterior triangle of the neck, anterior view.

2. Use string to connect the origin of each suprahyoid muscle with its insertion on the hyoid bone (Figure 11.7a). Visualize each muscle's path and how, as a group, these muscles elevate the hyoid.
3. Identify the origins of the infrahyoid muscles.
   - Sternothyroid originates at the manubrium of the sternum.
   - Sternohyoid originates at the manubrium and the medial end of the clavicle.
   - Thyrohyoid originates from the thyroid cartilage. Although the thyroid cartilage is not demonstrated on the skeleton, you can approximate its position just inferior to the hyoid bone.
   - Omohyoid (superior belly) originates from the medial end of the clavicle. (The inferior belly will be discussed with the posterior triangle.)
4. Use string to connect the infrahyoid muscles' origins and insertions. Visualize the path of each muscle and how, as a group, they depress the hyoid bone.
5. On the skeleton, identify the origins of the deep muscles in the anterior triangle.
   - **Longus capitis** originates from the transverse processes of vertebrae C3 through C6 and inserts on the inferior aspect of the occipital bone.
   - **Longus colli** originates from the bodies of vertebrae C1 through T3 and inserts on the transverse processes of vertebrae C3 through C6.
6. Use string to connect the origins and insertions of these deep muscles. Visualize the path of the longus capitis and how it flexes the head. Longus colli consists of several overlapping muscle bundles and flexes the neck.
7. Use an anatomical model, chart, or atlas to identify the muscles of the anterior triangle listed above and verify the following:
   - The mylohyoid and geniohyoid muscles form the muscular floor of the mouth.
   - Both digastric and omohyoid have two bellies connected by a central tendon. The central tendon of the digastric connects it to the hyoid bone. The central tendon of the omohyoid connects it to the medial end of the clavicle.
   - The superior belly of the omohyoid is located in the anterior triangle. The inferior belly of this muscle is found in the posterior triangle.
   - The sternothyroid is the only superficial muscle in the anterior triangle that is not attached to the hyoid bone.
   - The longus capitis and longus colli are deep to the other muscles of the anterior triangle.
8. Place two fingers on your laryngeal prominence (Adam's apple), then move your fingers up about 1 to 1.5 cm and apply gentle pressure to feel the hyoid bone. As you swallow, feel the hyoid bone being elevated and then depressed to its original position. When the hyoid is elevated, suprahyoid muscles act as prime movers and the infrahyoids act as antagonists. When the hyoid is depressed, the roles reverse.
9. The longus capitis and longus colli flex the head and neck, respectively. From anatomical position, observe while your lab partner flexes his or her head and neck anteriorly.

QUESTIONS TO CONSIDER

1. When the anterior triangle muscles act directly on the hyoid bone, they will have an indirect action on the larynx, which is enclosed by the thyroid cartilage. For example, when the suprahyoid muscles elevate the hyoid bone during swallowing, these same muscles will indirectly elevate the larynx. Based on the muscle anatomy you just studied, explain why the hyoid bone and larynx move simultaneously.

___

___

___

2. Although the primary action of the longus capitis is flexion of the head, it may also assist in flexion of the neck. Examine the position of the longus capitis (Figure 11.7b) on an anatomical model, chart, or atlas and explain why the longus capitis could assist in neck flexion.

___

___

___

Muscles of the Posterior Triangle

The posterior triangle muscles include the **splenius**, **levator scapulae**, **scalenes**, and the **inferior belly of the omohyoid** (Figures 11.2b, 11.7, and 11.8; also Figures 11.11 and 11.14 later in the exercise). The splenius muscles cover the posterior and lateral surfaces of the neck. The levator scapulae, although located in the posterior triangle, elevates the scapula and is usually considered to be a shoulder muscle. The scalenes connect cervical vertebrae to the first and second ribs, and their rib attachments are important during forced inspiration. Only the inferior belly of the omohyoid is located in the posterior triangle. The bony attachments, innervations, and actions of these muscles are described in Table 11.2C.

1. On the skeleton, locate the position of the **splenius.** It originates from the spinous processes of vertebrae C7 through T6 and the ligamentum nuchae, which connects the spinous processes of the cervical vertebrae (not present on a skeleton). Use string to trace the path of **splenius capitis** from its origin to its insertion on the mastoid process.
2. Next, run the string from the origin of the splenius to the transverse processes of vertebra C1 through C4. These are the insertions for the **splenius cervicis**, and also the origins of the **levator scapulae**.

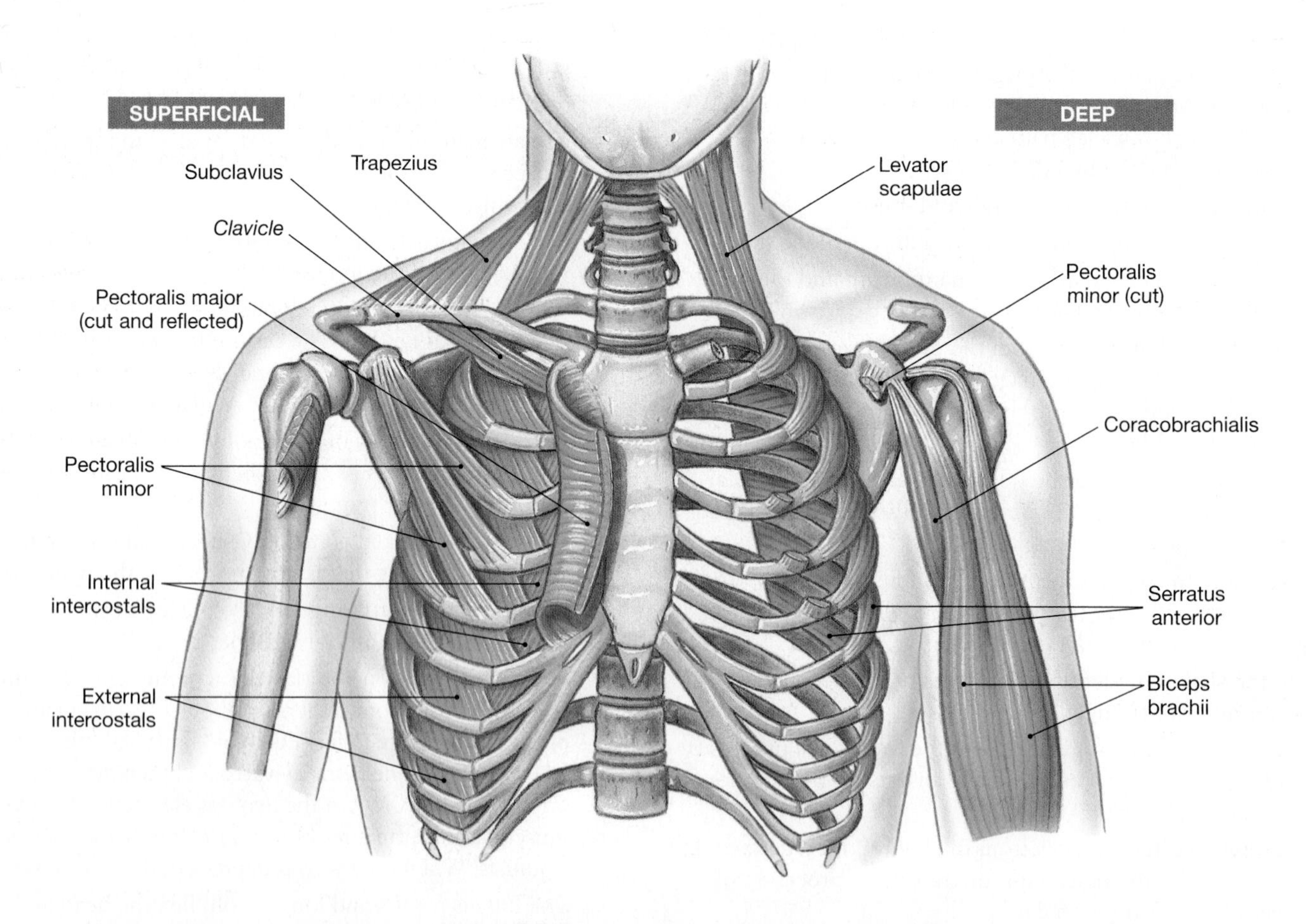

Figure 11.8 Muscles associated with the thoracic cage, anterior view. The intercostal muscles are located in the spaces between the ribs and are vital for respiration.

3. Use string to follow the path of levator scapulae from its origin inferiorly and laterally to its insertion on the superiomedial border of the scapula.
4. The three scalene muscles (posterior, middle, and anterior) originate from the anterior regions of ribs 1 and 2. Use string to trace their path as they travel superiorly to insert on the transverse processes of C4 through C6.
5. Identify the superior margin of the scapula. The inferior belly of the omohyoid originates here and travels medially. It is connected to the superior belly of the omohyoid by a central tendon that anchors the muscle to the medial end of the clavicle. Trace this path with string.
6. Use an anatomical model, chart, or atlas to identify the muscles of the posterior triangle and verify the following:
   - The inferior belly of the omohyoid is in the posterior triangle, but the superior belly is an infrahyoid muscle in the anterior triangle. The muscle as a whole acts as an infrahyoid muscle by depressing the hyoid bone.
   - Portions of the **brachial plexus**, the network of nerves that supply the upper extremity, travel through the posterior triangle. Because of their superficial position, the nerves are vulnerable to injury as they extend through this region.
   - The scalenes are important muscles of respiration. During deep inspiration, they elevate the superior ribs to allow more air to enter the lungs.
7. The splenius and scalenes laterally flex the neck. Have your lab partner perform this action, bending the neck from side to side, while you observe. The splenius also extends the neck and head. Have your lab partner start with his or her head and neck bent forward, then observe as the head and neck are straightened (extended).

WHAT'S IN A WORD The omohyoid is named according to where the muscle inserts. The first part of the muscle's name, *omo,* is from Greek and means "shoulder." The inferior belly of the omohyoid inserts on the superior margin of the scapula. The second part of the name, *hyoid,* refers to the hyoid bone, where the superior belly inserts.

The word *levator* is derived from Latin and means "lifter." The levator scapulae "lifts" or elevates the scapula.

The word *splenius* is derived from Greek and means "bandage." The fibers of the splenius muscles travel obliquely and wrap around the posterior surface of the neck like a bandage.

QUESTIONS TO CONSIDER 1. Explain why the origins and insertions of the scalenes are reversed when they act on the cervical vertebrae to laterally flex the neck, compared to when they act on the ribs during forced inspiration.

____

____

____

____

2. In addition to its other functions, the splenius also rotates the head, as when you shake your head to say "no." As the head rotates, it pivots around a vertical axis that travels through the midsagittal plane of the body. Perform this action on a skeleton and identify the joint at which it occurs.

____

Muscles of the Thorax

The muscles of the thorax, or chest, are primarily involved in respiration. Intercostal muscles run between the ribs and move them and the sternum. The diaphragm separates the thorax from the abdominal cavity below. The coordinated action of these muscles allows air to move in and out of the lungs during breathing.

ACTIVITY 11.3 Identifying Respiratory Muscles of the Thorax

Intercostal Muscles

The intercostal spaces, between the ribs, contain three layers of **intercostal muscles**. Functionally, the superficial layer, the **external intercostal muscles**, and the intermediate layer, the **internal intercostal muscles**, are the most significant (Figures 11.8 and 11.9b). The deepest layer includes the **transversus thoracis muscle**, anteriorly; the **innermost intercostal muscles**, laterally; and the **subcostal muscles**, posteriorly. The skeletal attachments, innervations, and functions of these muscles are summarized in Table 11.3.

1. On the skeleton, identify the intercostal spaces. The intercostal muscles that fill these spaces attach to adjacent ribs.
2. Along the lateral aspect of one intercostal space, tape a small piece of string to connect adjacent ribs. Make sure the string is traveling in a vertical plane to represent the direction of the muscle fibers in the innermost intercostal muscle.
3. In the same intercostal space, attach a second piece of string to the inferior rib. Tape the string to the superior rib by allowing it to pass superiorly and anteriorly from the first attachment. The second string represents the direction of the internal intercostal muscle.
4. Again, in the same intercostal space, attach a third piece of string to the superior rib and allow it to travel inferiorly and anteriorly to the inferior rib. This string represents the direction of the external intercostal muscle.
5. Use an anatomical model, chart, or atlas to identify the three intercostal muscle layers. For each layer of muscle, observe the direction of the muscle fibers and confirm that they match the arrangement of your strings. In addition, verify the following:
   - In the intercostal spaces between the tips of the ribs and the sternum (where the costal cartilage is located), a connective tissue membrane replaces the external intercostal muscles.

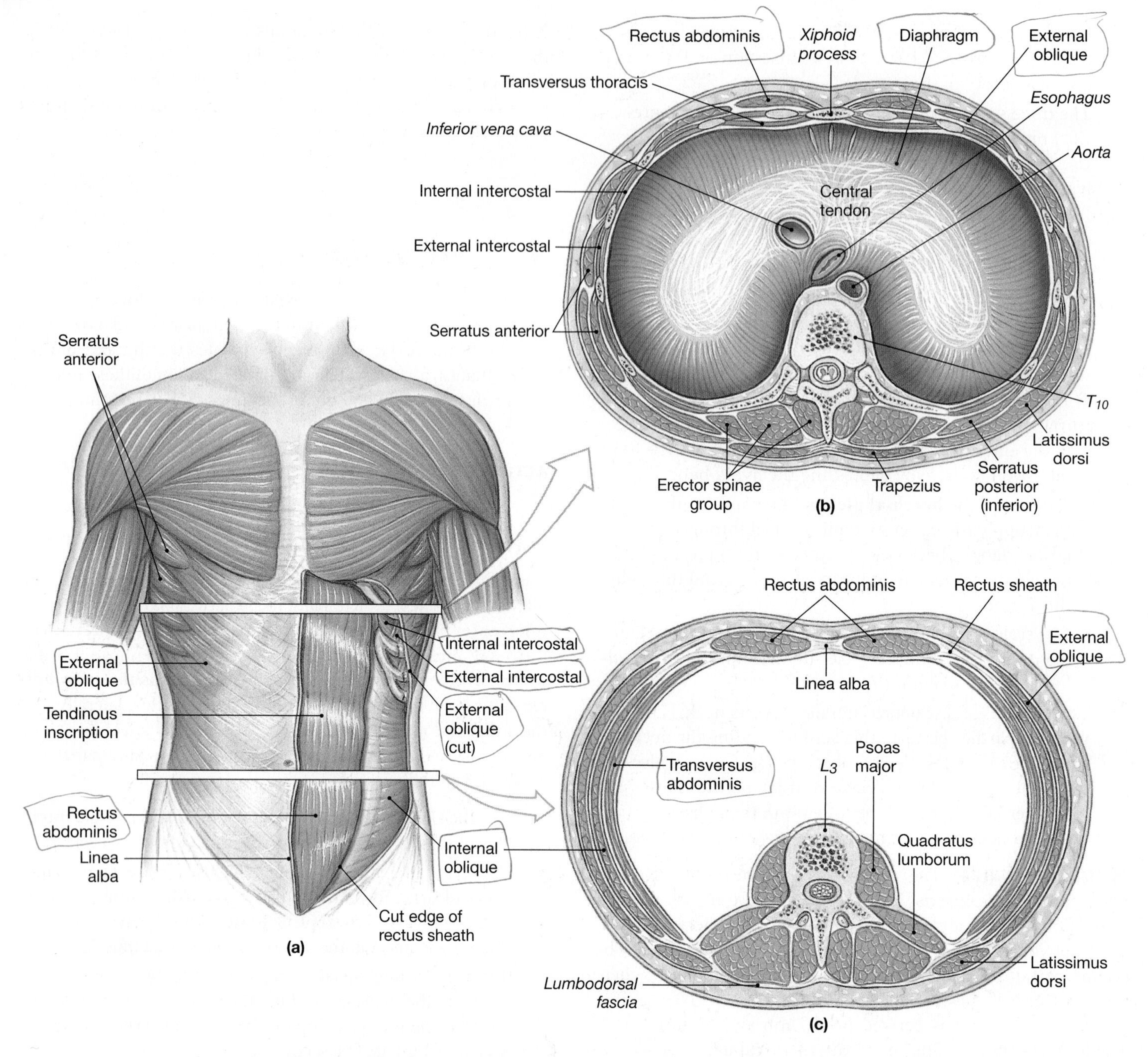

Figure 11.9 Muscles of the thorax and abdomen. a) Superficial thoracic and abdominal muscles, anterior view; **b)** cross section at the level of the diaphragm; **c)** cross section at the level of the umbilicus.

- Posteriorly, in the intercostal spaces between the angles of the ribs and the vertebral column, another connective tissue membrane replaces the internal intercostal muscles.
- Among the deepest layer of muscles, the transversus thoracis muscle is limited to the back of the sternum and the anterior intercostal spaces, the innermost intercostal muscles are found only in the lateral spaces, and the subcostal muscles are located more posteriorly. Note that the fibers of the subcostal muscles span more than a single intercostal space.

6. Place the index fingers from each hand on your ribs on both sides of the thorax and breathe normally. From the neutral position, you will feel your ribs being elevated by the external intercostal muscles as you inspire quietly. When you expire quietly, you will feel your ribs being depressed. This action occurs passively (without the assistance of muscle activity). However, during forced expiration the ribs will be depressed even more by the internal intercostal muscles. The innermost intercostal, transversus thoracis, and subcostal muscles are synergists.

Table 11.3 Respiratory Muscles of the Thorax

Muscle	Origin	Insertion	Action	Innervation
1. Diaphragm	Internal surfaces of xiphoid process and inferior six ribs; anterior surfaces of vertebrae L1, L2, and L3	Central tendon of the diaphragm	Increases volume of thoracic cavity by pulling central tendon inferiorly	Phrenic nerves (C3–C5)
2. External intercostals	Inferior border of superior rib	Superior border of inferior rib	Elevates the ribs during normal inspiration	Intercostal nerves (branches of thoracic spinal nerves)
3. Internal intercostals	Superior border of inferior rib	Inferior border of superior rib	Depresses the ribs during forced expiration	As above
4. Innermost intercostals	As above	As above	Assists internal intercostals in depressing the ribs	As above
5. Transversus thoracis	Internal surface of sternum	Various costal cartilages	As above	As above
6. Subcostal	Internal surface of inferior rib, near the angle	Internal surface of rib 2 or 3 superior to origin, near the angle	As above	As above

7. With your fingers on your ribs, breathe normally for a few cycles. When ready, forcefully expire. You will feel additional depression of the ribs during the forced expiration.

QUESTION TO CONSIDER The external and internal intercostal muscles are attached to the superior and inferior ribs that border the intercostal spaces. Based on their functions, determine the origins and insertions of the external and internal intercostal muscles in each intercostal space.

___

___

___

___

The Diaphragm

The **diaphragm** is a musculotendinous partition that separates the thoracic and abdominal cavities (Figure 11.9b). It has a concave surface that faces the abdominal cavity and a convex surface that faces the thoracic cavity. The diaphragm serves as the primary respiratory muscle. Its attachments, innervation, and specific functions are described in Table 11.3.

1. On the skeleton, identify the origins of the diaphragm: the xiphoid process, anteriorly; the costal margin of the rib cage, anterolaterally; and the anterior surfaces of the bodies of vertebrae L1 through L3, posteriorly. The origin of the diaphragm follows the circumference of the torso at the border between the thoracic and abdominopelvic cavities (Figure 11.9b). The diaphragm's insertion, the central tendon, cannot be identified on the skeleton.
2. Using an anatomical model, chart, or atlas, identify the diaphragm and verify the following:

- The diaphragm separates the thoracic cavity from the abdominopelvic cavity.
- Superiorly, the diaphragm forms two domes that face the thoracic cavity. The right dome is somewhat higher and rests on the liver; the left dome rests on the stomach.
- The domes are formed by the **central tendon of the diaphragm**, which serves as the insertion for the muscle (Figure 11.9b).
- The diaphragm has three major openings (Figure 11.9b) that allow for the passage of vital structures. The most inferior opening, the **aortic opening (hiatus)**, is located anterior to the body of the 12th thoracic vertebra and transmits the descending aorta. The **esophageal opening (hiatus)** lies at the level of the 10th thoracic vertebra and allows the esophagus to enter the abdominal cavity to reach the stomach. The **caval opening**, located at the 8th thoracic vertebra, is the most superior of these openings and serves as a passageway for the inferior vena cava as it enters the thoracic cavity to reach the heart.

WHAT'S IN A WORD The word *diaphragm* is derived from Greek and means "partition or wall." The diaphragm forms a wall that separates the thoracic and abdominopelvic cavities.

CLINICAL CORRELATION

A congenital diaphragmatic hernia occurs about once in every 2200 births. It is typically caused by a structural defect in the posterolateral aspect of the diaphragm and usually develops on the left side. As a result, abdominal contents (e.g., portions of the small intestine) will be displaced into the thoracic cavity. Normal lung development could be inhibited and life-threatening breathing difficulties could arise.

QUESTION TO CONSIDER The aortic hiatus does not pass *through* the diaphragm. Instead, it is a passageway that is *posterior* to the diaphragm. Explain why this distinction is significant. (*Hint:* Consider what structure travels through the aortic hiatus and what would happen to that structure if it passed through the diaphragm.)

____

____

____

____

Muscles of the Abdominal Wall

Although the inferior ribs and pelvic bones shelter some abdominopelvic organs, most of the abdominal wall lacks bony reinforcement. The muscles of the abdominal wall fill this gap to form a strong, multilayered enclosure that protects vulnerable abdominal viscera from injury. Most of the anterior and lateral aspects of the abdominal wall are covered by four pairs of muscles.

- The long, vertical **rectus abdominis** muscles cover the anterior abdominal wall (Figures 11.9a and 11.10).
- Three layers of broad, flat muscles cover the anterolateral aspect of the abdominal wall. From superficial to deep they are the **external oblique**, **internal oblique**, and **transversus abdominis** (Figures 11.9a and c; Figure 11.10).

The bony attachments, innervations, and actions of the abdominal wall muscles are described in Table 11.4.

ACTIVITY 11.4 Identifying Abdominal Wall Muscles

1. On the skeleton, identify the bony attachments of the rectus abdominis. This muscle originates on the pubic symphysis and pubic crest and inserts on the xiphoid process of the sternum and costal cartilages of ribs 5 through 7.
2. Attach a piece of string between these bony attachments to represent the orientation of the rectus abdominis muscles and the location of the **linea alba**, a midline fibrous cord that separates the two muscles (Figures 11.9a and 11.10).
3. On the skeleton, identify the inner surfaces of costal cartilages 7 through 12, and the iliac crest. These are the skeletal origins of the innermost flat muscle layer, transversus

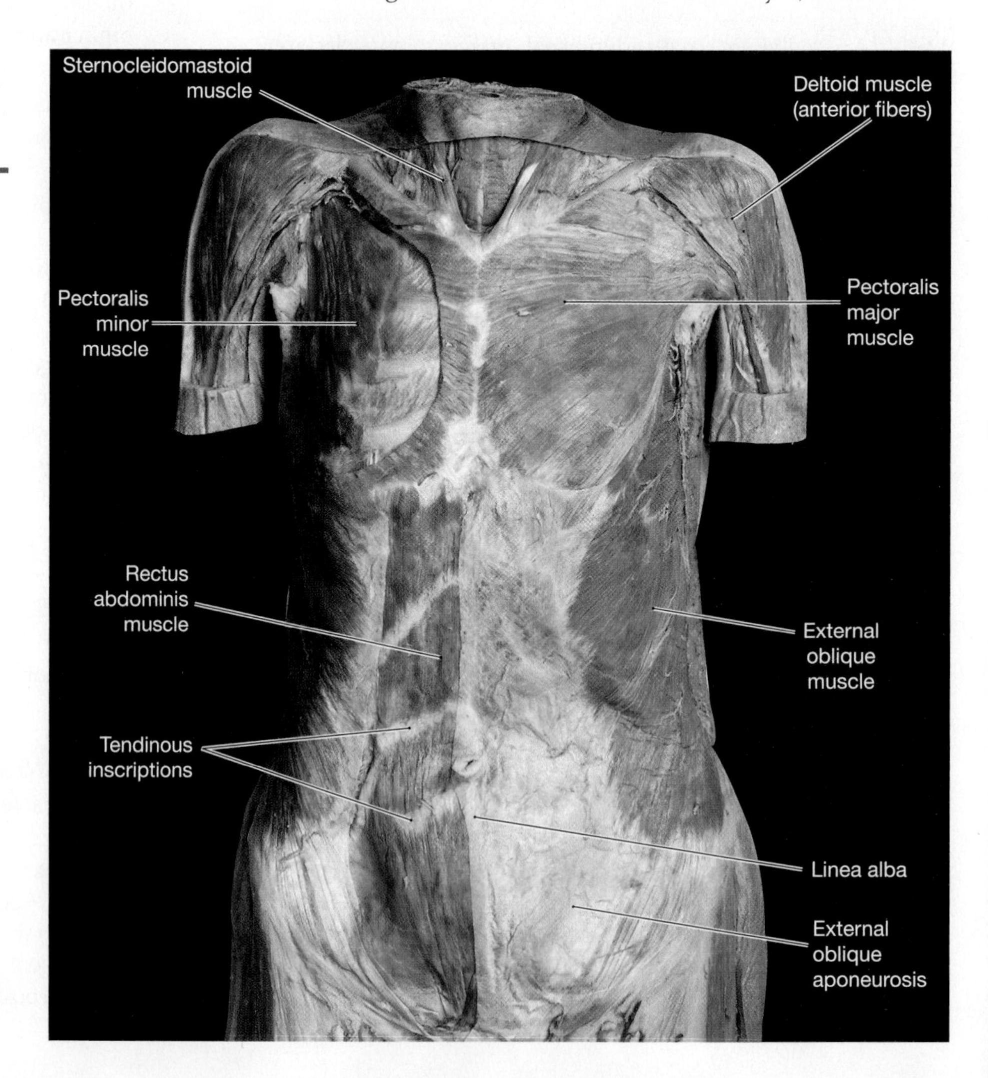

Figure 11.10 Cadaver dissection of thoracic and abdominal muscles, anterior view. A superficial dissection is illustrated on the right in the photograph; a deep dissection on the left.

Table 11.4 Muscles of the Abdominal Wall

Muscle	Origin	Insertion	Action	Innervation
1. Rectus abdominis	Pubic crest; pubic symphysis	Xiphoid process; costal cartilages of ribs 5–7	Flexion of vertebral column; stabilizes pelvis during walking; increases intraabdominal pressure during forced expiration	Intercostal nerves from T7–T12
2. External oblique	External surfaces of inferior eight ribs	Linea alba; pubic crest; pubic tubercle; iliac crest	Rotation and lateral flexion of trunk; increases intraabdominal pressure during forced expiration	As above
3. Internal oblique	Thoracolumbar fascia; iliac crest; inguinal ligament	Linea alba; pubic crest; inferior three or four ribs	As above	Intercostal nerves from T7–T12; branches from spinal nerve L1
4. Transversus abdominis	Thoracolumbar fascia; inguinal ligament; costal cartilages of inferior six ribs; iliac crest	Linea alba; pubic crest	Increases intraabdominal pressure during forced expiration	As above

abdominis. This muscle also originates from the **lumbodorsal (thoracolumbar) fascia**, a broad sheet covering deep back muscles in the thoracic and lumbar regions (see Figures 11.14 and 11.15 later in the exercise).

4. Attach a piece of string to one of the skeletal origins of the transversus abdominis. Guide the string anteriorly and horizontally approximating the abdominal wall, until it reaches the string representing the linea alba. This path represents the orientation of the muscle fibers in transversus abdominis. The linea alba is the muscle's insertion.
5. On the skeleton, identify the iliac crest. The anterior two thirds marks the skeletal origin of the middle flat muscle layer, the internal oblique. Like transversus abdominis, internal oblique also originates from the thoracolumbar fascia.
6. Attach a piece of string to the iliac crest. Guide the string along a circular and diagonal path passing anterosuperiorly toward the internal oblique's insertion: the inferior borders of ribs 10 through 12 and the linea alba. The circular diagonal path of the string represents the orientation of the internal oblique muscle fibers. Compare the orientation of the internal oblique with the transversus abdominis. In the space provided, describe the different paths of these two muscles.

____

____

____

____

7. On the skeleton, identify ribs 5 through 12. The outer surfaces of these ribs are the skeletal origin of the outermost flat muscle layer, the external oblique.
8. Attach a piece of string to the external surface of one rib, then pass the string anteroinferiorly toward the external oblique's insertion: the linea alba, the anterior half of the iliac crest, and the pubic tubercle. This path represents the orientation of the external oblique muscle fibers.
9. Compare the orientation of the external oblique with the other two muscle layers. Note that the muscle fibers of the internal and external obliques both travel diagonally but at right angles to each other.
10. Use an anatomical model, chart, or atlas to identify the muscles of the abdominal wall and confirm the following (Figures 11.9a and c, Figure 11.10):
    - The **rectus sheath**, formed by the aponeuroses of the three oblique muscles, forms a fascial enclosure for the rectus abdominis.
    - The linea alba is formed by the interlacing fibers of the rectus sheath and separates the two rectus abdominis muscles.
    - The flat muscles form a strong, three-layered muscular wall, reinforced by muscle fibers in the three layers traveling in different directions.
11. From the anatomical position, have your lab partner bend forward at the waist while you observe. This action occurs at the intervertebral discs between thoracic and lumbar vertebrae. Rectus abdominis is the prime mover for flexion of the thoracic and lumbar regions of vertebral column. The external and internal oblique muscles assist the erector spinae muscles in lateral flexion and rotation of the torso. Observe your lab partner as he or she performs these actions.
12. Sit or stand quietly with your fingers resting on the anterior wall of your abdomen. Breathe normally for a short period and when you are ready, expire forcefully after a normal expiration. During the forced expiration, you will feel rectus abdominis contract. Contractions of the abdominal wall muscles push abdominal organs up against the diaphragm which, in turn, increases internal thoracic pressure and helps force air out of the lungs.

WHAT'S IN A WORD The names of the abdominal muscles provide clues about their orientation. For example, *rectus* is from Latin and means "straight." The muscle fibers of rectus abdominis travel in a straight vertical path. The word *oblique* is also derived from Latin and means "diagonal or slanting." The fibers of the external and internal oblique muscles travel diagonally across the anterolateral abdominal wall.

QUESTIONS TO CONSIDER 1. Speculate on the significance of the muscle fibers in the three flat muscle layers traveling in different directions.

____

____

____

2. The external and internal oblique muscles laterally flex the vertebral column to the same side. What muscles would act as antagonists for lateral flexion of the vertebral column to the left side?

____

____

____

3. The external oblique rotates the vertebral column to the opposite side; the internal oblique rotates the vertebral column to the same side. What muscles would act as antagonists for rotation of the vertebral column to the left side?

____

____

____

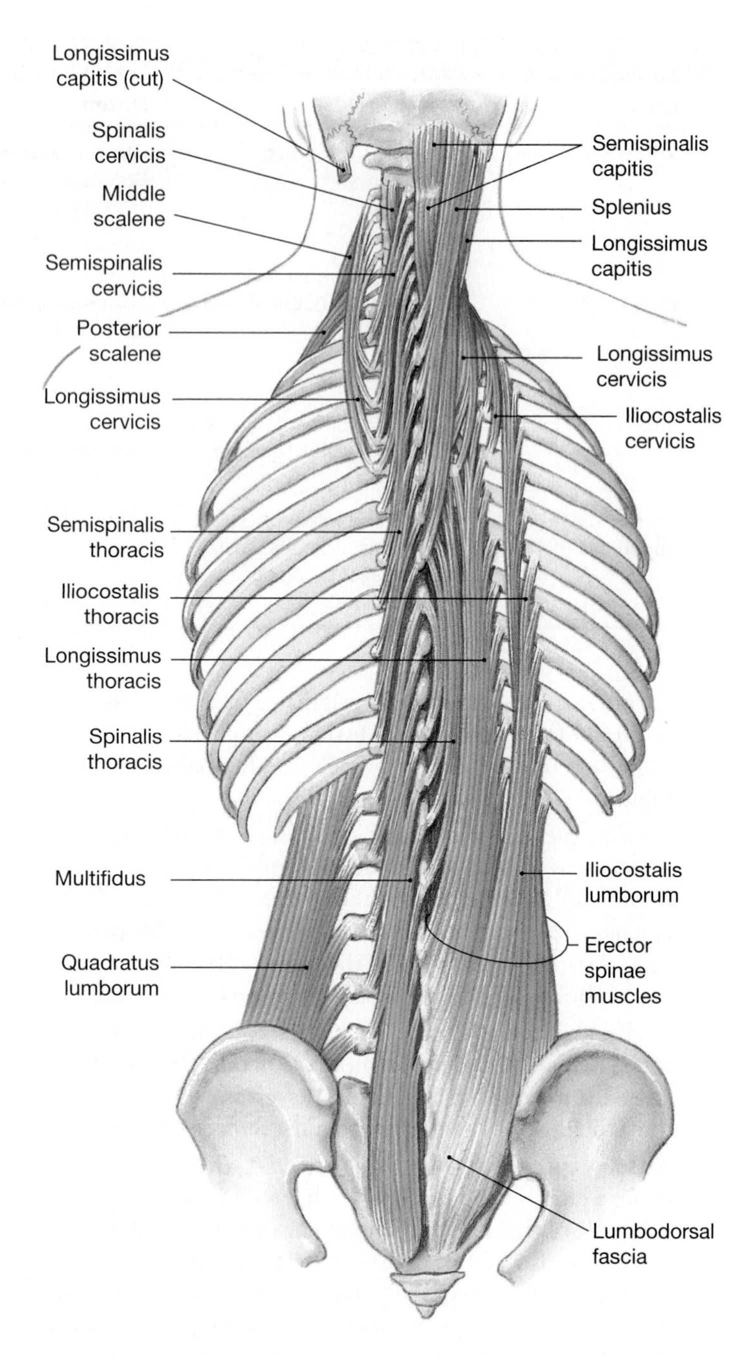

Figure 11.11 Deep muscles of the back. The erector spinae muscles are illustrated on the right. Small, deeper muscles are shown on the left.

Deep Back Muscles

The deep (intrinsic) back muscles are important because of their actions on the vertebral column and head, and their role in maintaining normal posture. They include the **erector spinae muscles**, the **splenius muscles**, and the **transversospinalis muscles**.

The erector spinae muscles are comprised of three muscular bands that travel vertically from the sacrum to the posterior surface of the skull, along each side of the vertebral column, From lateral to medial, the three muscles are **iliocostalis**, **longissimus**, and **spinalis** (Figure 11.11). Iliocostalis and longissimus are sometimes referred to as the **sacrospinalis**.

Each muscle in the complex can be subdivided according to the vertebral region in which it travels. Accordingly, the terms *lumborum* (iliocostalis only), *thoracis*, *cervicis*, and *capitis* are added to the names to identify the specific subdivisions.

The deep layer of intrinsic back muscles includes a number of short muscles (**semispinalis**, **multifidus**, **rotatores**) that are located in the groove formed by the spinous and transverse processes of vertebrae (Figure 11.11). Collectively, these muscles are called the **transversospinalis**.

The bony attachments, innervations, and principal actions of the deep back muscles are summarized in Table 11.5.

ACTIVITY 11.5 Identifying Deep Back Muscles

1. On the skeleton, identify the posterior aspect of sacrum, iliac crest, spinous processes of T11 through T12 and all

Table 11.5 Deep Back Muscles

Muscle Group/Muscle	Origin	Insertion	Action	Innervation
1. Splenius a. Splenius capitis b. Splenius cervicis	Spinous processes of inferior cervical and superior thoracic vertebrae	Mastoid process, occipital bone, and transverse processes of superior cervical vertebrae	Extension of the head and neck; lateral flexion and rotation of the head and neck	Cervical spinal nerves
2. Erector spinae a. Iliocostalis b. Longissimus c. Spinalis	Thoracolumbar fascia; posterior surface of sacrum; iliac crest; spinous processes of T11–T12 and L1–L5	*Iliocostalis:* angles of ribs and transverse processes of C4–C6 *longissimus:* transverse processes of thoracic and cervical vertebrae, tubercles of inferior nine ribs, and mastoid process on skull; *spinalis:* spinous processes of thoracic and cervical vertebrae	Extension of vertebral column and head; lateral flexion of vertebral column and head; maintains normal curvature of vertebral column	Cervical, thoracic, and lumbar spinal nerves
3. Transversospinalis a. Semispinalis b. Multifidus c. Rotatores	Transverse processes of vertebrae	Spinous processes of more superior vertebrae	Extension of vertebral column; rotation of vertebral column; stabilizes the vertebral column	As above

five lumbar vertebrae. These skeletal structures are the bony origins of the erector spinae muscles.

2. Take several pieces of string of different lengths and attach them to these origins. Extend the strings to the various insertions of the erector spinae.
   - *Iliocostalis:* angles of the ribs and transverse processes of C4 through C6
   - *Longissimus:* transverse processes of thoracic and cervical vertebrae, tubercles of the inferior nine ribs, and the mastoid process
   - *Spinalis:* spinous processes of thoracic and cervical vertebrae
3. Observe the arrangement of the strings and note that these muscles extend the vertebral column and head and are vital for maintaining the normal curvatures of the vertebral column and the relative positions of individual vertebrae.
4. Examine the connections of the transversospinalis. Attach several pieces of string to the transverse processes of various vertebrae and extend each string up by one to three vertebrae, connecting it to a spinous process. Notice that these muscles are in bony grooves formed by the serial arrangement of transverse and spinous processes.
5. Use an anatomical model, chart, or atlas to identify the deep back muscles and verify the following (Figure 11.11):
   - As a group, the erector spinae muscles stretch from the pelvis, inferiorly, to the posterior surface of the head, superiorly.
   - The erector spinae muscles are vital for supporting the body and maintaining erect posture.
   - The transversospinalis muscles are located deep to the erector spinae in bony grooves on each side of the vertebral column.
6. Bend forward (flex) at the waist. As you slowly straighten (extend) the spine, concentrate on feeling the pull in your back, along the vertebrae. You are feeling the collective actions of the deep back muscles.

CLINICAL CORRELATION

Sudden or excessive rotation or extension of the vertebral column can result in back strain. Back injuries usually occur in the lumbar region (lower back pain), perhaps during participation in sports or by incorrectly lifting heavy objects. The disproportionate force tears muscle fibers in the erector spinae and damages associated back ligaments. Often, the injured muscles go into spasm, causing pain and loss of function. Back injuries can be prevented by warming up properly before participating in physical activities and by using the strong lower extremity muscles while keeping the back straight when lifting heavy objects.

QUESTIONS TO CONSIDER

1. Acting together, the erector spinae muscles are the primary extensors of the vertebral column and the head. The transversospinales are synergists for this action. Identify a muscle that acts as an antagonist.

____

2. Acting on one side only, the erector spinae laterally flexes and rotates the vertebral column and head to the ipsilateral side. What muscles in the abdominal wall and neck act as synergists for these actions?

____

____

3. Comment on the following statement: It is important to keep the deep back muscles in balance with the abdominal wall muscles. (Hint: Consider the negative impact of having strong deep back muscles but weak abdominal muscles.)

____

____

Muscles of the Pelvic Floor

The **pelvic diaphragm** (Figure 11.12) is a funnel-shaped muscular layer that supports the organs in the pelvic cavity and closes the outlet of the bony pelvis. It is composed of two muscles: **levator ani** and **coccygeus**. The rectum, urethra, and vagina (in females) pass through this muscular sheet.

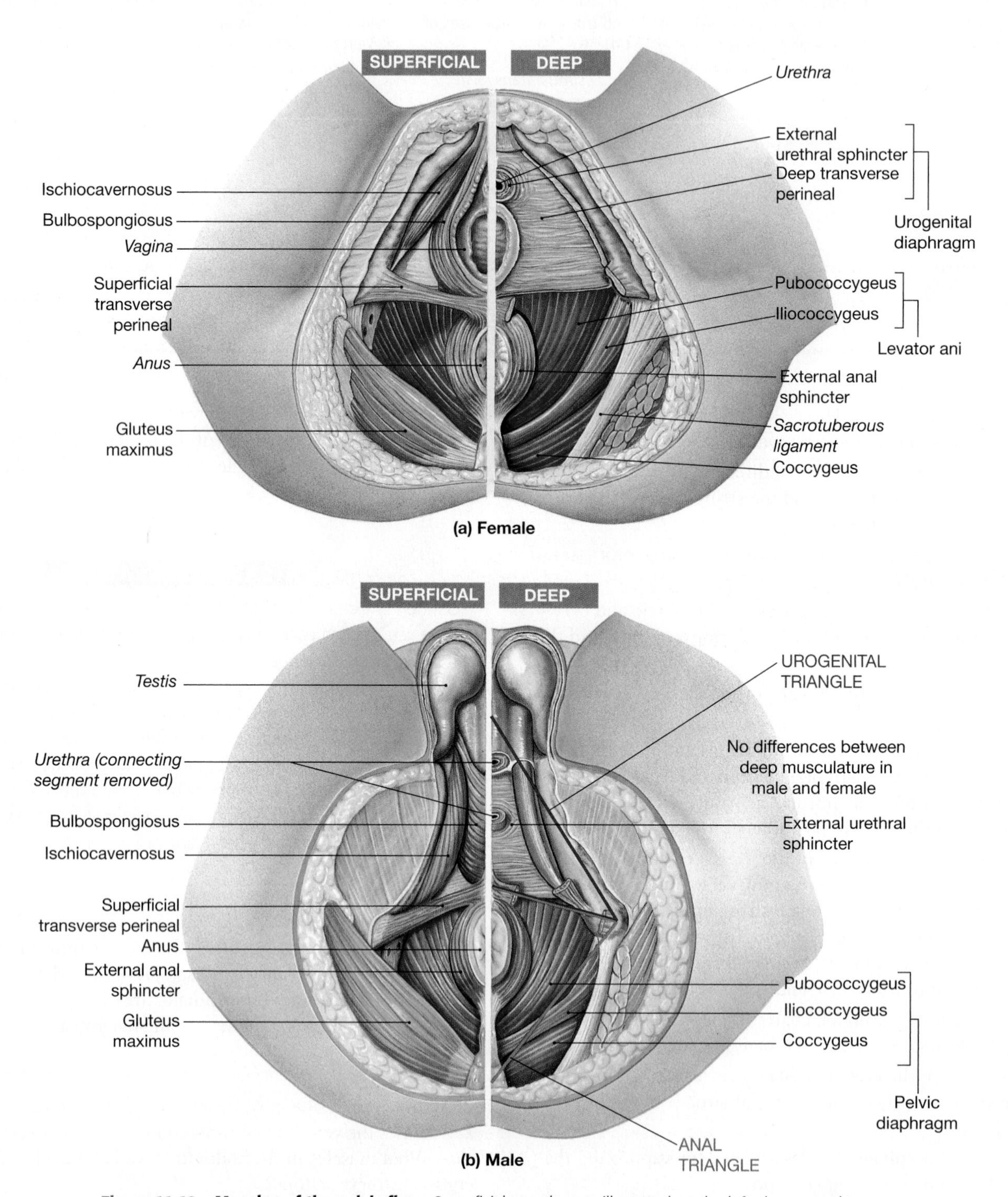

Figure 11.12 Muscles of the pelvic floor. Superficial muscles are illustrated on the left, deep muscles on the right. **a)** Female; **b)** male.

Just anterior to the pelvic diaphragm, filling a small gap, is the **urogenital diaphragm**. It includes the **external urethral sphincter**, which surrounds the urethral opening (Figure 11.12). Contraction of this voluntary muscle closes the urethra so that urination can be delayed.

The **bulbospongiosus** and **ischiocavernosus** are located just inferior to the urogenital diaphragm. They assist in maintaining erection of the penis and clitoris. Posterior to these muscles and surrounding the anus is the **external anal sphincter.** Contraction of this voluntary muscle allows the delay of defecation (Figure 11.12).

The bony attachments, innervations, and functions of the pelvic floor muscles are described in Table 11.6.

ACTIVITY 11.6 Identifying Pelvic Floor Muscles

1. On the skeleton, identify the bony margins of the diamond-shaped **perineum.**
   - Two ischial tuberosities
   - Pubic symphysis
   - Coccyx
2. Position a ruler to connect the ischial tuberosities. This divides the perineum into two **perineal triangles.**
   - The anterior **urogenital triangle** contains the external genitalia.
   - The posterior **anal triangle** contains the anal opening.
3. Use an anatomical model, chart, or atlas to identify the muscles of the pelvic floor and verify the following:
   - The levator ani and coccygeus muscles are attached to the bony margins of the perineum, forming the pelvic diaphragm that supports pelvic viscera.
   - The larger levator ani muscle occupies both perineal triangles and is composed of three muscle pairs: **pubococcygeus, puborectalis,** and **iliococcygeus.**
   - Coccygeus is triangular, posterior to the levator ani, and in the anal triangle.
   - The urethra, vagina (in females), and anus open through the pelvic diaphragm.
   - Sphincter muscles surround the urethra and anal openings.
   - Muscles associated with external genitalia, the bulbospongiosus and ischiocavernosus which assist in the erection, are present in both sexes.
   - Many perineal muscles attach to the **perineal body (central tendon)**, in the midline of the perineum. This arrangement provides much of the support for the organs in the pelvis.

Table 11.6 Muscles of the Pelvic Floor

Muscle Group/Muscle	Origin	Insertion	Action	Innervation
1. Pelvic diaphragm a. Levator ani: Pubococcygeus Puborectalis Iliococcygeus	Broad origin that extends from the pubis to ischial spine	Internal surface of coccyx; blends in with levator ani on opposite side	Supports pelvic organs; elevates and retracts anus; flexes coccygeal joints	Spinal nerve S4; inferior rectal nerve (S2–S3)
b. Coccygeus	Ischial spine	Sacrum and coccyx	Flexes coccygeal joints	Spinal nerves S4 and S5
2. External urethral sphincter	Rami of pubis and ischium	Central tendon (perineal body)	*Males:* closes urethra; compresses prostate and bulbourethral glands *Females:* closes urethra; compresses vagina and greater vestibular glands	Pudendal nerve (S2–S4)
3. Bulbospongiosus	Central tendon (perineal body); in males, midline raphe (band of connective tissue) along ventral surface of penis	Corpus cavernosum of penis or clitoris	*Males:* compresses base of penis; stiffens penis; ejects urine or semen *Females:* compresses and stiffens clitoris; constricts vaginal opening	As above
4. Ischiocavernosus	Ischial tuberosities	Crus of corpora cavernosum at the root of penis or clitoris	Compresses and stiffens penis (males) or clitoris (females)	As above
5. External anal sphincter	Coccyx	Surrounds anal opening	Closes anal opening	As above

QUESTION TO CONSIDER Speculate on potential complications that could develop if the levator ani muscles were damaged or weakened.

____

____

____

Muscles of the Upper Extremity

The muscles of the upper extremity act at the shoulder, elbow, and wrist as well as the joints in the hand. In the arm and forearm, the muscles are grouped into compartments based on their location, and the muscles within a single compartment have similar functions. All muscles of the upper extremity are innervated by branches of the **brachial plexus**, a network of nerves that receives contributions from spinal nerves C5 to T1.

Muscles of the Shoulder

The muscles of the shoulder assist in stabilizing the scapula during arm movements. They also perform a number of movements on the scapula, which increase the range of motion at the shoulder, and are responsible for many actions of the arm. The shoulder muscles are divided into anterior and posterior groups. The bony attachments, innervations, and actions of the shoulder muscles are described in Table 11.7.

Anterior Shoulder Muscles

The anterior shoulder muscles include the **pectoralis major**, **pectoralis minor**, and **serratus anterior** (Figures 11.10 and 11.13). These muscles form the bulk of the muscular wall of the anterior thorax and act directly on the scapula or humerus.

ACTIVITY 11.7 Identifying Anterior Shoulder Muscles

1. On the skeleton, identify the clavicle. The medial half of this bone and the anterior surface of the sternum are the origins of pectoralis major. From this large attachment site, the muscle passes laterally and converges at its insertion on the humerus, lateral to the intertubercular groove.
2. After you have identified these attachments, run string between them to verify that the pectoralis major crosses the shoulder joint.
3. Identify the anterior surfaces of ribs 3, 4, and 5, which are the origins of the deeper pectoralis minor. Then identify the coracoid process on the scapula, which is its insertion.
4. Use string to connect the origins and insertion of this muscle, noting that pectoralis minor does not directly cross the shoulder joint. This muscle moves the scapula (depression and inferior rotation), but not the arm.
5. Serratus anterior originates from the external surfaces of ribs 1 through 8. Attach a piece of string to one of these ribs and extend it to the anterior surface of the medial border of the scapula. Note that the string for the serratus anterior must pass between the scapula and the rib cage to reach its destination. Similar to the pectoralis minor, serratus anterior acts directly on the scapula (protraction and superior rotation).
6. Use an anatomical model, chart, or atlas to identify the anterior shoulder muscles and verify the following:
   - Pectoralis major crosses the shoulder joint and acts directly on the arm.
   - Pectoralis minor and serratus anterior do not cross the shoulder joint and move only the scapula.
   - Pectoralis minor lies deep to pectoralis major.
   - Serratus anterior passes between the scapula and the rib cage to reach its insertion.

WHAT'S IN A WORD The origin of serratus anterior on the ribs resembles the cutting edge of a serrated knife. This is the basis for the muscle's name, *serratus*.

CLINICAL CORRELATION

The serratus anterior keeps the scapula in position against the thoracic wall. If this muscle is paralyzed, the medial border of the scapula moves laterally and posteriorly, especially when the individual leans on the hand against a wall. This gives the scapula the appearance of a wing and, thus, this condition is called a winged scapula.

Serratus anterior also performs superior rotation of the scapula, an action required to abduct the arm above the horizontal. Consequently, if the muscle is not functioning, a student will not be able to "raise the hand" if he or she wishes to be called upon in class to answer a question.

Posterior Shoulder Muscles

Similar to the anterior shoulder muscles, all muscles of the posterior shoulder act directly on the scapula or humerus. This group includes the following:

- **Latissimus dorsi** and **trapezius**, the two large superficial back muscles

Table 11.7 Muscles that Act at the Shoulder Joint

Muscle Group/Muscle	Origin	Insertion	Action	Innervation
A. Anterior shoulder				
1. Pectoralis major	Medial half of clavicle; sternum; costal cartilages of ribs 1–6	Lateral lip of intertubercular groove on humerus	Flexion, medial rotation, and adduction of the arm	Medial and lateral pectoral nerves (C5–C8; T1)
2. Pectoralis minor	External surfaces of ribs 3–5	Coracoid process of scapula	Depression and inferior rotation of scapula	Medial pectoral nerve (C8, T1)
3. Serratus anterior	External surfaces of ribs 1–8	Medial border of scapula	Protraction and superior rotation of scapula; holds scapula against thoracic wall	Long thoracic nerve (C5–C7)
B. Posterior shoulder				
1. Latissimus dorsi	Spinous processes of vertebrae T6–T12; thoracolumbar fascia; iliac crest; inferior three ribs	Intertubercular groove of humerus	Extension, adduction, and medial rotation of arm	Thoracodorsal nerve (C6–C8)
2. Trapezius	Posterior surface of occipital bone; ligamentum nuchae; spinous processes of thoracic vertebrae	Lateral third of clavicle; acromion process; spine of scapula	*Superior fibers:* elevate scapula *Middle fibers:* retract scapula *Inferior fibers:* depress scapula	Accessory nerve (cranial nerve XI)
3. Deltoid	Lateral third of clavicle; acromion process; spine of scapula	Deltoid tuberosity of humerus	*Anterior fibers:* medial rotation and flexion of arm *Middle fibers:* abduction of arm *Posterior fibers:* lateral rotation and extension of arm *Anterior and posterior fibers:* act together to adduct the arm	Axillary nerve (C5–C6)
4. Levator scapulae	Transverse processes of vertebrae C1–C4	Medial border of scapula near superior angle	Elevates scapula; acts with pectoralis minor to inferiorly rotate scapula	Dorsal scapular nerve (C4–C5)
5 Rhomboids (major and minor)	*Minor:* ligamentum nuchae; spinous processes of vertebrae C7 and T1 *Major:* spinous processes of vertebrae T2–T5	Medial border of scapula	Same as levator scapulae; also retracts scapula	Dorsal scapular nerve (C4–C5)
6. Teres major	Scapula, near inferior angle	Medial lip of intertubercular groove	Same as latissimus dorsi	Lower subscapular nerve (C6–C7)
7. Rotator cuff muscles				
a. Supraspinatus	Supraspinous fossa of scapula	Greater tubercle of humerus	Abduction of arm	Suprascapular nerve (C4–C6)
b. Infraspinatus	Infraspinous fossa of scapula	Greater tubercle of humerus	Lateral rotation of arm	As above
c. Teres minor	Lateral border of scapula	Greater tubercle of humerus	Lateral rotation of arm	Axillary nerve (C5–C6)
d. Subscapularis	Subscapular fossa of scapula	Lesser tubercle of humerus	Adduction and medial rotation of arm	Upper and lower subscapular nerves (C5–C7)

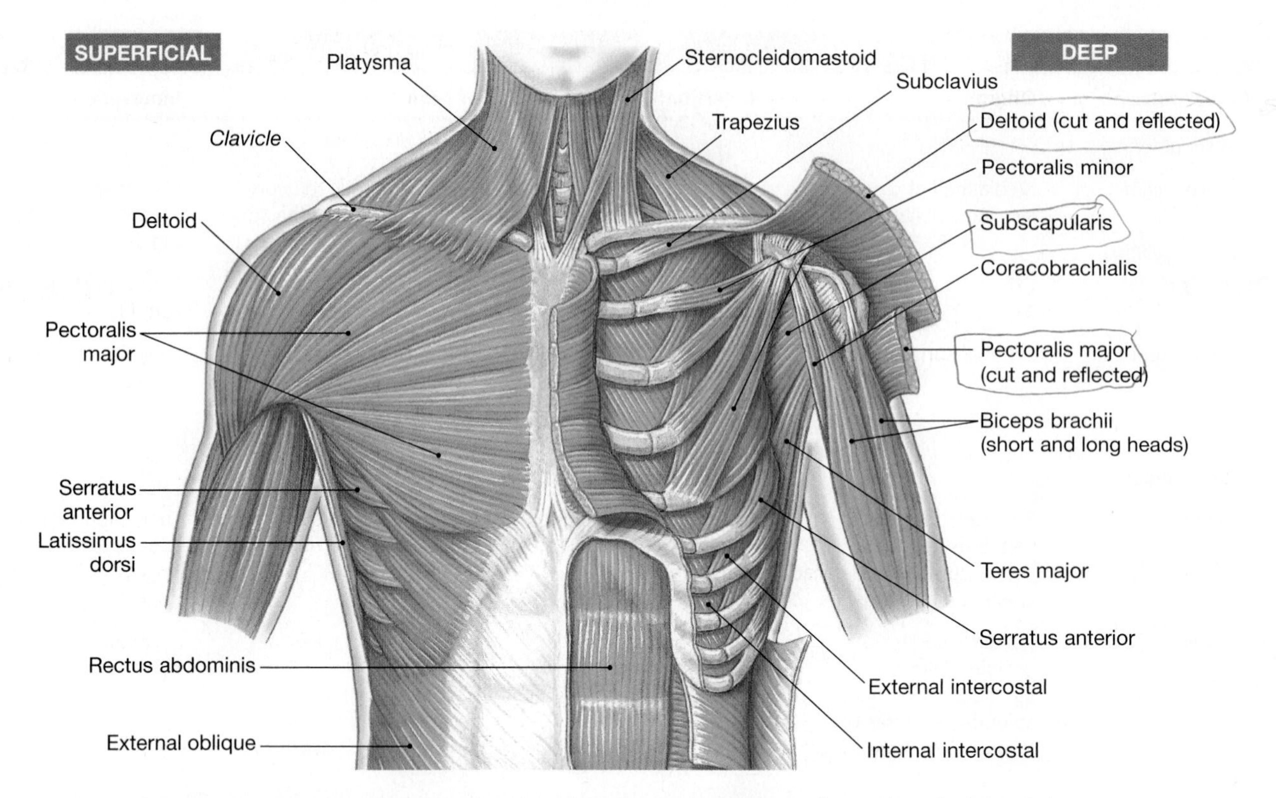

Figure 11.13 Muscles of the anterior shoulder. Superficial muscles are illustrated on the left and deep muscles on the right.

- **Levator scapulae** and **rhomboids**, located deep to the trapezius
- **Deltoid, teres major**, and the **rotator cuff muscles** that surround the shoulder joint (Figures 11.14 and 11.15).

ACTIVITY 11.8 Identifying Posterior Shoulder Muscles

1. On the skeleton, locate the extensive origin of the latissimus dorsi, which arises from the thoracolumbar fascia (not on the skeleton), the spinous processes of T7 through T12, the iliac crest, and the inferior second or third ribs.
2. Identify the muscle's insertion at the intertubercular groove.
3. Use string to connect the various points of origin with the insertion and note that the muscle passes medial to the shaft of the humerus before inserting on the intertubercular groove.
4. Identify the midline origin of trapezius that spans from the occipital bone down along the ligamentum nuchae to the spinous processes of the thoracic vertebrae.
5. Locate the insertion of trapezius onto a V-shaped region that includes the lateral third of the clavicle, the acromion process, and the scapular spine.
6. Deltoid originates where trapezius inserts. From here, it passes over the shoulder and tapers to insert on the deltoid tuberosity. Identify these attachments.
7. Locate the transverse and spinous processes of cervical and thoracic vertebrae. These are the origins of three muscles that elevate, retract, and inferiorly rotate the scapula—the levator scapulae (transverse processes of C1 to C4), rhomboid minor (spinous processes of C7 to T1), and rhomboid major (spinous processes of T2 to T5).
8. Trace the path of levator scapulae and the rhomboids from their origins to their insertions on the scapula's medial border (Figures 11.14 and 11.15).
9. Locate the origin of teres major at the inferior angle of the scapula, then attach a string from there to the insertion at the intertubercular groove on the humerus. This illustrates the position of the teres major.
10. Observe that teres major and latissimus dorsi have similar insertions and, thus, similar functions (Figure 11.14 and Table 11.7B).
11. Identify the spine of the scapula. This bony landmark separates the two fossae on the posterior surface of the scapula. The supraspinous fossa, superior to the spine, is the origin of **supraspinatus.** The infraspinous fossa, inferior to the spine, is the origin of **infraspinatus. Teres minor** also originates on the posterior aspect of the scapula along the lateral margin of the bone. Locate the greater tu-

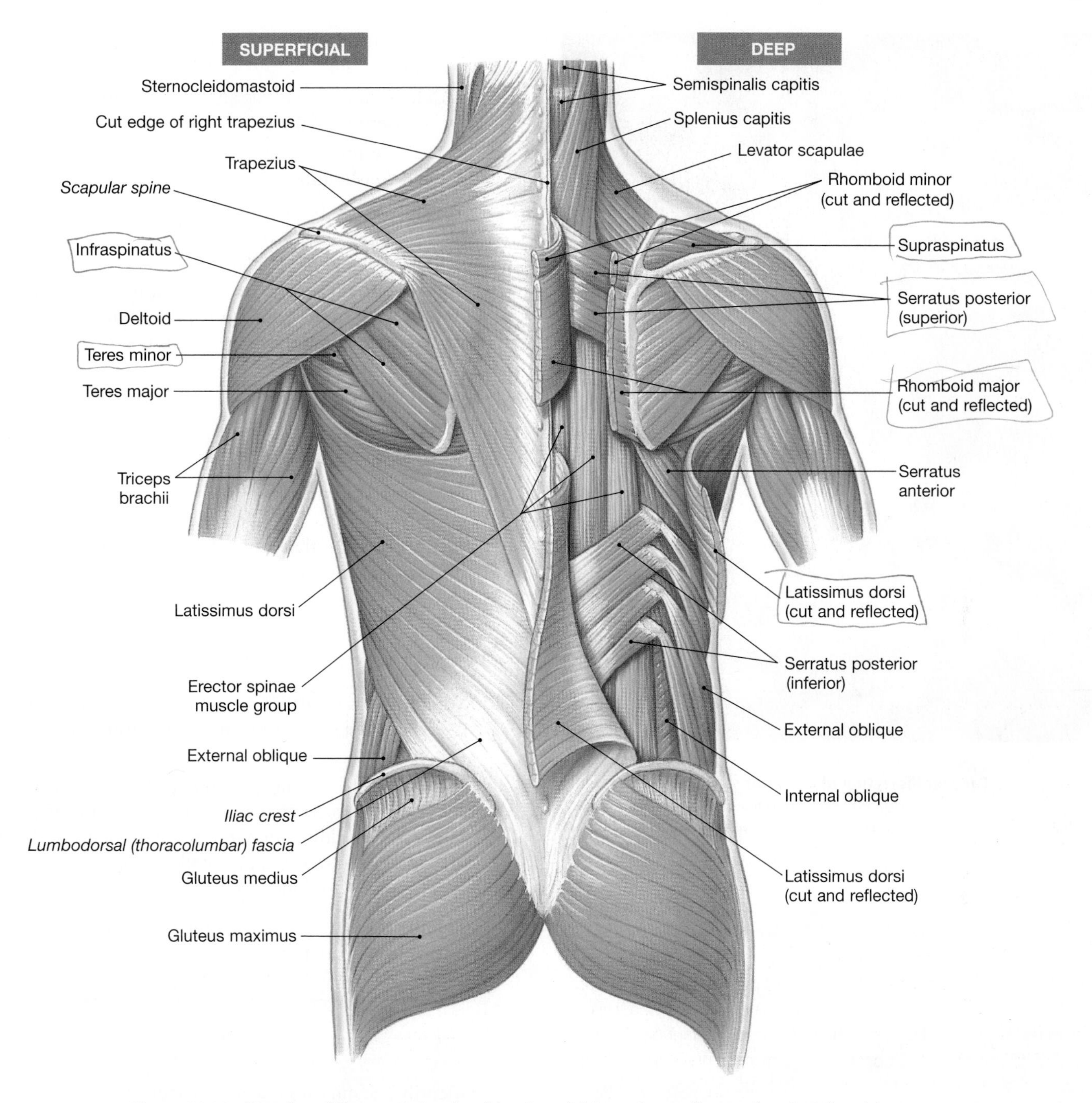

Figure 11.14 Muscles of the posterior shoulder. Superficial muscles are illustrated on the left and the deep muscles on the right.

bercle of the humerus, near the shoulder joint. These three muscles insert on this surface.

12. The anterior surface of the scapula is called the subscapular fossa, which is the origin for **subscapularis**. From its origin, subscapularis inserts onto the lesser tubercle of the humerus, just medial to the larger greater tubercle.
13. The four muscles just described—supraspinatus, infraspinatus, teres minor, and subscapularis—connect the scapula to the greater and lesser tubercles of the humerus. With four pieces of string, mark off the origins and insertions of these muscles. Notice that the insertions on the tubercles of the humerus form a cuff around the shoulder joint. Functionally, these muscles serve as lateral and medial rotators of the arm (Table 11.8B). Consequently, they are referred to as the rotator cuff muscles.

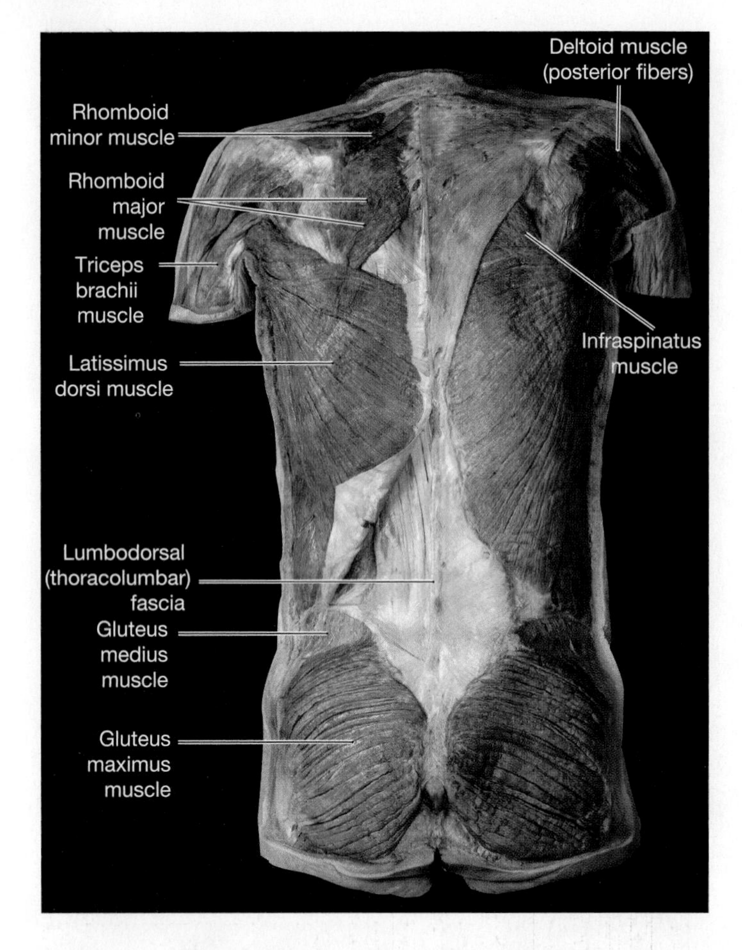

Figure 11.15 Cadaver dissection of posterior shoulder muscles. On the left side, the trapezius has been removed to reveal deeper muscles.

CLINICAL CORRELATION

The term *rotator cuff* refers to a reinforcing connective tissue sheath around the shoulder that is formed by the inserting tendons of the supraspinatus, infraspinatus, teres minor, and subscapularis muscles. The rotator cuff strengthens the shoulder joint on all sides except its inferior aspect. Consequently, most shoulder dislocations occur when the head of the humerus is displaced inferiorly. The rotator cuff can be stretched or torn if it is exposed to excessive rotational forces (torque). The violent twisting actions that occur while throwing a baseball or football make rotator cuff injuries common among baseball pitchers and football quarterbacks.

14. Use an anatomical model, chart, or atlas to identify the posterior shoulder muscles and verify the following:
    - Latissimus dorsi and trapezius are relative large superficial back muscles that both have large origins, but taper to relatively small insertions.
    - Levator scapulae and the rhomboids connect the vertebral column to the scapula and lie deep to the trapezius.
    - Latissimus dorsi and teres major have similar insertions and functions.
    - Deltoid drapes over the shoulder. Its origin is similar to the insertion of the trapezius.
    - The inserting tendons of the rotator cuff muscles reinforce the shoulder joint.
15. Watch your lab partner flex the arm by raising it anteriorly. Pectoralis major is the prime flexor of the arm, with deltoid acting as asynergist.
16. Watch your lab partner extend the arm by swinging it posteriorly. Latissimus dorsi is the prime extensor of the arm, with teres major and deltoid acting as synergists.

    Identify muscles that are antagonists of pectoralis major during arm flexion.

    ____

    ____

    Identify muscles that are antagonists of latissimus dorsi during arm extension.

    ____

    ____

17. Watch your lab partner abduct the arm by swinging it laterally. Deltoid is the prime mover for arm abduction, with supraspinatus assisting.
18. Watch your lab partner return to anatomical position from the abducted position, by moving the arm toward the body. This is adduction of the arm and the prime movers are latissimus dorsi and pectoralis major; teres major and subscapularis are synergists.

    Identify the muscles that are antagonists of the deltoid during arm abduction.

    ____

    ____

    Identify the muscles that are antagonists of latissimus dorsi and pectoralis major during arm adduction.

    ____

    ____

19. Watch as your lab partner rotates the arm. Notice that during medial rotation, the anterior surface of the humerus rotates toward the midline of the body and during lateral rotation, it rotates away. For medial rotation, the prime movers are latissimus dorsi and subscapularis, and synergists are pectoralis major, teres major, and deltoid. For lateral rotation, deltoid is the prime mover and infraspinatus and teres minor are synergists.

Identify the muscles that are antagonists of latissimus dorsi and subscapularis during medial rotation of the arm.

____

____

Identify the muscles that are antagonists of the deltoid during lateral rotation of the arm.

____

____

20. Some shoulder muscles connect the scapula to bones of the axial skeleton (ribs and vertebrae), and produce movements through the sternoclavicular and acromioclavicular joints. Work with your lab partner to demonstrate the following actions of the scapula, starting in anatomical position.
    - Abduct the arm parallel to the floor, then move it anteriorly. As you do this action, the scapula is protracted. Now move your arm posteriorly as far as you can. During this action, the scapula is retracted. Protraction is done by serratus anterior and retraction by the rhomboids and trapezius.
    - Abduct your arm past the parallel position as if you were "raising your hand." To do this, the scapula is rotated superiorly by serratus anterior. Returning your arm to anatomical position requires inferior rotation of the scapula by pectoralis minor and levator scapulae.
    - Watch the scapula move as your partner shrugs his or her shoulders. Scapular elevation is carried out by trapezius, levator scapulae, and the rhomboids. Scapular depression is done by pectoralis minor and trapezius.

WHAT'S IN A WORD Some muscles are named according to their general shape. Consider the following examples.

- The symbol for the Greek letter, delta (Δ), is a triangle. The deltoid muscle is the triangle-shaped muscle that drapes over the shoulder.
- The trapezius has the general shape of a trapezoid. Its name is derived from Greek and means "trapezoid shape."
- The word *rhomboid* is from Greek, and refers to a parallelogram. Both rhomboid muscles are shaped like this geometric form.

QUESTIONS TO CONSIDER 1. Consider the following muscles.

- Pectoralis minor, serratus anterior, and the rhomboids connect the scapula to bones of the axial skeleton.
- Deltoid and supraspinatus muscles originate on the pectoral girdle and insert on the humerus. They abduct the arm.

If you lost the function of pectoralis minor, serratus anterior, and the rhomboids,

a. How would this affect the stability of the scapula?

____

____

b. How would this affect your ability to abduct your arm?

____

____

2. Pectoralis minor and serratus anterior act as respiratory muscles during forced inspiration.

a. Describe the origins and insertions of these muscles when they act during respiration.

____

____

____

____

b. Speculate on what their function might be as respiratory muscles.

____

____

3. From a functional standpoint, explain why the deltoid is an extremely important muscle.

____

____

____

4. Review the insertions of the rotator cuff muscles on the humerus and explain why inferior dislocations of the shoulder joint are the most common.

____

____

Muscles of the Arm

Muscles of the arm are divided into anterior and posterior compartments. The anterior compartment muscles are flexors of the arm and forearm (the flexor compartment) and are innervated by the musculocutaneous nerve. The posterior compartment muscles are extensors of the arm and forearm (the extensor compartment) and are innervated by the radial nerve. The specific actions and bony attachments of the arm muscles are summarized in Table 11.8.

Table 11.8 Muscles of the Arm

Muscle Group/Muscle	Origin	Insertion	Action	Innervation
A. Anterior compartment				
1. Biceps brachii	*Long head:* superior margin of glenoid fossa of scapula *Short head:* coracoid process of scapula	Radial tuberosity of radius; fascia of medial forearm via bicipital aponeurosis	Flexion and supination of forearm; flexion of arm	Musculocutaneous nerve (C5–C6)
2. Brachialis	Anterior surface of humerus	Coronoid process of ulna	Flexion of forearm	As above
3. Coracobrachialis	Coracoid process of scapula	Medial surface of humerus	Flexion and adduction of arm	Musculocutaneous nerve (C5–C7)
B. Posterior compartment				
Triceps brachii	*Long head:* inferior margin of glenoid fossa of scapula *Lateral head:* posterior surface of humerus *Medial head:* posterior surface of humerus, inferior to origin of lateral head	Olecranon process of ulna	Extension of forearm; long head extends and adducts arm	Radial nerve (C6–C8)

Anterior Arm Compartment

The anterior compartment of the arm includes three muscles (Figure 11.16a). The most superficial muscle is the **biceps brachii**. The two deep muscles are the **brachialis** and the **coracobrachialis**.

ACTIVITY 11.9 Identifying Anterior Arm Muscles

1. On a skeleton, locate the scapular origins of the two heads of biceps brachii.
   - Superior margin of the glenoid fossa (**long head**)
   - Coracoid process (**short head**)
2. Identify the muscle's insertion on the radial tuberosity.
3. Connect the origins and insertion with string and notice that biceps brachii crosses both the shoulder and elbow and, consequently, acts at both joints.
4. Identify the origin of brachialis from the distal anterior surface of the humerus.
5. Next, identify its insertion on the coronoid process of the ulna.
6. Run a string between these attachments and verify that brachialis crosses only the elbow joint.
7. Locate the attachments of coracobrachialis. As its name implies, it originates from the coracoid process on the scapula, and inserts onto the medial shaft of the humerus.
8. Run string between these attachments to verify that coracobrachialis crosses only the shoulder joint.
9. Use an anatomical model, chart, or atlas to identify the anterior arm muscles and verify the following (Figure 11.16a):
   - Biceps brachii is the most superficial muscle of the anterior arm and has two heads of origin: a long head and a short head.
   - Brachialis lies deep to the biceps brachii.
   - Coracobrachialis is a slender muscle deep and medial to the biceps brachii.
10. Biceps brachii and brachialis are the principal flexors of the forearm. Place your hand on the anterior surface of your lab partner's arm while he or she flexes the forearm. You will feel biceps brachii contract.
11. Biceps brachii is also the prime mover for forearm supination. Place your hand across your partner's biceps and feel the change as he or she supinates the forearm. Coracobrachialis is a synergist during adduction and flexion of the arm. Identify the muscles that the coracobrachialis assists during the following:
    - Adduction of the arm: ____________________
    - Flexion of the arm: ____________________

QUESTIONS TO CONSIDER 1. Flexion of the forearm is not the same as flexion of the arm. Briefly compare these two movements.

____

____

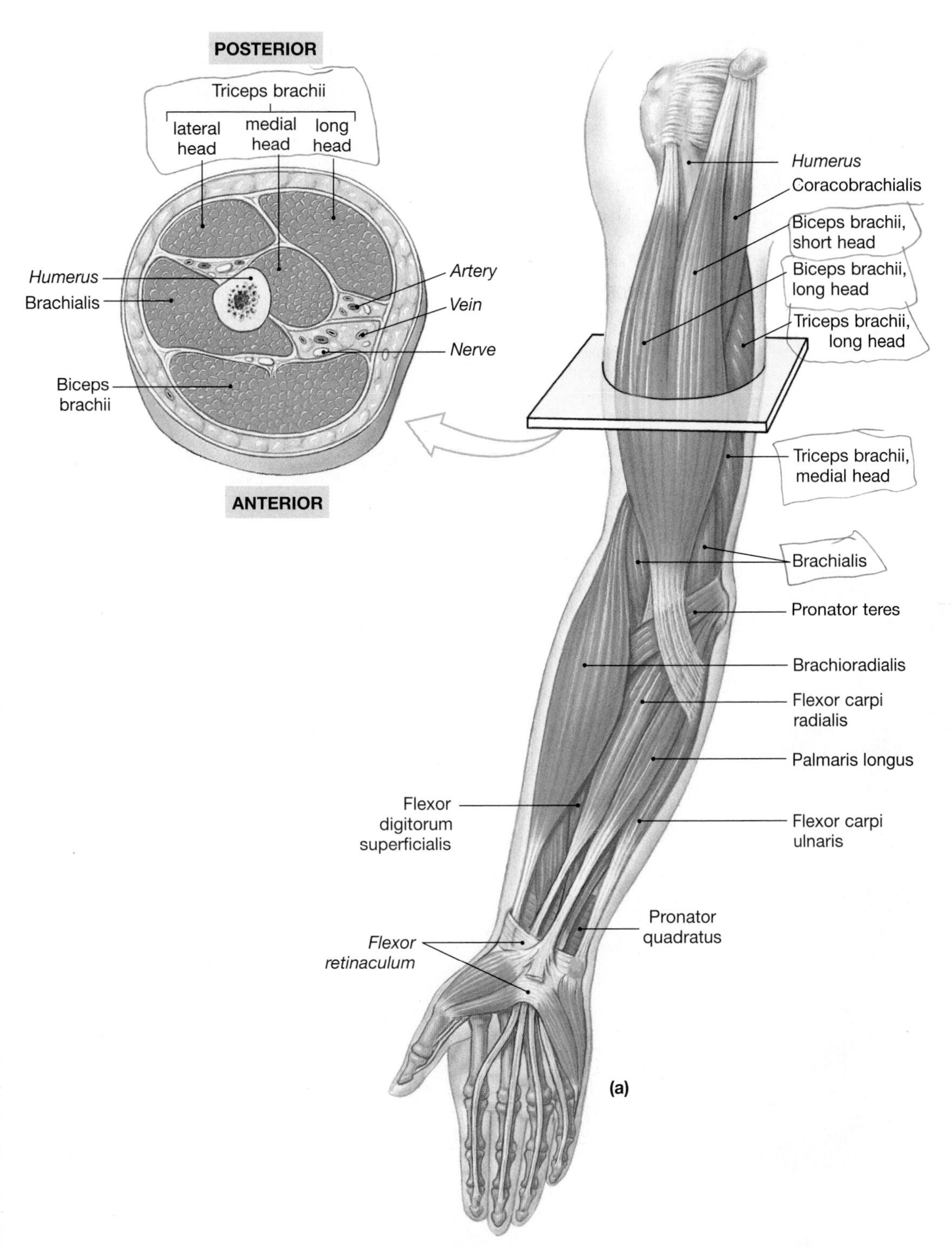

Figure 11.16 Muscles of the arm and superficial muscles of the forearm. a) Anterior view with an accompanying cross section of the arm muscles. *(continues)*

2. The brachialis does not act at the shoulder and the coracobrachialis does not act at the elbow joint. Can you explain why?

____

____

Posterior Arm Compartment

The **triceps brachii** (Figure 11.16b) is the only muscle in the posterior compartment of the arm. The muscle has three heads of origin that are attached to the scapula and humerus, and a common insertion on the ulna.

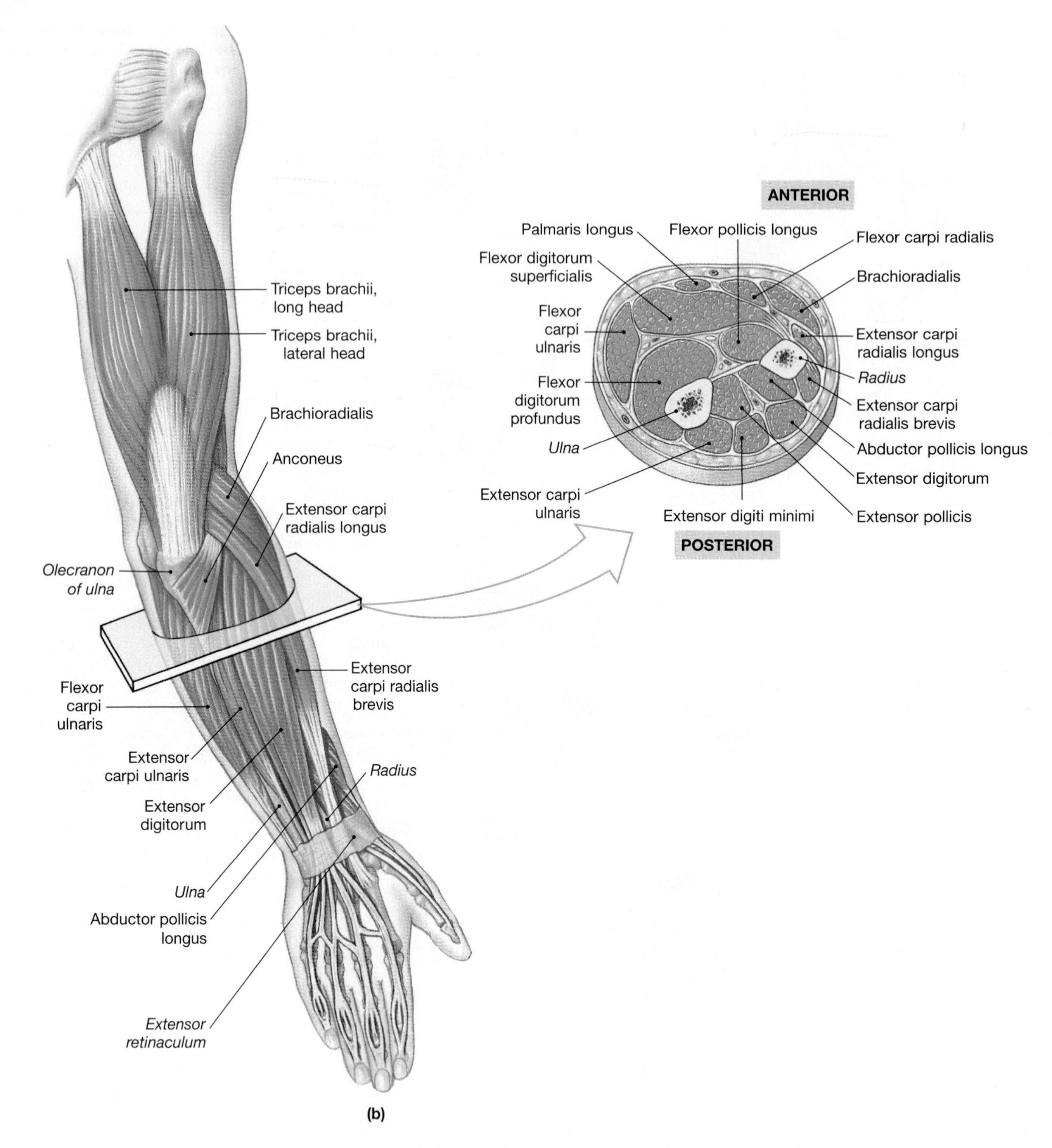

Figure 11.16 **Muscles of the arm and superficial muscles of the forearm *(continued)*. b)** Posterior view with an accompanying cross section of the forearm muscles.

ACTIVITY 11.10 Identifying Triceps Brachii

1. On a skeleton, identify the three origins of the triceps brachii.
   - **Long head**—inferior margin of the glenoid fossa of the scapula
   - **Lateral head**—proximal end of the posterior humeral shaft
   - **Medial head**—distal end of the posterior humeral shaft
2. Identify the muscle's insertion on the olecranon process of the ulna.
3. Connect these attachments with string and verify that the triceps brachii crosses the shoulder and the elbow, and therefore performs actions at both joints.
4. Using an anatomical model, chart, or atlas, identify the triceps brachii and verify the following (Figure 11.16b):
   - The triceps brachii is the only posterior muscle of the arm.
   - The three heads of the triceps brachii have a common tendon of insertion attached to the olecranon process of the ulna.
5. Triceps brachii is the prime extensor of the forearm. Perform and observe this action with your lab partner. Identify the muscles that act as antagonists of the triceps brachii.

   ____

6. Triceps brachii also acts at the shoulder as a synergist during adduction and extension of the arm. Identify the muscles that triceps brachii assists during the following:
   - Adduction of the arm: ______________________________
   - Extension of the arm: ______________________________

QUESTIONS TO CONSIDER 1. Explain why the force of gravity will allow you to extend your forearm even if the triceps brachii is not functioning.

____

____

2. Explain how extension of the forearm differs from extension of the arm.

____

____

Muscles of the Forearm

Similar to the muscles of the arm, the muscles of the forearm are divided into an anterior and a posterior compartment. The principal actions of anterior compartment muscles are flexion of the wrist and fingers, and pronation of the forearm. Most of these muscles are innervated by the median nerve (the exceptions to this rule are indicated in Table 11.9). Posterior compartment muscles extend the wrist and fingers, and supinate the forearm. All these muscles are supplied by the radial nerve. The specific actions and bony attachments of the forearm muscles are described in Table 11.10.

Anterior Forearm Compartment

The anterior forearm compartment contains three muscular layers: the superficial layer contains four muscles, the intermediate layer has one, and the deep layer has three (Figures 11.16a, 11.17a and b). The superficial and intermediate muscles have a common origin at the medial epicondyle of the humerus. The latter bony landmark is often called the **common flexor origin**.

Many muscles in the anterior compartment have long tendons that travel through the **carpal tunnel** to reach their insertions on the hand. The floor of the carpal tunnel is the concave surface formed by the carpal bones. The roof is formed by the **flexor retinaculum**, a connective tissue band that stretches over the bony concavity (Figure 11.18a).

CLINICAL CORRELATION

Many long tendons from muscles of the anterior forearm pass through the carpal tunnel to reach their insertions on the fingers. The median nerve also passes through the carpal tunnel and supplies the skin over roughly the lateral half of the hand and several small hand muscles, including muscles that move the thumb. Individuals who perform tasks involving repeated flexion of the wrists and fingers may develop carpal tunnel syndrome. This condition is characterized by compression of the median nerve caused by inflammation and swelling of the tendons passing through the carpal tunnel. Symptoms of carpal tunnel syndrome include tingling sensations in the skin covering the thumb, index finger, middle finger, and ring finger, and difficulty opposing and flexing the thumb. The partial loss of thumb function means fine motor activities that depend on normal thumb function will be impaired. Usually, rest and anti-inflammatory drugs can diminish the symptoms of carpal tunnel syndrome. In severe cases, however, surgically cutting the flexor retinaculum to relieve pressure (carpal tunnel release) might be necessary.

ACTIVITY 11.11 Identifying Anterior Forearm Muscles

1. On the skeleton, identify the medial epicondyle of the humerus. This is the common flexor origin for all the superficial and intermediate muscles of the anterior forearm. Some muscles have additional origins on the proximal shafts of the ulna and radius.
2. Attach a string to the common flexor origin and extend it inferolaterally to midway along the lateral aspect of the radial shaft. The string marks the position of **pronator teres**. Realize that this muscle has a second head of origin on the coronoid process of the ulna. The primary action of pronator teres (pronation of the forearm) occurs at the proximal radioulnar joint.

Table 11.9 Muscles of the Forearm

Muscle group/muscle	Origin	Insertion	Action	Innervation
A. Anterior compartment				
Superficial muscles				
1. Pronator teres	Medial epicondyle of humerus; coronoid process of ulna	Lateral surface of radius	Pronation and flexion of the forearm	Median nerve (C6–C7)
2. Flexor carpi radialis	Medial epicondyle of humerus	Base of metacarpal of the second digit	Flexion and abduction of the wrist	As above
3. Palmaris longus	As above	Flexor retinaculum; palmar aponeurosis	Flexion of the wrist	Median nerve (C7–C8)
4. Flexor carpi ulnaris	Medial epicondyle of humerus; olecranon process of radius	Medial carpal bones; metacarpal of fifth digit	Flexion and adduction of the wrist	Ulnar nerve (C7–C8)
Intermediate muscle				
5. Flexor digitorum superficialis	Medial epicondyle of humerus; coronoid process of ulna; anterior surface of radius	Middle phalanges of medial four digits	Flexion of wrist and fingers	Median nerve (C7, C8, T1)
Deep muscles				
6. Flexor digitorum profundus	Medial and anterior surfaces of ulna; interosseous membrane	Distal phalanges of medial four digits	Flexion of wrist and fingers	*Medial part:* ulnar nerve (C8, T1) *Lateral part:* median nerve (C8, T1)
7. Flexor pollicis longus	Anterior surface of radius; interosseous membrane	Distal phalanx of thumb	Flexion of the wrist and thumb	Median nerve (C8, T1)
8. Pronator quadratus	Anterior surface of distal ulna	Anterior surface of distal radius	Pronation of the forearm	As above

(continues)

3. Attach three additional pieces of string to the common flexor origin and extend them toward the wrist joint.
   - Attach the first string to the base of the second metacarpal bone to represent **flexor carpi radialis.** Note that the tendon for this muscle passes through the carpal tunnel to reach its insertion.
   - Attach the second string to the base of the fifth metacarpal bone to represent **flexor carpi ulnaris.** Realize that this muscle has a second head of origin on the olecranon process of the ulna. Its tendon of insertion does not pass through the carpal tunnel.
   - Run the third string between the first two to approximate the path of the **palmaris longus.** This muscle does not insert onto a bone. Rather, it attaches to the flexor retinaculum, which forms the roof of the carpal tunnels.

WHAT'S IN A WORD The term *carpi* refers to the wrist; any muscle with *carpi* in its name will act at the wrist joint.

4. Examine the positions of the strings. The muscles cross both the elbow joint and the wrist joint, and therefore will have functions at both locations. However, their primary actions will occur at the wrist joint.
5. Verify that each of the four medial digits of the hand (the fingers, but not the thumb) have three phalanges. The intermediate muscle of the anterior forearm, **flexor digitorum superficialis**, inserts on the middle phalanx of each finger. This muscle originates on the common flexor tendon, the proximal end of the ulna, and the anterior surface of the radial shaft. It descends through the anterior forearm and its four tendons pass through the carpal tunnel to reach their insertions. Verify on the skeleton that, between its origin and insertion, flexor digitorum superficialis crosses several joints, including the elbow, wrist, metacarpophalangeal (knuckles), and proximal interphalangeal (fingers).
6. Use string to mark the path of the **pronator quadratus**, a deep, rectangular muscle in the anterior forearm that orig-

Table 11.9 (continued)

Muscle group/muscle	Origin	Insertion	Action	Innervation
B. Posterior compartment				
Superficial muscles				
1. Brachioradialis	Lateral supracondylar ridge of humerus	Lateral surface of distal radius	Flexion of the forearm	Radial nerve (C5–C7)
2. Extensor carpi radialis longus	As above	Metacarpal of the second digit	Extension and abduction of the wrist	Radial nerve (C6, C7)
3. Extensor carpi radialis brevis	Lateral epicondyle of humerus	Metacarpal of the third digit	Extension and abduction of the wrist	Radial nerve (C7, C8)
4. Extensor digitorum	As above	Extensor expansions of medial four digits	Extension of wrist and fingers	As above
5. Extensor digiti minimi	As above	Extensor expansion of the fifth digit	Extension of the fifth digit	As above
6. Extensor carpi ulnaris	Lateral epicondyle of humerus; posterior surface of ulna	Metacarpal of the fifth digit	Extension and adduction of the wrist	As above
7. Anconeus	Lateral epicondyle of humerus	Olecranon process of ulna	Very weak extension of the forearm	Radial nerve (C7, C8, T1)
Deep muscles				
8. Supinator	Lateral epicondyle of humerus; proximal ulna	Posterior, lateral, and anterior surfaces of proximal radius	Supination of the forearm	Radial nerve (C5, C6)
9. Abductor pollicis longus	Posterior surfaces of radius and ulna; interosseous membrane	Metacarpal of thumb	Abducts and extends thumb	Radial nerve (C7, C8)
10. Extensor pollicis longus	Posterior surface of ulna and interosseous membrane	Distal phalanx of thumb	Extends thumb	As above
11. Extensor pollicis brevis	Posterior surface of radius and interosseous membrane	Proximal phalanx of thumb	Extends thumb	As above
12. Extensor indicis	Posterior surface of ulna and interosseous membrane	Extensor expansion of second digit	Extends the second digit	As above

inates at the distal end of the ulna and inserts on the distal end of the radius.

7. Locate the attachments of **flexor digitorum profundus**, which extends from the shaft of ulna to the distal phalanges of the four medial digits (fingers).
8. Locate the attachment of **flexor pollicis longus**, which travels from the shaft of the radius to the distal phalanx of the thumb.
9. Observe that these latter two muscles do not cross the elbow joint, and their tendons of insertion pass through the carpal tunnel at the wrist joint.

WHAT'S IN A WORD The term *pollicis* refers to the thumb; any muscle with *pollicis* in its name will act on the thumb.

10. Use an anatomical model, chart, or atlas to verify the following (Figures 11.16a, 11.17a and b):
    - The medial epicondyle of the humerus is a common origin for several muscles.
    - Many of the muscles cross two or more joints.
    - Flexor carpi radialis and the muscles that insert on digits—flexor digitorum superficialis, flexor digitorum profundus, and flexor pollicis longus—have tendons that pass through the carpal tunnel.
    - Flexor digitorum profundus travels deep to the flexor digitorum superficialis.
    - Palmaris longus is a relatively small muscle with a long tendon. It is not functionally important and, in fact, is commonly missing on one or both sides.

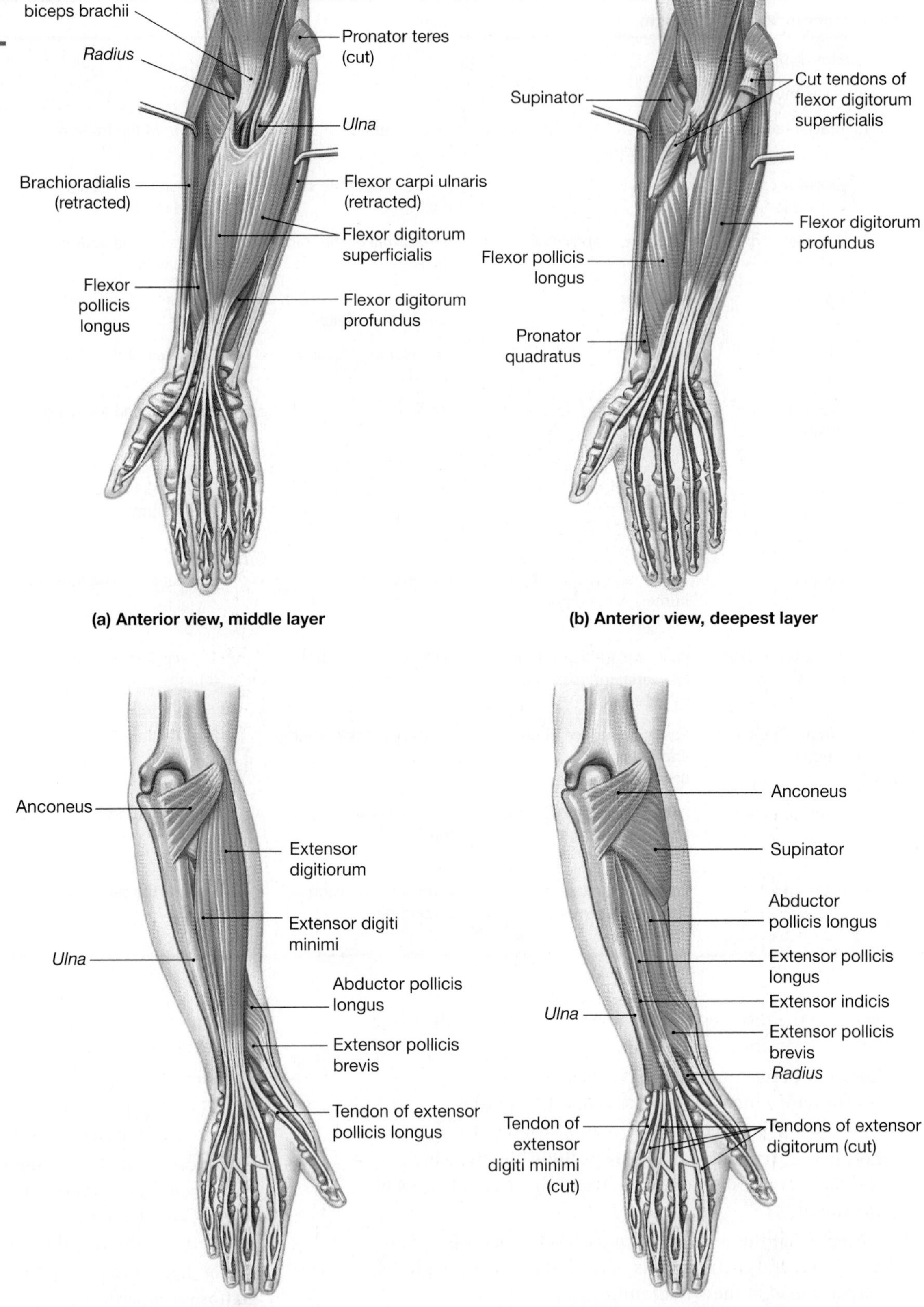

Tendon of biceps brachii
Radius
Pronator teres (cut)
Ulna
Brachioradialis (retracted)
Flexor carpi ulnaris (retracted)
Flexor digitorum superficialis
Flexor pollicis longus
Flexor digitorum profundus
(a) Anterior view, middle layer
Supinator
Cut tendons of flexor digitorum superficialis
Flexor digitorum profundus
Flexor pollicis longus
Pronator quadratus
(b) Anterior view, deepest layer
Anconeus
Extensor digitiorum
Extensor digiti minimi
Ulna
Abductor pollicis longus
Extensor pollicis brevis
Tendon of extensor pollicis longus
(c) Posterior view, middle layer
Anconeus
Supinator
Abductor pollicis longus
Extensor pollicis longus
Extensor indicis
Ulna
Extensor pollicis brevis
Radius
Tendon of extensor digiti minimi (cut)
Tendons of extensor digitorum (cut)
(d) Posterior view, deepest layer

Figure 11.17 Deep muscles of the forearm. a) and **b)** Anterior views; **c)** and **d)** posterior views.

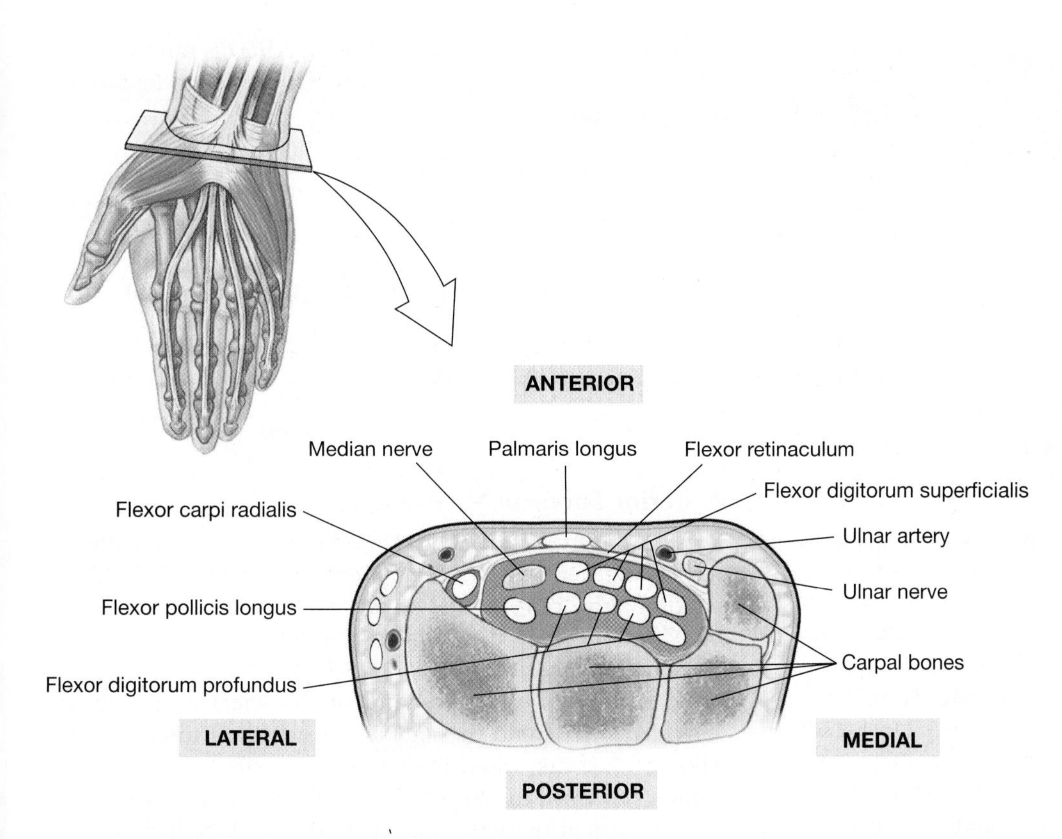

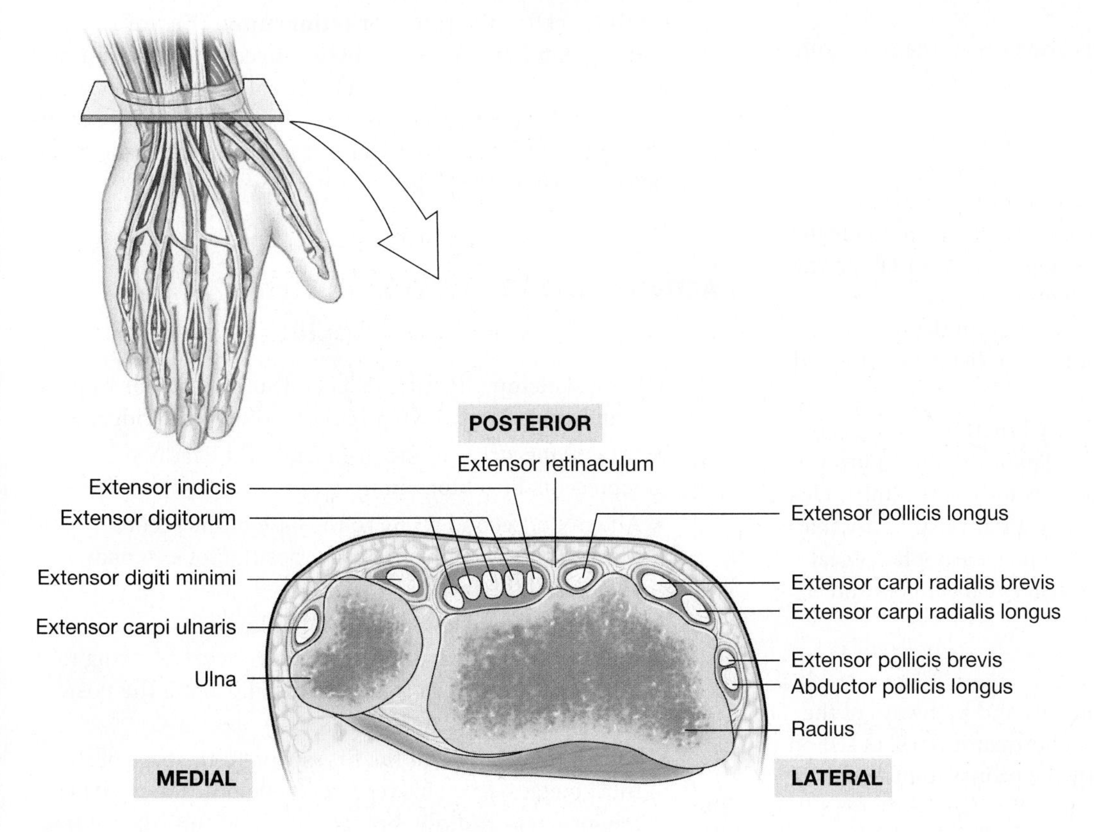

Figure 11.18 Structures passing through the tunnels formed by bones and retinacula at or near the wrist.
a) Cross section of the carpal tunnel along the proximal row of carpal bones of the right wrist. The tendons of several anterior forearm muscles and the median nerve pass through the carpal tunnel on their way to the hand.
b) Cross section of the tunnels formed by the extensor retinaculum at the distal ends of the right radius and ulna, just proximal to the wrist. The tendons of several posterior forearm muscles pass deep to the extensor retinaculum as they travel to the hand.

CLINICAL CORRELATION

The palmaris longus tendon is often used to replace damaged tendons or ligaments around joints during reconstructive surgery. For example, "Tommy John surgery" (named after former New York Yankees pitcher Tommy John) has become an increasingly more common operation to salvage the careers of professional baseball pitchers. During this procedure, the tendon of the palmaris longus is often used to repair the ulnar (medial) collateral ligament of the elbow.

11. The pronator quadratus is the prime mover for forearm pronation, while the pronator teres acts as a synergist for this action. Demonstrate this action on your own forearm.

    The muscles that function as antagonists of pronator quadratus will perform what action?

    ____

12. From the anatomical position, have your lab partner flex his or her wrists while you observe. The prime movers for wrist flexion are the flexor carpi ulnaris, flexor carpi radialis, and flexor digitorum superficialis. The palmaris longus is a relatively weak muscle with an insignificant functional role.

    Flexor digitorum profundus also crosses the wrist joint. Speculate on its functional role in wrist flexion.

    ____

13. In addition to wrist flexion, the flexor carpi radialis abducts the wrist, and the flexor carpi ulnaris adducts the wrist. From the anatomical position, have your lab partner perform these actions while you observe. Notice that during abduction, the hand is bent away from the body in the coronal plane. During adduction, the wrist is bent toward the body.
14. Flexion of the digits is the primary function of the flexor digitorum superficialis (fingers), flexor digitorum profundus (fingers), and the flexor pollicis longus (thumb). Flex your fingers and thumb by making a tight fist. Verify that when you perform this action, the metacarpophalangeal (knuckles) joints and the interphalangeal (fingers and thumb) joints are flexed.

QUESTIONS TO CONSIDER

1. Review the attachments and pathways of the pronator quadratus and pronator teres. Based on your observations, speculate on why the pronator quadratus is the more powerful pronator.

____

____

____

2. Explain why the flexor digitorum superficialis can assist other muscles in flexing the forearm, but the flexor digitorum profundus cannot.

____

____

3. If you lost all function of flexor carpi radialis, flexor carpi ulnaris, and palmaris longus, would you still have the ability to flex your wrist? Explain.

____

____

Posterior Forearm Muscles

The posterior compartment of the forearm contains seven superficial muscles and five deep muscles (Figures 11.16b, 11.17c and d). Many of these muscles have a common origin at the lateral epicondyle of the humerus, which is often called the **common extensor origin**.

Most muscles in the posterior compartment have long tendons that insert onto the digits. To reach their insertions, the tendons must first travel through tunnels, similar to the carpal tunnels described earlier. The floors of these tunnels are formed by the bones at the wrist and the roof by a thick band of connective tissue called the **extensor retinaculum** (Figure 11.18b).

A strong band of connective tissue covers the posterior and lateral surfaces of the proximal phalanges of the medial four digits (the fingers). These coverings are called the **extensor expansions** (Figure 11.19b). They serve as the insertion sites for muscles that extend the fingers.

ACTIVITY 11.12 Identifying Posterior Forearm Muscles

1. On the skeleton, identify the lateral supracondylar ridge of the humerus. Attach two pieces of string to the ridge.
   - Attach one string to the distal end of the radius. This represents **brachioradialis.**
   - Attach the second string to the base of the second metacarpal. This illustrates the position of **extensor carpi radialis longus.**
2. Identify the lateral epicondyle of the humerus, called the common extensor origin. Attach three pieces of string to it. Allow all the strings to extend distally along the posterior forearm.
   - Attach the free end of the first string to the base of the third metacarpal. This represents the position of the **extensor carpi radialis brevis.** Note that this muscle travels medially to the extensor carpi radialis longus.
   - Attach the second string to the base of the fifth metacarpal. This illustrates the position of the **extensor carpi ulnaris**, the most medial muscle in the posterior forearm.

- Attach the third string to the proximal phalanx of the fifth digit. This represents the **extensor digiti minimi.**

3. **Extensor digitorum**, a superficial muscle, also arises at the lateral eipcondyle and travels medial to extensor carpi ulnaris and extensor digiti minimi. In the hand, its tendon splits into four branches that insert on the extensor expansions of the fingers at the proximal phalanges. Identify these bones on the skeleton, and visualize this muscle by extending string between its origin and insertions.
4. Locate the posterior surface of the proximal ulna, just distal to the olecranon process. The **supinator** originates here and from the lateral epicondyle. It inserts on the posterior, lateral, and anterior surfaces of the proximal radius, wrapping around the bone. Use string to follow its path on the skeleton.
5. Identify the distal halves of the posterior radial and ulnar shafts. Four deep posterior forearm muscles originate here and act on the index finger and thumb. On the skeleton, use string to approximate the positions of these muscles by identifying their insertion sites.
   - **Abductor pollicis longus** = base of metacarpal for the thumb
   - **Extensor pollicis brevis** = base of proximal phalanx of the thumb
   - **Extensor pollicis longus** = base of distal phalanx of the thumb
   - **Extensor indicis** = extensor expansion of the index finger
6. Use an anatomical model, chart, or atlas to identify the muscles of the posterior forearm, and verify the following (Figures 11.16b, 11.17c and d):
   - As in the anterior forearm, many posterior forearm muscles cross two or more joints.
   - All superficial muscles of the posterior forearm originate on or near the lateral epicondyle of the humerus.
   - The inserting tendons of all muscles that extend the wrist and fingers pass deep to the extensor retinaculum.
   - Brachioradialis and supinator are the only posterior forearm muscles that do not insert in the hand.
   - With the exception of the supinator, the deep muscles of the posterior forearm originate on middle and distal regions of the ulnar and/or radial shafts and insert on bones of the thumb or index finger.
7. Place your forearm in the midprone position. From this position, flex your forearm. The brachioradialis is a strong forearm flexor from this position.
8. Watch your lab partner extend his or her wrists. The prime mover for wrist extension is extensor digitorum. Extensor carpi radialis longus, extensor carpi radialis brevis, and extensor carpi ulnaris act as synergists.

   Identify the antagonist muscles for wrist extension.

   ____

9. In addition to wrist extension, the extensor carpi radialis longus and brevis muscles abduct the wrist, and the extensor carpi ulnaris adducts the wrist. Recall that wrist abduction and adduction are also functions of muscles in the anterior forearm. Perform these actions while considering the positions of these muscles.

   Identify the anterior forearm muscle that does the following:

   - Abducts the wrist ______________________________
   - Adducts the wrist ______________________________

10. From the closed fist position, open your hand to expose the palmar (anterior) surface. This action requires you to extend your fingers from the flexed position. Recall that three posterior forearm muscles extend the fingers. Extensor digitorum extends all four fingers, extensor indicis extends the index finger, and extensor digiti minimi extends the pinky.
11. Stand in the anatomical position and perform the following thumb actions.
    - Bend your thumb, in the coronal plane, toward the midline of the body. This is thumb flexion, a function of flexor pollicis longus in the anterior forearm muscle.
    - From the flexed position, move your thumb, in the coronal plane, away from the midline of the body (back to anatomical position and beyond). This is thumb extension, which is a function of extensor pollicis longus and brevis in the posterior forearm.
    - From the anatomical position, move your thumb anteriorly (away from the body) in a sagittal plane. This is thumb abduction, which is a function of the abductor pollicis longus, a posterior forearm muscle.
    - From the abducted position, move your thumb posteriorly (toward the body) in a sagittal plane. This is thumb adduction, which is a function of an intrinsic hand muscle that will be described later.

QUESTIONS TO CONSIDER

1. The extensor carpi radialis longus and extensor carpi radialis brevis abduct the wrist; the extensor carpi ulnaris adducts the wrist. If you lost the function of all these muscles, abduction and adduction of the wrist would still be possible. Explain why.

____

____

2. Compare the actions of supination and pronation of the forearm. Speculate on which action is more powerful. (Hint: When you use a screwdriver with your right hand to turn a screw into a piece of wood, what action are you performing?)

____

____

3. The muscles that flex your wrist and fingers are on the anterior surface of the forearm and hand; the muscles that extend your wrist and fingers are on the posterior surface. With your wrist extended, make a fist by tightly flexing your fingers. Next, with your wrist flexed, make a fist once again. Based on the assumption that making a fist (flexion of the fingers) is the specific action of interest, answer the following questions.

a. Identify one muscle that could act as a prime mover.

____

b. What is the antagonistic action?

____

c. Identify one muscle that could act as an antagonist.

____

d. In which case, with the wrist extended or flexed, was the action of making a fist easier?

____

e. Based on your answer for (d), identify one muscle that acts as a synergist by making it easier to make a fist.

____

f. What is the action performed by the synergist identified in (e)?

____

g. Identify one muscle that could act as both an antagonist and a synergist.

____

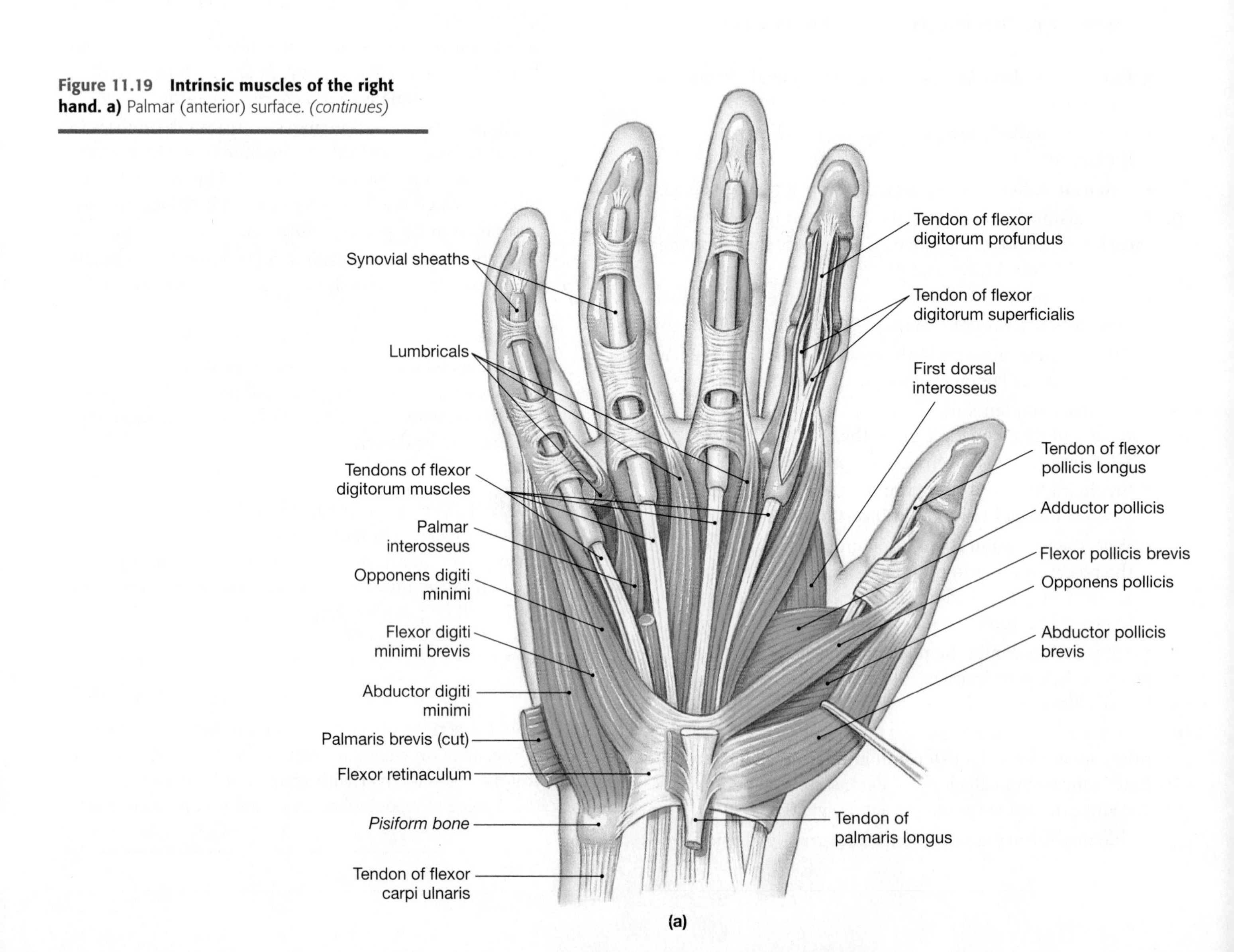

Figure 11.19 Intrinsic muscles of the right hand. a) Palmar (anterior) surface. *(continues)*

Muscles of the Hand

The muscles in this group (Figure 11.19) perform fine motor skills with the hand. Activities such as writing with a pen, grasping the handle of a coffee mug, buttoning a shirt, snapping the fingers, or sewing a hem line on a dress would not be possible without the use of these muscles. The bony attachments, innervations, and functions of the intrinsic hand muscles are described in Table 11.10.

ACTIVITY 11.13 Identifying Intrinsic Hand Muscles

1. On the skeleton, review the bones of the wrist and hand: carpals, metacarpals, and phalanges. The instrinsic muscles of the hand originate and insert on these bones.
2. Identify the lateral carpals—the scaphoid and trapezium. Three of the four muscles of the thenar eminence—**flexor pollicis brevis**, **abductor pollicis brevis**, and **opponens pollicis**—originate on these bones and the flexor retinaculum.
3. Identify the proximal phalanx of the thumb. Flexor pollicis brevis and abductor pollicis brevis insert on the lateral side of the base of this bone.
4. Identify the first metacarpal (the thumb). Opponens pollicis inserts on the lateral side of this bone.
5. Identify the second and third metacarpal bones. The fourth muscle of the thenar eminence, **adductor pollicis**, originates from the shafts and bases of these bones as well as adjacent carpal bones. It inserts on the medial side of the base of the thumb's proximal phalanx.
6. Use string to follow the paths of the muscles of the thenar eminence.
7. Identify the two most medial carpal bones—the pisiform and the hamate. The three muscles of the hypothenar eminence originate from these bones. **Abductor digiti minimi** originates from the pisiform; **flexor digiti minimi** and **opponens digiti minimi** originate from the hook of the hamate and flexor retinaculum.
8. Identify the proximal phalanx of the fifth digit. Abductor digiti minimi and flexor digiti minimi insert on the medial side of the base of this bone.

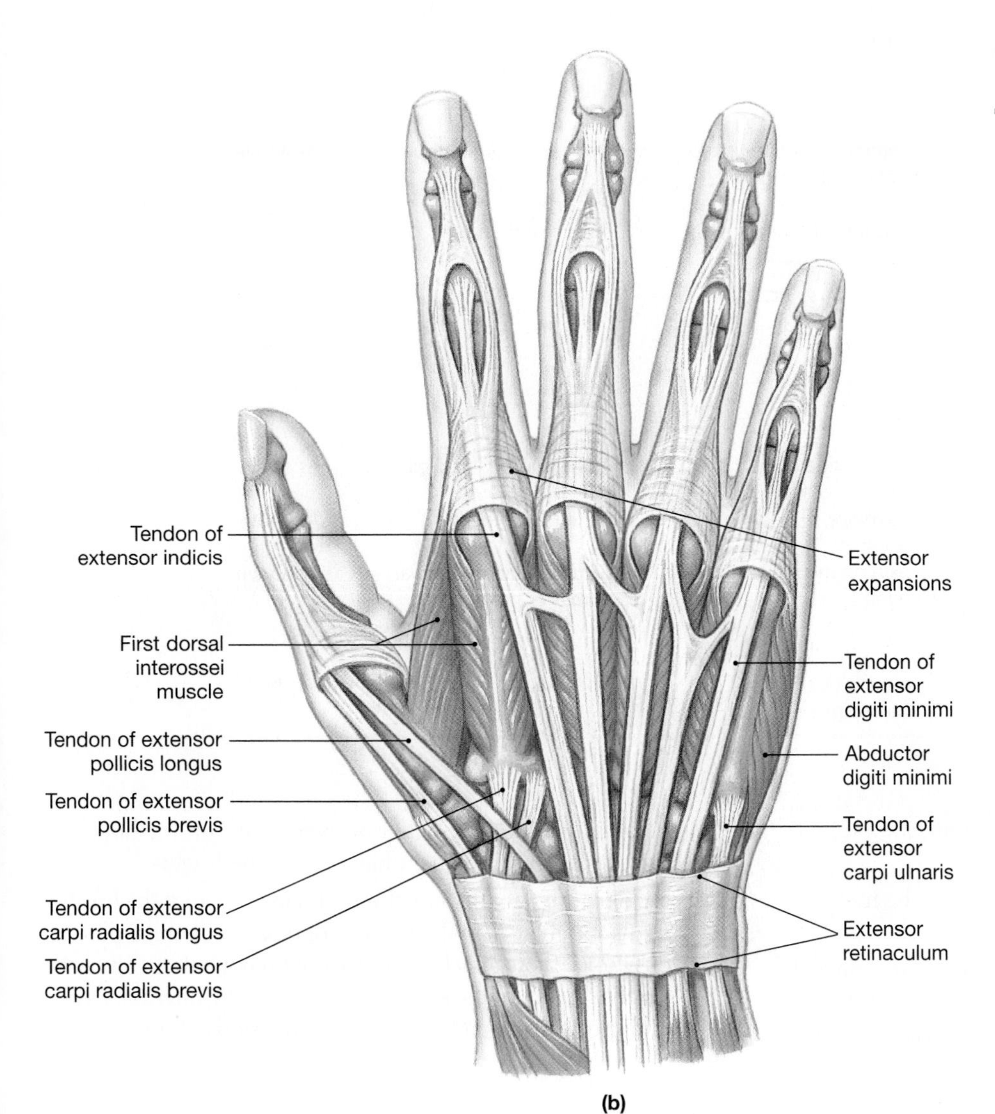

Figure 11.19 Intrinsic muscles of the right hand *(continued).* b) Dorsal surface.

Table 11.10 Intrinsic Muscles of the Hand

Muscle Group/Muscle	Origin	Insertion	Action	Innervation
1. Lumbricals (4)	Tendons of flexor digitorum profundus	Lateral sides of extension Expansions of medial four digits	Flexion at metacarpophalangeal (MP) joints and extension at interphalangeal (IP) joints of medial four digits	*Lumbricals 1 and 2:* median nerve (C8, T1) *Lumbricals 3 and 4:* ulnar nerve (C8, T1)
2. Dorsal interossei (4)	Sides of two adjacent metacarpal bones	Extensor Expansions and proximal phalanges of digits 2, 3, and 4	Abduction of digits 2, 3, and 4; flexion of digits 2, 3, and 4 at MP joints; extension of digits 2, 3, and 4 at IP joints	Ulnar nerve (C8, T1)
3. Palmar interossei (3)	Palmar surfaces of metacarpals of digits 2, 4, and 5	Extensor Expansions and proximal phalanges of digits 2, 4, and 5	Adduction of digits 2, 4, and 5; flexion of digits 2, 4, and 5 at MP joints; extension of digits 2, 4, and 5 at IP joints	As above
4. Muscles of the thenar eminence				
Superficial group				
a. Abductor pollicis brevis	Flexor retinaculum; lateral carpal bones	Lateral side of proximal phalanx of thumb	Abduction of thumb; opposition of thumb	Median nerve (C8, T1)
b. Flexor pollicis brevis	As above	As above	Flexion of thumb; opposition of thumb	As above
Deep group				
a. Opponens pollicis	As above	Lateral side of metacarpal bone of the thumb	Opposition of the thumb	As above
b. Adductor pollicis	*Oblique head:* bases of second and third metacarpal bones and adjacent carpal bones *Transverse head:* palmar surface; third metacarpal	Medial side of proximal phalanx of thumb	Adduction of thumb; flexion of thumb	Ulnar nerve (C8, T1)
5. Muscles of the hypothenar eminence				
Superficial group				
a. Abductor digiti minimi	Pisiform bone	Medial side of proximal phalanx of fifth digit	Abduction of fifth digit	As above
b. Flexor digiti minimi	Hook of the hamate; flexor retinaculum	As above	Flexion of fifth digit	As above
c. Palmaris brevis	Palmar aponeurosis	Skin on medial side of the hand	Wrinkles skin on medial border of palm	As above
Deep muscle				
Opponens digiti minimi	Hook of the hamate; flexor retinaculum	Medial side of fifth metacarpal	Opposition of fifth digit	As above

9. Identify the fifth metacarpal. Opponens digiti minimi inserts on its medial side.
10. Use string to trace the paths of the muscles of the hypothenar eminence.
11. Examine the spaces between the metacarpals. Four **dorsal interossei** and three **palmar interossei** originate from the sides of the metacarpal shafts and fill these spaces. The interossei insert on the extensor expansions on the dorsal surfaces of the proximal phalanges of the fingers.
12. Note that the final group of muscles in the hand, the **lumbricals**, originates on the tendons of flexor digitorum profundus. Since these attachments are not on bone they cannot be demonstrated on the skeleton. Similar to the interossei, the lumbricals insert on the extensor expansions of the fingers.

WHAT'S IN A WORD The term *interossei* is derived from Latin and means "between the bones." The name reflects the position of these muscles between the metacarpal bones.

Each lumbrical is a small, slender muscle that is similar in appearance to an earthworm. The word *lumbrical* is derived from Latin and means "earthworm."

13. Use an anatomical model, chart, or atlas to identify the intrinsic muscles of the hand and to verify the following (Figure 11.19):
    - These muscles are small and attach to bones of the wrist and hand.
    - The four muscles of the thenar eminence act exclusively on the thumb.
    - The three muscles of the hypothenar eminence act exclusively on the pinky finger.
14. The main reason the hand is able to perform fine motor skills is because of the opposable thumb. From anatomical position, perform thumb opposition while your lab partner observes. During this action, the thumb is rotated medially so that it can touch another digit. Opponens pollicis is the prime mover for this action; abductor pollicis brevis is a synergist.
15. To demonstrate the importance of thumb opposition, write your name by holding the pen in the normal fashion between your thumb and index finger. Write your name again, but hold the pen without using your thumb.

    In addition to opposition, the muscles of the thenar eminence also flex, abduct, and adduct the thumb. Identify the muscles, either from the hand or the forearm, that can perform the following actions.
    - Flex the thumb

    ___

    - Abduct the thumb

    ___

    - Adduct the thumb

    ___

16. Abduction and adduction of the fingers are carried out by the dorsal (abduction) and palmar (adduction) interossei. Hold the palm of your hand in front of your face and imagine a vertical midline running through the middle finger. When the fingers are abducted, they move away from the midline; when they are adducted, they move toward the midline. Perform these actions.
17. The interossei muscles and lumbricals flex the metacarpophalangeal (knuckle) joints and extend the interphalangeal (finger) joints. Both forearm and intrinsic hand muscles perform flexion and extension of the fingers. Identify the muscles from both groups that can carry out the following functions.
    - Flexion of the fingers

    ___

    ___

    - Extension of the fingers

    ___

    ___

18. The hypothenar muscles abduct, flex, and oppose the pinky finger. Attempt to oppose the pinky finger. Note that this action is not as efficient (and not as important) as thumb oppostion.

QUESTION TO CONSIDER The median nerve supplies the intrinsic hand muscles that oppose, abduct, and flex the thumb. The ulnar nerve supplies the intrinsic hand muscles that oppose the pinky finger, abduct and adduct the fingers, and adduct the thumb. Consider two possible hand injuries—an injury to the median nerve and an injury to the ulnar nerve. Speculate on which type of injury would be more serious.

___

___

___

Muscles of the Lower Extremity

The muscles of the lower extremity act at the hip, knee, and ankle joints as well as the joints in the foot. They are innervated by nerves that are derived from two spinal nerve plexuses: the **lumbar plexus** that receives contributions from spinal nerves L1 through L4, and the **sacral plexus** that is composed of spinal nerves L4 and L5, and S1 through S4.

Muscles of the Gluteal Region

The gluteal region extends from the iliac crest, superiorly, to the inferior border of the gluteus maximus, inferiorly (Figure 11.20a). The muscles are divided into superficial and deep groups. The superficial group includes the three gluteal muscles: **gluteus maximus**, **gluteus medius**, and **gluteus minimus** (Figures 11.20a and b). Collectively, they form a mass of muscle tissue known as the buttock. These muscles are important extensors and abductors of the thigh. They also stabilize the pelvis and thigh while standing erect. The deep group of gluteal muscles (Figure 11.20b) includes **piriformis**, **obturator internus**, **superior gemellus**,

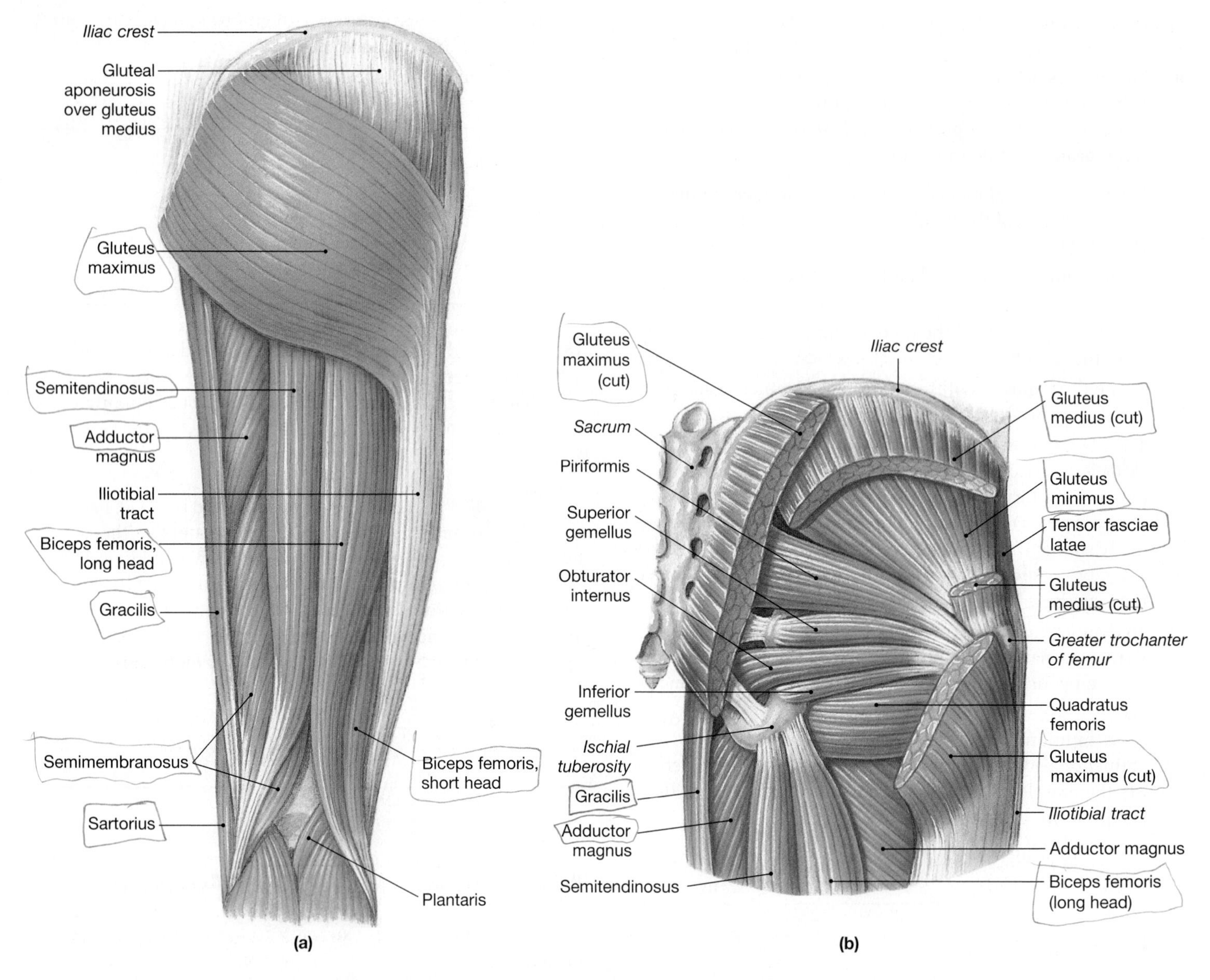

Figure 11.20 Muscles of the gluteal region and posterior thigh. a) Superficial gluteal muscles and posterior thigh muscles, posterior view; **b)** the gluteus maximus and gluteus medius have been cut to reveal the deep gluteal muscles.

inferior gemellus, and **quadratus femoris**. The piriformis abducts the thigh; the other muscles in this group laterally rotate it. The bony attachments, innervations, and specific actions of the gluteal muscles are described in Table 11.11.

ACTIVITY 11.14 Identifying the Gluteal Muscles

1. On the skeleton, identify the posterior end of the iliac crest. Place two fingers on this bony surface, then pass your fingers inferiorly onto the outer surface of the ilium and then the back of the sacrum and coccyx. These locations mark the origin of the massive gluteus maximus.
2. From this origin, gluteus maximus passes inferolaterally and inserts on the gluteal tuberosity of the femur. Locate this on the posterior surface of the femur, lateral to the lesser trochanter. Gluteus maximus also inserts on the **iliotibial tract**, a thick band of deep fascia that runs along the lateral thigh.
3. Identify the ischial tuberosity. The gluteus maximus forms a thick protective pad that passes over the ischial tuberosity and cushions the bone when you sit. The ischial tuberosity can be palpated at the inferior border of the gluteus maximus.
4. Gluteus medius lies deep to gluteus maximus. Locate its origin on the external surface of the ilium along an arc just

Table 11.11 Muscles of the Gluteal Region

Muscle Group/Muscle	Origin	Insertion	Action	Innervation
A. Superficial group				
1. Gluteus maximus	Lateral surface of ilium; dorsal surface of sacrum and coccyx	Iliotibial tract; gluteal tuberosity of femur	Extension and lateral rotation of thigh; powerful Extensor for climbing and running; stabilizes pelvis and thigh while standing erect	Inferior gluteal nerve (L5, S1, S2)
2. Gluteus medius	Lateral surface of ilium	Greater trochanter of femur, lateral surface	Abduction and medial rotation of thigh; steadies the pelvis during walking	Superior gluteal nerve (L5, S1)
3. Gluteus minimus	Lateral surface of ilium, inferior to origin of gluteus medius	Greater trochanter of femur, anterior surface	As above	As above
B. Deep group				
1. Piriformis	Anterior surface of sacrum	Greater trochanter of femur, superior surface	Abducts thigh	Branches of spinal nerves S1 and S2
2. Obturator internus	Internal surface of obturator membrane and adjacent bones	Greater trochanter of femur, medial surface	Laterally rotates thigh	Nerve to obturator internus (L5, S1)
3. Superior gemellus	Ischial spine	As above	As above	As above
4. Inferior gemellus	Ischial tuberosity	As above	As above	Nerve to quadratus femoris (L5, S1)
5. Quadratus emoris	As above	Intertrochanteric crest of femur	As above	As above

inferior to the iliac crest. From there, the muscle travels inferolaterally to the greater trochanter. Trace that path on the skeleton. The gluteus minimus lies deep and has attachments similar to gluteus medius.

CLINICAL CORRELATION

Most of the gluteus medius lies deep to gluteus maximus. However, in the superolateral part of the buttock, gluteus medius extends beyond the lateral border of gluteus maximus (Figure 11.20a). The exposed portion of the gluteus medius is called the safe area, because intramuscular injections can be made without causing injury to the sciatic nerve.

5. Locate the greater trochanter on the femur. This large landmark is the insertion site for all but one of the deep gluteal muscles.
6. Attach a piece of string to the anterior surface of the sacrum, extend it laterally through the greater sciatic notch, and attach it to the superior tip of the greater trochanter. This marks the postion of the piriformis. Notice that the piriformis is partly within the pelvic cavity and partly within the gluteal region.
7. Locate the **obturator foramen**. In life, it is almost completely covered by the **obturator membrane**. The obturator internus muscle originates on the inner surface of the obturator membrane and the surrounding bones.
8. Attach a string to the internal surface of a bone rimming the obturator foramen. Pass the string laterally, through the lesser sciatic notch, and attach to the greater trochanter. Note the position of the string and verify the following:
   - Obturator internus lies partly within the pelvic cavity and partly within the gluteal region.
   - To insert on the greater trochanter, obturator internus must turn anteriorly shortly after it passes through the lesser sciatic notch.
9. Attach one string to the ischial spine and a second to the ischial tuberosity. Extend each string laterally and attach the free ends to the greater trochanter. Notice that the first string, attached to the ischial spine, passes superior to the string representing the obturator internus. The first string represents the superior gemellus. The second string passes inferior to the "obturator internus" and represents the position of the inferior gemellus.

10. Attach a string to the ischial tuberosity and extend it laterally to the femur. Attach the free end to the intertrochanteric crest of the femur. This illustrates the path of quadratus femoris muscle.

WHAT'S IN A WORD The buttock mass is formed by the gluteal muscles. The term *gluteus* comes from the Greek word *gloutos*, which means "buttocks."

The word *piriformis* is derived from Latin and means "pear shaped." The piriformis muscle has a form that resembles a pear.

The superior gemellus and inferior gemellus are roughly the same size and shape. Their name (*gemellus*) originates from Latin and means "twin."

11. Use an anatomical model, chart, or atlas to identify the superficial and deep gluteal muscles and to verify the following features (Figure 11.20).
    - Gluteus maximus is the most superficial muscle in the gluteal region and forms a bulky, protective pad for the ischial tuberosity and adjacent pelvic regions.
    - Gluteus maximus covers all but the superolateral corner of the deeper gluteus medius.
    - Gluteus minimus lies deep to gluteus medius and they have similar attachments.
    - Piriformis is a reliable anatomical landmark for identifying the sciatic nerve, which emerges from the muscle's inferior border.
    - All deep gluteal muscles insert on or near the greater trochanter.
    - Piriformis and obturator internus originate in the pelvic cavity and insert in the gluteal region.
12. Perform lateral and medial rotation of the thigh while your lab partner observes. Verify that during lateral rotation, the anterior surface of the femur rotates away from the midline of the body; during medial rotation, it rotates toward the midline.
13. The gluteus maximus is a prime mover for thigh extension. Perform this action by standing from the seated position. The gluteus medius is the prime mover for thigh abduction. Gluteus minimus and piriformis are synergists for this action. Perform this movement.
14. The gluteus medius and gluteus minimus are important muscles for locomotion because they stabilize the pelvis during walking and running. Stand in antomical position and palpate the left superolateral corner of the gluteal region (the "safe area"). Raise your right foot off the ground. As you do this, you can feel the left gluteus medius contracting to prevent the right side of the pelvis from sagging. When the right limb is raised, the unsupported right side of the pelvis tends to tilt inferiorly. The left gluteal muscles steady the pelvis by pulling the left side inferiorly. This prevents the right side from sagging, allowing the right foot to clear the ground while walking.

QUESTION TO CONSIDER Explain how the gluteal muscles, which act at the hip joint, are similar to the rotator cuff muscles, which act at the shoulder joint.

____

____

____

Muscles of the Thigh

The thigh contains three main muscle groups.

- The **anterior thigh muscles** act primarily to flex the thigh and extend the leg.
- The **medial thigh muscles** act primarily to adduct and flex the thigh.
- The **posterior thigh muscles** extend the thigh and flex the leg.

Anterior Thigh Compartment

The anterior thigh is dominated by four muscles, known collectively as the **quadriceps femoris**. Other muscles in this compartment include the **iliopsoas**, **sartorius**, and **tensor fasciae latae** (Figure 11.21). The bony attachments, innervations, and actions of these muscles are described in Table 11.12A.

ACTIVITY 11.15 Identifying Anterior Thigh Muscles

1. On the skeleton, attach string to the anterior superior iliac spine. Pass the string anteromedially and attach the free end to the pubic tubercle. The string illustrates the position of the **inguinal ligament**, the superior border of the anterior thigh.
2. Locate the bodies of T12 through L5. Attach a string to the lateral surface of any of them. Attach another piece of string to the iliac fossa. Pass both strings deep to the "inguinal ligament" and attach the free ends to the lesser trochanter on the femur. The strings illustrate the position of the iliopsoas muscle. This muscle has two parts: the **psoas major** that originates on the vertebral bodies, and the **iliacus** on the iliac fossa. Note that the iliopsoas is partly within the abdominopelvic cavity and partly within the anterior thigh (Figure 11.21b).
3. The four muscles that comprise the quadriceps femoris—**rectus femoris**, **vastus lateralis**, **vastus medialis**, and **vastus intermedius**—cover nearly all the anterior, lateral, and medial aspects of the femur. Locate the origins of the following muscles.
    - *Rectus femoris:* anterior inferior iliac spine
    - *Vastus lateralis:* lateral lip of the linea aspera
    - *Vastus medialis:* medial lip of the linea aspera
    - *Vastus intermedius:* anterior and lateral sides of the femur

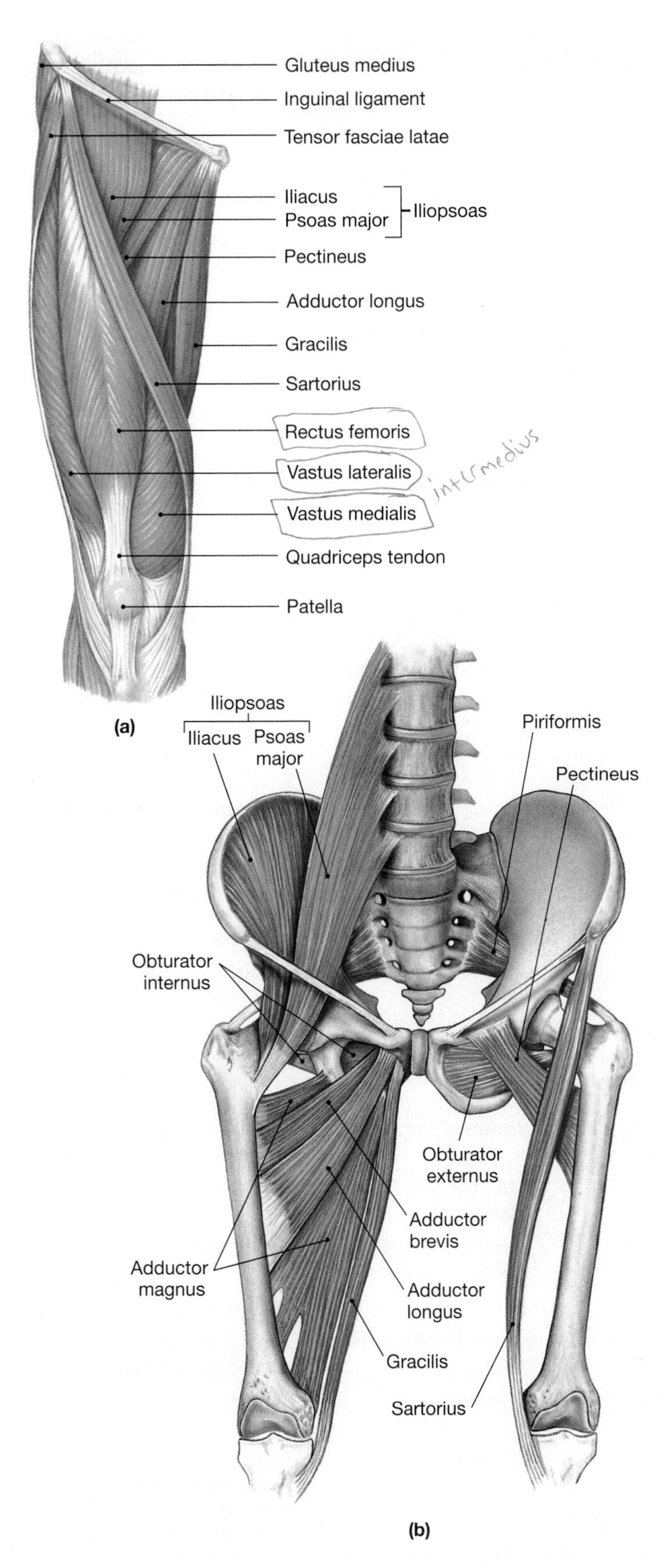

Figure 11.21 Muscles of the anterior and medial thigh. a) Anterior view of the right thigh; **b)** anterior view of the iliopsoas and medial thigh muscles.

4. Attach strings to the four origins, as described. Allow all four strings to converge at the patella and attach all of them to this bone. These represent the four muscles of the quadriceps femoris. Note that rectus femoris crosses the hip joint, but the other three do not. All four muscles give rise to the common **quadriceps tendon**, which attaches to the patella.
5. Attach a piece of string from the patella to the tibial tuberosity, which is the insertion for the quadriceps femoris. The string represents the **patellar ligament**.
6. Note that the quadriceps tendon and patellar ligament cross the patella before attaching to the tibia. This provides a mechanical advantage when the quadriceps femoris extends the knee.
7. Locate the anterior superior iliac spine. It is the origin for sartorius and tensor fasciae latae.
8. From its origin, trace the path of sartorius inferomedially across the thigh and to its insertion on the medial surface of the proximal end of the tibia. Identify the insertion of the sartorius on the skeleton. Note that this muscle crosses both the hip and knee joints.
9. The tensor fasciae latae travels inferiorly and attaches to the iliotibial tract, not seen on the skeleton. Tensor fascia latae crosses the hip joint and, by virtue of its attachment to the iliotibial tract, the knee joint as well.
10. Use an anatomical model, chart, or atlas to identify and examine the muscles of the anterior thigh and verify the following (Figure 11.21):
    - Sartorius is the most superficial muscle in the anterior thigh.
    - Iliopsoas is formed by two separate muscles, the psoas major and iliacus, which originate in the abdominopelvic cavity and insert in the anterior thigh.
    - The quadriceps femoris group contains four muscles—rectus femoris and the three vastus muscles—that comprise the bulk of the anterior thigh.
    - Vastus intermedius lies deep to rectus femoris and is intermediate to vastus lateralis and vastus medialis.
    - The combination of the quadriceps tendon, patella, and patellar ligament allows the quadriceps femoris to insert on the tibial tuberosity.
11. Flex your thigh. Iliopsoas is the principal flexor (prime mover) of the thigh, with rectus femoris, sartorius, and tensor fasciae latae acting as synergists. Identify a gluteal muscle that is an antagonist of the iliopsoas.

    ____

12. Sit on a lab stool with your legs and feet hanging freely, then extend your leg so that it is elevated to a horizontal position. The quadriceps femoris muscles are the prime movers for extension of the leg.

Table 11.12 Muscles of the Thigh

Muscle Group/Muscle	Origin	Insertion	Action	Innervation
A. Anterior thigh muscles				
1. Iliopsoas				
a. Iliacus	Iliac fossa; anterior surface of sacrum	Lesser trochanter of femur	Flexion of the thigh	Femoral nerve (L2, L3)
b. Psoas major	Bodies of vertebrae T12–L5 nerves	As above	As above	Branches of lumbar spinal (L1–L3)
2. Sartorius	Anterior superior iliac spine of ilium	Medial surface of tibia, near the knee joint	Flexion, abduction, and lateral rotation of thigh; flexion of leg	Femoral nerve (L2, L3)
3. Quadriceps femoris				
a. Rectus femoris	Anterior inferior iliac spine of ilium	Patella by the quadriceps tendon and tibial tuberosity by the patellar ligament	Extension of leg; flexion of thigh	Femoral nerve (L2–L4)
b. Vastus lateralis	Greater trochanter and lateral lip of linea aspera of femur	As above	Extension of leg	As above
c. Vastus medialis	Intertrochanteric line and medial lip of linea apsera	As above	As above	As above
d. Vastus intermedius	Anterior and lateral surfaces of femoral shaft	As above	As above	As above
4. Tensor fasciae latae	Anterior superior iliac spine of ilium	Iliotibial tract	Flexes, abducts, and medially rotates thigh; steadies trunk on thigh while standing	Superior gluteal nerve (L4, L5)
B. Medial thigh muscles				
1. Obturator externus	External surface of obturator membrane and adjacent bones	Trochanteric fossa of femur	Laterally rotates thigh	Obturator nerve (L3, L4)
2. Pectineus	Superior ramus of pubic bone	Pectineal line of femur	Adduction and flexion of thigh	Femoral nerve (L2, L3); may receive a branch from obturator nerve
3. Adductor longus	Body of pubic bone	Linea aspera of femur	Adduction and flexion of thigh	Obturator nerve (L2–L4)
4. Adductor magnus	*Adductor part:* inferior ramus of pubic bone; ramus of ischium	*Adductor part:* gluteal tuberosity, linea aspera, and medial supracondylar line of femur	*Adductor part:* same as adductor longus	As above
	Hamstring part: ischial tuberosity of ischium	*Hamstring part:* adductor tubercle of femur	*Hamstring part:* extension of thigh	Tibial division of sciatic nerve (L4)
5. Adductor brevis	Body and inferior ramus of pubic bone	Pectineal line and linea aspera of femur	Adduction and flexion of thigh	Obturator nerve (L2–L4)
6. Gracilis	Body and inferior ramus of pubic bone	Medial surface of tibia, near the knee joint	Adduction of thigh; flexion and medial rotation of leg	Obturator nerve (L2–L3)
C. Posterior thigh muscles				
1. Semitendinosus	Ischial tuberosity of ischium	Medial surface of tibia, near the knee joint	Extension of thigh; flexion of leg	Tibial division of sciatic nerve (L5, S1, S2)
2. Semimembranosus	As above	Posterior surface of medial condyle of tibia	As above	As above
3. Biceps femoris	*Long head:* as above *Short head:* linea aspera and lateral supracondylar line of femur	Lateral side of head of fibula	As above	*Long head:* as above *Short head:* common fibular (peroneal) division of sciatic nerve (L5, S1, S2)

13. From the sitting position, cross your legs. Working with your lab partner, verify that during this action, all of the following movements are performed.
    - Flexion of the thigh
    - Abduction of the thigh
    - Lateral rotation of the thigh
    - Flexion of the leg

    Sartorius is a synergist for all of these actions, which will cross the legs in a position that was once used by tailors. For this reason, sartorius is called the "tailor's muscle."

QUESTIONS TO CONSIDER 1. The iliopsoas can also flex the trunk when sitting up from a supine position. In this situation, the origins and insertions of the muscles are reversed. Can you explain why?

___

___

___

2. The primary function of the quadriceps femoris is extension of the leg, but it also assists in hip flexion. Explain why the quadriceps can act at the knee as well as the hip.

___

___

Medial Thigh Compartment

The medial thigh muscles (Figure 11.21) are commonly referred to as the adductors of the thigh or the groin muscles. The principal actions of most muscles in this group are adduction and flexion of the thigh. They are most active during activities that move the thighs together, such as riding a horse. They also function as synergists during the gait cycle and play a role in maintaining erect posture. Specific actions, bony attachments, and innervations of the medial thigh muscles are described in Table 10.12B.

CLINICAL CORRELATION

A groin pull (groin injury) refers to some degree of damage to one or more medial thigh muscles. Typically, the injury presents as a strain, stretch, or tear of the muscles near their origins on the pelvis. Groin injuries often occur among athletes who play sports that require quick running starts, such as sprinting, soccer, stealing bases in baseball, running downfield to catch a pass in football, or executing a fast break in basketball.

ACTIVITY 11.16 Identifying Medial Thigh Muscles

1. On the skeleton, identify the body of each pubic bone (pubis) and verify that they articulate at the symphysis pubis. Identify the superior and inferior rami that extend posterolaterally from the body of the pubis. **Pectineus**, **adductor longus**, **adductor brevis**, and **adductor magnus** originate along bony surfaces on the body of the pubis and its two rami.
2. On the femur, identify the pectineal line, the linea aspera the medial and lateral supracondylar ridges, and the adductor tubercle. Place your index finger on the pectineal line just below the lesser trochanter, where it begins. Move your finger inferiorly along the pectineal line, linea aspera, and medial supracondylar ridge until you reach the adductor tubercle. This path marks the insertions for pectineus, adductor longus, adductor brevis, and adductor magnus.
3. Attach a string to the body of the pubis. Extend the string posteriorly and inferiorly so that it passes medial to the femur, and attach the free end to the linea aspera. The string illustrates the general orientation of the four medial thigh muscles. If the lower extremity on your skeleton is hanging freely, pull on the string to move the femur. Note that by pulling the string, you can adduct and flex the thigh.
4. Adductor magnus has two parts. The adductor part has skeletal attachments and functions typical of the other adductor muscles. The hamstring part is similar to the muscles of the posterior thigh. It originates at the ischial tuberosity and inserts on the adductor tubercle. Compare the bony attachments of the adductor and hamstring portions of the adductor magnus.
5. Attach a string to the body of the pubis. Extend it inferiorly along the medial aspect of the femur and attach it to the medial side of the tibia at its proximal end. The string represents the **gracilis**, which crosses both the hip and the knee.
6. Locate the obturator foramen and recall that the obturator membrane covers it. **Obturator internus**, a gluteal muscle, originates from the *internal* surface of the membrane and surrounding bones. **Obturator externus** originates from the *external* surface of the membrane and surrounding bones, and passes laterally to insert in the trochanteric fossa at the base of the greater trochanter. Identify the fossa on the skeleton.

WHAT'S IN A WORD The medial thigh muscles include adductor brevis, adductor longus, and adductor magnus. In Latin, the words *brevis, longus,* and *magnus* mean "short," "long," and "large." These terms are used regularly to compare muscles that are located in the same region and share a common function.

7. Use an anatomical model, chart, or atlas to identify the muscles of the medial thigh and verify the following (Figure 11.21b):
    - Most medial thigh muscles originate on the pubis and insert on the femur.
    - The gracilis is the most medial muscle in the group and the only one that crosses the knee.
    - Adductor magnus is the largest and deepest muscle of the medial thigh, and has two distinct functional parts: the adductor part and the hamstring part.

- Obturator externus, which originates on the external side of the obturator membrane, is the counterpart of the obturator internus, which originates on the internal side of the membrane.

8. Sit on a lab stool with your feet hanging freely and your thighs separated, as if you were sitting on a horse. With your hands on your medial thighs, attempt to draw them together against the resistance of the seat. As you perform this action (adduction of the thighs), you will feel your medial thigh muscles contract.

   Identify the gluteal muscles that are antagonists for thigh adduction.

   ____

   ____

QUESTION TO CONSIDER You are at a nightclub in which the main attraction is a mechanical bull. You decide to try your luck with "bull riding." Although it is a wild ride, you surprise yourself by staying on for 10 seconds. Shortly thereafter, you develop tightness and pain in your medial thighs. Explain why you are experiencing this discomfort.

____

____

____

Posterior Thigh Compartment

The three long muscles in the posterior thigh are called the "hamstrings." These muscles—the **semitendinosus**, **semimembranosus**, and **biceps femoris**—share a common origin on the ischial tuberosity (Figure 11.20b), deep to the gluteus maximus, and they cross both the hip and knee joints (Figure 11.20a). The bony attachments, innervations, and actions of the posterior thigh muscles are described in Table 11.13C.

WHAT'S IN A WORD The name *hamstrings* is derived from the practice of butchers in 18th-century England inserting hooks into these muscles to hang the carcasses of pigs in their shops.

ACTIVITY 11.17 Identifying Posterior Thigh Muscles

1. On the skeleton, identify the origins for the two heads of biceps femoris.
   - Ischial tuberosity (long head)
   - Distal end of the linea aspera on the femur (short head)
2. Identify the head of the fibula, which is where both heads insert, by a common tendon.
3. Use a string to trace the path of both heads of the biceps femoris.
4. The ischial tuberosity is also the origin for the two medial hamstring muscles: the semitendinosus and semimembranosus. These two muscles travel inferiorly and insert on the medial condyle of the tibia (semimembranosus) and the medial surface of the tibial shaft, just inferior to the condyle (semitendinosus). Identify these insertion sites on the skeleton, and use string to trace the paths of these muscles.
5. Verify that all these muscles cross both the hip and the knee.
6. Use an anatomical model, chart, or atlas to identify the posterior thigh muscles and to verify the following (Figure 11.20):
   - The common origin for the hamstrings is the ischial tuberosity.
   - All posterior thigh muscles act at both the hip and the knee.
   - On the medial side of the posterior thigh, the semimembranosus is deep to the semitendinosus.
   - The biceps femoris has two heads of origin and a single tendon of insertion.
7. The posterior thigh muscles are the prime movers for extension of the thigh and flexion of the leg. Perform these actions while considering the locations of these muscles. The posterior thigh muscles are assisted by gluteus maximus and adductor magnus (hamstring portion) during thigh extension, and by the gastrocnemius in the posterior leg (described later), during leg flexion.
8. Identify the muscles that act as antagonists of the posterior thigh muscles during the following:
   - Extension of the thigh

     ____

   - Flexion of the leg

     ____

CLINICAL CORRELATION

Hamstring injuries are common among individuals who participate in sports that involve running, jumping, or quick starts and stops. Hamstring pulls can vary in severity, ranging from small tears of a few muscle fibers to an avulsion (forcible separation) of tendinous attachments from the ischial tuberosity. These injuries can be painful and slow to heal.

QUESTIONS TO CONSIDER 1. Explain how flexion of the leg differs from flexion of the forearm.

____

____

2. When the thighs and legs are flexed (e.g., when you are seated), the semimembranosus and semitendinosus can extend the trunk (e.g., when you stand from a seated position). Describe the origins and insertions of these muscles during trunk extension.

____

____

____

Muscles of the Leg

The leg is naturally divided into three muscular compartments, each sharing the same general function and innervated by the same branch of the sciatic nerve. The anterior compartment contains muscles that dorsiflex the foot and extend the toes. They are innervated by the deep fibular (peroneal nerve). The lateral compartment muscles evert the foot and are supplied by the superficial fibular (peroneal) nerve. In the posterior compartment, the muscles plantarflex the foot and flex the toes. They are innervated by the tibial nerve. The specific actions and bony attachments of the leg muscles are described in Table 11.13.

Anterior Leg Compartment

The anterior compartment of the leg contains four muscles that originate on the bones in the leg (tibia or fibula) and insert on bones in the foot (Figure 11.22a). These muscles have relatively long tendons that travel through tunnels, similar to the carpal tunnel at the wrist. The floors of these tunnels are formed by the tarsal bones of the foot and the roofs by thick bands of connective tissue that form the **superior** and **inferior extensor retinaculum** (Figure 11.22a and Figure 11.23c later in the exercise).

ACTIVITY 11.18 Identifying Anterior Leg Muscles

1. On the skeleton, observe the bones of the leg from an anterior view. Place a finger on the lateral condyle of the tibia and run your finger distally along the tibial shaft, about halfway along its length. The region you have outlined is the origin of **tibialis anterior**.
2. Use string to trace the path of tibialis anterior from its origin to its insertion on the medial cuneiform and base of the first metatarsal. Note that this muscle covers much of the anterior surface of the tibial shaft.
3. Once again, place a finger on the lateral condyle of the tibia. From this location, move your finger inferiorly along the superior three fourths of the shaft of the fibula. This area of bone represents the origin of **extensor digitorum longus**.
4. Use string to trace the path of the extensor digitorum longus from its origin to its insertion, via a split tendon, on the middle and distal phalanges of the lateral four toes. Verify that this muscle is lateral to the tibialis anterior.
5. Attach string to the anterior surface of the fibula, about halfway along its length. Extend the free end of the string to the distal phalanx of the great toe. This marks the position of **extensor hallucis longus.** Note that this muscle is between tibialis anterior and extensor digitorum longus.
6. The **fibularis (peroneus) tertius** is a small muscle that is actually an extension of extensor digitorum longus. Trace its path as it arises from the distal third of the fibula and inserts on the base of the fifth metatarsal. This is a weak muscle and is sometimes missing.

WHAT'S IN A WORD The term *hallucis* refers to the great toe; any muscle with *hallucis* in its name will act on the great toe.

7. Use an anatomical model, chart, or atlas to identify the anterior leg muscles (Figure 11.22a) and to verify the following:
   - These muscles cover the anterior surfaces of the tibia and fibula.
   - These muscles have long tendons of insertion held in place by connective tissue bands called the extensor retinacula.
   - The extensor hallucis longus is slightly deeper to tibialis anterior and extensor digitorum longus.
8. Sit on a lab stool so that your feet are hanging freely. From this position, dorsiflex your feet. Dorsiflexion is a bending action at the ankles during which the feet and toes are directed superiorly. Note that when you dorsiflex the feet, the toes are usually in the extended position. Toe extension is a function of extensor digitorum longus (lateral four toes) and extensor hallucis longus (the great toe).

CLINICAL CORRELATION

The tibialis anterior is a strong dorsiflexor of the foot, particularly when the foot is elevated during the gait cycle. Paralysis of tibialis anterior causes the foot to remain plantar flexed and to drag along the ground when the foot is raised while walking. This condition is called "foot drop."

9. In addition to being a strong dorsiflexor, tibialis anterior also inverts the foot. From the sitting position, invert your right foot. Verify that the sole of the foot is turned medially during this action.

QUESTION TO CONSIDER The inserting tendons of the anterior leg muscles are held in position by the superior and inferior retinacula. If the retinacula were absent, what effect would this have on the function of these muscles?

____

____

____

Table 11.13 Muscles of the Leg

Muscle Group/Muscle	Origin	Insertion	Action	Innervation
A. Anterior compartment				
1. Tibialis anterior	Lateral condyle and lateral surface of tibia; interosseous membrane	Medial cuneiform bone and first metatarsal	Dorsiflexion and inversion of the foot	Deep fibular (peroneal) nerve (L4–L5)
2. Extensor digitorum longus	Lateral condyle of tibia; medial surface of fibula; interosseous membrane	Middle and distal phalanges of digits 2–5	Dorsiflexion of the foot; extension of the toes	Deep fibular (peroneal) nerve (L5–S1)
3. Extensor hallucis longus	Anterior surface of fibula; interosseous membrane	Distal phalanx of great toe	Dorsiflexion of the foot; extension of the great toe	As above
4. Fibularis (peroneus) tertius	Anterior surface of fibula; interosseous membrane	Fifth metatarsal	Dorsiflexion and eversion of the foot	As above
B. Lateral compartment				
1. Fibularis (peroneus) longus	Head and lateral surface of fibula	First metatarsal and medial cuneiform	Eversion of the foot; fibular	Superficial (peroneal) nerve (L5, S1, S2)
2. Fibularis (peroneus) brevis	Lateral surface of fibula	Fifth metatarsal	As above	As above
C. Posterior compartment				
Superficial group				
1. Gastrocnemius	*Lateral head:* lateral condyle of femur *Medial head:* popliteal surface of femur, just superior to medial condyle	Calcaneus by the calcaneal (Achilles) tendon	Plantar flexion of foot; flexion of the leg	Tibial nerve (S1, S2)
2. Soleus	Head of fibula; posterior surface of fibula; medial aspect of tibia	As above	Plantar flexion of foot	As above
3. Plantaris	Lateral supracondylar line of femur	As above	Assists in plantar flexion of foot and flexion of leg	As above
Deep group				
4. Popliteus	Lateral condyle of femur; lateral meniscus in knee joint	Posterior surface of tibia	Flexion of leg; medial rotation of leg to unlock knee prior to leg flexion	Tibial nerve (L4, L5, S1)
5. Flexor digitorum longus	Posterior surface of tibia	Distal phalanges of digits 2–5	Plantar flexion of foot; flexion of toes	Tibial nerve (S2, S3)
6. Flexor hallucis longus	Posterior surface of fibula; interosseous membrane	Distal phalanx of great toe	Plantar flexion of foot; flexion of great toe	As above
7. Tibialis posterior	Posterior surfaces of tibia and fibula; interosseous membrane	Navicular, cuboid, and cuneiform bones; second, third, and fourth metatarsals	Plantar flexion and inversion of foot	Tibial nerve (L4–L5)

Lateral Leg Compartment

The two muscles in the lateral leg compartment travel along the lateral side of the fibula and insert on bones in the foot. Similar to the anterior compartment muscles, the inserting tendons travel through tunnels, formed by bone and connective tissue, to reach their insertions (Figure 11.22b).

ACTIVITY 11.19 Identifying Lateral Leg Muscles

1. On the skeleton identify the head and superior two thirds of the lateral surface of the fibula. This region is the origin of the **fibularis (peroneus) longus**. Attach a piece of string anywhere along its length. Extend the string inferiorly and

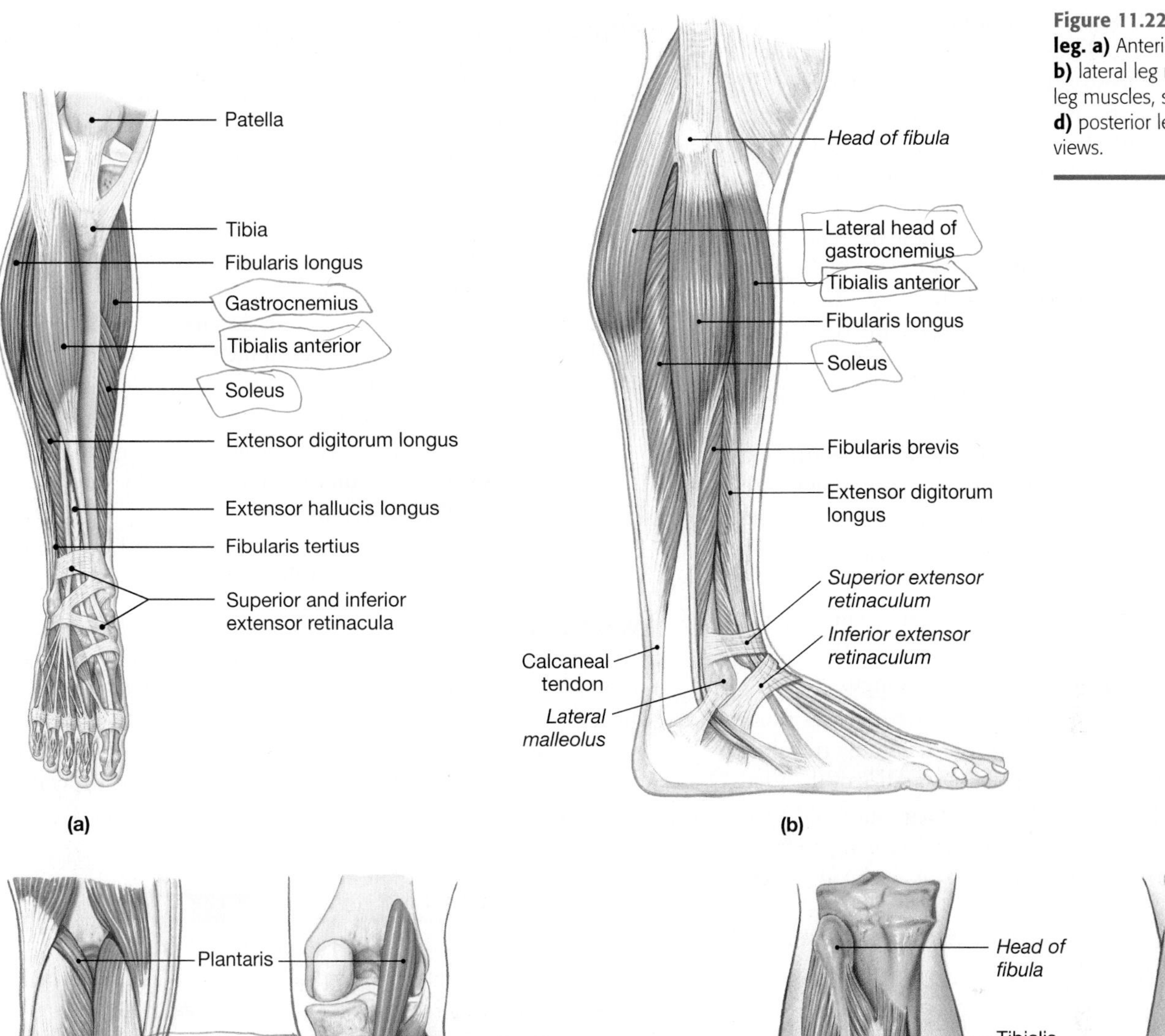

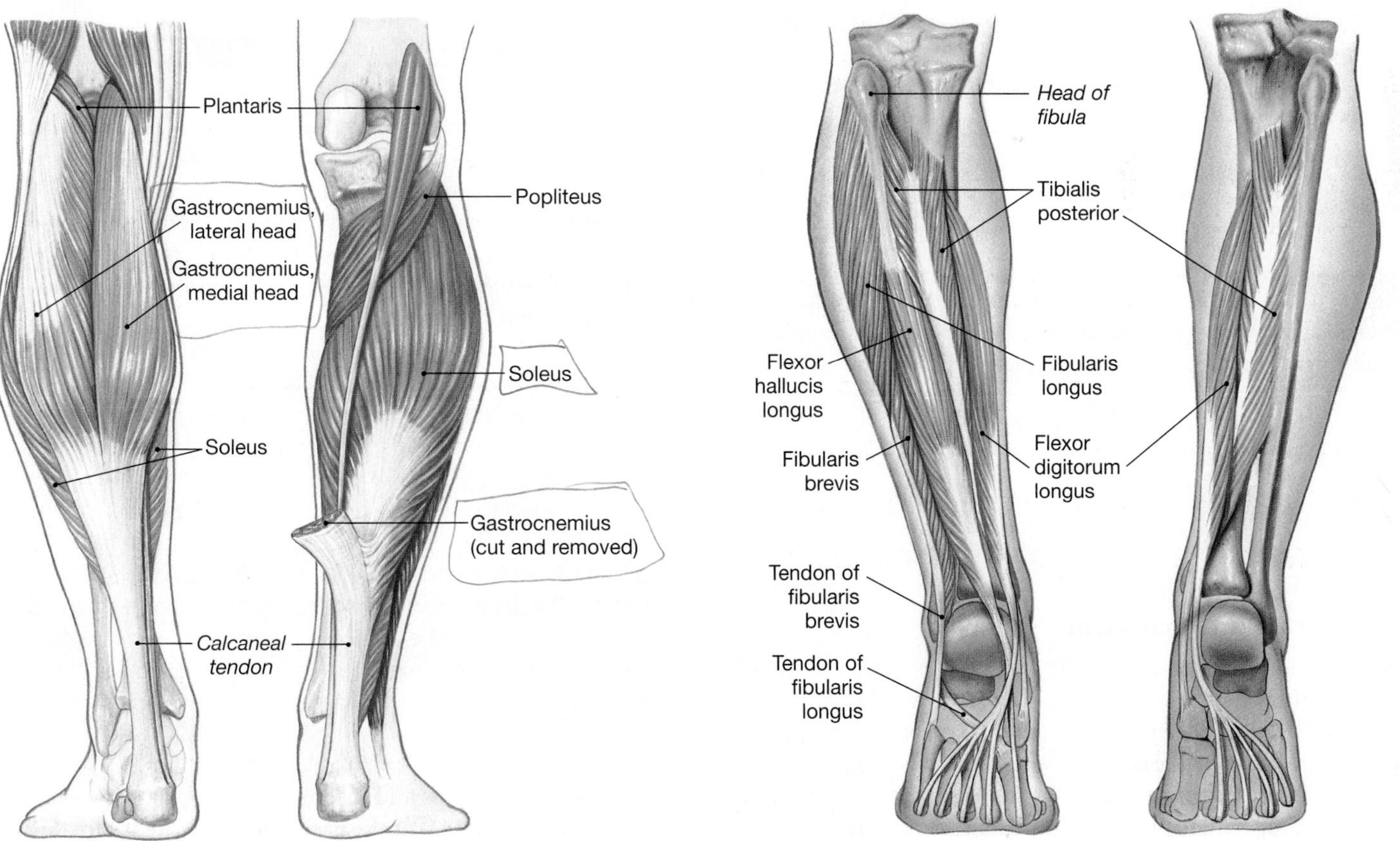

Figure 11.22 Muscles of the leg. a) Anterior leg muscles; **b)** lateral leg muscles; **c)** posterior leg muscles, superficial views; **d)** posterior leg muscles, deep views.

allow it to pass lateral to the calcaneus and posterior to the lateral malleolus, and then bend anteriorly toward the cuboid. At the cuboid, allow the string to curve onto the sole of the foot and travel medially. Finally, on the medial side of the foot, attach the free end of the string to either the medial cuneiform or the metatarsal of the great toe. The string illustrates the position of the fibularis longus and its tendon of insertion. Observe how the tendon wraps under the foot to provide support for the lateral longitudinal and transverse arches of the foot.

2. Attach a second piece of string along the inferior half of the lateral surface of the fibula. As before, pass the string lateral to the calcaneus and posterior to the lateral malleolus. As the string bends anteriorly toward the cuboid, attach the free end to the base of the metatarsal of the little toe. This illustrates the position of **fibularis (peroneus) brevis.**
3. Use an anatomical model, chart, or atlas to identify the lateral leg muscles and verify the following (Figure 11.22b):
   - Both muscles in the lateral leg compartment originate on the lateral side of the fibular shaft.
   - Fibularis brevis lies deep to fibularis longus.
   - The inserting tendon of fibularis longus passes from the lateral side to the medial side of the foot by traveling along the sole, supporting the two arches of the foot.
4. The principal function of the lateral leg muscles is foot eversion. Sit on a lab stool with your feet hanging freely. Evert your foot and verify that the sole of the foot turns laterally during this action.

   Identify the muscles that are antagonists of the lateral compartment muscles.

____

____

QUESTION TO CONSIDER If the two muscles in the lateral leg compartment were paralyzed, the foot would remain in an abnormal inverted position. Explain why this would occur.

____

____

____

Posterior Leg Compartment

The posterior compartment of the leg contains three superficial and four deep muscles (Figures 11.22c and d). The superficial muscles are commonly referred to as "calf muscles" and have a common insertion on the posterior surface of the calcaneus (Figure 11.22c). Three of the four deep muscles have long tendons of insertion (Figure 11.22d) that reach the foot by passing through a tunnel covered by a connective tissue band, the flexor retinaculum.

ACTIVITY 11.20 Identifying Posterior Leg Muscles

1. On the skeleton identify the two heads of origin for the **gastrocnemius**, the massive superficial muscle of the posterior leg.
   - Lateral condyle of the femur (lateral head)
   - Popliteal surface of the femur, just superior to the medial condyle (medial head)
2. Deep to the gastrocnemius is the **soleus.** Locate its origins on the head of the fibula and on the posterior surfaces of the proximal ends of the fibula and tibia.
3. Identify the posterior surface of the calcaneus. This region is commonly referred to as the heel. Both the gastrocnemius and soleus insert on this bony surface by way of the **calcaneal (Achilles) tendon**.
4. Use string to trace the paths of these muscles from their origins to their insertion on the calcaneus.

WHAT'S IN A WORD The name *gastrocnemius* is derived from two Greek words. The word *gaster* means "stomach," and *kneme* means "leg." Taken literally, *gastrocnemius* means "stomach of the leg."

The name *soleus* is derived from the Latin word *solea*, meaning "sole fish." The soleus muscle resembles this fish species in appearance.

CLINICAL CORRELATION

Injuries to the calcaneal tendon and associated structures are common among athletes. Calcaneal bursitis is an inflammation of the deep calcaneal bursa positioned between the calcaneal tendon and the posterior surface of the calcaneus. The injury is caused by excessive friction on the bursal membrane as the tendon slides over it. It results in pain along the posterior aspect of the heel. Calcaneal bursitis is common among long-distance runners and tennis players.

Calcaneal (Achilles) tendonitis is a common running injury. It is caused by tissue inflammation leading to small tears in the calcaneal tendon. The injury causes pain while walking and is extremely slow to heal. Individuals with a history of calcaneal tendonitis are vulnerable to a much more severe injury—a ruptured calcaneal tendon. This injury can occur if a plantar flexed foot is suddenly dorsiflexed. Typically, there is a sudden pain in the calf and an audible snap as the tendon ruptures. A fleshy lump forms in the calf due to the shortening of the gastrocnemius and soleus, and a gap can be felt where the tendon once was. When the calcaneal tendon is ruptured, the individual cannot use the limb. The foot can be dorsiflexed, but plantar flexion is difficult.

5. Review the origins and insertions of the gastrocnemius and soleus. Note that the gastrocnemius crosses the knee and ankle, but soleus crosses only the ankle.

6. Return to the lateral condyle, where the lateral head of the gastrocnemius originates. A small, superficial muscle, the **plantaris**, originates on the posterior surface of the femur just superior to the lateral condyle. The plantaris has a very long tendon that travels between gastrocnemius and soleus and inserts on the posterior surface of the calcaneus with the calcaneal tendon.
7. Attach a piece of string to the lateral condyle of the femur. Extend the string inferomedially and attach the free end to the posterior surface of the tibia, just inferior to the tibial condyles. The string marks the position of the **popliteus**, a deep muscle in the posterior compartment of the leg. This muscle also originates on the fibrous capsule and lateral meniscus of the knee joint.
8. Examine the posterior surfaces of the tibia and fibula. These are the origins for the remaining three deep muscles in the posterior leg—**tibialis posterior**, **flexor hallucis longus**, and **flexor digitorum longus**.
9. Attach a string to the posteromedial aspect of the tibia and extend it inferiorly to pass posterior to the medial malleolus. Allow the string to curve anteriorly onto the sole of the foot and attach the free end to the navicular or any of the adjacent bones. This illustrates the path of tibialis posterior.
10. Attach a second piece of string to the posterior surface of the fibula. Extend the string inferomedially so that it travels posterior to the distal end of the tibia. Allow the string to curve onto the medial side of the calcaneus and then the sole of the foot. Attach the free end to the distal phalanx of the great toe. This string represents the position of flexor hallucis longus. Notice that on the sole of the foot, the string travels along the central axis of the medial longitudinal arch of the foot.
11. Attach a third piece of string to the posterior surface of the tibia. Extend the string inferiorly so that it passes posterior to the medial malleolus and onto the medial side of the calcaneus. From here, allow the string to curve onto the sole of the foot and attach to the distal phalanx of any of the lateral four toes. This string represents the position of flexor digitorum longus. The tendon branches to insert on the distal phalanges of all four lateral toes.
12. Examine the relative positions of the three strings ("muscles") as described. Note that the origin of tibialis posterior is intermediate to the origins of flexor hallucis longus and flexor digitorum longus.
13. On the leg, verify that the origin of flexor hallucis longus is *lateral* to the origin of flexor digitorum longus. On the foot, observe that the tendons of these two muscles cross so that the insertion of flexor hallucis longus is *medial* to the insertion of flexor digitorum longus. Also, note the paths of all three tendons as they travel along the sole of the foot and realize that they all (especially flexor hallucis longus) help support the medial longitudinal arch of the foot.
14. Use an anatomical model, chart, or atlas to identify the muscles of the posterior leg and to verify the following (Figure 11.22):
    - Gastrocnemius and soleus are the largest muscles in the posterior leg. The soleus lies deep to the gastrocnemius.
    - The calcaneal (Achilles) tendon is the insertion tendon for gastrocnemius, soleus, and plantaris.
    - Gastrocnemius acts at both the knee and ankle, whereas soleus acts only at the ankle.
    - The popliteus is unusual in that part of its origin is on structures within the knee joint.
    - On the leg, the flexor hallucis longus is *lateral* to flexor digitorum longus; however, on the sole of the foot, the tendon of flexor hallucis longus is *medial* to the tendon of flexor digitorum longus.
    - The long inserting tendons of flexor hallucis longus, flexor digitorum longus, and tibialis posterior help support the medial longitudinal arch.
15. Although popliteus weakly flexes the leg, its more important function is to "unlock" the knee. Stand in an erect position with both feet firmly on the floor. In this position, the knees are fully extended and in the "locked" position. Flex your right leg by lifting it off the floor. When you perform this movement, popliteus "unlocks" the knee by rotating the femur *laterally* so that the leg can be flexed.
16. Sit on a lab stool and extend your right leg so that it is parallel to the floor. Your right foot is off the floor, but in this position it is fully extended and "locked." Flex your right leg so that it rests on the floor or hangs freely. When you perform this movement, the popliteus "unlocks" the knee by rotating the tibia *medially.*
17. Gastrocnemius and soleus are the prime movers for foot plantar flexion. Sitting on a lab stool with your feet hanging freely, plantar flex your feet.

    Identify leg muscles that will act as synergists for plantar flexion.

    ____

    ____

    Identify leg muscles that will be antagonists of the gastrocnemius and soleus during plantar flexion.

    ____

    ____
18. Stand in an erect position with your feet firmly on the floor. Lift your heels off the floor by standing on your toes. Your feet are plantar flexed. Have your lab partner assume a crouched position, as if he or she is in the starting blocks and about to start a race, then push off with the back leg to start the "race." The push-off action requires plantar

flexion of the foot. (*Warning:* Do not perform the "starting block" action unless there is enough space available and your instructor consents to the activity.)

19. The flexor hallucis longus (big toe) and flexor digitorum longus (lateral four toes) flex the toes. Note that when you plant flex the feet, the toes are usually in the flexed position.
20. An important action of tibialis posterior is foot inversion. Identify the muscles that are antagonists of tibialis posterior.

____

____

QUESTIONS TO CONSIDER

1. Explain why the gastrocnemius can flex the leg but soleus cannot.

____

____

2. The tibialis posterior inverts the foot. If you lost the function of the tibialis posterior, would foot inversion still be possible? Explain.

____

____

3. The popliteus can "unlock" the knee prior to flexion by either rotating the femur laterally or rotating the tibia medially. Describe and compare the origin and insertion of the popliteus when it rotates the femur versus when it rotates the tibia.

____

____

____

Muscles of the Foot

The muscles in the foot (Figure 11.23) are organized into four layers along the plantar surface; the first layer is the most superficial and the fourth layer is the deepest. Although there are many parallels between the musculature of the foot and the hand, the foot cannot perform the very specific, fine motor skills characteristic of the hand. The bony attachments, innervations, and functions of these muscles are described in Table 11.14.

ACTIVITY 11.21 Identifying Intrinsic Foot Muscles

1. On the skeleton, identify the plantar surface of the calcaneus, which is the origin of the three muscles in the first layer (Figure 11.23a).
2. From this origin, use a piece of string to trace the path of these muscles to their insertions on the phalanges as follows:
   - **Abductor hallucis** travels medially and inserts on the proximal phalanx of the great toe.
   - **Abductor digiti minimi** travels laterally and inserts on the proximal phalanx of the little toe.
   - **Flexor digitorum brevis** travels between these two muscles and inserts on the middle phalanges of the four lateral toes.
3. **Quadratus plantae**, in the second layer, inserts on the tendon of flexor digitorum longus (Figure 11.22b). Attach a string to the medial side of the calcaneus. Pass the string diagonally across the plantar surface of the foot toward the little toe and attach it to the distal phalanx. This represents the flexor digitorum longus tendon.
4. Attach one end of another piece of string to the plantar surface of the calcaneus, and run it directly forward until it meets the "flexor digitorum longus tendon" string. This represents quadratus plantae.
5. The four **lumbricals** are also in the second layer (Figure 11.22b). They originate on the separate tendons of flexor digitorum longus and insert on the extensor expansions of the four lateral toes, similar to the lumbricals in the hand. Visualize this on the skeleton.
6. The third muscle layer contains three small muscles acting on either the great toe or little toe (Figure 11.23b).
   - Identify the plantar surfaces of the cuboid and the lateral cuneiform bones. **Flexor hallucis brevis** originates from these bones. The muscle passes anteriorly to insert on the base of the proximal phalanx of the great toe. Trace that path.
   - Identify the second through fourth metatarsals. **Adductor hallucis** has two heads that originate from the bases (oblique head) and the heads (transverse head) of these bones. The two muscle heads converge and insert on the lateral side of the proximal phalanx of the great toe. Use string to follow these paths.
   - Identify the fifth metatarsal. **Flexor digiti minimi brevis** originates at its base, travels anteriorly, and inserts on the proximal phalanx of the same toe.
7. Identify the five metatarsal bones and the spaces between them. The muscles in the fourth layer, four **dorsal interossei** and three **plantar interossei**, originate on these bones and run between them, inserting on the proximal phalanges of the lateral four digits.
8. Use an anatomical model, chart, or atlas to identify the intrinsic foot muscles (Figures 11.23) and verify the following:
   - The muscles on the sole of the foot are arranged in four layers.
   - The organization of many intrinsic foot muscles is similar to comparable muscles in the hand.
   - The dorsal surface of the foot has two small muscles: **extensor hallucis brevis** and **extensor digitorum brevis** (Figure 11.23c), which are functionally insignificant.

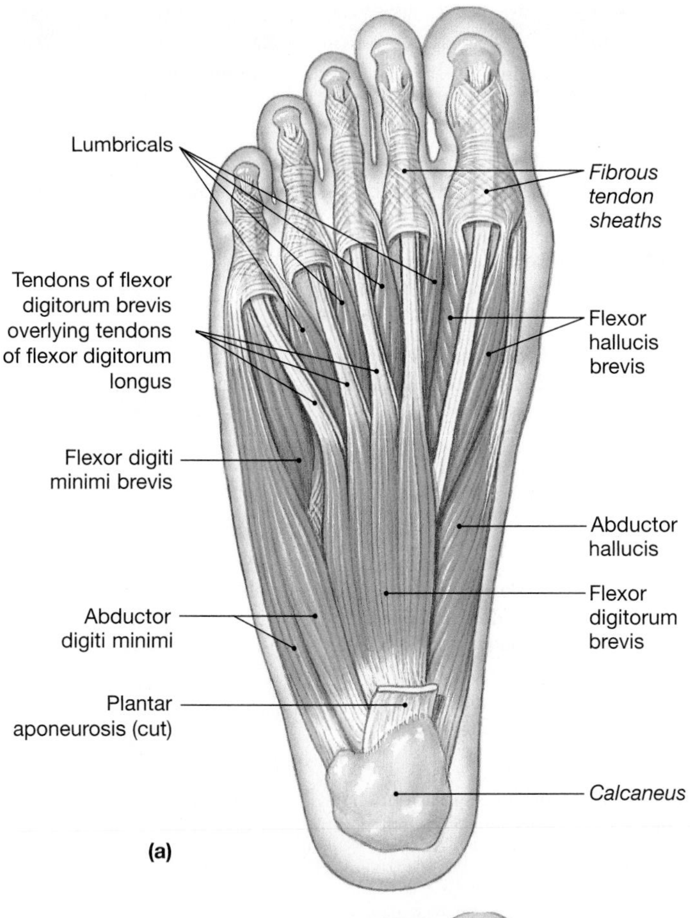
Lumbricals
Tendons of flexor digitorum brevis overlying tendons of flexor digitorum longus
Flexor digiti minimi brevis
Abductor digiti minimi
Plantar aponeurosis (cut)
Fibrous tendon sheaths
Flexor hallucis brevis
Abductor hallucis
Flexor digitorum brevis
Calcaneus
(a)

Figure 11.23 Intrinsic muscles of the foot. a) Superficial muscles, plantar view; **b)** deep muscles, plantar view; **c)** dorsal view.

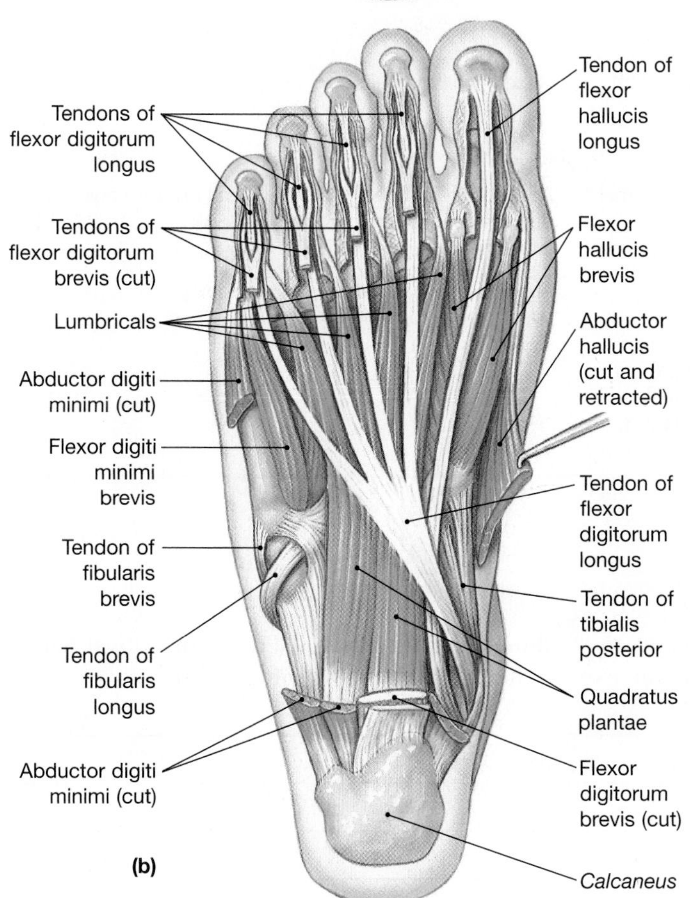
Tendons of flexor digitorum longus
Tendons of flexor digitorum brevis (cut)
Lumbricals
Abductor digiti minimi (cut)
Flexor digiti minimi brevis
Tendon of fibularis brevis
Tendon of fibularis longus
Abductor digiti minimi (cut)
Tendon of flexor hallucis longus
Flexor hallucis brevis
Abductor hallucis (cut and retracted)
Tendon of flexor digitorum longus
Tendon of tibialis posterior
Quadratus plantae
Flexor digitorum brevis (cut)
Calcaneus
(b)

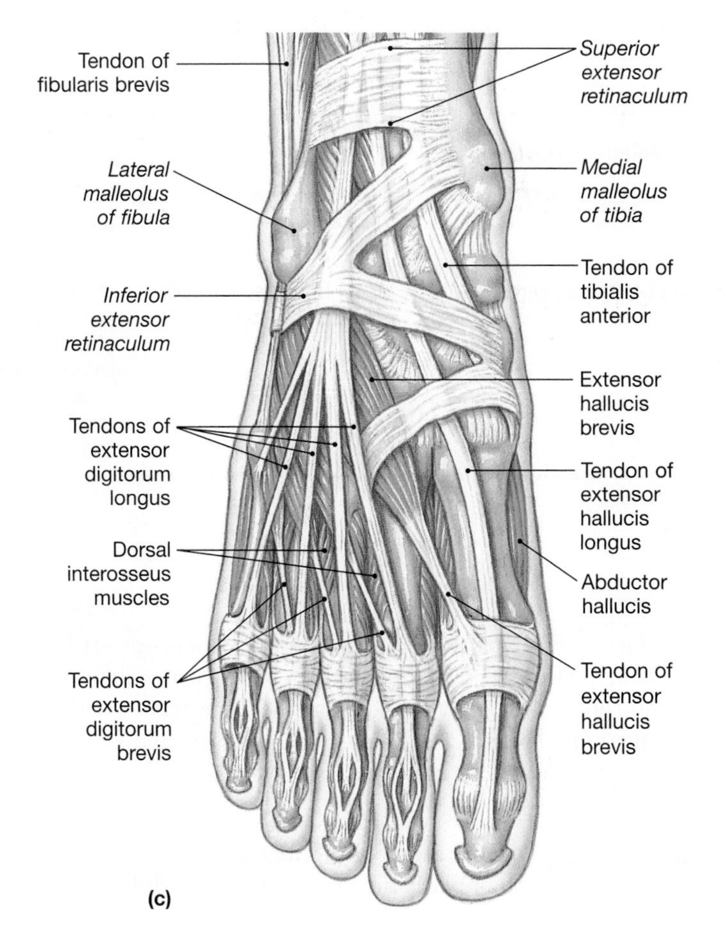
Tendon of fibularis brevis
Lateral malleolus of fibula
Inferior extensor retinaculum
Tendons of extensor digitorum longus
Dorsal interosseus muscles
Tendons of extensor digitorum brevis
Superior extensor retinaculum
Medial malleolus of tibia
Tendon of tibialis anterior
Extensor hallucis brevis
Tendon of extensor hallucis longus
Abductor hallucis
Tendon of extensor hallucis brevis
(c)

Table 11.14 Intrinsic Muscles of the Foot

Muscle Group/Muscle	Origin	Insertion	Action	Innervation
A. First muscle layer				
1. Abductor hallucis	Tuberosity of calcaneus	Medial aspect of proximal phalanx of great toe	Abducts and flexes great toe; supports medial longitudinal arch	Medial plantar nerve (S2, S3)
2. Flexor digitorum brevis	As above	Middle phalanges of digits 2–5	Flexes lateral four digits; supports medial and lateral longitudinal arches	As above
3. Abductor digiti minimi	As above	Lateral aspect of proximal phalanx of fifth digit	Abducts and flexes small toe; supports lateral longitudinal arch	Lateral plantar nerve (S2, S3)
B. Second muscle layer				
1. Quadratus plantae (flexor accessorius)	Inferior surface of calcaneus	Tendon of flexor digitorum longus	Flexes digits 2–5	As above
2. Lumbricals	Tendons of flexor digitorum longus	Medial aspect of extensor; expansions of digits 2–5	Flexion of proximal phalanges and extension of middle and distal phalanges of digits 2–5	*Medial one:* medial plantar nerve (S2, S3) *Lateral three:* lateral plantar nerve (S2, S3)
C. Third muscle layer				
1. Flexor hallucis brevis	Cuboid and lateral cuneiform bones	Proximal phalanx of great toe	Flexes great toe	Medial plantar nerve (S2, S3)
2. Adductor hallucis	*Oblique head:* bases of metatarsals 2–4 *Transverse head:* plantar ligaments of metatarsophalangeal joints	Lateral aspect of proximal phalanx of great toe	Adducts great toe; supports transverse arch	Lateral plantar nerve (S2, S3)
3. Flexor digiti minimi brevis	Fifth metatarsal	Proximal phalanx of fifth digit	Flexes small toe	As above
D. Fourth muscle layer				
1. Plantar interossei (3)	Medial aspect of metatarsals 3–5	Medial aspect of proximal phalanges of digits 3–5	Adducts lateral three digits; stabilizes forefoot while standing	Lateral plantar nerve (S2, S3)
2. Dorsal interossei (4)	Medial and lateral aspects of adjacent metatarsals	*Medial one:* medial aspects of proximal phalanx of second digit *Lateral three:* lateral aspects of proximal phalanges of digits 2, 3, and 4	Abducts lateral four digits; stabilizes forefoot while standing	As above

9. Have your lab partner sit on a lab stool with his or her feet hanging freely and barefoot. Watch your partner abduct and adduct the great toe. Perform the same actions with the thumb. Notice the difference in precision and efficiency between the hand and foot.

The hand is capable of finer motor skills than the foot partly due to the functional differences between the thumb and the great toe. From a functional standpoint, explain the difference and why it is significant.

____

____

____

____

Exercise 11 Review Sheet

Gross Anatomy of the Muscular System

Name ______________________

Lab Section ______________________

Date ______________________

1. Describe the basic anatomical organization of a whole muscle.

____

____

____

2. Brachialis flexes the forearm at the elbow. The muscle is attached to the distal half of the humerus and to the coronoid process of the ulna. Which attachment is the origin and which is the insertion? Explain.

____

____

Questions 3–6: Define the following terms.

3. Prime mover:

____

4. Antagonist:

____

5. Synergist:

____

6. Fixator:

____

Questions 7–10: For the following actions, list the muscles that would act as prime movers, antagonists, and synergists.

	Action	Prime Mover(s)	Antagonist(s)	Synergist(s)
7. Abduction of the arm				
8. Medial rotation of the arm				
9. Extension of the leg				
10. Flexion of the knee				

Questions 11–24: Identify the labeled muscles.

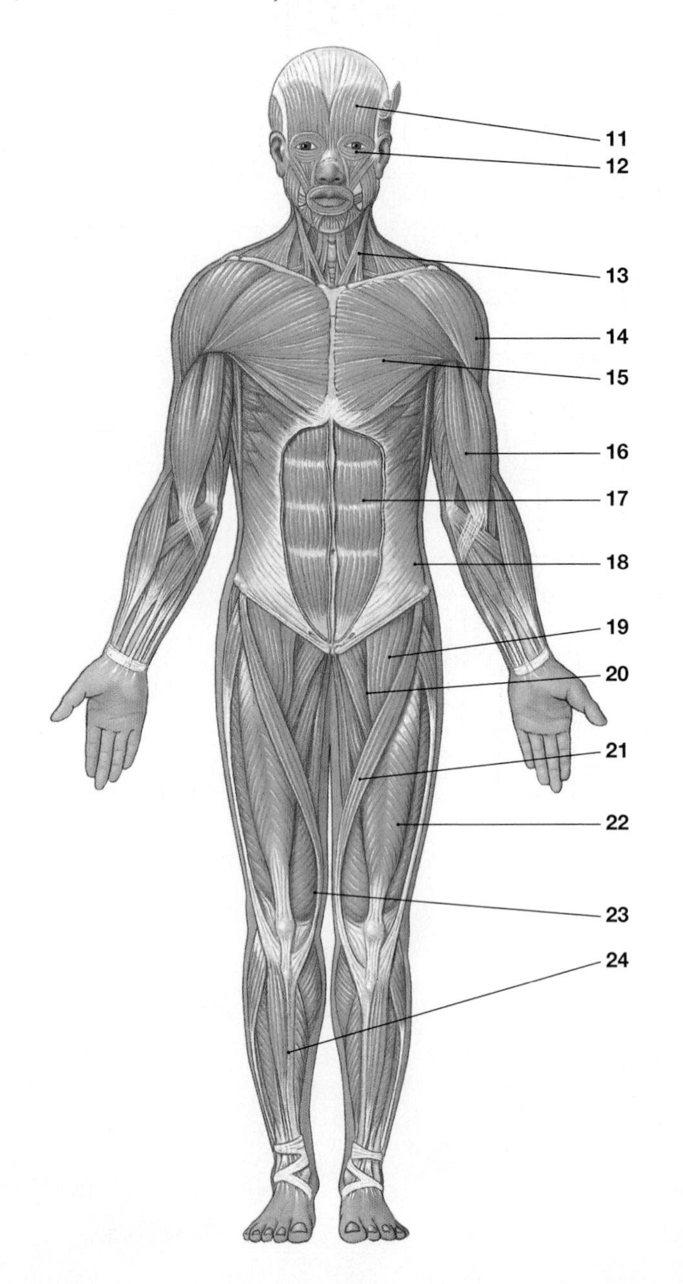

11. ______________________

12. ______________________

13. ______________________

14. ______________________

15. ______________________

16. ______________________

17. ______________________

18. ______________________

19. ______________________

20. ______________________

21. ______________________

22. ______________________

23. ______________________

24. ______________________

Questions 25–38: Identify the labeled muscles.

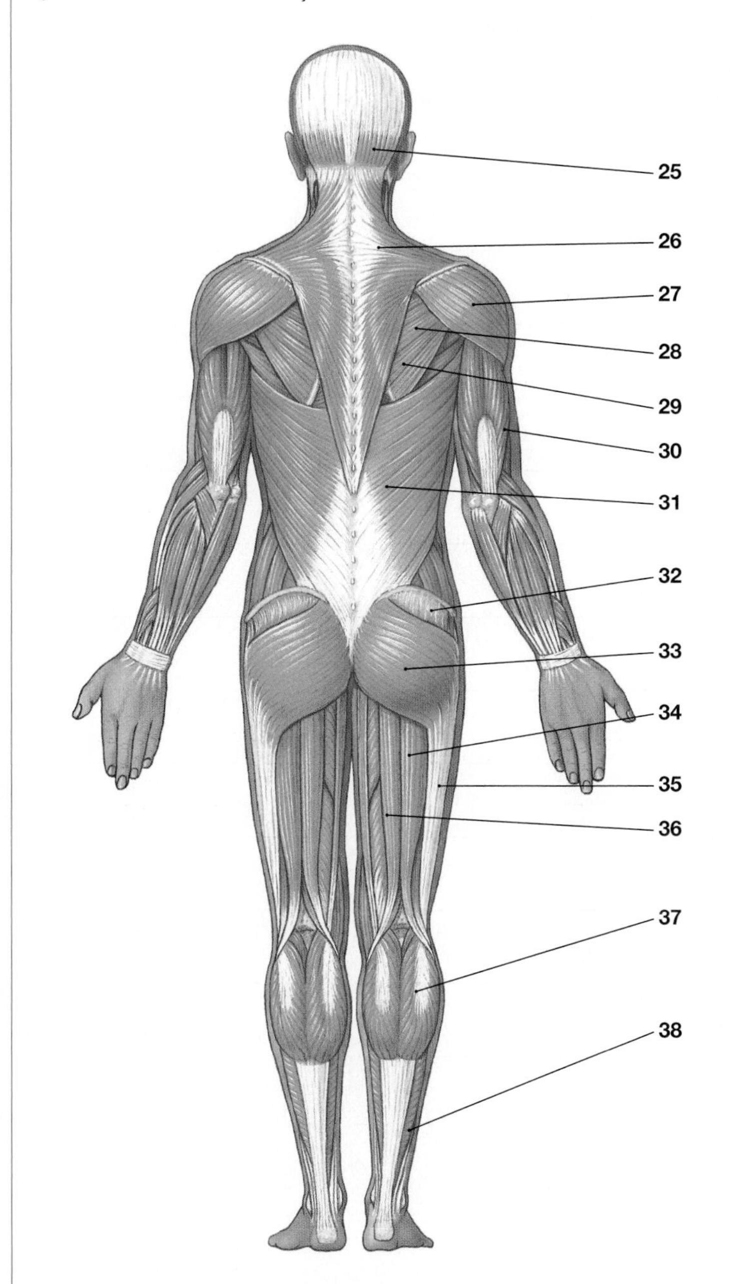

25. ______________________________

26. ______________________________

27. ______________________________

28. ______________________________

29. ______________________________

30. ______________________________

31. ______________________________

32. ______________________________

33. ______________________________

34. ______________________________

35. ______________________________

36. ______________________________

37. ______________________________

38. ______________________________

Questions 39–42: The vast majority of skeletal muscles are named according to one or more distinguishing features that describe their structure or function. Complete the table by providing examples of muscles whose names are derived from the listed criteria.

Criterion	Example
39. Number of origins a. biceps (5 two origins)	
b. triceps (5 three origins)	
c. quadriceps (5 four origins)	
40. Primary action of muscle a. flexor/extensor	
b. abductor/adductor	
c. supination/pronation	
41. Shape of muscle a. serratus (serrated edge)	
b. rhombus (diamond-shaped)	
c. quadratus (square-shaped)	
42. Size of muscle a. maximus/minimus	
b. longus/brevis	
c. major/minor	

Exercise 12

Physiology of the Muscular System

Laboratory Objectives

On completion of the activities in this exercise, you will be able to:

- Explain the difference between isotonic and isometric muscle contractions.
- Understand the interactions between muscles that contract concentrically and eccentrically during body movements.
- Appreciate the relationship between the contraction speed and load (resistance to contraction).
- Explain the motor unit.
- Explain the functional significance of motor unit recruitment.
- Understand how muscle fatigue affects normal muscle activity.
- Evaluate the range of motion of muscles at various joints.

Materials

- Dumbbells of various weights, or similar objects that are easy to hold
- Goniometers
- Floor mats
- Biopac Student Lab System

When a skeletal muscle contracts, it typically exerts a force, called the **muscle tension**, on an object. The object in turn exerts an opposing force, known as the **load**, which resists the muscle tension. When a muscle acts on an object, the type of contraction depends on the relationship between the muscle tension and load. For example, during an **isotonic contraction**, the tension exceeds the load; the contracting muscle changes in length and a movement is initiated (Figures 12.1a and b). During **isometric contractions**, the muscle tension is unable to overcome the load created by the object. The contracting muscles will not change in length and the object will not move (Figures 12.1c and d). Although isometric contractions do not perform work, they are important for maintaining posture and for stabilizing joints where movements are not desired.

Skeletal muscles are innervated by **motor nerve fibers (motor axons)** that originate in the brain or spinal cord and travel along some cranial and all spinal nerves to reach their destinations (Figure 12.2). When a nerve fiber penetrates a muscle, it divides into numerous terminal branches. Each branch forms a **neuromuscular junction** with one muscle fiber (the neuromuscular junction is discussed in Exercise 9). Each nerve fiber, its terminal branches, and all the muscle fibers it innervates represent a **motor unit** (Figure 12.2). The size of motor units can vary from as little as 4 or 5 muscle fibers to as many as 2000 muscle fibers. Typically, muscles that are used for highly specialized motor skills, such as the intrinsic hand muscles, have motor units with fewer than 10 muscle fibers. Conversely, large weight-bearing muscles, such as the gluteus maximus, contain motor units with hundreds of muscle fibers.

The contraction of muscle fibers in a motor unit is determined by the motor unit's **threshold stimulus**, which is the minimum stimulus strength required to initiate a contraction. If the threshold stimulus is reached, all the muscle fibers in a motor unit contract simultaneously. A whole muscle contains a mosaic of motor units with varying threshold stimuli. The muscle fibers of motor units are not bunched together, but dispersed throughout the muscle and mixed with muscle fibers of other motor units (Figure 12.2). Thus, when a motor unit is stimulated, it causes a generalized contraction of the entire muscle. If the stimulus strength increases, additional motor units are turned on and the contraction is more forceful.

The special arrangement and the properties of motor units allow a muscle to perform **graded contractions**, which refers to the ability to contract in a smooth, sustained manner and at various levels of strength. For example, when rowing a boat on a river, the muscles of your upper extremities must contract with a higher level of coordination and greater force if you are rowing against the current rather than with the current. To overcome the greater load that exists against the current, two adjustments must be made.

- The frequency of nerve stimulation to each motor unit must increase to maintain a period of smooth, sustained muscle contraction, known as **tetanus.**
- The strength of the stimulus must increase to activate (recruit) additional motor units so that the muscles can contract with greater force. This is called **recruitment of motor units.**

CLINICAL CORRELATION

The disease tetanus has no relationship to normal muscle function but gets its name from the primary effect on its victim: sustained and forceful contractions of skeletal muscles. The disease is caused by the bacterium *Clostridium tetani,* which grows rapidly in anaerobic (no oxygen) environments. Thus, deep puncture wounds (e.g., a deep cut caused by a nail) provide an excellent opportunity for the disease to develop. Once the *Clostridium* bacteria are established, they multiply rapidly and release a powerful toxin that inhibits the regulation of motor neuron activity. The outcome is sustained and abnormally forceful contractions of skeletal muscles. One common symptom of tetanus, which occurs early during infection, is difficulty in opening the mouth. For this reason, the disease is sometimes called **lockjaw.**

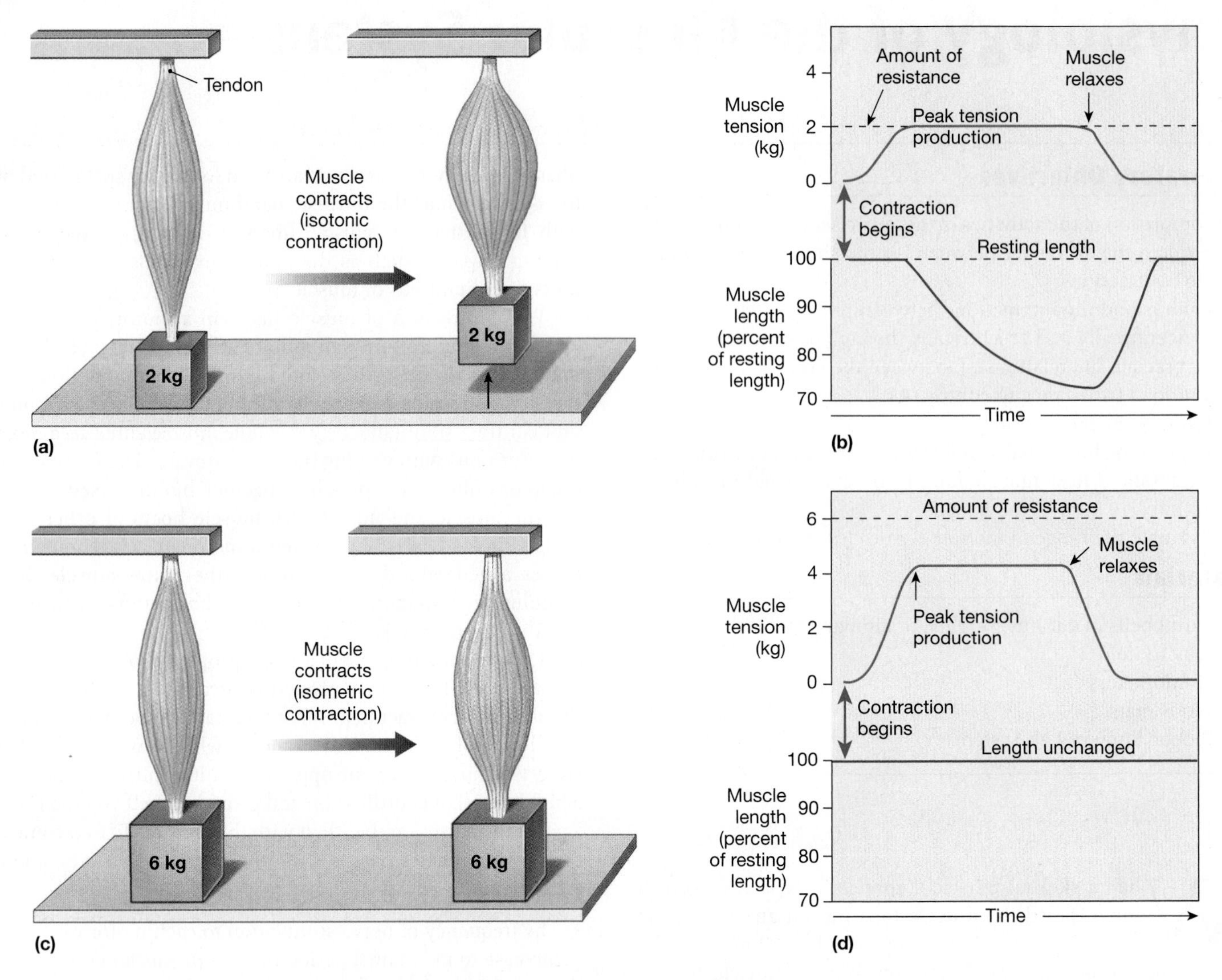

Figure 12.1 Comparison of isotonic and isometric muscle contractions. a) and b) Muscle tension is greater than the load and the weight is lifted off the surface. The muscle reaches a peak tension and remains constant during the period of contraction. During contraction the length of the muscle changes. **c) and d)** Muscle tension is less than the load and the weight is not lifted off the surface. The muscle reaches a peak tension during the period of contraction but the muscle length does not change.

Symptoms of tetanus, such as difficult swallowing and muscle spasms and stiffness, usually begin to appear about 2 weeks after the initial infection. People can recover from tetanus, but those with severe cases may die from respiratory failure if the diaphragm is affected. In the United States and other developed countries, tetanus is not considered a serious public health problem because proper immunization practices (tetanus shots) prevent infection. However, booster shots are required periodically.

Although muscle contractions are typically evaluated by the forces they generate, muscle flexibility is also important. Muscle flexibility refers to the range of motion (ROM) at a particular joint. The degree of flexibility depends on the potential of muscles and tendons to lengthen when movement occurs at a joint. Individuals with above average flexibility have a greater potential for musculotendinous elongation and exhibit a greater ROM at most joints. Activities such as figure skating, gymnastics, ballet dancing, and yoga require an exceptionally high level of ROM at most joints. On the other hand, long-distance running, basketball, bicycling, and horseback riding require only a normal ROM.

Types of Muscle Contractions

Muscles can perform two types of isotonic contractions: **concentric** and **eccentric**. During concentric contractions (Figures 12.1a and b), muscles shorten in length and perform

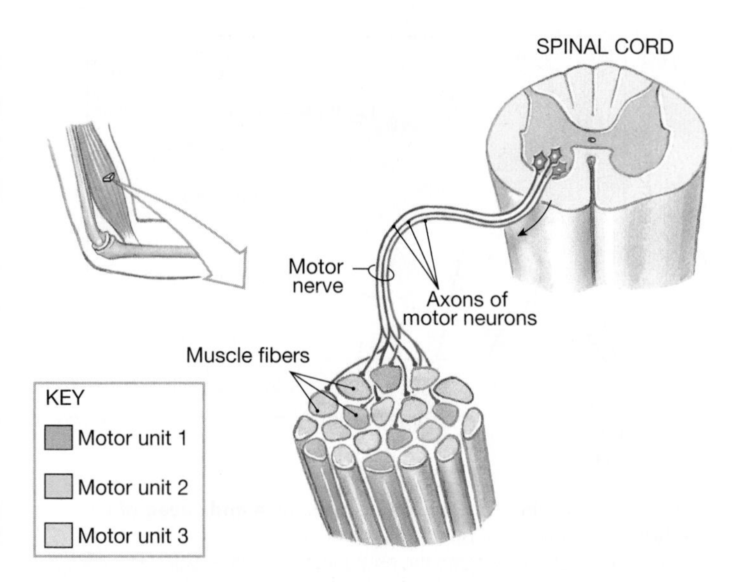

Figure 12.2 The components of a motor unit. The muscle fibers of a motor unit are not clustered together, but intermingled with the muscle fibers of other motor units.

work by moving an object from one position to another. During eccentric contractions, muscles increase in length but do not typically cause an object to be moved. Instead, they act in coordination with other muscles that contract concentrically. For example, lifting a book off a desk involves flexion of the forearm. While the anterior arm muscles (biceps brachii and brachialis) contract concentrically to perform the action, your posterior arm muscle (triceps brachii) contracts eccentrically. Placing a book on a desk requires extension of the forearm. To carry out this action, the triceps brachii will contract concentrically while the anterior arm muscles contract eccentrically to control the action. As this simple example demonstrates, nearly all muscular activities (e.g., lifting, walking, running, climbing) involve both concentric and eccentric contractions.

If you attempt to push a parked automobile with its emergency brake engaged, you can feel muscles in your upper and lower extremities tighten as they contract, but the car will not budge. Your effort will be in vain because your muscles are contracting isometrically—the load generated by the car is greater than the tension produced by your muscles (Figures 12.1c and d).

WHAT'S IN A WORD The term *isotonic* is derived from Greek and means "equal tension" (*isos,* "equal"; *tonos,* "tension"). The name refers to the state of a contracting muscle as it shortens to overcome a constant load (tension), such as lifting a bottle of water off a table.

The term *isometric* also has Greek origins and means "of equal dimensions" (*isos,* "equal"; *metron,* "measure"). The name refers to the condition of a muscle that contracts but does not change in length.

ACTIVITY 12.1 Demonstrating Isotonic and Isometric Contractions

Isotonic Muscle Contractions

1. Stand in an upright position with your feet flat on the floor.
2. While keeping your feet flat on the floor, slowly and carefully lower your torso by performing a deep knee bend.
3. Remain in the crouched position for a few seconds.
4. Slowly elevate your torso, returning to the upright position.
5. During the deep knee bend, identify the thigh muscles that are performing the following:

   a. Concentric isotonic contractions: ________________

   ____

   b. Eccentric isotonic contractions: ________________

   ____

6. During the return to an upright position, identify the thigh muscles that are performing the following:

   a. Concentric isotonic contraction: ________________

   ____

   b. Eccentric isotonic contraction: ________________

   ____

Isometric Muscle Contractions

1. Stand in the threshold of a doorway and place the palms of your hands flat along each side frame.
2. Attempt to push the sides of the threshold away from you.
3. Alternatively, place the palms of your hands flat against a wall in the laboratory and attempt to push the wall away from you.
4. Identify muscles that are contracting as you perform one of the above actions.

   ____

   ____

5. Explain why the muscles you identified in step 4 are contracting isometrically.

   ____

   ____

QUESTIONS TO CONSIDER 1. Describe the relationship between the muscle tension and load during an isotonic contraction.

____

____

2. In general, describe the changes (if any) that occur in the lengths of the sarcomeres of a muscle that is performing the following:
- Concentric isotonic contraction:

____

- Eccentric isotonic contraction:

____

- Isometric contraction:

____

3. At what step during the deep knee bend are your thigh muscles contracting isometrically? Briefly explain your answer.

____

____

____

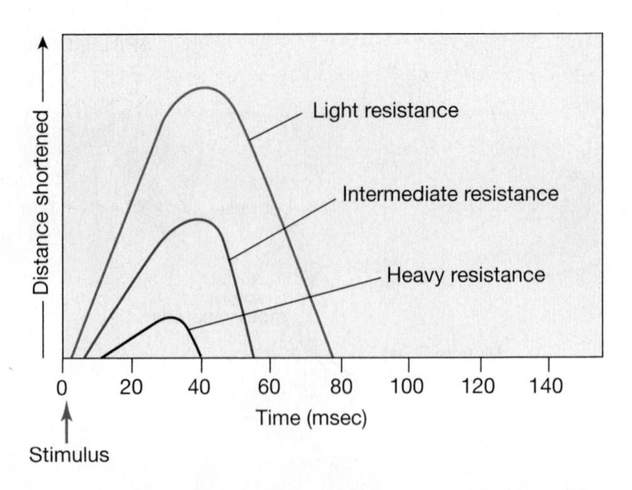

Figure 12.3 Graphic comparison of resistance and speed of contraction. As the load increases, it will take a longer period of time for the muscle to begin shortening and the amount of shortening will be less.

Resistance Versus Contraction Speed

The speed at which a muscle contracts is inversely related to the resistance or load. When a muscle begins to contract, tension increases gradually. To move an object, muscle tension must exceed the load, and it will take longer for tension to exceed the load of a heavier object than a lighter object (Figure 12.3). Thus, you can lift a light object much more quickly than a heavy object.

ACTIVITY 12.2 Examining Resistance Versus Contraction Speed

1. Obtain a series of three or four dumbbells (or other objects that are easy to hold) of various weights.
2. Rest the forearm of your dominant side on the laboratory bench with your palm directed superiorly.
3. From this position, grasp the lightest dumbbell (or other object) and lift it with moderate force by flexing your forearm. Do not use maximum force, and keep your elbow on the bench surface as you lift.
4. Repeat step 3 with the other dumbbells, in ascending weight order. Be sure you apply the same level of force with each lift as you used in the previous step, and keep your arm and forearm in the same relative positions as before.
5. Describe your results in terms of speed of contraction and increasing resistance.

____

____

QUESTIONS TO CONSIDER 1. While performing this activity, why is it important that your arm and forearm remain in the same position?

____

____

2. Explain why your thigh and leg muscles will contract more slowly while running up a steep hill compared to along a level surface.

____

____

Motor Unit Summation

The strength of a muscle contraction depends on the number of motor units that are stimulated. If the muscle contracts slowly and with little force, relatively few motor units have been activated. If the muscle receives more frequent nerve impulses, additional motor units are activated and the muscle contracts faster and with more power. The activation of increasing numbers of motor units is called **multiple motor unit summation** or **motor unit recruitment**.

ACTIVITY 12.3 Demonstrating Multiple Motor Unit Summation

1. Obtain a 2.25-kg (about 5-lb) dumbbell or other object of similar weight that is easy to hold.

2. Rest the forearm of your dominant side comfortably on the laboratory bench with your palm directed superiorly.
3. Grasp the dumbbell (or other object) and flex your forearm with the minimum amount of force (muscle tension) required to lift it. Keep your elbow on the bench surface at all times.
4. Repeat step 3 several times, increasing the force of contraction with each lift. On the last turn, lift the dumbbell by applying peak muscle tension.
5. Describe your results in terms of multiple motor unit summation.

___

___

___

QUESTIONS TO CONSIDER 1. How does the recruitment of motor units affect contraction speed? Provide some examples in your answer.

___

___

___

2. Explain why recruitment of motor units is important for overcoming relatively large loads (e.g., lifting a pen compared to lifting a heavy anatomy and physiology textbook).

___

___

___

3. Even when at rest, some motor units are active, so the muscles are contracted to some degree at all times. The state of contraction, at rest, is called **muscle tone**. Consider situations in which you are standing upright but relaxed and not moving, or sitting at your desk at rest. Speculate on how these conditions would change if your muscles lacked tone.

___

___

___

Muscle Fatigue

Muscle fibers can contract as long as adequate energy is available. Over time, however, the metabolic production of energy cannot keep up with the demand and metabolic wastes, such as lactic acid, accumulate in the sarcoplasm. As a result, the muscle will no longer generate a force. This breakdown in function is called **muscle fatigue**.

CLINICAL CORRELATION

Virtually all the energy (ATP) used by muscles at rest or during light to moderate exercise is derived from **aerobic respiration,** a process that occurs in the mitochondria and requires an adequate supply of oxygen. As long as oxygen is available, aerobic metabolism will proceed normally to supply muscle cells with enough ATP to function normally. If activity becomes more intense and continues for an extended period, however, a point will be reached when oxygen supply will not keep pace with the demand. If this occurs, muscle metabolism will switch to an anaerobic process called **glycolysis.** Aerobic respiration produces much more ATP than glycolysis, but it is rather slow because it involves a large number of biochemical steps. Glycolysis is fast and efficient, but produces only a small amount of ATP. It serves as a "quick fix" when muscle cells need energy immediately, such as running a 100-m sprint. If you rely too heavily on glycolysis for your energy needs, there is a price to pay. The primary waste product of glycolysis is lactic acid. If enough oxygen is available, muscle cells are able to convert lactic acid to pyruvic acid, which can be used to produce ATP. Lactic acid that is released into the bloodstream is transported to the liver where it is converted to glucose. Under conditions of intense muscle activity, there is an oxygen deficiency. As a result, lactic acid tends to accumulate in muscle cells and muscle fatigue sets in. The oxygen deficiency leads to what is called the **oxygen debt,** which is the amount of oxygen required for muscles to recover from the period of intense activity. Recovery activities include conversion of lactic acid to pyruvic acid (in muscle cells) or glucose (in the liver) and replenishment of energy reserves in muscle cells. The oxygen debt is paid back by a period of deep breathing for a period of time after the intense exercise. This explains why sprinters are breathing heavily immediately after a race.

ACTIVITY 12.4 Demonstrating Muscle Fatigue

1. Obtain a 2.25-kg (about 5-lb) dumbbell or another object of similar weight that is easy to hold.
2. Stand in an erect position and hold the dumbbell in your dominant hand, by your side.
3. With the dumbbell in hand, abduct your arm so that your upper extremity is parallel to the floor. The time when you begin this exercise is noted as "time 0" in Table 12.1.
4. Keep your upper extremity parallel to the floor for as long as you can.
5. When your upper limb begins to sag and you are laboring to keep your upper limb parallel to the floor, change the position of your arm and forearm in one of two ways.
    - Rotate your forearm so that your hand turns 90°.
    - Move your arm anteriorly so that it is projecting forward.
6. After making a position change, attempt to reposition your upper limb so that it is parallel to the floor once again. Have your lab partner note the time when you perform this maneuver in Table 12.1.

Table 12.1 Muscle Fatigue Exercise: Comparison Between Dominant and Nondominant Upper Extremities

	Start time of exercise (sec)	Time of position change (min/sec)	End time of exercise (min/sec)
Dominant side	0		
Nondominant side	0		

7. Keep your upper limb parallel to the floor for as long as you can. When you can no longer maintain the position, lower the dumbbell and rest. Have your lab partner note the time when you end the exercise in Table 12.1.
8. Repeat steps 1 through 5 with your nondominant upper limb. Have your lab partner record the appropriate times in Table 12.1.

QUESTIONS TO CONSIDER 1. Did you note any differences between your dominant and nondominant upper limbs? Provide an explanation for these differences, if any.

____

____

2. When you changed the position of your upper limb, speculate on why you were able to hold the dumbbell in position for an extended period of time.

____

____

3. Explain why, after a period of time, you were unable to keep the dumbbell in the original position.

____

____

4. One factor that causes muscle fatigue is the accumulation of lactic acid in the sarcoplasm of muscle cells. How will this impair normal muscle function? (Hint: Consider the pH inside muscle cells.)

____

____

Muscle Flexibility

The two types of flexibility are static and dynamic. For **static flexibility**, the degree of difficulty in performing the range of motion (ROM) is not considered. For **dynamic flexibility**, resistance to motion is a consideration and the ease or difficulty of a joint moving through the ROM is assessed.

Dynamic flexibility is more important for evaluating athletic ability or the physical condition of a joint (as with evaluating a joint for arthritis), but it is difficult to accurately measure in a student laboratory setting without specialized equipment. Thus, in this activity, you will be performing methods to measure the static flexibility of various joints.

To measure static flexibility, you will use an instrument called a **goniometer**. A goniometer is used like a protractor to measure the angle formed by a moving body part at a joint during a range of motion evaluation. It has two arms, one stationary and one moving, that are joined together and extend from a central pivot location (the fulcrum). To measure the ROM of a joint, the fulcrum of the goniometer is placed over the joint and the two arms are placed in line with the moving body part. When the desired movement is made, the stationary arm remains in its original position; the moving arm is swept across an arc that reflects the movement of the body part. The angle that is formed by the two arms of the goniometer at the fulcrum is the range of motion. For example, to measure the ROM during flexion at the knee, the fulcrum of the goniometer is placed over the lateral aspect of the knee joint and the arms are aligned with the long axis of the leg. This represents the starting position. The ROM can be measured as the angle made by the leg when it is flexed from the extended position.

If a goniometer is not available, you can use a visual method to estimate the range of motion. Have your lab partner carefully observe you during the ROM evaluations and make visual comparisons with the appropriate diagram in Figure 12.4.

Conduct the ROM exercises in this activity, either by measuring with a goniometer or by visual estimation. If you use a goniometer, you must maintain the instrument's original position at all times while measuring range of motion. Only the moving arm of the goniometer should change position during movements. Collect data for both the right and left sides of the body and record your results in Table 12.2.

ACTIVITY 12.5 Testing Muscle Flexibility

Shoulder Abduction and Adduction

1. From anatomical position, abduct your arm until it is parallel to the floor (Figure 12.4a). This is your start position. If using a goniometer, your lab partner should position the fulcrum over the anterior aspect of your shoulder and the arms should be aligned with the long axis of your arm.

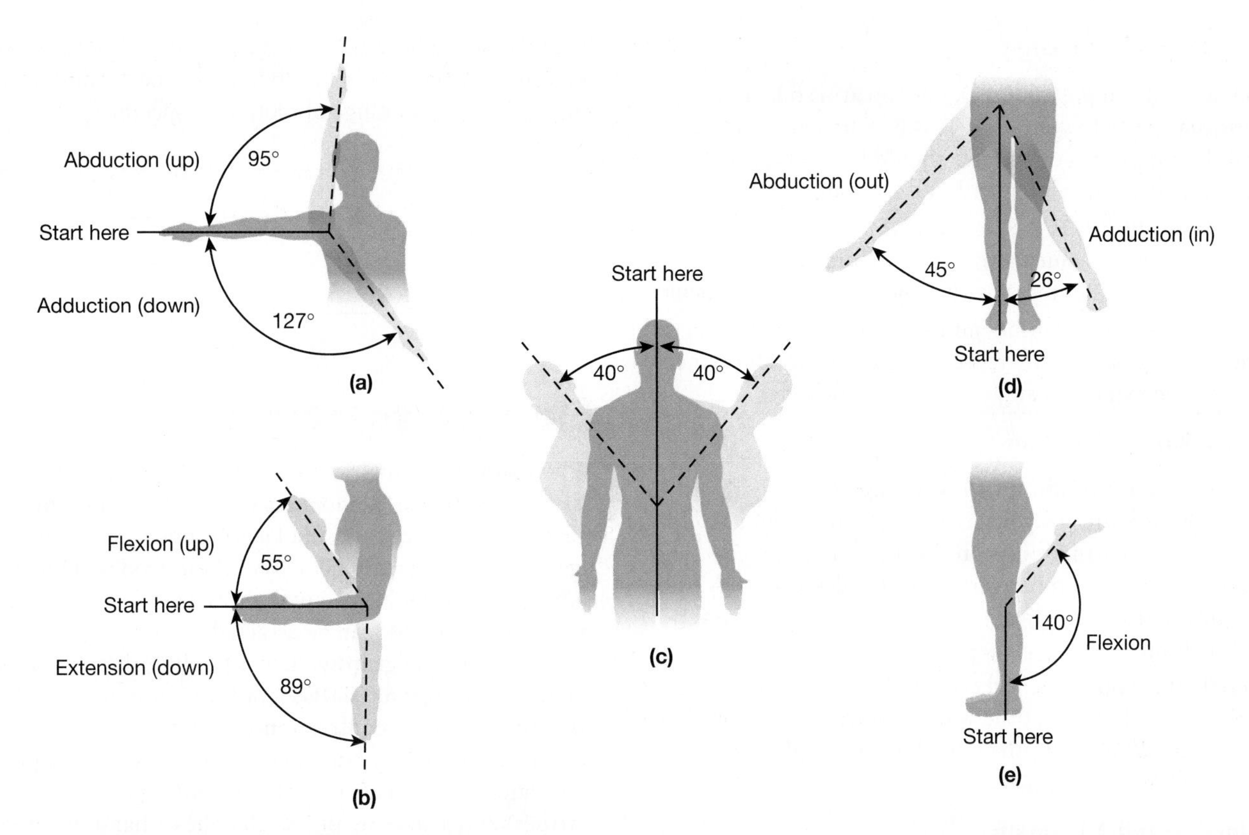

Figure 12.4 Evaluating the range of motion at various joints. a) Abduction–adduction of the shoulder; **b)** flexion–extension of the elbow; **c)** lateral flexion of the lumbar vertebrae; **d)** abduction–adduction of the hip; **e)** flexion of the knee.

Table 12.2 Flexibility Tests: Measuring Range of Motion (ROM)

Joint	Movement	Average ROM	Measured ROM (mark with "x" for visual estimation) Right	Measured ROM Left	Rating (below average, average, or above average) Right	Rating Left
Shoulder	Abduction	95°				
	Adduction	127°				
Elbow	Flexion	55°				
	Extension	89°				
Lower back	Lateral flexion	40°				
Hip	Abduction	45°				
	Adduction	26°				
Knee	Flexion	140°				

2. Abduct your arm as much as possible so that your hand is above your head (Figure 12.4a).
3. Have your lab partner adjust the moving arm of the goniometer and measure the angle produced.
4. Return your arm and the goniometer to the start position, then adduct your arm as much as possible so that it crosses anterior to your torso (Figure 12.4a).
5. Your lab partner should measure and record the adduction angle.

Elbow Flexion and Extension

1. From anatomical position, flex one forearm so it is perpendicular (90°) to your arm. This is your start position. The fulcrum of the goniometer should be placed over the lateral aspect of the elbow with the arms aligned with the long axis of the forearm.
2. Flex the forearm as much as possible (Figure 12.4b) and have your lab partner measure the angle with the goniometer.
3. Return to the start position, and then extend your forearm as much as possible (Figure 12.4b). Have your lab partner measure the extension angle with the goniometer.

Lower Back Range of Motion

1. Stand in an erect position with your knees slightly flexed, and place the palms of your hands on your anterior thighs. Have your lab partner position the fulcrum of the goniometer over the lumbar vertebrae at waist level with the arms directed superiorly along the vertical axis of the vertebral column.
2. Laterally flex your vertebral column by bending sideways (in the coronal plane) at the waist (Figure 12.4c). Your lab partner should measure the lateral flexion angle with the goniometer.

Hip Abduction and Adduction

1. Stand in an erect position. Have your lab partner place the fulcrum of the goniometer over the anterior aspect of the hip joint with the arms aligned with the long axis of the thigh.
2. Abduct your thigh as much as possible by moving your lower limb away from the midline of the body in a coronal plane (Figure 12.4d).
3. Your lab partner should measure the abduction angle with the goniometer.
4. Adduct the thigh back to the start position. From here, adduct the thigh even more by swinging the lower limb as much as possible to the contralateral side of the body (Figure 12.4d). Have your lab partner measure the adduction angle with the goniometer.

Knee Flexion

1. Stand in an erect position.
2. Have your lab partner place the fulcrum of the goniometer over the lateral aspect of the knee joint with the arms aligned with the long axis of the leg.
3. Flex your leg as much as possible.
4. Your lab partner should measure the flexion angle with the goniometer (Figure 12.4e).

QUESTIONS TO CONSIDER

1. Identify sources of experimental error when measuring the range of motion with a goniometer.

____

____

____

2. Speculate on how below-average range of motion could have a negative effect on athletic ability. Provide examples by discussing the role of range of motion in specific sports.

____

____

____

____

Electromyography

When a skeletal muscle contracts, electric impulses are generated along the cell membranes of muscle fibers. This activity, in turn, creates changes in the electric potential (changes in voltage) in overlying regions of skin. If electrodes are placed on the skin, the changes in the cutaneous electrical activity caused by muscle contractions can be detected and recorded by a technique called **electromyography**. The recording that is made is called an **electromyogram (EMG)**. An EMG provides insight into the underlying muscle contractions.

In the following activity, you will use the Biopac Student Laboratory System to record the EMG generated by your lab partner's forearm as he or she clenches a hand dynamometer. A **dynamometer** (*dyno* means "power," *meter* means "measure") measures the force exerted by the contracting forearm muscles. By comparing the force recordings with the EMG, you will be able to correlate motor unit recruitment with strength of muscle contraction. Additionally, you will examine and quantify the phenomenon of fatigue. Upon completion of the activity, you will be able to make comparisons of fatigue-resistance and maximum clench strength between the dominant and nondominant forearms.

ACTIVITY 12.6 Recording EMG Using the BIOPAC System

BIOPAC Systems, Inc.

Setup and Calibration

1. Have your lab partner remove all jewelry, watches, or other metal objects.
2. Position three electrodes on your lab partner's dominant forearm as shown in Figure 12.5. Attach electrode leads (SS2L) as illustrated.
3. In the same manner, position three electrodes on your lab partner's nondominant forearm but do not attach electrode leads at this time.
4. Attach a hand dynamometer (SS25L/LA) to Channel 1, the electrode leads (SS2L) to Channel 3, and headphones (OUT1) to the back of the Acquisition Unit, and then turn on the unit.
5. Start the Biopac Student Laboratory System, choose Lesson 2 (LO2-EMG-2), and click **OK**. When prompted, enter a unique filename and click **OK**.

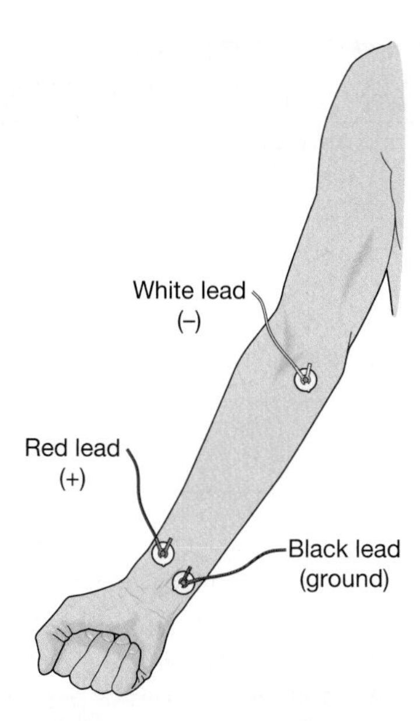

Figure 12.5 Electrode placement for forearm EMG recording. The electrode lead colors should match the positions illustrated in the diagram.

6. Begin the calibration procedure by setting the dynamometer down and clicking the **Calibrate** button. A pop-up window instructs you to remove any force from the transducer so that the computer can establish a zero-force level. With the dynamometer at rest on the table, click **OK.**
7. Next, as instructed by the pop-up window, have your lab partner pick up the dynamometer with the dominant hand; if using the SS25L, hold it close to, but not touching, the metal crossbar and keep the crossbar parallel to the floor, not at an angle (Figure 12.6). Your partner will need to hold the dynamometer in the same position for all measurements for each arm. When ready, click **OK.**
8. In the next pop-up window, click the **Calibrate** button to continue calibration. Instruct your partner to wait 2 seconds, squeeze the dynamometer with a maximum clench for about 4 seconds, and then relax. This procedure tells the computer what a maximum clench looks like. The Calibration procedure will automatically stop after 8 seconds.
9. If the recordings in both channels have a zero baseline and positive signals representing the maximum squeeze of the dynamometer, you are ready to record. If not, then check the electrodes and connections and click **Re-do Calibration.**

Recording Data

In this activity, you will record two segments of data on your lab partner's dominant arm. Segment 1 will be a series of clenches with increasing force. Segment 2 will be a prolonged clench to investigate muscle fatigue. You will click on the **Suspend** button to automatically insert a marker between the two segments. You will then repeat the experiment and collect similar data using your partner's nondominant arm (segments 3 and 4). Read this entire section first so that you will be familiar with the activities before you begin recording data.

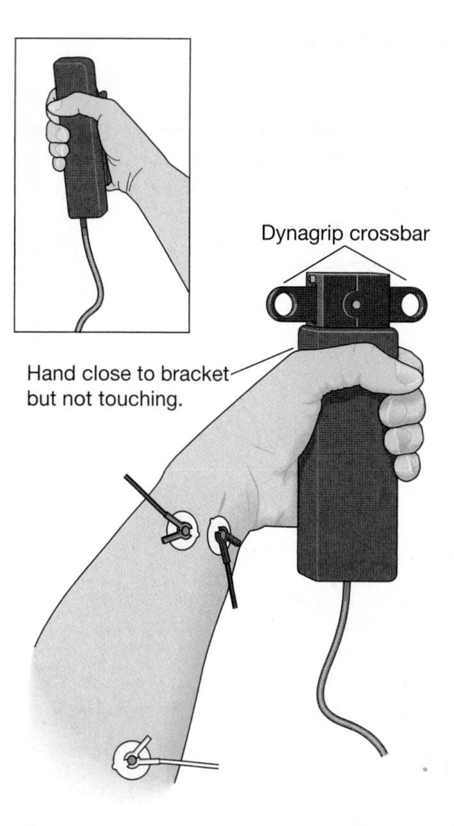

Figure 12.6 Proper hand position for using dynamometer. The hand should be close to, but not touching, the metal crossbar. Hand placement must be consistent throughout the experiment.

1. Click **Record** to begin the recording. For Segment 1, have your lab partner perform a series of clenches using the dynamometer with increasing degrees of force.
   - The first clench should be one of 5 kg. Your partner should increase the force applied by 5 kg for each of the subsequent clenches until he or she can clench no harder. This will result in clenches of 5 kg, 10 kg, 15 kg, and so on until a maximum is reached.
   - To reach a particular target force, your partner should watch the dynamometer tracing on the screen and compare it to the force markings shown on the vertical axis to the right.
   - For each clench, your partner should squeeze until the target force is reached, hold for 2 seconds at that level, then release. After waiting for 2 seconds, the next clench may begin.
   - To keep track of the time, note that each vertical line on the screen is a 2-second interval.
2. When you complete the series of clenches (Segment 1), click **Suspend** to pause the recording. The resulting data should resemble Figure 12.7.
3. Click **Resume** to begin the second segment, a marker labeled "continued clench at maximum force" will automatically be inserted.
4. For Segment 2, have your lab partner clench the dynamometer with maximum force without watching the computer screen and hold this clench until the force

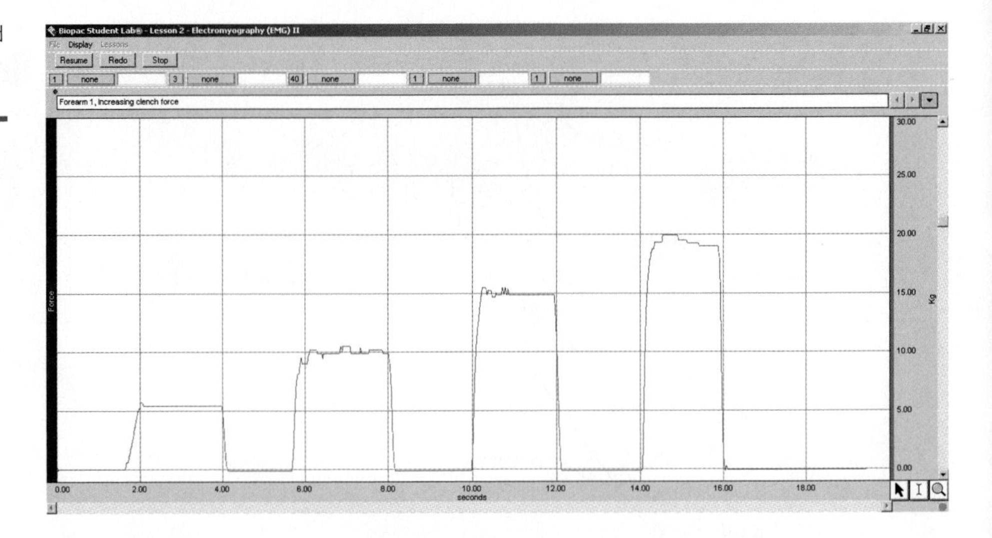

Figure 12.7 Motor unit recruitment. The recorded data illustrate four successive clenches of increasing force.

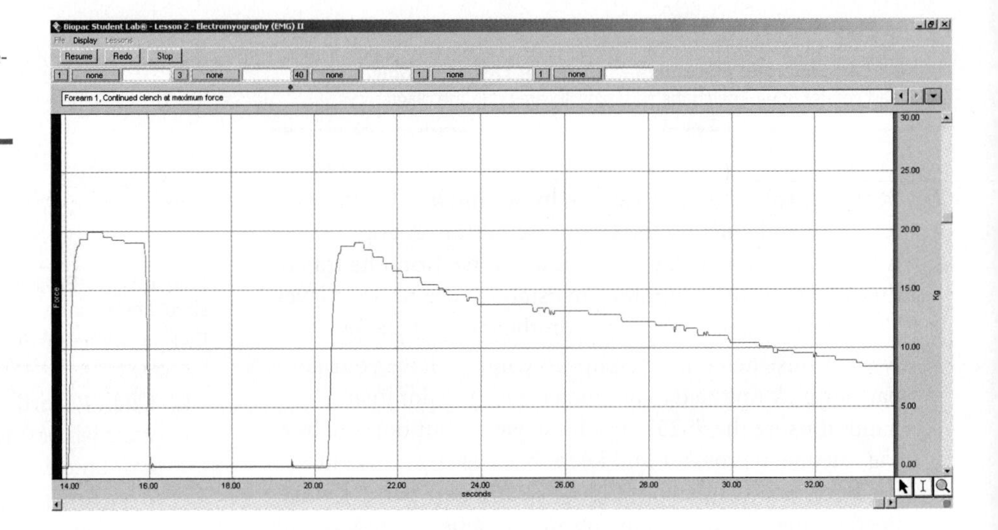

Figure 12.8 Muscle fatigue. The recorded data illustrate a prolonged clench. The force has dropped below 50% of the initial maximum by the end of the recording.

tracing decreases by more than 50% (e.g., below 12 kg for a maximum force of 24 kg). Since your lab partner should not be watching the screen, you should notify him or her when the tracing has fallen consistently below 50% of the initial maximum.

5. Click **Suspend** to pause the recording. Your partner may then release his or her grip and relax. The resulting data should appear similar to that illustrated in Figure 12.8.
6. When you have completed Segment 2, you may listen to an "auditory version" of the EMG. While wearing the headphones, click the **Listen** button and have your lab partner clench and relax with the dynamometer. Note that the frequency of the auditory signal corresponds with the intensity of the muscle contractions. This data will not be recorded.
7. Click on **Stop** when you are ready to continue with the next experiment. A pop-up window will appear asking if you are sure you want to stop recording.
   - Clicking **No** will bring you back to the **Resume** or **Stop** options, providing one last chance to redo the fatigue recording for Forearm 1 (dominant arm).
   - Clicking **Yes** will end the recording session and automatically save your data with an added file extension "1-L02," which indicates forearm 1 for Lesson 2. This extension will be added onto your previously entered filename (e.g., it would appear as SmithEMG02–1–L02).
8. Disconnect the electrode leads from your lab partner's dominant arm and reconnect them in the same manner to the nondominant arm.
9. At the next pop-up window, click **Forearm 2**, calibrate the computer as described previously, and repeat the entire recording experiment (steps 1–4) with the nondominant forearm.
10. When you have finished recording from the nondominant arm, click **Done**.

Data Analysis

1. Enter the **Review Saved Data** mode and select the data file for Lesson 2. This file has the extension "L02."
2. When opened, you should see a graph window consisting of three channels: Ch 1 Force (kg), Ch 3 Raw EMG (mV), and Ch 40 Integrated EMG (mV).

Table 12.3 Segment 1 Data: Motor Unit Recruitment

	Forearm 1 (Dominant)		Forearm 2 (Nondominant)	
Clench #	Force at peak [Ch 1] mean (kg)	Int. EMG [Ch 40] mean (mV)	Force at peak [Ch 1] mean (kg)	Int. EMG [Ch 40] mean (mV)
1				
2				
3				
4				
5				
6				
7				
8				
% increase from first to last clench				

Table 12.4 Segment 2 Data: Fatigue

Forearm 1 (Dominant)			Forearm 2 (Nondominant)		
Maximum clench force	50% of max clench force	Time to fatigue	Maximum clench force	50% of max clench force	Time to fatigue
Ch 1 value	Calculate	Ch 1 delta T	Ch 1 value	Calculate	Ch 1 delta T

- The Force tracing comes from the dynamometer.
- The Raw EMG represents electrical activity detected by the electrodes.
- The Integrated EMG is like a "contour" tracing of the shape and intensity of the Raw EMG. It is similar to taking the absolute value of the Raw EMG and reduces some of the "noise" that makes taking measurements from the raw signal difficult.

Since you will not need to view the Raw EMG, you may hide that channel by clicking on the Ch 3 box while simultaneously holding down the **Ctrl** key (PC) or **Option** key (Mac).

3. To analyze the data from Segment 1 of the dominant arm, set up the first two pairs of measurement boxes as follows:

   Ch 1 mean
   Ch 40 mean

4. Begin with Channel 1 (dynamometer force). Using the I beam cursor tool, highlight the "plateau" of the first clench. The plateau should be a relatively flat region of the tracing in the middle of the clench, representing the "hold" period of the clench-hold-release cycle. Record the mean values for channels 1 and 40 in the **Forearm 1** column of Table 12.3. Repeat the same measurement for each of the successive clenches. Note that the mean force measurements in Ch 1 should approximate the 5-10–15-kg target forces.
5. Calculate the percentage increase that occurred in Force and Integrated EMG activity from the weakest clench to the strongest clench. Record the result in the final row of Table 12.3.
6. Scroll along the bottom of the data window to the beginning of Segment 2, which includes the recording of muscle fatigue. Segment 2 should begin after the first marker that appears just above the waveforms.
7. Change the measurement boxes as follows:

   Ch 1 value
   Ch 1 delta T (ΔT)

8. In Ch 1, use the I beam cursor tool to click on a point of maximal clench force immediately following the start of Segment 2. Note the resulting value displayed in the Ch 1 measurement box and record in Table 12.4. Calculate 50% of that value and record in the table.
9. Examine the tracing in Ch 1 and make an eyeball approximation of the point at which the tracing is 50% down from the maximum clench point. Then use the I beam cursor to click on points near this region, noting the value displayed in the Ch 1 measurement box until you locate the point

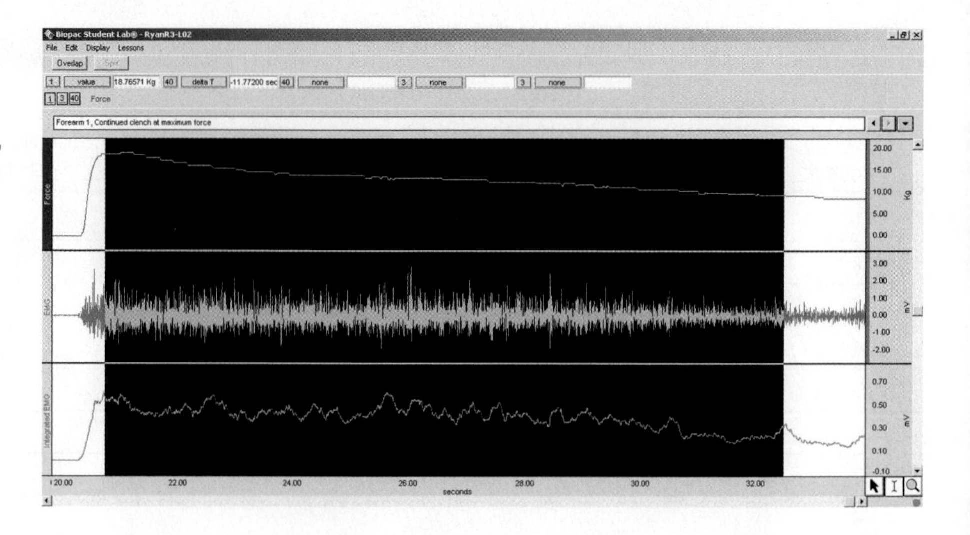

Figure 12.9 Measuring delta T for muscle fatigue. The selected area extends from the maximum clench force immediately following the start of Segment 2 to a force equal to 50% of the maximum.

that equals the 50% value you calculated in step 8. When you have located this point, click and drag backwards to the beginning of the clench. Release the mouse when you have selected this region (Figure 12.9).

10. Note the value for delta T in Ch 1 and record in Table 12.4. This value represents the time it took for the clench force to fatigue to 50% of maximum. You may disregard any + or − sign for the delta T value.
11. Return to **Review Saved Data** and select the data file for the nondominant forearm. This file has the extension "2-L02."
12. Repeat the entire analysis (steps 2–10) for Forearm 2 and enter the data in the appropriate columns of Tables 12.3 and 12.4.

QUESTIONS TO CONSIDER

1. What does increased EMG activity physiologically represent?

___

___

___

2. Describe the relationship between Force and EMG activity. Explain how this relates to motor unit recruitment.

___

___

___

___

3. Consider the percentage increase in Force and EMG activity between the weakest and strongest clenches (Table 12.3). What can you conclude about the relative strength of the dominant and nondominant forearms?

___

___

___

4. Consider the Force and Fatigue data for both forearms (Table 12.4). What might you conclude about the strength and aerobic capacity of each forearm? What factors might contribute to these differences?

___

___

___

___

Exercise 12 Review Sheet

Physiology of the Muscular System

Name ______________________

Lab Section ______________________

Date ______________________

1. Describe the relationship between muscle tension and load.

____

____

2. From a functional standpoint, discuss the difference between isotonic and isometric muscle contractions.

____

____

3. Discuss the similarities and differences between concentric and eccentric muscle contractions.

____

____

4. Discuss the relationship between muscle function and motor unit size.

____

____

5. The muscle fibers in a motor unit are dispersed throughout a muscle and intermingle with muscle fibers of other motor units. What is the functional significance of such an arrangement?

____

____

6. Discuss how recruitment of motor units contributes to a muscle's ability to carry out graded contractions.

____

____

Exercise 13

Histology of Nervous Tissue

Laboratory Objectives

On completion of the activities in this exercise, you will be able to:

- Explain the general organization of the nervous system.
- Recognize the important microscopic structures of a neuron.
- Explain the general structure and function of multipolar, bipolar, and unipolar neurons.
- Understand the functions of the neuroglia.
- Explain how an impulse is transmitted at a synapse.
- Identify important microscopic structures in the cerebrum and cerebellum.
- Identify important microscopic structures of the spinal cord.
- Understand how ventral and dorsal spinal roots give rise to a spinal nerve.
- Describe the organization of a peripheral nerve.

Materials

- Compound light microscopes
- Prepared microscope slides of the following nerve tissues:
  - Ox spinal cord smear
  - Cerebrum
  - Cerebellum
  - Spinal cord cross section with dorsal and ventral roots
  - Peripheral nerve cross section
- Anatomical models of a nerve cell and a synapse
- Colored drawing pencils

The nervous system provides the body with a means for maintaining homeostasis and ensuring that bodily functions are carried out efficiently. It performs this function by conducting electric impulses along nerve fibers to target organs. As a result of this electrical activity, the nervous system is able to control and integrate the activities of all the organs and organ systems. Neural responses occur quickly but have only short-term effects.

The nervous system is responsible for three basic functions.

- *Reception.* With the assistance of sensory receptors, it is able to detect changes that occur inside the body and in the surrounding environment.
- *Integration.* It is able to interpret and integrate sensory input by storing the information as memory and producing thoughts.
- *Response.* It is able to respond to the sensory input by initiating muscular contractions or glandular secretions.

The two major divisions of the nervous system (Figure 13.1) are the **central nervous system (CNS)** and the **peripheral nervous system (PNS).** The central nervous system consists of the brain and the spinal cord. It is the control center for all nervous system function. All sensory information must be delivered to the CNS if it is to be detected and integrated. The CNS produces all motor impulses to muscles and glands. The peripheral nervous system comprises all nerves that connect the brain and spinal cord to muscles, glands, and receptors. Nerves connected to the brain are called **cranial nerves**; those connected to the spinal cord are **spinal nerves.** Cranial and spinal nerves convey sensory information from receptors (e.g., pain receptors in the skin) to the CNS, and transmit motor information from the CNS to **effector organs** (e.g., muscles and glands). Thus, the PNS can be divided into a **motor (efferent) division** with **motor (efferent) neurons**, and a **sensory (afferent) division** with **sensory (afferent) neurons.**

The motor, or efferent, division consists of two parts (Figure 13.1).

1. The **somatic nervous system** contains efferent neurons extending from the CNS to skeletal muscle (voluntary actions).
2. The **autonomic nervous system** contains efferent neurons extending from the CNS to smooth muscle, cardiac muscle, and glands (involuntary actions).

The sensory, or afferent, division contains three components (Figure 13.1).

1. Afferent neurons that receive stimuli from **somatic sensory receptors** that detect **general sensations** (touch, pressure, temperature, pain, and body position) in the skin, skeletal muscles, and joints.
2. Afferent neurons that receive stimuli from **visceral sensory receptors** that detect sensations in internal organs.
3. Afferent neurons that receive stimuli from **special sensory receptors** that detect **special sensations** (smell, taste, vision, hearing, equilibrium).

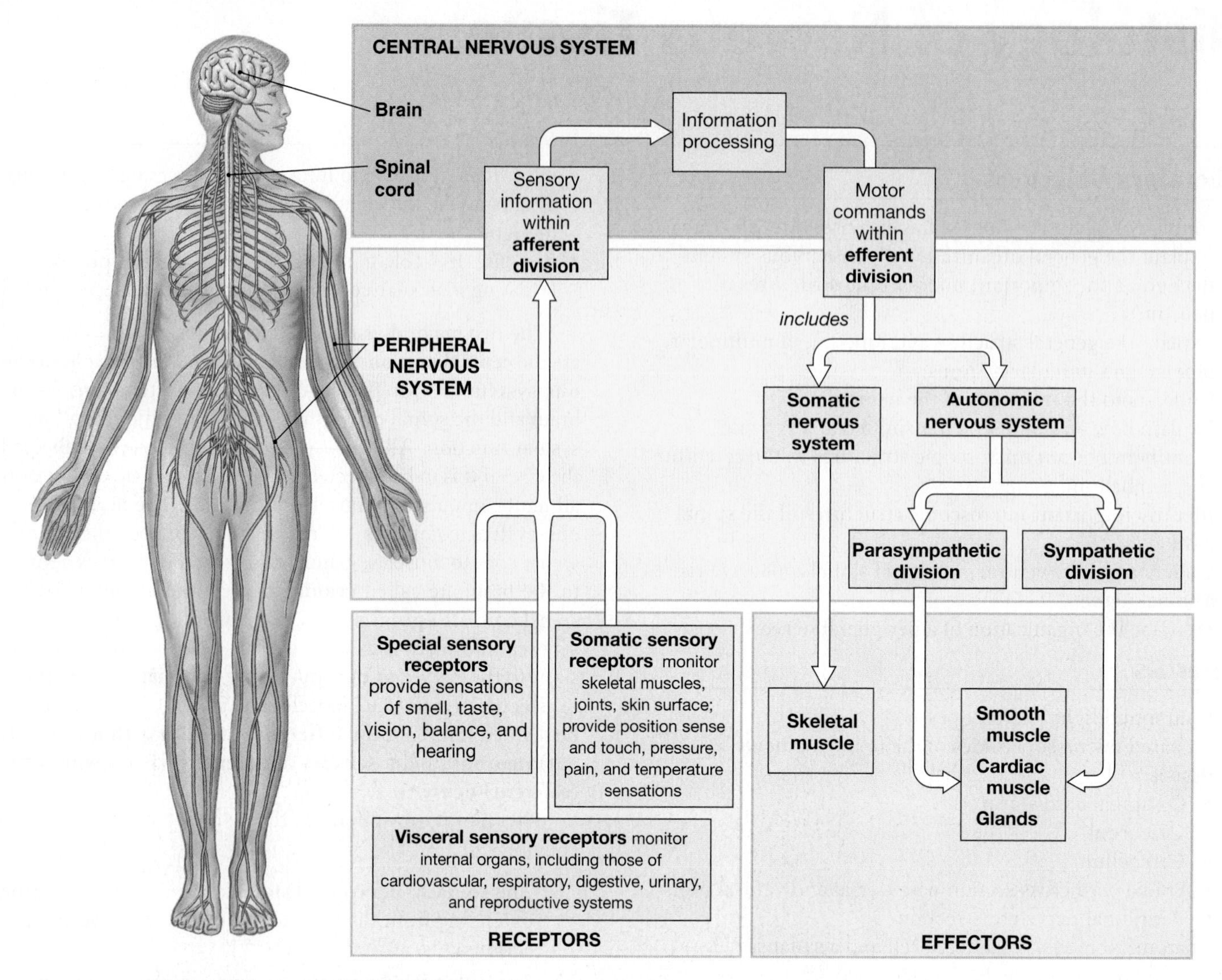

Figure 13.1 General organization of the nervous system. The central nervous system (CNS) includes the brain and spinal cord. The peripheral nervous system (PNS) includes the peripheral nerves (cranial and spinal nerves).

Cell Types in Nervous Tissue

The structural and functional unit of the nervous system is the **neuron**, or **nerve cell** (Figure 13.2). A typical neuron consists of a cell body that contains a large, round nucleus with a well-defined nucleolus. In the surrounding cytoplasm, the most prominent organelles include mitochondria, the Golgi apparatus, and clusters of free ribosomes and the rough endoplasmic reticulum (RER), known as **chromatophilic (Nissl) bodies**. The supporting cytoskeleton includes **neurofilaments** (intermediate filaments) and **neurotubules** (microtubules). In the CNS, neuron cell bodies are found in the surface layer of the cerebrum (**cerebral cortex**) and cerebellum (**cerebellar cortex**), and in deeper clusters called **nuclei** (singular = **nucleus**; not to be confused with the nucleus of a cell). In the PNS, neuron cell bodies are organized in clusters called **ganglia** (singular = **ganglion**).

WHAT'S IN A WORD The term *chromatophilic* is derived from Greek and means "color loving." The chromatophilic bodies in nerve cells readily stain with biological dyes and typically appear as dark blue or purplish structures.

Extending from the cell body are two types of cell processes: **dendrites** and **axons.** It is possible for neurons to possess many dendrites, but they usually have only one axon. The dendrites, along with the cell body, serve as contacts to receive impulses from other neurons. Usually, axons give off several collateral branches, each ending with a dilated region known as the **axon terminal** or **synaptic knob**. Axons can be quite variable in length. For example, the axons of the bipolar neurons in the retina of the eyeball are quite short; however, the axons of motor neurons that travel from the lumbar spinal cord to the foot can be greater than 1 meter (3 to 4 feet).

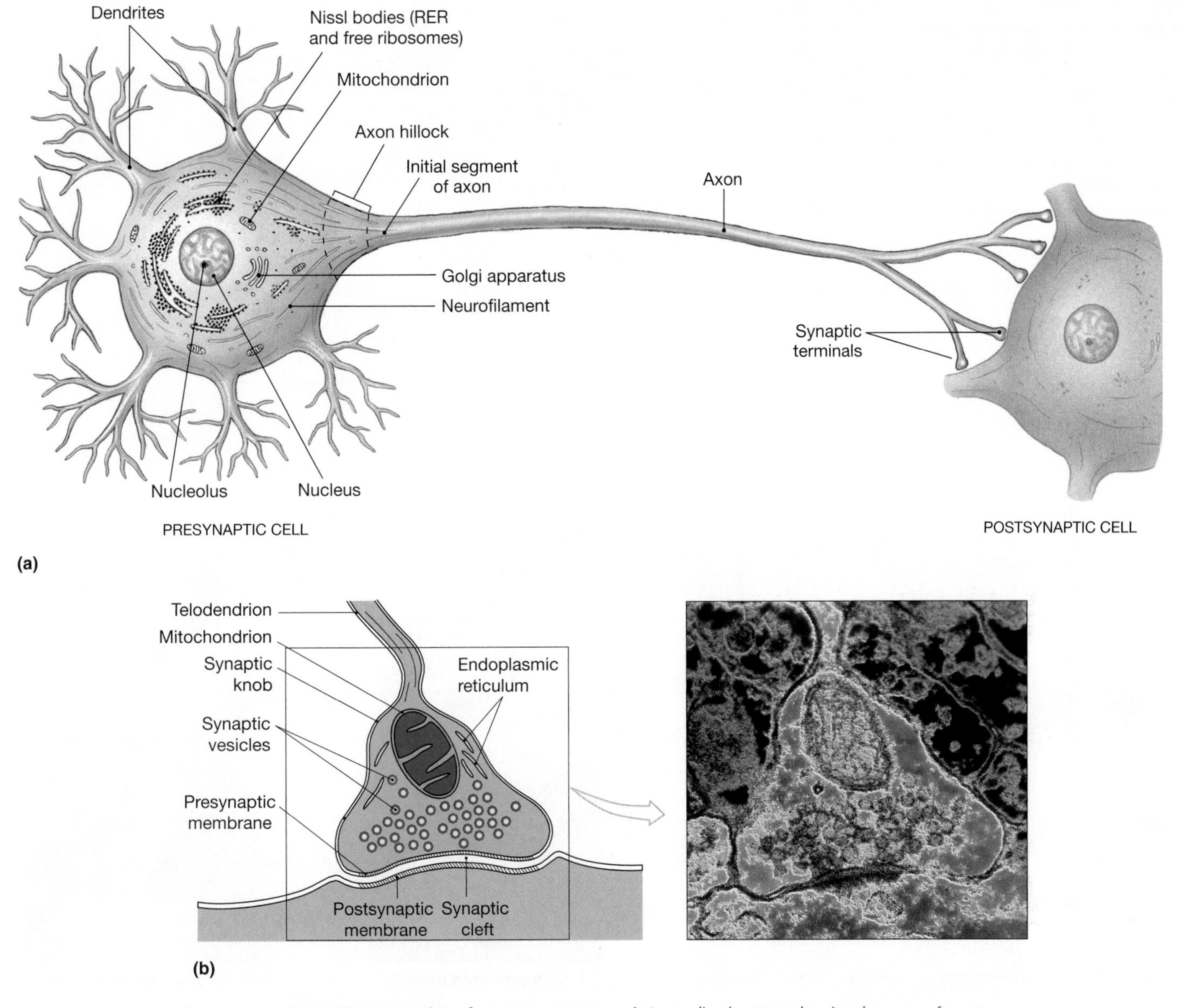

Figure 13.2 Synaptic connections between neurons. a) Generalized pattern showing the axon of a presynaptic cell forming synapses with the dendrites and cell body of a postsynaptic cell. **b)** Enlargement of a synapse between two neurons. The synapse is formed by the synaptic knob of a presynaptic neuron and a dendrite or the cell body of a postsynaptic neuron. Neurotransmitter is released by exocytosis from the presynaptic membrane, diffuses across the synaptic cleft, and attaches to receptors on the postsynaptic membrane.

Some axons are surrounded by a **myelin sheath** (Figure 13.3), which consists of several circular layers of fused cell membrane. Because of its high lipid content, the myelin sheath provides an insulating covering for nerve fibers, much like the insulation around electric wires in your home. The myelin sheath is produced by two types of cells. In the peripheral nervous system, it is produced by **neurolemmocytes (Schwann cells)**, and in the central nervous system by **oligodendrocytes**. Nerve fibers that are surrounded by myelin are called **myelinated fibers**; those without myelin are known as **unmyelinated fibers**. At regular intervals along a myelinated axon are small gaps in the myelin sheath. These interruptions are called **nodes of Ranvier**. The effect of the myelin sheath on the conduction of an electric impulse (action potential) will be studied later in this exercise.

Neurons are classified according to their function. As mentioned earlier, sensory (afferent) neurons conduct nerve impulses from sensory receptors, located in the body wall and internal organs (viscera), to the central nervous system. Motor (efferent) neurons conduct nerve impulses from the central

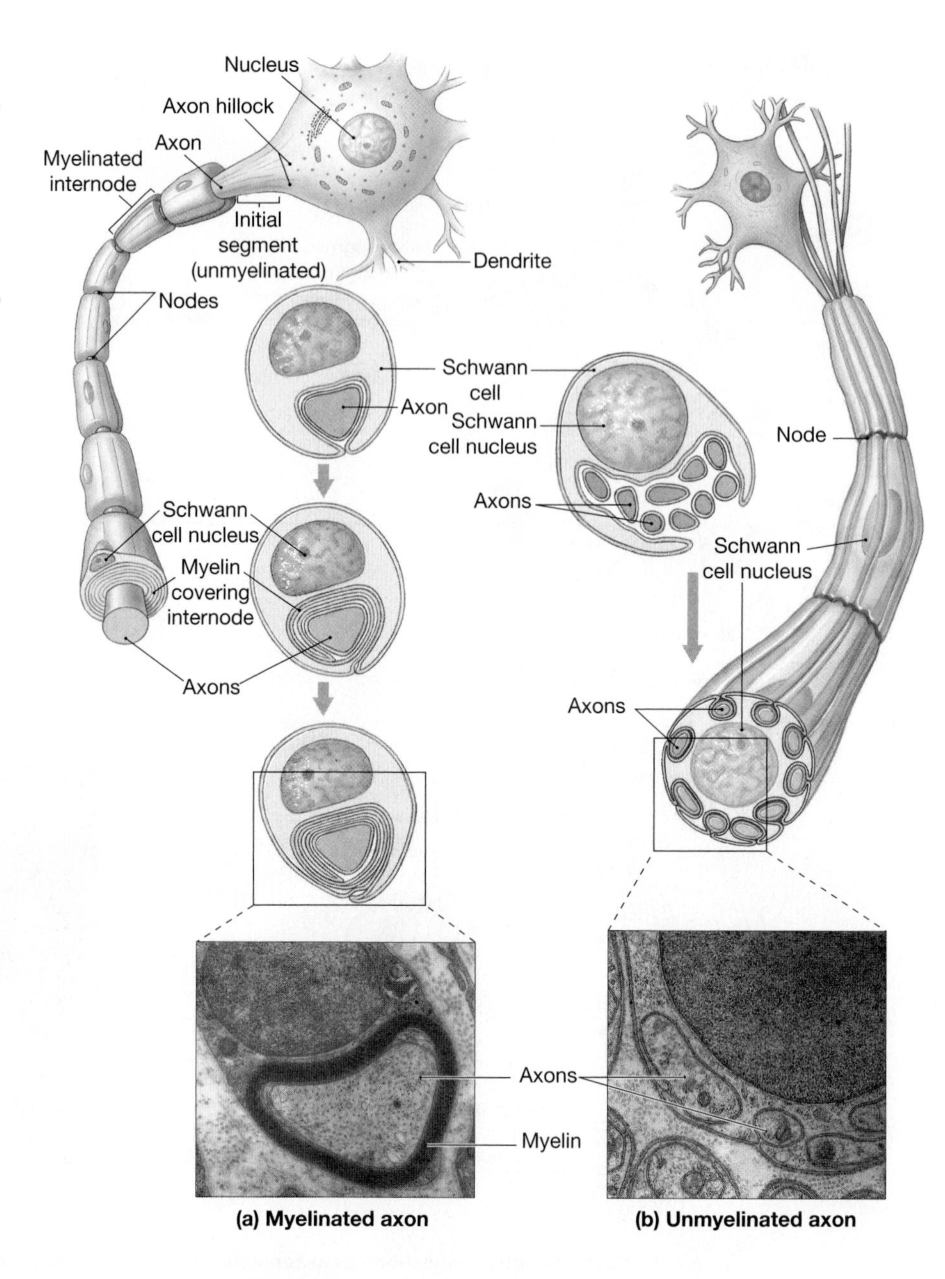

Figure 13.3 Myelinated and unmyelinated axons in the peripheral nervous system. a) A myelinated axon, showing the development of the myelin sheath by a Schwann cell. Several Schwann cells are needed to form the myelin sheath around each axon. **b)** A group of unmyelinated axons, illustrating their relationship with a Schwann cell. Although the myelin sheath is not formed, these axons are protected, to some degree, by Schwann cells.

nervous system to muscles or glands. A third type, **interneurons** or **association neurons**, form links between sensory and motor neurons. Interneurons also act as relay stations for the transmission of impulses from one part of the brain or spinal cord to another. The vast majority of neurons in the central nervous system are interneurons.

Neurons are also classified by the number of cell processes they possess (Figure 13.4). **Multipolar neurons** make up over 99% of nerve cells and include motor neurons and interneurons. They contain many cell processes—typically one axon and multiple dendrites. A small number of multipolar neurons contain only dendrites and are called **anaxonic neurons** (*anaxonic* means "no axon"). **Bipolar neurons** contain two cell processes: one axon and one dendrite. These cells are specialized sensory neurons; their locations are restricted to the retina of the eye, the inner ear, and the olfactory (sense of smell) epithelium in the upper nasal cavity. **Unipolar neurons** are the typical sensory neurons that transmit impulses from peripheral sensory receptors to the central nervous system. They possess only one cell process that divides into two branches: a peripheral branch that acts as a dendrite by transmitting impulses from a sensory receptor toward the cell body, and a central branch that acts as an axon by transmitting impulses to the central nervous system. When unipolar neurons first form, they have two distinct processes that fuse together as the cells mature. For this reason, they are sometimes called **pseudounipolar neurons** (*pseudo* means "false").

Other cells in the nervous system, the **neuroglia**, or **glial cells**, provide support and protection for neurons and other structures. Glial cells are more numerous than neurons. They have relatively small cell bodies from which cytoplasmic processes extend. The number of processes and the complexity of their branching patterns vary between the different cell types. Neuroglia are found in both the central nervous system

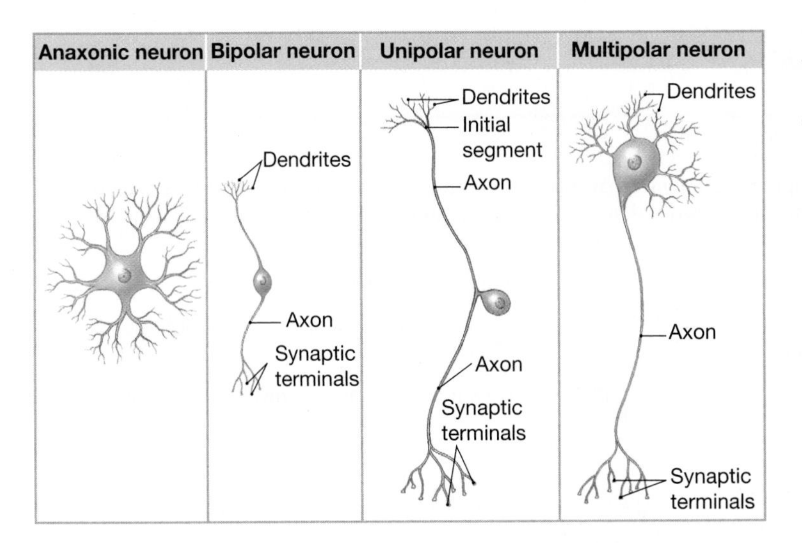

Figure 13.4 Four types of neurons. Multipolar neurons function as motor neurons and interneurons. Bipolar and unipolar neurons function as sensory neurons. The function of anaxonic neurons is not clearly understood.

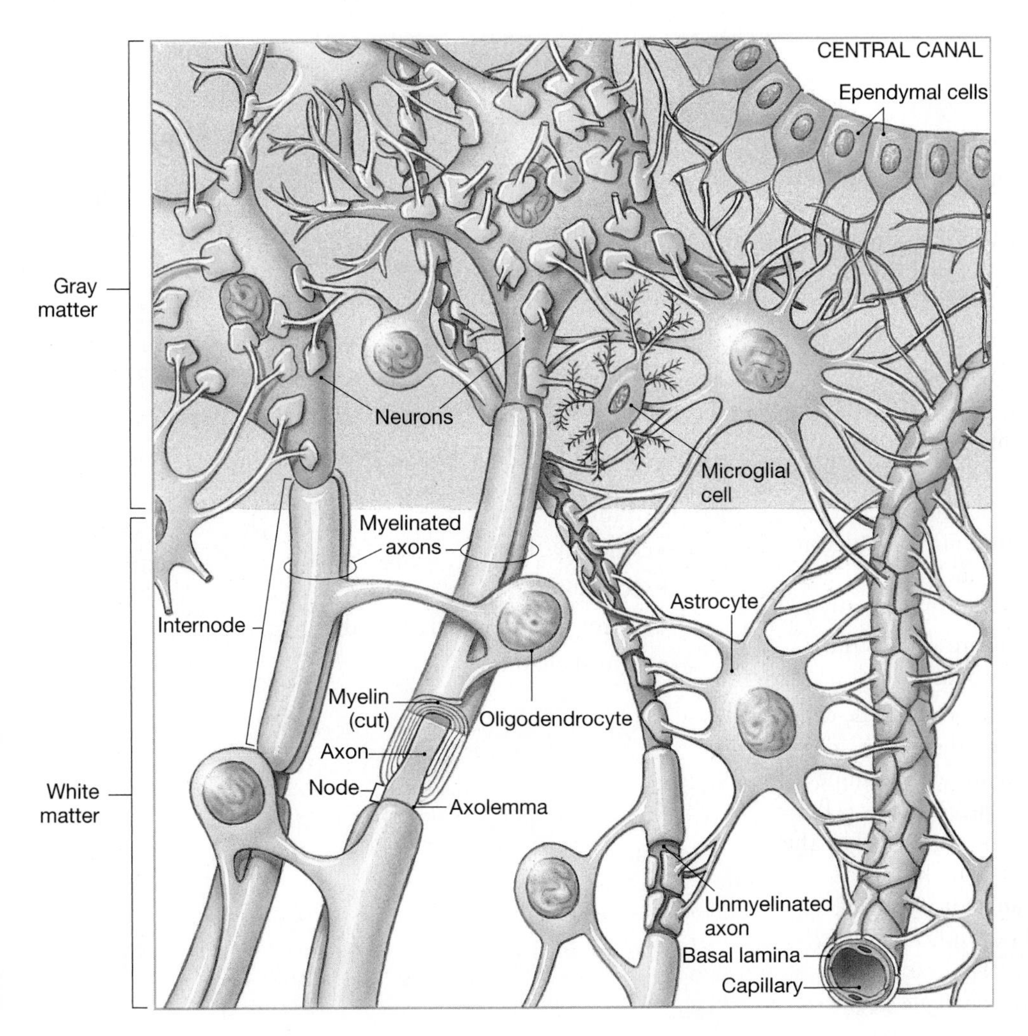

Figure 13.5 Nervous tissue in the central nervous system. The close relationship between glial cells and neurons is demonstrated.

(Figure 13.5) and the peripheral nervous system. The specific functions of each type are described in Table 13.1.

WHAT'S IN A WORD The word *glial* is derived from the Greek word *glia*, which means "glue." The name reflects the general function of the glial cells: sustaining and protecting neurons.

Because multipolar neurons are the most numerous, it is common to study the general structure of a neuron by observing these cells. The ox spinal cord is an excellent specimen, because the neurons are relatively large and structures are easy to identify with the light microscope.

Table 13.1 Neuroglia Cell Types

Cell type	Function
I. Neuroglia in the CNS	
1. Astrocytes	The cell processes of these cells form a supporting network that connects neurons to blood vessels. Astrocytes help to form the blood-brain barrier, which prevents the passage of potentially harmful substances from the blood to brain tissue. They also have a role in regulating levels of oxygen, carbon dioxide, and nutrients.
2. Microglial cells	These cells act as phagocytes. They protect the CNS from disease-causing microorganisms and clear away cellular debris.
3. Ependymal cells	These cells are modified epithelial cells that line the ventricles of the brain and central canal of the spinal cord. They facilitate the circulation of cerebrospinal fluid in the ventricles.
4. Oligodendrocytes	These cells produce the myelin sheath around axons in the CNS. Each oligodendrocyte can form myelin sheaths around several axons.
II. Neuroglia in PNS	
1. Satellite cells	These cells surround neuron cell bodies in peripheral ganglia and regulate levels of oxygen, carbon dioxide, and nutrients.
2. Neurolemmocyte (Schwann cell)	These cells produce the myelin sheath around axons in the PNS. Each neurolemmocyte forms a myelin sheath around only one axon.

ACTIVITY 13.1 Identifying Major Components of Multipolar Neurons

1. Obtain a prepared slide of an ox spinal cord smear, or a similar slide that demonstrates multipolar neurons (Figure 13.6).
2. View the slide under low power. As you scan the slide, note several large neurons distributed throughout the tissue specimen.
3. Move the slide so that a neuron is positioned in the center of the field of view. Switch to high power.
4. Identify the **cell body** of the neuron. The most prominent structure in this region is the large round nucleus. Identify the darkly stained nucleolus within the nucleus.
5. Notice that the cytoplasm in the cell body is filled with a darkly stained, granular substance. This material is the rough ER or the chromatophilic (Nissl) bodies.
6. Identify the many cell processes that extend from the cell body. One cell process is an axon, and all the other processes are dendrites. It is often difficult to determine whether a process is an axon or a dendrite. In a few cells, however, the presence of the axon can be confirmed by the identification of the **axon hillock**. The axon hillock is connected to the **initial segment** of an axon, which is where the axon arises from the cell body. It is relatively easy to identify because it has a distinctive funnel shape and lacks the dark-staining Nissl bodies that are prominent in the cell body and dendrites.
7. Attempt to identify bundles of filaments that travel through the cell body and extend into the dendrites and the axon. The filaments comprise the cytoskeleton and include neurotubules and neurofilaments. On your slide, you will not be able to differentiate between the two filament types.
8. Attempt to identify the much smaller cell bodies of the glial cells, which outnumber the neurons. If the quality of your slide is not good, the cell bodies will be difficult to detect, but the nuclei should be visible.
9. If available, study a model of the neuron and compare its structures with your miscroscopic observations.
10. In the space provided, draw a region of the spinal cord smear as you view it with the microscope and label the following structures.
    - Cell body
    - Dendrites
    - Axon
    - Nucleus
    - Nucleolus
    - Chromatophilic (Nissl) bodies
    - Cytoskeleton

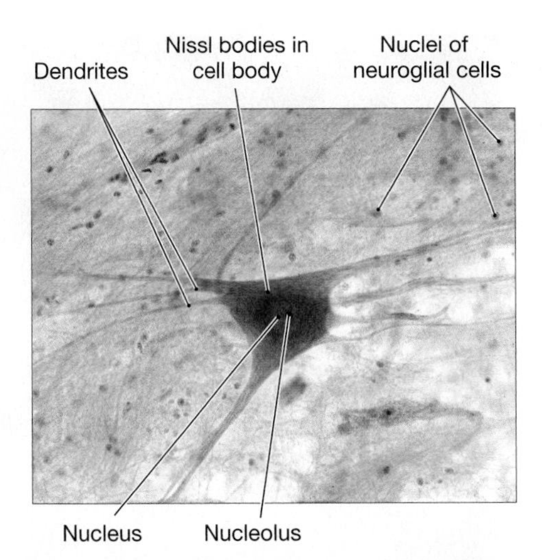

Figure 13.6 Multipolar neuron from an ox spinal cord smear. Multipolar neurons comprise the vast majority of neurons in the central nervous system.

QUESTION TO CONSIDER From your microscopic observations, you likely noticed that the cell bodies of neurons have an abundance of chromatophilic (Nissl) bodies. Provide a functional explanation for this structural feature.

____

____

The Synapse

Axons are capable of generating impulses (**action potentials**) and transmitting them at special junctions known as **synapses.** A synapse is usually formed by the **presynaptic membrane** on an axon terminal and the **postsynaptic membrane** on an effector cell (Figure 13.2b). An effector cell could be another neuron (neuroneuronal junction), a muscle cell (neuromuscular junction), or a gland cell (neuroglandular junction). Usually, a narrow extracellular space, the **synaptic cleft**, separates the presynaptic and postsynaptic membranes.

At a typical synapse, arrival of an action potential promotes the entry of calcium ions into the synaptic knob. The presence of calcium promotes secretory vesicles to fuse with the presynaptic membrane and to release their neurotransmitter by exocytosis. The neurotransmitter diffuses across the synaptic cleft and binds to receptors on the postsynaptic membrane.

ACTIVITY 13.2 Examining the Synapse

1. If available in the laboratory, examine a model of a synapse. Alternatively, study figures of synapses in a histology atlas, your textbook, or Figure 13.2b in this exercise.
2. Identify the following structures.
   - The *axon,* also known as the *nerve fiber,* conducts *action potentials* toward the synapse. The cytoplasm of the axon (the **axoplasm**) contains numerous - *neurofibrils* and *neurotubules,* components of the cytoskeleton. The axon is part of the *presynaptic neuron.*
   - The *synaptic knob* is the dilated termination of the axon. It contains numerous **mitochondria** and **secretory vesicles** filled with **neurotransmitter** molecules.
   - The *presynaptic membrane* is the cell membrane located at the end of the synaptic knob.
   - The *postsynaptic membrane* is the portion of the **postsynaptic neuron's** cell membrane adjacent to the synaptic cleft.
   - The *synaptic cleft* is the narrow space that separates the presynaptic and postsynaptic membranes.
3. Consider the mechanism for the transmission of an impulse from one neuron to another. Number the processes listed below in the order that they occur during synaptic transmission.

   _______ Neurotransmitter diffuses across the synaptic cleft.

   _______ Nerve impulse travels along the axon to the synaptic knob.

   _______ Secretory vesicles containing neurotransmitter fuse with the presynaptic membrane.

   _______ Neurotransmitter binds to receptors on the postsynaptic membrane.

   _______ Neurotransmitter released from the presynaptic membrane by exocytosis.

   _______ Action potential promotes the entry of calcium ions into the synaptic knob.

QUESTIONS TO CONSIDER 1. Synaptic knobs contain numerous mitochondria. Speculate on a reason why these organelles are strategically located in this region of a neuron.

____

____

2. A neuron's cytoskeleton, particularly the neurotubules, plays an important role in transporting materials from the cell body to the axon terminals. Speculate on why this process, known as **axoplasmic transport**, is a vital activity for sustaining normal function at synapses.

____

____

The Cerebrum

The cerebrum consists of two **cerebral hemispheres**. Similar to most regions of the brain, each cerebral hemisphere has two types of nervous tissue: **gray matter** and **white matter**. The gray matter consists of nerve cell bodies, dendrites, unmyelinated axons, and glial cells. It is located along the surface (cerebral cortex) and in deeper regions (nuclei). Immediately deep to the cerebral cortex is the white matter, which contains myelinated axons and glial cells. Bundles of axons in the white matter form fiber tracts that connect various brain regions.

ACTIVITY 13.3 Examining Microscopic Structure of the Cerebrum

1. Obtain a prepared slide of the cerebrum.
2. View the slide under low power. As you scan the slide, note two distinct regions (Figure 13.7a).
   - Along the surface is the layer of gray matter known as the cerebral cortex.
   - Deep to the cerebral cortex is the white matter.
3. In the cerebral cortex, identify the relatively large cell bodies of multipolar neurons and the nuclei of the much smaller glial cells, whose cell bodies are difficult to see. Note that the glial cells outnumber the neurons (Figure 13.7).
4. Move the slide so that a region of gray matter is in the center of the field of view. Switch to high power (Figure 13.7b). In the gray matter, you will see darkly stained cells throughout the field of view.
5. Scan the slide until you find a multipolar neuron, which has a relatively large cell body with the dark-staining Nissl bodies distributed throughout the cytoplasm. Depending on the section, the large nucleus with a distinct nucleolus may be identified.
6. Identify the many cell processes that extend from the cell body. One cell process is an axon, and all others are dendrites. If you are viewing a slide stained with hematoxylin and eosion (h & e), you might be able to confirm the presence of the axon by identifying the axon hillock.
7. Move the slide to a deeper area where very few cell bodies are present. This is the white matter, which contains mostly myelinated nerve fibers. You should be able to identify the nuclei of glial cells. Among the glial cells present here are the oligodendrocytes that produce the myelin sheath around axons in the central nervous system.
8. In the space provided, draw a region of the cerebrum as you view it with the microscope and label the following structures.
   - Gray matter in the cerebral cortex
   - Multipolar neurons
   - Glial cell nuclei (in both gray matter and white matter)
   - White matter deep to the cerebral cortex
   - Myelinated axons

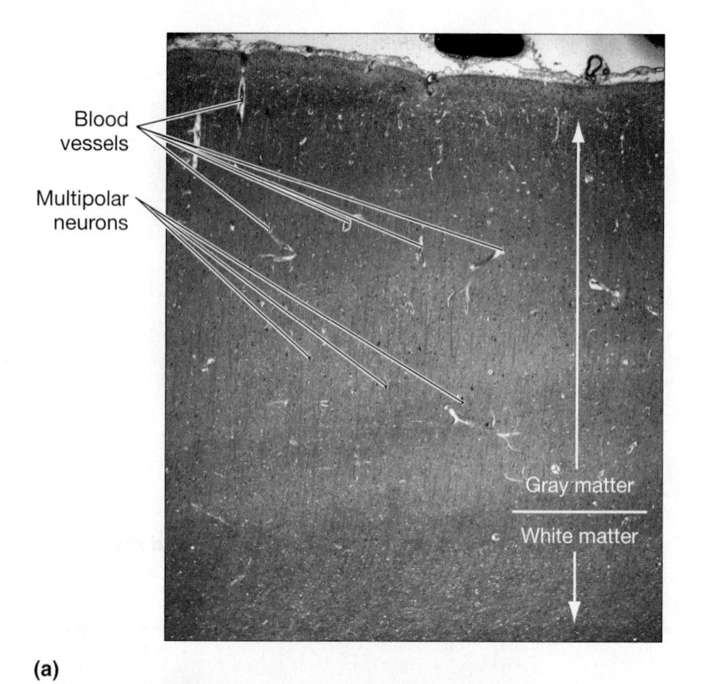

(a)

Multipolar neurons

Glial cells

(b)

Figure 13.7 Light micrographs of nervous tissue in the cerebrum. **a)** Low-power view illustrating the relationship between the gray matter in the cerebral cortex and the deeper white matter (LM × 200); **b)** high-power view of the cerebral cortex illustrating multipolar neurons and the much smaller glial cells (LM × 200).

9. If available in your laboratory, obtain a slide of the cerebrum stained with silver. View the slide under low power. The slide has been specifically prepared to highlight nerve cells by staining them black against a light background (Figure 13.8). As you scan the slide, you will immediately notice the black-stained network of nerve cell processes.

In the white matter, you will see only nerve fibers, but if you move to an area of gray matter, you will see the cell bodies and numerous cell processes of multipolar neurons.

10. Place a region of gray matter containing several neurons in the center of the field of view and switch to high power. At this magnification, you can clearly see cell processes extending from all sides of the cell bodies (Figure 13.8).

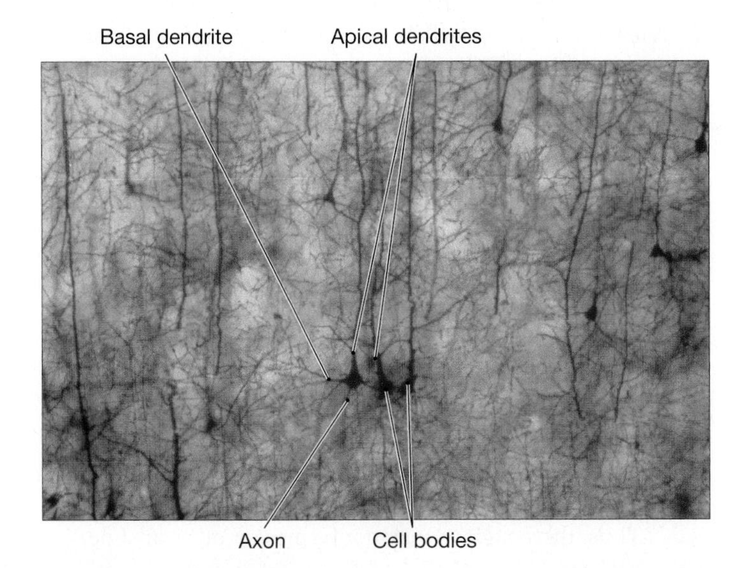

Figure 13.8 Multipolar neurons in the cerebral cortex, stained with silver. With this special preparation, the cell processes extending from the cell body and spatial arrangement of the neurons can be fully appreciated.

11. Slowly turn the fine adjustment knob of the microscope. Notice that the cells that you were viewing move out of focus but others at different depths of field come into focus. You can truly appreciate the three-dimensional quality of the structure you are viewing.

The Cerebellum

The arrangement of nervous tissue in the cerebellum is similar to what is seen in the cerebrum. The cerebellum consists of two **cerebellar hemispheres.** Each hemisphere contains a surface layer of gray matter, the **cerebellar cortex** (Figure 13.9), and a deep region of white matter called the **arbor vitae.** Deep to the arbor vitae are additional regions of gray matter known as the **cerebellar nuclei.**

ACTIVITY 13.4 Examining Microscopic Structure of the Cerebellum

1. Obtain a prepared slide of the cerebellum.
2. View the slide under low power. As you scan the slide, identify the outer layer of gray matter, the cerebellar cortex, and the deeper layer of white matter (Figure 13.9). This pattern is similar to what you saw in the cerebrum.
3. Try to identify the three layers of gray matter.
   - The relatively thick superficial **molecular layer** contains interneurons that interconnect the other neurons in the cerebellar cortex.

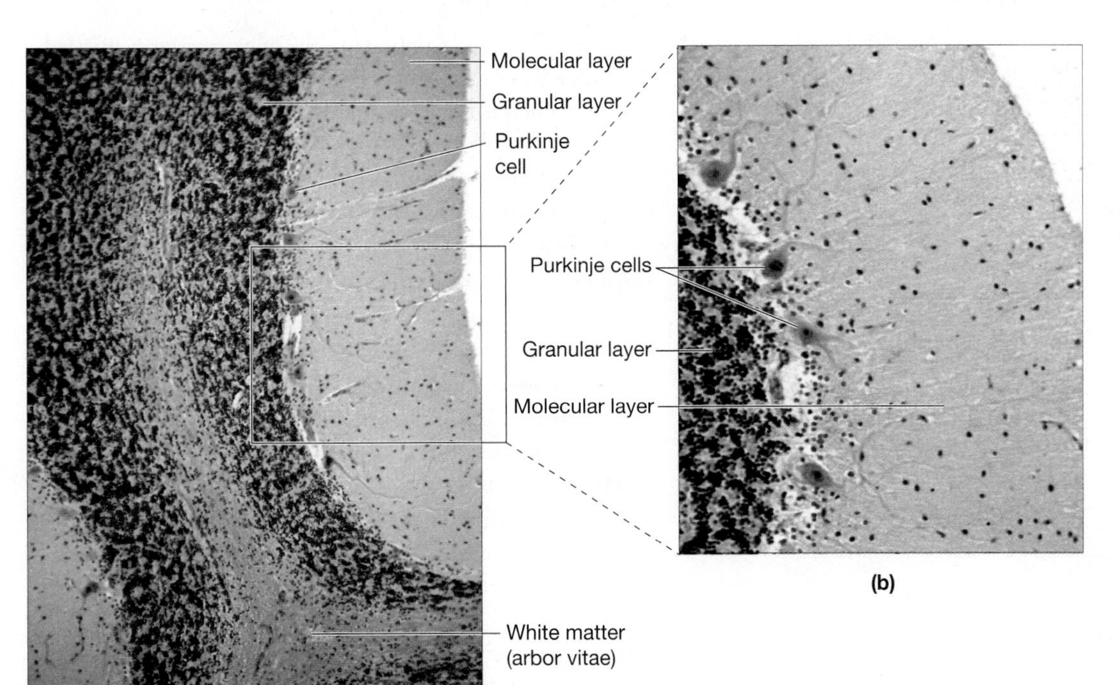

Figure 13.9 Light micrographs of nervous tissue in the cerebellum. a) Low-power view illustrating the relationship between the gray matter in the cerebellar cortex and the deeper white matter (arbor vitae) (LM × 100); **b)** higher power view of the cerebellar cortex. The three cellular layers including the large Purkinje cells can be seen (LM × 200).

- The thin **Purkinje cell layer** is occupied by multipolar neurons whose axons serve as a major output channel, transmitting information from the cortex to deeper cerebellar nuclei.
- The deep **granular layer** includes neurons that receive most of the information coming to the cerebellum from other brain areas.

4. Switch to high power. Within the molecular and granular layers, identify the cell bodies and cell processes (dendrites and unmyelinated axons) of small neurons.
5. The distinguishing structural feature in the cerebellum is the single row of large, flask-shaped **Purkinje cells** in the Purkinje cell layer (Figure 13.9b). Identify these multipolar cells on your slide.
6. In the space provided, draw a region of the cerebellum that you observed and label the following structures:
   - Cerebellar cortex
   - Purkinje cells
   - White matter

QUESTIONS TO CONSIDER 1. Compare the microscopic structure of the cerebrum with that of the cerebellum and identify similarities and differences (e.g., cell types and shapes, layering, or structural patterns of nervous tissue).

a. Similarities:

____

____

b. Differences:

____

____

2. From your light microscopic observations of the cerebrum and cerebellum, you likely noticed that virtually all the neurons are multipolar. Provide an explanation for why this is the case. (Hint: Consider the general function of these neurons.)

____

____

The Spinal Cord

Unlike the cerebrum and cerebellum in the brain, the spinal cord lacks a cortical layer of gray matter. Instead, all the gray matter is deep and completely surrounded by the more superficial white matter. The gray matter is easy to identify because of its unique butterfly shape (Figures 13.10a and b).

ACTIVITY 13.5 Examining Microscopic Structure of the Spinal Cord

1. Obtain a slide of spinal cord tissue, viewed in cross section.
2. View the slide under low power. At this magnification you can clearly identify the butterfly-shaped gray matter, surrounded by peripheral regions of white matter (Figures 13.10a and b).
3. Carefully observe the gray matter. On each side, identify the following parts (Figures 13.10a and b).
   - The **posterior horn** receives sensory fibers from spinal nerves.
   - The **anterior horn** contains the cell bodies of *lower* motor neurons. They are multipolar cells that contribute motor fibers to spinal nerves.
   - The **lateral horns** contain the cell bodies of autonomic motor neurons. They are restricted to thoracic and lumbar spinal cord levels (sympathetic neurons) and portions of the sacral spinal cord (parasympathetic neurons). Thus, these regions may not be present on your slide.
   - The **gray commissure** is a narrow band of tissue that connects the gray matter on each side. The nerve fibers of interneurons pass through this area to relay impulses from one side of the cord to the other. The **central canal**, which contains cerebrospinal fluid, passes through the gray commissure.
4. On each side of the spinal cord, the white matter is divided into three distinct columns (**funiculi**; singular = **funiculus**) of tissue. Each is named according to its relative position. On your slide, observe the white matter and identify the following regions (Figures 13.10a and b).
   - Posterior (dorsal) column of white matter (posterior funiculus)
   - Anterior (ventral) column of white matter (anterior funiculus)
   - Lateral column of white matter (lateral funiculus)

   The white matter columns contain two types of fiber tracts that connect the spinal cord and the brain: The **ascending tracts** transmit sensory information while the **descending tracts** convey motor information.
5. If available, observe an anatomical model of the spinal cord in cross section. Correlate the structures illustrated on the model with what you have just observed under the microscope.
6. Once again, view the spinal cord cross section under low power. Place a region of the anterior horn in the center of the field of view and switch to high power. At this magnification you can identify the cell bodies of multipolar motor neurons (Figure 13.10c). These cells are also called

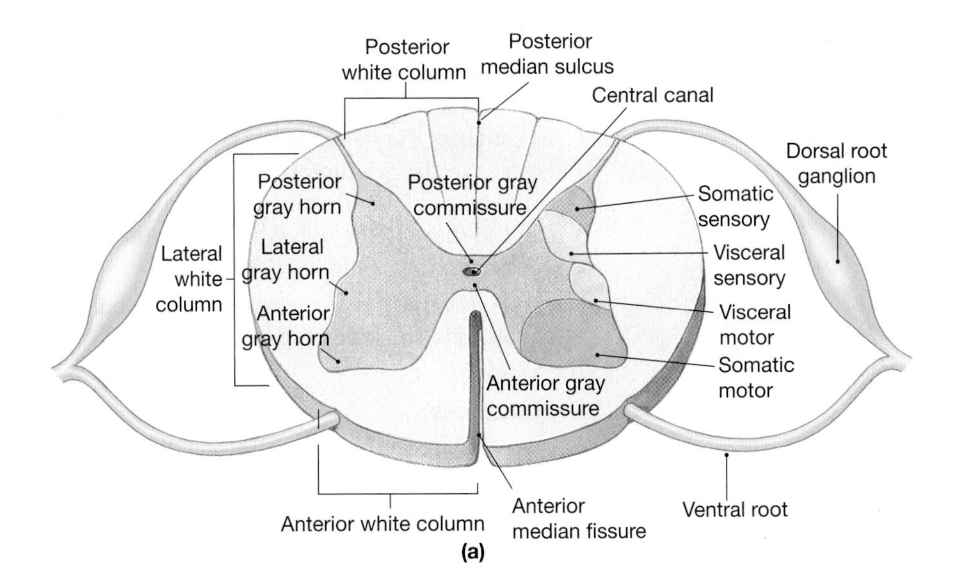

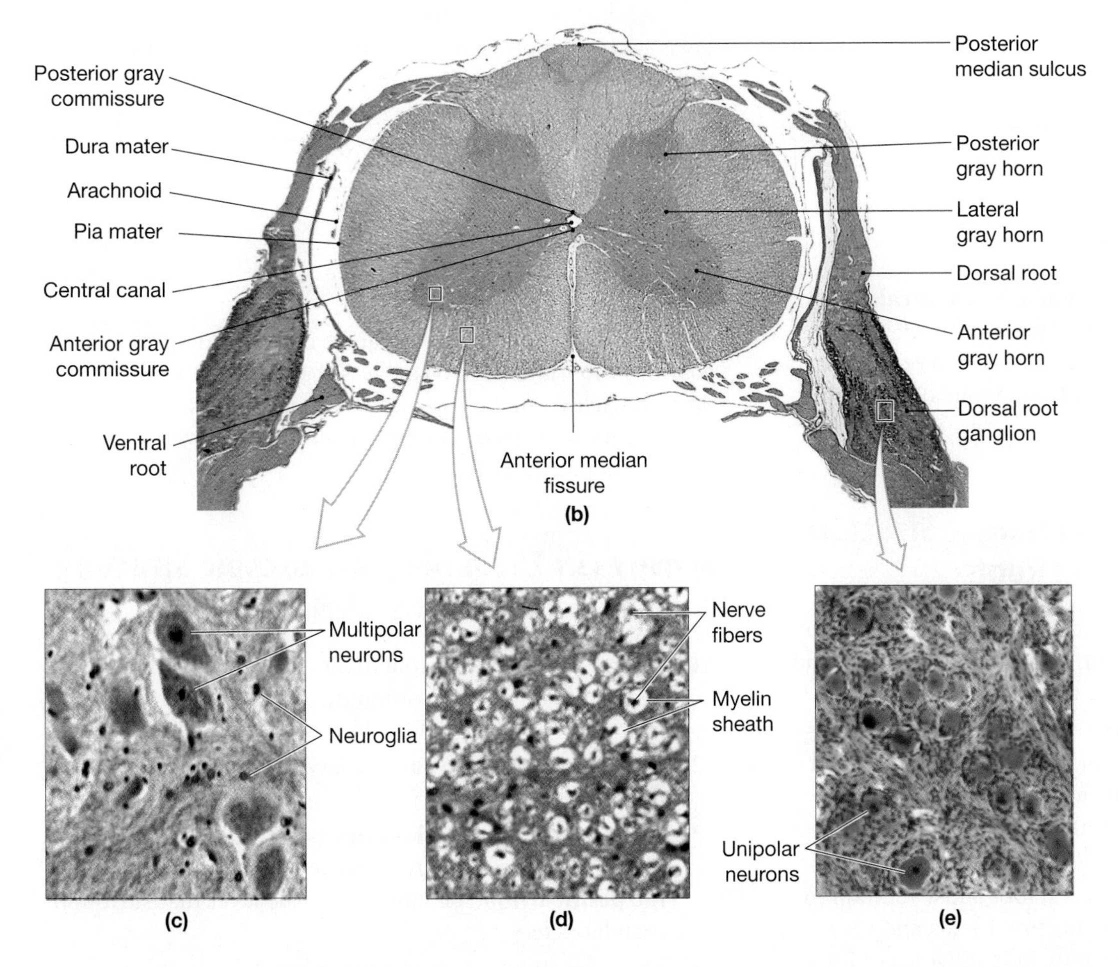

Figure 13.10 Microscopic structure of the spinal cord. a) Cross-sectional diagram, illustrating the deep gray matter horns and the superficial white matter columns; **b)** light micrograph of a spinal cord cross section with dorsal and ventral roots (LM × 40); **c)** high power of the gray matter in the anterior horn, illustrating multipolar motor neurons and glial cells (LM × 200); **d)** high power of the white matter, illustrating myelinated axons (LM × 200); **e)** high power of a dorsal root ganglion, illustrating unipolar sensory neurons (LM × 200).

lower motor neurons. They give rise to axons that travel in spinal nerves to reach skeletal muscles.

7. Switch back to low power. Place a region of white matter in the center of the field of view and switch the microscope to high power. At this magnification you can see numerous cross sections of darkly stained nerve fibers surrounded by clear spaces. The clear spaces represent the myelin sheaths. However, myelin is no longer present in this prepared section (Figure 13.10d).
8. In the space provided, draw a cross section of the spinal cord that you observed under the microscope and label the following structures.
   - Posterior horn
   - Anterior horn

- Lateral horn
- Gray commissure
- Central canal
- Posterior column of white matter
- Anterior column of white matter
- Lateral column of white matter

QUESTION TO CONSIDER Compare the organization of gray and white matter in the spinal cord versus the cerebrum and cerebellum.

____

____

Spinal Nerve Roots

Each spinal nerve is formed by the union of a **ventral root** and a **dorsal root**. The ventral root transmits motor nerve fibers from the anterior horn of the spinal cord to a spinal nerve. The dorsal root transmits sensory nerve fibers from a spinal nerve to the posterior horn of the spinal cord.

ACTIVITY 13.6 Examining Microscopic Structure of Spinal Nerve Roots

1. Obtain a slide of the spinal cord cross section that also illustrates the spinal nerve roots. At low power, position the slide so that the spinal cord is in the center of the field of view.
2. Move the slide to either side of the spinal cord and identify the dorsal root and ventral root. Along the course of the dorsal root is a dilated region called the **dorsal root ganglion** (see step 4). The ventral root does not have a comparable structure.
3. Locate the position where a dorsal root and a ventral root merge to form a spinal nerve (Figures 13.10a and b). Switch to high power and identify individual nerve fibers traveling along the nerve roots.
4. Switch back to low power and identify the dorsal root ganglion once again. Within the ganglion you will see numerous cell bodies of unipolar sensory neurons. These neurons contain one cell process that divides into two branches. One branch sends nerve fibers along a spinal nerve to reach peripheral sensory receptors. The other branch sends nerve fibers along a dorsal root to reach the posterior horn of the spinal cord.
5. Position your slide so that the dorsal root ganglion is in the center of the field. Switch to high power and examine the cells in the ganglion (Figure 13.10e). In some cells, you will be able to identify the nucleus and nucleolus. Attempt to identify the initial portion of the single cell process on some cells.
6. Return to your previous drawing of the spinal cord cross section. Add the following structures.
   - Ventral root
   - Dorsal root
   - Dorsal root ganglion with the cell bodies of unipolar neurons
   - Beginning of a spinal nerve

QUESTION TO CONSIDER Explain what would happen if all the ventral roots that supply the spinal nerves to the lower extremities were severed.

____

____

Peripheral Nerves

A typical peripheral nerve (cranial nerves and spinal nerves) contains thousands of nerve fibers that are organized into bundles (fascicles) and enclosed by connective tissue coverings. Some fibers are surrounded by a myelin sheath, while others are unmyelinated.

ACTIVITY 13.7 Examining Microscopic Structure of Peripheral Nerves

1. Obtain a slide of a peripheral nerve cross section and view it under low power. Position the slide so that the nerve is in the center of the field of view (Figure 13.11).
2. Identify the **epineurium**, the layer of connective tissue that surrounds the entire nerve.
3. Identify several **nerve bundles (nerve fasicles)**. Each bundle contains numerous **nerve fibers (axons)**. Identify the **perineurium**, a connective tissue layer that surrounds each fascicle.
4. Move the slide so that one nerve bundle is in the center of the field of view. Switch to high power so that you can identify individual nerve fibers. The fibers will appear as darkly stained circles. Identify myelinated nerve fibers. On the slide, these fibers are surrounded by a clear space. Due to the special preparation of the tissue, the myelin sheath is absent. Attempt to identify the **endoneurium**, the connective tissue layer that covers each nerve fiber.

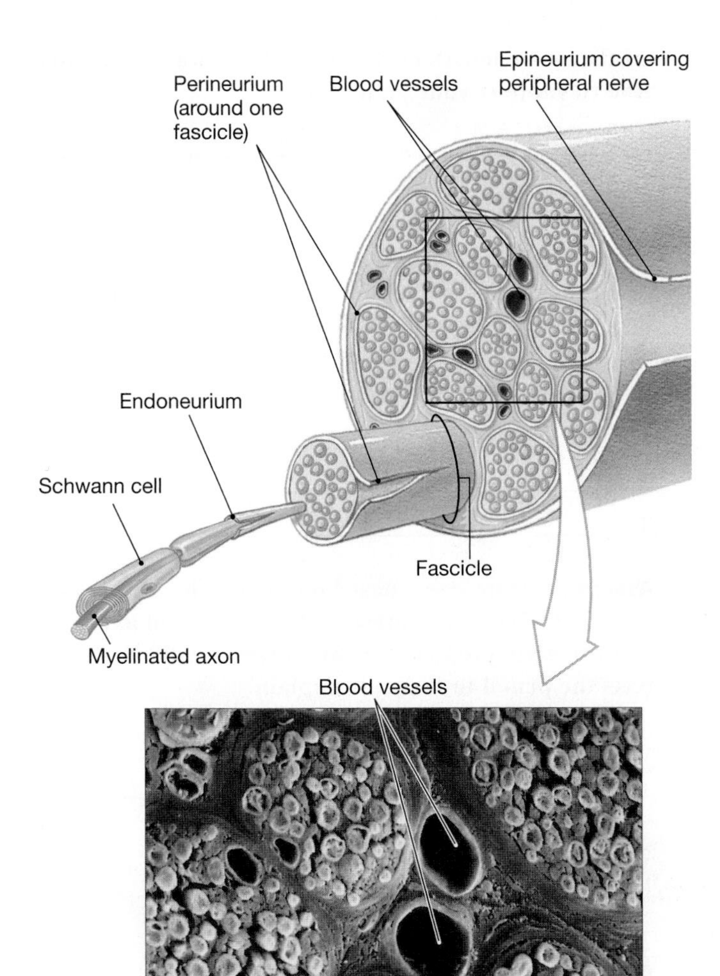

Figure 13.11 Cross section of a peripheral nerve. Each nerve fiber is surrounded by the endoneurium. Nerve fascicles are surrounded by the perineurium. The entire nerve is surrounded by the epineurium.

5. In the space provided, draw a cross section of a peripheral nerve and label the following structures.
   - Epineurium
   - Perineurium
   - Endoneurium
   - Nerve bundle (fascicle)
   - Nerve fiber (axon)
   - Clear space around some axons, where myelin sheath would be found

QUESTION TO CONSIDER How does the organization of a peripheral nerve compare with the organization of a skeletal muscle? (Hint: See Exercises 10 and 11 to review the levels of organization of a skeletal muscle.)

____

____

Propagation of Action Potentials

An action potential is generated at the initial segment of an axon. When this occurs, it creates a local current that depolarizes the adjacent segment of the axonal membrane, resulting in the generation of a second action potential. This process repeats itself to create a rapid succession of action potentials that move along the axon toward the synaptic knob. Thus, action potentials are propagated along the entire length of the axon.

As mentioned earlier, some nerve fibers are surrounded by a myelin sheath (Figure 13.3). The presence or absence of myelin will influence the way the action potentials are propagated. Action potentials can be generated along the entire length of an unmyelinated axon. Myelinated axons generate action potentials only at the nodes of Ranvier, where myelin is lacking. The insulating property of the myelin sheath inhibits the passage of sodium and potassium ions across the membrane, and it is that movement that creates the action potential. Thus, action potentials are not produced along the myelinated segments of the axon. Instead, they are transmitted by skipping from node to node. This type of transmission, known as **saltatory propagation**, is much faster than what occurs in an unmyelinated axon.

CLINICAL CORRELATION

Multiple sclerosis (MS) is a neurological disease that typically occurs in young adults. It is caused by the destruction of the myelin sheath around motor and sensory axons in the brain and spinal cord. The result is a progressive loss of motor and sensory function. Typical symptoms include muscle paralysis, impaired vision, loss of balance, slurred speech, and skin numbness. Multiple sclerosis is an autoimmune disease, which means that the affected individual's own immune system mistakenly attacks and destroys the myelin sheath. The severity and intensity of the disease varies among those who are affected.

Recent investigations suggest that MS could develop when the immune system confuses the myelin sheath proteins for viral proteins with similar amino acid sequences and attacks them. Treatment with interferons, antiviral proteins that are produced by cells in the immune system, has been successful in mitigating the symptoms of MS. However, there is currently no cure for the disease.

ACTIVITY 13.8 Demonstrating the Significance of the Myelin Sheath

1. Organize two groups of students. The first group will represent axon A and should include nine or ten students; the second group will represent axon B and should include four or five students.
2. Mark off a space in the laboratory or adjacent hallway that is 20 feet long.
3. The two groups should arrange themselves so that they form two parallel rows. The first person in each group represents the initial segment of an axon. They should be standing side by side at the zero mark. The last person in each group represents a synaptic knob. They should be standing side by side at the 20-foot mark. All other members in both groups should position themselves between the first and last persons so that there is an equal distance between each individual.
4. The first person in each group should hold a pencil. When the instructor gives the command to start, the students should hand the pencil down the line to the end. Members of axon A should be able to pass the pencil down the line without walking forward. Members of axon B will probably have to walk to the next person to pass the pencil.
5. **Form a Hypothesis** **Before you begin, predict which group (axon A or axon B) will be first to pass the pencil the entire length of the "axon." Provide an explanation to support your prediction.**

   ____

   ____

6. After completing this activity for the first time, verify your results by repeating the exercise one or two more times. Record the results in the table provided.

Trial	Faster "Axon" (A or B)
1	
2	
3	

7. **Assess the Outcome** **Based on the results of this activity, which group represented a myelinated axon? Did you correctly predict that this "axon" would be first to pass the pencil to the end? Explain.**

   ____

   ____

QUESTIONS TO CONSIDER

1. With regard to activity 13.8: What process does the passing of the pencil represent?

____

____

2. Do the individuals in the myelinated axon represent myelinated segments or nodes of Ranvier? Explain.

____

____

3. Earlier, you learned that multiple sclerosis is caused by damage to the myelin sheath around axons.
   - How will multiple sclerosis affect the transmission of action potentials along axons?
   - With regard to this activity, to demonstrate the effects of multiple sclerosis, would you add or subtract members of the myelinated axon? Explain.

   ____

   ____

Exercise 13 Review Sheet

Histology of Nervous Tissue

Name ______________________________

Lab Section ______________________________

Date ______________________________

1. Compare the central nervous system and the peripheral nervous system by discussing the components found in each.

____

____

2. Describe how nerve impulses are transferred from one neuron to the next at a synapse.

____

____

3. What is a myelin sheath? How does the myelin sheath affect the conduction of impulses along a nerve fiber?

____

____

Questions 4–7: Fill in the blanks in the following table.

Types of Neurons

Unipolar/Bipolar/Multipolar	Sensory/Motor/Interneuron	Afferent/Efferent/Neither
4. Unipolar		
5.	Motor	
6.		Neither
7. Bipolar		

Questions 8–12: Match the cell type in column A with the appropriate function in column B.

A	B
8. Ependymal cells ______	a. help to form the blood-brain barrier.
9. Astrocytes ______	b. produce the myelin sheath in the PNS.
10. Schwann cells ______	c. line the walls of ventricles in the brain.
11. Microglial cells ______	d. serve as afferent neurons.
12. Oligodendrocytes ______	e. produce the myelin sheath in the CNS.
	f. protect the CNS from disease-causing microorganisms.

Questions 13–14: Explain the difference between the following pairs of terms.

13. White matter and gray matter

____

____

14. Ganglion and nucleus

____

____

Questions 15–24: In the following diagram, identify structures by labeling with the color that is indicated.

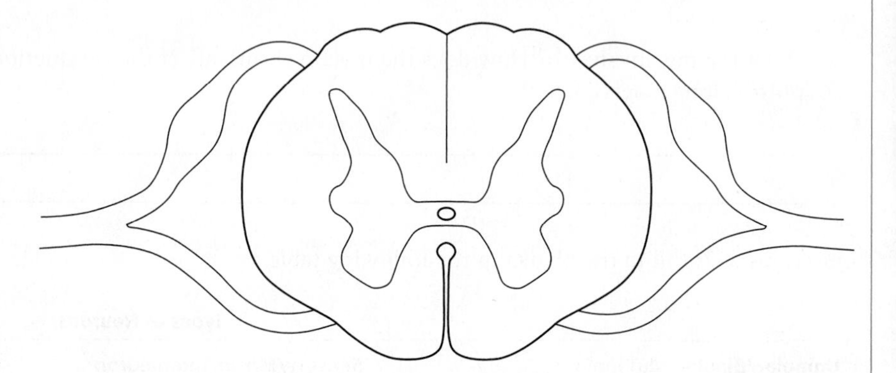

Gray matter:

15. Anterior horn = **green**
16. Posterior horn = **red**
17. Lateral horn = **yellow**
18. Gray commissure = **black**

White matter:

19. Lateral columns = **blue**
20. Anterior column = **purple**
21. Posterior column = **brown**
22. Dorsal root ganglion = **red**
23. Dorsal root = **yellow**
24. Ventral root = **red**

Exercise 14

The Brain and Cranial Nerves

Laboratory Objectives

On completion of the activities in this exercise, you will be able to:

- Locate the major gross anatomical structures of the human brain.
- Locate the gross anatomical structures of a sheep brain and compare its structure with that of the human brain.
- Identify the cranial nerves by name and number on a model or diagram.
- Evaluate the functions of the cranial nerves.

Materials

- Anatomical models or figures of the human brain
- Colored pencils
- Preserved sheep brains
- Dissecting trays
- Dissecting tools
- Dissecting gloves
- Protective eyewear
- Face mask (optional)
- Human skulls
- Pipe cleaners
- Small flashlight
- Cotton swabs
- Samples of coffee, pepper, and aromatic spices
- 10% table salt (NaCl) solution
- 10% sucrose (sugar) solution
- 10% quinine solution
- Biopac Student Lab System

The Brain

The human brain can be divided anatomically by various means. Based on its embryological origins, it can be divided into five major regions: telencephalon (anterior forebrain), diencephalon (posterior forebrain), mesencephalon (midbrain), metencephalon (anterior hindbrain), and myelencephalon (posterior hindbrain). Another way to divide it is by its adult structures, as follows:

- Cerebrum
- Diencephalon (thalamus, hypothalamus, pineal gland)
- Cerebellum
- Brainstem (midbrain, pons, medulla oblongata)

This laboratory exercise will focus on the organization of the adult brain, but both systems are linked together in Table 14.1.

The Cerebrum

The **cerebrum** is the largest and most prominent structure in the human brain (Figures 14.1 and 14.2). It is responsible for our ability to receive and interpret sensations, understand and form intellectual thoughts, store memory, develop emotions and personality, and initiate voluntary motor activities.

The cerebrum is divided into two **cerebral hemispheres** that are connected by a bridge of nerve fibers called the **corpus callosum** (Figure 14.2). Each hemisphere is divided into five lobes. The **frontal**, **parietal**, **occipital**, and **temporal lobes** are named after the cranial bones that cover them (Figure 14.1). A fifth lobe, the **insula**, is deep to the temporal lobe.

The surface of each hemisphere is a thin layer (2–5 mm) of **gray matter** called the **cerebral cortex**. The cortex has a complex folding pattern that gives rise to the ridges of tissue known as **gyri** (singular = **gyrus**). Each gyrus is separated by a groove (Figures 14.1 and 14.2). A shallow groove is called a **sulcus** (plural = **sulci**) and a deep groove is a **fissure**.

Deep to the cortex is an area of **white matter**, which makes up the bulk of the cerebral tissue (Figure 14.3). The white matter contains myelinated fiber tracts that connect the following:

- Cerebral regions within the same hemisphere (**association fibers**)
- Cerebral regions from both hemispheres (**commissural fibers**, such as the corpus callosum)
- Cerebrum with other parts of the brain (**projection fibers**)

The **basal nuclei** (caudate nucleus, putamen, globus pallidus) are regions of gray matter deep within the cerebral hemispheres (Figure 14.3). These structures monitor voluntary movements initiated in the primary motor area by regulating the intensity and degree of muscular contractions and making adjustments in muscle tone.

CLINICAL CORRELATION

Neurons located in a midbrain nucleus called the **substantia nigra** project nerve fibers to the basal nuclei where they release the neurotransmitter, **dopamine.** If the substantia nigra is damaged, dopamine levels are reduced, and the basal nuclei become less active. The result is a gradual and systemic increase in muscle tone, and symptoms associated with **Parkinson's disease** can develop: muscle tremors when at rest and difficulty initiating and controlling voluntary movements. While walking, people with Parkinson's disease have a shuffling gait and tend to be bent forward. It is not clear how Parkinson's develops, but recent research suggests that exposure to pesticides might increase the risk of acquiring the disease.

Treatment for Parkinson's disease includes administration of L-dopa, a precursor of dopamine, in combination with other drugs. Recent studies using transplants of fetal substantia nigra tissue have shown promising results. However, the use of fetal tissue for this purpose is highly controversial, and whether research progresses in this direction remains questionable.

Table 14.1 Major Structures of the Adult Human Brain

Brain region	Structures	Functions
1. Telencephalon (anterior forebrain)	a. Lateral ventricles	Produces cerebrospinal fluid.
	b. Cerebrum	Interprets sensory impulses; controls voluntary motor activity; regulates muscle tone; stores information in memory; develops emotional and intellectual thoughts.
2. Diencephalon (posterior forebrain)	a. Third ventricle	Produces cerebrospinal fluid.
	b. Thalamus	Relay station for sensory impulses to cerebral cortex.
	c. Hypothalamus	Regulates autonomic nervous system; controls the pituitary gland; controls body temperature; controls food intake and thirst; receives sensory information from internal organs; regulates sleep–wake cycles.
	d. Pineal gland	Secretes melatonin, a hormone which assists the hypothalamus in the regulation of sleep–wake cycles.
3. Mesencephalon (midbrain *)	a. Cerebral aqueduct	Connects third and fourth ventricles.
	b. Cerebral peduncle	Relays motor impulses from cerebrum to spinal cord; relays sensory information from spinal cord to thalamus.
	c. Superior colliculi	Controls reflexes related to visual stimuli.
	d. Inferior colliculi	Controls reflexes related to auditory stimuli.
4. Metencephalon (anterior hindbrain)	a. Part of fourth ventricle	Produces cerebrospinal fluid.
	b. Cerebellum	Regulates motor activity so that muscle contractions are smooth, coordinated, and timely; maintains posture and regulates muscle tone.
	c. Pons *	Contains fiber tracts that connect the cerebrum to the cerebellum, medulla oblongata, and spinal cord; contains the origins of several cranial nerves; contains centers for regulating breathing rate.
5. Myelencephalon (posterior hindbrain)	a. Part of fourth ventricle	Produces cerebrospinal fluid.
	b. Medulla oblongata *	Contains centers for regulating heart rate, breathing rate, and blood vessel diameter; contains the origins of several cranial nerves; contains reticular formation that maintains wakefulness and filters out unnecessary sensory information.

* The midbrain, pons, and medulla oblongata form the brainstem.

Figure 14.1 Lateral view of the human brain. The major lobes in the left cerebral hemisphere are illustrated. The insula, not shown in the figure, is deep to the temporal lobe.

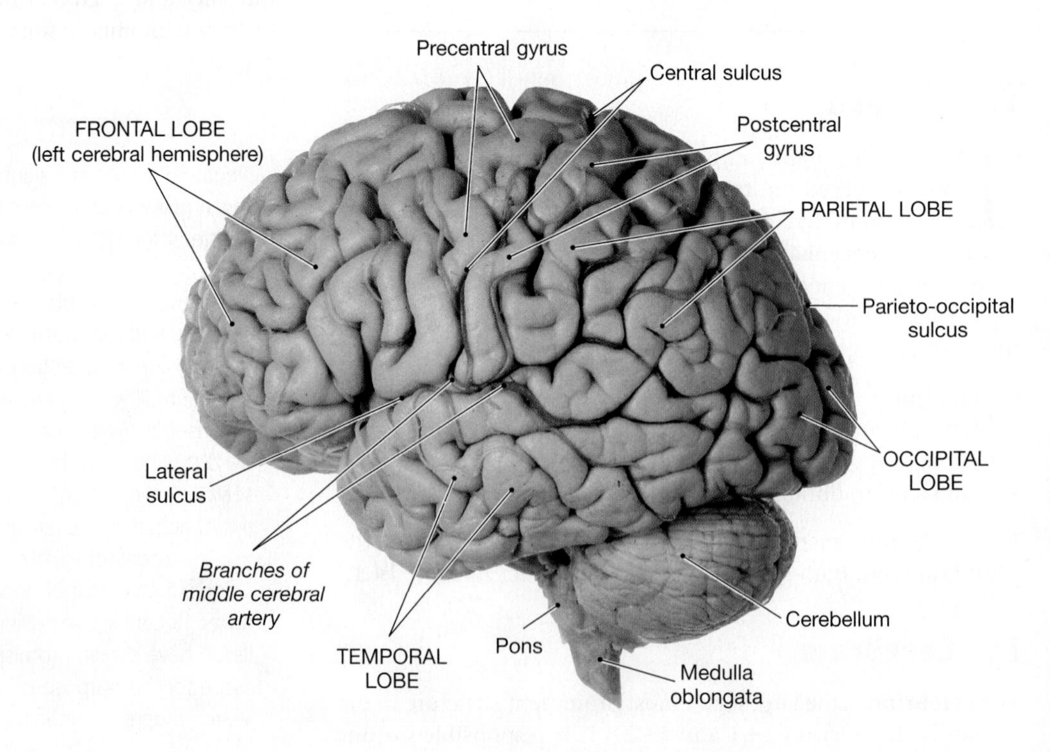

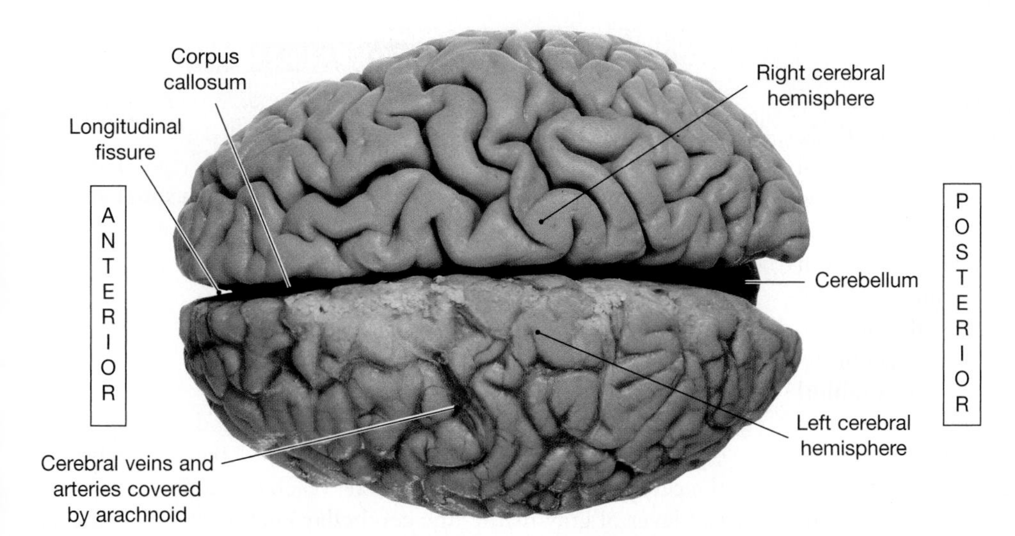

Figure 14.2 Superior view of the human brain. The two cerebral hemispheres are separated by the longitudinal fissure. On the surface of the cerebrum, note the numerous gyri separated by sulci.

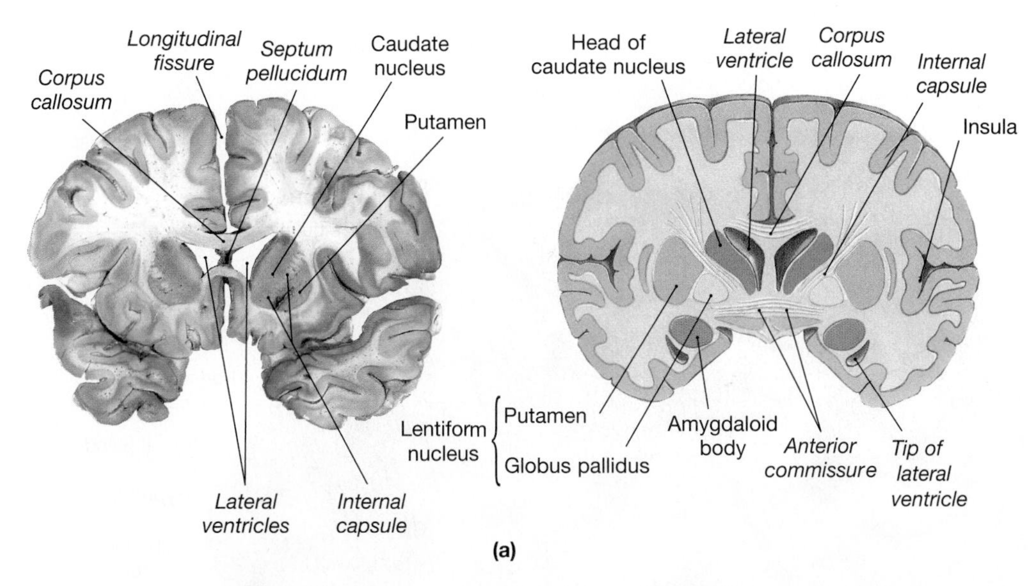

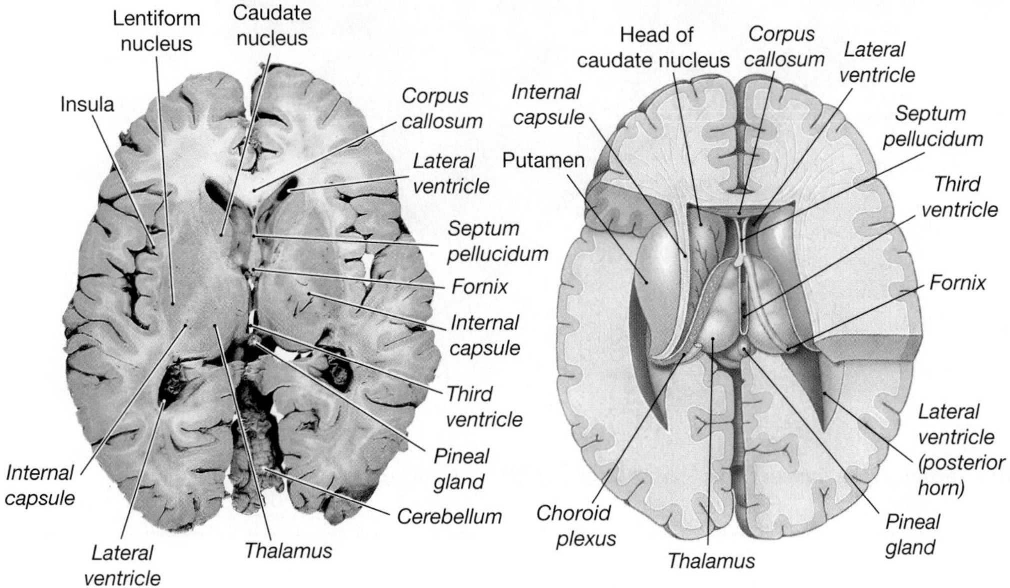

Figure 14.3 Sectional views of the human brain. a) Dissection and corresponding diagram of a coronal section; **b)** dissection and corresponding diagram of a transverse section.

The Diencephalon

Located centrally between the two cerebral hemispheres is an area called the **diencephalon**, composed primarily of the **thalamus**, **hypothalamus**, and **pineal gland** (Figure 14.4). The thalamus is medial to the basal nuclei. It is a large, oval-shaped mass of gray matter. Nearly all incoming information to the cerebral cortex first passes through the thalamus. Thus, it serves as a sensory relay station and has important roles in regulating motor activities, emotions, learning and memory, and visceral functions.

The hypothalamus is inferior to the thalamus. It is connected to the **pituitary gland** by a stalk of tissue called the **infundibulum**. The hypothalamus produces numerous hormones that influence the function of the pituitary gland. In addition, the hypothalamus has an important role in regulating homeostatic functions including heart rate, blood pressure, respiration, digestion, body temperature, water balance, and food intake.

The **pineal gland** is a small endocrine gland found posterior to the thalamus. It secretes melatonin, a hormone that assists the hypothalamus in regulating day–night cycles and sleeping patterns.

CLINICAL CORRELATION

Melatonin production increases in darkness and declines during daylight hours. During winter, when the amount of daylight decreases at higher latitudes, daily melatonin output increases. This change is believed to be the physiological basis for **seasonal affective disorder (SAD).** Symptoms of SAD include disruption of normal sleeping and eating patterns and adverse mood changes.

The Cerebellum

The **cerebellum** is posterior to the brainstem and inferior to the occipital lobe of the cerebrum (Figure 14.4). Like the cerebrum, the cerebellum has two hemispheres, each containing a superficial layer of gray matter, the **cerebellar cortex**, and a deep area of white matter. Deep within the white matter is another area of gray matter, the **cerebellar nuclei.**

The cerebellum is linked to other regions of the brain and the spinal cord by three pairs of nerve fiber tracts known as

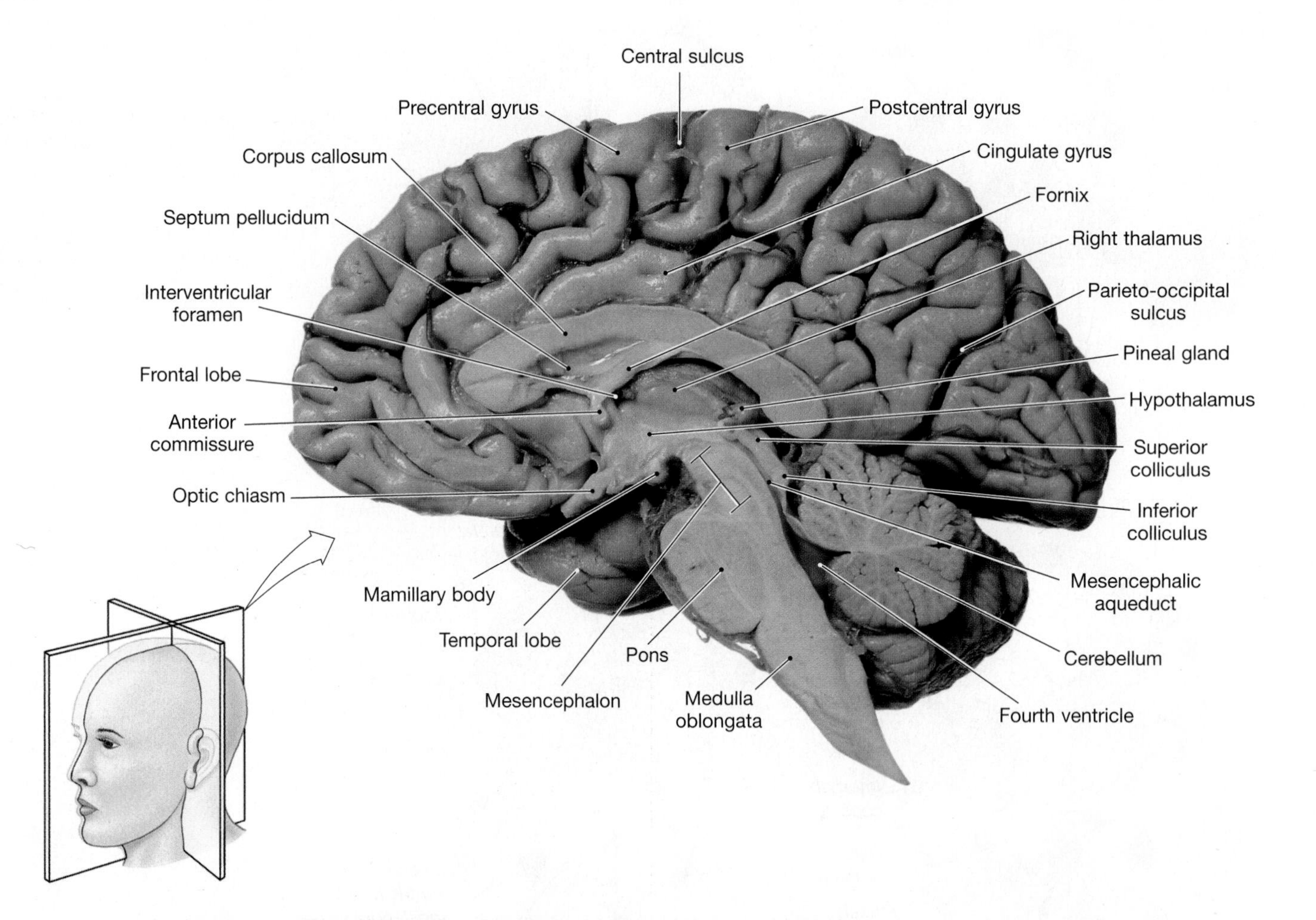

Figure 14.4 Midsagittal section of the human brain. The relationship of the cerebrum with other structures is clearly seen.

cerebellar peduncles. Via these connections, the cerebellum receives information from sensory receptors in skeletal muscles and joints (proprioceptors) that keep it informed of the current body position. It also receives information from motor pathways from the primary motor area for initiating voluntary movements. The cerebellum compares this information and makes adjustments to produce smooth, well-coordinated muscle contractions.

The Brainstem

The brainstem consists of three parts: the **midbrain**, **pons**, and **medulla oblongata**. The midbrain (mesencephalon) is inferior and posterior to the thalamus (Figure 14.4). The **cerebral peduncles**, located along the anterior aspect of the midbrain, contain both motor and sensory pathways that connect the cerebrum and thalamus to other regions in the brainstem and to the spinal cord. On the posterior surface are four mounds of tissue called the **corpora quadrigemina**. This structure contains two **superior colliculi**, which are centers for visual reflexes, and two **inferior colliculi**, which are centers for auditory reflexes.

WHAT'S IN A WORD The term *colliculus* is derived from Latin and means "mound." The collective name for the inferior and superior colliculi, *corpora quadrigemina*, is also Latin and means, literally, "bodies of quadruplets." More loosely, it translates to "four bodies"; it refers to the four mounds (bodies) of tissue that comprise the inferior and superior colliculi.

The pons (Figures 14.4 and 14.5) is the mound of tissue on the anterior surface of the brainstem. It contains nuclei for several cranial nerves, respiratory control centers, and nerve fiber tracts that connect the spinal cord, cerebellum, and other brain regions.

The medulla oblongata (Figures 14.4 and 14.5) is directly continuous with the spinal cord. All ascending and descending pathways that connect the brain and spinal cord must pass through this region. The medulla oblongata also contains nuclei for several cranial nerves and control centers for respiration, heart rate, and blood pressure.

The Cranial Meninges

The brain and spinal cord are covered by three connective tissue layers: **dura mater**, **arachnoid**, and **pia mater**. These layers are known collectively as the **meninges** (Figure 14.6). Along with the cranial bones, vertebrae, and cerebrospinal fluid, the meninges protect the brain and spinal cord from trauma and injury.

The Cranial Nerves

Twelve pairs of cranial nerves originate from the nervous tissue of the brain (Figure 14.5). Some cranial nerves are mixed nerves that contain both motor and sensory nerve fibers, while others are totally sensory. The cranial nerves only supply structures in the head and neck, except for the vagus nerve, which also supplies organs in the thorax and abdomen.

The motor components of the cranial nerves are composed of motor neurons whose cell bodies are located in motor nuclei inside the brain. These neurons transmit impulses along their axons to target structures such as skeletal muscles in the head and neck (e.g., muscles of facial expressions), glands in the head (e.g., salivary glands), and involuntary muscles in the walls of thoracic and abdominal organs (e.g., cardiac muscle in the heart or smooth muscle in the stomach).

The sensory components of cranial nerves consist of sensory neurons whose cell bodies are found in sensory ganglia outside the brain. These neurons transmit nerve impulses from peripheral sensory organs to the brain. The sensory neurons are connected to general sensory organs such as pressure, pain, and temperature receptors in the skin, and to special sensory organs such as taste receptors of the tongue, and photoreceptor cells in the retina of the eye.

Sheep Brain Dissection

The sheep brain is anatomically similar to the human brain, so it serves as an excellent model for dissection. At various stages in the dissection you will be asked to compare the structures that you identify in the sheep brain with equivalent structures in the human brain. Have anatomical models of the human brain readily available for this purpose. If models are not accessible, you may also use diagrams or photographs of the human brain in an anatomical atlas, your textbook, or this exercise.

The Sheep Meninges

Dissection of the meningeal coverings must be completed before making observations of the brain. Proceed with care and patience to avoid damaging important structures.

ACTIVITY 14.1 Dissecting the Sheep Meninges

1. Obtain a sheep brain and a set of dissecting tools.
2. The dura mater is the tough, opaque outer covering of the brain. It may have already been removed during commercial preparation. If not, carefully cut it away without damaging underlying structures by lifting the dura mater away from the surface of the brain with forceps and cutting the membrane with a pair of scissors. Begin at the posterior end of the brain and cut just to the left or right of the midsagittal plane along the superior surface.

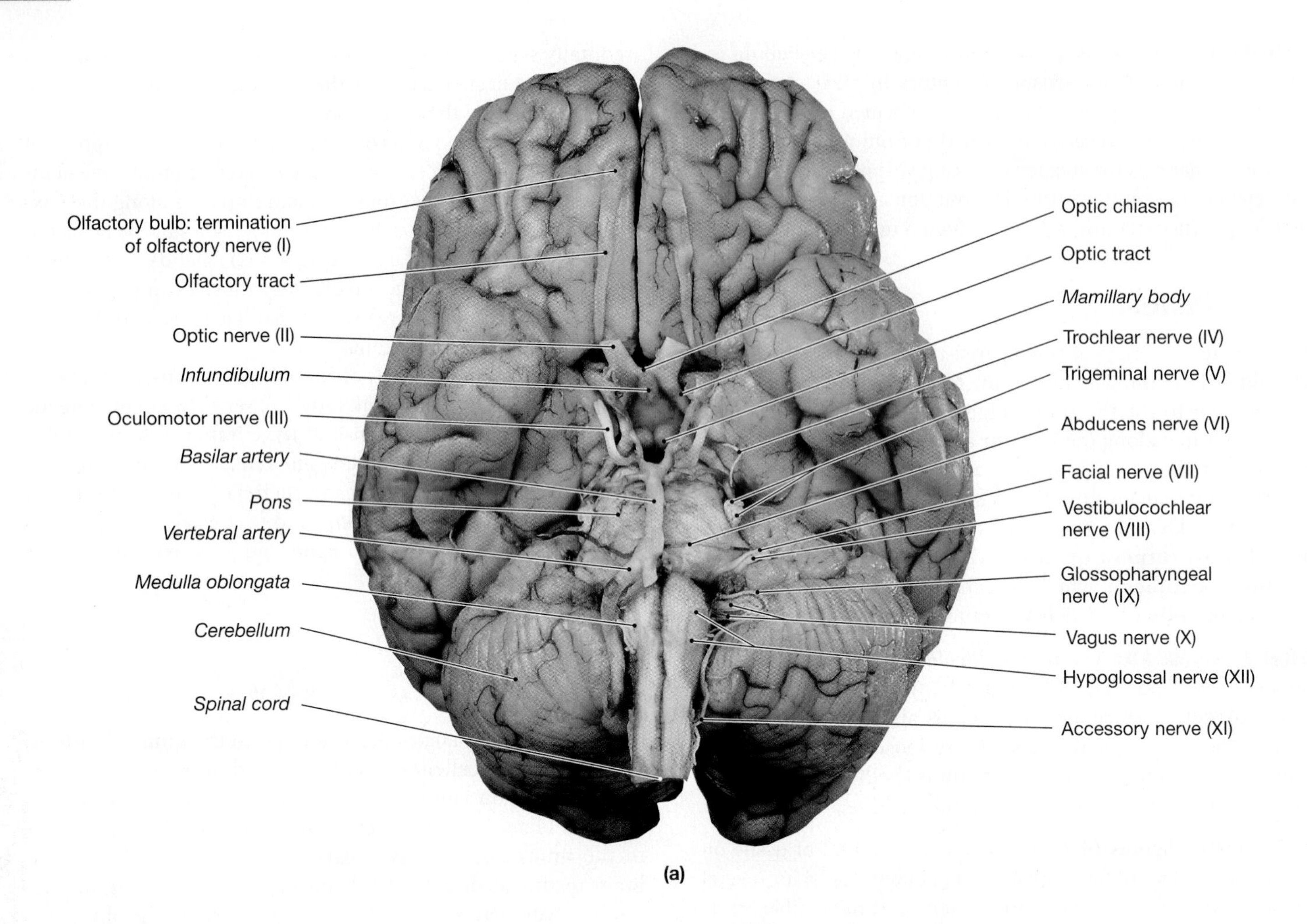

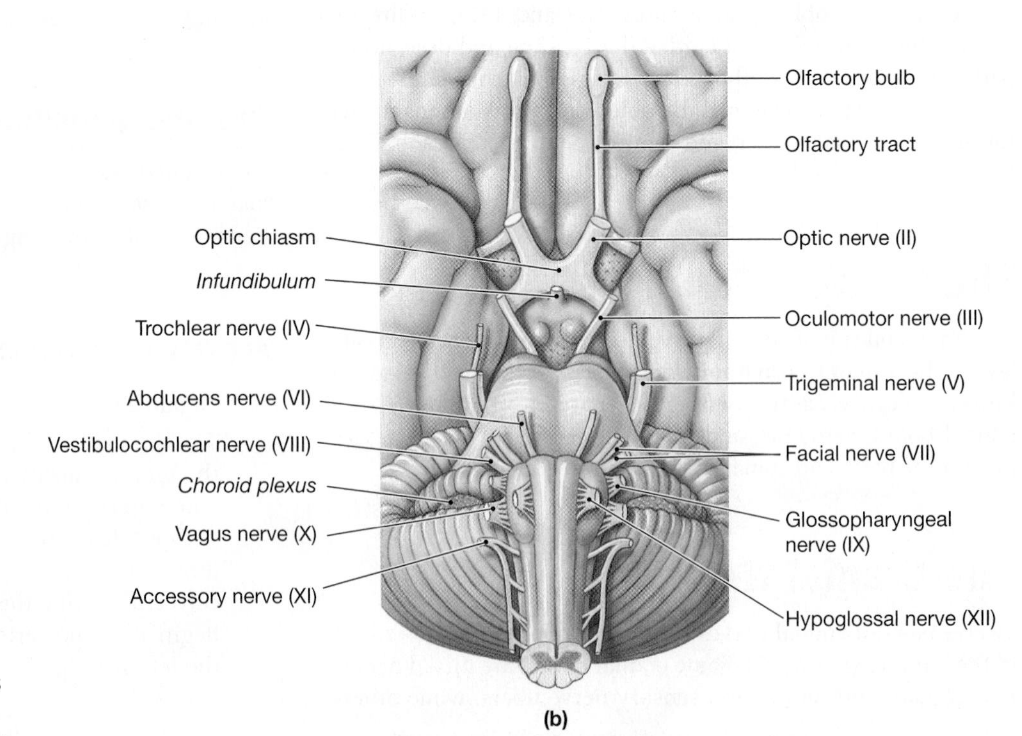

Figure 14.5 Inferior surface of the brain illustrating the origins of the cranial nerves. a) Brain dissection; **b)** corresponding diagram of the region where the cranial nerves originate.

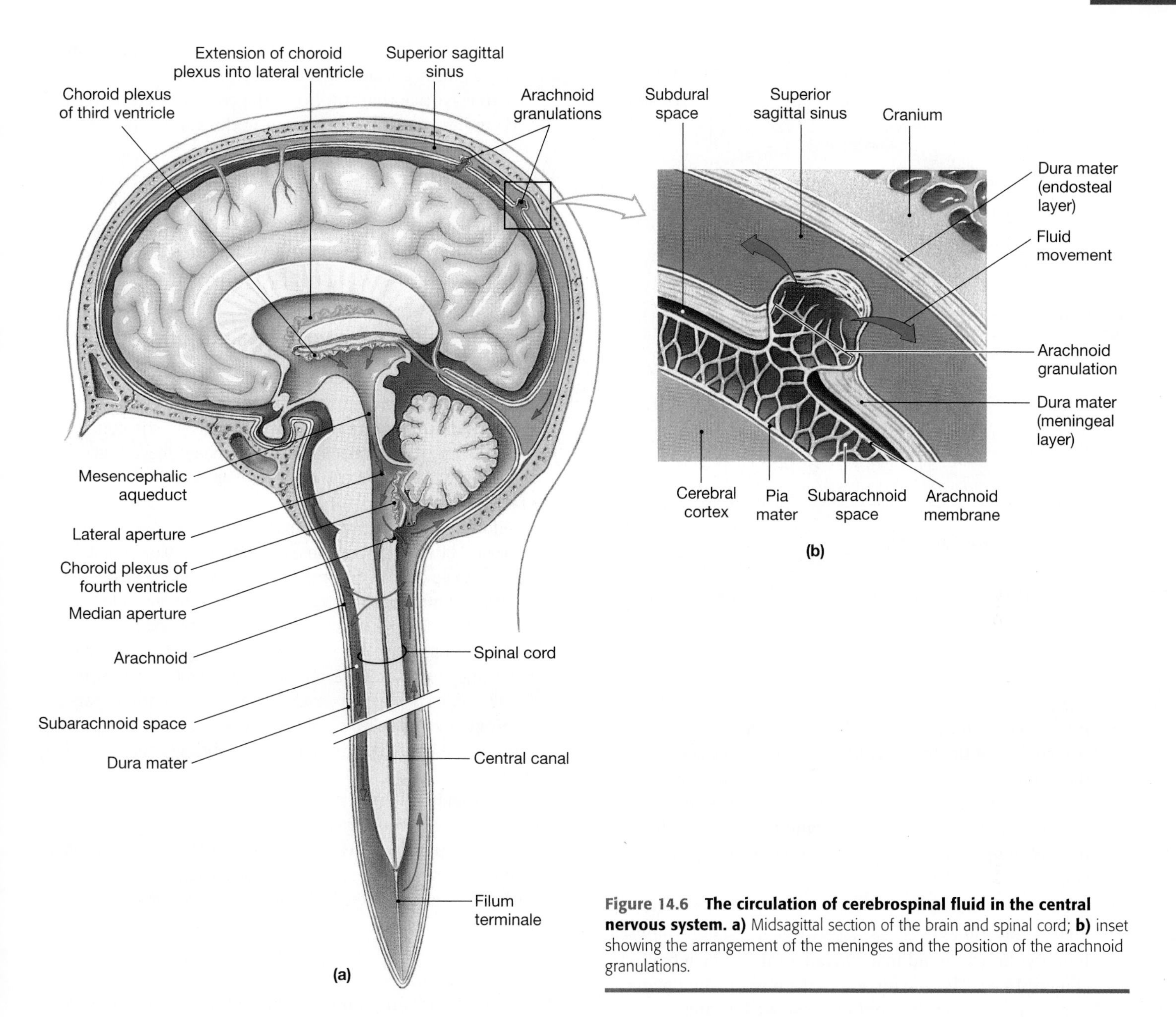

Figure 14.6 The circulation of cerebrospinal fluid in the central nervous system. a) Midsagittal section of the brain and spinal cord; **b)** inset showing the arrangement of the meninges and the position of the arachnoid granulations.

3. As you proceed, attempt to identify three dural partitions (Figure 14.7).
   - The **falx cerebri** folds into the **longitudinal fissure** and separates the cerebral hemispheres.
   - The **falx cerebelli** separates the cerebellar hemispheres.
   - The **tentorium cerebelli** separates the cerebrum (occipital lobe) and cerebellum.
4. Once you have identified the dural partitions, you will have to make at least one cut across each of them with your scissors. This will allow for easier removal of the dura mater.
5. The dura mater has two layers: an outer layer that is attached to the inside of the cranial bones and an inner layer that forms the brain covering. You cannot identify these layers except at specific locations where they separate to form **dural sinuses**. Along the midsagittal plane, verify that the two layers of the dura mater separate to form the **superior sagittal sinus**. This sinus serves to drain venous blood from the brain to the internal jugular vein (Figures 14.6 and 14.7).

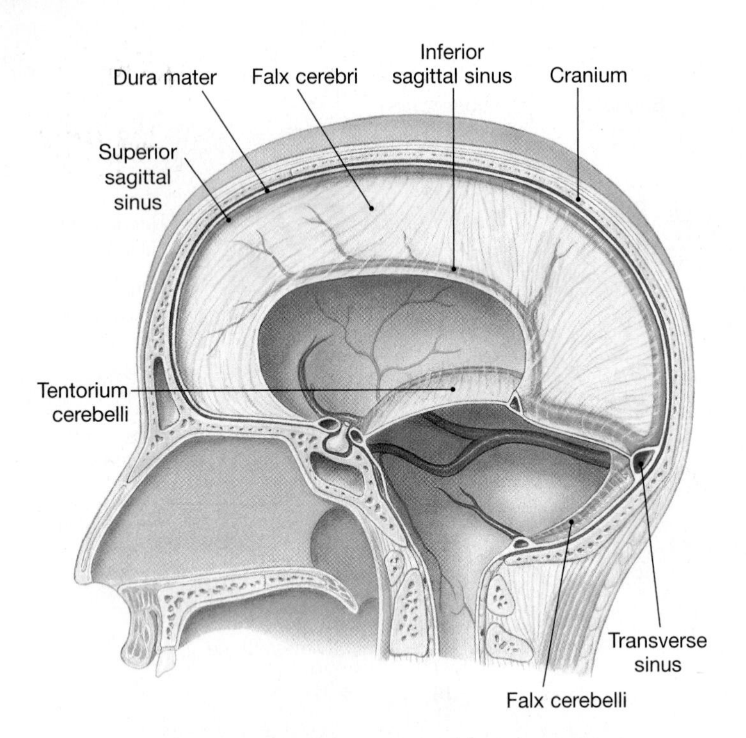

Figure 14.7 The cranial cavity with the brain removed. The positions of the dural partitions and dural sinuses are shown.

6. It is very difficult to remove the dura mater from the inferior surface of the brain without damaging the origins of the cranial nerves. Do not be concerned if this occurs; however, attempt to save the two **optic nerves (cranial nerve II)** and the **optic chiasm**, where some nerve fibers from each optic nerve cross over to the opposite side (Figures 14.5 and 14.8c).
7. On a model or diagram of the human cranial meninges, identify the dura mater, dural partitions, and dural sinuses.
8. Identify the arachnoid in a diagram of the human brain (Figure 14.6). The arachnoid lies deep to the dura mater. Typically, this layer does not preserve well during commercial preparation and cannot be identified on your sheep brain.
9. Locate the **subarachnoid space** deep to the arachnoid layer (Figure 14.6). This space is filled with **cerebrospinal fluid**, which provides a protective, shock-absorbing layer for the brain. Traveling through the space are major blood vessels that supply the brain.

WHAT'S IN A WORD The name *arachnoid* is derived from Greek and means "like a spider." The middle meningeal layer is called the arachnoid because slender filamentous strands that traverse the subarachnoid space and connect the arachnoid to the deeper pia mater take on the appearance of a spider's web.

10. Identify the pia mater in a diagram of the human brain (Figure 14.6). It is the thin, translucent layer that is closely attached to the surface of the brain. The somewhat glossy surface of the sheep brain is the thin pia mater. Notice that it follows the complicated pattern of convolutions and grooves along the surface of the cerebrum.
11. Carefully lift a small portion of the pia mater away from the surface of the brain with a pair of forceps. Note that in the area from which you removed the pia mater, the brain's surface has been exposed and the glossy appearance is absent.

CLINICAL CORRELATION

The dura mater is associated with two potential spaces: the **epidural space** between the dura mater and the cranial bones, and the **subdural space** (Figure 14.6b) between the dura mater and arachnoid. The term *potential space* means that a true space does not really exist, but one could form as a result of injury (e.g., a blow to the head).

Epidural and **subdural hematomas** can form when meningeal blood vessels are torn as a result of a blow to the head. In an epidural hematoma, blood collects in the epidural space when **meningeal arteries,** traveling along the outer dural layer, rupture. The pterion (the temple) is a common region for an epidural hematoma to form because the anterior branch of the middle meningeal artery travels superficially through this region.

A subdural hematoma develops when blood from damaged **meningeal veins** collects in the subdural space. These typicallly form when a blow to the head causes the brain to shake inside the skull.

Epidural hematomas are more dangerous because arterial blood pressure is much greater than venous blood pressure. Thus, the hematoma spreads quickly over a relatively short period of time. In both cases, however, an expanding hematoma may exert fatal pressure on the brain unless it is promptly recognized and surgically treated.

QUESTION TO CONSIDER Identify three special features of the meninges around the sheep brain that are similar to the meninges around the human brain.

____

____

Surface Anatomy of the Sheep Brain

A logical way to begin your study of brain anatomy is to first identify and learn some of the prominent external structures. This activity will provide you with a good overview of the brain's structure before any incisions are made to examine the internal anatomy. In addition, the dominance of the cerebral hemispheres and the complex system of convolutions on their surfaces will be clearly demonstrated.

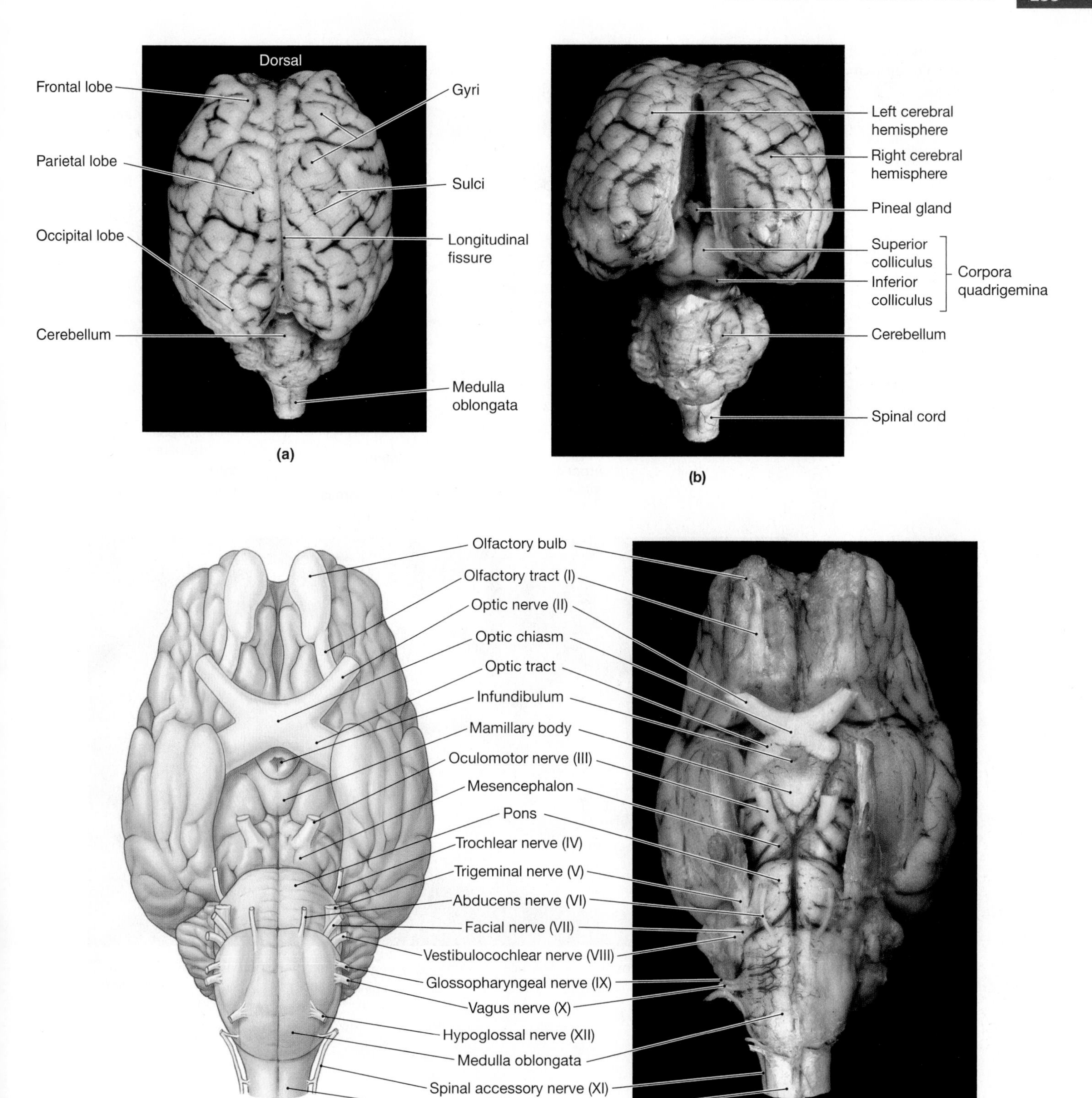

Figure 14.8 Various views of the sheep brain. a) Superior view; **b)** superior view with midbrain structures exposed; **c)** inferior view; *(continues)*

Figure 14.8 Various views of the sheep brain. *(Continued)*
d) Midsagittal section.

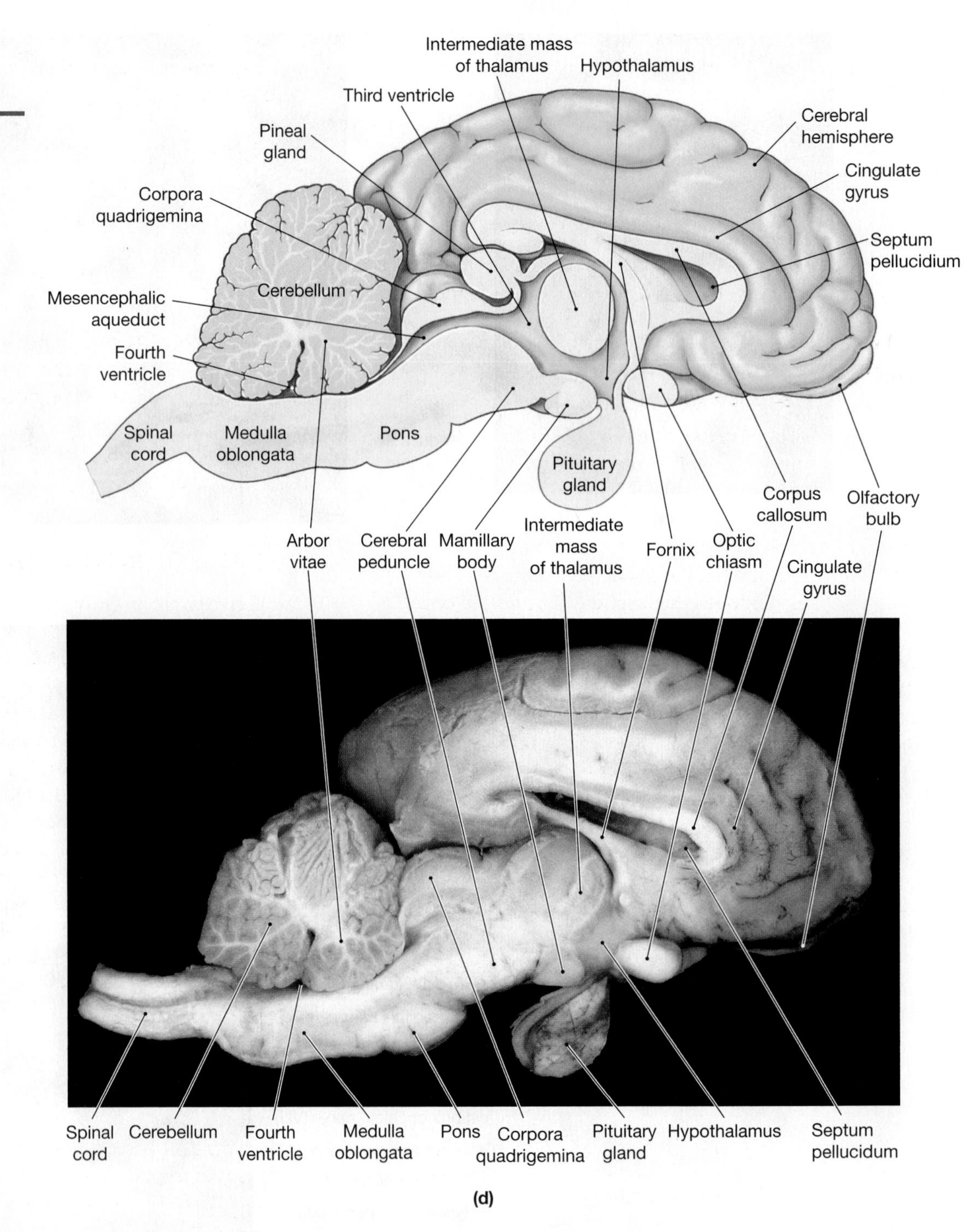

ACTIVITY 14.2 Dissecting Surface Structures on the Sheep Brain

1. Identify the cerebrum, which consists of two cerebral hemispheres. Note that the hemispheres are separated by the longitudinal fissure (Figure 14.8a).
2. Notice that the surfaces of the hemispheres are characterized by a complex system of raised gyri (singular = gyrus). The gyri are separated by shallow sulci or deep fissures.
3. Identify the following cerebral lobes.
   - The frontal lobes are located at the anterior end of the cerebrum.
   - The parietal lobes are just posterior to the frontal lobes. In each hemisphere, the frontal lobe is separated from the parietal lobe by the central sulcus. This sulcus is difficult to identify on the sheep brain.
   - The occipital lobes are located at the posterior tip of the cerebrum. Each occipital lobe is separated from its cor-

responding parietal lobe by the **parieto-occipital sulcus.** Like the central sulcus, the parieto-occipital sulcus is hard to identify.

- The temporal lobes are located along the lateral aspects of each hemisphere. Identify the lateral fissure, which separates the temporal lobe from the frontal lobe.

4. Obtain a model or diagrams of the human brain (Figures 14.1 and 14.2). Examine the complex pattern of the gyri and grooves. Identify the cerebral lobes, longitudinal fissure, and lateral sulcus. Attempt to identify the central sulcus and the parieto-occipital sulcus. As in the sheep brain, these two sulci are difficult to find in the human brain.
5. Carefully separate the two cerebral hemispheres of the sheep brain with your hands and look down into the longitudinal fissure. By doing this, you will identify the corpus callosum, which contains myelinated nerve fibers that connect the hemispheres.
6. Identify the cerebellum, just posterior to the cerebrum (Figure 14.8). Note that the cerebellum, like the cerebrum, is divided into two hemispheres, the cerebellar hemispheres.
7. Gently separate the cerebrum and cerebellum at the **transverse fissure** and identify two pairs of prominent tissue elevations on the roof of the midbrain. The anterior pair are called the superior colliculi (singular = colliculus), and the posterior pair are the inferior colliculi. Collectively, all four of these tissue elevations are called the corpora quadrigemina (Figure 14.8b).
8. Observe the inferior surface of the brain (Figure 14.8c) and identify the following structures.
   - The **olfactory bulbs** are relatively large, ovoid masses on the inferior surface of the frontal lobe. They give rise to **olfactory nerves (cranial nerve I)**, which pass through the cribriform plate of the ethmoid bone to enter the nasal cavity. If a sheep skull is available, identify the cribriform plate. Otherwise, identify the cribriform plate on the human skull.
   - The **olfactory tracts** leave the olfactory bulbs and pass posteriorly to the olfactory areas in the temporal lobes.
   - The **optic chiasm**, where the two **optic nerves** converge, is anterior to the **pituitary gland**. At this location, some fibers of each optic nerve cross to the opposite side.
   - The pituitary gland is located posterior to the optic chiasm. On most commercial preparations of the sheep brain, the pituitary is absent. If this is the case on your dissection, identify this structure on the inferior surface of the human brain model.
   - On the brainstem, the elevated mass of tissue on the ventral surface is the pons.
   - The medulla oblongata is the column of tissue that extends posteriorly from the pons and is directly continuous with the spinal cord.
9. On a model or in diagrams of the human brain, identify the structures you found on the sheep brain (Figures 14.1, 14.2, and 14.5).

QUESTIONS TO CONSIDER 1. After reviewing the general function of the cerebrum (Table 14.1), explain why this structure comprises the vast majority of the brain's total mass.

____

____

2. As you have observed, the cerebral cortex is characterized by a complex series of tissue folds or gyri. Speculate on the functional advantage for this structural pattern.

____

____

Sectional Anatomy of the Sheep Brain

To fully appreciate its complex anatomical details, the brain must be examined from various planes of section. The midsagittal section (Figures 14.4 and 14.8d) is a classic view for studying brain anatomy. From this perspective, you will be able to identify most external brain features as well as several deep structures that cannot be seen from the surface. This view also offers the best perspective for understanding how the adult brain structures are organized and how they form from the five brain vesicles that develop in the fetus (Table 14.1).

The coronal and transverse sections (Figure 14.3) are particularly useful for observing the organization of the gray matter and white mater in the cerebrum. From these views the position of the surface gray matter in the cerebral cortex in relation to the white matter and deep gray matter nuclei (i.e., the basal nuclei) is clearly seen.

ACTIVITY 14.3 Examining a Midsagittal Section of the Sheep Brain

1. Place the sheep brain on the dissecting tray with the dorsal surface up. Using a scalpel or small knife, make an incision along the midsagittal plane of the brain. Cut along the longitudinal fissure in an anterior to posterior direction, staying as close to the midline as possible. When you reach the posterior end of the cerebrum, continue your midsagittal incision through the cerebellum and along the length of the brainstem.
2. Examine the midsagittal section of the sheep brain (Figure 14.8d) and identify the structures described herein. Have a model or diagram of the human brain available so that you can make comparative observations (Figure 14.4).
3. From this view, the medial surface of the cerebral hemisphere can be seen. From anterior to posterior, identify the frontal, parietal and occipital lobes.

4. Locate the corpus callosum, which is the arching band of white matter located just inferior to the cerebral hemisphere. Identify the **cingulate gyrus**, which is located on the medial surfaces of the frontal and parietal lobes and curves over the corpus callosum. This is a part of the **limbic system**, where emotions and other related behaviors are regulated.
5. Just inferior to the corpus callosum is a thin membrane called the **septum pellucidum**. This membrane forms a barrier between the two lateral ventricles. Break the septum pellucidum (it may already be broken) to reveal the cavity of the lateral ventricle.
6. Identify the thalamus, a large circular region inferior to the corpus callosum. The right and left thalami form the lateral walls of the **third ventricle**. In a midsagittal section, the third ventricle is not intact because the incision that you made passes through this cavity.
7. The region just inferior to the thalamus is the hypothalmus. If present on your dissection, identify the pituitary gland and note that it is connected to the hypothalamus by the infundibulum.
8. Locate the **mamillary body**, a nucleus in the hypothalamus. The mamillary body is a component of the limbic system. It serves as a control center for motor reflexes associated with eating.
9. Identify the **fornix**, a gently curving fiber tract that runs between the corpus callosum and the thalamus. The fornix connects the mamillary body to the **hippocampus**, which is another limbic system structure, important for memory. You will not be able to identify it because it is located deep within the temporal lobe.
10. Identify the pineal body, a relatively small structure that is wedged between the thalamus and midbrain.
11. Immediately posterior to the cerebrum is the cerebellum. Note the presence of the gray matter along the outer surface of the cerebellum. This gray matter layer is the cerebellar cortex. A region of white matter, the **arbor vitae**, is deep to the cortex.

WHAT'S IN A WORD In a midsagittal section, the highly branched pattern of the cerebellar white matter can be clearly seen. Because it resembles the branches of a tree, the white matter in the cerebellum is often referred to as the *arbor vitae*, which is the Latin term for "tree of life."

12. Identify the three parts of the brainstem: the midbrain, pons, and medulla oblongata.
13. The midbrain is directly posterior to the thalamus and hypothalamus. Superiorly, the midbrain is dominated by the superior and inferior colliculi. A fiber tract, known as the cerebral peduncle, passes inferiorly. Traveling between the colliculi and cerebral peduncle is the **cerebral (mesencephalic) aqueduct**, a narrow passageway that connects the third and fourth ventricles.
14. The pons is immediately posterior to the midbrain, followed by the medulla oblongata. The **fourth ventricle** is the cavity located between the cerebellum, superiorly, and the pons and medulla oblongata, inferiorly. Note that the fourth ventricle is continuous with the central canal of the spinal cord.

ACTIVITY 14.4 Examining a Coronal Section of the Sheep Brain

1. Obtain one half of the sheep brain from your midsagittal incision or a whole brain, if available.
2. Using a scalpel or small knife, make an incision along a coronal plane. Begin your incision by cutting through the parietal lobe. Be sure that your cut passes through the temporal lobes.
3. Examine your coronal section of the brain and identify the following structures. Have a model or diagram of the human brain available so that you can make comparative observations (Figure 14.3a).
   - The *cerebral cortex,* a thin layer of gray matter on the surface of the cerebrum
   - The *white matter,* deep to the cerebral cortex
   - The *basal nuclei,* a region of gray matter deep within the cerebrum
   - The two *lateral ventricles* deep within the cerebrum and the *septum pellucidum* that divides them (This relationship is best observed on a whole brain.)
   - The *third ventricle,* located between the *left* and *right portions of the thalamus* (This relationship is best observed on a whole brain.)
   - The *internal capsule,* a projection fiber tract that travels between the thalamus and basal nuclei
   - The *corpus callosum,* the main commissural fiber tract that travels between the two cerebral hemispheres

The Brain Ventricles and Cerebrospinal Fluid

The ventricles are a series of interconnected deep cavities (Figure 14.9) that are derived from the lumen (internal space) of the embryonic neural tube. The walls of the ventricles are lined by **ependymal cells**, a type of glial cell. The **choroid plexuses**, located along the roofs of all the brain ventricles, produce cerebrospinal fluid. Each choroid plexus is composed of capillary networks that are in close contact with ependymal cells. The cerebrospinal fluid (CSF) is formed when blood plasma is filtered through the walls of the capillaries, travels through the ependymal cells, and empties into the ventricles.

Cerebrospinal fluid (CSF) is drained by the **arachnoid granulations (arachnoid villi)**. These structures are extensions of the arachnoid that project into the dural sinuses (Figure 14.6b). The cerebrospinal fluid filters through the granulations and drains into the venous blood by entering one of the dural si-

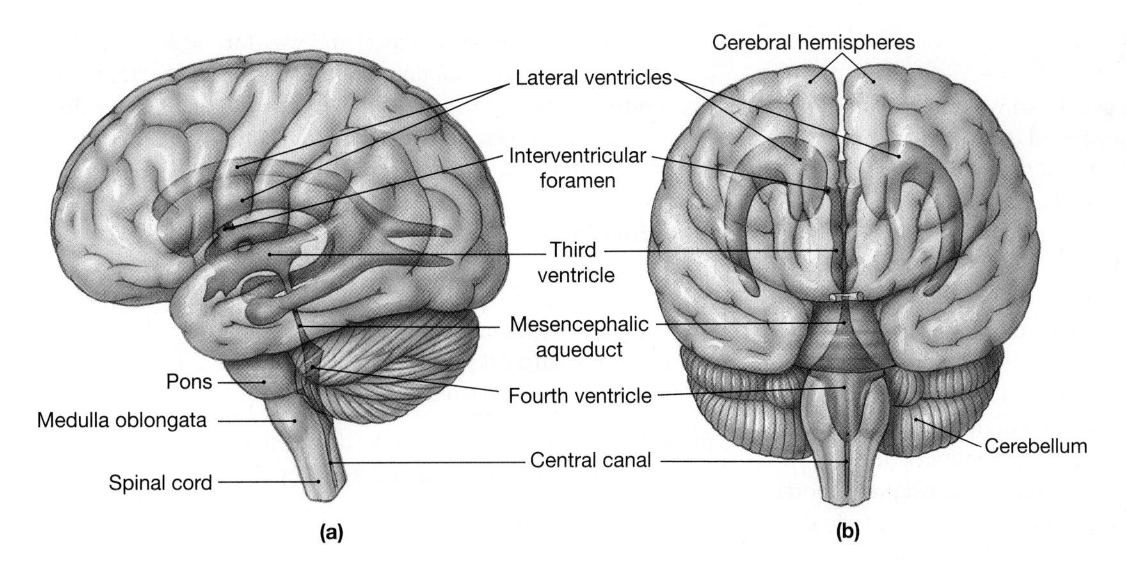

Figure 14.9 The brain ventricles. The deep ventricles are illustrated in relation to more superficial structures: **a)** lateral view; **b)** anterior view.

nuses. On average, about 500 ml of CSF circulates through the ventricles and subarachnoid space each day.

CLINICAL CORRELATION

The overaccumulation of cerebrospinal fluid in the brain ventricles or subarachnoid space is called **hydrocephalus** ("water on the brain"). Hydrocephalus can be caused by a blockage or reduction of normal drainage of CSF from the brain ventricles at the arachnoid granulations. This abnormality could be caused by a tumor or an infection such as meningitis. As CSF accumulates, the increase in fluid pressure could cause brain damage. If hydrocephalus occurs in newborns and infants, the developing cranium becomes much larger than normal. The condition can be more serious in adults because the cranial bones have stopped growing, and thus, there is no way to compensate for the rise in pressure. Hydrocephalus is typically treated by implanting a shunt (a length of plastic tubing) that diverts cerebrospinal fluid from a brain ventricle to either the peritoneal cavity or a superficial vein in the neck. With proper treatment, the prognosis for newborns with hydrocephalus is very good.

The three-dimensional qualities of the ventricles can be fully appreciated if you examine them from a variety of views and sections (Figures 14.3, 14.4, 14.9).

ACTIVITY 14.5 Examining Brain Ventricles and CSF Flow

1. With the aid of anatomical models or diagrams, carefully study the organization of the four brain ventricles: the two lateral ventricles, the third ventricle, and the fourth ventricle (Figures 14.6a and 14.9). Note that the two lateral ventricles have irregular bends and folds that reach into all the cerebral lobes. Also observe that the ventricles and their connecting channels form a continuous pathway that leads directly into the central canal of the spinal cord.
2. Beginning with the choroid plexus along the roof of the right lateral ventricle, follow the pathway that a drop of CSF would take to reach the central canal of the spinal cord. List all structures, in the correct order, that the drop of CSF would encounter.

 ____

 ____

3. Beginning once again at the choroid plexus in the right lateral ventricle, trace the pathway that a drop of CSF would take to reach the subarachnoid space. In what structure would this pathway take a different direction, compared to the pathway described in step 2?

 ____

 List all the new structures that the drop of CSF would encounter.

 ____

 ____

QUESTION TO CONSIDER Because the subarachnoid space is filled with cerebrospinal fluid, the brain is surrounded by an aqueous (mostly water) liquid layer. Explain why this arrangement provides protection for the brain.

____

____

Electroencephalography

Electrical activity in the brain is generated when nerve impulses are sent along axons and transferred to other neurons at synapses. In a living individual, this activity is always present, even while sleeping, and the electric field that is produced can be recorded and measured by placing electrodes on the scalp.

An **electroencephalogram (EEG)** is a recording of the brain's electrical activity. An EEG, and therefore the results of this activity, vary according to the physiological, emotional, and mental state of the individual being recorded. The wave patterns produced by an EEG generally fall into one of four basic types, each characterized by a particular combination of *frequency* (the number of wave patterns per time period) and *amplitude* (the size of the waves). Generally speaking, as an individual becomes more alert, waves become less synchronized and their amplitude decreases. **Alpha waves** are detected in healthy adults who are awake but resting, with their eyes closed. These low-frequency, high-amplitude waves are created by synchronization of the electrical activities throughout the cerebral cortex. Higher frequency, lower amplitude **beta waves** are identified in individuals who are experiencing a stressful situation or concentrating on performing a specific task. They also occur during rapid eye movement (REM) or deep sleep, when the eyes tend to move back and forth. **Theta** and **delta waves** typically occur during periods of sleep other than REM sleep, but they may also occur if a brain tumor or some other disorder is present. Theta waves can also be detected during periods of intense frustration in otherwise healthy individuals.

In the following activity, you will use the Biopac Student Lab System to record your lab partner's EEG through electrodes placed on the occipital region of the brain. You will examine the four EEG rhythms and investigate the effects of visual attention and mental stress on EEG waveforms.

ACTIVITY 14.6 Electroencephalography Using the Biopac Student Lab System

BIOPAC Systems, Inc.

Setup and Calibration

1. Position three electrodes on your lab partner's scalp as shown in Figure 14.10. Part your lab partner's hair tightly and firmly press the disposable electrode in the middle of each part. Attach electrode leads (SS2L) as illustrated. After the electrodes are attached, a swim cap or self-adhering wrap can be used to help keep them in place.
2. Have your lab partner lie on his or her back with the head tilted comfortably to one side. Instruct your partner to minimize movement throughout the experiment.
3. Attach the electrode leads (SS2L) to Channel 1 and turn on the Acquisition Unit.
4. Start the Biopac Student Lab System, choose Lesson 3 (L03-EEG-1), and click **OK.** When prompted, enter a unique file name and click **OK.**
5. Click on the **Calibrate** button and wait for the computer to adjust the signal. Calibration automatically stops after 8 seconds. If the recording is relatively flat without any large fluctuations or spikes, you are ready to record. If not, then check the electrodes and connections and click **Redo Calibration.**

Recording Data

You will record three segments of data in this exercise:

- 10 seconds with your lab partner's eyes closed
- 10 seconds with eyes open
- 10 seconds with eyes reclosed

You will insert markers into the recording to designate the start of each segment by pressing either the **F9** key (PC) or **Esc** key (Mac).

1. Have your lab partner lie quietly and breathe slowly with his or her eyes closed.
2. Click on the **Record** button.
3. After 10 seconds, ask your lab partner to open his or her eyes and try not to blink. Immediately insert a marker.
4. After 10 seconds with the eyes open and not blinking, ask your lab partner to close his or her eyes. Immediately insert a marker.
5. Once you have collected the final 10 seconds with the eyes closed, click on the **Stop** button.
6. You will see the raw EEG on the screen. To calculate each of the four rhythms from the raw EEG, click once on the **alpha** button, the **beta** button, the **delta** button, and the **theta** button. As you click, data for each of these waveforms will be added to the file.
7. Only after you have added all four waveforms should you click on the **Done** button.

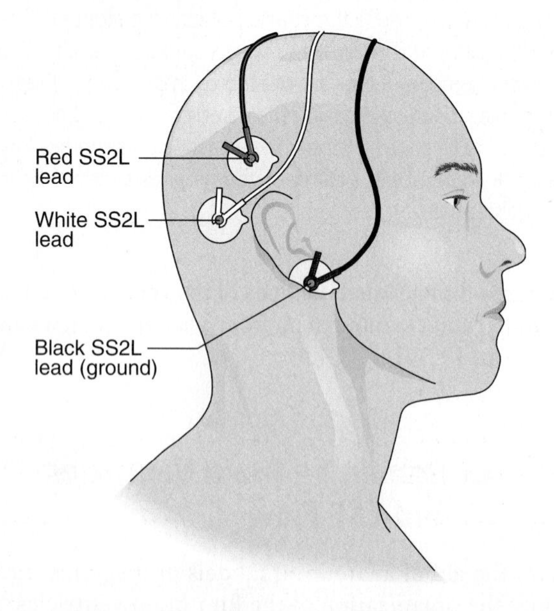

Figure 14.10 Placement of electrodes for occipital lobe EEG recording. The electrode lead colors should match the positions illustrated in the diagram. The ground electrode may also be placed on facial skin just inferior to the earlobe.

8. When prompted by the computer, select **Record another lesson**, click **OK**, choose Lesson 4 (**L04-EEG-2**), and click **OK**. When prompted, type a unique file name and click **OK**.

   In this next experiment you will record four segments of data:
   - 10 seconds relaxed with eyes closed
   - 20 seconds performing mental math with eyes closed
   - 10 seconds recovering from hyperventilation with eyes closed
   - 10 seconds relaxed with eyes open

   You will click on the **Suspend** button to automatically insert a marker in between segments.
9. This lesson requires the same electrode placement and setup as Lesson 3. Therefore, repeat the "Setup and Calibration" steps as previously described.
10. Have your lab partner lie quietly and breathe slowly with his or her eyes closed.
11. Click on the **Record** button and collect data for 10 seconds. Click on the **Suspend** button. *[Note: The "alpha thermometer" on the right side of the screen provides a visual reference for observing EEG activity in real time.]*
12. Make sure your lab partner keeps his or her eyes closed. Give your lab partner a mental math problem and click on the **Resume** button. The problem should be challenging and require approximately 20 seconds to complete. (Example: Take the number 2 and double it, double it again, double again, divide by 4, multiply by 15, divide by 10, multiply by 12.)
13. Collect data as your lab partner completes the problem while remaining lying down with eyes closed. After 20 seconds, or once he or she has solved the problem, click on the **Suspend** button.
14. Ask your lab partner to hyperventilate by taking quick, deep breaths for 2 minutes while remaining lying down with closed eyes. Immediately after your partner has hyperventilated, click on the **Resume** button and record for 10 seconds while he or she is recovering. Then click on the **Suspend** button.
15. Have your lab partner remain lying down but with open eyes. Click on the **Resume** button and record for 10 seconds. Then click on the **Suspend** button and then **Done**.
16. Your lab partner may now remove the electrodes.

Data Analysis

1. Enter the **Review Saved Data** mode and select the appropriate data file for Lesson 3, which ends in-L03.
2. Hide Channel 1 (raw EEG) by clicking on the Channel 1 box while simultaneously holding down the **Ctrl** key (PC) or **Option** key (Mac).
3. The activity level, or amplitude, of alpha and beta waves can be determined by measuring the standard deviation (stddev) of the data points. Channel 2 displays alpha wave activity and Channel 3 displays beta wave activity. To compare the amplitudes, set the channel measurement boxes as follows:

   Ch 2 stddev
   Ch 3 stddev
   Ch 4 stddev
   Ch 5 stddev
4. Using the I-beam cursor tool, select the region corresponding to the first "eyes closed" segment (from time 0 to the first marker) as shown in Figure 14.11. Record the standard deviations for the alpha and beta waveforms in Table 14.2.
5. Repeat step 4 for the region corresponding to the "eyes open" segment (from the first event marker to the second) and the region corresponding to the "eyes reclosed" segment (from the second event marker to the end). Record the standard deviations for all four waveforms in Table 14.2.

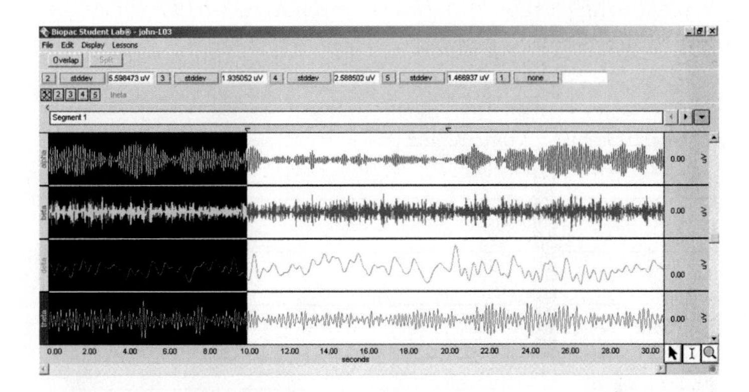

Figure 14.11 Measuring the amplitude of EEG waves. The selected area (from 0 to first event marker) represents the data gathered when the eyes were closed. The illustration shows a data window with Channels 2 (alpha waves), 3 (beta waves), 4 (delta waves), and 5 (theta waves) displayed.

Table 14.2 EEG Amplitude Measurements (stddev. uV) from Lesson 3

Rhythm	Channel	Eyes closed	Eyes open	Eyes reclosed
Alpha	Ch 2			
Beta	Ch 3			
Delta	Ch 4			
Theta	Ch 5			

6. To compare the frequency of the four EEG waveforms, change the measurement boxes as follows:

   Ch 2 Freq

   Ch 3 none

7. Using the magnifying glass tool, zoom in on the waveforms so that you can clearly identify individual cycles. Then, using the I-beam cursor tool, select a single peak-to-peak cycle (Figure 14.12) from the alpha wave. Record the frequency in Table 14.3. Take two additional single-cycle measurements from the alpha wave and calculate the mean frequency.
8. Repeat step 7 for three single cycles from each of the beta, delta, and theta waves. Record the data in Table 14.3.

QUESTIONS TO CONSIDER 1. What changes do you observe in the amplitude of the alpha and beta waves when the eyes are closed and open (Table 14.2)?

____

____

2. How do the frequencies of the four waves you measured compare to the normal ranges presented in Table 14.3?

____

____

9. Enter the **Review Saved Data** mode and select the appropriate data file for Lesson 4, which ends in-L04.
10. Set the channel measurement box to **Ch 41** and **mean.** Channel 41, labeled **alpha RMS** (root-mean-squared), is an indicator of the magnitude of alpha activity.
11. Using the I-beam cursor tool, select the region corresponding to the first "eyes closed, resting" segment (from time 0 to the eyes-open marker) as shown in Figure 14.13. Record the mean in Table 14.4.
12. Repeat step 11 for the region corresponding to the "mental math" segment (from the second to third markers), the region corresponding to the "after hyperventilation" segment (from the third to fourth markers), and the region corresponding to the "eyes open" segment (from the fourth marker to the end). Record the means in Table 14.4.
13. Calculate the change in alpha activity that occurs from rest (eyes closed; segment 1) to each experimental condition (segments 2–4). To do so, subtract the value in row 1 of Table 14.4 from each of rows 2 through 4. Record in Table 14.5. Designate either an increase in activity (+), a decrease (−), or no change (=).

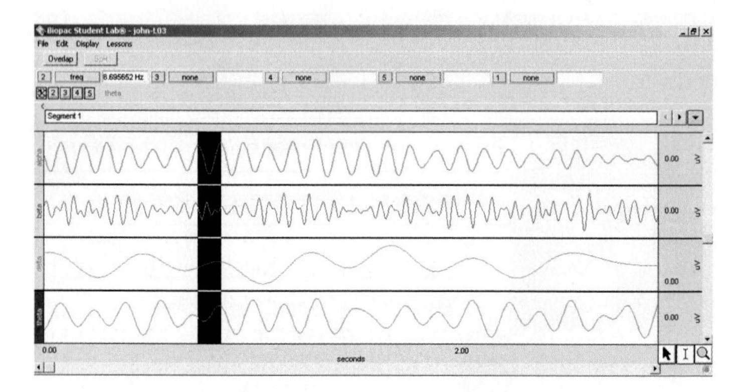

Figure 14.12 Measuring the frequency of EEG waves. The illustration shows EEG waves during a portion of the first 10-second interval when the eyes were closed. The selected area (from peak to peak) represents a single cycle of the alpha wave.

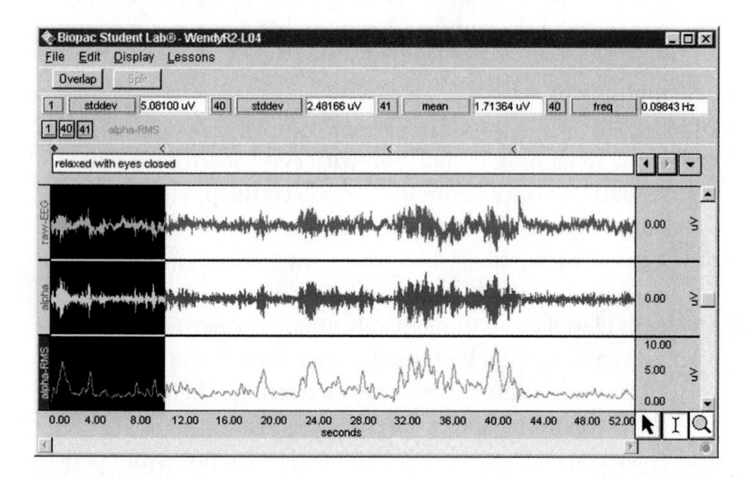

Figure 14.13 Measuring the amplitude of alpha rms waves. The selected area (from 0 to first eyes open marker) represents the data gathered when the eyes were closed.

Table 14.3 EEG Frequency Measurements (Freq. Hz) from Lesson 3

Rhythm	Channel	Cycle 1	Cycle 2	Cycle 3	Mean	Normal
Alpha	Ch 2					8–13
Beta	Ch 3					13–30
Delta	Ch 4					1–5
Theta	Ch 5					4–8

Table 14.4 Mean EEG Amplitude Measurements (uV) from Lesson 4

Segment	Condition	Mean [41-Alpha rms]
1	Eyes closed	
2	Mental math	
3	After hyperventilation	
4	Eyes open	

Table 14.5 Change in Mean EEG Amplitude from Resting, Eyes Closed

Segment	Experimental condition	Difference (experimental–"eyes closed")	Summary (+, −, =)
2	Mental math		
3	After hyperventilation		
4	Eyes open		

QUESTIONS TO CONSIDER

1. Which conditions produced the highest and lowest alpha activity?

2. Hyperventilation causes pH levels of the blood and cerebrospinal fluid to rise. These changes affect the electrical activity of the brain. How does recovery from hyperventilation affect alpha activity?

3. How does the level of concentration by the subject affect the alpha activity?

4. If the subject found the mental math too challenging and gave up on finishing the calculation, how might this be indicated in the alpha activity?

Cranial Nerve Origins

There are 12 pairs of cranial nerves; the first 2 pairs originate from the forebrain and the other 10 pairs arise from the brainstem (Figure 14.5). They exit various foramina in the skull (Figure 14.14) to reach their destinations, and have special names that describe their general distribution or functions. They are also identified by Roman numerals in the order of their cranial origins, from anterior to posterior (Figure 14.5). The functions of the 12 cranial nerves are described in Table 14.6.

ACTIVITY 14.7 Identifying the Origins of the Cranial Nerves

Examine the inferior surface of the brain (Figure 14.5) on an anatomical model or diagram and note the following structural features of the cranial nerves.

1. The olfactory nerves (cranial nerves I) originate in the olfactory epithelium in the roof of the nasal cavity. The nerve fibers of the olfactory nerves pass through small holes in the cribriform plate of the ethmoid bone (Figure 14.14) and synapse with neurons in the olfactory bulbs on the inferior surface of the frontal lobes (Figure 14.15). From the olfactory bulbs, nerve fibers travel within the olfactory tracts (Figure 14.5) to reach the olfactory areas in the temporal lobes.
2. The optic nerves (cranial nerves II) originate from cells in the retinas of the eyes. From there, each nerve passes through the optic canal (Figure 14.14) to reach the optic chiasm, where some fibers of each nerve cross over to the contralateral side before traveling to and synapsing in the thalamus. The thalamus gives rise to the optic radiations, which project to the visual areas in the occipital lobes (Figure 14.16).
3. Cranial nerves III through XII originate from various regions of the brainstem. They arise in ascending order, from anterior to posterior (Figure 14.5).
   - The **oculomotor nerves (cranial nerves III)** and the **trochlear nerves (cranial nerves IV)** originate from the midbrain.

Figure 14.14 Superior view of the cranial cavity. The openings through which the cranial nerves exit the skull are identified.

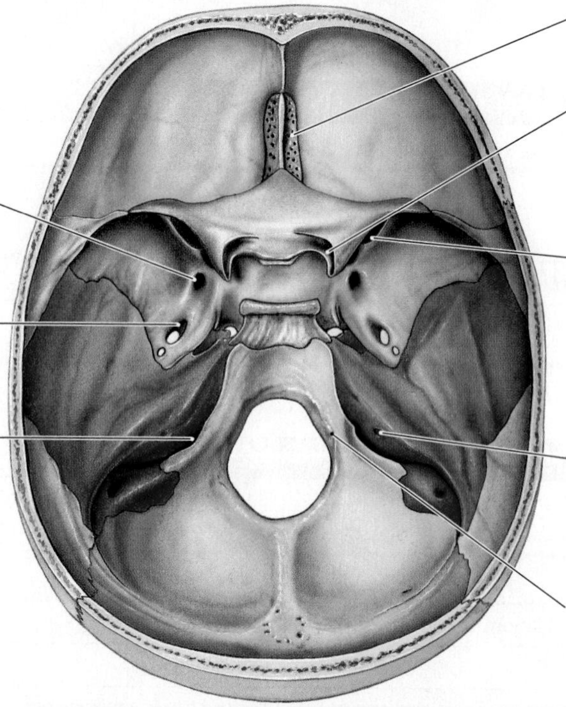

Cribriform plate
Transmits the olfactory (I) nerve

Optic canal
Transmits the optic (II) Nerve

Foramen rotundum
Transmits the Trigeminal (V) nerve, maxillary division

Superior orbital fissure
Transmits the
Oculomotor (III) nerve
Trochlear (IV) nerve
Trigeminal (V) nerve, ophthalmic division
Abducens (VI) nerve

Foramen ovale
Transmits the Trigeminal (V) nerve, mandibular division

Jugular foramen
Transmits the
Glossopharyngeal (IX) nerve
Vagus (X) nerve
Accessory (XI) nerve

Internal acoustic meatus
Transmits the
Facial (VII) nerve
Vestibulocochlear (VIII) Nerve

Hypoglossal canal
Transmits the Hypoglossal (XII) nerve

Table 14.6 Summary of the Cranial Nerves

Nerve	Exiting foramen	Region of origin	Function
I. Olfactory	Cribiform plate	Forebrain	Sensory fibers for olfaction (sense of smell) from the nasal epithelium.
II. Optic	Optic canal	Forebrain	Sensory fibers for vision from the retina.
III. Oculomotor	Superior orbital fissure	Midbrain	Voluntary motor fibers to four extrinsic eye muscles (inferior rectus, medial rectus, superior rectus, inferior oblique); autonomic fibers to ciliary body muscles and pupil constrictors.
IV. Trochlear	Superior orbital fissure	Midbrain	Voluntary motor fibers to one extrinsic eye muscle (superior oblique).
V. Trigeminal			
1. Ophthalmic	Superior orbital fissure	Pons	Sensory fibers from cornea, skin of nose, forehead, anterior scalp.
2. Maxillary	Foramen rotundum	Pons	Sensory fibers from mucous membrane in nasal cavity, upper teeth and gums, palate, upper lip and skin of cheek.
3. Mandibular	Foramen ovale	Pons	Sensory fibers from the tongue, lower teeth and gums, anterior two thirds of tongue, skin of chin and lower jaw; voluntary motor fibers to muscles of mastication.
VI. Abducens	Superior orbital fissure	Pons	Voluntary motor fibers to one extrinsic eye muscle (lateral rectus).
VII. Facial	Enters internal acoustic meatus; exists via stylomastoid foramen	Pons	Voluntary motor fibers to muscles of facial expression; autonomic fibers to lacrimal gland, and sublingual and submandibular salivary glands; sensory (taste) fibers to anterior two thirds of tongue.
VIII. Vestibulocochlear	Internal acoustic meatus	Pons-medulla border	Sensory fibers from inner ear for hearing and equilibrium.
IX. Glossopharyngeal	Jugular foramen	Medulla oblongata	Voluntary motor fibers to one muscle in pharynx; autonomic fibers to parotid salivary gland; general sensory and taste fibers from posterior one third of tongue.
X. Vagus	Jugular foramen	Medulla oblongata	Voluntary motor fibers to muscles of pharynx and larynx; autonomic fibers to thoracic and abdominal organs.
XI. Accessory	Jugular foramen		
1. Cranial portion		Medulla oblongata	Voluntary motor fibers to muscles of larynx and soft palate (with vagus nerve).
2. Spinal portion		Spinal cord (C1–C5)	Voluntary motor fibers to trapezius and sternocleidomastoid.
XII. Hypoglossal	Hypoglossal canal	Medulla oblongata	Voluntary motor fibers to the tongue.

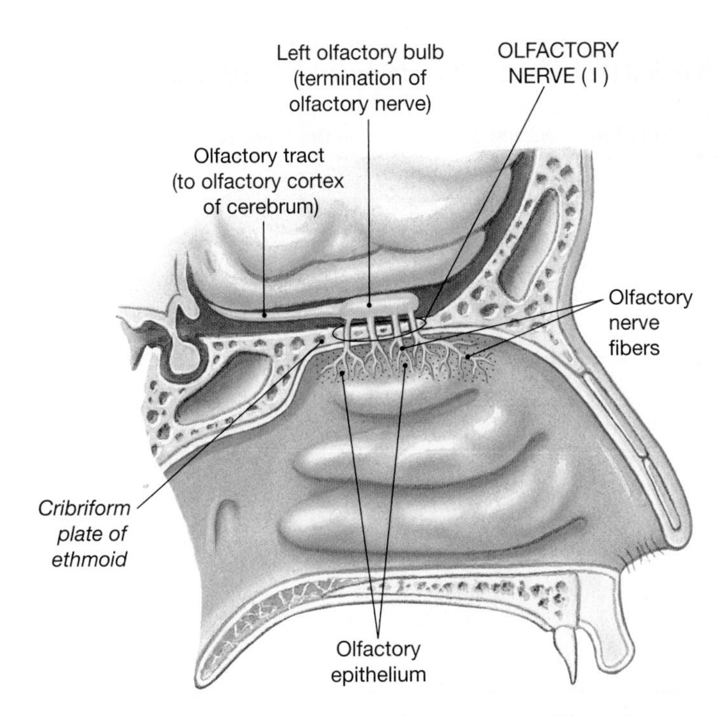

Figure 14.15 Midsagittal section of the nasal cavity illustrating the origin of the olfactory (I) nerve. Branches of the olfactory nerve pass through the cribriform plate to reach the olfactory bulbs.

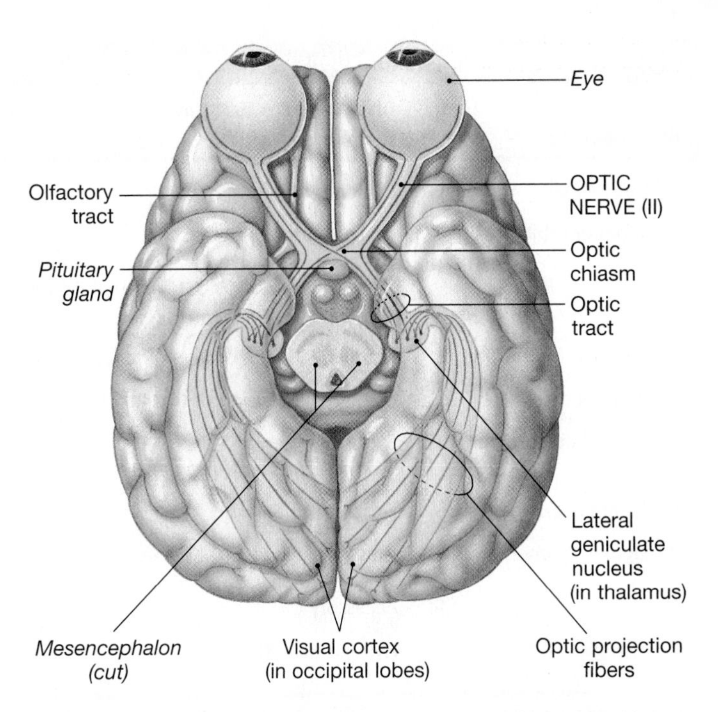

Figure 14.16 Inferior view of the brain, illustrating the optic nerve and visual pathway. The two optic nerves merge at the optic chiasm, where some fibers from each nerve cross over to the opposite side.

- The **trigeminal nerves (cranial nerves V)**, **abducens nerves (cranial nerves VI)**, and the **facial nerves (cranial nerves VII)** originate from the pons.
- The **vestibulocochlear nerves (cranial nerves VIII)** originate from the pons-medulla border.
- The **glossopharyngeal nerves (cranial nerves IX)**, **vagus nerves (cranial nerves X)**, **accessory nerves (cranial nerves XI)**, and **hypoglossal nerves (cranial nerves XII)** originate from the medulla.

QUESTION TO CONSIDER Fractures in the base of the skull can often result in damage to one or more cranial nerves. Reason why the cranial nerves would be vulnerable under these circumstances.

____

____

Cranial Nerve Passageways

To reach the structures that they innervate, the cranial nerves must first leave the cranial cavity by passing through an opening (foramen, fissure, or canal) in the skull. For example, identify the cribriform plate of the ethmoid bone (Figure 14.14) and note the numerous small holes that dot its surface. These openings lead to the roof of the nasal cavity. Fibers of the olfactory nerve travel through these holes to reach the olfactory bulbs. Within the cranial cavity, the olfactory bulbs rest on the cribriform plate.

ACTIVITY 14.8 Identifying the Openings for the Cranial Nerves

1. Obtain a human skull and observe the bony openings from inside the cranial cavity. Identify the openings, listed in Table 14.7 and illustrated in Figure 14.14, that transmit cranial nerves II through XII.
2. Demonstrate the exit of each cranial nerve from the skull by gently pushing a pipe cleaner through the appropriate opening.
3. Observe the location of the pipe cleaner as it emerges from the other side of the opening. Record its position in the space provided in Table 14.7.
4. As you perform this exercise, note the following features.
   - Some openings in the skull transmit several cranial nerves (Figure 14.14). For example, the glossopharyngeal (IX), vagus (X), and accessory (XI) nerves (see Figures 14.21, 14.22, and 14.23 later in the exercise) pass through the jugular foramen (Figure 14.14) to innervate structures in the neck. The vagus nerve also supplies autonomic nerves to thoracic and abdominal organs (see Figure 14.22 later in the exercise).
   - As you might predict, the bony openings lead to regions of the head or neck that contain structures innervated by the cranial nerves passing through them. To verify this concept, correlate the region where your pipe cleaner travels with the structures that are innervated by the cranial nerve in question. These structures are listed in Table 14.6.

Table 14.7 Identifying the destinations of the Cranial Nerves

Bony opening	Cranial nerves that travel through the opening	Position of pipe cleaner as it exits from the other side of the opening
1. Optic canal	Optic (II) nerve	____________
2. Superior orbital fissure	Oculomotor (III) nerve Trochlear (IV) nerve Ophthalmic division of trigeminal (V) nerve Abducens (VI) nerve	____________
3. Foramen rotundum	Maxillary division of trigeminal (V) nerve	____________
4. Foramen ovale	Mandibular division of trigeminal (V) nerve	____________
5. Internal acoustic meatus	Facial (VII) nerve Vestibulocochlear (VIII) nerve	____________
6. Stylomastoid foramen	Facial (VII) nerve	____________
7. Jugular foramen	Glossopharyngeal (IX) nerve Vagus (X) nerve Accessory (XI) nerve	____________
8. Hypoglossal canal	Hypoglossal (XII) nerve	____________

- The facial (VII) nerve has a rather complex pathway. It enters the **internal acoustic meatus** with the vestibulocochlear (VIII) nerve (Figure 14.14). From this passageway, branches of the facial nerve travel through canals in the temporal bone (these canals are deep and cannot be identified) to reach the lacrimal (tear) gland, the mucous membrane in the nasal cavity, and salivary glands in the oral cavity. Another branch exits the skull by passing through the **stylomastoid foramen** (located between the styloid process and mastoid process on the inferior surface of the skull) to innervate the muscles of facial expression (see Figure 14.19 later in the exercise). The stylomastoid foramen is too small to insert a pipe cleaner. However, if you position a pipe cleaner at the foramen and bend it anteriorly, you will observe that the pipe cleaner is directed to the facial area of the skull.

QUESTION TO CONSIDER The superior orbital fissure, the internal acoustic meatus, and the jugular foramen all transmit two or more cranial nerves. Explain why.

____

____

Cranial Nerve Function

The normal function of the cranial nerves can be evaluated by performing a simple battery of tests. Working with your laboratory partner, carry out the following analysis of the selected cranial nerves and record your results in Table 14.8. For many of these tests, only a portion of the nerve's function will be evaluated.

ACTIVITY 14.9 Testing the Functions of Cranial Nerves

1. *Olfactory (I) nerve.* The olfactory nerve (Figure 14.15) provides us with the ability to smell substances. Have your lab partner close his or her eyes and one nostril. Bring samples of coffee, pepper, and other spices close to the open nostril, one at a time, and ask your partner to identify them. Perform the same test with the other nostril, but pass the substances in a different order. Correct identification of each substance indicates normal functioning of the olfactory nerve.
2. The optic nerve (Figure 14.16) begins in the retina of the eyes and is responsible for vision. Your ability to read the printed text in this manual indicates normal function of the optic nerve.
3. *Oculomotor (III), trochlear (IV), and abducens (VI) nerves.* These three nerves are motor nerves and innervate the six extrinsic muscles that perform voluntary eye movements (Figure 14.17). The muscles innervated by each nerve are as follows:
   - *Oculomotor:* inferior oblique, superior rectus, inferior rectus, medial rectus
   - *Trochlear:* superior oblique
   - *Abducens:* lateral rectus

   Have your lab partner sit comfortably on a laboratory seat, with his or her head facing forward and one eye covered. Place your finger about 30 cm (1 foot) in front of your partner's open eye, and move it in all directions. With the head still, your partner should be able to follow your finger with normal eye movements. Perform the same test with the other eye.

Table 14.8 Functional Assessment of the Cranial Nerves

Cranial nerve	Test performed	Results	Assessment of nerve function
Olfactory (I)			
Oculomotor (III)			
Trochlear (IV)			
Abducens (VI)			
Oculomotor (III) (autonomic function)			
Trigeminal (V)			
Facial (VII)			
Vestibulocochlear (VIII)			
Glossopharyngeal (IX)			
Accessory (XI)			
Hypoglossal (XII)			

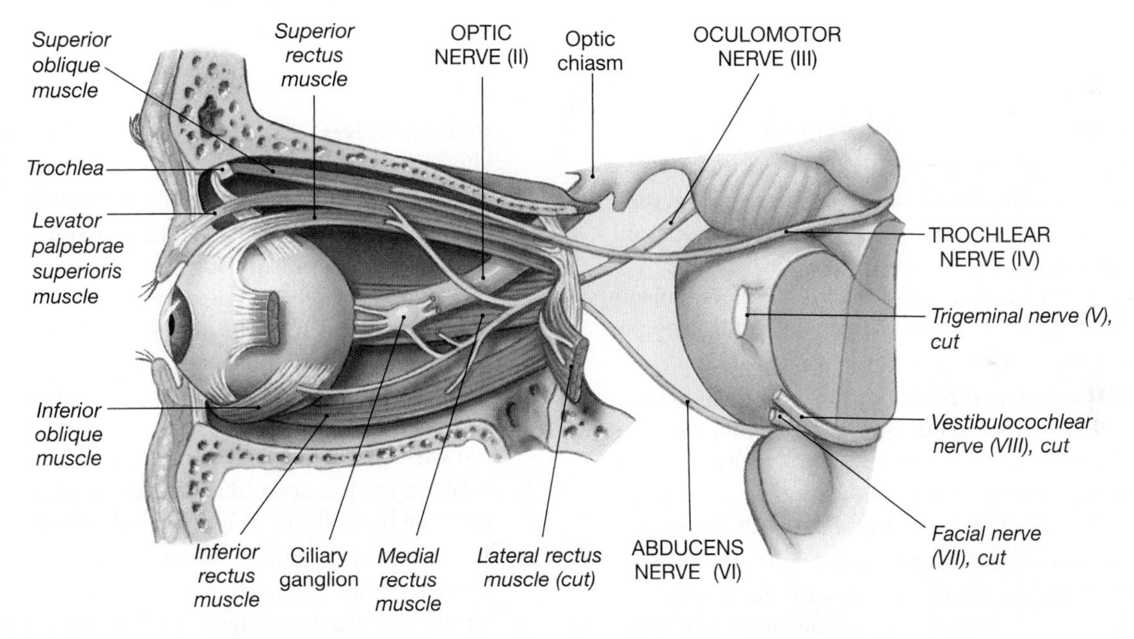

Figure 14.17 The cranial nerves that innervate the extrinsic eye muscles. The oculomotor (III) nerve controls four muscles. The trochlear (IV) and abducens (VI) nerves each control one muscle.

The oculomotor (III) nerve also contains parasympathetic nerve fibers that cause the constrictor muscles in the iris to contract. When this occurs, the diameter of the pupil decreases and the amount of light entering the eye is reduced. To test this function, shine a flashlight at an angle (not directly) into your lab partner's eyes and observe the constriction of the pupils. When the light is removed, the pupils will dilate back to normal.

4. *Trigeminal (V) nerve.* The trigeminal nerve is both motor and sensory. It innervates the muscles of mastication (chewing) and transmits general sensory information (touch, pressure, pain, temperature) from the face to the brain (Figure 14.18).
   - To test for normal motor function, place your hand under your partner's mandible and ask your partner to open the mouth against your resistance. The ability to depress the mandible indicates normal motor function of the muscles of mastication.
   - To test for sensory function, have your partner close his or her eyes. Take your finger and gently press on various areas of your partner's face. As you apply gentle pressure on the mandibular, maxillary, and ophthalmic regions,

Figure 14.18 The three divisions of the trigeminal (V) nerve. The trigeminal nerve is entirely sensory except for motor branches to the muscle of mastication.

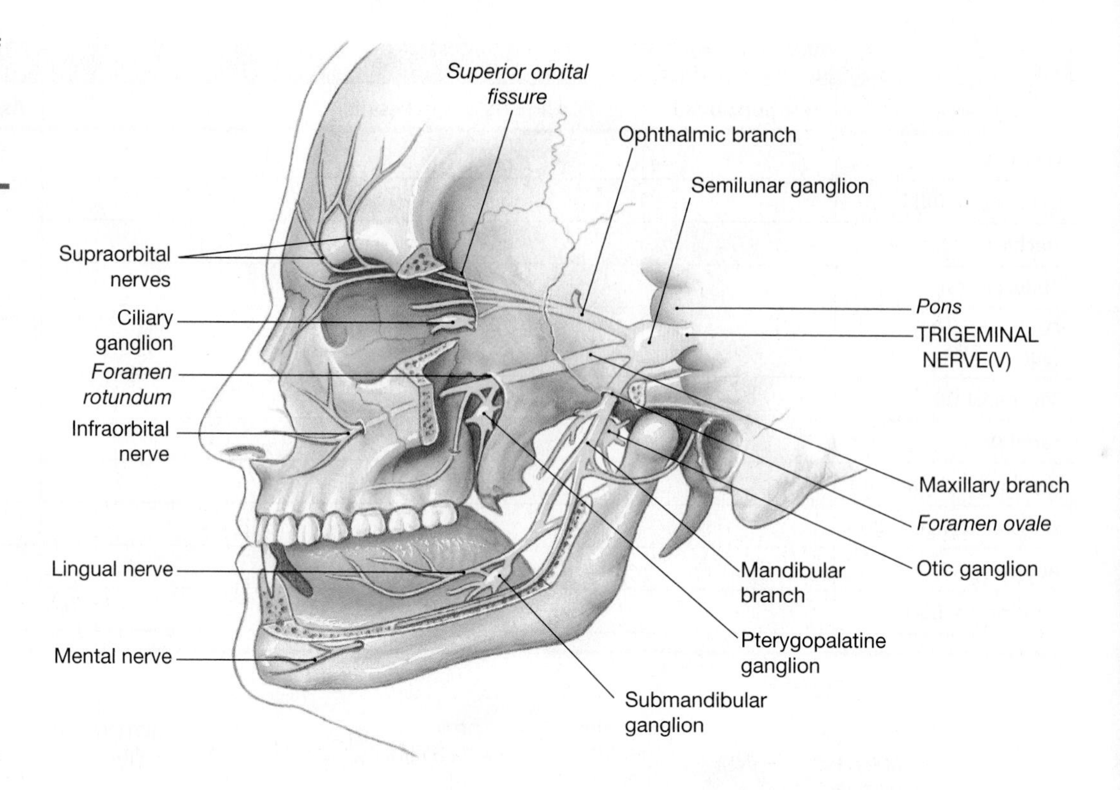

your partner should be able to detect from where the pressure sensation is originating.

5. *Facial (VII) nerve.* The facial nerve is both motor and sensory. Its motor fibers control the actions of the muscles of facial expression and the secretions from the lacrimal (tear) gland and salivary glands. Its sensory fibers transmit taste sensations from the anterior two thirds of the tongue to the brain (Figure 14.19).
   - To test for normal motor function, have your lab partner perform various facial expressions (e.g., smiling, puffing the cheeks, raising the eyebrows). Check to see if normal function occurs on both sides of the face. For example, when smiling, are both corners of the mouth elevated?
   - To test for normal sensory function, have your partner close the eyes and open the mouth with the tongue slightly protruded. Dip separate cotton swabs in a 10% table salt solution and a 10% sucrose solution. Gently place the swabs onto the tip of your partner's tongue in any order and ask him or her to identify each taste. If taste sensation is normal, your partner should be able to differentiate between salty and sugary tastes.
6. *Vestibulocochlear (VIII) nerve.* The vestibulocochlear nerve transmits sensory information from receptors in the inner ear to the brain (Figure 14.20). The cochlear division of the nerve is responsible for our ability to hear sounds. As part of our equilibrium (balance) sense, the vestibular division allows individuals to detect linear (along a straight line) and angular (turning) movements. To assess the function of the vestibular division, have your lab partner sit, with eyes closed and ears plugged, on a stool equipped with wheels. Slowly roll the stool, either forward or backward, and ask your partner to identify the direction of movement. Next, slowly move the stool forward for a short distance and then turn to the right or left. Ask your partner to identify the direction of the turn.
7. *Glossopharyngeal (IX) nerve.* The glossopharyngeal nerve is both motor and sensory. It innervates one pharyngeal muscle (stylopharyngeus) that has a role in swallowing. It also supplies sensory neurons for taste to the posterior third of the tongue and general sensory neurons to the pharynx, posterior tongue, and tympanic membrane or eardrum (Figure 14.21). To test the nerve's taste function, dip a cotton swab in a 10% quinine solution, and gently place the swab on the posterior third of the tongue. The ability to taste the bitter flavor of quinine indicates normal function.
8. *Vagus (X) nerve.* Among its many functions, the vagus nerve supplies voluntary motor fibers to muscles of the larynx and pharynx (Figure 14.22) used for talking and swallowing. Your ability to speak normally to your lab partner indicates normal function for the vagus nerve.
9. *Accessory (XI) nerve.* The accessory nerve is motor only and innervates the trapezius and sternocleidomastoid muscles (Figure 14.23).
   - Ask your laboratory partner to shrug his or her shoulders. This action, which requires elevation of the scapulae, is performed by the trapezius muscles.
   - Have your partner turn (rotate) his or her head to the right and look up at the same time. As the action is being performed, feel (by palpation) or observe the contraction of the sternocleidomastoid on the left side.
10. *Hypoglossal (XII) nerve.* The hypoglossal nerve is motor only and innervates the muscles that move the tongue (Figure 14.23). To evaluate the function of this nerve,

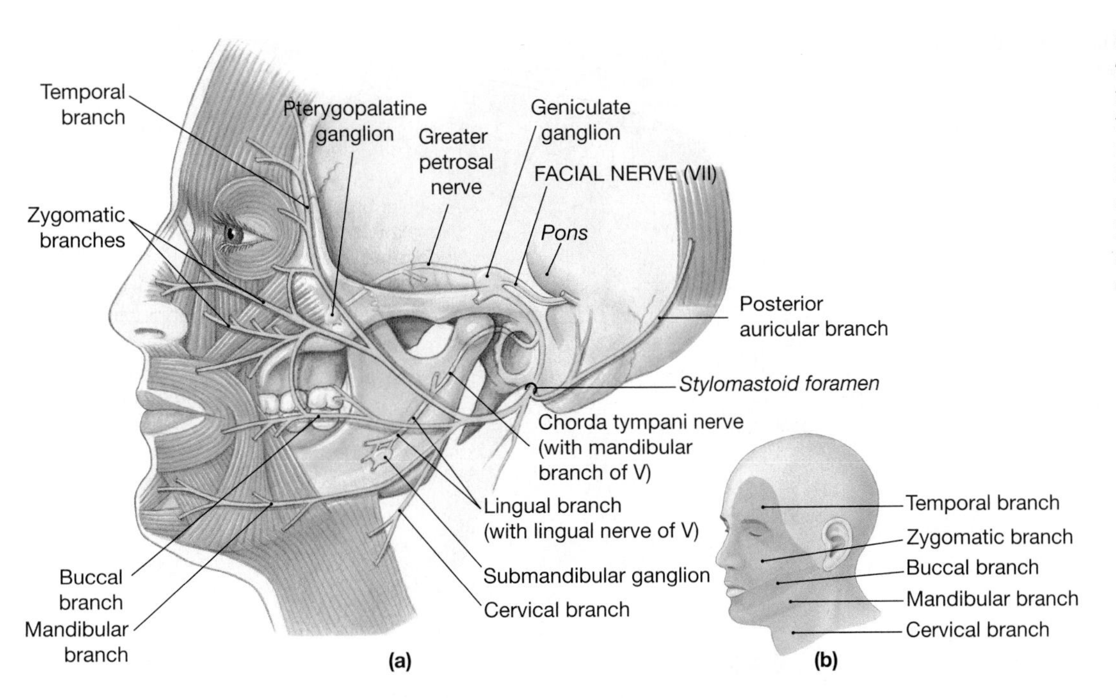

Figure 14.19 The branches of the facial (VII) nerve. a) Motor fibers of the facial nerve innervate all the muscles of facial expression. A major sensory component includes branches to taste receptors on the anterior two thirds of the tongue; **b)** the five major superficial branches of the facial nerve.

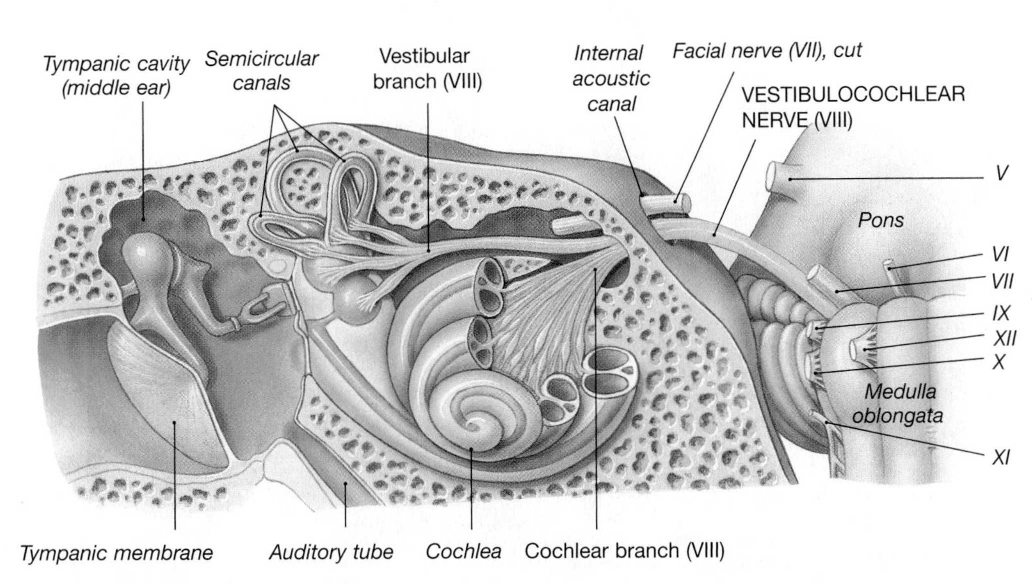

Figure 14.20 The distribution of the vestibulocochlear (VIII) nerve. This nerve is entirely sensory. It supplies sensory receptors in the inner ear for hearing and equilibrium.

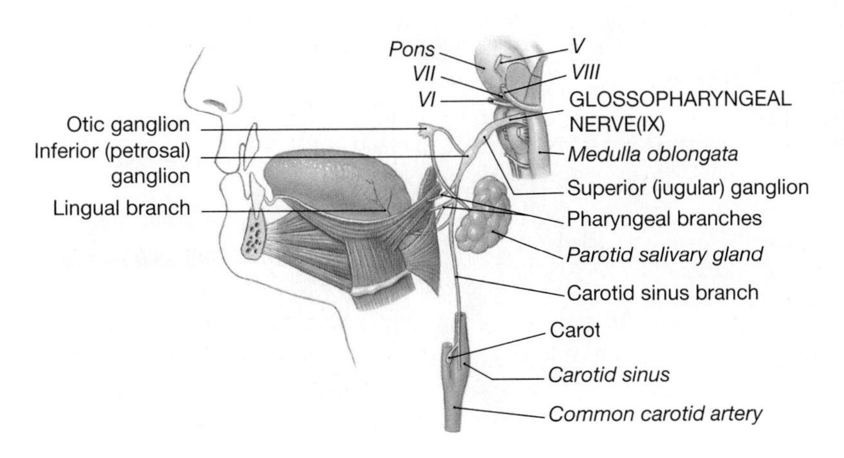

Figure 14.21 The branches of the glossopharyngeal (IX) nerve. This nerve contributes motor fibers to the stylopharyngeus muscle in the pharynx, general sensory fibers to the pharynx, sensory fibers to taste receptors on the posterior one third of the tongue, and autonomic fibers to the parotid salivary gland.

Figure 14.22 The distribution of the vagus (X) nerve. The vagus nerve is the only cranial nerve that supplies structures outside the head and neck.

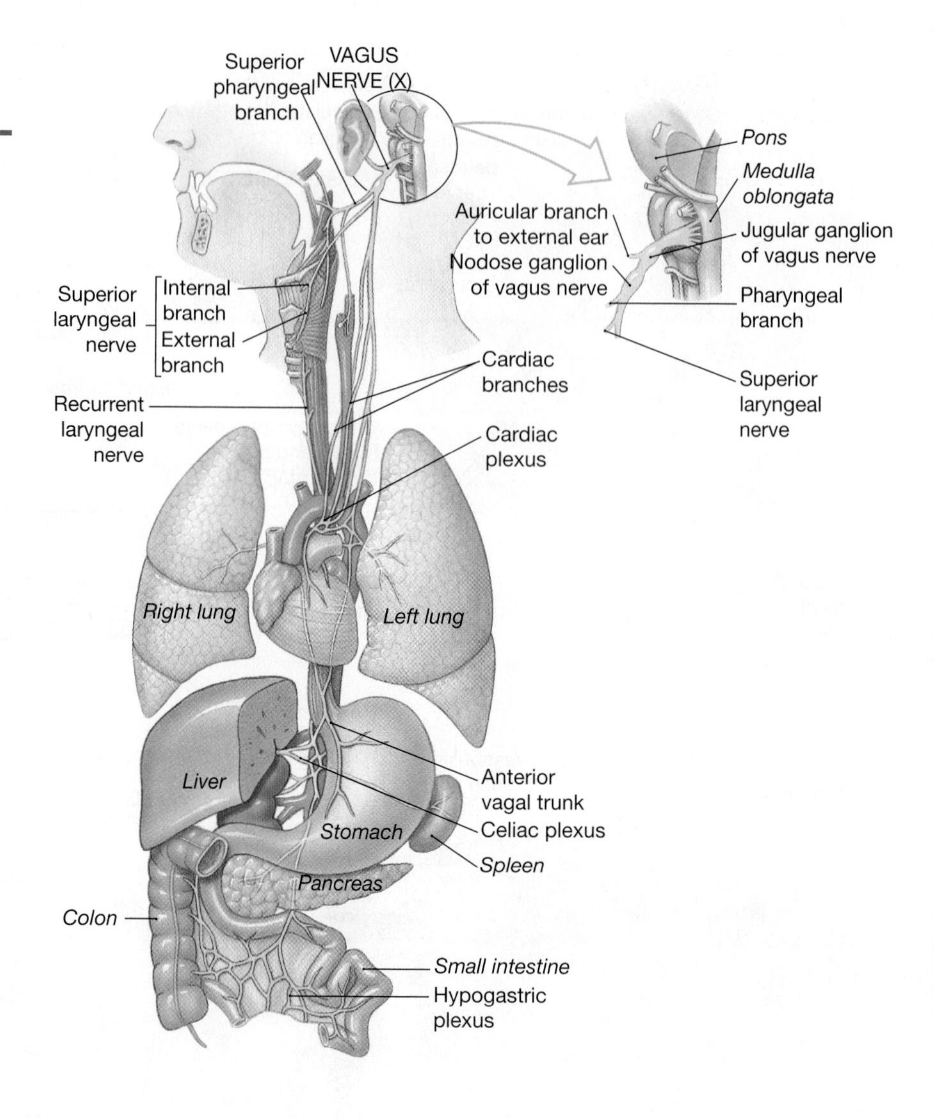

Figure 14.23 The branches of the accessory (XI) and hypoglossal (XII) nerves. The accessory nerve innervates laryngeal muscles (with the vagus nerve), the trapezius, and the sternocleidomastoid. The hypoglossal nerve innervates the muscles that move the tongue.

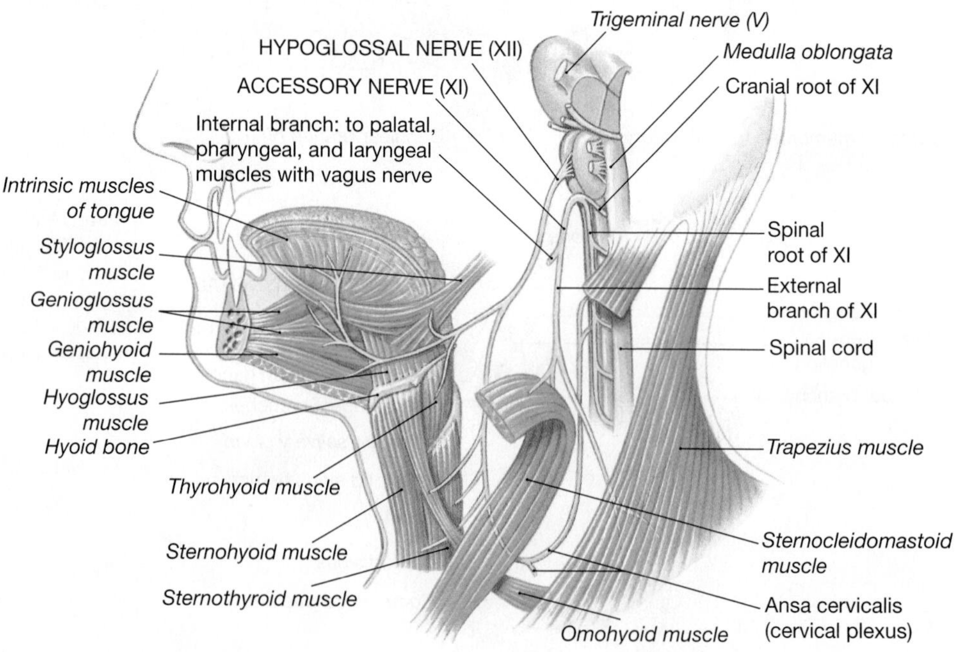

have your laboratory partner protrude his or her tongue. If both hypoglossal nerves are functioning normally, the tongue will move straight out when protruded. If the tongue deviates to one side, it could be caused by damage to the hypoglossal nerve on that side.

QUESTIONS TO CONSIDER

1. Three cranial nerves innervate the extrinsic muscles of the eye. The most severe loss of eye movement will result from damage to which of these cranial nerves? Explain.

____

____

2. Sensory fibers from the glossopharyngeal (IX) nerve supply taste receptors on the tongue. If these sensory fibers are severed, taste sensation will not be completely lost. Explain why.

____

____

3. You have a dental carie (a cavity) in your lower left second molar. Your dentist injects the anesthetic, novocaine, to eliminate any possible pain during treatment. Which cranial nerve is affected by the novocaine? Explain.

____

____

Exercise 14 Review Sheet

The Brain and Cranial Nerves

Name ______________________________

Lab Section ______________________________

Date ______________________________

Questions 1–4: Match the brain region in column A with the appropriate structures in column B. (Use all the structures in column B; some regions in column A will have multiple answers.)

A	B	
1. Cerebrum ______________	a. Thalamus	f. Medulla oblongata
2. Diencephalon ______________	b. Midbrain	g. Basal nuclei
3. Cerebellum ______________	c. Frontal lobe	h. Pons
4. Brainstem ______________	d. Cerebellar cortex	i. Corpus callosum
	e. Hypothalamus	

5. Describe how the three meningeal layers are arranged around the brain. Draw a diagram that illustrates your written answer.

6. Describe the anatomical relationship of gray matter and white matter in the cerebrum.

____

____

Questions 7–12: In the following diagram, identify the structures by labeling with the color that is indicated.

7. Temporal lobe = **green**
8. Frontal lobe = **red**
9. Occipital lobe = **blue**
10. Parietal lobe = **yellow**
11. Brainstem = **brown**
12. Cerebellum = **purple**

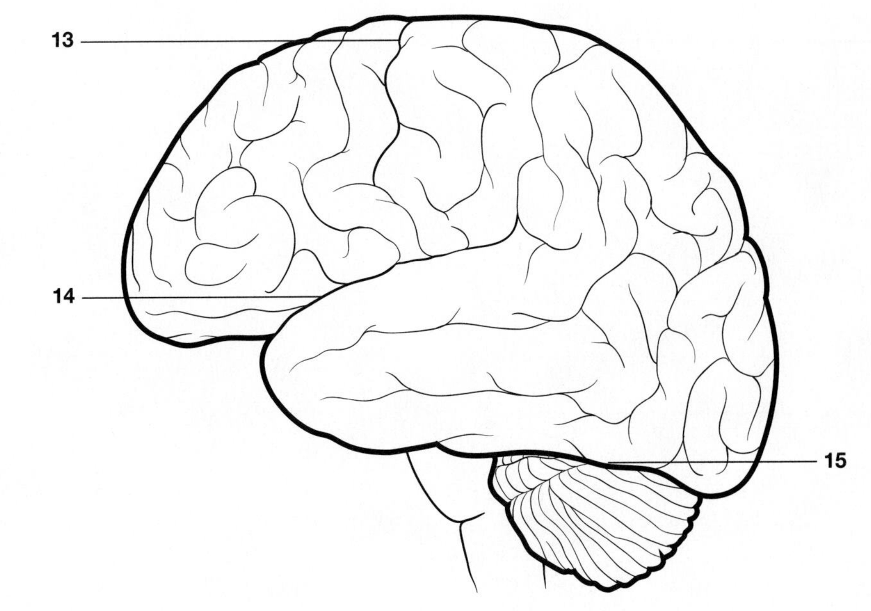

Questions 13–15: In the previous diagram, identify the labeled grooves.

13. ______________________________

14. ______________________________

15. ______________________________

Questions 16–27: In the following diagram, identify the structures by labeling with the color that is indicated.

16. Parietal lobe = **yellow**
17. Frontal lobe = **red**
18. Occipital lobe = **blue**
19. Cerebellum = **green**
20. Thalamus = **orange**
21. Corpus callosum = **brown**
22. Fornix = **black**
23. Hypothalamus = **purple**
24. Pons = **yellow**
25. Pituitary = **blue**
26. Medulla oblongata = **yellow**
27. Midbrain = **brown**

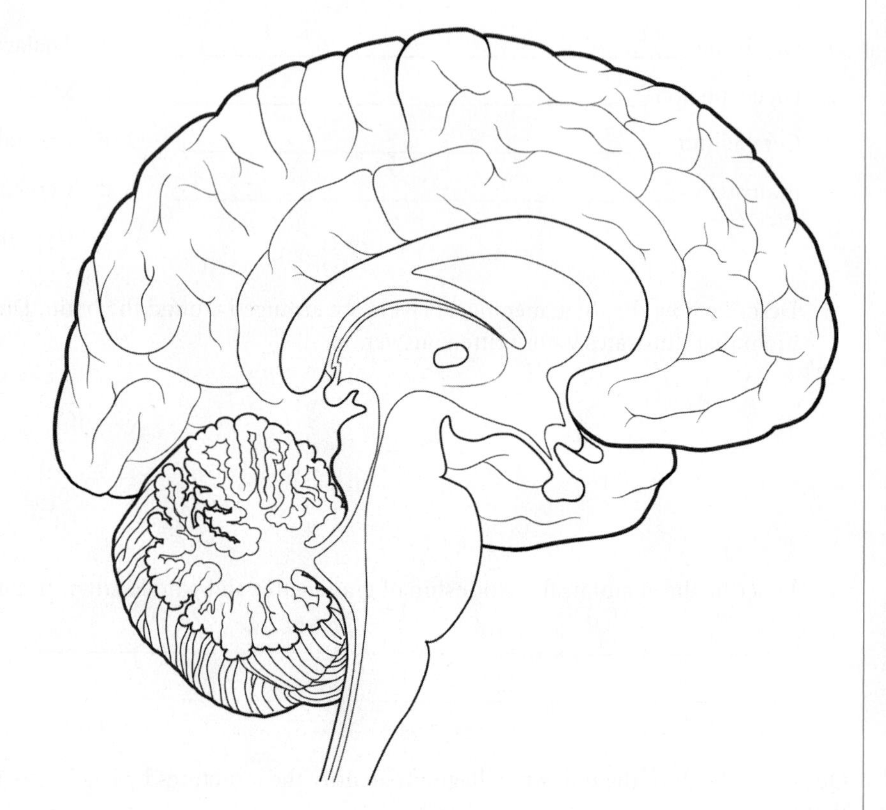

Questions 28–33: In the following diagram, identify the labeled cranial nerves.

28. ______________________

29. ______________________

30. ______________________

31. ______________________

32. ______________________

33. ______________________

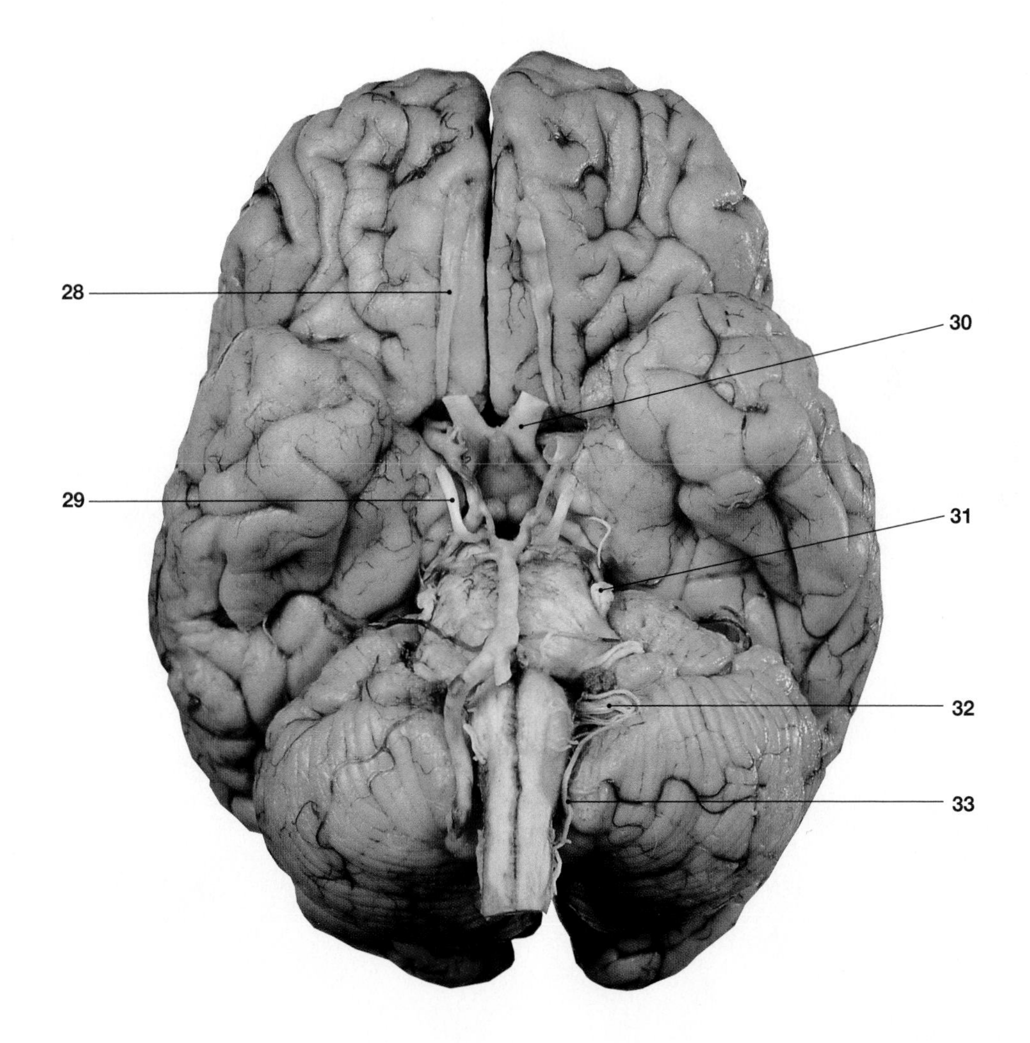

Exercise 15

The Spinal Cord and Spinal Nerves

Laboratory Objectives

On completion of the activities in this exercise, you willl be able to:

- Explain the organization of the spinal meninges.
- Identify the gross anatomical structures of the spinal cord.
- Describe the organization and distribution of spinal nerves.
- Assess the function of spinal nerves using a two-point discrimination test.
- Assess the function of spinal nerves by testing general sensory function in dermatomes.

Materials

- Anatomical model of the entire spinal cord and vertebral column
- Anatomical model of a spinal cord cross section showing its relationship with the spinal meninges and vertebrae
- Diagrams of the spinal cord and spinal nerves from anatomical atlases
- Colored pencils
- Measuring compasses with a millimeter scale
- Tuning forks

The spinal cord begins at the foramen magnum of the skull, where it is continuous with the medulla oblongata of the hindbrain. It ends at or near the intervertebral disc between the L1 and L2 vertebrae, where it forms a cone-shaped structure known as the **conus medullaris** (Figure 15.1a). Throughout its course, the spinal cord is enclosed and protected by the **vertebral canal**, a bony tunnel formed by the **vertebral foramina** of successive vertebrae (Figure 15.2).

On average, the spinal cord is approximately 2 cm (0.75 in) in diameter. However, the diameter is greater in the cervical and lumbar regions, where spinal nerves to the upper and lower extremities originate. These expanded areas of the spinal cord are called the **cervical** and **lumbar enlargements**.

The spinal cord is divided into 31 segments: 8 cervical, 12 thoracic, 5 lumbar, 5 sacral, and 1 coccygeal. Each segment gives rise to a pair of **spinal nerves** (Figure 15.1a) that pass through openings between successive vertebrae, called **intervertebral foramina**, to reach peripheral body regions (Figure 15.2).

The spinal nerves are identified according to the spinal cord segment from which they originate. They are all classified as **mixed nerves** because they contain both afferent (sensory) and efferent (motor) nerve fibers. Each spinal nerve is formed by the union of a dorsal root and a ventral root (Figures 15.1b and 15.2). The ventral root conveys motor fibers from the spinal cord to the nerve; the dorsal root transmits sensory fibers from the nerve to the spinal cord.

The Spinal Meninges

The spinal cord is protected by the same three meningeal layers as the brain: the **dura mater**, **arachnoid**, and **pia mater**. These layers are continuous with the cranial meninges at the foramen magnum of the skull. Blood vessels within the spinal meninges give off branches that deliver blood to the spinal cord. Like the brain, the spinal cord is surrounded by a shock-absorbing fluid. The **subarachnoid space**, between the arachnoid and pia mater, is filled with cerebrospinal fluid.

ACTIVITY 15.1 Identifying Spinal Meninges and Associated Structures

1. Obtain an anatomical model or a diagram from an anatomy atlas that illustrates a cross section of the spinal cord, with the surrounding spinal meninges (Figure 15.2b).
2. Note that the basic organization of the dura mater, arachnoid, and pia mater, including the presence of a fluid-filled subarachnoid space around the spinal cord, is similar to that of the brain.
3. Note the following special structural features of the spinal meninges that differ from the cranial meninges.
   - Around the spinal cord, only one dural layer exists. It is a continuation of the inner dural layer around the brain.
   - The spinal dura mater is not attached to the internal surfaces of covering bones as it is around the brain. Instead, a true space, called the **epidural space**, exists between the dura mater and the vertebrae (Figure 15.2b). This space is filled with fat tissue, blood vessels, and nerves; the fat tissue provides a protective padding for the spinal cord.

CLINICAL CORRELATION

Anesthetics can be injected into the epidural space, resulting in a temporary loss of function (both sensory and motor) to the spinal nerves at and adjacent to the area of injection. This procedure, known as an **epidural block,** is difficult to do in most areas of the spinal cord because the epidural space is quite narrow. To reduce the pain of labor and childbirth, an epidural block can be performed, with relatively little risk, by injecting anesthetics into the epidural space between the L3 and L4 vertebrae.

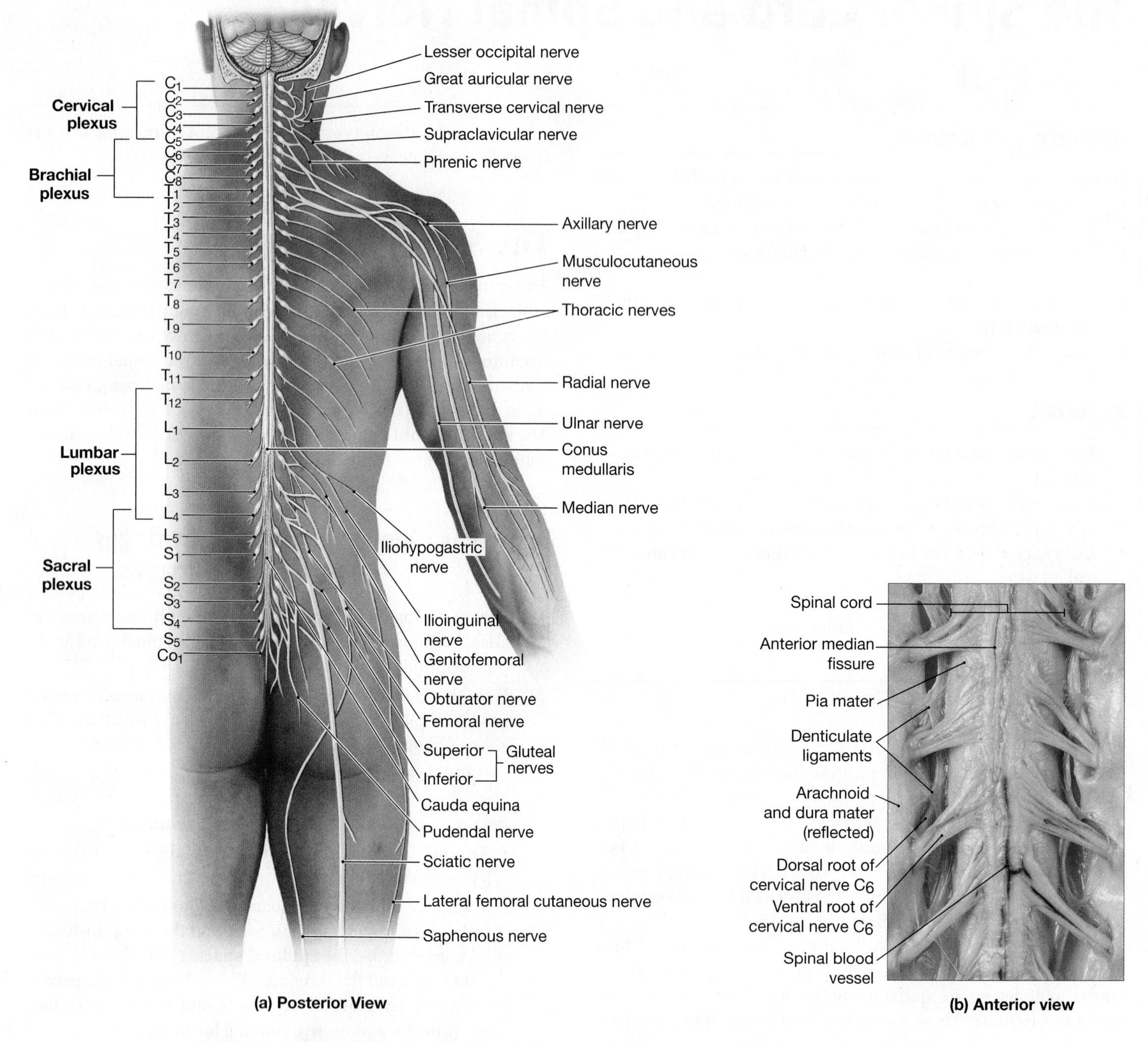

Figure 15.1 The adult spinal cord with spinal nerves. a) Posterior view of the entire spinal cord, illustrating the segmental pairs of spinal nerves and the formation of spinal nerve plexuses; **b)** anterior view of the cervical spinal cord, with the dorsal and ventral roots of cervical spinal nerves.

- The spinal cord ends between vertebrae L1 and L2. However, the dura mater and arachnoid extend beyond this termination to between vertebrae S2 and S3. Thus, the subarachnoid space ends as a blind sac filled with cerebrospinal fluid (Figure 15.3).
- The spinal pia mater has two special features that serve to secure the spinal cord in its position. The **filum terminale** is a thin strand of pia mater that extends inferiorly from the termination of the spinal cord (conus medullaris) to the coccyx. The **denticulate ligaments** (Figures 15.1b and 15.2b) are lateral extensions of pia mater that connect the spinal cord along its entire length to the dura mater. These structures serve as anchors to prevent excessive movement of the spinal cord.

WHAT'S IN A WORD The name *filum terminale* is derived from Latin and means "terminal thread." The name aptly describes

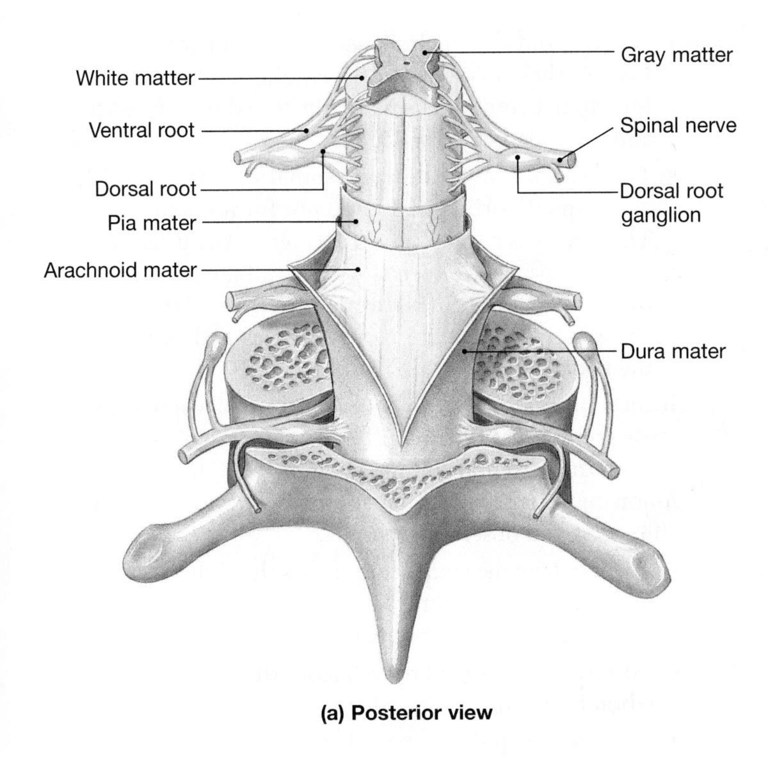

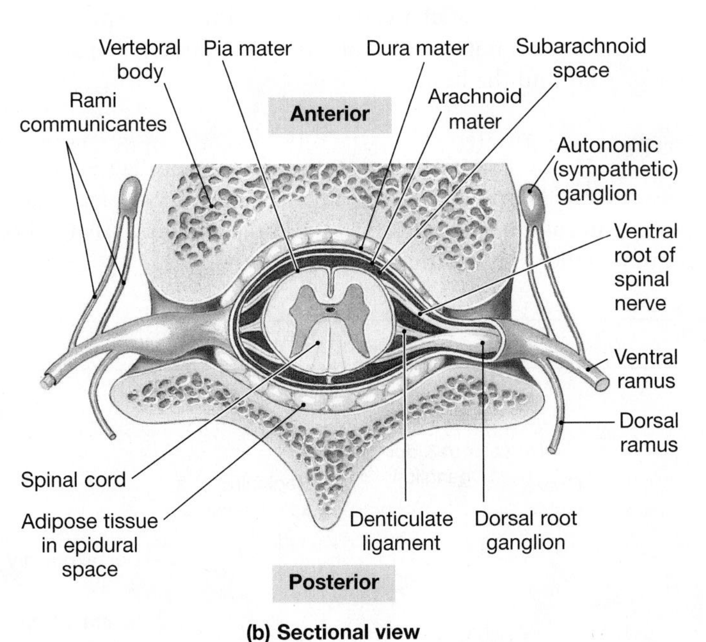

Figure 15.2 A segment of the spinal cord. The relationship of the spinal cord with a vertebra and the spinal meninges is illustrated. The formation of spinal nerves and their exits from intervertebral foramina are also represented. **a)** Posterior view; **b)** cross section.

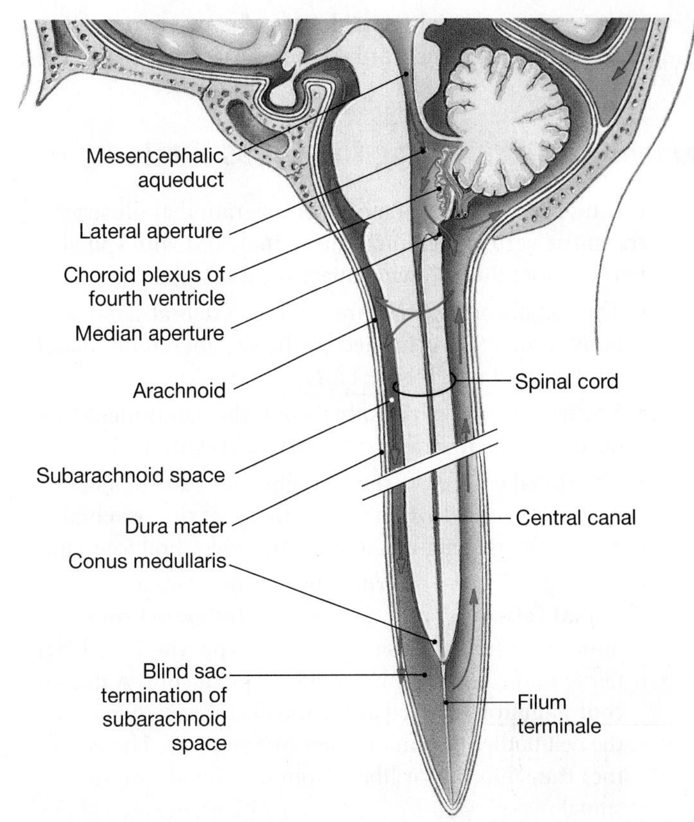

Figure 15.3 Sagittal section of the brainstem, cerebellum, and spinal cord with the surrounding meningeal layers. Although the spinal cord ends between vertebrae L1 and L2, the arachnoid and dura mater extend to between vertebrae S2 and S3. Inferior to the termination of the spinal cord, the subarachnoid space is a blind sac filled with cerebrospinal fluid.

this structure: a thin strand of pia mater that extends from the termination of the spinal cord to the coccyx.

The word *denticulate* is derived from the Latin word *dentatus*, which means "toothed." The denticulate ligaments resemble rows of small teeth on each side of the spinal cord. ■

QUESTION TO CONSIDER A **lumbar puncture (spinal tap)** is a procedure during which a needle is inserted between adjacent lumbar vertebrae into the subarachnoid space to collect a sample of cerebrospinal fluid. Explain why a lumbar puncture is done inferior to the second lumbar vertebra.

____

____

The Spinal Cord and Spinal Nerves

The spinal cord serves as an information pathway that connects the brain and the spinal nerves and as an integration center for spinal reflexes (reflexes will be studied in Exercise 16). As the spinal cord descends from its origin at the base of the skull, it is protected by the vertebral column. However, the length of the spinal cord is much shorter than the length of the vertebral column, terminating between vertebrae L1 and L2. This length discrepancy affects the exiting pathways of spinal nerve roots as they emerge from the spinal cord segments to form spinal nerves.

ACTIVITY 15.2 Examining the Gross Anatomy

1. Examine an anatomical model or diagram that illustrates the entire vertebral column and spinal cord with spinal nerves. Note the following anatomical relationships.
   - The spinal cord passes through the vertebral canal, a bony tunnel that is formed by the vertebral foramina of all the vertebrae (Figure 15.2).
   - The termination of the spinal cord, the conus medullaris, occurs between vertebrae L1 and L2 (Figure 15.1a).
   - The spinal cord gives rise to segmental pairs of spinal nerves (Figures 15.1 and 15.2) that exit the vertebral column by passing through the intervertebral foramina.
   - Each spinal nerve is formed by two nerve roots: the **dorsal (afferent) root** and **ventral (efferent) root.** The dorsal root transmits sensory fibers from the spinal nerve to the posterior horn of the spinal cord. A **dorsal root ganglion**, located along the dorsal root, contains the cell bodies of unipolar sensory neurons. The ventral root transmits motor fibers from the spinal cord to a spinal nerve.
   - During development, the spinal cord grows at a slower rate than the vertebral column and ends between the first and second lumbar vertebrae. Cervical and thoracic spinal nerve roots pass more or less laterally as they form spinal nerves that exit the vertebral column. Lumbar, sacral, and coccygeal spinal nerve roots, however, must travel inferiorly within the vertebral canal to reach their respective intervertebral foramina. These inferior spinal nerve roots are called the **cauda equina** (Figure 15.1a).

WHAT'S IN A WORD The phrase "cauda equina" is Latin for "the tail of a horse." The lumbar, sacral, and coccygeal spinal nerve roots are given this name because they resemble a horse's tail as they travel inferiorly to form spinal nerves.

   - Soon after spinal nerves emerge from the intervertebral foramina, they divide into several branches or rami (Figure 15.4). The **dorsal primary rami** supply the deep back muscles and the skin on the dorsal surface of the trunk. The **ventral primary rami** supply all structures of the extremities, the skin along the lateral and ventral surfaces of the trunk, and all skeletal muscles except the deep back muscles. The **meningeal rami** pass back through the intervertebral foramina to supply the vertebrae and meninges. The **rami communicantes** are branches along thoracic and lumbar spinal nerves that are associated with the sympathetic division of the autonomic nervous system.
   - Although the spinal nerves arise from the spinal cord in a segmental fashion, the ventral rami of most form complex networks so that the emerging peripheral nerves contain components from several spinal nerve roots. Each of these nerve networks is called a **plexus.** The four spinal cord plexuses are illustrated in Figure 15.5 and described in Table 15.1.
   - The ventral rami of thoracic spinal nerves (T2–T12, and a small portion of T1) do not form plexuses. These nerves supply the structures in the intercostal spaces (spaces between the ribs) and are called **intercostal nerves** (Figure 15.1a). The inferior intercostal nerves also supply the muscles and skin along the abdominal wall.
2. Inspect an anatomical model or diagram of a spinal cord cross section that illustrates the relationship of the spinal cord to the vertebral column (Figure 15.2). Identify the following features and compare them with the relationships that you observed in step 1.
   - Observe how the spinal cord passes through the vertebral foramen of the vertebra. Note that the vertebral foramina, in series, form the vertebral canal, described in step 1.
   - Confirm that a spinal nerve passes through an intervertebral foramen.
   - Identify the epidural space between the dura mater and the bone of a vertebra. Recall that the epidural space is not present (or exists as a potential space) around the brain.

QUESTION TO CONSIDER Examine an anatomical model or diagram of a spinal cord that illustrates the anatomical relationships of the epidural space. From the anatomy, what risk(s) can you identify in performing an epidural block?

____

____

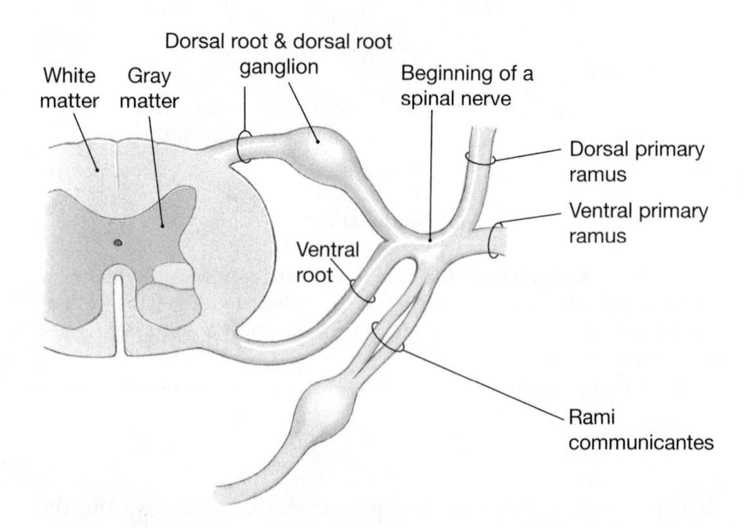

Figure 15.4 Cross section of the spinal cord and a spinal nerve. The spinal nerve is formed by the dorsal and ventral roots and gives off several branches (rami).

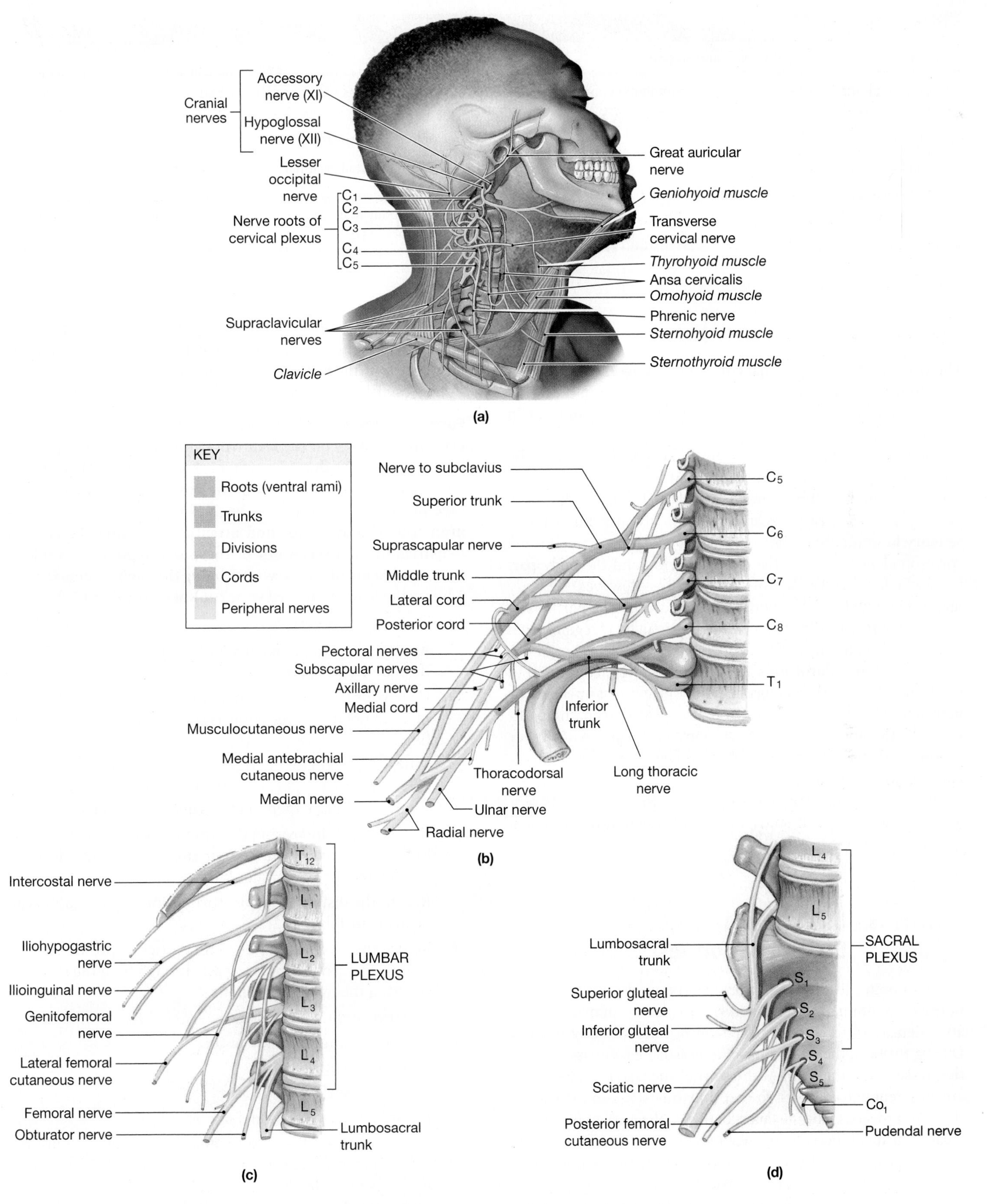

Figure 15.5 The organization of the four spinal nerve plexuses. Each plexus gives rise to peripheral nerves that contain nerve fibers originating from several spinal nerves: **a)** Cervical plexus; **b)** brachial plexus; **c)** lumbar plexus; **d)** sacral plexus.

Table 15.1 Spinal Nerve Plexuses

Plexus	Spinal nerve contributions	Distribution
1. Cervical plexus	C1–C4; with a small contribution from C5	Portions of skin and muscles of the head, neck, and shoulders; the phrenic nerve supplies the diaphragm.
2. Brachial plexus	C5–C8; T1	All structures of the upper extremities.
3. Lumbar plexus	L1–L4	Anterolateral abdominal wall, external genitalia, and part of the lower extremity
4. Sacral plexus	L4–L5; S1–S4	The buttocks, perineum, and part of lower extremity; includes the sciatic nerve, the largest nerve in the body that supplies all the muscles in the posterior thigh, leg, and foot.

Spinal Nerve Function

The white matter in the spinal cord contains **ascending** and **descending pathways**. These pathways serve as communication links between spinal nerves and the sensory and motor areas in the brain.

General sensations reach the **primary sensory area** in the postcentral gyrus (located in the parietal lobe of the cerebrum) via the ascending pathways in the spinal cord (Figure 15.6). The primary sensory area is organized as a body map known as a **sensory homunculus**. The homunculus illustrates a one-to-one correspondence between the cortical region and the body part from which it receives sensory information. However, the body map is constructed disproportionately because large portions of the sensory area are devoted to relatively small body regions such as the hands and fingers (Figure 15.6).

The **primary motor area** in the precentral gyrus (located in the frontal lobe of the cerebrum) is the origin of **upper motor neurons**, whose nerve fibers travel along the descending pathways. In the anterior horn these neurons synapse with **lower motor neurons**, whose nerve fibers travel along spinal nerves and synapse with skeletal muscles.

In the following activities, we focus on the spinal cord's sensory function. You will study the motor function of the spinal cord in Exercise 16.

ACTIVITY 15.3 Using a Two-Point Discrimination Test

The two-point discrimination test can be used not only to test normal functioning of spinal nerves, but also to compare the relative density of receptors in the skin of selected body regions. During this test, the two points of a measuring compass are gently pressed onto the skin. At the start of the test, the two points are close enough together so that only one sensation is felt. As the test proceeds, the distance between the two points is increased in small increments until two distinct touch or pressure sensations are felt. The ability to feel two distinct sensations is referred to as two-point discrimination.

The relative densities of skin receptors on various skin regions can be determined by performing a series of two-point discrimination tests. The distance between the compass points is inversely related to the density of receptors. Accordingly, the shorter the distance between the two compass points at which two distinct sensations can be felt, the greater the density of skin receptors.

Form a Hypothesis **You will be performing two-point discrimination tests on the following skin regions: (1) tip of the index finger, (2) dorsal surface of the hand, (3) posterior surface of the neck, (4) side of the nose, and (5) anterior surface of the arm. Before you begin, examine the sensory homunculus (Figure 15.6) and attempt to predict the relative densities of skin receptors in these five regions by ranking them on a scale of 1 to 5, with 1 being the highest density and 5 being the lowest. Record your rankings in Table 15.2, in the column labeled "Predicted."**

1. Arrange the two points of a measuring compass so that they are 1 mm apart.
2. Have your laboratory partner close his/her eyes and gently place the points of the compass on the tip of his/her index finger.
3. With your laboratory partner's eyes remaining closed, increase the distance between the two compass points slightly and place them on the same area of the finger.
4. Repeat step 3, increasing the distance between the points by small increments, until your laboratory partner feels two distinct points on the skin.
5. Record the distance at which two-point discrimination can be made in Table 15.2.
6. Return the compass points to their original positions (1 mm apart) and repeat steps 1 through 5 on the skin, covering the following regions.
   - Dorsal surface of the hand
   - Posterior surface of the neck
   - Side of the nose
   - Anterior surface of the arm
7. Record all your results in Table 15.2.
8. Review your test results and rank the five skin regions in terms of relative density of sensory receptors. Record your rankings in Table 15.2 (in the column labeled "Actual") on a scale of 1 to 5, with 1 being the highest density and 5 being the lowest.

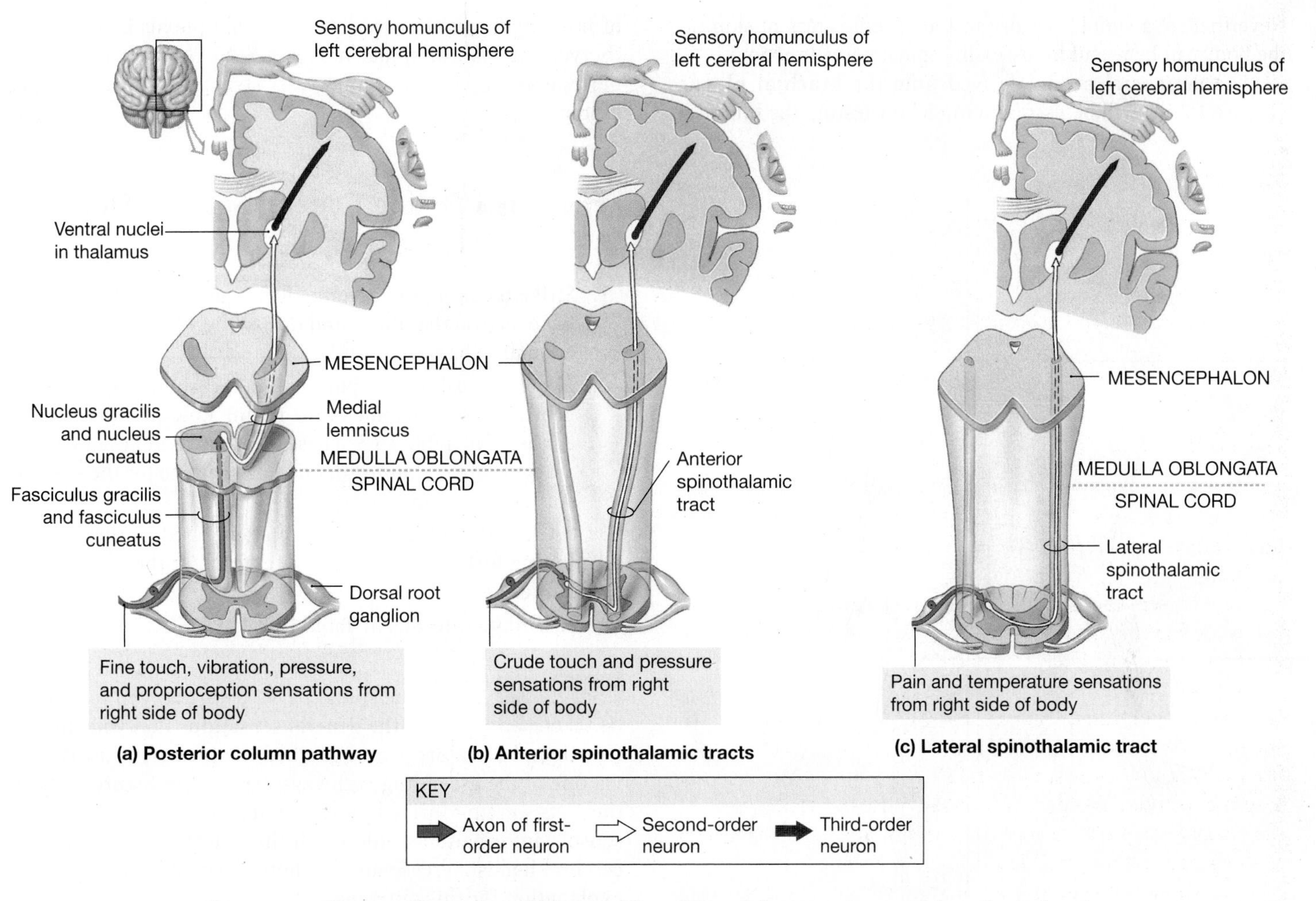

Figure 15.6 The ascending sensory pathways. The pathways deliver general sensory information to the primary sensory area in the postcentral gyrus in the parietal lobe. The primary sensory area is organized into a body parts map (sensory homunculus) and the sensory nerve fibers cross over to the opposite side in the brainstem **a)** or spinal cord **b)** and **c)**.

Table 15.2 Two-Point Discrimination Test

Area of skin	Distance between compass points at which two-point discrimination can be made (mm)	Relative density of sensory receptors (ranked 1 to 5)	
		Predicted	Actual
Tip of index finger			
Dorsal surface of the hand			
Posterior surface of the neck			
Side of the nose			
Anterior surface of the arm			

QUESTION TO CONSIDER Compare your predictions of relative receptor densities with your actual experimental results. How accurate were your predictions? Explain.

____

____

Dermatomes

The surface of the body can be divided into sensory cutaneous regions, each innervated by a single pair of spinal nerves. These cutaneous regions are called **dermatomes** (Figure 15.7). There is considerable overlapping of adjacent dermatomes and the dermatomal body maps can vary significantly between individuals.

Nevertheless, a simple sensory test at specific areas of skin on the body can be useful for assessing spinal nerve function.

The peripheral nerves derived from the **brachial plexus** (Figure 15.5b) will be used as a model for testing the function of individual spinal nerves. The brachial plexus is formed by the ventral rami of spinal nerves C5–T1. It gives rise to the peripheral nerves that innervate structures in the upper extremity.

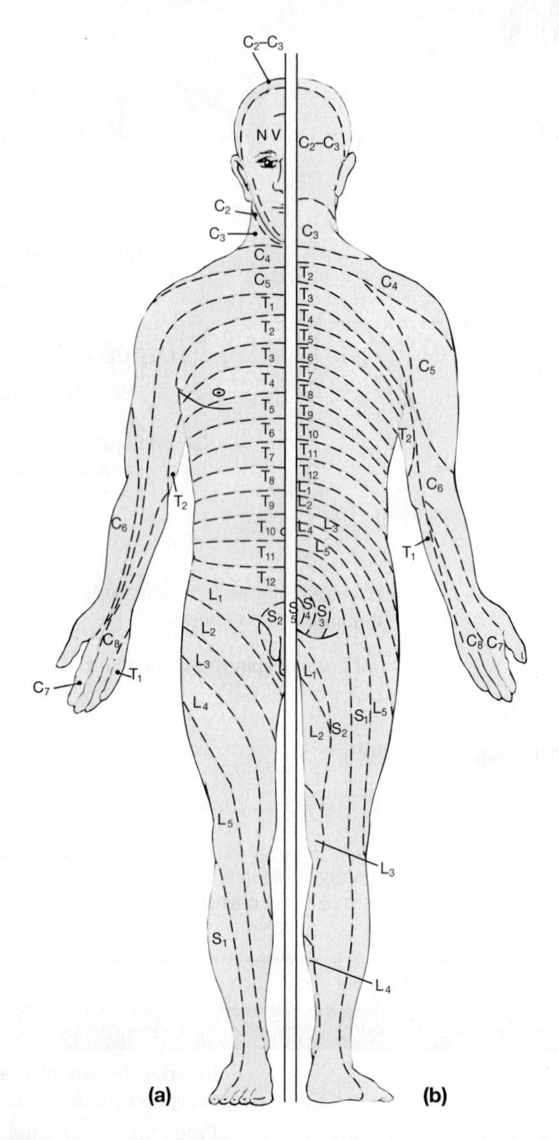

Figure 15.7 Generalized organization of dermatomes on the human body. There is considerable overlap between adjacent dermatomes and a great deal of variation in their distribution between individuals. **a)** Anterior view; **b)** posterior view.

ACTIVITY 15.4 Testing General Sensory Function in Dermatomes

1. Strike the prongs of a tuning fork against the edge of your lab bench so that the sound due to the vibration is loud enough to be easily heard.
2. Using your laboratory partner as a test subject, quickly but gently press the handle of the tuning fork on the skin areas listed in Table 15.3. These areas are used because they are typically within the defined dermatome for the specific spinal nerve being tested.
3. The ability to feel the vibration generated by the tuning fork indicates normal function of the spinal nerve.
4. Record your results in Table 15.3.

QUESTIONS TO CONSIDER

1. For the spinal nerve tests that you just performed, the general sensations that you studied (touch and vibratory sense) reached the primary sensory area via one of the ascending pathways depicted in Figure 15.6. If the tests were performed on the right side of the body, the sensory information would reach the sensory area in the left cerebral hemisphere. Examine Figure 15.6 and provide an explanation for this outcome.

____

____

2. Review your results of the two-point discrimination test. Speculate on a reason why the sensory homunculus is arranged disproportionately (Figure 15.6). That is, why is such a large portion of the sensory area devoted to the hands and fingers, compared to other larger regions of the body?

____

____

Table 15.3 Functional Assessment for Spinal Nerves of the Brachial Plexus

Spinal nerve	Dermatomal area	Result of test
C5	Skin covering anterior and lateral aspect of deltoid muscle	
C6	Skin covering the thenar eminence in the hand	
C7	Skin covering the anterior surface of the distal end of the middle finger	
C8	Skin covering the hypothenar eminence in the hand	
T1	Skin covering the superomedial aspect of the arm	

Exercise 15 Review Sheet

The Spinal Cord and Spinal Nerves

Name ______________________________

Lab Section ______________________________

Date ______________________________

1. Describe how the three meningeal layers are arranged around the spinal cord. Discuss similarities and differences with the meningeal layers around the brain. Draw a diagram that illustrates your written answer.

2. Describe the various branches (rami) of a spinal nerve.

Questions 3–8: Define the following terms.

3. Cervical and lumbar enlargements:

4. Conus medullaris:

5. Filum terminale:

6. Cauda equina:

7. Spinal nerve plexus:

8. Intercostal nerves:

9. Study the anatomy of the spinal nerve plexuses in Figure 15.5. Explain why a peripheral nerve, such as the sciatic nerve (Figure 15.5d), can contain nerve fibers from several spinal nerves.

____

____

____

10. If all the ventral roots of spinal nerves to the right lower extremity were damaged, what would be the outcome?

____

____

Exercise 16

Human Reflex Physiology

Laboratory Objectives

On completion of the activities in this exercise, you will be able to:

- Describe the organization of the motor homunculus.
- List the fiber connections between the cerebellum and other parts of the central nervous system.
- Describe the function of the cerebellum.
- Indicate why reflexes are important to day-to-day functioning.
- Assess basic functions of coordination and balance.
- Describe the function of each element of the reflex arc.
- Perform spinal reflex tests for the upper and lower extremities.

Materials

- Reflex hammers
- Chalk
- Labeling or masking tape

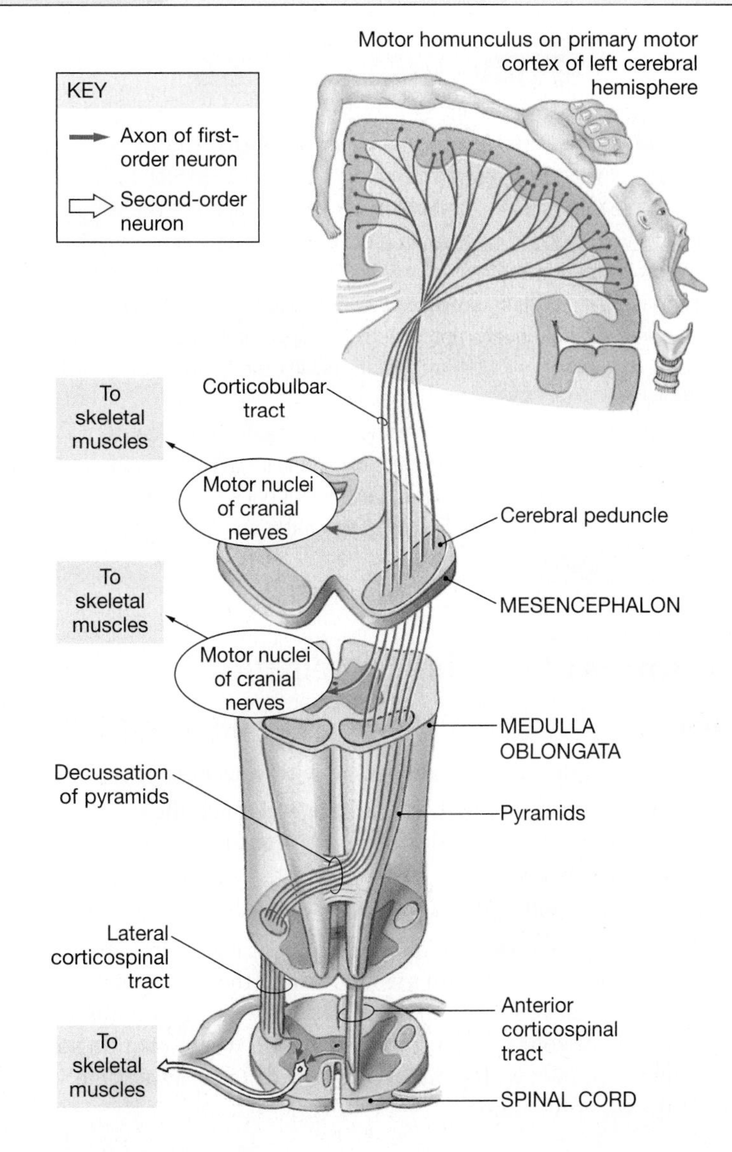

Figure 16.1 Descending motor pathways. The pathways begin in the primary motor area in the precentral gyrus of the frontal lobe and travel to the brainstem or spinal cord. Note the body parts map (motor homunculus) in the primary motor area and the crossing over of the motor fibers.

Motor neurons in the precentral gyrus of the frontal lobe initiate all voluntary motor function. This region, known as the **primary motor area**, can be divided into body regions represented by a **motor homunculus** that illustrates the correspondence between the cortical regions and the body parts it controls (Figure 16.1). The organization of the motor homunculus is related to the number of motor units that are present in muscles, not the size of the body region. For example, the muscles of the hands and face contain many small motor units that make complex and specific motor skills possible, so the motor areas that control these body regions are relatively large. Conversely, the motor area that controls the more expansive torso of the body is relatively small, because the muscles in this area contain larger but fewer motor units that are designed for less specific functions.

The descending motor pathways, which originate in the primary motor area, descend through the brainstem and spinal cord before distributing motor fibers to cranial and spinal nerves. Since most descending pathways cross over to the opposite side either in the brainstem or spinal cord, the motor area on the right side of the brain directs motor function on the left side of the body, and vice versa.

Coordinated Movement and Balance

The cerebellum interacts with the motor areas by ensuring that all movements are smooth and well coordinated. It receives inputs for motor commands from the cerebrum, for equilibrium (balance) from the inner ear, for visual stimuli from the retinas of the eyes, and for proprioception (body position in space) from receptors called proprioceptors in muscles and joints. After analyzing this input, the cerebellum sends information that fine-tunes the motor activity back to the cerebrum. Thus, normal cerebellar function gives us the ability to maintain normal balance and posture, and allows us to perform highly specialized motor functions (e.g., playing a musical instrument, typing on a keyboard, climbing a ladder) gracefully and efficiently.

WHAT'S IN A WORD The term *proprioception* is derived from two Latin words: *proprius,* meaning "one's own," and *capio,* meaning "to take." Proprioception is a general sensory awareness of the position and movement of one's own body parts.

CLINICAL CORRELATION

Damage to the cerebellum, due to trauma or stroke, can result in **ataxia,** a condition characterized by uncoordinated and shaky movements and an inability to maintain balance. A person with ataxia may have balance problems so severe that he or she will not be able to stand or sit in an erect position. Voluntary movements are performed very slowly and require a great deal of concentration. In addition, timing and coordination of movements is impaired so that an attempt to grab an object (e.g., reaching for a coffee cup or a pen) often results in moving the hand beyond the target.

Cerebellar function is also temporarily impaired by alcohol and some drugs. For this reason, police officers will ask drivers who are suspected of driving under the influence of alcohol or drugs to walk in a straight line or to touch the tip of the nose with the index finger.

ACTIVITY 16.1 Assessing Coordination

Test 1

1. Stand in the anatomical position with your eyes open.
2. Move your right hand to your face and touch the tip of your nose with the index finger. Repeat this action with your left hand.
3. Repeat steps 1 and 2 with your eyes closed.
4. In the space provided, record and analyze your test results. Note how accurate you are in touching the tip of your nose when your eyes are open versus closed. If you missed the tip of your nose, record the position where your finger made contact with your face and the distance away from the tip of the nose.

____

____

____

____

Test 2

1. Stand in the anatomical position with your eyes open.
2. Abduct your upper extremities so that they are parallel to the floor.
3. With each index finger extended, move your upper extremities forward and attempt to bring the fingertips together.
4. Repeat steps 1 through 3 with your eyes closed.
5. In the space provided, record and analyze your test results. Note the accuracy in your ability to touch your fingertips with your eyes open versus closed. If your fingertips do not make contact, record the distance between the two fingers as they pass by each other.

____

____

____

____

QUESTIONS TO CONSIDER

1. Discuss the importance of visual stimuli in performing the tests in Activity 16.1.

____

____

2. When your eyes are closed, what other sensory stimuli do you rely on to successfully complete the exercises? __________

____

____

3. When your eyes are closed, is it more or less difficult to perform the exercises? Explain. __________

____

____

4. Compare and discuss your results with the results of other students in the laboratory. __________

____

____

ACTIVITY 16.2 Assessing Balance

Test 1

1. With tape, mark off a line along the floor that is 6 m (20 ft) in length.
2. With your eyes open, walk slowly along the tape, heel to toe, without losing your balance. Have your lab partner record the time that it takes to walk the entire length of the tape. __________
3. Repeat step 2 several times, increasing your speed for each turn. Record the time after each turn. Stop the test when you no longer can maintain your balance.

____

____

4. Repeat steps 2 and 3 with your eyes closed. Record the time at each turn. (*Important: For safety, two people should*

remain to the sides of the subject to act as spotters while he or she performs this activity.)

5. Compare and discuss your results with the results of other students in the laboratory.

Test 2

1. Draw a vertical line on a chalkboard, or position a piece of tape in a vertical line on a wall in the laboratory.
2. Stand in an erect position, adjacent to the chalk or tape line, with your upper limbs by your sides and your eyes open. Remain standing for at least 2 minutes.
3. Your laboratory partner will time the exercise and record any body movements that you make during the test. Use the chalk or tape line as a reference position for recording movements.
4. When the 2-minute period is completed, sit comfortably for a short time.
5. When you are rested and comfortable, stand in an erect position, adjacent to the vertical reference line, but this time close your eyes. Remain standing for 2 minutes.

   Form a Hypothesis Before you begin, make a prediction about the results.
6. Your laboratory partner should, once again, note any body movements that you make during the test.
7. **Assess the Outcome Make comparisons of the results from the two tests. In particular, contrast the number and type of movements that were made and note any evidence that a loss of balance had occurred when the eyes were open versus closed. Did your results support or refute your hypothesis?**

8. Compare and discuss your results with those of others in the laboratory.

QUESTION TO CONSIDER Do you believe that your abilities to perceive and interpret visual stimuli are more important for performing coordinated movements or for maintaining balance? Explain.

ACTIVITY 16.3 Assessing Coordinated Movement and Balance

1. Stand in an upright position, with your upper limbs by your sides and your eyes open.
2. When you are relaxed, raise one leg off the ground and touch your heel to the anterior surface (the shin) of your other leg.
3. Move the heel inferiorly (toward the floor), but try to keep it in contact with your other leg.
4. During the test, your lab partner will evaluate your performance by observing how fluid your movements are and how well you maintain your balance.
5. Repeat steps 1 through 4 from the other side.
6. In the space provided, record and analyze your test results.
7. Compare and discuss your results with other students in the laboratory.

QUESTION TO CONSIDER Discuss why well-coordinated muscle contractions and maintenance of balance and posture are important for performing complex actions (e.g., dancing, running, climbing).

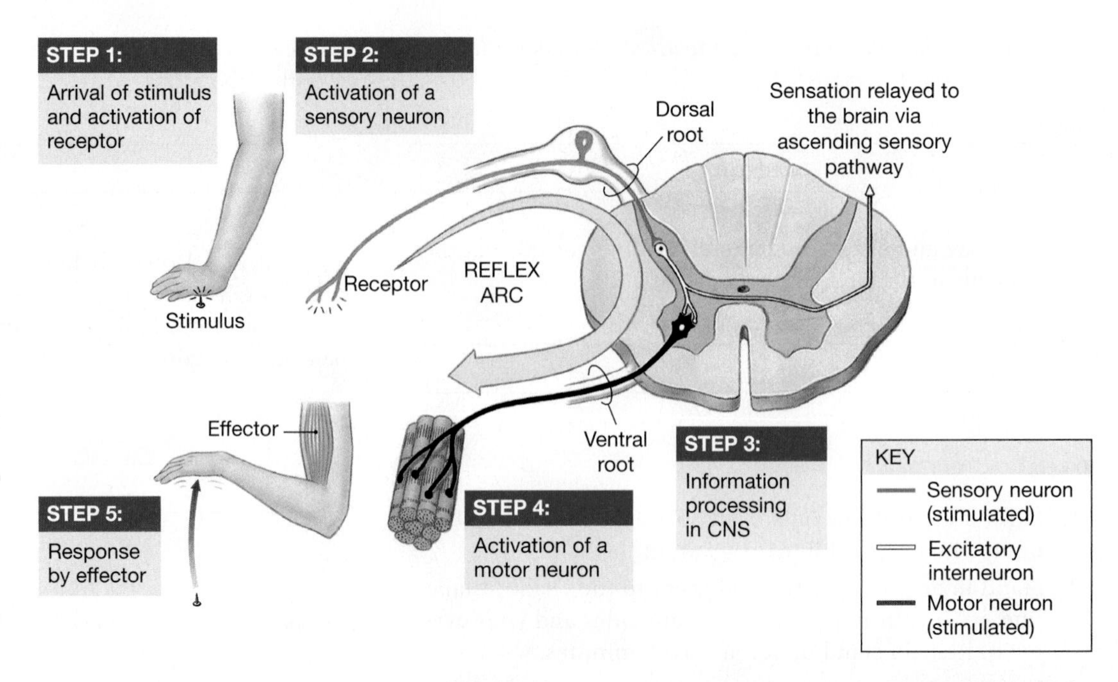

Figure 16.2 The components of a reflex arc. Step 1: A sensory receptor responds to a stimulus. **Step 2:** A sensory neuron transmits sensory impulses to the CNS. **Step 3:** An integration center in the CNS (either the spinal cord or the brainstem) receives and transfers sensory information to a motor neuron. **Step 4:** A motor neuron transmits motor impulses to an effector organ. **Step 5:** An effector organ receives motor impulses and acts in response to the stimulus. Effector organs can be muscle fibers (skeletal, smooth, or cardiac muscle) or glands.

The Reflex Arc

A **reflex** is an automatic, involuntary response to a change that can occur inside or outside the body. Many reflexes play a critical role in protecting the body by regulating homeostasis. The neuronal pathway of a reflex, known as a **reflex arc**, contains five elements (Figure 16.2).

- **Sensory receptor** that responds to a stimulus
- **Sensory neuron** that delivers sensory impulses to the CNS
- **Integration center** in the CNS (either the spinal cord or the brainstem) where the sensory information is received and transferred to motor neurons
- **Motor neuron** that delivers motor impulses from the CNS to an effector organ
- **Effector organ** that receives motor impulses and acts in response to the stimulus; effector organs can be muscle fibers (skeletal, smooth, or cardiac muscle) or glands

Reflexes are classified according to the location of the integration center, the type of effector organ that is activated, and the number of synapses between the sensory and motor neurons in the reflex arc. Table 16.1 identifies the various types of reflexes based on these criteria.

Table 16.1 Classification of Nerve Reflexes

Criterion	Type of reflex
A. Location of integration center	
1. Brainstem	Cranial reflex
2. Spinal cord	Spinal reflex
B. Type of effector organ	
1. Skeletal muscle	Somatic reflex
2. Cardiac muscle, smooth muscle, glands	Visceral reflex
C. Number of synapses between sensory and motor neurons in the reflex arc	
1. One synapse (sensory neuron synapses directly with motor neuron)	Monosynaptic reflex
2. Two or more synapses (one or more interneurons intervene between the sensory and motor neurons)	Polysynaptic reflex

Assessing the Function of Spinal Reflexes: The Reflex Arc

The spinal reflexes that you will observe in the laboratory are examples of **stretch reflexes** (Figure 16.3a). They involve the activity of receptors called **muscle spindles** that are embedded between muscle fascicles in skeletal muscles. When a muscle is stretched, muscle spindles are elongated. As a result, a reflex muscle contraction will occur, and excessive stretching of the muscle will be prevented.

Other spinal reflexes include the **deep tendon reflex**, **withdrawal reflex** (Figure 16.3b), and **crossed extensor reflex** (Figure 16.3c). These are described in Table 16.2.

CLINICAL CORRELATION

Stretch reflexes play a critical role in reducing the possibility of muscle damage due to overextension. They are also important for maintaining normal upright posture. Accordingly, they are very active in the muscles of the lower extremities. In a clinical setting, stretch reflex tests are performed to check normal function of spinal nerves.

With a lab partner, perform the reflex tests described in the following activities. Conduct the tests on both the right and left sides. For most of these tests, the reflex contractions are very subtle, so, careful observations are required to see the results.

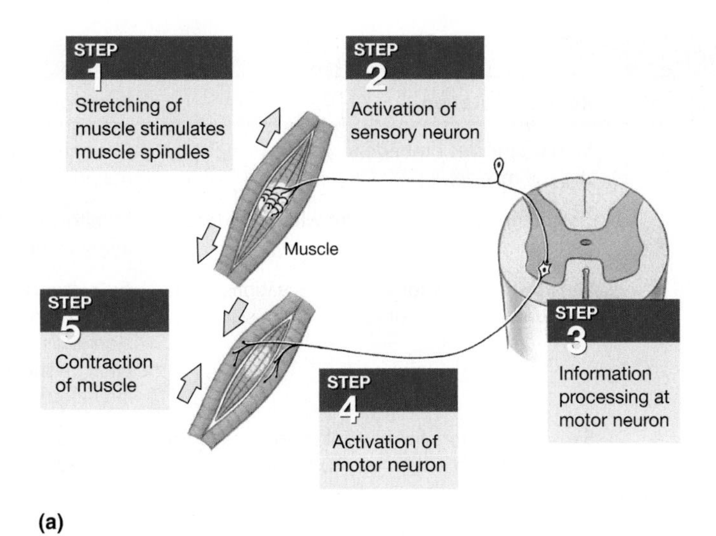

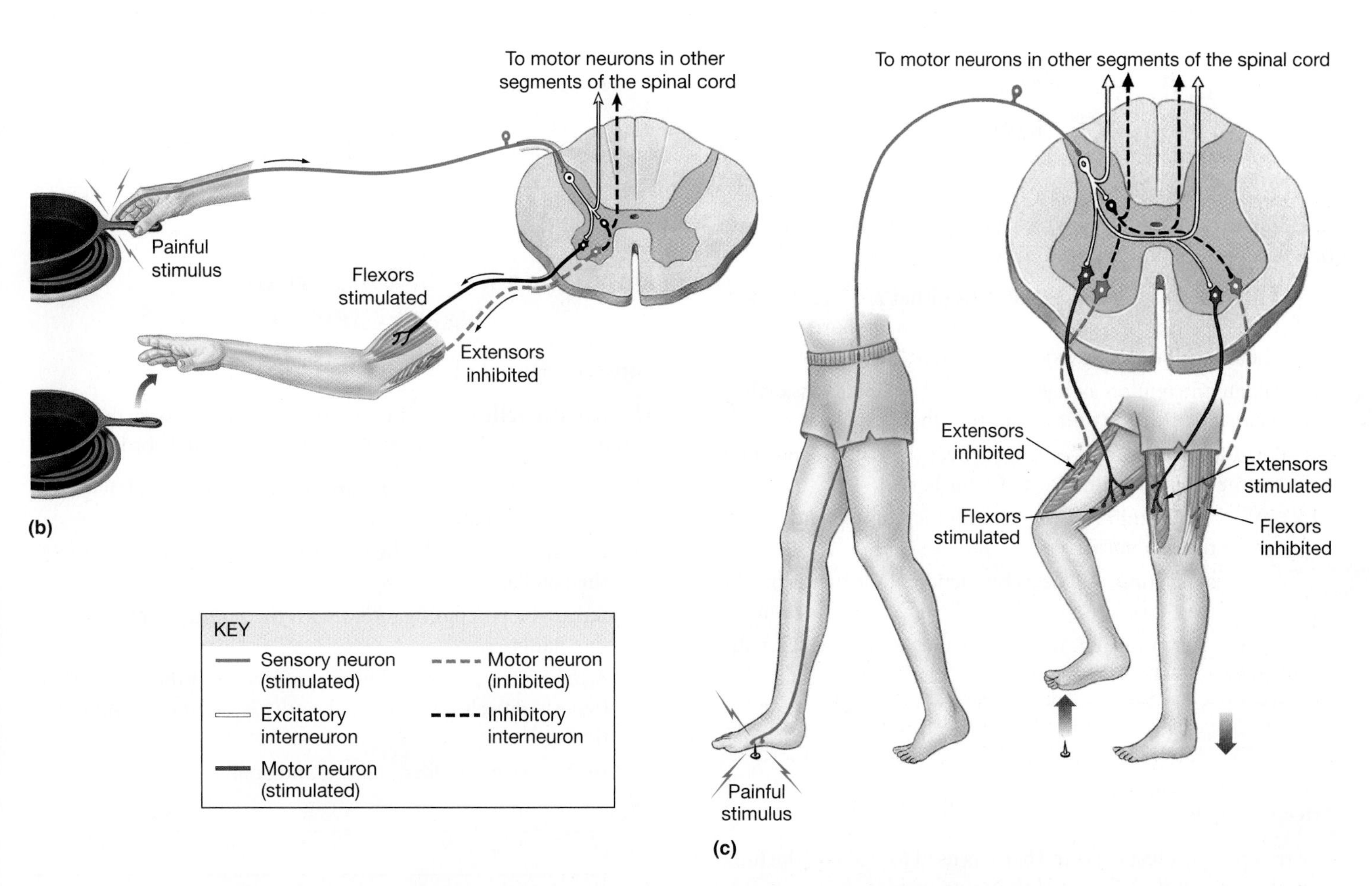

Figure 16.3 Types of spinal reflexes. a) Stretch reflex; **b)** withdrawal (flexor) reflex; **c)** crossed extensor reflex. The stretch reflex is monosynaptic. Withdrawal and crossed extensor reflexes are polysynaptic.

Table 16.2 Types of Spinal Reflexes

Type of reflex	Monosynaptic/polysynaptic	Receptor/location	Function
Stretch reflex	Monosynaptic	Muscle spindles embedded within a whole skeletal muscle.	Prevents excessive stretching of muscle; maintenance of normal upright posture.
Deep tendon reflex	Polysynaptic	Golgi tendon organs, located within tendons.	Monitors muscle tension; protects tendons from injury by preventing overcontraction of muscles.
Withdrawal (flexor) reflex	Polysynaptic	Sensory receptors for general sensations (pain, pressure, touch, temperature) in the skin.	Typically acts upon an extremity; causes flexion of a body part, resulting in movement away from a negative (e.g., pain, hot) stimulus; inhibits antagonistic extensor muscles (reciprocal inhibition).
Crossed extensor reflex	Polysynaptic	Sensory receptors for general sensations (pain, pressure, touch, temperature) in the skin.	Occurs simultaneously with the withdrawal reflex. When withdrawal reflex causes flexion of a body part, the crossed extensor reflex promotes extension of the corresponding body part on the contralateral side.

ACTIVITY 16.4 Reflex Tests for the Upper Extremity

Biceps Reflex

The biceps reflex test (Figure 16.4a) is used to assess the function of level C5 and C6 spinal nerves.

1. With your partner seated, take his or her arm and locate the biceps brachii tendon. This can be seen more easily when the elbow is flexed and the muscle is contracted.
2. Palpate the tendon as it passes over the cubital fossa (the shallow depression anterior to the elbow joint).
3. Position your thumb (or finger) over the tendon and stabilize the elbow with the rest of your hand.
4. Strike your thumb (or finger) with the pointed end of a rubber reflex hammer.
5. A normal response will be a contraction of the biceps brachii muscle; you should notice a slight flexion of the forearm.
6. In the space provided, record and analyze your test results.

____

____

Triceps Reflex

The triceps reflex test (Figure 16.4b) is used to evaluate the function of the C7 and C8 levels of the spinal cord.

1. Palpate your lab partner's triceps brachii muscle in the posterior arm.
2. Follow the muscle inferiorly until you can feel the triceps tendon just superior to the elbow joint.
3. Strike the triceps tendon with the pointed end of the reflex hammer and observe the response.
4. A normal response will be a contraction of the triceps brachii muscle; you should notice a slight extension of the forearm.
5. In the space provided, record and analyze your test results.

____

____

ACTIVITY 16.5 Reflex Tests for the Lower Extremity

Patellar ("Knee Jerk") Reflex

The patellar reflex test (Figure 16.4c) is used to evaluate the function of the L2, L3, and L4 levels of the spinal cord.

1. Have your lab partner sit on the laboratory bench with his or her legs hanging freely.
2. On one leg, identify the patellar ligament just inferior to the patella.
3. Strike the patellar ligament with the pointed end of the reflex hammer.
4. A normal response will be a contraction of the quadriceps femoris muscles; you should notice a slight extension of the leg.
5. In the space provided, record and analyze your test results.

____

____

Ankle Jerk Reflex

The ankle jerk reflex test (Figure 16.4d) assesses the function of the S1 and S2 levels of the spinal cord.

1. Have your lab partner sit on the laboratory bench with his or her legs hanging freely.
2. On one side, identify the calcaneal (Achilles) tendon as it passes posterior to the ankle joint.

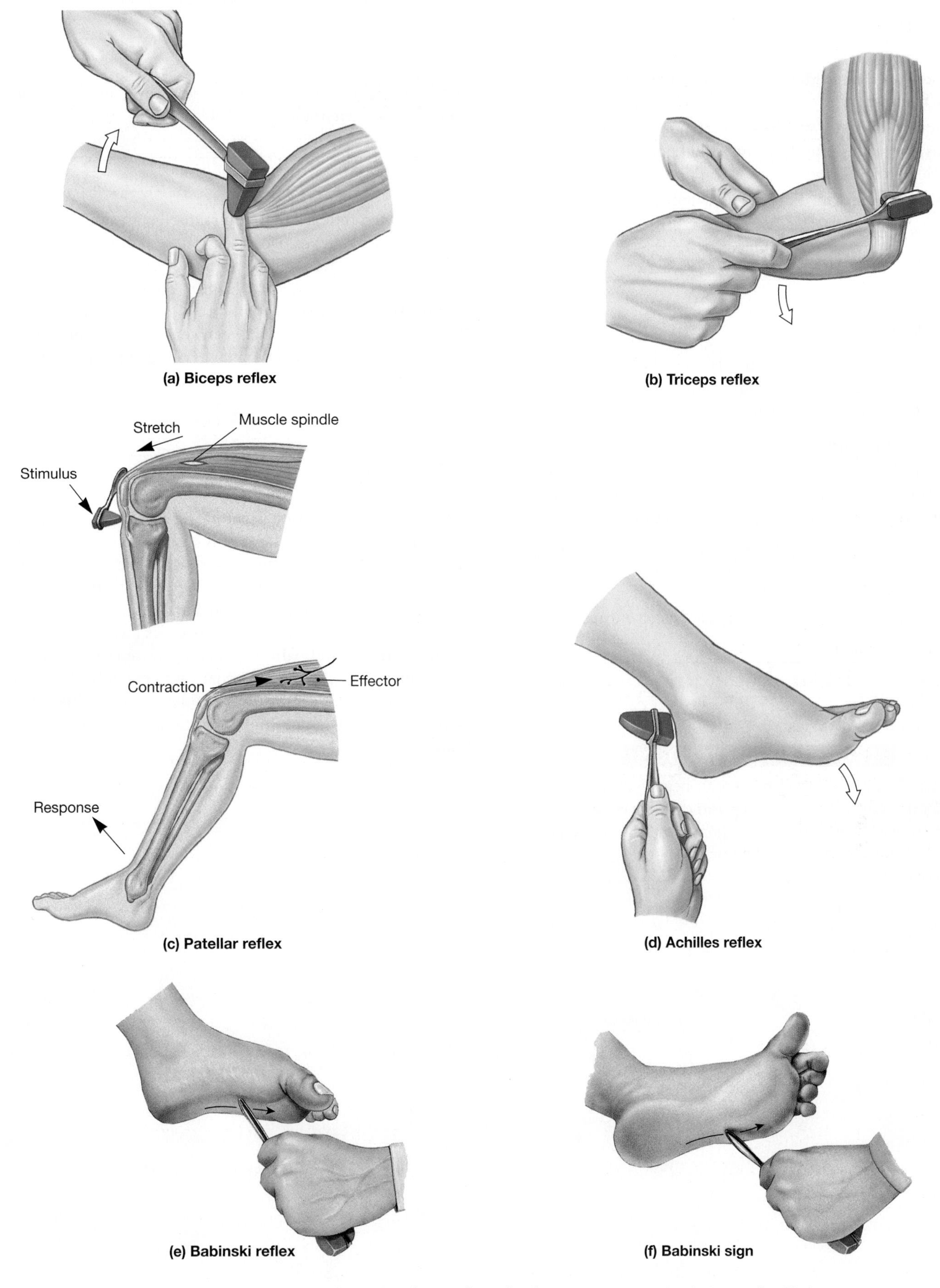

Figure 16.4 Various types of stretch reflexes. All stretch reflexes are monosynaptic. **a)** Biceps reflex; **b)** triceps reflex; **c)** patellar (knee jerk) reflex; **d)** Achilles (ankle jerk) reflex; **e)** Babinski (plantar) reflex and **f)** the Babinski sign.

3. Strike the calcaneal tendon with the broad side of the reflex hammer.
4. A normal response will be a contraction of the gastrocnemius and soleus muscles; you should notice a slight plantar flexion of the foot.
5. In the space provided, record and analyze your test results.

Babinski Reflex

The Babinski test, known also as the **plantar reflex** (Figure 16.4e), examines the function of the S1 and S2, and to a lesser extent, the L4 and L5 levels of the spinal cord.

1. Have your partner remove his or her shoes.
2. Run the handle of the reflex hammer from the heel along the lateral margin of the foot to the tarsometatarsal joints; then move across the plantar surface toward the great toe.
3. A normal response is plantar flexion of the foot. An abnormal response is extension of the great toe and abduction (spreading) of the other toes.
4. In the space provided, record and analyze your test results.

CLINICAL CORRELATION

For infants, in which myelination of nerve fibers is not complete, an abnormal plantar reflex, called a **positive Babinski's sign** (Figure 16.4e), is common. In older children and adults, however, it could indicate damage to the descending (motor) tracts in the spinal cord.

QUESTIONS TO CONSIDER

1. Classify stretch reflexes by considering the criteria listed in Table 16.1.

2. The stretch reflexes that you studied are examples of monosynaptic reflexes. Deep tendon, withdrawal, and crossed extensor reflexes are all polysynaptic. Examine the reflex arcs for monosynaptic (Figure 16.3a) and polysynaptic (Figure 16.3b) reflexes. In which type of reflex will the transfer of neural information between sensory and motor neurons be faster? Provide an explanation to support your answer.

3. Spinal reflexes operate without the assistance of higher brain centers. However, for a polysynaptic reflex, you can become aware of the stimulus that initiated the reflex. For example, if you accidentally touch a hot frying pan, a withdrawal reflex will cause your hand to move away from the pan, but you will also have conscious awareness of the hot temperature and pain. Conscious awareness of a sensation requires the sensory information to reach the primary sensory area in the cerebrum. Speculate on how this would be possible. (Hint: Study the neuronal pathways in Figures 16.3b and c.)

Exercise 16 Review Sheet

Human Reflex Physiology

Name ____________________

Lab Section ____________________

Date ____________________

1. Discuss the relationship between the organization of the motor homunculus and the number of motor units in the muscles of a body region.

2. What is the significance of the crossing over of the descending motor pathways?

3. Describe the type of neural inputs that the cerebellum receives and the type of neural outputs that it sends out.

4. Based on your answer to question 3, describe the function of the cerebellum.

5. Describe the anatomical components of a reflex arc.

6. Discuss the various ways that reflexes can be classified.

7. Review the functions of spinal reflexes that are described in Table 16.2. Discuss the functional relationship between a flexor reflex and a crossed extensor reflex.

Exercise 17

Special Senses

Laboratory Objectives

On completion of the activities in this exercise, you will be able to:

- Explain the events leading to the perception of a sensation.
- Differentiate between general sensation and special sensation.
- Identify the gross anatomical features of the sensory organs and accessory structures for the special senses: olfaction, gustation, vision, hearing, and equilibrium.
- Describe the microscopic anatomy (histology) of the olfactory epithelium, taste buds, retina, and cochlea.
- Describe the neural pathways for the special senses.
- Understand and explain the physiology of the sensory organs and accessory structures for the special senses: olfaction, gustation, vision, hearing, and equilibrium.
- Dissect a cow eye and describe its anatomy.
- Perform simple tests to assess the function of the special senses.

Materials

- Anatomical models of the following:
  - Midsagittal section of the head
  - Human brain
  - Human eye
  - Human ear
- Human skulls
- Compound light microscopes
- Prepared microscope slides of the following:
  - Olfactory epithelium
  - Tongue, with taste buds
  - Retina
  - Cochlea
- Watch or clock with second hand
- Vials of wintergreen oil
- Vials of peppermint oil
- Vials of other substances with strong, distinct odors (if wintergreen and peppermint oils are not available)
- Absorbent cotton balls
- Paper towels
- Granulated sugar (sucrose)
- 10% granulated sugar solution
- Table salt (NaCl)
- 10% table salt solution
- Phenylthiocarbamide (PTC) test strips
- Cubes of various food items (food types to remain unknown to students)
- Toothpicks
- Pipe cleaners or slender probes
- Dissecting tools
- Dissecting gloves
- Dissecting trays
- Protective eyewear
- Face masks (optional)
- Fresh or preserved cow eyes
- Dissecting pins
- Metric rulers
- Pencils
- Flashlights
- Snellen charts
- Astigmatism charts
- Ishihara color plates
- Tuning forks
- Colored pencils

Sensory receptors are extensions of the nervous system. They contain specialized cells or cell processes that relay specific information (a stimulus) from inside or outside the body to the central nervous system. To perceive a sensation, the following must occur.

- A stimulus strong enough to initiate a response in the nervous system must be detected by a sensory receptor.
- The receptor must convert the stimulus to an action potential, a process called **transduction**.
- The action potential must be conducted from the receptor to the CNS along a neural pathway.
- The action potential must reach a specific region of the brain, where the stimulus is perceived and interpreted.

Action potentials that travel to the cerebral cortex are consciously interpreted as sensations. If the action potential terminates at a nucleus in the spinal cord or brainstem and does not reach the cerebral cortex, a reflex response will be initiated, but conscious perception of the sensation will not occur.

Sensations can be categorized into two groups: general senses and special senses. **General senses** include touch, pressure, vibration, pain, temperature, proprioception (perception of position and movement of body parts), and chemical and fluid pressure (i.e., blood pressure). Receptors for these sensations are located throughout the body and are relatively simple in structure. **Somatic sensations** are detected by peripheral receptors in skin, muscles, tendons, and joints. **Visceral sensations** are detected by deep receptors in the walls of internal organs and blood vessels.

Special senses include **olfaction** (smell), **gustation** (taste), **vision**, **hearing**, and **equilibrium** (body orientation). Receptors for these sensations are localized to specific regions in the head and have a complex structure. Unlike the general senses, special sensory information must travel along complex pathways to reach the CNS.

General sensations were studied in the previous two laboratory exercises. In this exercise, you will study the special senses.

Olfaction

Sensory receptors for olfaction are located in the **olfactory epithelium** that covers the superior portion of the nasal cavity (Figures 17.1a and 17.2). The olfactory epithelium consists of three cell types (Figures 17.1b and c).

- **Olfactory receptor cells** are bipolar neurons that are stimulated by chemical substances (**odorants**) in the nasal cavity.
- **Supporting cells** are columnar epithelial cells that surround the olfactory receptors.
- **Basal cells** are interspersed between the bases of the supporting cells and divide regularly to produce new olfactory receptor cells.

CLINICAL CORRELATION

Olfactory receptor cells have relatively short lifespans (about 1 month) and are constantly being replaced by the division and differentiation of basal cells. This replacement of olfactory cells is a rare example of nerve cell turnover in adult humans.

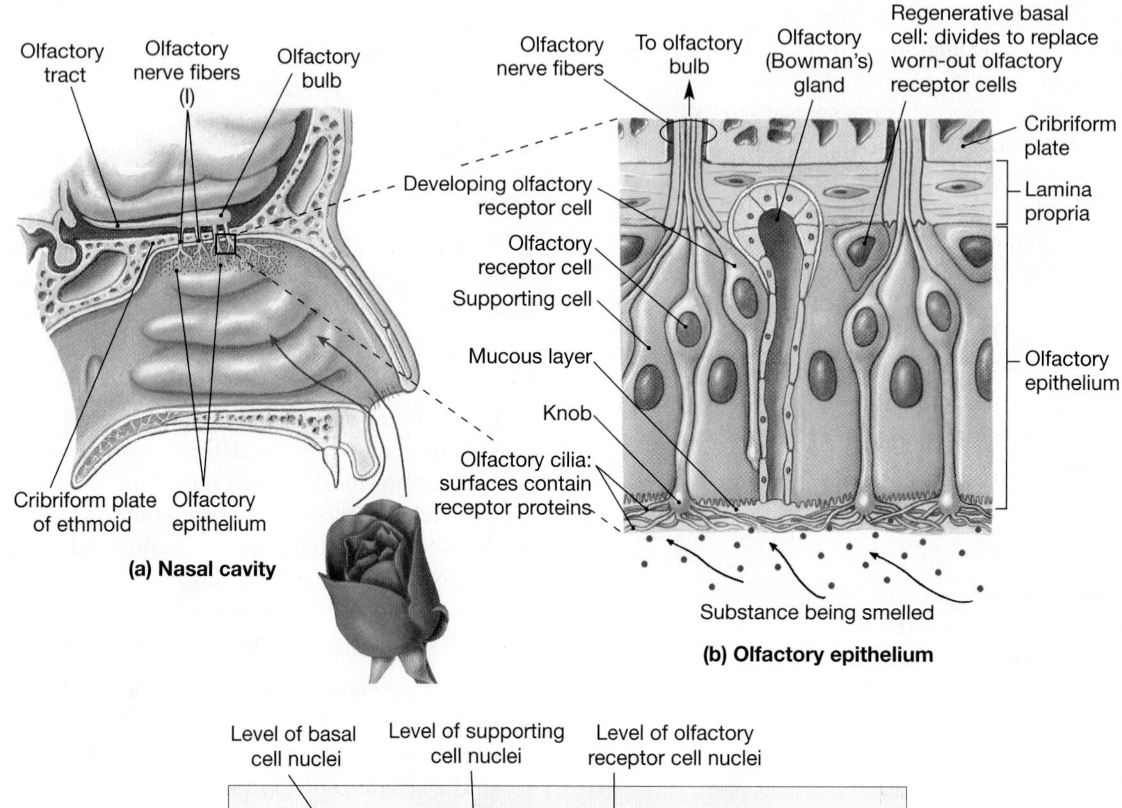

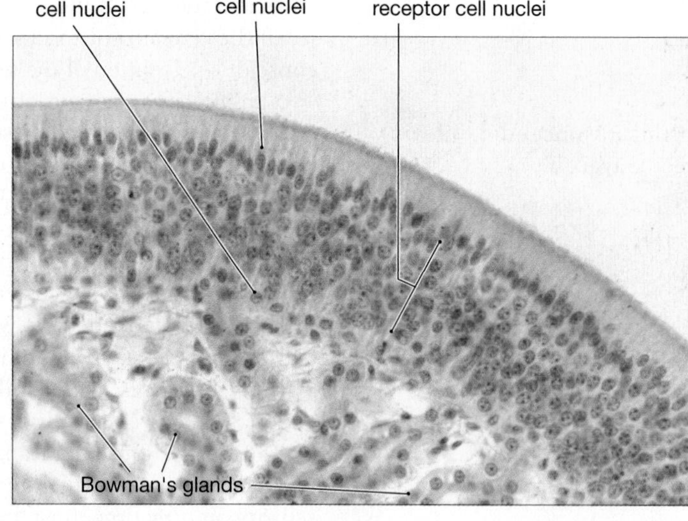

Figure 17.1 The olfactory epithelium. a) Midsagittal section of the nasal cavity showing the location of the olfactory epithelium; **b)** diagram; **c)** light micrograph of the microscopic structure of the olfactory epithelium (LM × 500).

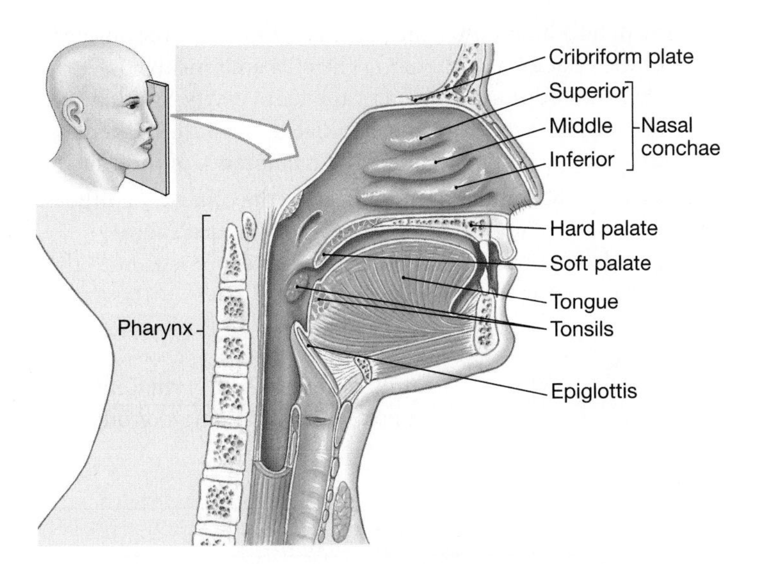

Figure 17.2 Midsagittal section of the head. The olfactory epithelium covers the superior portion of the nasal cavity, including the superior concha and the inferior surface of the cribriform plate. Taste buds are located on the superior surface of the tongue, soft palate, epiglottis, and pharyngeal wall.

WHAT'S IN A WORD The word *olfaction* is derived from the Latin word *olfacio*, which means "to smell." Structures with the terms *olfaction* or *olfactory* in their names have functions related to the sense of smell.

ACTIVITY 17.1 Examining the Anatomy of Olfactory Structures

Gross Anatomy

1. Obtain a midsagittal section of the head (Figure 17.2) and locate the nasal cavity. Identify the **superior nasal concha** and the inferior surface of the **cribriform plate**. These structures and the superior portion of the **perpendicular plate** are covered by the mucous membrane that contains the olfactory epithelium.
2. Obtain a skull and remove the calvaria (skullcap) to view the cranial cavity (Figure 17.3a). Identify the numerous **olfactory foramina**, which are narrow openings that pass through the cribriform plate. These passageways transmit the unmyelinated axons of the **olfactory nerve (cranial nerve I)**. The axons originate from the cell bodies of olfactory receptor cells in the olfactory epithelium.

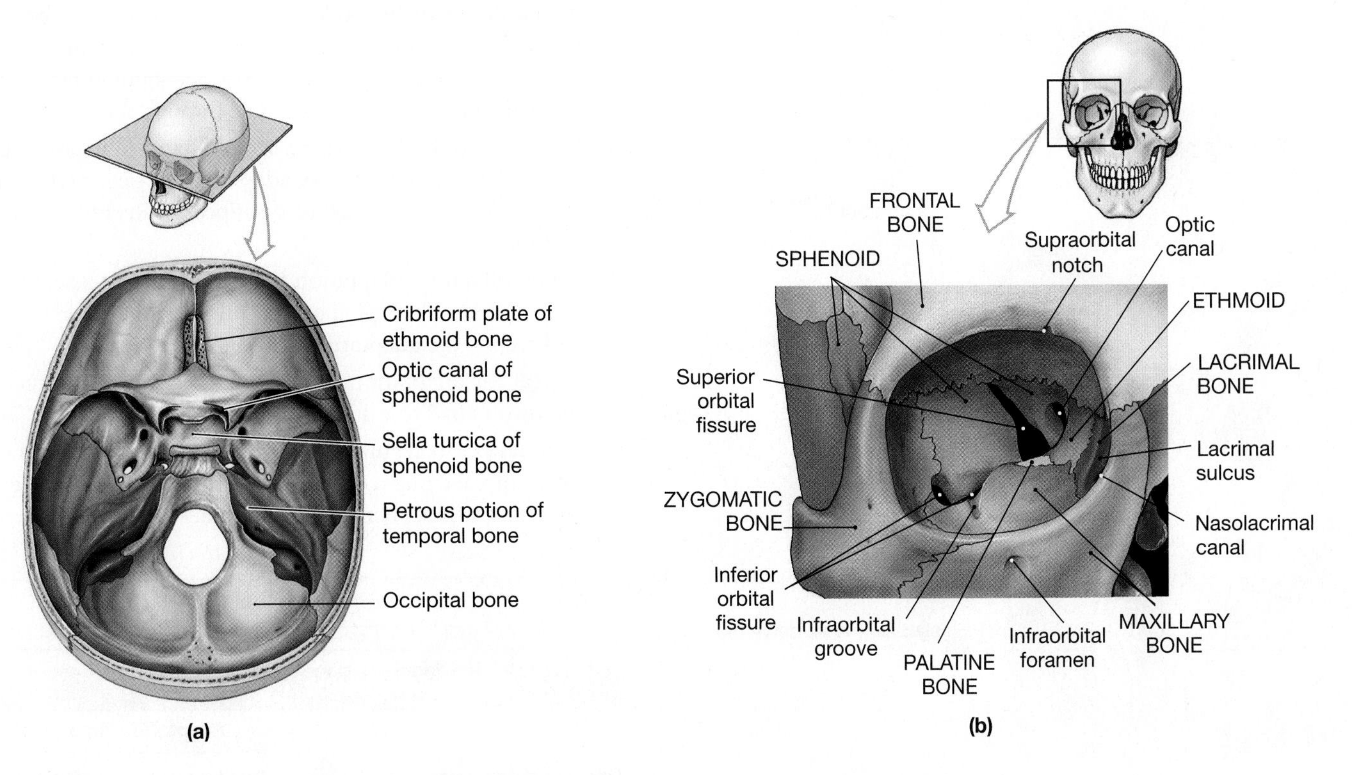

Figure 17.3 Structures on the skull associated with the special sensory receptors. a) Structures in the cranial cavity linked to the senses of smell, vision, hearing, and equilibrium; **b)** the bones that contribute to the formation of the orbital cavities.

3. Obtain a model of the brain and view its inferior surface (Figure 17.4). Identify the two **olfactory bulbs** on the inferior surface of the frontal lobes. The olfactory bulbs are regions of gray matter, where olfactory nerve axons synapse with multipolar neurons. Identify the **olfactory tracts** that project posteriorly from the olfactory bulbs, sending axons to the **olfactory area** in the **temporal lobe**.
4. Place the brain into the cranial cavity of the skull (do not expect an exact fit, but do the best you can). Notice that the olfactory bulbs rest on the superior surface of the cribriform plate.

Microscopic Anatomy

1. Obtain a prepared microscope slide of the olfactory epithelium.
2. View the slide under low power. The three cell types in the epithelium—olfactory receptor cells, supporting cells, and basal cells—are not distinguishable on your slide. However, you can identify the approximate levels of the cell nuclei for each type (Figure 17.1c), as follows:
   - Both the olfactory receptor cells and supporting cells are columnar and extend from the surface to the base of the epithelium. However, the nuclei of the receptor cells are typically located more centrally in the epithelium, while the nuclei of supporting cells are closer to the surface.
   - The basal cells are small, rounded, or cone-shaped cells. On your slide, the nuclei of these cells form a single layer along the base of the epithelium.

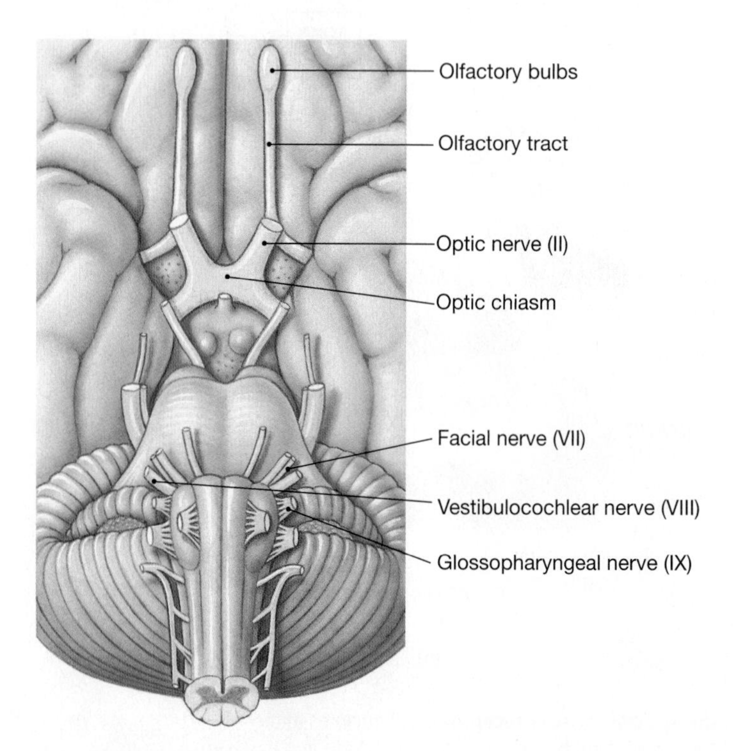

Figure 17.4 Inferior surface of the brain. Cranial nerves linked to the special senses are illustrated.

3. Switch to high power and identify **cilia** along the surface of the olfactory epithelium. The cilia are sensitive to chemical odorants that enter the nasal cavity. Stimulation of olfactory receptor cells is initiated when these chemicals bind to receptors along the membranes of the cilia.
4. Identify the lamina propria, deep to the olfactory epithelium. This connective tissue layer contains **olfactory glands** (Figures 17.1b and c) that secrete mucus onto the surface ot the olfactory epithelium.

QUESTION TO CONSIDER Why would sniffing, rather than normal inhalation, tend to increase the stimulation of olfactory receptor cells?

____

____

ACTIVITY 17.2 Testing Olfactory Adaptation

Adaptation refers to a decreasing sensitivity to a stimulus over time. For olfaction, this means that your sensitivity to an odor diminishes during a period of continual stimulation.

1. Have your lab partner sit quietly with both eyes closed and the left nostril squeezed shut.
2. Place several drops of wintergreen oil (or any other substance with a strong, distinct odor) on a small ball of absorbent cotton and hold it under your lab partner's open right nostril. During the test period, your lab partner should inhale normally through the nose but exhale through the mouth.
3. Use a watch or clock with a second hand to measure the time required for olfactory adaptation to occur (the odor is significantly diminished or disappears). Record the result in Table 17.1.
4. When olfactory adaptation has occurred, test a second substance, such as peppermint oil, with the same nostril and record the adaptation time in Table 17.1.
5. Repeat the adaptation tests for the left nostril and record the results in Table 17.1.
6. Compare your results with other students in your laboratory. For each substance, describe the variation in the adaptation times between individuals.

____

____

Table 17.1 Olfactory Adaptation

	Adaptation time (sec.)	
Odorant substance	**Right nostril**	**Left nostril**
1.		
2.		

QUESTIONS TO CONSIDER

1. Explain why you can smell a second substance after adapting to a first substance.

____

____

2. Explain why the olfactory adaptation time for one substance can be relatively short, but for another substance it can be much longer.

____

____

3. For any odorant substance, the olfactory adaptation time can vary between individuals. Suggest a reason why this is true.

____

____

Gustation

Sensory receptors for gustation (taste) are located in **taste buds**, primarily on the superior surface of the tongue but also on the soft palate, pharynx, and epiglottis. On the tongue, taste buds are located in the epithelium of surface projections known as **papillae** (singular = **papilla**; Figures 17.5a and b). The **fungiform** and **filiform papillae** are scattered over the entire anterior two thirds of the tongue. Each fungiform papilla contains between 5 and 10 taste buds. Although they are more numerous, the filiform papillae lack taste buds. The 9 to 12 **circumvallate papillae** are aligned in a V shape on the posterior aspect of the tongue. Each one contains over 100 taste buds.

Taste buds are ovoid structures that resemble a flower bud. They are embedded in the surrounding epithelium but, at the apex, a **taste pore** opens to the oral cavity (Figure 17.5c). Taste buds are composed of three different cell types. The **basal cells** divide regularly to produce new **supporting cells**.

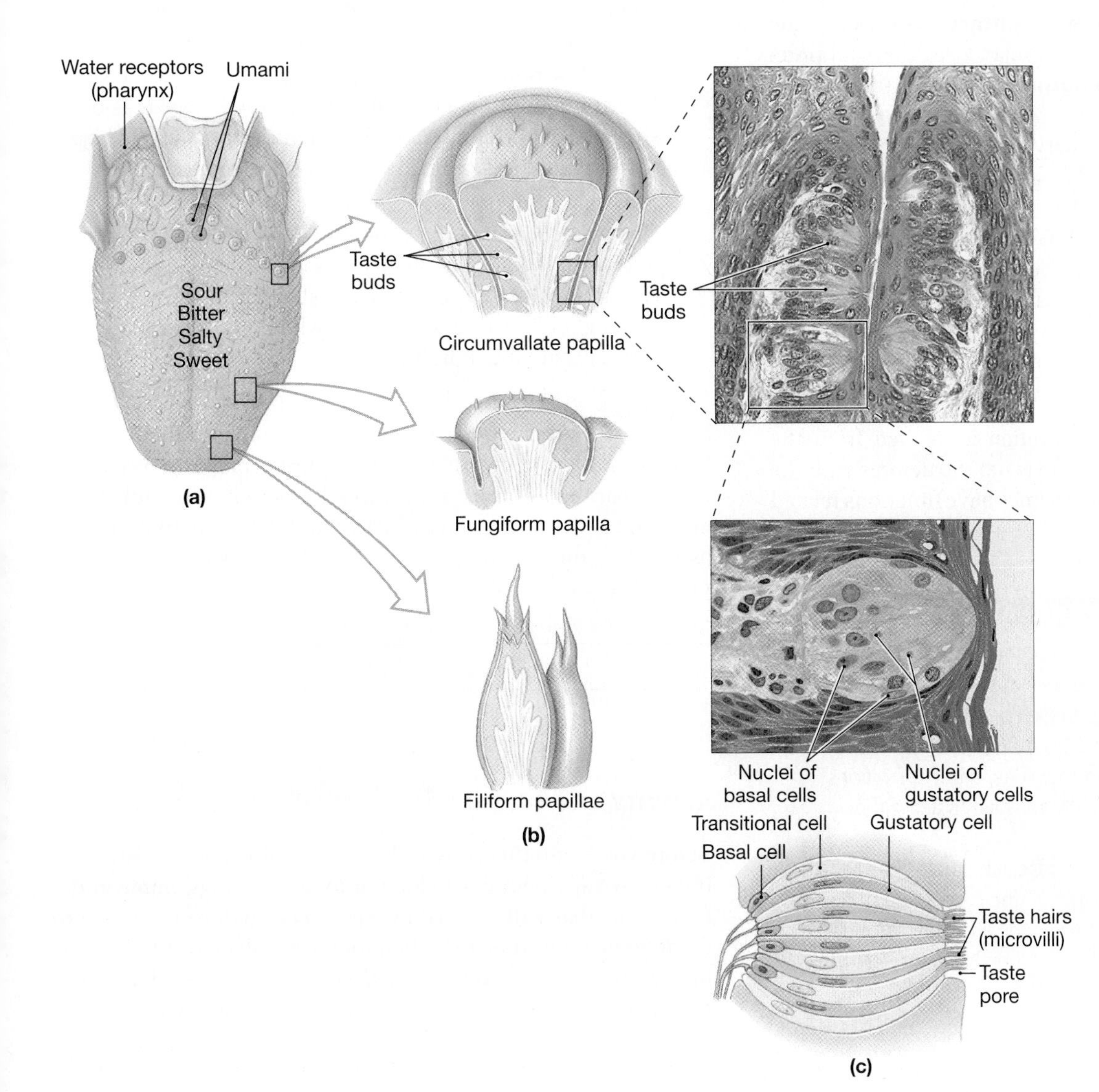

Figure 17.5 Taste buds on the tongue. a) The superior surface of the tongue, showing the relative positions of the papillae; **b)** enlargements of the three types of papillae on the tongue's surface; **c)** light micrographs and corresponding diagram that illustrates the structure of taste buds; corresponding diagram that illustrates the structure of taste buds (LM × 300).

The supporting cells, in turn, differentiate into **gustatory cells**, which have a lifespan of about 10 days.

Basal cells are oval or rounded cells at the base of a taste bud. Supporting cells and gustatory cells are similar to each other in appearance. Both are elongated and taper at both ends, and extend the entire length of the taste bud. In addition, both cells possess microvilli, known as **taste hairs**, that extend through the taste pore. Gustatory cells act as taste receptors. However, all three cell types synapse with sensory nerve fibers, and it is unclear whether basal and supporting cells also have a gustatory function.

Taste buds are innervated by one of three cranial nerves.

- The **facial nerve (cranial nerve VII)** supplies the taste buds on the anterior two thirds of the tongue.
- The **glossopharyngeal nerve (cranial nerve IX)** innervates the posterior one third of the tongue.
- The **vagus nerve (cranial nerve X)** innervates the taste buds on the soft palate, pharyngeal wall, and epiglottis.

Gustatory cells, when stimulated by dissolved chemicals in foods and fluids, transmit electrical impulses at synapses with afferent axons from cranial nerves VII, IX, or X. Taste information is relayed to the **gustatory (solitary) nucleus** in the medulla oblongata. From there it is relayed to the **thalamus** and finally to the **primary gustatory cortex** in the parietal lobe.

Humans can detect five primary taste sensations: **sweet**, **sour**, **salty**, **bitter**, and **umami**. The gustatory sensation *umami* is associated with the taste of beef or chicken broth. The detection of monosodium glutamate (MSG), a flavor enhancer added to some foods, is due to the sensitivity of some taste buds to the umami sensation. Recent research has indicated the existence of taste receptors for water, particularly in the pharyngeal taste buds. These receptors appear to have a role in water balance and blood pressure regulation.

WHAT'S IN A WORD The word *gustation* is derived from the Latin word *gustacio*, which means, "to taste." Structures with the terms *gustation* or *gustatory* in their names have functions related to the sense of taste.

CLINICAL CORRELATION

It was once believed that the surface of the tongue could be divided into regions based on differences in sensitivities to the various taste stimuli. According to these so-called tongue maps, sweet sensations are perceived by taste buds on the anterior tip of the tongue, sour sensations on the sides, bitter sensations along the posterior surface, and salty sensations along the edges.

More recent evidence from taste research studies has shown that tongue maps are incorrect. Although there could be minor sensitivity differences, in general, taste buds are capable of detecting all taste sensations, regardless of their locations on the surface of the tongue.

ACTIVITY 17.3 Examining the Anatomy of Gustatory Structures

Gross Anatomy

1. Obtain a midsagittal section of the head (Figure 17.2) and locate the **oral cavity** and the **pharynx**. Identify the structures that contain taste buds (Figures 17.2 and 17.5a and b): the **tongue**, the **soft palate**, the **wall of the pharynx**, and the **epiglottis**.
2. Look in a mirror and stick your tongue out as far as possible to observe the surface papillae.
   - Circumvallate papillae form an inverted V near the back of the tongue.
   - Fungiform papillae, on the anterior two thirds of the tongue, contribute to the rough, uneven surface that you see.
   - Note the lack of papillae on the posterior one third of the tongue.

Microscopic Anatomy

1. Obtain a microscope slide that is specially prepared to illustrate taste buds in the epithelium of the tongue.
2. Under low power, locate taste buds embedded in the surface epithelium. In a typical slide, they will appear as oval-shaped cell clusters stained light pink and contrast sharply with the darker staining epithelial cells (Figure 17.5c).
3. Switch to high power and observe a taste bud more closely. Depending on the quality of your slide, you may be able to identify the nuclei of cells in the taste bud (Figure 17.5c). The nuclei in basal and transitional (supporting) cells are darkly stained while the nuclei in gustatory cells are much lighter.

QUESTION TO CONSIDER Taste buds are embedded in the epithelium of circumvallate and fungiform papillae. What type of papillae were you most likely observing in your microscope slide of the tongue? Explain. (Hint: Qualitatively assess the population density of taste buds on your slide preparation.)

____

____

ACTIVITY 17.4 Testing Gustatory Sensations

Before you begin this activity, be sure to inform your instructor if you have any allergies to food or food flavorings. Some of the food items that will be used in this activity will be unknown to the food taster so it is important for your instructor to have this information before you begin. If you do have an allergy to a food item that will be used, do not perform the activity.

Assessing Taste Bud Stimulation

1. Dry the superior surface of your tongue with a clean paper towel.
2. Place a few crystals of granulated sugar on the tip of your tongue. Have your lab partner note the time.
3. Alert your lab partner by raising your hand as soon as you can taste the sugar. Have your partner record the time of stimulation in Table 17.2.
4. Rinse your mouth with water and dry your tongue with a clean paper towel.
5. Repeat steps 2 through 4, but this time use a drop of 10% granulated sugar solution in place of the sugar crystals.

   Form a Hypothesis **Before you begin step 5, make a prediction of the results.**

   ____

6. Record the time of stimulation in Table 17.2.
7. Did it take a longer, shorter, or the same length of time to detect the sweet taste of the sugar solution compared with the sugar crystals? Explain your results.

   ____

   ____

8. Repeat steps 1 through 6, using crystals of table salt; then repeat again using a 10% salt solution. Be sure to use a clean paper towel each time you dry the surface of your tongue. Record your results in Table 17.2.
9. In terms of the length of time to detect taste, how do your results with salt compare with your earlier results with sugar? Explain.

   ____

   ____

Investigating the Link Between Taste and Inheritance

1. The ability to detect the bitter taste of phenylthiocarbamide (PTC) is an inherited trait that is present in about 70% of the population. If you assume that your laboratory class reflects the general population, predict the number of PTC tasters and nontasters in your class and record the result in Table 17.3.
   - Predicted # PTC tasters = 0.70 × total class population
   - Predicted # PTC nontasters = 0.30 × total class population
2. Obtain a specially prepared paper strip flavored with PTC.
3. Place the paper strip on your tongue and chew it so that PTC can mix with your saliva. Do not swallow the paper.
4. Can you detect the bitter taste of PTC? ______________
5. Tally the actual numbers and calculate the percentages of PTC tasters and nontasters in your class. Record your results in Table 17.3.
   - % PTC tasters = (# PTC tasters ÷ total class population) × 100
   - % PTC nontasters = (# PTC nontasters ÷ total class population) × 100
6. Are the predicted results in close agreement with the actual results? If not, suggest a reason for the disparity.

   ____

   ____

CLINICAL CORRELATION

Research has shown that the inherited ability to detect PTC is strongly correlated with the ability to taste other bitter substances. In many mammals, the ability to perceive bitter tastes is believed to be a warning against eating poisonous substances. In humans, our understanding of the genetic basis of PTC detection is being used by the food industry to improve the taste of some foods.

Table 17.2 Assessing the Stimulation of Taste Buds

Test substance		Time of stimulation (sec)
1. Table sugar	a. Solid crystals	
	b. 10% solution	
2. Salt	a. Solid crystals	
	b. 10% solution	

Table 17.3 Percentage of Phenylthiocarbamide Tasters in the Laboratory Class and the General Population

		Predicted class population		Actual class population	
	General population (%)	Number	%	Number	%
PTC tasters	70%		70%		
PTC nontasters	30%		30%		

Examining the Effect of Olfaction on Gustation

1. Obtain small cubes of three to five different types of food that have been selected by your lab instructor. All the cubes should be the same size. The identity of the food types should never be revealed to your lab partner (the food taster) during this entire experiment. The food handler should wear gloves and use toothpicks to manipulate the food cubes.

   Form a Hypothesis **Before you begin the following procedure, make a prediction on the effect your sense of smell will have on your ability to taste food.**

   ____

   ____

2. Have your lab partner dry his or her tongue with a paper towel, and then sit in a relaxed position with the eyes closed and the nostrils pinched shut.
3. Without revealing its identity, place the first food cube on your lab partner's dry tongue.
4. Ask your lab partner to identify the food item in the following sequence of conditions.
   - Immediately after placing the cube on the tongue, with eyes and nostrils closed
   - After chewing the cube, with eyes and nostrils remaining closed
   - After opening the nostrils, but with eyes still closed
5. Record your results in Table 17.4 using the following protocol.
   - For each food item, place one plus (+) sign to indicate the condition under which the food was correctly identified.
   - If the food was incorrectly identified, place a minus (−) sign under the appropriate condition.
   - If, after a correct identification, the taste was enhanced under subsequent conditions, place two (or three) plus (+) signs in the appropriate space.
6. Repeat steps 2 through 5 for the other types of food. Be sure your lab partner rinses his or her mouth with water and dries the tongue with a clean paper towel between each taste test.
7. Record all results in Table 17.4, using the protocol as described.

   Assess the Outcome **Did your prediction agree with your experimental results? Explain.**

   ____

   ____

QUESTION TO CONSIDER In the previous activity, you examined the relationship between olfactory and gustatory sensations. Given what you have learned, explain why your sense of taste is reduced when you have a cold and your nose is blocked.

____

____

Vision

The eyes are spherical structures, approximately 2.5 cm (1 in) in diameter, that function as the visual sense organs of the body. They are located within the bony orbits of the skull, where they are cushioned by a protective layer of fat. Each eye is composed of three layers (tunics) of tissue (Figure 17.6a): the outer **fibrous tunic**, the middle **vascular tunic (uvea)**, and the inner **sensory** or **neural tunic (retina)**.

The sensory receptors for vision are located in the retina, which consists of an outer **pigmented layer** and an inner **nervous layer**. In the pigmented layer, **melanocytes** produce the pigment **melanin**, which prevents reflection or scattering of light waves when they reach the retina. The nervous layer contains three types of neurons each in their own cellular layer (Figure 17.7a). **Photoreceptor cells** form the outermost layer.

Table 17.4 Identification of Food Items

Food item	Conditions for identification: On tongue before chewing; and nostrils closed eyes	Chewing with eyes and nostrils closed	Chewing with nostrils open and eyes closed
1.			
2.			
3.			
4.			
5.			

The two types of photoreceptive cells, the **rods** and **cones**, are sensitive to light stimuli. The rods are important for our ability to see during low light conditions and for peripheral vision. They are highly concentrated along the periphery of the retina and tend to decrease in number toward the center. The cones are important for visual acuity (viewing objects in sharp focus) and color vision. They are most concentrated in central regions of the retina and tend to diminish in number toward the periphery. **Bipolar cells** are located in the middle layer and form synapses with the photoreceptive cells. **Ganglion cells**, which form the innermost layer, synapse with the bipolar cells. The axons of the ganglion cells form the **optic nerve** (Figure 17.7b)

Light enters the eye (Figure 17.6b) and, upon reaching the retina, must first penetrate the ganglion and bipolar cell layers before reaching the photoreceptor cells (Figure 17.7a). When photoreceptor cells are stimulated, they transfer the impulses to bipolar cells, which in turn form synapses with ganglion cells. **Horizontal cells** are located at the level where photoreceptor cells synapse with bipolar cells and **amacrine cells** are found where bipolar cells synapse with ganglion cells (Figure 17.7a). These two cell types form lateral connections between cells in the three retinal layers. Their actions improve visual contrast by allowing the retina to make adjustments for various intensities of light stimuli.

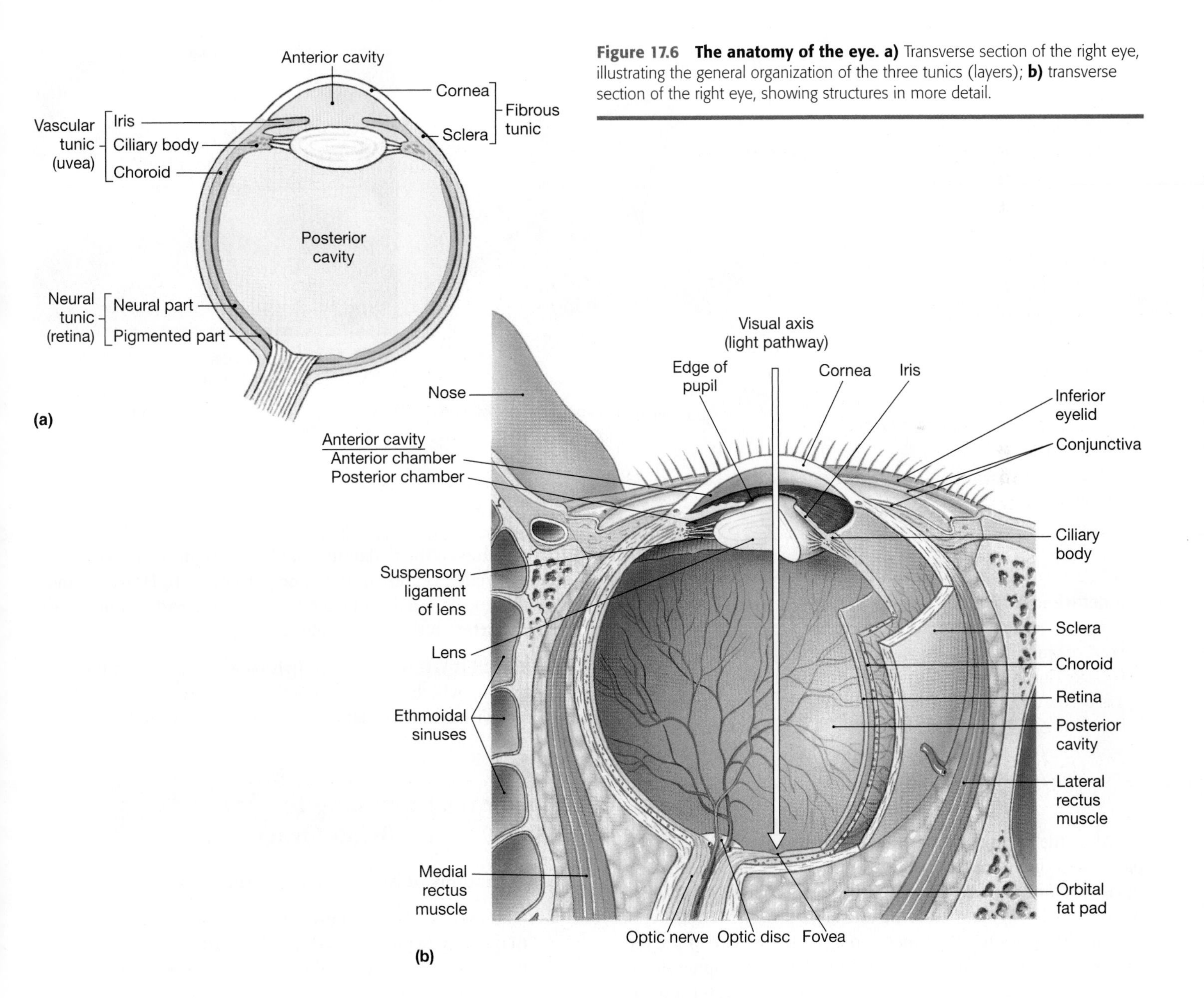

Figure 17.6 The anatomy of the eye. a) Transverse section of the right eye, illustrating the general organization of the three tunics (layers); **b)** transverse section of the right eye, showing structures in more detail.

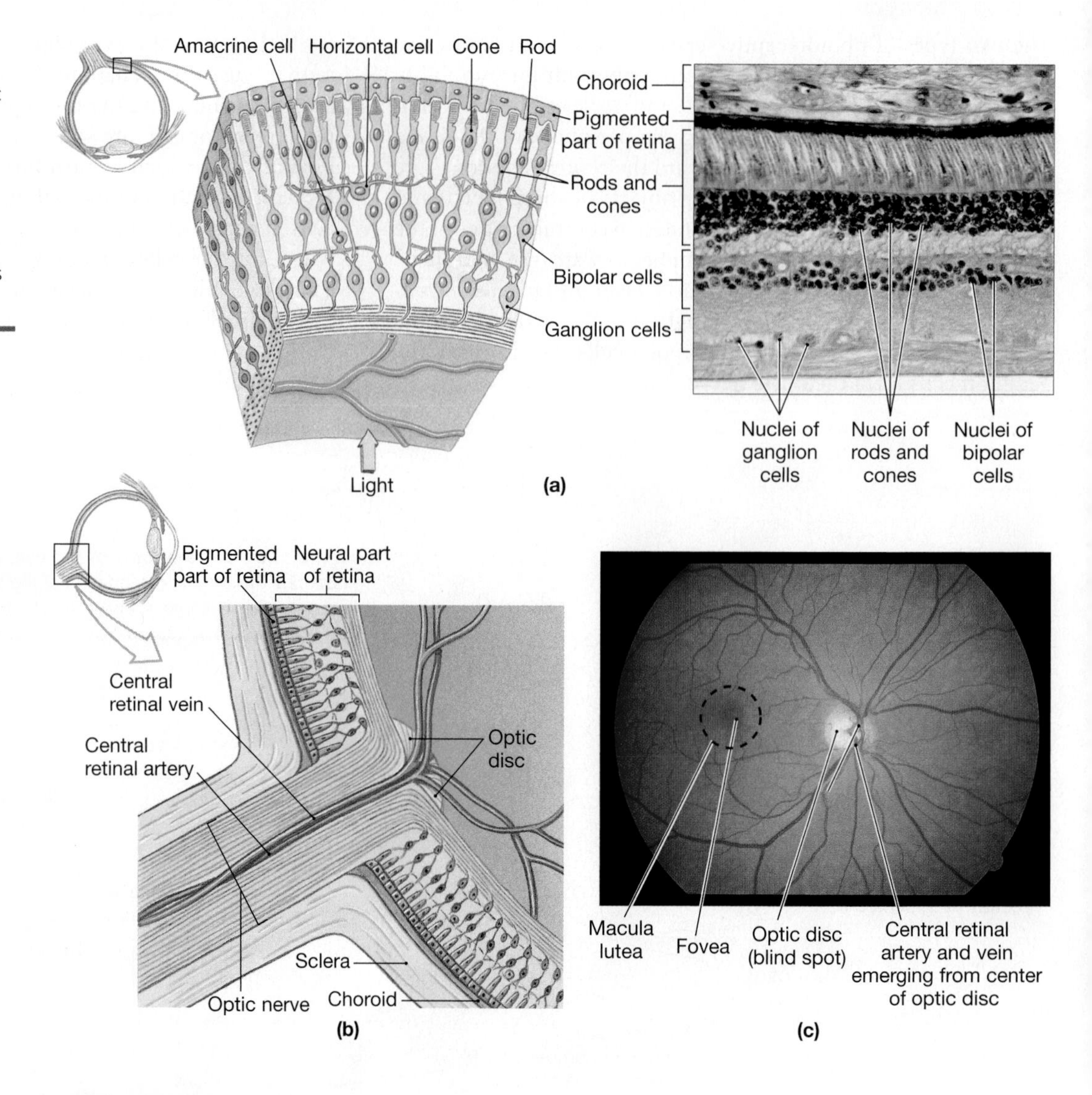

Figure 17.7 The structure of the retina. a) Diagram and light micrograph illustrating the light microscopic structure of the retina. The cellular arrangement of the nervous layer is highlighted (LM × 400); **b)** the optic nerve exiting the eyeball at the optic disc; **c)** photograph of the retina as viewed through the pupil. The optic disc, macula lutea, and fovea centralis are shown.

CLINICAL CORRELATION

A **detached retina** is a separation between the pigmented and nervous layers of the retina. It can be caused by a blow to the head or progressive degeneration of the retina due to disease or old age. Fluid can accumulate between the separated layers. If this occurs, the nervous layer bulges forward, causing distorted or loss of vision in the affected visual field. A detached retina can be repaired by laser surgery or cryosurgery (applying extreme cold to the damaged area).

After the optic nerves exit the eyeballs (Figure 17.7b), they pass through the **optic canals** (Figure 17.3) and meet at the **optic chiasm** (Figure 17.4). At the optic chiasm, optic nerve fibers, which originate from the medial half of the retinas, cross over to the opposite side. Fibers from the lateral half of the retinas remain on the same side (Figure 17.8). The optic chiasm gives rise to the **optic tracts**, which project to the **lateral geniculate bodies** in the thalamus. The lateral geniculates relay some fibers to reflex centers in the **superior colliculi.** However, most visual information is sent along the **optic radiations** to the **visual cortex** in the occipital lobes (Figure 17.8).

WHAT'S IN A WORD The word *vision* is derived from the Latin word *visio*, which means "to see." Structures with the terms *vision* or *visual* in their names have functions related to the ability to see.

ACTIVITY 17.5 Examining the Anatomy of Vision Structures

Gross Anatomy of Accessory Eye Structures

1. Obtain a skull and remove the calvaria to view the cranial cavity (Figure 17.3a). At the border between the anterior and middle cranial fossae, identify the optic canals in the sphenoid bone. Pass a pipe cleaner or slender

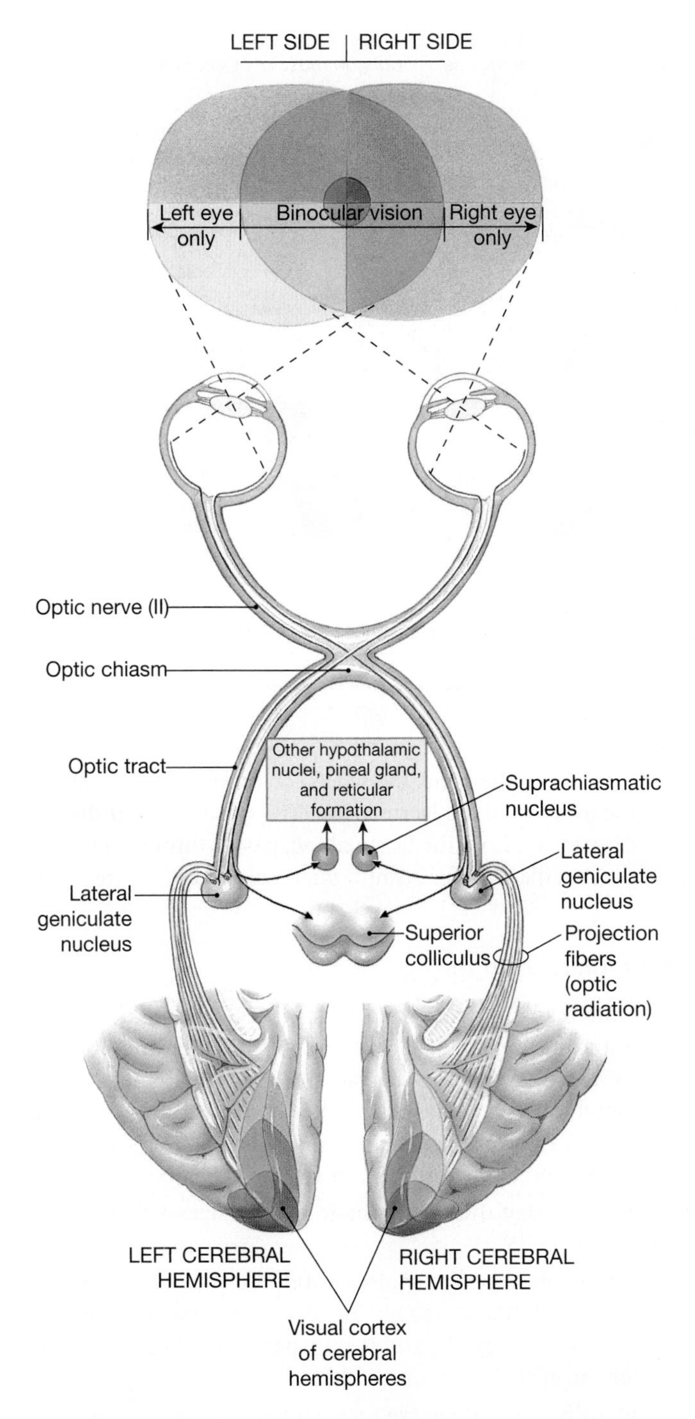

Figure 17.8 The neural pathway for vision. At the optic chiasm, there is partial crossing over of optic nerve fibers. Consequently, the visual cortex on each side of the brain receives information from both visual fields.

probe anteriorly into one of the optic canals. *Where does the pipe cleaner emerge?*

____

____

2. Obtain a model of the brain and view its inferior surface (Figure 17.4). Identify the optic chiasm, where the two optic nerves converge and partial crossing over of nerve fibers occurs.
3. In the cranial cavity of the skull, locate the **sella turcica** of the sphenoid bone (Figure 17.3a). Place the brain in the cranial cavity (do not expect an exact fit, but do the best you can) and notice that the optic chiasm is positioned just superior to the sella turcica.
4. View the skull from an anterior view and locate the **orbital cavities.** Seven bones contribute to the formation of the orbits: **frontal, sphenoid, zygomatic, maxilla, lacrimal, ethmoid,** and **palatine** (Figure 17.3b). Attempt to identify these bones in the orbits. This is easiest on high-quality skulls with well-marked sutures and little or no damage.

CLINICAL CORRELATION

A traumatic injury to the eye can cause a "blowout" fracture to the orbit. The bones most vulnerable to injury are the very thin ethmoid and lacrimal along the medial orbital wall and the maxilla on the inferior wall. Fractures to these bones may also involve the adjacent paranasal sinuses.

5. Observe your lab partner's face or look at yourself in a mirror, and identify the following accessory structures of the eye (Figure 17.9a).
   - The **eyebrows** are bands of thick hair that arch across the superior margins of the orbits and protect the eyeballs from perspiration and direct sunlight.
   - The **eyelids (palpebrae)** are fleshy coverings composed mostly of skeletal muscle and skin. Blinking the eyelids keeps the eye surface moist at all times. Each eyelid contains modified sebaceous glands called **tarsal (Meibomian) glands** that prevent the eyelids from sticking together while blinking.
   - The **palpebral fissure** is the space between the two eyelids. Through the fissure, you can see only the anterior one sixth of the eyeball, which includes the transparent **cornea** that covers the **iris** and **pupil**, and a small fraction of the **sclera** (white of the eye).
   - The two eyelids converge medially and laterally at the **medial canthus** and **lateral canthus.**
   - The **lacrimal caruncle** is the small reddish mass of soft tissue at the medial canthus. Glands in this structure produce mucus that often accumulates while sleeping.
   - **Eyelashes** are the thick hairs emerging from the margins of both eyelids that filter out airborne particulate matter.
   - The **conjunctiva** is a thin, protective mucous membrane that lines the inner surface of the eyelids (**palpebral conjunctiva**) and curves onto the anterior surface of the eye (**bulbar** or **ocular conjunctiva**) to cover the sclera. The shiny surface that you can observe on the sclera is the bulbar conjunctiva.

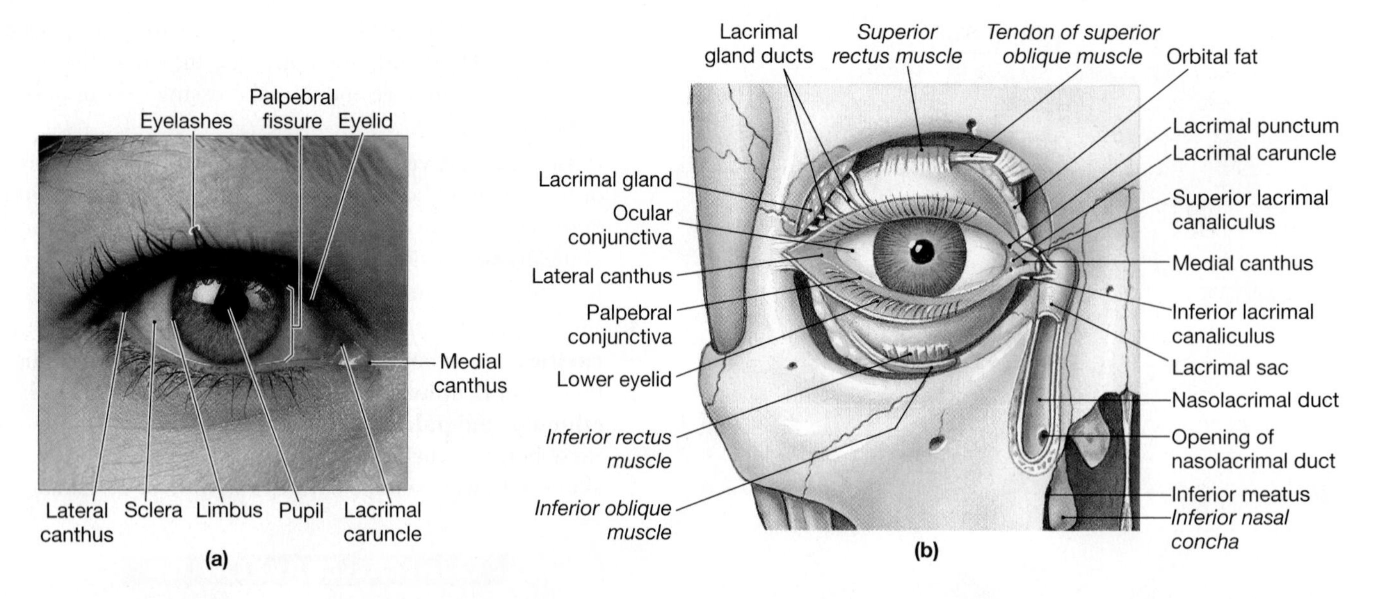

Figure 17.9 Accessory structures of the eye. a) Photograph of the right eye, illustrating surface structures; **b)** diagram of a deep dissection of the right orbital cavity, illustrating the lacrimal apparatus.

CLINICAL CORRELATION

Bacterial infections, physical trauma, or chemical irritations can lead to inflammation and damage to various eye structures. Infections of the tarsal glands can lead to the growth of a small lump or cyst, known as a **chalazion.** Similar infections to the sebaceous ciliary glands may result in the formation of a pus-filled sac or swelling, called a **sty.** Damage and swelling of the conjunctiva is referred to as **conjunctivitis.** A common outcome of this condition is redness over the surface of the sclera due to the dilation of blood vessels. For this reason, conjunctivitis is also called **pinkeye.**

6. On the skull, locate a small depression in the frontal bone, just inside the superolateral margin of the orbital cavity. This depression marks the location of the **lacrimal gland** (Figure 17.9b), which produces watery, alkaline **tears.** The lacrimal gland and its accessory structures are known as the **lacrimal apparatus** (Figure 17.9b). The accessory structures regulate the movement of tears across the surface of the eye and drain it into the nasal cavity. Tears contain an antibacterial enzyme called **lysozyme.** Thus, in addition to cleaning and lubricating the surface of the eyeball, this fluid protects the eye from many bacterial infections.
7. On your lab partner or yourself, once again locate the lacrimal caruncle at the medial canthus. When you blink, tears are swept across the surface of the eye to the lacrimal caruncle, where they drain into two small pores called **lacrimal puncta** (singular = **punctum**). From there, tears flow through two small canals to reach the **lacrimal sac** (Figure 17.9b). On a skull, locate the lacrimal bone on the medial wall of the orbit (Figure 17.3b). A small groove traveling along the bone, known as the **lacrimal fossa**, marks the location of the lacrimal sac. The **nasolacrimal duct** receives tears from the lacrimal sac, passes through a canal in the maxilla, and drains into the nasal cavity (Figure 17.9b).

Gross Anatomy of the Eye

1. Obtain an anatomical model of the eye and identify its outermost layer, the fibrous tunic. Note that the fibrous tunic has two parts. The vast majority is a tough, opaque membrane known as the sclera, or white of the eye. It consists of tightly bound elastic and collagen fibers and provides protection and support for the eyeball.
2. Observe that the six **extrinsic eye muscles** attach to the sclera.
3. The smaller anterior portion of the fibrous tunic is the cornea. Unlike the sclera, the cornea is more delicate, lacks blood vessels, and is transparent to allow light to enter the eyeball (Figure 17.6b).
4. Identify the optic nerve (cranial nerve II) as it exits the eye along the posterior surface of the sclera (Figures 17.6b and 17.7b).
5. Remove a portion of the fibrous tunic to reveal the middle tissue layer, the vascular tunic (Figure 17.6b). Observe that this layer is composed mostly of the **choroid**, a thin layer of tissue that contains numerous small blood vessels. The choroid is densely populated with melanocytes. The melanin produced by these cells prevents incoming light waves from being reflected back out of the eye.
6. Identify the two anterior extensions of the choroid: the ciliary body and the iris (Figure 17.6b). The ciliary body

is a thickened ring of tissue, composed mostly of smooth muscle that surrounds the lens. Identify the **suspensory ligaments** that connect the ciliary body to the lens. The iris is composed of pigment cells that are responsible for eye color and two layers of smooth muscle that surround a central opening known as the pupil. Contraction of smooth muscle in the iris causes dilation or constriction of the pupil, and thus, regulates the amount of light entering the eye.

7. Identify the **lens.** Note that it has a biconvex shape and, like the cornea, it is transparent (Figure 17.6b). The lens is composed of several tightly packed layers of fibrous proteins, arranged like the layers of an onion. It refracts (bends) incoming light waves and focuses them onto the retina. Thus, it has a function similar to the lens of a camera, which focuses light onto photographic film. As described earlier, the lens is held in its position by the suspensory ligaments of the ciliary body (Figure 17.6b). This arrangement allows the ciliary body to control the change in the shape of the lens while focusing on near or distant objects.

CLINICAL CORRELATION

A **cataract** occurs when the lens thickens and becomes less flexible. As a result, the normally transparent lens becomes clouded and vision is blurred. Cataracts can be corrected by removing the damaged lens and replacing it with an artificial implant.

8. Identify the innermost tissue layer of the eye, the neural tunic or retina. The retina contains three layers of nerve cells (Figure 17.7a): the photoreceptive cells (rods and cones), bipolar cells, and ganglion cells. Axons of the ganglion cells form the optic nerve (Figure 17.7b).
9. Identify the **optic disc** along the posterior surface of the retina (Figures 17.7b and c). This marks where the optic nerve exits the eyeball. The optic disc is also called the **blind spot** because it lacks photoreceptor cells. Therefore, light that is focused on this region cannot be seen.
10. Locate the **macula lutea**, just lateral to the blind spot. In the center of the macula lutea is a small region known as the **fovea centralis** (Figures 17.7b and c). These two structures contain the highest concentrations of cone cells. Thus, the sharpest vision is attained when images are focused on these retinal regions. The fovea centralis is unique because it contains only cones and is the area of keenest vision.

WHAT'S IN A WORD In Latin, the term *macula lutea* means "yellow spot." The name refers to the unique region ("spot") on the retina where the concentration of photoreceptive cone cells is the greatest. The macula lutea is occasionally referred to as the *macula retinae* ("spot on the retina"). The term *fovea centralis,* also derived from Latin, means "central pit." It is a shallow depression in the center of the macula lutea.

11. Identify the two major cavities of the eyeball (Figure 17.6).
    - The larger of the two, the **posterior cavity (vitreous chamber)**, extends from the lens, anteriorly, to the retina, posteriorly. It is filled with a jellylike substance called the **vitreous humor.**
    - The **anterior cavity** extends from the cornea, anteriorly, to the lens, posteriorly. Note that the anterior cavity is subdivided into the **anterior chamber**, between the cornea and iris, and the **posterior chamber**, between the iris and lens. The anterior cavity is filled with a watery fluid called **aqueous humor.**

CLINICAL CORRELATION

The vitreous humor that fills the posterior cavity of the eye helps to maintain normal intraocular pressure and holds the retina firmly against the choroid.

In the anterior cavity, aqueous humor is produced by the ciliary body in the posterior chamber and is drained by veins in the wall of the anterior chamber. If normal drainage is blocked, intraocular pressure can increase and cause damage to the retina and optic nerve. This condition, called **glaucoma,** can be controlled in its early stages by administering eye drops that enhance the rate of fluid drainage.

Dissection of the Cow Eye

1. Obtain dissecting instruments, dissecting gloves, a face mask, and safety glasses.
2. Obtain a preserved or fresh cow eye and place it on a dissecting tray.
3. Notice the thick protective layer of fat on the external surface of the eye (Figure 17.10a). Carefully remove the fat, without damaging other structures.
4. Identify the optic nerve (cranial nerve II) as it exits from the posterior aspect of the eyeball. It will appear as a solid white cord.
5. Identify the sclera, or white of the eye, and notice that it comprises most of the external fibrous tunic. On the sclera, locate the attachments for the extrinsic muscles of the eye (Figure 17.10b). The shiny membrane that covers the anterior portion of the sclera is the bulbar (ocular) conjunctiva.
6. Locate the cornea, which is the anterior portion of the fibrous tunic (Figure 17.10b). Normally, the cornea is transparent, but if your eye is preserved, it will be opaque.
7. Hold the posterior portion of the eyeball securely on the dissecting tray. With a scalpel, make an incision about 6 mm (0.25 in) posterior to the cornea. The sclera is a relatively thick, fibrous layer, so be sure to apply enough pressure to cut through the wall. Be aware that aqueous humor from the anterior cavity could squirt out during this procedure.
8. Insert scissors into the initial incision and make a circular cut around the cornea, always remaining about 6 mm posterior to the corneal margin.

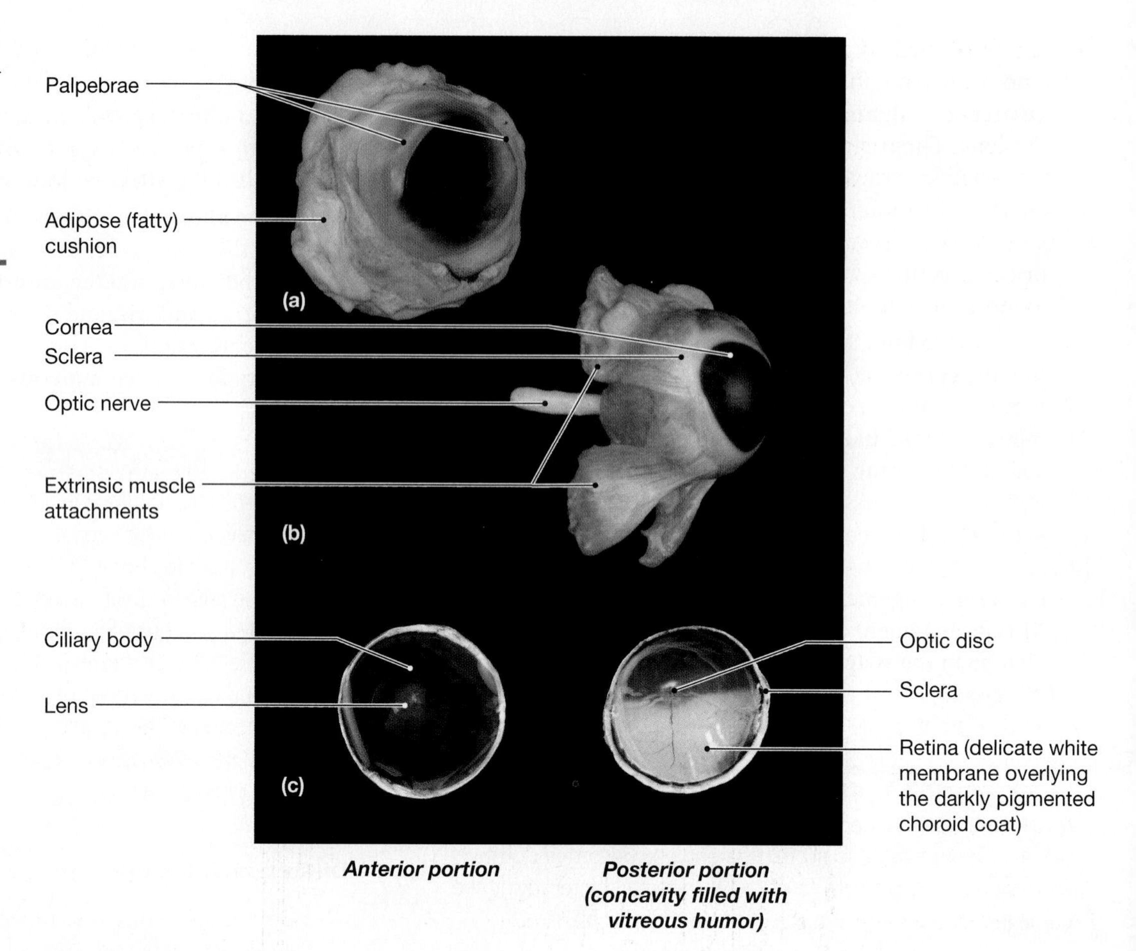

Figure 17.10 Dissection of the cow eye. a) The whole eye with adipose tissue attached; **b)** the whole eye with adipose tissue removed to illustrate the extrinsic eye muscles attached to the sclera; **c)** the cut into anterior and posterior portions to demonstrate internal structures.

9. Lift away the anterior portion of the eyeball. The aqueous humor will escape from the anterior cavity, but the vitreous humor should remain in the posterior cavity (Figure 17.10c).
10. Identify the following structures in the anterior portion of the eyeball:
    - The centrally located lens is a biconvex disc (Figure 17.10c). Like the cornea, it is normally transparent but will be opaque if your eye is preserved.
    - The ciliary body appears as a thick black ring around the lens (Figure 17.10c).
    - Gently lift one side of the lens with a blunt probe and attempt to identify the delicate suspensory ligaments. Notice that the suspensory ligaments connect the lens to the ciliary body.
    - Carefully remove the lens and identify the iris. The iris is heavily pigmented and will probably appear very dark or black. The opening through the center of the iris is the pupil.
11. Identify the following structures in the posterior portion of the eyeball.
    - The vitreous humor is the thick, gel-like substance that fills the posterior cavity.
    - Remove the vitreous humor and identify the retina, which appears as a whitish or yellowish layer. The retina can be easily separated from the underlying choroid, but is attached to the posterior surface of the eyeball at the optic disc (blind spot). Notice that the optic disc is the location where the optic nerve exits the eye.
12. Remove a portion of the retina to expose the deeply pigmented choroid. The choroid appears iridescent due to the presence of a membrane called the **tapetum lucidum**. This unique structure reflects some light back onto the retina, thus improving vision at night or during other periods when light intensity is low. The tapetum lucidum is not present in the human eye.

Microscopic Anatomy of the Retina

1. Obtain a prepared microscope slide of the retina.
2. View the retina with the low-power objective lens on your microscope. If possible, adjust the position of your slide so that the field of view is similar to what is depicted in Figure 17.7a. Thus, the choroid will be at the top of the field of view.
3. Identify the pigmented layer of the retina (Figure 17.7a). It appears as a darkly stained band deep to the choroid.

4. Locate the nervous layer of the retina, which contains the three bands of nerve cells. The photoreceptive cell layer, consisting of the rods and cones, is just deep to the pigmented layer of the retina. This is followed by a row of bipolar cells, and finally, a band of ganglion cells. The axons of the ganglion cells form the optic nerve. The cell types in the nervous layer can best be identified by locating their nuclei, which are organized into three distinct bands. (Figure 17.7a).
5. As you view the slide, realize that light first contacts the retina at the ganglion cell layer (Figure 17.7a). Light must pass through these cells and the bipolar cells to reach and stimulate the rods and cones in the photoreceptive cell layer.

QUESTIONS TO CONSIDER 1. The concentration of rods increases as you move from the center to the periphery of the retina. On the other hand, the concentration of cones increases as you move from the periphery to the center where the macula lutea and fovea centralis are located. If you compare the functions of rods and cones, how do you account for the different distribution patterns of these cells?

____

____

____

2. Compare the anatomy of the human and cow eyes and identify any differences in their structure.

____

____

____

ACTIVITY 17.6 Performing Visual Tests

Demonstrating the Blind Spot

1. Hold Figure 17.11 about 46 cm (18 in) from your eyes.
2. Close your left eye and focus your right eye on the plus (+) sign.
3. Move the figure slowly toward your face until the black dot disappears.
4. Explain why the dot becomes invisible. ______________________________

____

____

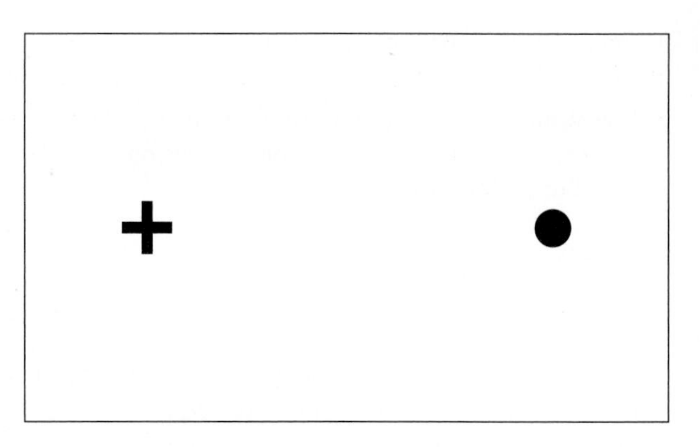

Figure 17.11 Demonstrating the blind spot. Follow the procedure in Activity 17.6 to illustrate how images focused on the optic disc cannot be seen.

Demonstrating Binocular Vision and Depth Perception

Partial crossing over of nerve fibers from the optic nerves occurs at the optic chiasm. Thus, the visual area in each cerebral hemisphere receives and interprets visual information originating from both eyes (Figure 17.8). Alone, each eye has a different view, but together, the left and right visual fields overlap (Figure 17.8), providing humans (and other primates) with **binocular vision**. The information from overlapping visual fields is translated by the visual cortex into **three-dimensional vision**, and the ability to locate objects in space (**depth perception**).

1. Hold a pencil vertically, at arm's length, in front of your face. Focus on the pencil with both eyes. To be able to see the pencil clearly, make sure that there is a sharp contrast between it and the background of your visual field.
2. Close your right eye completely but continue to look at the pencil with your left eye. Then, close your left eye and open your right, always focusing on the pencil.
3. Repeat step 2 several times. Do you notice any change in the relative position of the pencil when you switch back and forth between the visual fields of the left and right eyes? Provide an explanation for your results. ______________________________

____

____

4. Once again, hold a pencil vertically at arm's length, in front of your face. If you are right handed, hold the pencil with your left hand; if you are left handed, hold the pencil with your right hand.
5. With both eyes open, quickly place the tip of the index finger of your free hand on the top of the pencil. Repeat this action two or three times.
6. Place your free hand back to the side of your body.
7. Close your left eye and again place the tip of your index finger on the top of the pencil. As before, perform this action quickly, not slowly and thoughtfully. Repeat this action two or three times.
8. Repeat step 7 with your right eye closed and left eye open.

9. Compare the results for each test by addressing the following questions.
   - Was it more or less difficult to place the tip of your finger on the pencil when only one eye was open compared to both eyes open?

   ____

   ____

   - How accurate were you in placing the tip of your finger on the pencil when both eyes were open compared to only one eye open?

   ____

   ____

   - Did you notice any differences in your ability to accurately perform the task when only the right eye was open compared to only the left eye open?

   ____

   ____

   - How do you account for any differences in the degree of difficulty or accuracy that you have noted? If you did not notice any differences, provide an explanation for that result, as well.

   ____

   ____

Near-Point Accommodation

As discussed earlier, smooth muscle contractions in the ciliary body can create changes in the shape of the lens to focus on near or distant objects. For example, when focusing on distant objects, the lens is flat; when focusing on near objects, it is more spherical (Figure 17.12). Thus, the lens exhibits some degree of elasticity. As a person ages, the elasticity of the lens decreases and focusing on close objects becomes more difficult (this explains why many older people need reading glasses). This condition is called **presbyopia**. The elasticity of the lens can be tested by measuring the **near-point accommodation**, the distance from the eyes at which an object begins to be blurred or distorted. Perform the following simple test to measure your near-point accommodation.

WHAT'S IN A WORD The term *presbyopia* is derived from two Greek words: *presbys,* which means "old man," and *ops,* which means "eye." Thus, *presbyopia* means "eyes of an old man" or "old vision."

Form a Hypothesis Before you begin the following procedure, refer to Table 17.5 and predict the near-point accommodation for you and your lab partner. Record this information in Table 17.6.

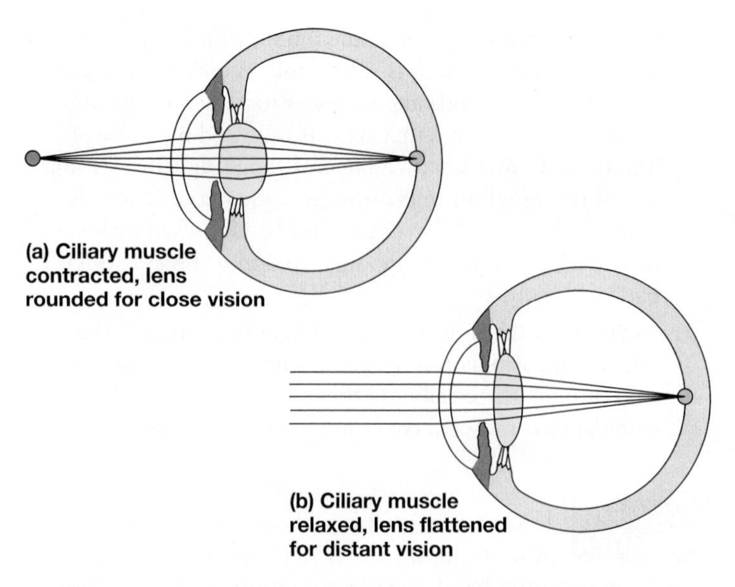
(a) Ciliary muscle contracted, lens rounded for close vision
(b) Ciliary muscle relaxed, lens flattened for distant vision

Figure 17.12 Visual accommodation. The lens can change its shape so that objects viewed at various distances can be properly focused on the retina. **a)** The condition of the lens for close vision; **b)** the condition of the lens for distant vision.

Table 17.5 Near-Point Accommodation Distances at Various Ages

Age (yrs)	Accommodation distance (cm)
10	7.5
20	9.0
30	11.5
40	17.2
50	52.5
60	83.3

1. Hold a dissecting pin at arm's length in front of your eyes.
2. Close one eye and slowly move the pin toward your face until the image just begins to become blurred or distorted.
3. Hold the pin at this position and have your lab partner measure the distance, in centimeters, between the pin and your eye. Repeat the procedure to confirm your results and record the data in Table 17.6.
4. Repeat the procedure for the other eye. Record the data in Table 17.6.

Assess the Outcome Compare your results with those of your lab partner and one or two other students in the laboratory and consider the following:

- For each subject, was the actual near-point accommodation about the same for both eyes?

____

____

Table 17.6 Near-Point Accommodation

Subject	Age	Near-point accommodation (cm)		
		Predicted Value	Actual Value, Left Eye	Actual Value, Right Eye
1.				
2.				
3.				
4.				

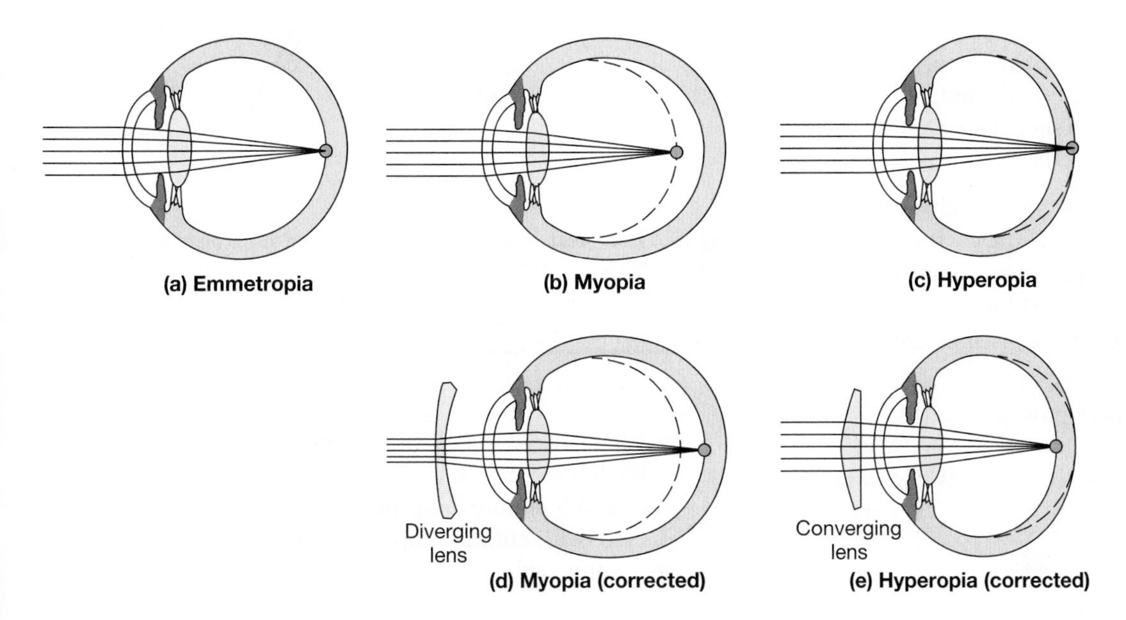

Figure 17.13 The function of the lens during normal vision and visual abnormalities. a) In normal vision (emmetropia), the lens focuses the image on the retina; **b)** myopia (nearsightedness) is the inability of the lens to focus distant objects on the retina. In this situation, objects are focused in front of the retina; **c)** hyperopia (farsightedness) is the inability of the lens to focus close objects on the retina. In this situation, objects are focused behind the retina; **d)** and **e)** both myopia and hyperopia can be treated with corrective lenses.

- For each subject, compare the actual near-point accommodation (for both eyes) with the predicted value and decide whether it is normal for your age.

____

____

Visual Acuity

Visual acuity refers to an individual's ability to see objects clearly at various distances (sharpness of vision). Visual acuity is generally tested by using a **Snellen chart.** The chart contains rows of letters printed in progressively smaller sizes on a white background. Individuals with normal vision should be able to clearly see the letters of a given size from a specific distance. That distance is printed at the end of each line. You can test your own visual acuity by performing the following test. If you wear glasses, perform the test without the glasses and then repeat the test with your glasses. You should not remove contact lenses in the laboratory, but make a note that you are wearing them.

1. Stand 20 ft from a Snellen chart.
2. Cover one eye and read each row of letters from top to bottom until you can no longer read the figures; have your lab partner check for accuracy.
3. Check the line that was last read accurately. If "20/20" is printed at the end of the line, then you have normal vision (**emmetropia**; Figure 17.13a). If the ratio is less than one, it indicates that you have difficulty seeing distant objects; you are **nearsighted** or **myopic** (Figure 17.13b). A ratio of 20/40, for example, indicates that objects you can clearly see at 20 ft are clearly seen by a person with normal vision at 40 ft. If the ratio is greater than one, it suggests that your vision is better than normal. A ratio of 20/15, for example, indicates that objects you can clearly see at 20 ft are clearly seen by a person with normal vision at only 15 ft.

4. Record your results in the space provided.
5. Repeat the test for the other eye, and then with both eyes. Record your results in the space provided.

Visual acuity	Without glasses	With glasses
Right eye:	____________	____________
Left eye:	____________	____________
Both eyes:	____________	____________

Test for Astigmatism

If there are defects in the curvature of the lens or cornea, light entering the eye will not be focused at a single point on the retina. This condition, known as **astigmatism**, can cause blurred vision. You can test for astigmatism in your own eyes by performing the following test. If you wear glasses, perform the test without them and then repeat the test with them. You should not remove contact lenses in the laboratory, but make a note that you are wearing them.

1. With one eye closed, view the astigmatism chart in Figure 17.14. Focus on the center of the chart.
2. If all the radiating lines are in sharp focus, there is no astigmatism. If some of the lines are blurred or appear lighter than others, some astigmatism is apparent.
3. Repeat the test with the other eye.
4. Record your results in the space provided.

Astigmatism	Without glasses (+/−)	With glasses (+/−)
Presence in right eye:	____________	____________
Presence in left eye:	____________	____________

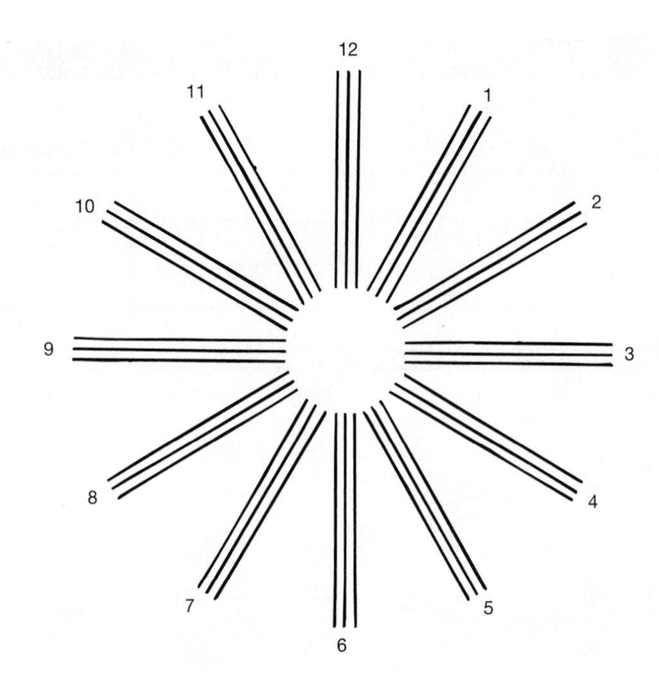

Figure 17.14 Test chart for astigmatism. The test is performed with one eye closed and the test eye focused on the center of the chart. If all the radiating lines are in sharp focus, there is no astigmatism. If some of the lines are blurred or appear lighter than others, some astigmatism is apparent.

CLINICAL CORRELATION

Lasik eye surgery is a procedure during which a laser is used to reshape the curvature of the cornea. The technique is designed to correct the refraction of light by the cornea so that the lens can properly focus an image onto the retina. Lasik surgery is becoming an increasingly acceptable alternative to wearing glasses or contact lenses.

Demonstrating the Pupillary Reflex

1. Have your lab partner shield his or her left eye from any light that is directed to the right side of the face. This can be accomplished by closing the left eye and holding a piece of cardboard or index card on the side of the nose to serve as a barrier between the eyes.
2. Shine a flashlight, at an angle (not directly), into the open right eye for about 5 to 10 seconds at a distance of 20 cm (about 8 in). In the space provided, describe the change that occurs in the size of the right pupil.
3. Wait 3 to 5 minutes and repeat step 2 with the left eye open and the right eye protected from the light. In the space provided, describe the change that occurs in the size of the left pupil.
4. Wait 3 to 5 minutes and shine a flashlight, at an angle, on the face with both eyes open. Describe the results in the space provided.

	Result
Right eye open, left eye closed:	____________
Left eye open, right eye closed:	____________
Both eyes open:	____________

5. After the pupils of both eyes have returned to normal size under normal room light conditions, have your lab partner move into a dark room (one without windows, if available) for 3 to 5 minutes.
6. When your lab partner merges from the dark room, immediately observe the conditions of his or her pupils. Describe any change that you observe. ____________

____

____

7. Describe the actions of the smooth muscle layers in the iris (Figure 17.15) that account for the results that you observed. ____________

____

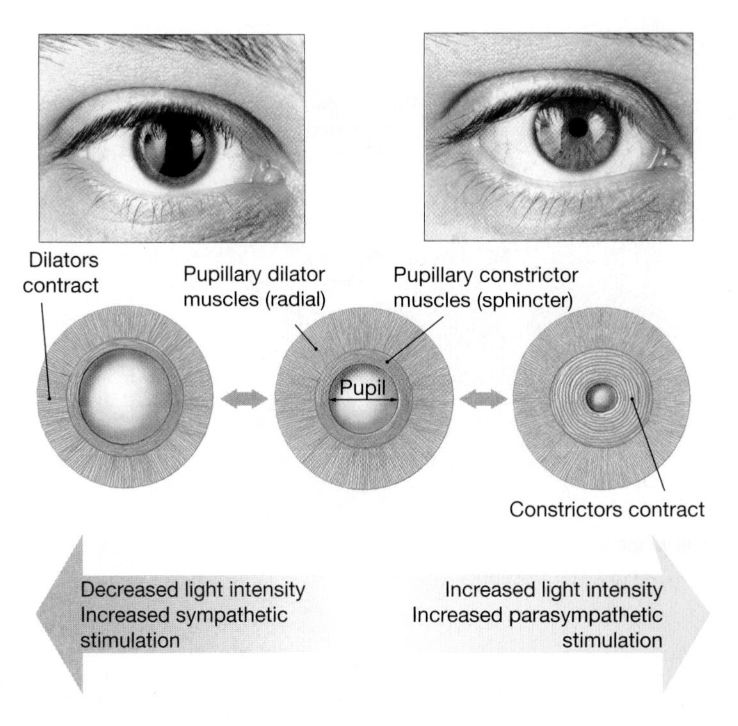

Figure 17.15 Smooth muscle layers in the iris. Contraction of the inner circular muscle layer constricts the pupil. Contraction of the outer radial muscle layer dilates the pupil.

Test for Color Blindness

The photoreceptor cells, known as cones, are responsible for color vision. In the retina, the three types of cones are each able to absorb light at specific wavelengths: Red cones absorb red light, blue cones absorb blue light, and green cones absorb green light. The lack of one or more types of cone can cause **color blindness**.

CLINICAL CORRELATION

Color blindness is a **sex-linked genetic trait,** which means that the gene for color blindness is located on the X chromosome. For a female to be color blind, she must possess the gene for color blindness on both X chromosomes. Males have only one X chromosome. Therefore, only one color-blind gene is needed for the trait to be expressed. The most typical pattern of inheritance for color blindness is for a mother who carries the gene, but has normal vision, to pass the trait to her sons. The vast majority of color-blind individuals are males. Expression of the gene in females is relatively rare.

You can test for color blindness by viewing **Ishihara color plates**.

1. Have your lab partner hold Ishihara color plates about 80 cm (30 in) in front of you. Be sure that you are viewing them in a brightly lit room.
2. Within 2 or 3 seconds, report to your partner what you see in the plates.
3. Have your lab partner record your responses and compare them with those in the plate book.
4. Inability to correctly identify the figures in the plates indicates some degree of color blindness.
5. Repeat this procedure to test for color blindness in your lab partner.
6. Are any students in your laboratory color blind?

____

QUESTION TO CONSIDER On a scale of 1 to 5, with 1 being "poor" and 5 being "excellent," how would you rate your eyesight? Based on the results of the visual tests that you have just completed, provide supporting evidence for your rating by writing a summary that describes the general condition of your eyesight.

	Poor				Excellent
Eyesight rating	1	2	3	4	5

Summary

____

____

____

____

Hearing and Equilibrium

The ear is a sensory organ for hearing and equilibrium (balance). Most of its components are located within the **petrous portion** of the temporal bone (Figure 17.3a) and are divided into three regions (Figure 17.16): the **external ear**, **middle ear (tympanic cavity)**, and **inner ear**.

The inner ear, also known as the **labyrinth**, is the location of the sensory organs for hearing and equilibrium. It consists of two primary compartments: the **bony labyrinth** and the **membranous labyrinth** (Figure 17.17). The bony labyrinth is a maze of bony passageways that course through the temporal bone. The membranous labyrinth is a network of membranous capsules contained within the bony labyrinth. Both compartments are filled with fluid; the bony labyrinth contains **perilymph** and the membranous labyrinth contains **endolymph**. Thus, the inner ear has a "tube within a tube" architecture, with the inner membranous labyrinth floating in the fluid contained in the bony labyrinth (Figure 17.17b). The fluids in the labyrinth provide a medium for the conduction of sound waves for hearing and for the detection of changes in body position and movement for equilibrium.

The labyrinth is composed of three areas (Figure 17.17a). The **vestibule** is the central area and is subdivided into the

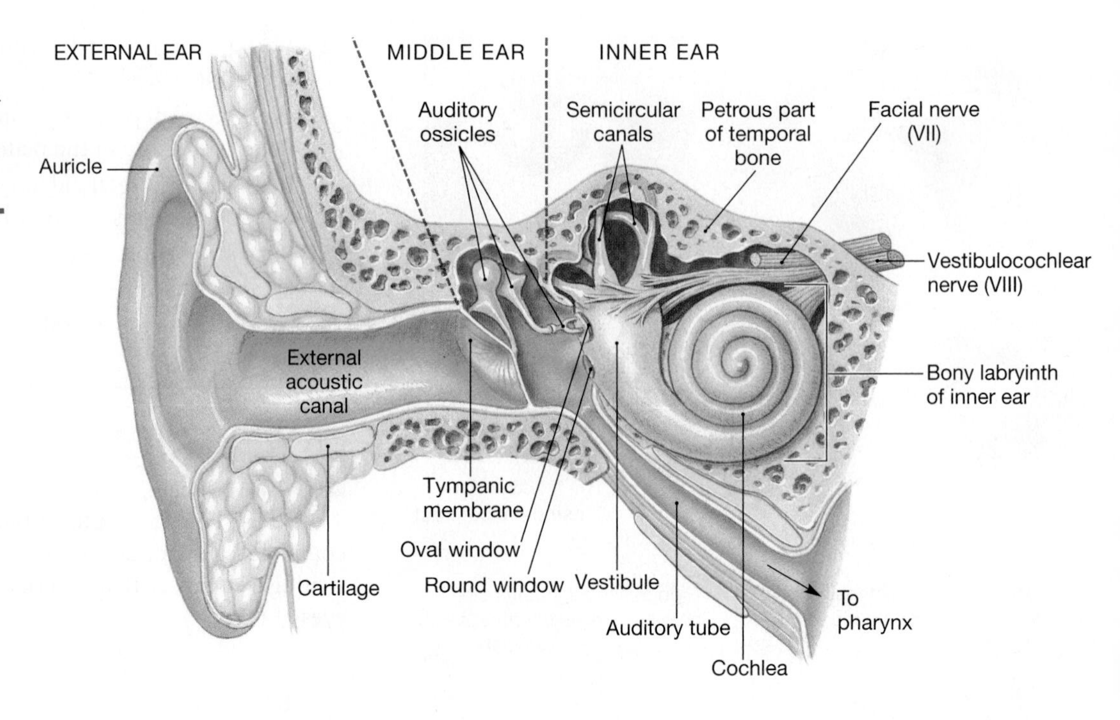

Figure 17.16 The anatomy of the ear. The ear is divided into three regions: external ear, middle ear, and inner ear. The middle and inner ear regions are located within the petrous portion of the temporal bone.

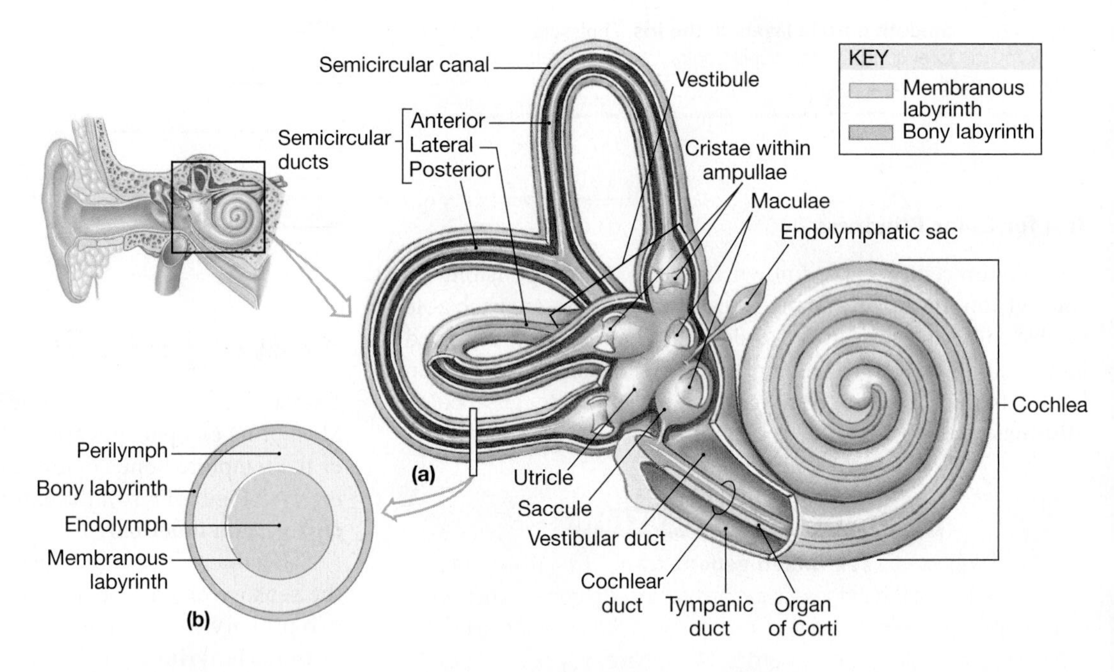

Figure 17.17 The structure of the inner ear. a) The entire inner ear complex includes the semicircular canals, vestibule, and cochlea. In all three subdivisions, the bony labyrinth encloses the membranous labyrinth. **b)** Cross section of a semicircular canal, illustrating the "tube within a tube" architecture of the inner ear.

utricle and **saccule**. This region contains sensory receptors, called **maculae** (singular = **macula**), that detect changes in linear movements and respond to gravitational forces (**static equilibrium**). Walking forward (or backward) along a straight path or jumping off a diving board into a pool are examples of body position changes detected by the vestibule. The **semicircular canals** consist of three semicircular tubes connected to the posterior wall of the vestibule. The semicircular canals lie at right angles to each other along the three planes of movement. Thus, there is an anterior, posterior, and lateral semicircular canal. This portion of the labyrinth contains sensory receptors known as **cristae** (singular = **crista**) that detect changes in angular acceleration or deceleration (**dynamic equilibrium**). Making a turn while driving a car or performing a gymnastics vault are examples of body position changes detected by the semicircular canals.

The **cochlea** is a snail-shaped structure that coils 2 1/2 times around a central axis of bone. It is attached to the anterior wall of the vestibule. This portion of the labyrinth contains the sensory receptors for hearing, known as the **spiral organ** or the **organ of Corti**.

Sensory **hair cells** are found in all receptor organs of the inner ear (maculae, cristae, and spiral organ). All hair cells contain long microvilli, called **stereocilia**, along their free surfaces. The

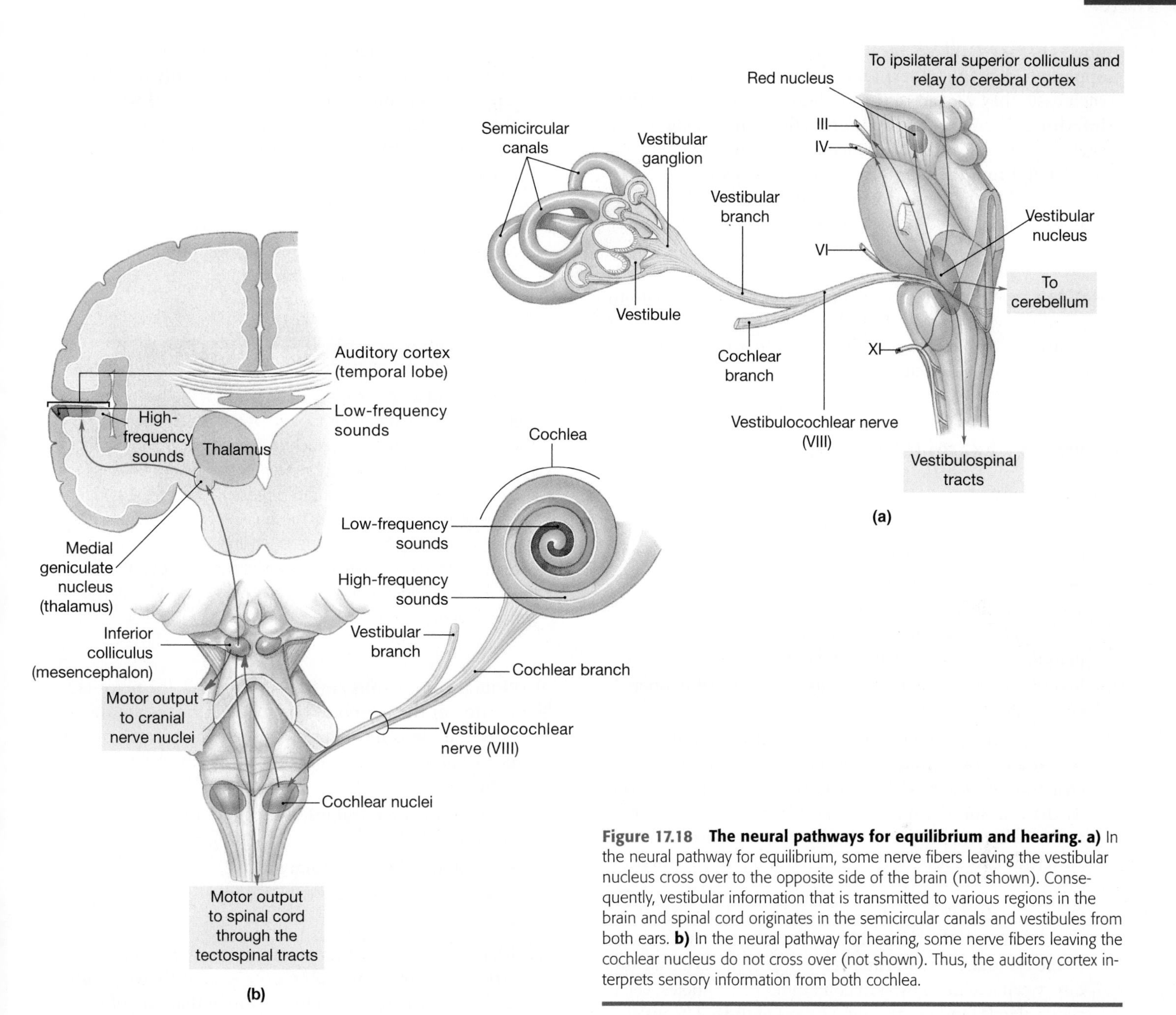

Figure 17.18 The neural pathways for equilibrium and hearing. a) In the neural pathway for equilibrium, some nerve fibers leaving the vestibular nucleus cross over to the opposite side of the brain (not shown). Consequently, vestibular information that is transmitted to various regions in the brain and spinal cord originates in the semicircular canals and vestibules from both ears. **b)** In the neural pathway for hearing, some nerve fibers leaving the cochlear nucleus do not cross over (not shown). Thus, the auditory cortex interprets sensory information from both cochlea.

hair cells in the maculae of the vestibule also possess a **kinocilium**, which is a long cilium (recall from your study of cell biology that microvilli and cilia differ in structure). The hair cells are stimulated when the stereocilia (and kinocilia in the vestibule) bend as a result of a mechanical stimulus. The stimulus could be gravitational forces in the vestibule, turning movements in the semicircular canals, or sound waves in the cochlea.

The sensory receptors in the labyrinth are supplied by the **vestibulocochlear nerve (cranial nerve VIII)**. This nerve consists of the following two divisions.

- In the **vestibular division** (Figure 17.18a), neurons, located in the **vestibular ganglia**, receive sensory information from receptors in the vestibule and semicircular canals, and send impulses to the **vestibular nuclei** in the medulla oblongata. From there, crossed and uncrossed fiber tracts travel to the **cerebellum**, to the **cerebral cortex** via relays in the **thalamus**, to **motor nuclei for cranial nerves** that control reflex eye, head, and neck movements, and to **lower motor neurons** in the spinal cord along the **vestibulospinal tracts**. The information sent to these locations is designed to monitor balance, equilibrium, and muscle tone.
- The **cochlear division** (Figure 17.18b) receives impulses from receptors in the spiral organ and sends axons to the **cochlear nuclei** in the medulla oblongata. Some auditory

fibers that leave the cochlear nuclei cross over to the opposite side while others remain on the same side. In each case, they ascend to the midbrain and synapse in the **inferior colliculi**, where auditory reflexes involving skeletal muscles in the head, neck, and trunk are regulated. From the inferior colliculi, axons are transmitted to the **medial geniculate bodies** in the thalamus and relayed to the auditory areas in the temporal lobes.

WHAT'S IN A WORD Most scientific terms used in anatomy and physiology are derived from Latin and Greek words. The term *hearing* is an exception to this rule. It has its origins from an old English word, *heran,* which means "to hear."

ACTIVITY 17.7 Examining Ear Anatomy

Gross Anatomy

1. Obtain an anatomical model of the ear and identify its three divisions: the external ear, the middle ear (tympanic cavity), and the inner ear (Figure 17.16).
2. The model illustrates that the middle and inner ears are enclosed by bone. The bony chambers are formed by the petrous portion of the temporal bone.
3. Identify the temporal bone and its petrous portion on a skull (Figure 17.3a).
4. The external ear includes the **auricle (pinna)** and the **external auditory canal** (Figure 17.16). Identify these structures on the model. On yourself or a lab partner, verify that the auricle consists of a cartilagenous rim known as the **helix** and an inferior fleshy **lobule** or **earlobe.** The auricle directs sound waves into the external auditory canal.
5. On the model, observe that the external auditory canal travels through the temporal bone, connecting the auricle to the **tympanic membrane (eardrum).** The external auditory canal is lined by skin containing hair follicles, sebaceous glands, and **ceruminous (wax) glands.** The wax glands are modified apocrine sweat glands that produce **cerumen,** a waxy substance that keeps the eardrum soft and waterproof and traps foreign particles that might enter the external auditory canal.
6. Identify the tympanic membrane. Observe that it forms a partition between the external and middle ears (Figure 17.16). The tympanic membrane is a thin, translucent membrane of fibrous connective tissue. Sound waves traveling through the external auditory canal cause the tympanic membrane to vibrate. This action transfers the sound waves to the bony ossicles in the middle ear.
7. Locate the middle ear (tympanic cavity), which is an air-filled chamber lined by a mucous membrane (Figure 17.16). Identify the **three bony ossicles (malleus, incus, stapes)** within the middle ear. Verify that these small bones connect the eardrum to the **oval window** on the wall of the inner ear (Figure 17.16). The ear ossicles transmit and amplify sound waves across the tympanic cavity to the oval window.
8. Identify the **pharyngotympanic (auditory) tube** along the anterior wall of the tympanic cavity. Notice that this tube provides a direct communicating link between the middle ear and the **nasopharynx.**

CLINICAL CORRELATION

Equalization of air pressure on both sides of the tympanic membrane is functionally important for hearing. When atmospheric pressure is reduced, such as when traveling to higher elevations, the tympanic membrane bulges outward because the auditory tube is collapsed and the pressure inside the tympanic cavity is greater than outside. The bulging may be painful and can impair hearing because the flexibility of the membrane is reduced. The act of yawning or swallowing will open the auditory tube and equalize the air pressure on both sides of the tympanic membrane.

9. Note the numerous cavities throughout the petrous portion of the temporal bone. These spaces are the **mastoid air cells.** They are directly connected to the posterior wall of the tympanic cavity by several tubular connections.
10. Identify the three regions of the inner ear or labyrinth (Figures 17.16 and 17.17a).
    - Vestibule (the utricle and saccule)
    - Semicircular canals
    - Cochlea
11. Identify the oval window and the round window. These structures are located along the wall of the vestibule near the base of the cochlea. Notice that one of the bony ossicles, the stapes, is attached to the oval window.
12. Identify the vestibulocochlear nerve (cranial nerve VIII) as it exits the inner ear complex (Figure 17.16). Verify that this nerve consists of two divisions: the cochlear division and the vestibular division.

Microscopic Anatomy

1. Obtain a prepared microscope slide of a cross section of the cochlea.
2. View the slide under low power. The cochlea is a bony structure that resembles a snail. It contains three chambers that spiral for 2½ turns (Figure 17.19a). On the slide, identify a number of cross sections that illustrate

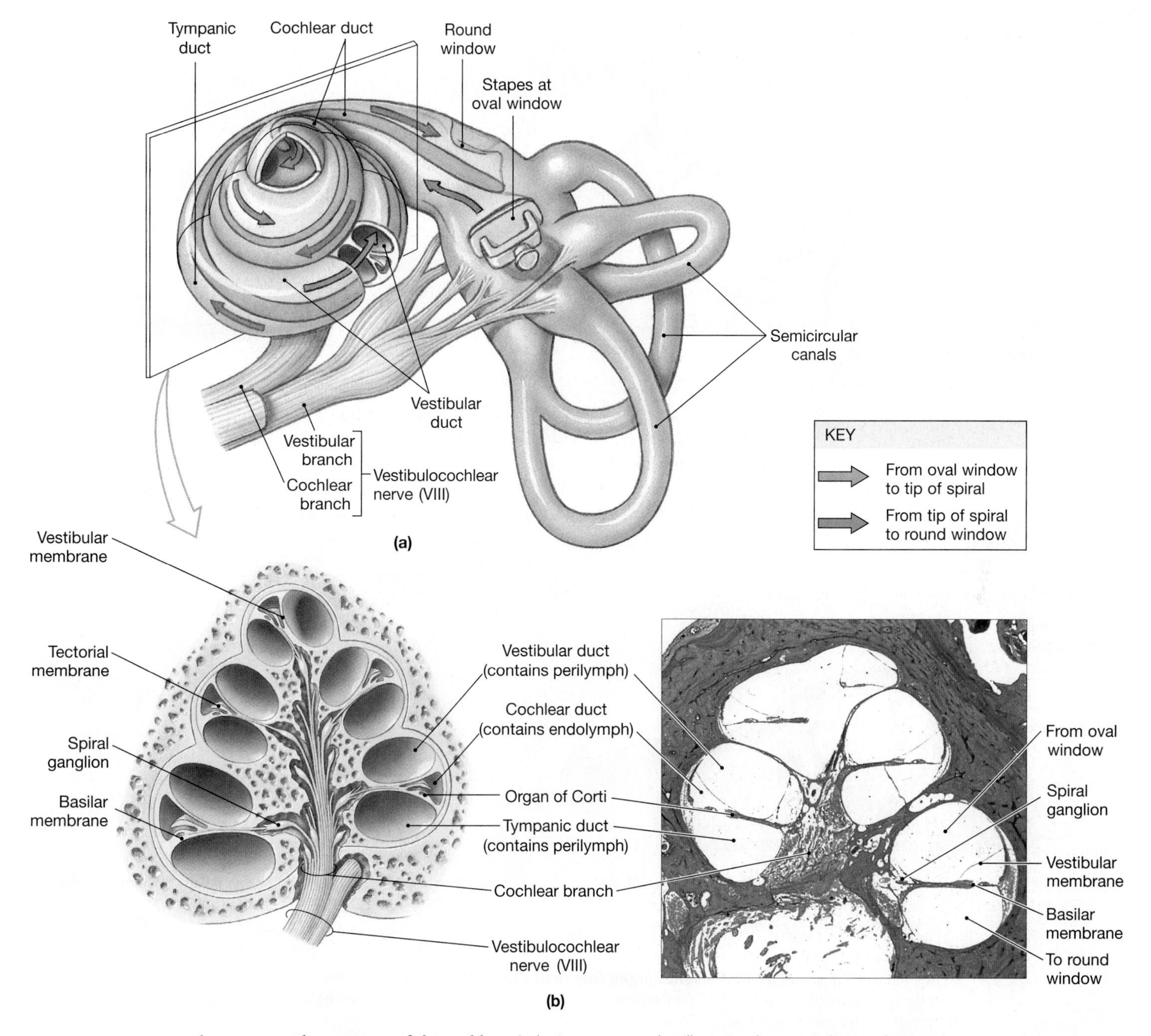

Figure 17.19 The structure of the cochlea. a) The inner ear complex, illustrating the 2½ spiraling revolutions of the three internal chambers of the cochlea. The arrows indicate the transmission of sound waves through the vestibular duct (purple) and tympanic duct (red); **b)** diagram and light micrograph of the cochlea in cross section (LM × 40).

the three cochlear chambers. In one of these cross sections, identify the three chambers (Figures 17.19b and 17.20a).

- The **cochlear duct** is the middle chamber and is a part of the membranous labyrinth. It contains the sensory organ for hearing, the spiral organ (organ of Corti). The cochlear duct is filled with a fluid called **endolymph**.
- The **vestibular duct** is a part of the bony labyrinth and is filled with a fluid called **perilymph**. It begins at the oval window and spirals to the apex of the cochlea.
- The **tympanic duct** is also a part of the bony labyrinth and is filled with perilymph. It begins at the apex of the cochlea, where it is continuous with the vestibular duct, and spirals to the base to end at the round window (Figure 17.19a).

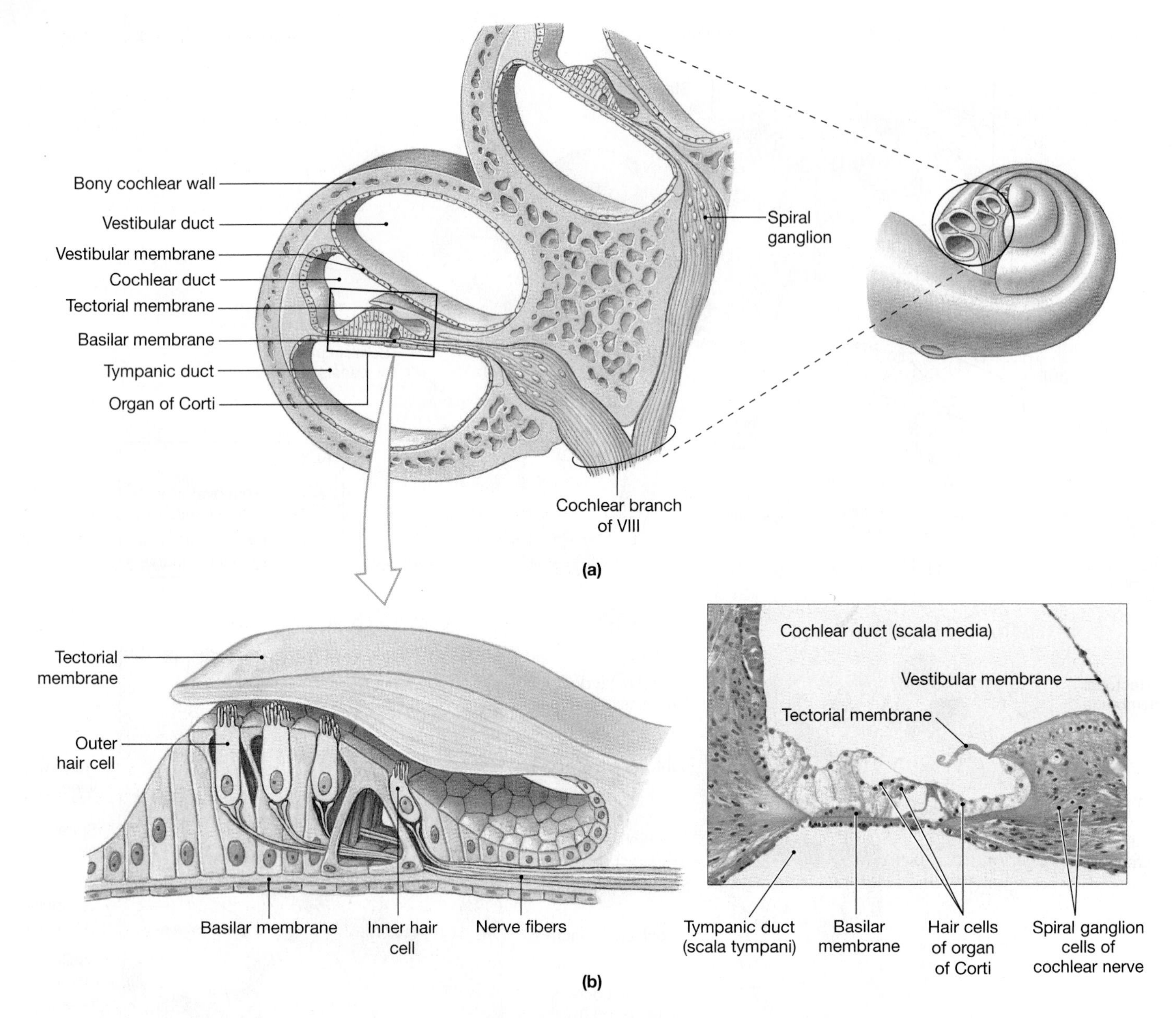

Figure 17.20 The structure of the spiral organ (organ of Corti). a) Cross section of the three internal chambers of the cochlea. The middle chamber, the cochlear duct, contains the spiral organ; **b)** diagram and light micrograph of the spiral organ (LM × 100).

3. Adjust the position of the slide so that the spiral organ in the cochlear duct is in the center of the field of view. Switch to high power and identify the following structures (Figure 17.20b).
   - The **vestibular membrane** separates the cochlear duct from the vestibular duct.
   - The **basilar membrane** separates the cochlear duct from the tympanic duct.
   - Hair cells of the spiral organ rest on the basilar membrane.
   - The **tectorial membrane** arches over the hair cells. The stereocilia of the hair cells are in contact with this membrane.
   - The **spiral ganglion** contains neuron cell bodies of the cochlear division of the vestibulocochlear nerve (cranial nerve VIII).

QUESTION TO CONSIDER Bacterial or viral infections that originate in the nose or throat can spread to the middle ear and mastoid air cells. Using your knowledge of the anatomy of the ear, explain how this could be possible.

___

___

ACTIVITY 17.8 Performing Hearing Tests

To obtain meaningful results, it is best to perform the following hearing tests in a quiet room or in an area of the laboratory where as much background sound as possible can be eliminated.

Hearing Acuity Test

Hearing acuity refers to an individual's ability to hear sounds clearly at various distances.

1. Carefully place a small amount of cotton at the entrance to the external auditory canal of the left ear.
2. While sitting quietly with your eyes closed, have your lab partner hold a watch with an audible ticking sound very close to your right ear.
3. Have your lab partner slowly move the watch away from your ear until you can no longer hear the ticking sound.
4. Measure the distance (in centimeters) at which the sound is no longer heard. Record the result in the space provided.
5. Repeat steps 1 through 4 for the left ear.

Distance (cm) at which ticking sound is inaudible

Right ear: ______________________

Left ear: ______________________

6. What conclusion can you make if the distance at which the ticking sound becomes inaudible:
   - Is the same for both ears?

   ____

   ____

   - Is greater in one ear compared to the other?

   ____

   ____

Test for Localizing Sound

When an individual has a hearing loss in one ear, sounds coming from some locations may be heard better than others. For example, if a person has better hearing acuity in his or her right ear, a sound originating closer to the right ear will be heard more clearly than one originating closer to the left ear. In such a case, a person might prefer to use a telephone with the right ear instead of the left.

1. Sit quietly with your eyes closed.
2. Have your lab partner hold a watch with an audible ticking sound close to one of your ears so it can be heard.
3. From this initial position, have your partner move the watch to various positions around your head so that it can always be heard.
4. After each move, indicate the new position of the watch to your lab partner.
5. Summarize your results by answering the following questions.
   - Was it more difficult to localize the sound when the watch was in a particular position? ______________

   ____

   ____

   - Do your results indicate the possibility of better hearing acuity in one ear over the other, or is your hearing acuity about the same in both ears?

   ____

   ____

Weber Test

The Weber test can be performed to determine whether an individual has **conductive deafness** or **sensorineural deafness.** Conduction deafness can occur if the transmission of sound waves is disturbed in the external acoustic canal, at the eardrum, or along the bony ossicles. Sensorineural deafness is the result of damage along the neural pathway for hearing.

1. Strike the prongs of a tuning fork with a rubber mallet or against the heel of your hand.
2. Place the end of the tuning fork handle on your forehead, midway between each ear.
3. If the tone of the fork is heard equally well in both ears, then you have equal hearing (possibly, equal loss of hearing) in both ears.
4. If you have sensorineural deafness, the tone will be detected in the normal ear only. You will be unable to hear the tone in the affected ear because the neural pathway for hearing is damaged and will be unable to send information to the auditory area in the temporal lobe (Figure 17.18b).
5. If you have conduction deafness, the tone will be heard in both ears, but it will be louder in the affected ear. In the normal ear, the cochlea receives background noise from the environment as well as vibrations from the tuning fork. The affected ear receives only the tuning fork vibrations, so they will sound louder.
6. Record the results of the Weber test in the space provided.

Weber test results (+/−)

	Normal hearing	Possible sensorineural deafness	Possible conduction deafness
Right ear:			
Left ear:			

Rinne Test

The Rinne test is used to determine whether sound conduction along the ear ossicles is impaired. Abnormal function of the bony ossicles is suggestive of a conductive hearing loss.

1. Strike the prongs of a tuning fork with a rubber mallet or against the heel of your hand.
2. Place the end of the tuning fork handle on the mastoid process on the right side of your skull. You should be able to hear the sound of the vibrating prongs by bone conduction.
3. When you can no longer hear the sound, place the tuning fork prongs near the entrance to the external auditory canal of your right ear.
4. If you can hear the sound once again, this time by air conduction, your hearing is not impaired. If you cannot hear the sound, a conductive hearing loss is possible.
5. Record your results in the space provided, as positive (+) if hearing is normal or negative (−) if there is a possible conductive hearing loss.
6. Strike the tuning fork again, but this time place the prongs near the entrance to the external auditory canal of your right ear so that you can first hear the sound by air conduction.
7. When you can no longer hear the sound, place the handle of the tuning fork on the right mastoid process to hear the sound by bone conduction.
8. If you cannot hear the sound by bone conduction, it indicates that your hearing is normal. If you can hear the sound, it indicates a possible conductive hearing loss in the right ear.
9. Record your results in the space provided, as positive (+) if hearing is normal or negative (−) if there is a possible conductive hearing loss.
10. Repeat steps 1 through 9 for the left ear.

Rinne test results (+/−)

	Tuning fork on mastoid process first and moved to external auditory canal	**Tuning fork near external auditory canal first and moved to mastoid process**
Right ear:	____________	____________
Left ear:	____________	____________

QUESTION TO CONSIDER On a scale of 1 to 5, with 1 being "poor" and 5 being "excellent," how would you rate your hearing ability? Based on the results of the hearing tests that you have just completed, provide supporting evidence for your rating by writing a summary that describes the general condition of your ability to hear.

	Poor				Excellent
Hearing rating	1	2	3	4	5

Summary

____

____

____

____

Exercise 17 Review Sheet

Special Senses

Name ______________________

Lab Section ______________________

Date ______________________

1. Identify differences and similarities between general senses and special senses.

Questions 2–4: Discuss the significance of the following structures with regard to olfactory sense.

2. Superior nasal concha

3. Cribriform plate

4. Olfactory bulbs

5. Describe the basic structure of a taste bud.

Questions 6–8: Discuss the significance of the following structures with regard to gustatory sense.

6. Circumvallate papillae

7. Wall of the pharynx

8. Cranial nerves VII (facial nerve) and IX (glossopharyngeal nerve)

9. Name the two components of the outer (fibrous) layer of the eye. How do these two structures differ?

____

____

10. Both the choroid and the pigmented layer of the retina contain melanin. In terms of eye function, why is this pigment important?

____

____

Questions 11–13: Describe the function of the following structures.

11. Iris:

____

____

12. Ciliary body:

____

____

13. Lens:

____

____

14. Describe the cellular arrangement of the nervous layer of the retina.

____

____

15. What is the functional difference between rods and cones?

____

____

16. The optic disc and the macula lutea are special regions of the retina. How do these two structures differ?

____

____

17. Compare the structure and function of the anterior and posterior cavities of the eye.

____

____

Questions 18–20: Discuss the significance of the following structures with regard to vision.

18. Optic chiasm

____

____

19. Eyelids

____

____

20. Lacrimal apparatus

____

____

21. Describe the pathway for sound waves through the three regions of the ear. (Hint: review Figures 17.16 and 17.19a)

____

____

22. Differentiate between the bony labyrinth and the membranous labyrinth.

____

____

23. Describe the functions of the three subdivisions of the labyrinth.

____

____

24. Which cranial nerve supplies the labyrinth? Identify the two divisions of this nerve and briefly describe the function of each.

____

____

Questions 25–27: Discuss the significance of the following structures with regard to hearing and equilibrium.

25. Hair cells

____

____

26. Petrous portion of the temporal bone

____

____

27. Perilymph and endolymph

____

____

Questions 28–35: In the following diagram, identify the structures by labeling with the color that is indicated.

28. Lens = **yellow**
29. Iris = **dark blue**
30. Sclera = **green**
31. Choroid = **red**
32. Retina = **orange**
33. Ciliary body = **brown**
34. Cornea = **purple**
35. Optic nerve = **light blue**

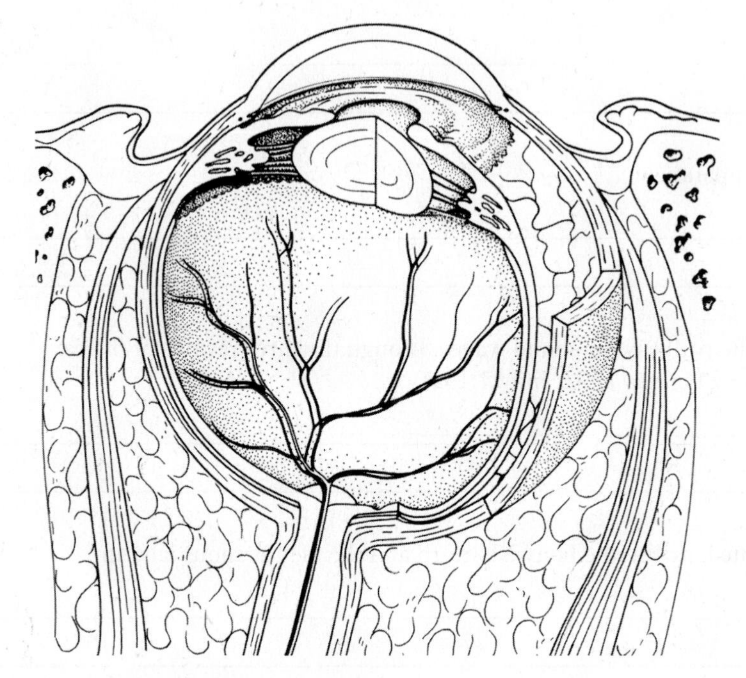

Questions 36–44: In the following diagram, identify the structures by labeling with the color that is indicated.

36. Semicircular canals = **red**
37. Cochlea = **green**
38. External acoustic canal = **blue**
39. Vestibule = **yellow**
40. Bony ossicles = **brown**
41. Vestibulocochlear nerve (VIII):
    a. Cochlear division = **purple**
    b. Vestibular division = **red**
42. Auditory tube = **orange**
43. Tympanic membrane = **pink**
44. Pinna = **green**

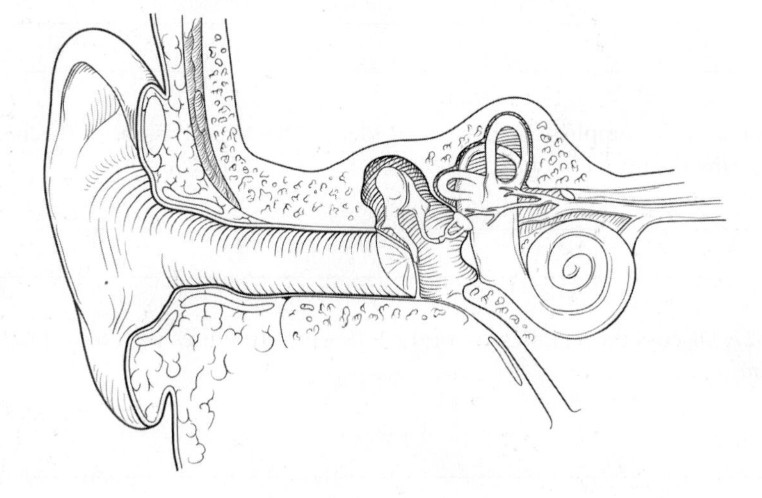

Exercise 18

The Endocrine System

Laboratory Objectives

On completion of the activities in this exercise, you will be able to:

- Describe the difference between an endocrine gland and an exocrine gland.
- Discuss how a hormone affects a target cell.
- Explain the functional relationship between the endocrine system and the nervous system.
- Identify the locations of the endocrine glands in the human body.
- Describe the anatomical relations of the endocrine glands to adjacent structures.
- List the hormones produced by the various endocrine glands and describe their functions.
- Describe the microscopic structure of the endocrine glands.

Materials

- Anatomical models:
  - Human brain
  - Human torso
- Human skulls
- Compound light microscopes
- Prepared microscope slides:
  - Pituitary gland
  - Thyroid gland
  - Parathyroid gland
  - Thymus
  - Pancreas
  - Adrenal gland
  - Ovary
  - Testis

The **endocrine system** consists of a diverse collection of organs and tissues that contain **endocrine glands.** These glands secrete chemicals known as **hormones** into nearby blood capillaries. Once in the circulatory system, hormones can be transported to **target cells** at some distant location. At the target cell, a hormone binds to a specific receptor. Once this occurs, the target cell will respond to the hormone's chemical message. Hormones can influence a target cell's metabolic activities by regulating the production of specific enzymes or critical structural proteins. This can be accomplished by promoting or inhibiting specific genes in the nucleus or by controlling the rate of protein synthesis. Hormones can also activate or deactivate an enzyme's activity by altering its three-dimensional structure.

Endocrine glands represent one of the two types of glands found in the body. The second type, **exocrine glands**, secrete substances into ducts, which transport the secretions into the lumina (internal cavities) of organs, into body cavities, or to the surface of the skin. Sweat glands and sebaceous glands, which you studied in Exercise 6, are examples of exocrine glands.

The endocrine system operates in conjunction with the nervous system to maintain homeostasis and to ensure that bodily functions are carried out efficiently. This functional relationship is sometimes expressed as a **neuroendocrine effect** in which nerve impulses can affect the release of hormones and, in turn, hormones can regulate the flow of nerve impulses.

As you learned earlier, the nervous system performs its functions by conducting electric impulses along nerve fibers and releasing neurotransmitters across synapses to nearby target cells. Neural responses occur quickly but do not last for long periods of time. Endocrine responses are not as rapid as neural responses but can persist for several hours to several days, and are equally effective in regulating physiological activities.

WHAT'S IN A WORD The word *hormone* is derived from the Greek word *hormao*, which means "to provoke" or "set in motion." Hormones released by endocrine glands influence their target organs by "setting in motion" or promoting a specific function.

Gross Anatomy of the Endocrine System

An overview of the endocrine system is illustrated in Figure 18.1. Some structures have an exclusively endocrine function. They include the **pituitary gland**, **pineal gland**, **thyroid gland**, **parathyroid glands**, and **adrenal glands**. In addition, the endocrine system includes several other organs that produce hormones but also perform nonendocrine functions. They include the **hypothalamus**, **thymus gland**, **pancreas**, **testes**, **ovaries**, **heart**, **stomach**, **small intestine**, and **kidneys**.

ACTIVITY 18.1 Examining Gross Anatomy of the Endocrine System

Endocrine Organs Located in the Head

1. Obtain a model of a midsagittal section of a human brain.
2. Identify the pineal gland, located along the roof of the third ventricle (Figure 18.2). **Melatonin**, the hormone secreted by this gland, is believed to control daily sleeping–waking patterns and other cyclical physiological processes (**circadian rhythms**).

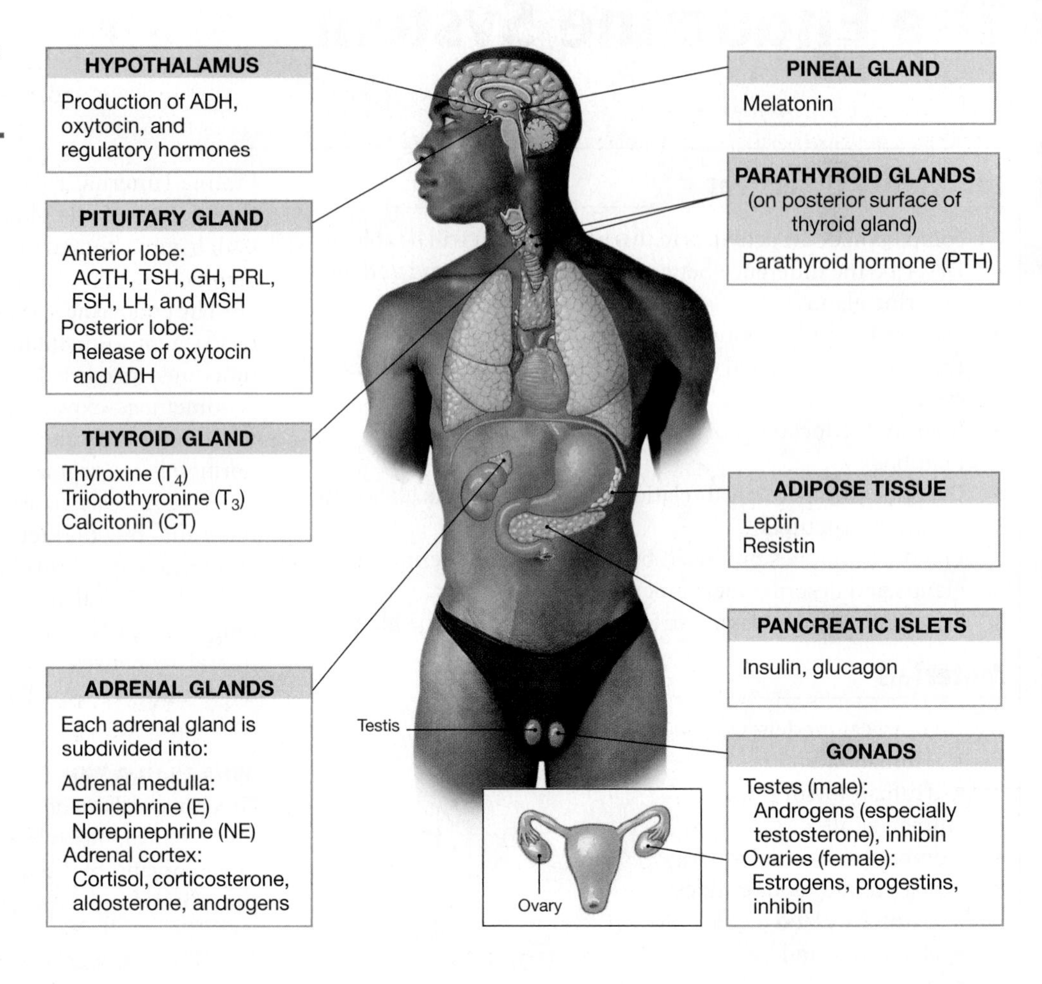

Figure 18.1 Overview of the endocrine system. The human endocrine system contains a diverse array of organs scattered throughout the body.

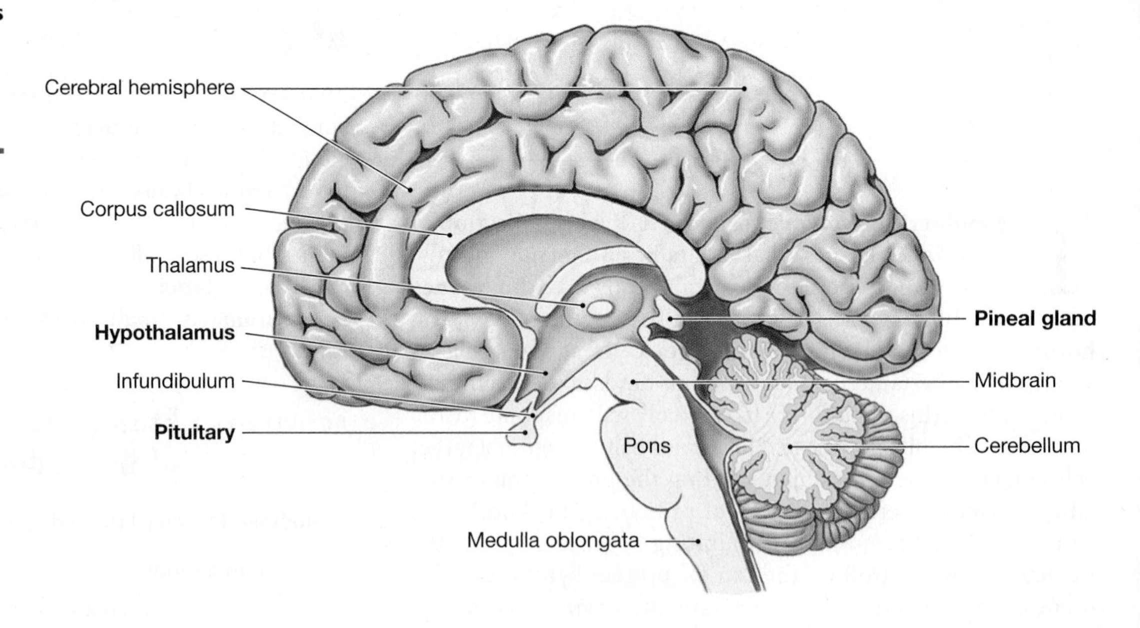

Figure 18.2 Endocrine structures in the brain. Cranial structures that produce hormones include the hypothalamus, pituitary gland, and pineal gland.

CLINICAL CORRELATION

Melatonin production and secretion increase during periods of darkness and decrease during periods of light. This fluctuation in activity is believed to be the underlying cause of **seasonal affective disorder.** This condition brings about unusual changes in mood, sleeping pattern, and appetite in some people living at high latitudes where periods of darkness are quite long during winter months.

3. Locate the hypothalamus, just inferior to the thalamus (Figure 18.2). It produces a number of **releasing hormones** that increase, and **inhibiting hormones** that reduce, the production and secretion of hormones in the anterior pituitary (Table 18.1). The hypothalamus also releases **antidiuretic hormone (ADH)**, which acts on the kidneys to reduce water loss, and **oxytocin**, which stimulates smooth muscle contractions, most notably labor contractions in the uterus and contractions in reproductive glands and ducts of both sexes during intercourse. ADH and oxytocin are transported along axons to the posterior lobe of the pituitary gland, where they are stored and eventually released.
4. Identify the pituitary gland, or **hypophysis**, which is directly connected to the hypothalamus by a stalk of tissue called the **infundibulum** (Figure 18.2). It lies within the **sella turcica** of the sphenoid bone (Figure 18.3a). Obtain a human skull and identify the sella turcica along the floor of the cranial cavity.

   The pituitary gland is divided into two distinct regions: the **anterior lobe** and the **posterior lobe** (Figure 18.3). The anterior lobe produces several hormones (Table 18.1) that regulate the activities of other structures, including other endocrine glands, throughout the body. The posterior lobe, as described earlier, stores and secretes the two hypothalamic hormones, ADH and oxytocin (Table 18.2).

WHAT'S IN A WORD The term *pituitary* is derived from the Latin word *pituita,* which means phlegm or thick mucous secretion. The Renaissance anatomist, Andreas Vesalius, gave the pituitary its name because he mistakenly thought that it produced a mucous secretion related to the throat. When the true function of the pituitary was determined, some 200 years later, it was given a new name, the *hypophysis,* which is the Greek word for "undergrowth". This is probably a better name for the gland since it describes its position suspended from the inferior surface of the hypothalamus.

Endocrine Organs Located in the Neck

1. Obtain a torso or head and neck model.
2. Locate the thyroid gland, which is composed of two elongated lobes located on each side of the **trachea**, just inferior to the **thyroid cartilage**. The **isthmus of the thyroid** travels across the anterior surface of the trachea and connects the two lobes (Figure 18.4a). The thyroid gland produces the hormones **thyroxine (T_4)** and **triiodothyronine (T_3)**, which regulate cell metabolism, general growth and development, and the normal development and maturation of the nervous system. It also produces **calcitonin** that reduces the levels of calcium ions in body fluids.

Table 18.1 Hormones Produced by the Anterior Pituitary

Hormone	Target	Effect	Hypothalamic regulatory hormone
Pars distalis			
Thyroid stimulating hormone (TSH)	Thyroid gland	Promotes secretion of thyroid hormones	Thyrotropin-releasing hormone (TRH)
Adrenocorticotropic hormone (ACTH)	Adrenal cortex	Promotes secretion of glucocorticoids	Corticotropin-releasing hormone (CRH)
Gonadotropins			
a. Follicle stimulating hormone (FSH)	Follicle cells in ovaries	Promotes estrogen secretion and follicle development	Gonadotropin-releasing hormone (GnRH)
	Sustentacular cells in testes	Promotes sperm maturation	
b. Luteinizing hormone (LH)	Follicle cells in ovaries	Promotes ovulation, corpus luteum formation, and progesterone secretion	
	Interstitial cells in testes	Promotes testosterone secretion	
Prolaction	Mammary glands	Stimulates milk production	Prolactin-releasing factor (PRF); prolactin-inhibiting hormone (PIH)
Growth hormone	All cells	Growth, protein synthesis, lipid mobilization, and catabolism	Growth hormone–releasing hormone (GH-RH); Growth hormone–inhibiting hormone (GH-IH)
Pars intermedia			
Melanocyte stimulating hormone (MSH)	Melanocytes	Increased melanin synthesis	Melanocyte stimulating hormone–inhibiting hormone (MSH-IH)

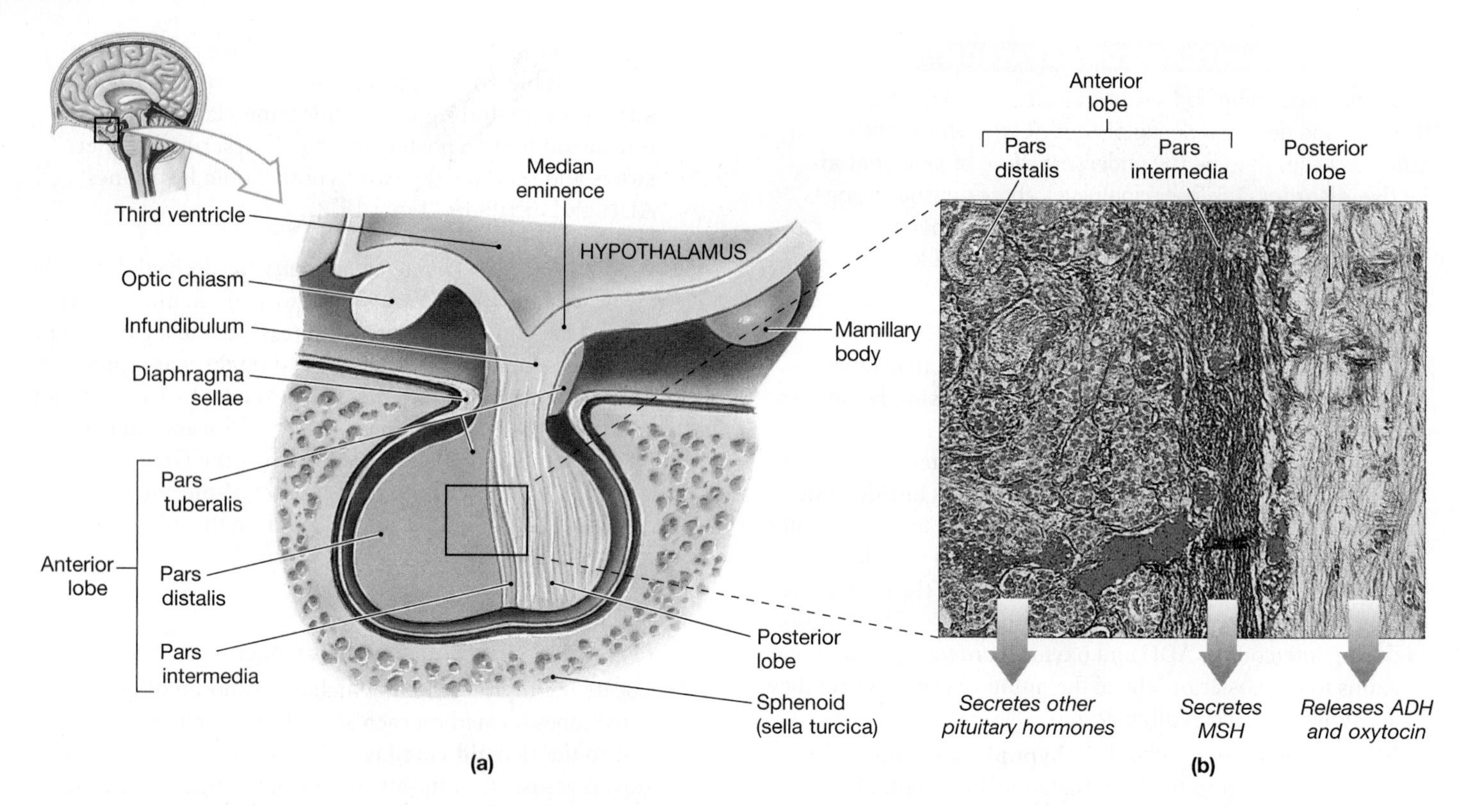

Figure 18.3 Anatomy of the pituitary gland. a) Diagram of the anterior and posterior lobes of the pituitary gland. Note that it is connected to the hypothalamus by the infundibulum. **b)** Light micrograph of the anterior and posterior lobes of the pituitary (LM × 100).

Table 18.2 Hormones Secreted by the Posterior Pituitary

Hormone	Target	Effect	Hypothalamic regulatory hormone
Antidiuretic hormone (ADH)	Kidneys	Reabsorption of water; elevation of blood volume and pressure	None; transported along axons from hypothalamus to posterior pituitary
Oxytocin (OT)	Uterus, mammary glands	Labor contractions; milk ejection	Same as above
	Ductus deferens, prostate gland	Contractions of ductus deferens and prostate gland	

3. Typically there are two pairs of parathyroid glands embedded on the posterior surfaces of the thyroid gland lobes (Figure 18.5a). These glands may not be illustrated on the models in your lab. If not, locate them in an illustration. They produce **parathyroid hormone (PTH)**, which opposes the action of calcitonin by increasing the concentration of calcium ions in body fluids.

Endocrine Organs Located in the Thoracic Cavity

1. Remove the anterior body wall from a torso model so that the contents of the thoracic cavity are exposed (Figure 18.6).
2. Note that the heart is located in the central region of the thoracic cavity, known as the **mediastinum**. If blood volume is elevated above normal, cardiac muscle cells in the heart wall secrete **natriuretic peptides**. These hormones act on the kidneys to promote the loss of sodium ions and water.

WHAT'S IN A WORD **Natriuretic peptides** promote **natriuresis**, or the excretion of sodium in the urine. The term *natriuresis* is derived from two Greek words: *natrium*, meaning "sodium" (the chemical symbol for sodium is Na), and *ouron*, meaning "urine".

3. The thymus gland is located just posterior to the sternum. If it is present on the models in your lab, observe how it covers the superior portion of the heart and extends superiorly into the base of the neck. The thymus produces a group of hormones called **thymosins** that promote the maturation of T-lymphocytes, a type of white blood cell

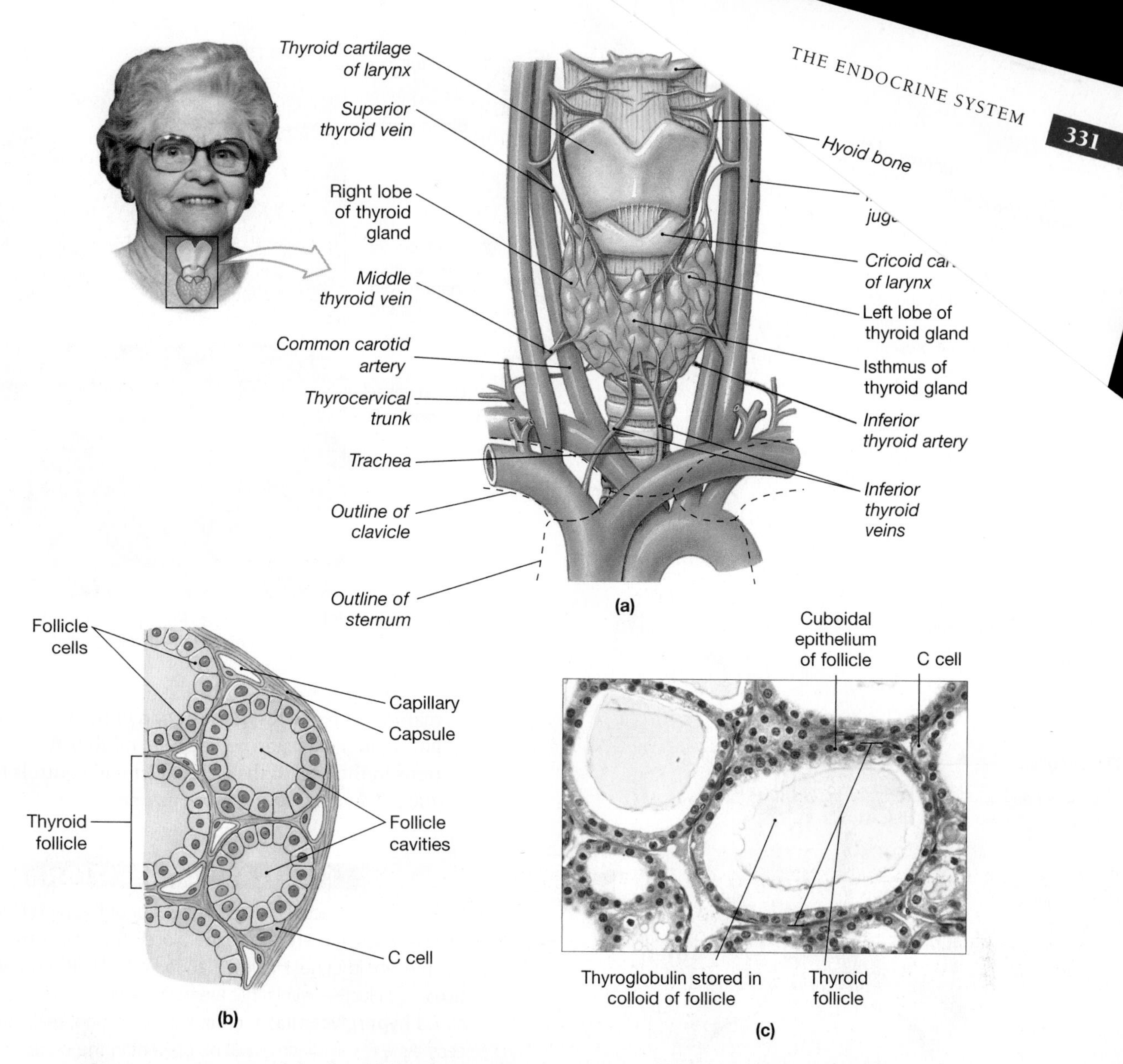

Figure 18.4 Anatomy of the thyroid gland. a) Diagram of the thyroid gland, illustrating its relationship with neighboring structures; **b)** diagram; and **c)** light micrograph, illustrating the microscopic structure of the thyroid gland (LM × 200).

that coordinates the body's immune response. The thymus is relatively large in newborns and young children. After puberty, the size of the thymus is gradually reduced, and the glandular tissue is replaced by fat and fibrous connective tissue. Exercise 23 presents the structure and function of the thymus in greater detail.

Endocrine Organs Located in the Abdominopelvic Cavity

1. On a torso model, remove the digestive organs from the abdominopelvic cavity to expose the structures along the posterior wall. The stomach and small intestine produce several hormones that are important for regulating digestive activities, which will be discussed later when you study the digestive system.
2. Locate the elongated pancreas that stretches across the posterior body wall between the duodenum (first part of the small intestine) and the spleen. The **head** of the pancreas is that portion which is nestled within the C-shaped curvature of the duodenum on the right side (Figure 18.7a). Moving to the left, the **body** of the pancreas is the main portion of the organ. It gives rise to an elongated **tail** that extends to the left toward the spleen. Although the pancreas is largely composed of exocrine glands that produce digestive enzymes, scattered throughout are regions of endocrine tissue

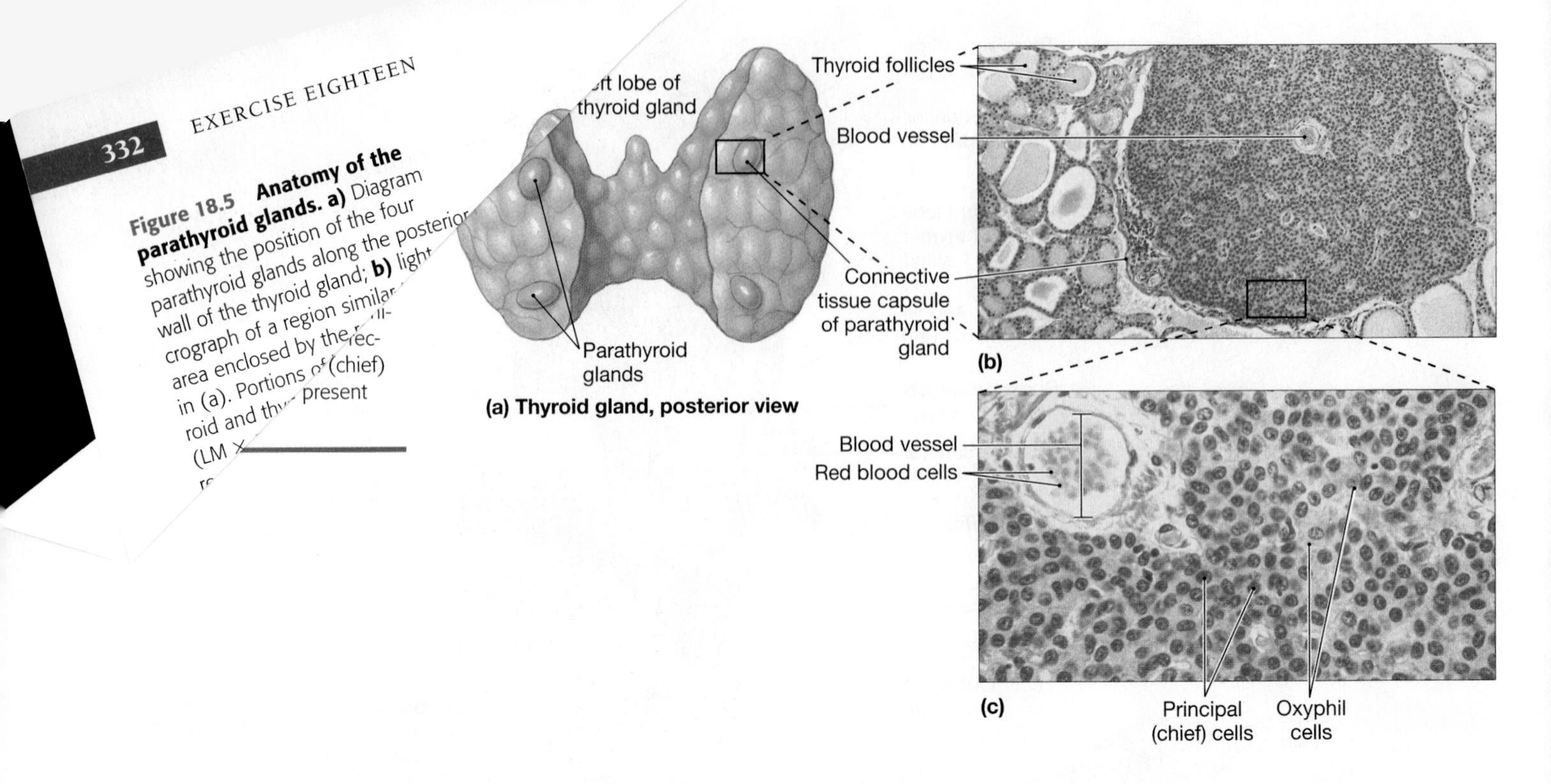

Figure 18.5 Anatomy of the parathyroid glands. a) Diagram showing the position of the four parathyroid glands along the posterior wall of the thyroid gland; **b)** light micrograph of a region similar to the area enclosed by the rectangle in (a). Portions of (chief) roid and thy... present (LM ×

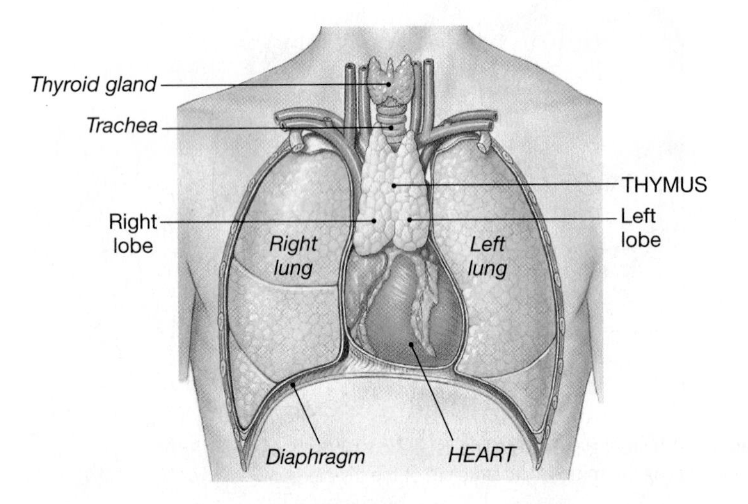

Figure 18.6 Endocrine structures in the thoracic cavity. The heart produces natriuretic peptides and the thymus produces thymosins.

known as the **pancreatic islets (islets of Langerhans**; Figures 18.7b and c). The two main hormones produced by the islet cells are **glucagon** and **insulin**, which regulate blood glucose levels. Glucagon elevates blood glucose levels by promoting the breakdown of glycogen, the synthesis of glucose from fats and proteins, and the release of glucose into the blood. Insulin lowers blood glucose levels by promoting glucose uptake into most cells. Additionally, in skeletal muscles and in the liver, insulin increases glucose storage by stimulating the production of glycogen. Two other hormones, **somatostatin** and **pancreatic polypeptide (PP)**, are also produced by the pancreatic islets. Somatostatin regulates the secretion of both insulin and glucagon. Pancreatic polypeptide inhibits muscular contractions in the wall of the gallbladder and controls the pancreatic production of digestive enzymes.

CLINICAL CORRELATION

Normally, any glucose that is filtered out of the blood by the kidneys is reabsorbed back into the blood. Thus, glucose is usually not present in urine. However, an individual with **diabetes mellitus** has glucose levels that are well above normal, a condition called **hyperglycemia,** and the kidneys cannot reabsorb the excess. As a result, glucose will be present in the urine. There are two main types of diabetes mellitus. Type I diabetes accounts for 5% to 10% of all cases in the United States. It usually develops in children or young adults and destroys the pancreatic cells that produce insulin. It can be treated by daily administration of insulin, supplemented by a carefully monitored dietary plan. Type II diabetes is far more common, making up 90% to 95% of all cases. In addition, a strong correlation exists between type II diabetes and obesity. People with type II diabetes produce normal amounts of insulin but cannot utilize the hormone effectively. This could be due to the production of defective insulin molecules or the lack of insulin receptors on target cells. Careful dietary control, weight reduction, and other lifestyle changes (e.g., regular exercise) are the best treatments for this form of the disease. Diabetes is a long-term, progressive disorder that has potentially serious systemic effects. It can contribute to blindness, heart disease, stroke, kidney failure, circulatory problems resulting in limb amputations, and nerve damage. It is also one of the leading causes of death in the United States.

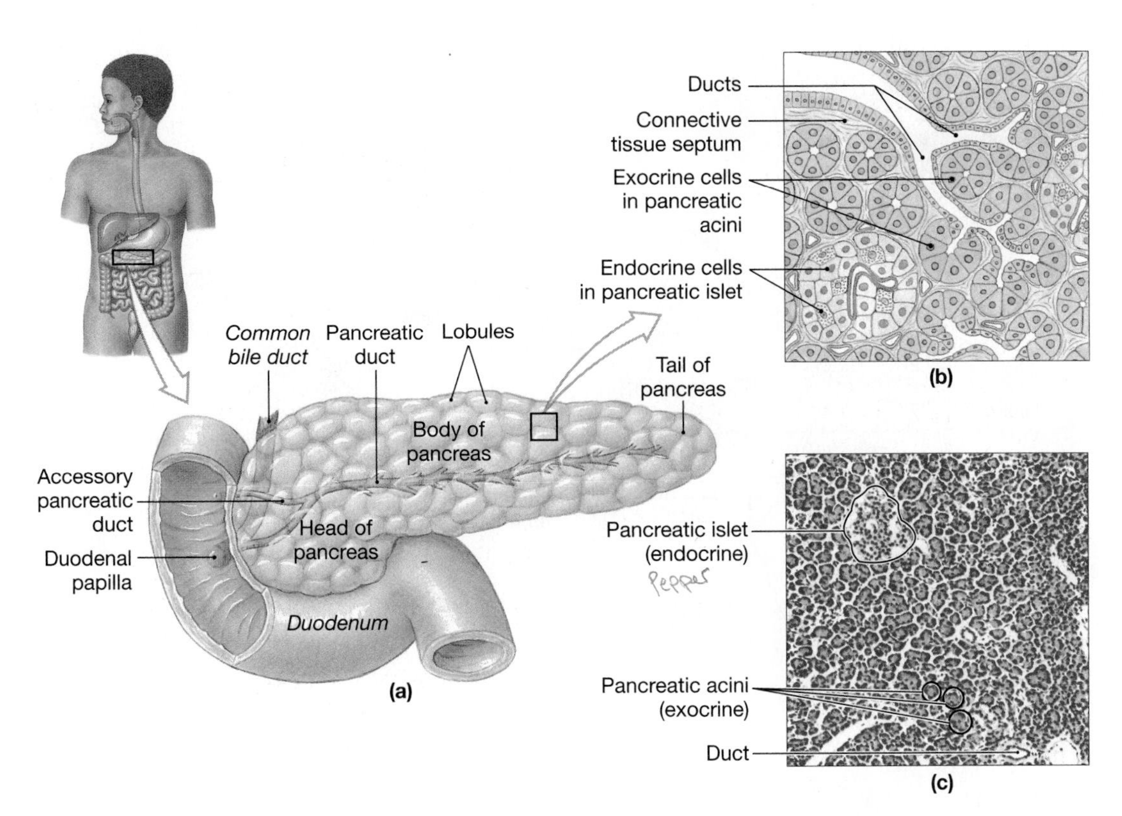

Figure 18.7 Anatomy of the pancreas. a) Diagram showing the relationship of the pancreas to the duodenum; **b)** diagram; and **c)** light micrograph illustrating the light microscopic structure of the pancreas. The pancreatic islets are the endocrine portions of the pancreas (LM × 100).

3. The adrenal (suprarenal) glands are pyramid-shaped structures resting on the superior margins of the kidneys (Figure 18.8a). Fibrous connective tissue attaches the adrenal glands to the connective tissue capsule that surrounds the kidneys. If possible, remove the anterior portion of one adrenal gland and observe its internal structure. Identify the inner **adrenal medulla** and the outer **adrenal cortex** (Figure 18.8b). The adrenal cortex produces three categories of hormones.
   - **Mineralcorticoids**, such as **aldosterone**, act on the kidneys to conserve water and sodium ions, and to secrete potassium ions.
   - **Glucocorticoids**, such as **cortisol**, act on many cells to conserve glucose by utilizing fatty acids and proteins as an energy source (**glucose-sparing effect**), and they function as anti-inflammatory agents by inhibiting cells in the immune system.
   - **Androgens (gonadocorticoids)** are male sex hormones that are produced in small quantities and converted to estrogens (female sex hormones) when they enter the blood. The function of adrenal androgens is not clear.

   The adrenal medulla releases two hormones, **epinephrine** and **norepinephrine**, in response to sympathetic nervous system activation, contributing to the fight-or-flight response. The effects include increased heart rate, blood pressure, and respiratory rate, and decreased digestive activity.

CLINICAL CORRELATION

Glucocorticoids are steroid hormones. Because of their inflammatory effects, these chemicals, or derivatives of them, are used in prescription and over-the-counter "steroid creams" to treat skin rashes such as poison ivy. In addition, many college and professional athletes are given **cortisone injections** to reduce the inflammation that occurs at an injured joint. These injections are effective in reducing injury-related pain, but they do little in repairing damaged tissue. Thus, an athlete who receives a series of cortisone injections might misinterpret a reduction in pain for complete recovery, return to normal activity prematurely, and possibly cause a more serious injury.

4. Locate the kidneys (Figure 18.8a). Although they are mostly involved with waste removal, they also have endocrine functions. Under the influence of parathyroid hormone, the kidneys release a hormone called **calcitriol**, which acts on the small intestine to increase absorption of calcium and phosphate. The kidneys also release **erythropoietin (EPO)**, which stimulates red blood cell production in bone marrow.
5. The gonads include the ovaries in females and the testes in males. On a female model, locate the ovaries along the lateral wall of the pelvic cavity (Figure 18.9a). They produce the female sex hormones called **estrogens**. On a male model, locate the testes. They originate in the abdominal cavity near the kidneys, but descend into the scrotum,

Figure 18.8 Anatomy of the adrenal gland. a) Diagram showing the relationship of the adrenal gland to the kidney and neighboring blood vessels; **b)** diagram showing the two regions of the adrenal gland—the inner adrenal medulla and the outer adrenal cortex; **c)** light micrograph illustrating the microscopic structure of the adrenal gland. Notice that the adrenal cortex is divided into three distinct zones, which are illustrated in the three insets at higher magnification (LM × 400).

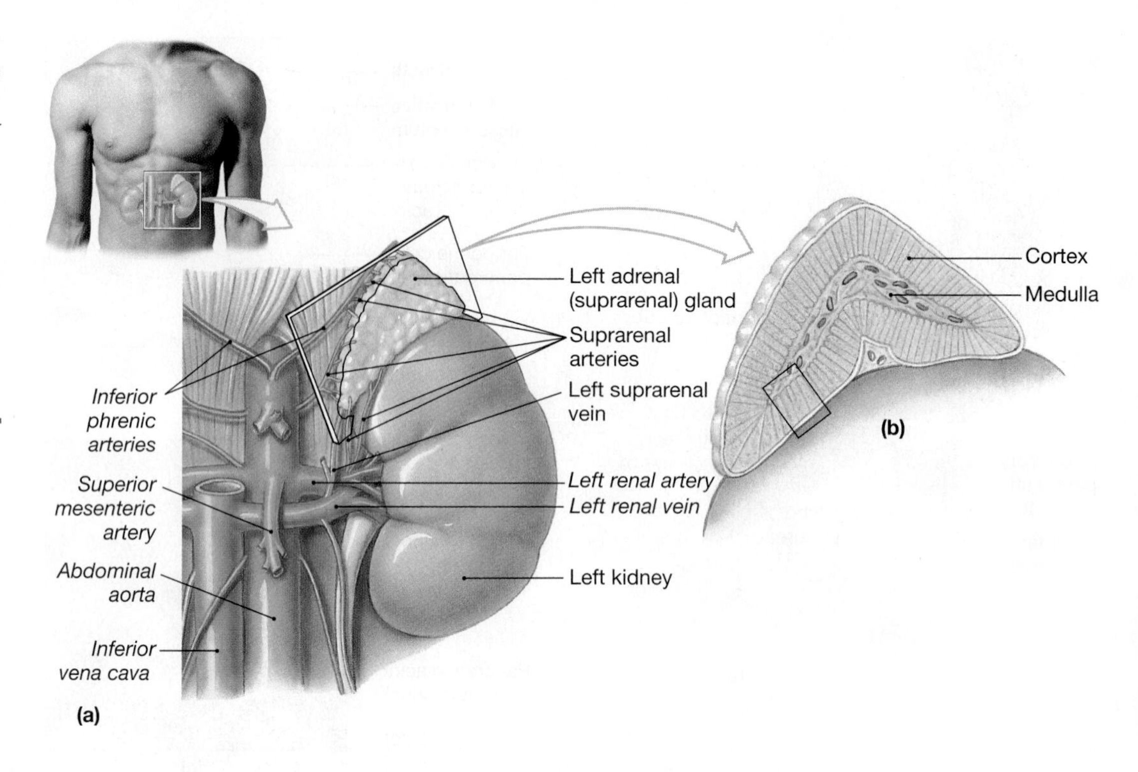

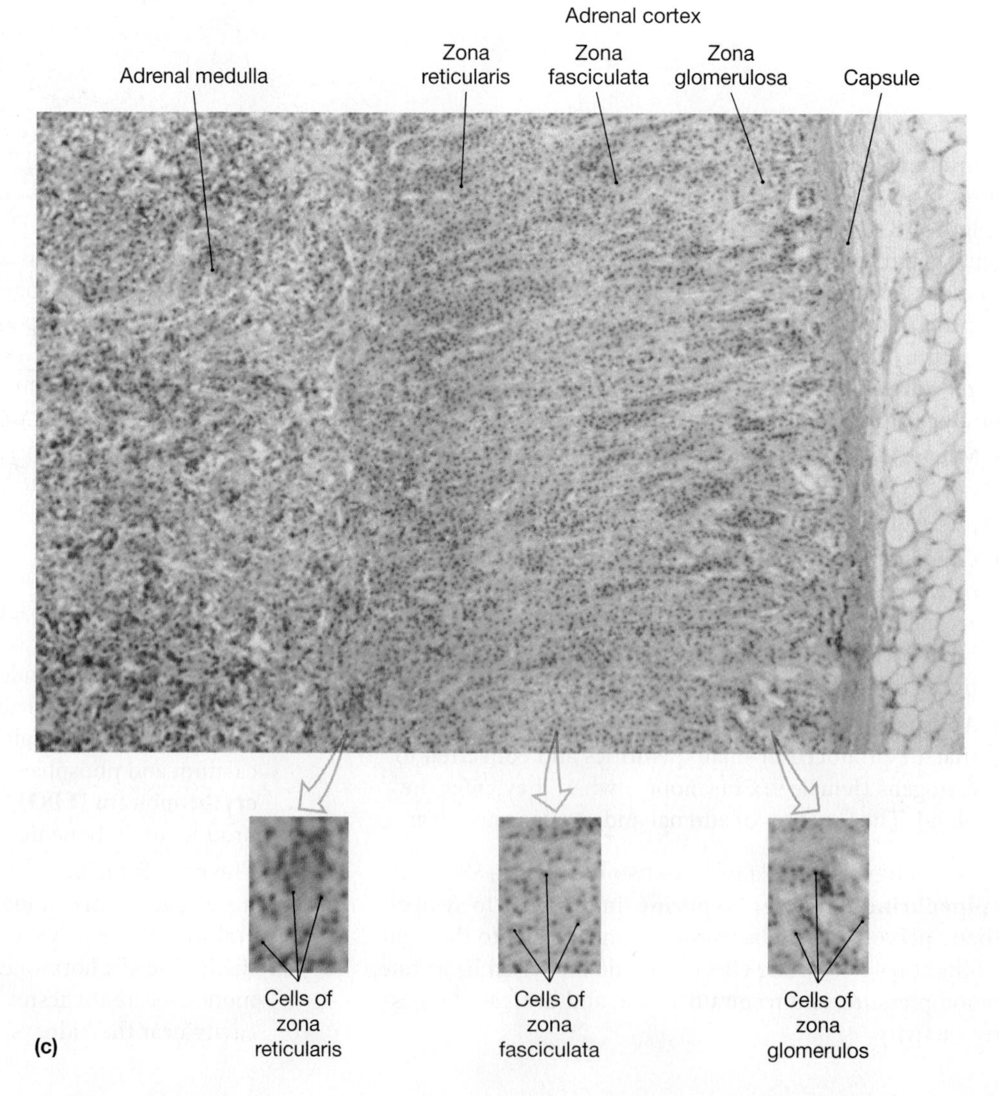

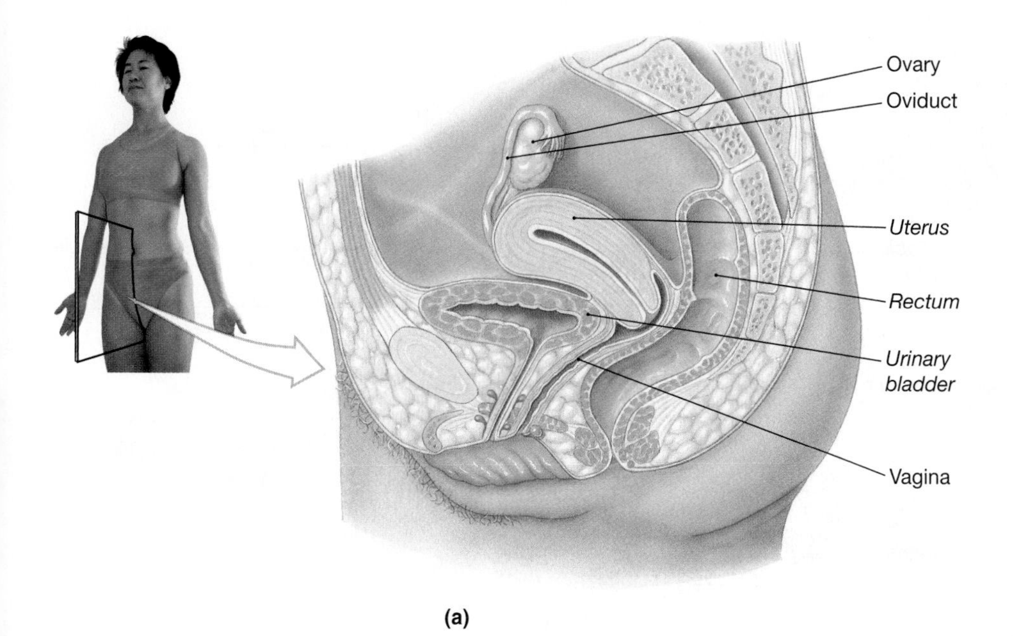

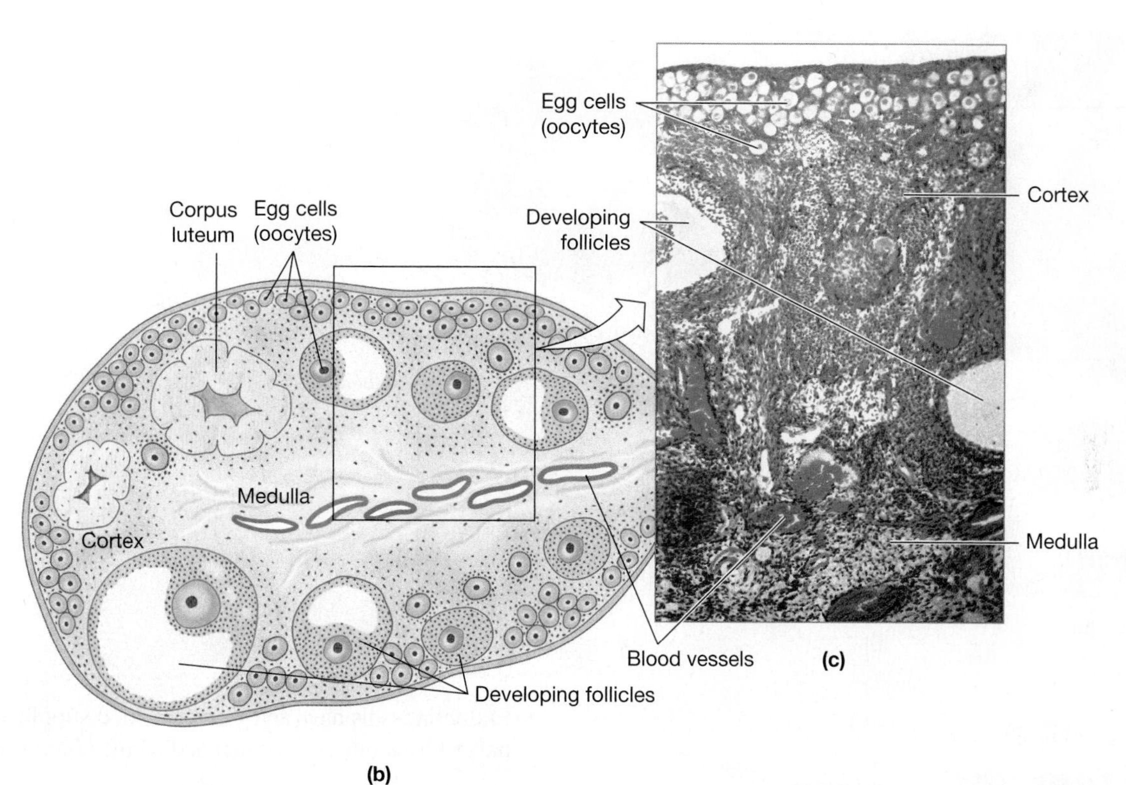

Figure 18.9 Anatomy of the ovary. a) Midsagittal section of the female pelvic cavity, showing the relationship of the ovary to neighboring structures; **b)** diagram illustrating the structure of the ovary in cross section; **c)** light micrograph illustrating the microscopic structure of the ovary. Notice the developing follicles in the cortex of the ovary. The follicular cells that surround the egg produce female sex hormones (LM × 40).

which is outside the body cavity (Figure 18.10a). The testes produce male sex hormones (**androgens**), of which **testosterone** is the most important. The sex hormones control the development and maturation of sex cells (egg and sperm), maintain accessory sex organs, and support secondary sex characteristics.

QUESTION TO CONSIDER Endocrine glands are surrounded by an extensive network of blood capillaries. Suggest a reason why this anatomical relationship is significant.

____

____

Figure 18.10 Anatomy of the testis. a) Midsagittal section of the male pelvic cavity. In the adult male, the testes are located in the scrotal sac, outside the body cavity. **b)** Low-power light micrograph showing cross sections of seminiferous tubules in a testis (LM × 100). **c)** High-power light micrograph and corresponding diagram of a single seminiferous tubule, similar to the tubule enclosed by the box in (b). The interstitial cells produce testosterone (LM × 200).

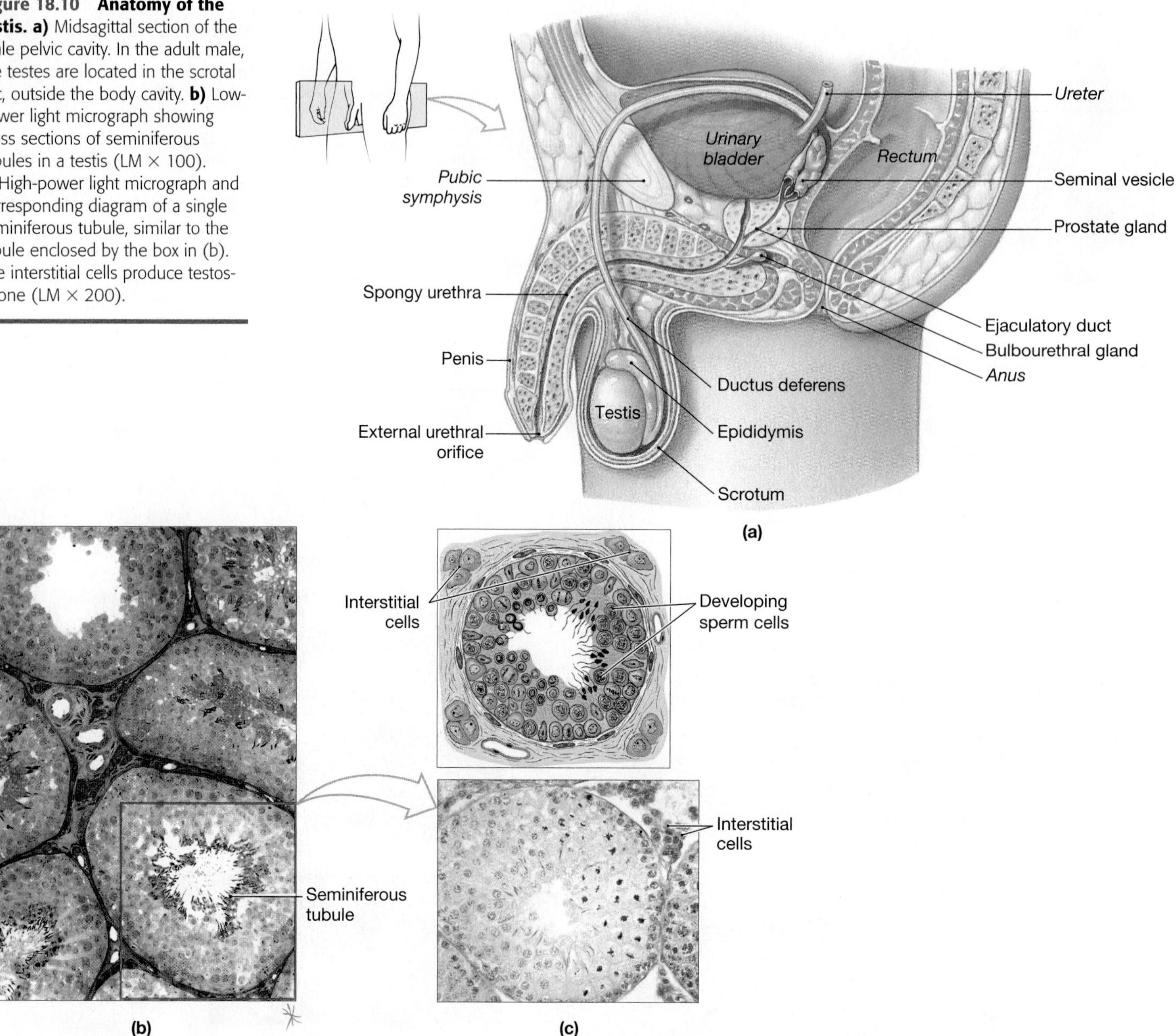

Microscopic Anatomy of the Endocrine System

The cells of endocrine glands possess the following common features.

- The cells are usually cuboidal or polyhedral (many sides) with a large, spherical nuclei.
- With the exception of the hypothalamus, all endocrine cells are derived from epithelial tissue.
- The cells are typically arranged in clusters, small islands (islets), or cords.
- Endocrine cells form glands that lack a system of ducts. Hormones are secreted directly into the surrounding tissue spaces and eventually gain entry into the blood circulation.
- Endocrine cells have an extensive blood supply, and all of them have at least one surface that is directly adjacent to a capillary.

As you study the microscopic anatomy of the various endocrine organs, be aware of these similarities as well as the unique features that characterize each structure.

ACTIVITY 18.2 Examining Microscopic Anatomy of Endocrine Organs

Pituitary Gland

1. Obtain a slide of the pituitary gland (hypophysis).
2. View the slide with the scanning or low-power objective lens. Depending on your slide preparation, adjacent brain and bone tissue may also be present.

- What region of the brain would you expect to see on your slide? ______________________________
- What skull bone would you expect to be present?

____

3. Move the slide so that the pituitary gland is centered. Identify the darker staining **anterior lobe (adenohypophysis)** and the lighter staining **posterior lobe (neurohypophysis** or **pars nervosa).** If possible, identify the infundibulum that connects the pituitary gland to the hypothalamus (Figure 18.3a).
4. Observe the anterior lobe of the pituitary with high power. The anterior lobe is a true endocrine gland because it contains several types of endocrine cells that produce and secrete hormones (Table 18.1). Identify the following regions of the anterior lobe (Figure 18.3b).
   - The **pars distalis** consists of glandular epithelial cells arranged in cords or clusters. Notice that these cells have a cuboidal shape and possess well-defined nuclei. As you scan the slide, you will see that the cells vary considerably in their staining properties. The different colors that you observe in these cells depend on the staining technique used to prepare your slide. Nevertheless, this variability reflects the fact that the pars distalis contains several cell types, each responsible for producing a specific hormone (Table 18.1).
   - The **pars intermedia** is a narrow band of tissue between the pars distalis and the posterior lobe. In the fetus, young children, and pregnant women, the cells in this region produce melanocyte-stimulating hormone (Table 18.1). In most adults, this region of the anterior pituitary is normally inactive.
   - The **pars tuberalis** is an extension of the anterior lobe that wraps around the infundibulum, then spreads along the inferior margin of the hypothalamus. If the infundibulum and hypothalamus are present on your slide, look for the pars tuberalis hugging their outside borders.

CLINICAL CORRELATION

Growth hormone (GH), secreted by the pars distalis, promotes protein synthesis in virtually all cells. It is particularly important for the growth and development of muscle, cartilage, and bone. Inadequate production (hyposecretion) of GH before puberty leads to a condition called **pituitary dwarfism.** People with this disorder have normal body proportions, but abnormally short bones due to reduced activity at the epiphyseal plates. Pituitary dwarfism can be successfully treated before puberty by administering synthetic GH.

Two other abnormalities are caused by excessive secretion (hypersecretion) of GH. **Gigantism** is the overproduction of GH before bone fusion. Individuals with this disorder have normal body proportions but excessively long limbs and can reach heights up to 8.5 ft. **Acromegaly** is caused by excessive GH production after bone fusion. In this condition, bones cannot lengthen, but instead become thicker and denser, particularly in the face, hands, and feet. Both gigantism and acromegaly are usually caused by a tumor in the pars distalis and can be treated by its surgical removal.

5. The **posterior lobe** of the pituitary gland, or **pars nervosa**, is actually an extension of the brain. It is not a true endocrine gland because it does not produce its own hormones. View the posterior lobe with high power (Figure 18.3b) and notice that most of this structure contains axons, which originate from neuron cell bodies in the hypothalamus. Antidiuretic hormone and oxytocin, produced in the hypothalamus, travel along these axons and are released from axon terminals in the posterior lobe.

Thyroid Gland

1. Obtain a slide of the thyroid gland.
2. View the slide with the scanning or low-power objective lens. Notice that the thyroid has a distinctive structure, consisting of numerous **thyroid follicles** of various sizes (Figures 18.4b and c).
3. Use the high-power objective lens to examine the thyroid follicles more closely. Notice that each follicle consists of a central **follicle cavity** surrounded by a single layer of cuboidal **follicle cells** (Figures 18.4b and c).
4. Inside the follicle cavities, identify a lightly staining material known as **colloid.** Follicle cells produce a globular protein known as **thyroglobulin** and secrete it into the colloid for storage. Thyroglobulin is later used to synthesize the thyroid hormones thyroxin (T_4) and triiodothyronine (T_3).
5. In the regions of connective tissue between the follicles, identify the **parafollicular cells (C cells)**, which produce calcitonin. They usually appear in small clusters and are characterized by their pale or lightly stained cytoplasm and large nuclei (Figures 18.4b and c).

Parathyroid Gland

1. Obtain a slide of the parathyroid gland.
2. View the slide with the scanning or low-power objective lens. Since the parathyroid glands are embedded in the posterior wall of the thyroid gland, your slide may display tissue from both structures (Figure 18.5b).
3. Center an area of parathyroid tissue and switch to high power. The darkly stained cells that fill the field are **principal (chief) cells**, which produce **parathyroid hormone**. If you look carefully, you should notice that these cells are arranged in a curvilinear fashion (Figure 18.5c).
4. A second cell type, the **oxyphil cells** are found only in human parathyroid glands. If you are viewing a human parathyroid, attempt to locate these cells (Figure 18.5c). They are larger, stain lighter, and are far fewer than the principal cells. The function of the oxyphil cells is unknown.

Pancreas

1. Obtain a slide of the pancreas.
2. View the slide with the low-power objective lens and identify the **pancreatic acini.** Each acinus contains a cluster of cuboidal cells **(pancreatic acinar cells)**, arranged around a central lumen (Figures 18.7b and c). The acinar cells are the exocrine portion of the pancreas, and produce digestive enzymes. Observe that the darkly stained acinar cells comprise the vast majority of the pancreas.
3. Scattered among the pancreatic acini, identify the islands of lighter staining cells. These are the pancreatic islets or islets of Langerhans (Figures 18.7b and c), which are the endocrine portion of the pancreas. The pancreatic islets contain four cell types. On your slides, you will probably be unable to identify the different cell types.

   However, each type is responsible for producing a specific hormone, as follows:
   - **Alpha cells** produce **glucagon.**
   - **Beta cells** produce **insulin.**
   - **Delta cells** produce **somatostatin.**
   - **F cells** produce **pancreatic polypeptide (PP).**

Adrenal Gland

1. Obtain a slide of the adrenal gland.
2. View the slide with the scanning or low-power objective lens and identify the outer **adrenal cortex** and the inner **adrenal medulla** (Figure 18.8b).
3. Center the adrenal cortex and switch to high power. Identify the following three cellular layers (Figure 18.8c).
   - The **zona gomerulosa** is the outermost layer and is covered by a connective tissue capsule (the capsule may not be present on your slide). It comprises 10% to 15% of adrenal cortical volume. Notice how the cells in this layer are arranged in small circular clusters. These cells produce **mineralcorticoids.**
   - The **zona fasciculata** is the middle layer and makes up 75% to 78% of the volume of the adrenal cortex. As you move the slide into this region, notice that the cells are larger and more lightly stained than those in the previous layer. The lighter staining is due to the large supply of lipids in the cytoplasm. Observe that the cells in this layer are organized into stacks or columns, rather than clusters. These cells manufacture **glucocorticoids.**
   - The **zona reticularis** is the smallest (7% to 10% of the total cortical volume) and innermost layer of the adrenal cortex. As you move into this layer, notice that the cells are more deeply stained and form an irregular, intersecting network. The cells in this layer produce a small amount of **androgens.**

WHAT'S IN A WORD The three zones of the adrenal cortex are named according to the organization of the cells in each layer. The term *glomerulosa* is derived from the Latin word *glomus*, which means "a ball". The cells in the zona glomerulosa are arranged in a spherical fashion. The term *fasciculata* has its origins from the Latin word *fasciculus*, which refers to "a bundle or cord". The name describes the columns of cells in the zona fasciculata. The word *reticularis* comes from the Latin term *reticulatus*, which means "netlike", and is suggestive of cell arrangment in the zona reticularis. ■

4. Switch back to low power, locate the adrenal medulla, and center this region in the field of view.
5. With high power, observe the cells in the adrenal medulla (Figure 18.8c). This region consists of loosely arranged polyhedral cells with large round nuclei. An extensive network of capillaries travels between the cells and a large **medullary vein** (or veins), which drains the entire adrenal gland, may be identified. The endocrine cells in the adrenal medulla resemble cells found in sympathetic ganglia, and their secretory activity is promoted by preganglionic sympathetic nerve fibers. The majority of the cells produce epinephrine, and a smaller number synthesize norepinephrine.

Ovary

1. Obtain a slide of the ovaries from a human or another mammalian species. The ovaries are the primary sex organs in the female.
2. View the slide with the scanning or low-power objective lens. Identify the two regions of the ovary (Figures 18.9b and c).
   - The outer **cortex** contains the ovarian follicles at various stages of development. Each follicle contains a developing egg cell, known as an **oocyte.**
   - The inner **medulla** is a region of loose connective tissue with numerous blood vessels, nerves, and lymphatics.
3. Scan the cortex and identify follicles at various stages of development (Figure 18.9c). In each developing follicle, identify the egg cell and the multiple layers of follicular cells that surround it. The follicular cells produce the female sex hormones known as **estrogens.**

Testis

1. Obtain a slide of the testes from a human or another mammalian species. The testes are the primary sex organs in the male.
2. View the slide with the scanning or low-power objective lens. Scan along the edge of the section and observe the fibrous connective tissue covering called the **tunica albuginea.** Connective tissue partitions derived from the tunica albuginea divide the testes into **lobules.**
3. Within each lobule of the testes are three or four **seminiferous tubules.** As you scan the slide under low power, the tubules can be observed throughout the field of view. Each tubule is surrounded by connective tissue and contains several layers of cells surrounding a central lu-

men (Figure 18.10b). Because of the plane of section, the lumen may not be evident in some tubule profiles.

4. Observe a seminiferous tubule under high power (Figure 18.10c). Most of the cells in the walls of the tubules are sperm cells in various stages of development. Collectively, these cells are called **spermatogenic cells**. As the sperm cells form, they move from the base to the lumen of the seminiferous tubules. Observe these various cells on the slide. Note that as you view the cells in the tubule walls, from base to lumen, their appearance changes progressively.
5. Scan the slide under high power and observe areas of connective tissue between seminiferous tubules. These interstitial areas contain the **interstitial (Leydig) cells** (Figure 18.10c), which produce the male sex hormone, testosterone.

QUESTION TO CONSIDER Based on your microscopic observations in the previous activity, identify structural similarities and differences in the various endocrine organs. Focus your attention on the arrangement and structure of the glandular cells in each structure.

Similarities:

____

____

Differences:

____

____

Exercise 18 Review Sheet

The Endocrine System

Name ______________________

Lab Section ______________________

Date ______________________

1. Discuss the differences between an endocrine gland and an exocrine gland.

____

____

2. Target cells respond to inputs from both the nervous system and the endocrine system. In general, how does a neural response differ from an endocrine response?

____

____

3. What is meant by a neuroendocrine effect?

____

____

4. Explain how the hypothalamus influences the function of the anterior lobe of the pituitary gland.

____

____

5. Describe the functional relationship between the hypothalamus and the posterior lobe of the pituitary.

____

____

Questions 6–12: Match the hormone in column A with its function in column B.

A	B
6. Insulin ______	a. Regulates cell metabolism
7. Oxytocin ______	b. Lowers blood glucose levels
8. Aldosterone ______	c. Promotes sperm development
9. Epinephrine ______	d. Elevates blood calcium levels
10. Testosterone ______	e. Promotes uterine contractions during labor
11. Thyroxine ______	f. Promotes egg development
12. Parathyroid hormone ______	g. Acts on the kidneys to conserve water and sodium
	h. Lowers blood calcium levels
	i. Promotes the fight-or-flight response
	j. Elevates blood glucose levels

Exercise 19
Blood Cells

Laboratory Objectives

On completion of the activities in this exercise, you will be able to:

- Describe the functions of blood.
- Describe the difference between the formed elements of blood and blood plasma.
- Identify all blood cell types when viewed with a light microscope.
- Perform a differential white blood cell count.
- Safely use a blood lancet to collect a blood sample.
- Safely determine your ABO and Rh blood type.
- Use the proper safety techniques to discard wastes that have been contaminated with blood.

Materials

- Prepared microscope slides of human blood smears
- Compound light microscopes
- Clean microscope slides
- Gloves
- Face masks
- Protective eyewear
- Wax labeling pencils
- Sterile blood lancets
- Sterile alcohol pads
- Warming tray
- Paper towels
- Containers for the disposal of biohazardous wastes
- Anti-A, anti-B, and anti-Rh blood typing solutions
- Simulated blood typing kits (an optional alternative to using human blood)

Blood is a highly specialized connective tissue that consists of various blood cells (**formed elements**) suspended in a fluid matrix (**blood plasma**). Its various functions include the transportation of substances, the regulation of various blood chemicals, and protection against infections and diseases.

The formed elements, comprising 45% of the total blood volume, include the **red blood cells (erythrocytes)**, **white blood cells (leukocytes)**, and **platelets** (Table 19.1). Red blood cells are flattened biconcave discs that lack nuclei and most organelles. Most of the cytoplasm is filled with the protein **hemoglobin**, which is used to transport oxygen and carbon dioxide.

WHAT'S IN A WORD The term *erythrocyte* is derived from two Greek words: *erythros*, meaning "red," and *kytos*, meaning "cell." Thus, red blood cells are called erythrocytes.

The term *leukocyte* is also derived from Greek and means "white cell" (*leukos* means "white").

The five white blood cell types are categorized into two major groups (Table 19.1). The **granulocytes** have distinct granules in their cytoplasm. They include the **neutrophils**, **eosinophils**, and **basophils**. The **agranulocytes** lack cytoplasmic granules, and include the **monocytes** and **lymphocytes**. White blood cells play essential roles in defending the body against infection (Table 19.1).

Platelets are not true cells, but cytoplasmic fragments derived from cells called **megakaryocytes**. Platelets have a critical role in the blood clotting mechanism and the repair of damaged blood vessels.

Plasma makes up the remaining 55% of the total blood volume. It is 90% water, but it also includes a wide variety of dissolved substances including gases, nutrients, hormones, waste products, ions, and proteins.

Blood performs a variety of essential functions that fall into three major categories.

1. *Transportation of substances*
   - Oxygen and nutrients are delivered to all body tissues for metabolism.
   - Cellular wastes are brought to the lungs and kidneys for elimination.
   - Hormones from endocrine glands are distributed to various target organs.
2. *Regulatory activities*
   - Body temperature is regulated by distributing heat throughout the body and shunting excess heat to the skin's surface for elimination.
   - Blood proteins act as buffers that function to maintain stable blood pH levels.
   - Various blood proteins and dissolved ions prevent dramatic fluctuations in blood volume by maintaining osmotic balance between the blood plasma and surrounding tissue fluids.
3. *Defensive activities*
   - White blood cells, antibodies, and various blood proteins protect the body from infections caused by bacteria, viruses, and other pathogens.
   - Platelets and various blood proteins protect the body from excessive blood loss by repairing damaged blood vessels and forming blood clots.

Table 19.1 Formed Elements in the Blood

Cell type	% total	Average # cells/µl	General description	Function
Erythrocytes (red blood cells; RBCs)	99.9% of all blood cells	4.4–6 million	Biconcave discs; lack nuclei and most other organelles; 7–8 µm diameter	Oxygen and carbon dioxide transport
Leukocytes (white blood cells, WBCs)	0.1% of all blood cells	6000–9000		
Granulocytes				
Neutrophils	40%–70% of WBCs	1800–7300	Multilobed nuclei; pale-staining granules; 10–14 µm diameter	Attack bacteria by phagocytosis
Eosinophils	1%–4% of WBCs	0–700	Bilobed nuclei; bright red or orange granules; 10–14 µm diameter	Attack parasitic worms; mitigate the effects of allergy and inflammation
Basophils	<1% of WBCs	0–150	Bilobed nuclei; blue-purple granules; 10–12 µm diameter	Enhance inflammatory response and tissue repair by releasing histamine and heparin
Agranulocytes				
Monocytes	4%–8% of WBCs	200–950	Kidney-shaped nuclei; pale-staining cytoplasm; 14–24 µm diameter; on average, the largest blood cells	Differentiate into macrophages—attack and destroy bacteria and viruses by phagocytosis
Lymphocytes	20%–45% of WBCs	1500–4000	Relatively large spherical nuclei; thin rim of pale-staining cytoplasm; size is variable, ranging from 5–17 µm in diameter	Regulate the immune response by direct cellular attack and by antibody production
Platelets		150,000–500,000	Cellular fragments of megakaryocytes	Repair of damaged blood vessels; involved in blood clotting 2–4 µm diameter

Formed Elements in Blood

Red Blood Cells

Red blood cells (erythrocytes) make up 99.9% of the formed elements. Thus, when you examine a slide of a normal human blood smear, you should expect the vast majority of the cells that you observe to be red blood cells. Careful and patient scanning of the slide will be required to locate and identify the various white blood cell types and platelets.

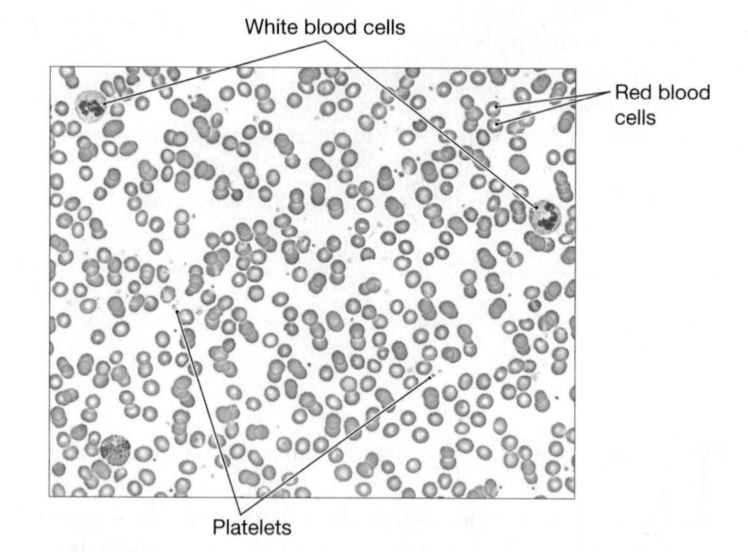

Figure 19.1 Human blood smear. The bulk of the formed elements in blood consists of red blood cells (erythrocytes). White blood cells and platelets are also shown.

ACTIVITY 19.1 Identifying Blood Cells

1. Obtain a slide of a normal human blood smear, prepared with Wright's stain.
2. View the slide under low power with a compound light microscope. Almost all the cells that you see in the field of view are the relatively small, pink-staining red blood cells (erythrocytes). You might also see a few larger and more darkly stained white blood cells (leukocytes) scattered among the erythrocytes (Figure 19.1). If white blood cells are not present in your field of view, carefully scan the slide until you identify them.
3. Examine red blood cells more closely under high magnification. They are uniquely shaped as **biconcave discs** (Figure 19.2), with a relatively thin central region and thick peripheral region. Many resemble doughnuts or lifesaver candies, because the thinner center is stained more lightly than the periphery. In addition, notice that mature erythrocytes lack a nucleus.
4. Identify any white blood cells in the field of view (Figure 19.1). Unlike the red blood cells, white blood cells contain distinct nuclei that are usually stained a deep blue or purple. In addition, most white blood cells are much larger than red blood cells.

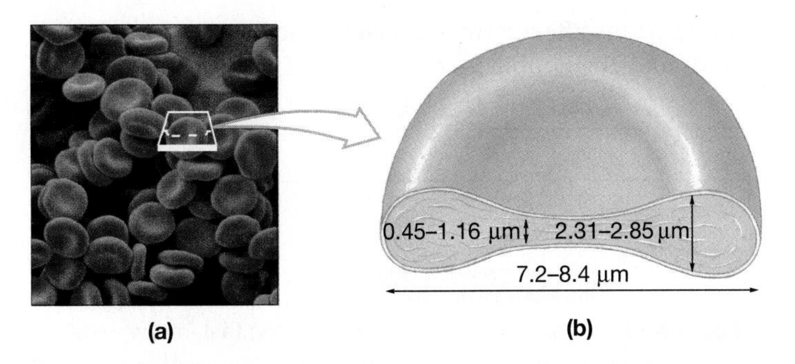

Figure 19.2 Structure of red blood cells. a) Electron microscopic view illustrating the three-dimensional structure of erythrocytes (LM × 800); **b)** sectional view illustrating the unique biconcave shape of a normal erythrocyte.

5. Attempt to identify small fragments of cell cytoplasm in your blood smear, known as platelets (Figure 19.1), which are important for blood clotting.

QUESTIONS TO CONSIDER 1. The vast majority of oxygen that is transported in the blood is bound to hemoglobin molecules in red blood cells. Given that another gas, carbon monoxide, has a much stronger affinity than oxygen to the same hemoglobin binding sites, describe how carbon monoxide poisoning would occur.

____

____

2. How does the lack of a nucleus help to explain why mature red blood cells have a relatively short life span (less than 120 days)?

____

____

CLINICAL CORRELATION

The **hematocrit** is a value that records the percentage of whole blood that is composed of cells (the formed elements). Since red blood cells comprise the vast majority (99.9%) of the formed elements, the hematocrit value is used to measure erythrocyte levels and thereby assess oxygen carrying capacity. The normal hematocrit range is 42 to 52 for men and 37 to 47 for women.

Another diagnostic measure for oxygen carrying capacity is **hemoglobin (Hb) concentration** in red blood cells. Hb concentration is measured in grams per deciliter (g/dl). Typically, a decline in the hematocrit value will also cause a decline in Hb concentration. Normal Hb concentrations are 14 to 18 g/dl in males and 12 to 16 g/dl in females.

Anemia is a condition characterized by a reduction in oxygen carrying capacity and a resulting decline in oxygen transport to cells and tissues. Anemia can occur when the hematocrit and/or the Hb concentration is lowered.

White Blood Cells

White blood cells play vital roles in defending the body from pathogens and foreign proteins. When these cells are activated, they are capable of migrating from blood vessels to surrounding tissues by a process called **diapedesis.** Neutrophils, eosinophils, and monocytes are able to ingest pathogens and debris from dead cells by phagocytosis. Basophils release chemicals (histamines and heparin) that enhance inflammation when tissue damage occurs. Lymphocytes are responsible for defenses against a specific pathogen. These activities include direct cellular attacks and the production of antibodies.

WHAT'S IN A WORD The word *diapedesis* is derived from Greek and means "leaping through." White blood cells can pass or "leap through" the walls of blood vessels to enter surrounding tissues where they destroy disease-causing organisms and ingest the remains of dead cells.

You have already discovered that white blood cells are not nearly as abundant as red blood cells. However, white blood cells are easy to identify because of their relatively large size and distinctive nuclei.

ACTIVITY 19.2 Identifying White Blood Cell Types

1. Under high power, carefully scan a slide of a normal human blood smear prepared with Wright's stain.
2. When you identify a white blood cell, use immersion oil and the oil immersion lens to observe the cell at a higher magnification and to identify the specific white blood cell type. For help, refer to Table 19.1 and Figure 19.3, and consider the following general features of white blood cells.
   - On average, white blood cells are approximately twice as large (10–12 μm in diameter) as red blood cells. Monocytes are the largest white blood cells, ranging from 14 to 24 μm in diameter. Lymphocytes range in size from 5 to 17 μm in diameter, and are often classified as small (5–8 μm), medium (9–12 μm), and large (13–17 μm).
   - The nuclei of white blood cells, which stain a deep blue or purple, have distinctive shapes. For example, a neutrophil has a multilobed nucleus, whereas in a basophil or an eosinophil, the nucleus is bilobed. The typical monocyte nucleus possesses a deep indentation on one side, giving it a kidney shape. In a lymphocyte, the nucleus is round and occupies the vast majority of the cell volume, leaving only a narrow rim of cytoplasm around the periphery.
   - Neutrophils are the most abundant white blood cells and will be the easiest to identify. Lymphocytes and monocytes are also relatively common. Eosinophils and especially basophils are quite rare and difficult to identify. If you think that you have found either of these cell types, verify the identification with your instructor. Be patient in your attempt to identify these cells, but do not be discouraged if you cannot find one. Perhaps your instructor will have better luck!

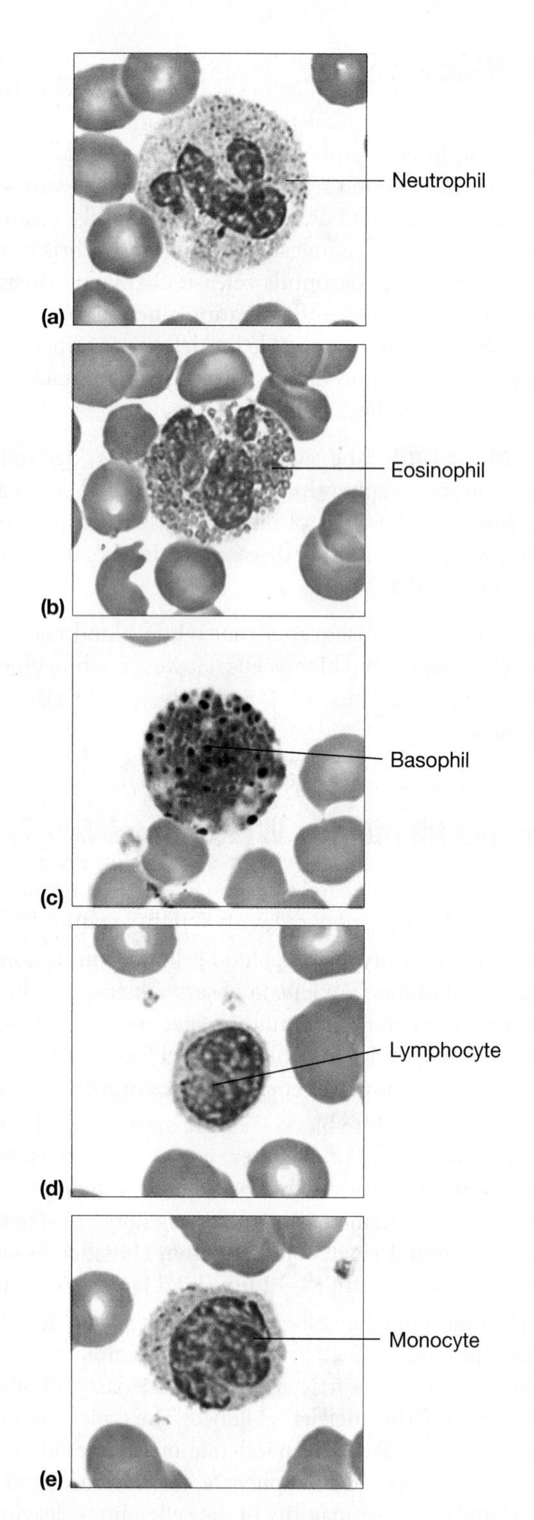

Figure 19.3 Structure of white blood cells. a) neutrophil; **b)** eosinophil; **c)** basophil; **d)** lymphocyte; **e)** monocyte.

QUESTION TO CONSIDER Over-the-counter antihistamine drugs are used to relieve the pain and discomfort of inflammation due to allergies, cold, and fever. However, overuse of these drugs could prolong your symptoms. Speculate on a reason for this effect. (Hint: Consider the function of basophils and eosinophils; see Table 19.1.)

____

____

CLINICAL CORRELATION

Leukemia is a cancerous disease of the blood that involves the uncontrolled propagation of abnormal white blood cells in the bone marrow. Leukemias can be classified according to the type of leukocyte involved. **Myeloid leukemias** involve granulocytes; **lymphoid leukemias** involve lymphocytes. For both categories, the disease can advance quickly (acute) or slowly (chronic). As leukemia progresses, abnormal leukocytes gradually replace normal cells, and bone marrow function is impaired. As a result, the production of red blood cells, normal white blood cells, and platelets declines significantly, leading to anemia, infections, and reduced blood clotting. One option to fight leukemia is a **bone marrow transplant.** In this procedure, the patient is exposed to a massive dose of radiation or chemotherapy to kill any abnormal and cancerous cells in the bone marrow. This exposure also kills the normal cells, so the individual is left highly susceptible to infections that could be fatal. Next, the patient is given healthy bone marrow tissue that, hopefully, will generate new populations of normal blood cells. Compatibility of blood and tissue types is a critical factor in transplant operations. If tissue rejection occurs, the donor's lymphocytes could attack and destroy the recipient's tissues, a condition that could cause death.

Differential White Blood Cell Count

White blood cells play vital roles in protecting us from infections and promoting inflammation in response to tissue damage and allergies. Under normal conditions, the blood will contain a certain percentage of each white blood cell type (Table 19.1). Any deviation in the normal percentage ranges could indicate an abnormal condition such as a bacterial, viral, or parasitic infection.

In a clinical laboratory, a **differential white blood cell count** is performed to determine the percentages of each white blood cell type. Although this procedure can be completed rather quickly using computers, the manual method described here will yield similar results.

ACTIVITY 19.3 Performing a Differential White Blood Cell Count

Form a Hypothesis **To complete your differential white blood cell count, you will be required to identify at least 100 white blood cells. Before you begin, use the information in Table 19.1 to predict the number of each white blood cell type that**

Table 19.2 Differential White Blood Cell Count

	Predicted results	Observed results	
Cell Type	**# of Cells**	**# of Cells**	**% of Total Leukocytes**
Neutrophil			
Eosinophil			
Basophil			
Monocyte			
Lymphocyte			
Total	100		100

you expect to identify. Record your prediction in the "Predicted Results" column in Table 19.2.

1. Obtain a slide of a human blood smear prepared with Wright's stain.
2. Move your slide to the upper left margin of the blood smear and focus under low power.
3. Switch to a high power and observe the upper left region of the blood smear. Find an area in which the blood cells are dispersed evenly throughout the field of view. For the most accurate results, the oil immersion lens should be used, although you can complete this activity with the highest dry objective lens.
4. Beginning at the upper left region of the blood smear, carefully scan the entire slide in the back-and-forth pattern illustrated in Figure 19.4. When you observe a white blood cell, identify the type and record it in Table 19.2.
5. After you have identified at least 100 white blood cells, convert the number of each cell type to a percentage of the total, using the following equation. Record these results in Table 19.2.

Percentage (%) = (# cells observed ÷ total # counted) × 100

Assess the Outcome **Examine the data that you have collected in Table 19.2. Do your predicted results agree with your observations? Provide an explanation for your observations.**

____

____

QUESTION TO CONSIDER **Infectious mononucleosis** is a disease believed to be caused by an infection of the Epstein-Barr virus. It is characterized by fever, sore throat, swollen lymph nodes, and an enlarged spleen. Discuss how a differential white blood cell count from an individual with this disease would differ from a normal count.

____

____

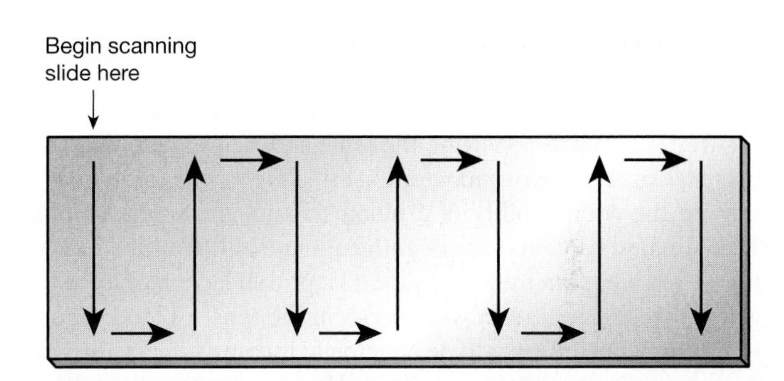

Figure 19.4 Method for performing a differential white blood cell count. Observations of a blood smear begin at the upper left corner of the microscope slide. The slide is scanned in a back-and-forth pattern as indicated by the arrows until the entire blood smear is viewed.

Blood Types

An individual's blood type is a genetically determined trait. It is based on the presence or absence of specific glycoprotein molecules, called **surface antigens** or **agglutinogens**, which are located on the cell membranes of erythrocytes. The immune system recognizes the surface antigens as being normal and will not attack them as a foreign substance.

Blood plasma contains **antibodies** or **agglutinins**, which are each genetically programmed to react with a specific surface antigen if it is present. Thus, if blood of two different types are mixed, an **antigen–antibody reaction** will occur and the result will be **agglutination**, or clumping of red blood cells.

In humans, there are over 50 blood groups, but most do not cause significant reactions when different types are mixed. In this activity, you will be studying two blood groups, the **ABO group** and the **Rh system**, which are significant for their antigenic reactions.

The ABO group is based on the presence or absence of two surface antigens on red blood cells, called A and B. Persons with **Type A** blood have surface antigen A on their red blood cell membranes, and persons with **Type B** blood have surface antigen B. If both surface antigens are present, the blood is **Type AB**;

Table 19.3 ABO Blood Groups

Blood type	Surface antigen	Antibody	Compatible donor	Incompatible donor
A	A	Anti-B	Type A, Type O	Type B, Type AB
B	B	Anti-A	Type B, Type O	Type A, Type AB
AB	A, B	None	Type AB, Type A, Type B, Type O	None
O	None	Anti-A, Anti-B	Type O	Type A, Type B, Type AB

if both surface antigens are absent, the blood is **Type O**. Blood plasma will contain antibodies for the surface antigen that is not present on red blood cells. Thus, Type A blood will have the **anti-B** antibody, and Type B blood will have **anti-A antibody**. Type AB blood will have neither antibody, but Type O blood will contain both anti-A and anti-B antibodies.

A summary of the ABO system is shown in Table 19.3 and Figure 19.5. When reviewing the table and figure, note that the presence or absence of surface antigens is the key factor in determining the compatibility of a blood transfusion. For example, Type A blood contains surface antigen A and anti-B antibodies. A type A recipient can receive any blood type that lacks surface antigen B, which could react with anti-B in the type A blood. Thus, for a blood transfusion, a Type A recipient is compatible with Type A or Type O donors (Figure 19.5). Using a similar rationale, a Type B recipient can receive blood from Type B or Type O donors.

Type O is a special blood type because both surface antigens A and B are absent. Thus, red blood cells in donor Type O blood will not agglutinate with antibodies that might be present in a recipient's blood. Because of this feature, Type O is the **universal donor** (Table 19.3; Figure 19.5). However, because Type O blood contains small levels of anti-A and anti-B antibodies, some agglutination can occur if it is mixed with different blood types. Type AB blood is unique because it lacks both anti-A and anti-B antibodies. Since there are no antibodies to cause agglutination, Type AB blood is referred to as the **universal recipient** (Table 19.3; Figure 19.5). However, antibodies that are present in Types A, B, and O blood could cause some agglutination in a Type AB recipient. The lesson to be learned from this information is that donor and recipient blood types can be different, but the best case scenario is always when the two blood types are the same.

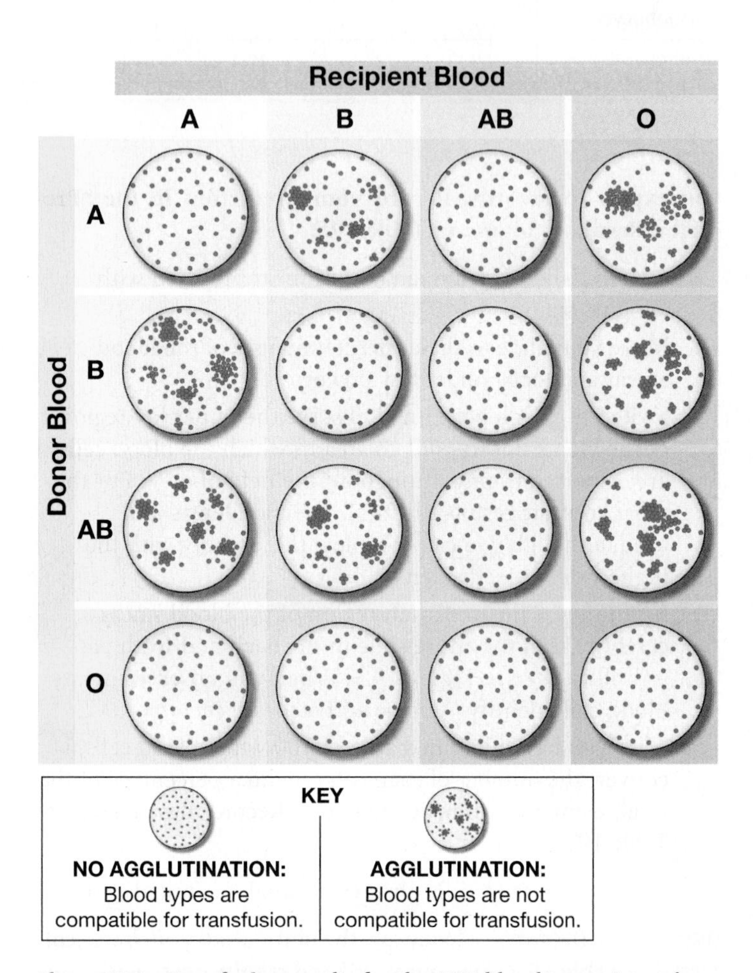

Figure 19.5 Transfusion results for the ABO blood types. A transfusion is compatible if the donor's blood lacks the antigens (A, B) that will react with antibodies (anti-A, anti-B) in the recipient's blood and cause agglutination. Notice that type O blood is the universal donor and Type AB is the universal recipient.

CLINICAL CORRELATION

An error in blood typing, leading to an incompatible blood transfusion, could prove to be a fatal mistake. Surface antigens on the red blood cells of the recipient blood will react with antibodies in the donor blood. As agglutination proceeds, some cells will swell and eventually rupture (**hemolysis**), releasing hemoglobin into the blood. As clumps of erythrocytes become trapped in capillary beds, blood delivery to cells and tissues in all parts of the body could be dramatically reduced, leading to multiple organ failure and possibly death.

The Rh system is named after the rhesus monkey, the primate in which the surface antigen, the **Rh factor**, was first discovered. For Rh blood typing, a plus (+) sign is used if Rh factor is present and a negative (−) sign is used if it is absent. For example, a person with Type A blood can be either A (+) or A (−), depending on whether the Rh factor is present or absent. Unlike the ABO group, a person who is Rh negative does not normally produce anti-Rh antibodies unless he or she is inadvertently exposed to Rh factor (e.g., through a blood transfusion).

CLINICAL CORRELATION

If an Rh (−) woman becomes pregnant, it is possible for the fetus to be Rh (+) if the father is Rh (+). During the pregnancy, fetal blood could leak from the placenta to the maternal bloodstream. As a result, the mother becomes sensitized to the Rh factor by producing anti-Rh antibodies. The fetus will not be affected because anti-Rh is not typically produced in a high enough concentration until after birth. However, potentially fatal consequences could result if there is a second pregnancy and the fetus is Rh (+). Antibodies from the maternal blood could cross the placenta, enter the fetal blood, and react with Rh factor. The resulting agglutination and hemolysis, known as **hemolytic disease of the newborn (HDN),** could be fatal. HDN can be easily prevented. If a mother is Rh (−), a drug called RhoGAM is given after the first pregnancy. RhoGAM prevents the production of anti-Rh antibodies by the mother and protects the second fetus from the dangers of fetal–maternal blood incompatibility.

Before you proceed with the next activity, please remember that you will be working with blood, a body fluid that may contain infectious organisms. Adhere to the following safety procedures when handling blood (or any other body fluid).

- Always assume that the blood with which you are working is infected with a disease-causing organism. This attitude will put you in the right frame of mind to work with extreme caution.
- Work with your own blood *only.* Under no circumstances should you collect or conduct experiments with the blood of another individual.
- Always wear gloves, safety eyewear, and a mask when working with blood. Never allow blood to come in contact with unprotected skin. Protective gear should be worn throughout the experiment and during cleanup.
- Mouth pipetting should never be done under any circumstances.
- If any part of your skin is accidentally contaminated with blood, disinfect the area immediately with a 70% alcohol solution for at least 30 seconds, followed by a 1-minute soap scrub and rinsing.
- Any blood that spills onto your work area should be disinfected with a 10% bleach solution or a commercially prepared disinfectant. The contaminated area should remain covered with the bleach/disinfectant for at least 30 minutes before wiping it off.
- Lancets, needles, and other sharp instruments (sharps) should be used only once. After use, disinfect all sharps in a 1:10 solution of household bleach and 70% alcohol for 30 minutes and then place in a puncture-proof container for disposal.
- All reusable glassware and other instruments should be disinfected in a 10% bleach solution for 30 minutes and then washed in hot, soapy water. If available, autoclaving this equipment before washing is recommended.
- Work areas should be cleaned with a 10% bleach solution or a commercially prepared disinfectant before and after any laboratory activity during which blood is used.
- As an alternative, simulated blood typing kits are available. These kits will allow you to conduct blood typing experiments without using human blood products.

ACTIVITY 19.4 Determining Your Blood Type

Form a Hypothesis **Before you begin the blood typing experiment, review the data in Table 19.4, which lists the distribution of blood types in the United States. Conduct a demographic survey of your class by counting the total number of students and the number of students in each population group. Record this information at the bottom of Table 19.5. Use the data tabulated in Table 19.4 to make a prediction of the distribution of ABO and Rh blood groups in your laboratory class. Record this information in the predicted columns in Table 19.5.**

1. Obtain a sterile glass microscope slide. With a wax pencil, draw a line that divides the slide into left and right sides. Label the left side "A" and the right side "B."
2. Obtain a second sterile glass microscope slide and label it "Rh."
3. Place a paper towel on your work space. This towel will be used to place blood collecting instruments prior to disposal.
4. Wash your hands thoroughly with warm water and soap and dry them completely with a paper towel.
5. You will be collecting a blood sample from yourself. Put on your protective eyewear and face mask. Place a surgical glove on the hand that will hold the lancet. The hand that will be used for collecting the blood sample should remain uncovered.
6. Use a sterile alcohol pad to clean the tip of your index finger on all sides. Place the used pad on the paper towel.

Table 19.4 Distribution of Blood Types in the United States

Incidence of blood types (%)					
Population Group	**O**	**A**	**B**	**AB**	**Rh (+)**
White	45	40	11	4	85
African American	49	27	20	4	95
Korean	32	28	30	10	100
Japanese	31	38	21	10	100
Chinese	42	27	25	6	100
Native American	79	16	4	1	100

7. Open a sterile blood lancet to expose the sharp tip. With a swift and deliberate motion, jab the lancet tip into the fingertip. After use, place the lancet on the paper towel until it can be properly discarded. The lancet should not be used more than once under any circumstances.
8. Squeeze a drop of blood to each half of the slide labeled A and B, and another drop to the slide labeled Rh.
9. Add a drop of anti-A serum to blood sample A. As you add the serum, keep the dropper from directly contacting the drop of blood. In the same way, add a drop of anti-B serum to blood sample B and a drop of anti-Rh serum to blood sample Rh.
10. Mix the blood samples and antisera with clean toothpicks. Be sure to use a different toothpick to mix each sample. Place the used toothpicks on the paper towel until they can be discarded safely.
11. Place both slides on a warming tray. Gently agitate the samples in a back-and-forth manner for 2 minutes.
12. Examine the blood samples for evidence of agglutination. Agglutination will occur when antibodies in the antiserum react with the corresponding surface antigen on red blood cells (Figure 19.6).

    For the ABO group, agglutination is possible when the following occur.

    - Type AB blood is exposed to either anti-A or anti-B serum (Figure 19.6a).
    - Type B blood is exposed to anti-B serum (Figure 19.6b).
    - Type A blood is exposed to anti-A serum (Figure 19.6c).
    - Type O blood will not agglutinate when exposed to either antiserum (Figure 19.6d).

    For determining the Rh status, Rh (+) blood will agglutinate when exposed to anti-Rh serum, but Rh (−) blood will not. The Rh agglutination is often difficult to observe with the unaided eye. If this is the case, use a microscope to observe the reaction.
13. Record your blood typing observations in Table 19.6 by writing a (+) or (−) for the presence or absence of agglutination.

Assess the Outcome **Collect the blood typing results from the rest of the class, and sort the data according to the various population groups that are represented. Calculate the incidence of blood types in each population group and record the data in the "actual" columns in Table 19.5. Did your predictions for blood type incidence agree with the actual results? Provide an explanation for your results. If the sample size in your class is small, discuss the limitations that this would have on your analysis.**

____

____

Table 19.5 Distribution of Blood Types in the Laboratory Class

Incidence of blood types (%)

Population Groups in the Class	O		A		B		AB		Rh(+)	
	Predicted	Actual	Predicted	Actual	Predicted	Actual	Predicted	Actual	Predicted	Actual
1.										
2.										
3.										
4.										
5.										

Total number of students in class ________________

Number of students in each population group:

Population group	# students
1.	
2.	
3.	
4.	
5.	

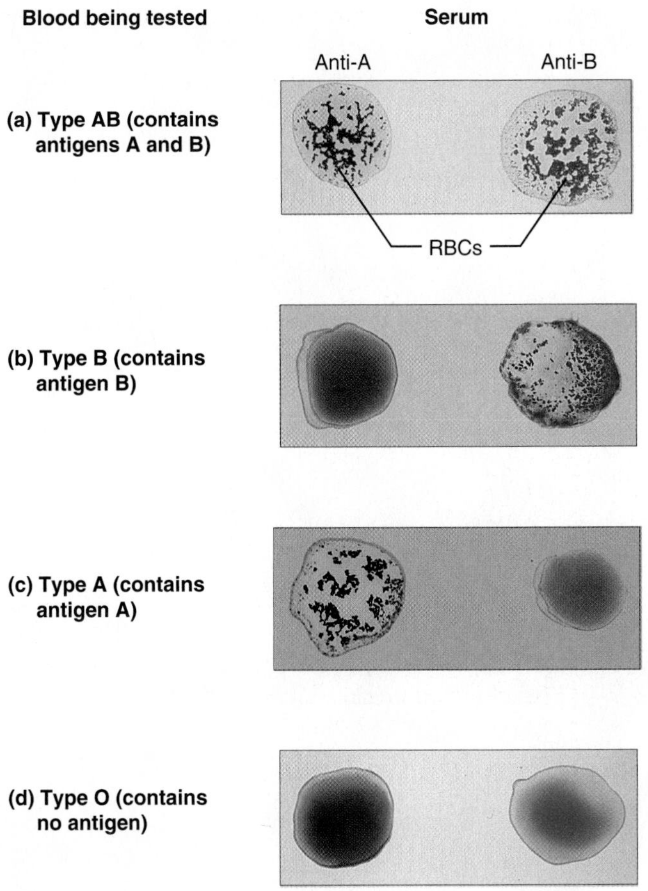

Blood being tested | Serum: Anti-A, Anti-B

(a) Type AB (contains antigens A and B)

RBCs

(b) Type B (contains antigen B)

(c) Type A (contains antigen A)

(d) Type O (contains no antigen)

Figure 19.6 ABO blood typing results. a) For Type AB blood, agglutination occurs when it is exposed to both anti-A and anti-B sera. **b)** For Type B blood, agglutination occurs when it is exposed to anti-B serum, but not to anti-A serum. **c)** For Type A blood, agglutination occurs when it is exposed to anti-A serum, but not to anti-B serum. **d)** For Type O blood, agglutination does not occur when it is exposed to either anti-A or anti-B serum.

QUESTIONS TO CONSIDER

1. While typing the blood of a patient who is about to undergo surgery, the medical technician determines that agglutination occurs when the blood is exposed to both anti-A and anti-B. Agglutination does not occur when the blood is exposed to anti-Rh. Based on the results of these tests, what is the blood type of the patient? Provide a brief explanation for your answer.

____

____

____

____

2. The destruction of fetal red blood cells can occur if anti-Rh antibody enters the blood of an Rh-positive fetus. For this potentially fatal situation to occur, the parents must have the following Rh blood typing characteristics.

- The mother is Rh (−) but possesses anti-Rh antibodies in her blood.
- The father is Rh (+).

Explain why these two conditions must be correct.

____

____

____

____

Table 19.6 Blood Typing Results

Name	Presence of agglutination (+/−)			
	Anti-A	Anti-B	Anti-Rh	Blood Type
1.				
2.				
3.				
4.				
5.				

Exercise 19 Review Sheet

Blood Cells

Name ____________________

Lab Section ____________________

Date ____________________

Questions 1–5: Define the following terms.

1. Formed elements

2. Blood plasma

3. Hemoglobin

4. Granulocytes versus agranulocytes

5. Diapedesis

6. Describe the basic functions of blood.

Questions 7–13: Match the cell type in column A with the correct description in column B.

A	B
7. Neutrophils ________	a. Large cells with kidney-shaped nuclei and no granules in the cytoplasm
8. Erythrocytes ________	b. Regulate the immune response
9. Monocytes ________	c. Involved in the repair of damaged blood vessels and blood clotting
10. Basophils ________	d. Biconcave discs that lack nuclei
11. Platelets ________	e. Contain bilobed nuclei with bright orange-red granules in the cytoplasm
12. Eosinophils ________	f. Enhance inflammatory response by releasing heparin and histamine
13. Lymphocytes ________	g. The most abundant white blood cell; contain multilobed nuclei; attack infectious agents by phagocytosis

14. Explain how the blood types in the ABO group are derived.

____

____

15. Explain why Type O blood is the universal donor and Type AB blood is the universal recipient.

____

____

Exercise 20

Gross Anatomy of the Heart

Laboratory Objectives

On completion of the activities in this exercise, you will be able to:

- Describe the anatomical relations of the heart with other structures in the thoracic cavity.
- Provide details on the arrangement of the connective tissue layers (the pericardium) that surround the heart.
- Identify important structures on the surface of the heart.
- Locate the major internal structures of the heart.
- Identify the tissue layers of the heart wall.
- Describe the coronary circulation.
- Describe the flow of blood through the heart.
- Dissect a sheep heart and compare its structure with the human heart.

Materials

- Anatomical models or figures of the human heart
- Plastic bags
- Cotton or cheesecloth
- Colored pencils
- Preserved sheep hearts
- Dissecting trays
- Dissecting tools
- Dissecting gloves
- Protective eyewear
- Face mask

The heart is a two-sided, double-pumping organ. The left side (the left pump) controls the flow of blood to all tissues and cells in the body, where oxygen and nutrients are delivered and metabolic wastes are taken away. The right side (the right pump) sends blood to the lungs, where oxygen stores in red blood cells are replenished and carbon dioxide, a metabolic waste, is released. To keep blood circulating throughout the body, the heart beats approximately 100,000 times and pumps between 7000 and 9000 liters of blood each day. By any standard, this is an arduous workload, but the fact that the heart can maintain this level of activity for decades, without stopping, is nothing short of remarkable.

In this exercise, you will examine the special anatomical features that reflect the enduring and efficient functioning of the heart. You will focus your examination on gross anatomical structure. If you would like to review the light microscopic structure of cardiac muscle, see Activity 10.2.

The Pericardium

The heart is enclosed by a membranous sac called the **pericardium** (Figure 20.1c). This structure consists of two parts. The outer **fibrous pericardium** is a tough, fibrous connective tissue layer that is fused to adjacent structures (the diaphragm, sternum, costal cartilages of ribs, thoracic vertebrae, and the great vessels emerging from the heart). The inner **serous pericardium** is a delicate serous membrane that forms a double-layered sac around the heart. It consists of the **parietal pericardium**, which covers the deep or inner surface of the fibrous pericardium, and the **visceral pericardium**, which forms the outer surface of the heart wall. The potential space between the parietal and visceral pericardial membranes is the **pericardial cavity** (Figure 20.1c). The cavity is filled with a watery fluid produced by the epithelial cells lining the serous pericardium. The fluid helps to reduce friction when the two serous membranes rub against each other as the heart beats.

CLINICAL CORRELATION

Inflammation of the pericardial membranes, known as **pericarditis,** increases the friction between the two membranes and causes an overproduction of fluid. As the fluid accumulates in the pericardial cavity it inhibits the normal movements of the heart wall and restricts cardiac output, leading to a condition called **cardiac tamponade.**

ACTIVITY 20.1 Examining the Organization of the Pericardium

1. Obtain a large, clear plastic bag and close off the open end.
2. With a heart model, push the inferior pointed tip (the apex of the heart) into the wall of the closed plastic bag. This action is similar to pushing a fist into the bag as illustrated in Figure 20.1c.
3. Notice that as you push the heart deeper into the closed bag, two layers of plastic, separated by a space, cover the organ (Figure 20.1c).
4. Continue pushing the heart into the bag until you reach the great vessels that are attached to the superior aspect of the heart (aorta, pulmonary trunk, superior vena cava).
5. The plastic layers represent the serous pericardium (Figure 20.1c), as follows:
   - The inner plastic layer that is in contact with heart wall represents the visceral pericardium.
   - The outer layer of plastic represents the parietal pericardium.
   - The space between the two plastic layers is the pericardial cavity.
   - Notice that the two layers of plastic are continuous with each other at the great vessels. In other words, the visceral pericardium is continuous with the

parietal pericardium where the great vessels are connected to the heart (Figure 20.1c).

6. Wrap a layer of cotton or cheesecloth over the outer plastic layer (the parietal pericardium). This layer represents the fibrous pericardium which is attached to the connective tissue layers that surround adjacent structures.

QUESTION TO CONSIDER What important function does the fibrous pericardium serve?

____

____

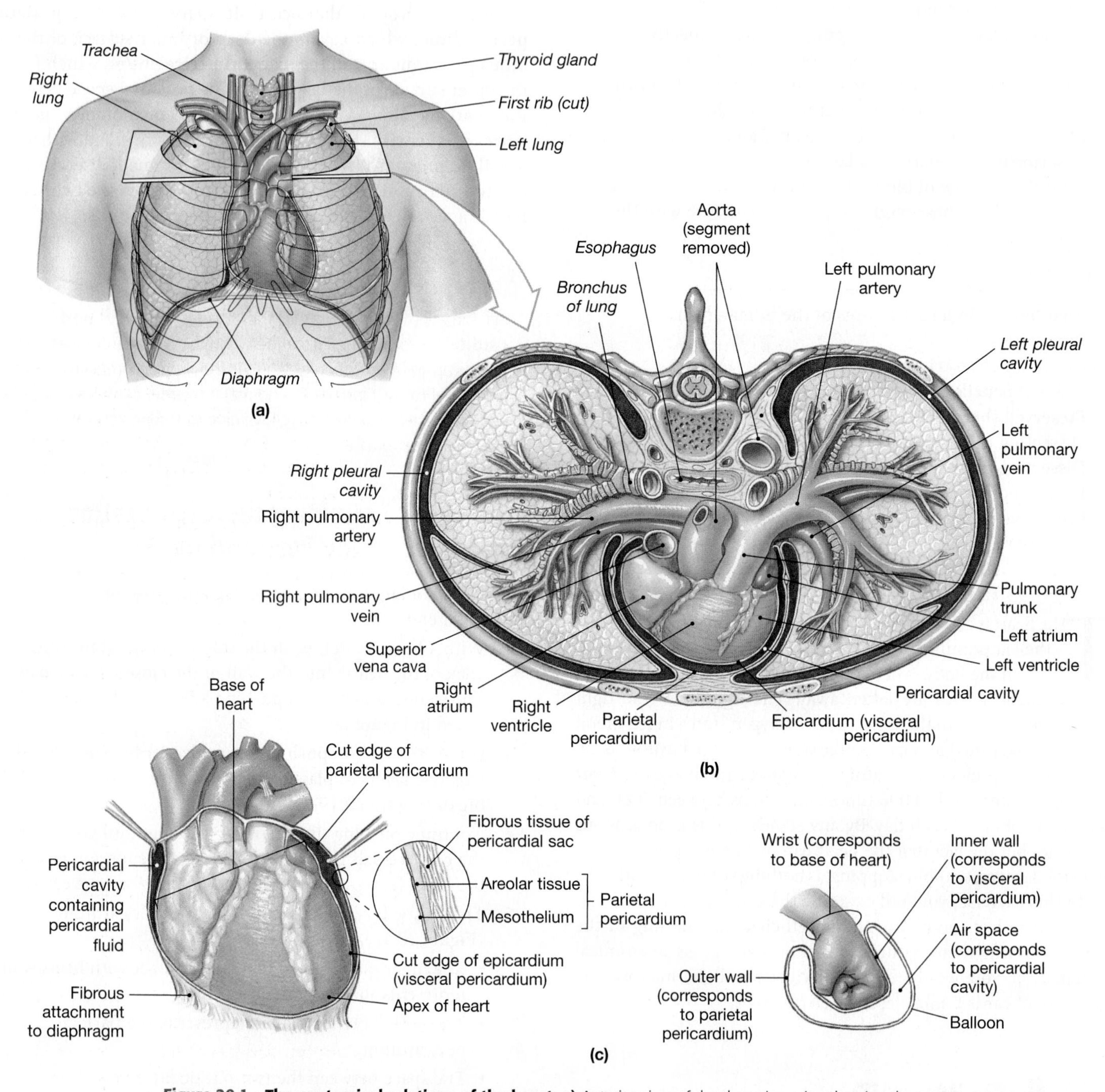

Figure 20.1 The anatomical relations of the heart. a) Anterior view of the thoracic cavity, showing the heart within the mediastinum and between the two lungs; **b)** transverse section of the thoracic cavity showing the position of the heart in relation to other structures; **c)** the relationship of the heart, pericardium, and pericardial cavity.

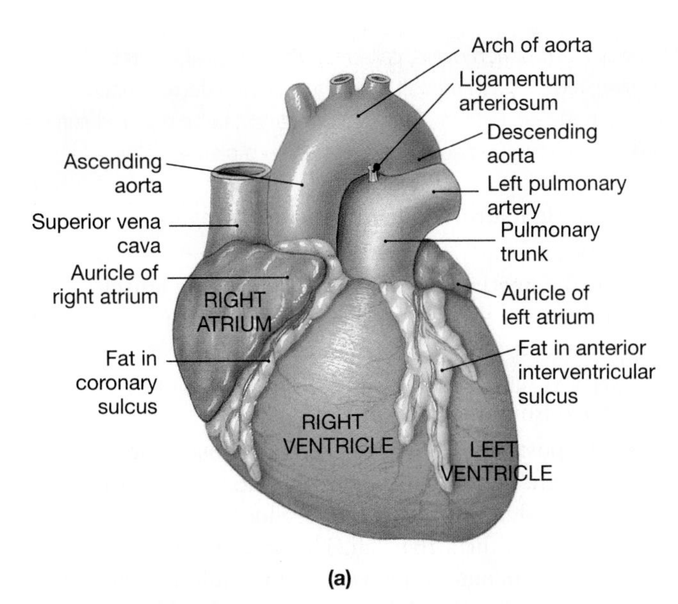

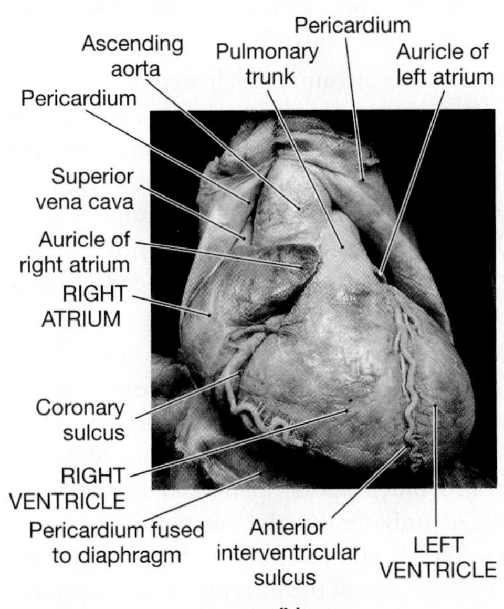

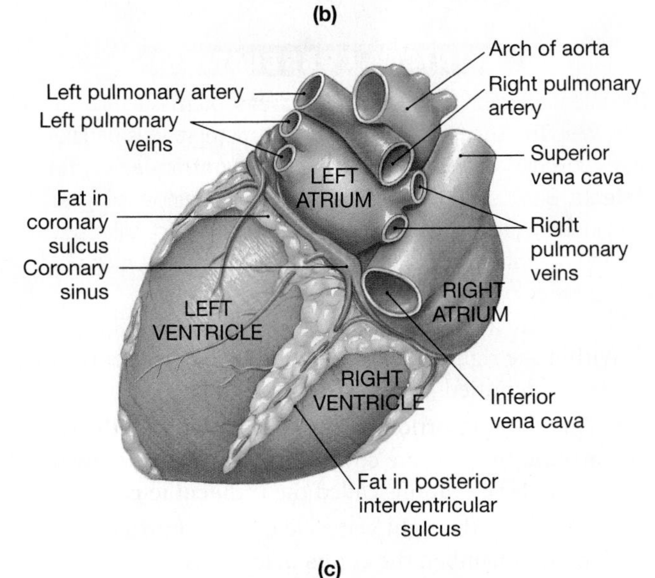

Figure 20.2 External anatomy of the heart. a) Diagram, and **b)** dissection, anterior view; **c)** diagram, posterior view.

Gross Anatomy of the Heart

The heart is a four-chambered organ that is shaped roughly like an inverted pear. On average, it is approximately 14 cm long and 9 cm wide, or slightly larger than a clenched fist. Its weight ranges from 230 to 280 grams in females and 280 to 340 grams in males. The heart and its surrounding pericardial cavity are located within the **mediastinum**, a centrally located area within the thoracic cavity. Two thirds of the organ is positioned to the left of the midline. It is bordered laterally by the pleural cavities, which surround the lungs, anteriorly by the sternum, and posteriorly by the esophagus and thoracic vertebrae (Figures 20.1a and b).

ACTIVITY 20.2 Examining the Gross Anatomy of the Human Heart

External Anatomy

1. Examine a model of the heart from an anterior view (Figures 20.2a and b) and make the following observations.
   - The heart is divided into left and right sides. Each side contains two chambers: a superior **atrium** that receives blood and an inferior **ventricle** that discharges blood. Identify the four heart chambers: **right atrium, right ventricle, left atrium** (best observed from a posterior view), and **left ventricle.**
   - The **apex of the heart** (Figure 20.1c) is formed by the inferior tip of the left ventricle. It is located at the level of the fifth intercostal space, 7 to 9 cm to the left of the median plane. Locate this position on a skeleton.
   - Extending off the main body of each atrium is a medial appendage known as an **auricle.** Locate the auricles of each atria.

WHAT'S IN A WORD The term *auricle* is derived from *auricular,* the Latin word for "external ear." The auricles of the atria were given that name because early anatomists noted their resemblance to the external ear.

   - Identify the **atrioventricular (coronary) groove (sulcus)**, which divides the atria, superiorly, from the ventricles, inferiorly.
   - From the atrioventricular groove, identify the **anterior interventricular groove** as it travels toward the apex along the anterior surface of the heart. It forms a border between the left and right ventricles.
   - Locate the **ascending aorta**, which receives blood from the left ventricle. The ascending aorta gives rise to the **arch of the aorta**. The aortic arch gives off three branches in the following order: the **brachiocephalic artery**, the **left common carotid artery**, and the **left subclavian artery**.

- Locate the **pulmonary trunk.** It is located anterior to the ascending aorta and receives blood from the right ventricle. The pulmonary trunk gives rise to the **right** and **left pulmonary arteries.**
- Locate the **superior vena cava** and **inferior vena cava** where they empty into the right atrium.

2. Examine a model of the heart from a posterior view (Figure 20.2c) and make the following observations.
   - Identify the relative positions of the four heart chambers. Notice that the left atrium is more easily identified from a posterior view.
   - The **base of the heart** is at the heart's posterior and superior aspects, and is formed primarily by the left atrium. It is located beneath the second pair of ribs and the **sternal angle** (junction of the sternal body and manubrium), and extends from T6 to T9 vertebrae. Locate this position on a skeleton.
   - Once again, identify the atrioventricular (coronary) groove as it continues along the posterior surface of the heart. Notice that it forms a complete circle around the heart, and forms a border between the atria and ventricles.
   - Locate the **posterior interventricular groove.** Like the anterior interventricular groove, the posterior groove descends toward the apex from the atrioventricular groove, and forms a border between the left and right ventricles.
   - Observe the four **pulmonary veins**—two on each side—as they enter the left atrium.
   - Once again, identify the superior and inferior vena cavae entering the right atrium.

Internal Anatomy

1. On the heart model, open the heart wall to expose the internal structures (Figure 20.3).
2. Identify the two superior chambers, the right and left atria. With one hand, place your thumb on one side and your index finger on the other side of the wall that separates the two atria. Your fingers are holding the **interatrial septum.** Identify the **fossa ovalis**, an oval depression along the interatrial septum within the right atrium.

CLINICAL CORRELATION

The wall that separates the right and left atria is called the **interatrial septum.** Inside the right atrium, along the interatrial septum, there is an oval depression called the **fossa ovalis** (Figure 20.3). This depression marks the site of the **foramen ovale,** an opening that connects the atria in the fetal heart. The foramen ovale has a valve that allows blood to travel from the right atrium to the left atrium but not in the reverse direction. This specialization in the fetal circulation allows most of the oxygenated blood coming from the placenta to bypass the nonfunctional lungs and the pulmonary circulation, and to pass directly to other vital organs via the systemic circulation. At birth, the foramen ovale closes when the valve fuses with the interatrial septum. Incomplete closure of the foramen ovalis, called an **atrial septal defect,** allows oxygenated blood in the left atrium to mix with deoxygenated blood in the right atrium. This malformation can be repaired surgically to prevent the two blood supplies from blending.

3. In the right atrium, identify the following distinct regions.
   - The anterior wall is defined by the rough surface formed by the **pectinate muscles.** The pectinate muscles continue onto the wall of the right auricle, the ear-shaped, muscular pouch that extends medially from the atrial wall (Figure 20.2a). This portion of the atrium is derived from the embryonic heart.
   - The posterior wall lacks pectinate muscles and is smooth. Verify that the openings for the superior vena cava and inferior vena cava are located along the smooth portion of the atrial wall. The posterior atrial wall is derived from embryonic veins. At the inferior end of the right atrium, the **right atrioventricular (AV) orifice** leads into the right ventricle.
4. Similar to the right atrium, the left atrium has two distinct regions with similar embryonic origins. The small anterior wall is dominated by the left auricle. Identify the pectinate muscles in this region. The posterior wall is smooth and relatively large. Verify that the four pulmonary veins drain into the left atrium from this region. Identify the left atrioventricular (AV) orifice, at the inferior end of the left atrium, leading into the left ventricle.
5. Identify the two inferior chambers, the right and left ventricles. Place the thumb and index finger of one hand on either side of the wall that separates the two ventricles. This structure is the **interventricular septum** (Figure 20.3). Notice that this wall is much thicker than the interatrial septum. Once again, identify the anterior and posterior interventricular grooves. Verify that these grooves delineate the anterior and posterior margins of the interventricular septum.

CLINICAL CORRELATION

The inferior portion of the interventricular septum is a thick muscular wall. The superior portion is a thin, membranous partition and, consequently, is a more likely site for **ventricular septal defects.** Because blood pressure in the left ventricle is higher than in the right ventricle, a ventricular septal defect will result in a left-to-right shunt of blood. This malformation, left unrepaired, can cause pulmonary disease and heart failure.

6. Within the right ventricle, the following distinct regions can be identified.
   - The inferior portion receives the blood from the right atrium. Its walls are covered by an irregular network of muscular elevations called the **trabeculae carneae.**
   - Superiorly, the right ventricle narrows into a cone-shaped chamber, the **conus arteriosus**, which leads to the pulmonary trunk. The wall of the conus arteriosus is smooth and, consequently, lacks trabeculae carneae.

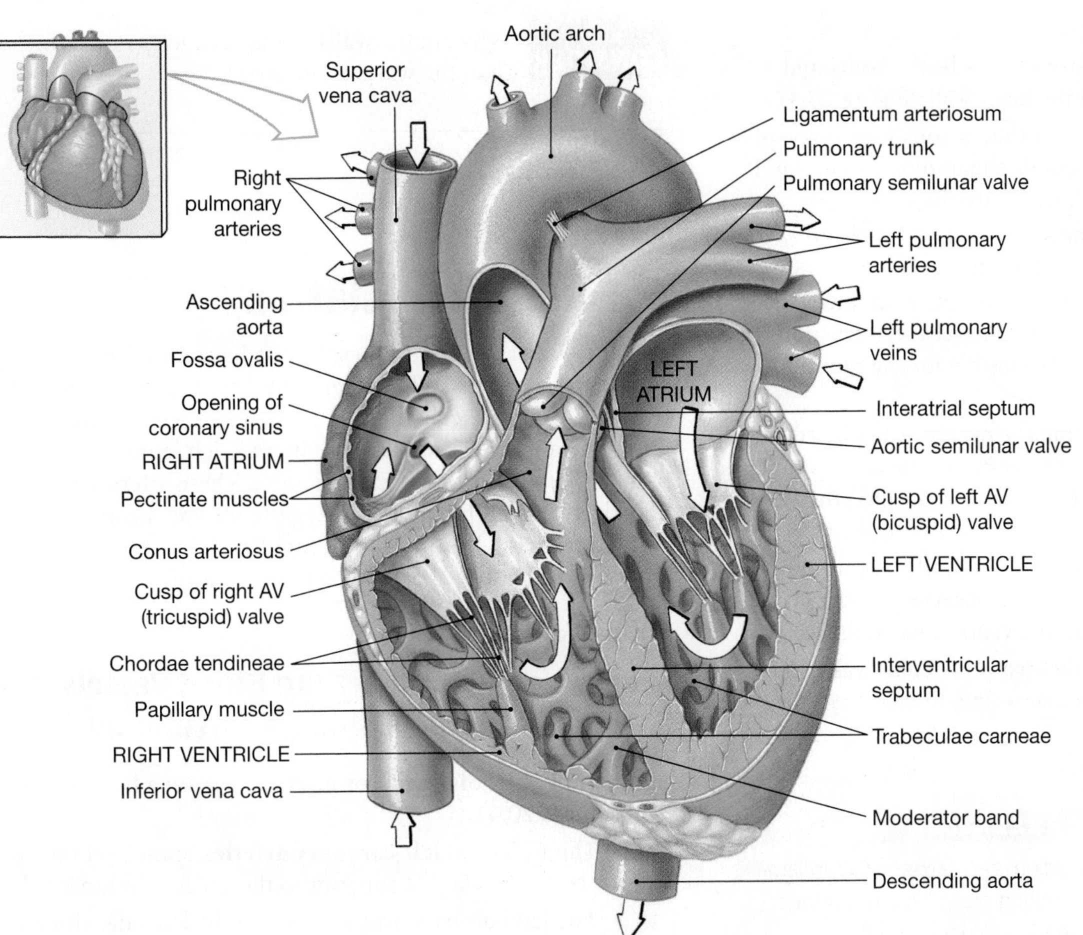

Figure 20.3 Internal anatomy of the heart. Internal structures of the heart chambers are revealed in a coronal section. The arrows indicate the direction of blood flow.

7. Inside the left ventricle, the arrangement of rough- and smooth-walled sections is similar to what is found in the right ventricle. The wall along the inferior portion of the left ventricle is rough, due to the presence of the trabeculae carneae. The **aortic vestibule** is the smooth-walled, superior region that leads to the aorta.
8. Identify the two pairs of **heart valves**, which are strategically located to regulate blood flow through the heart and into the great arteries (aorta and pulmonary trunk).
   - The two **atrioventricular (AV) valves** are positioned between the atria and ventricles at the atrioventricular orifices. These valves open and close as a result of pressure differences between the atria and ventricles during the pumping action of the heart. The right valve is called the **tricuspid valve** because of its three cusps. The left valve is named the **bicuspid (mitral) valve** because it possesses only two cusps. The cusps on the AV valves are membranous extensions of the endocardium that reach into the ventricular chambers. Fibrous cords, known as **chordae tendinae**, connect the inferior free margins of the cusps to **papillary muscles** located on the ventricular walls. The **chordae tendinae** and papillary muscles prevent the cusps from swinging back into the atria when the ventricles contract. As a result, backflow of blood into the atria is prevented.
   - The two **semilunar valves** can be identified at the junction of each ventricle to its respective great artery. The **pulmonary semilunar valve** is situated at the junction of the right ventricle and the pulmonary trunk, and the **aortic semilunar valve** is located at the junction of the left ventricle and the aorta (Figure 20.3). Each semilunar valve has three crescent-shaped cusps that are extensions of the great arterial walls. The operation of these valves is also controlled by changes in pressure, this time between the ventricles and great arteries. They open when the ventricles pump blood into the arteries. They close when the pumping action is complete to prevent backflow into ventricles.

WHAT'S IN A WORD The bicuspid valve is often referred to as the **mitral valve**, because when it is closed, the cusps resemble the tall pointed hat, with front and back peaks, worn by bishops and other members of the clergy. This ceremonial headdress is called a *mitre (miter)*.

The term *semilunar* means "half moon." The name refers to the half-moon shape of the cusps in the semilunar valves.

The Heart Wall

1. Observe the left ventricular wall on a heart model and identify the three layers of the heart wall (Figure 20.4).
   - The inner **endocardium** is a thin, serous membrane of connective tissue and a simple squamous epithelium. It lines the internal walls of the heart chambers.
   - The middle **myocardium** is the thickest layer and comprises the bulk of the heart wall. It is composed primarily of cardiac muscle fibers, separated by connective tissue containing capillaries and nerves. Contractions of cardiac muscle fibers are responsible for the pumping action of the heart.
   - Similar to the endocardium, the outermost **epicardium** is a serous membrane of connective tissue and a simple squamous epithelium. This layer is also called the *visceral pericardium*.
2. Examine the relative thickness of the walls surrounding the heart chambers (Figure 20.3). Observe that the atrial walls are much thinner than the ventricular walls.
3. Compare the thickness of the two ventricular walls and observe that the left ventricular wall is thicker than the right ventricular wall.

CLINICAL CORRELATION

Microbial infection of the endocardium can cause tissue inflammation or **endocarditis.** This condition often causes damage to the heart valves and could impede normal blood flow through the heart. In severe cases, blood clots can form along the walls of the ventricles. These clots can break off and travel to other blood vessels where they can bring about organ failure, heart attacks, or strokes.

Why are the walls of the ventricles much thicker than the walls of the atria?

____

____

Coronary Circulation

Like any other organ, the heart must have an adequate blood supply that delivers sufficient amounts of oxygen and nutrients to cardiac muscle cells and carries away carbon dioxide and other metabolic wastes. The right and left coronary arteries branch directly off the ascending aorta. These arteries and their branches deliver blood to all regions of the heart. Blood is drained from the heart wall by a number of cardiac veins which empty, directly or indirectly, into the right atrium.

ACTIVITY 20.3 Identifying the Blood Vessels of the Coronary Circulation

1. Identify the following coronary arteries on a heart model (Figure 20.5).
   - The **right** and **left coronary arteries** branch off the ascending aorta just superior to the aortic semilunar valve.
   - The left coronary artery travels to the left side, along the atrioventricular groove and posterior to the pulmonary trunk (Figure 20.5a). Soon after it emerges from behind the pulmonary trunk, it gives rise to two main branches: the **circumflex artery** and the **anterior interventricular artery** (Figure 20.5a).

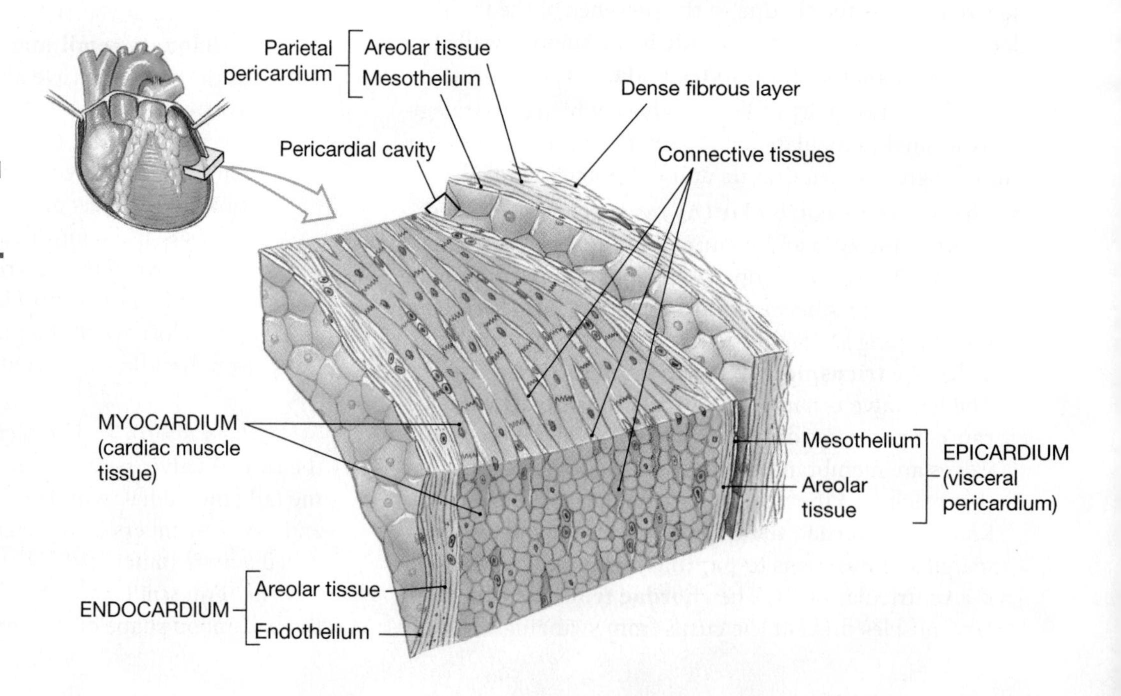

Figure 20.4 Organization of the pericardium and heart wall. The relationship of the heart wall (epicardium, myocardium, and endocardium) with the pericardium and pericardial cavity is illustrated. Note that the epicardium and the visceral pericardium are the same structure.

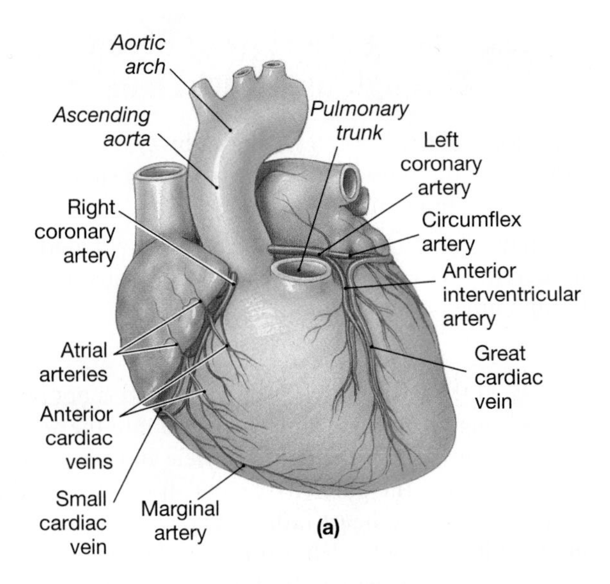

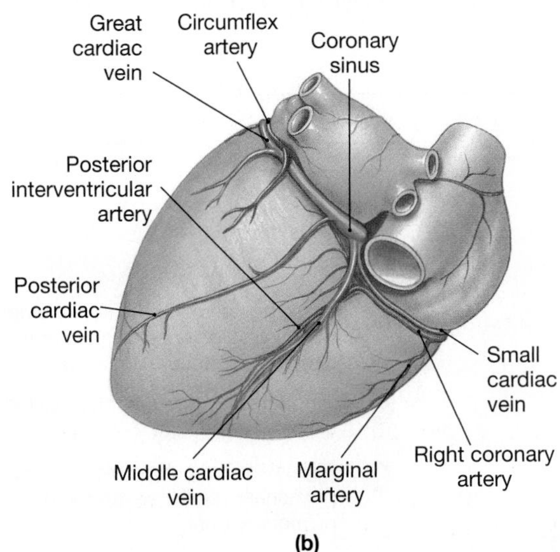

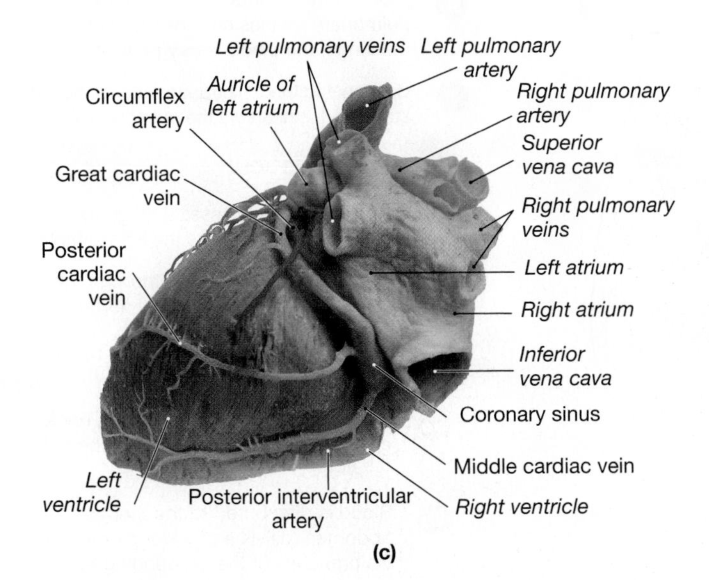

Figure 20.5 The coronary circulation. The major arteries and veins that supply and drain the heart wall are illustrated. **a)** Anterior view of the heart with the pulmonary trunk removed to reveal the left coronary artery; **b)** posterior view of the heart; **c)** dissection of the heart, posterior view.

- The circumflex artery travels along the atrioventricular groove. It curves around the left side (Figure 20.5a) and continues onto the posterior surface (Figures 20.5b and c).
- The anterior interventricular artery descends toward the apex along the anterior interventricular groove (Figure 20.5a).
- The right coronary artery travels to the right along the atrioventricular groove (Figure 20.5a). It curves around the right side and continues onto the posterior surface (Figure 20.5b).
- The right coronary artery and the circumflex artery form an **anastomosis** (a natural connection between two blood vessels) on the posterior surface of the heart.
- Just before the right coronary artery curves around to the posterior surface, it gives off a branch called the **marginal artery.** The marginal artery descends along the right margin of the right ventricle (Figure 20.5a).
- On the posterior surface of the heart, the right coronary artery gives off a second major branch, the **posterior interventricular artery**. This artery descends toward the apex along the posterior interventricular groove (Figures 20.5b and c).
- Near the apex of the heart, attempt to identify an anastomosis between the anterior and posterior interventricular arteries. This anastomosis may not be demonstrated on the models that are available in your laboratory.

2. Identify the following cardiac veins on the heart model (Figure 20.5).
   - On the posterior surface of the heart, the **coronary sinus** is a large dilated sac that runs along the atrioventricular groove and empties into the right atrium (Figures 20.5b and c). It drains most of the venous blood from the heart wall.
   - The **great cardiac vein** ascends along the anterior interventricular groove, running alongside the anterior interventricular artery (Figure 20.5a). At the atrioventricular groove it travels with the circumflex artery to the posterior surface, where it drains into the coronary sinus (Figures 20.5b and c).
   - The **middle cardiac vein** ascends along the posterior interventricular groove, traveling with the posterior interventricular artery. It drains into the coronary sinus as it approaches the base of the heart (Figures 20.5b and c).
   - The **small cardiac vein** runs alongside the marginal artery as it travels toward the atrioventricular groove (Figure 20.5a). At the atrioventricular groove, it travels with the right coronary artery to the posterior surface and drains into the coronary sinus (Figure 20.5b).
   - The **anterior cardiac veins** are small veins that travel a short distance along the anterior surface of the right ventricle. They drain directly into the right atrium (Figure 20.5a).

QUESTION TO CONSIDER During the previous activity, you identified two arterial anastomoses in the coronary circulation. What do you think is the functional significance of these arterial connections in the heart? (Hint: Consider what must occur if blood flow to a particular region of the heart is partially blocked.)

____

____

Blood Flow Through the Heart

The heart functions as a two-sided muscular pump that regulates two separate blood circulatory loops. The right side of the heart controls the **pulmonary circulation**, which is the flow of blood to and from the lungs. The left side of the heart controls the **systemic circulation**, which is the flow of blood to and from all body tissues (Figure 20.6). The sequence of events that defines one pumping cycle of the heart is known as the **cardiac cycle**. During the cardiac cycle, the two atria relax and contract together and the two ventricles relax and contract together. Thus, the flow of blood through pulmonary and systemic circuits is occurring simultaneously.

ACTIVITY 20.4 Tracing the Flow of Blood Through the Heart Chambers

1. On a model of the heart, trace the pathway of blood through the pulmonary circuit by reviewing steps one through five in Figure 20.6.
2. On a model of the heart, trace the pathway of blood through the systemic circuit by reviewing steps six through nine in Figure 20.6.

QUESTION TO CONSIDER If the left ventricle is damaged and is not pumping its normal volume of blood into the aorta, it will lag behind the pace of the right ventricle and eventually blood will back up in the pulmonary circulation. This condition is called **congestive heart failure**. Explain how restrictions in blood flow from the left ventricle and into the systemic circulation can have a negative effect on blood flow in the pulmonary circuit.

____

____

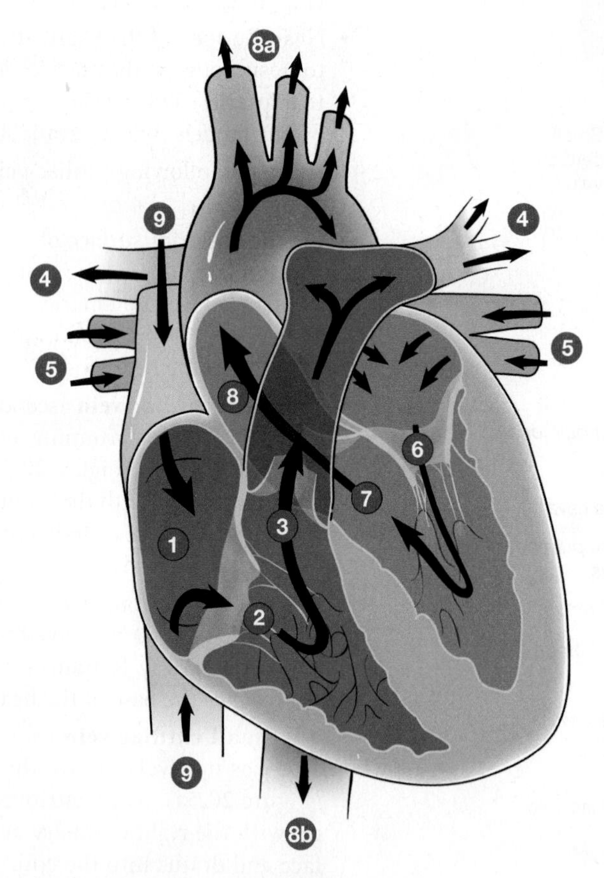

Pulmonary circulation:

1. Deoxygenated blood enters the right atrium via the venae cavae, coronary sinus, and anterior cardiac veins.
2. Blood passes through the tricuspid valve and enters the right ventricle.
3. Right ventricle pumps blood through the pulmonary semilunar valve into the pulmonary trunk.
4. Blood is transported to the lungs via the pulmonary arteries and their branches. In the lungs, the blood is oxygenated.
5. Oxygenated blood is transported to the left atrium of the heart via the pulmonary veins.

Systemic circulation:

6. Oxygenated blood passes through the bicuspid valve and enters the left ventricle.
7. Left ventricle pumps blood through the aortic semilunar valve into the aorta.
8. Oxygenated blood is transported by the aorta and its branches:
   - 8a Blood is distributed to the head, neck and upper extremities via branches of the aortic arch.
   - 8b Blood is distributed to the thorax, abdomen, pelvis and lower extremities via branches of the descending aorta.
9. Deoxygenated blood returns to the right atrium.

Figure 20.6 Overview of the pulmonary and systemic circulations. The right ventricle pumps blood into the pulmonary trunk to begin the pulmonary circulation. The left ventricle pumps blood into the ascending aorta to begin the systemic circuit.

The Sheep Heart

The sheep heart is remarkably similar to the human heart and thus represents an excellent model for studying cardiac structure. As you dissect, have models, illustrations, or photographs of the human heart readily available so that you can make structural comparisons.

ACTIVITY 20.5 Dissection of the Sheep Heart

Organization of the Pericardium

1. Identify the pericardial sac if it is present. This structure includes the fibrous pericardium and parietal pericardium. The fibrous pericardium is a thick outer layer of fibrous connective tissue and fat that encloses the heart. With a pair of scissors, cut along the pericardial sac for a short distance (about 2.5 cm or 1 inch) and fold it back to expose its inner surface. The thin, shiny layer along this surface is the parietal pericardium.
2. By making a cut in the pericardial sac, you have exposed the outside surface of the heart wall. Notice that the heart wall is covered by a thin, translucent membrane. With forceps, lift a portion of this membrane off the heart's surface. This is the visceral pericardium. Recall that the visceral pericardium and epicardium are the same structure. The space between the parietal pericardium and visceral pericardium is the pericardial cavity. Identify this space by placing a probe into it.

External Anatomy of the Heart

1. Carefully remove the pericardial sac to expose the entire heart. This can be accomplished by continuing your initial scissors cut toward the base of the heart and detaching the pericardial sac's attachments to the great vessels.
2. Observe the major sulci (grooves) that travel along the surface of the heart (Figures 20.7a and b). They can best be identified by the large amount of fat that is located along their paths.
   - The atrioventricular groove travels around the heart's circumference and separates the atria from the ventricles.

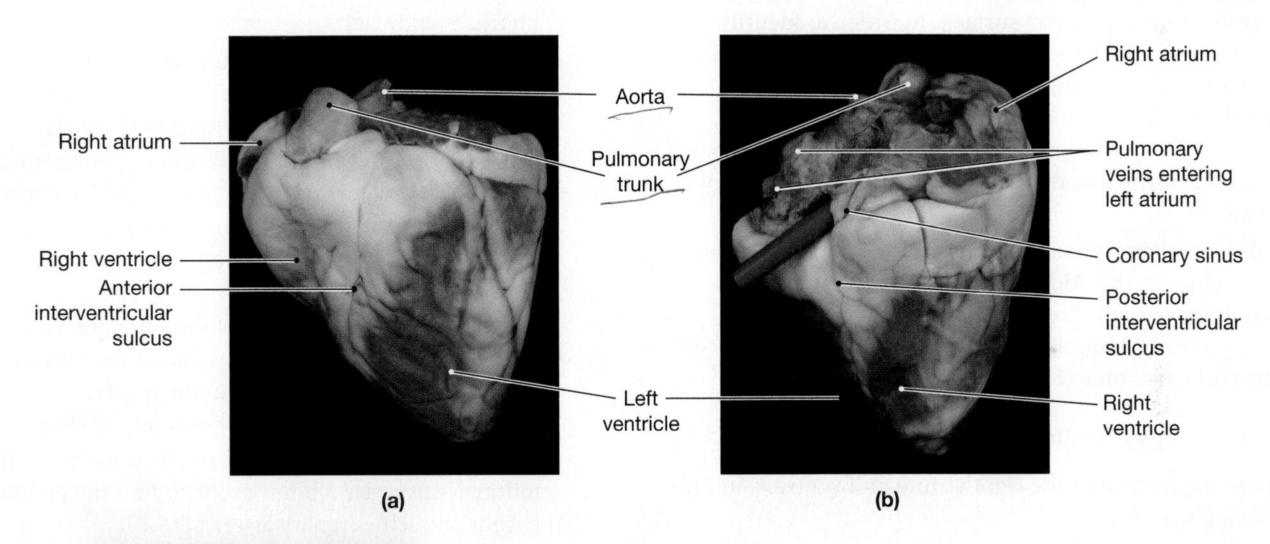

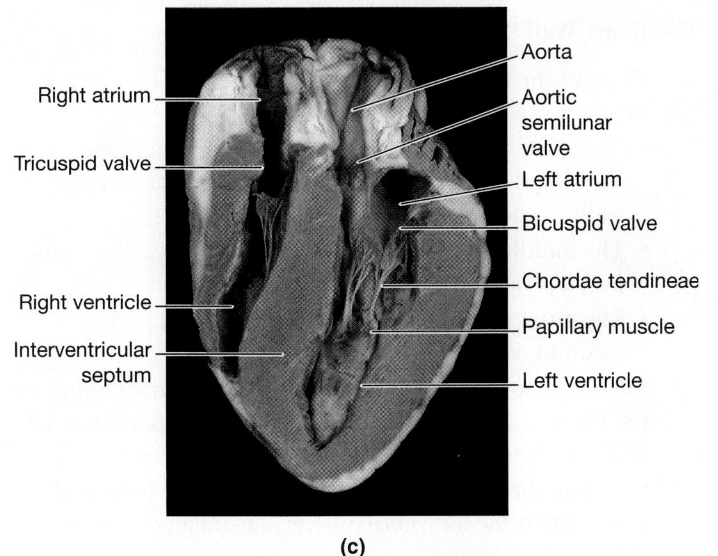

Figure 20.7 Anatomy of the sheep heart. a) Anterior view; **b)** posterior view; **c)** coronal section.

- The anterior interventricular sulcus (groove) travels between the left and right ventricles on the anterior surface.
- The posterior interventricular sulcus (groove) does the same on the posterior surface.

3. The atria on the sheep heart are quite small, and are comparable to the auricles on the human heart. Often, during commercial preparation of the heart, these chambers are partially removed, leaving the internal structures exposed. Consequently, the venae cavae leading into the right atrium and the pulmonary veins to the left atrium are usually absent. Identify the base of the heart by locating what remains of the atrial walls (Figure 20.7).
4. Hold the sheep heart so that the anterior surface is facing you (Figure 20.7a). At the inferior tip of the left ventricle, identify the apex of the heart. Identify the two great arteries—the pulmonary trunk and the aorta. From an anterior view, the pulmonary trunk is anterior to the aorta. Locate the anterior interventricular groove once again. Realize that this groove forms a boundary between the right and left ventricles (Figure 20.7a).
5. As described earlier, fat deposits are found along the major grooves on the heart's surface. In order to identify the blood vessels that travel along the grooves, it is necessary to remove this fat. With forceps, carefully strip away the fat from a small section along one of the grooves to verify the presence of blood vessels. Review the names of the blood vessels that travel in each groove (Figure 20.5).
6. Hold the sheep heart so that the posterior surface is facing you (Figure 20.7b). Along the atrioventricular groove, carefully remove the fat to reveal a thin walled, dilated blood vessel that empties into the right atrium. This vessel is the coronary sinus (Figure 20.7b).

Internal Anatomy of the Heart

1. Expose the interior of the right atrium and ventricle in the following manner.
   - Insert the blunt end of a pair of scissors into the superior vena cava. If the superior vena cava is not present, insert the scissors into the opening where the blood vessel drains into the right atrium.
   - Cut along the lateral margin of the right atrium.
   - Continue to cut along the lateral margin of the right ventricle to the apex. Be sure to cut through the entire thickness of the ventricular wall but avoid damaging internal structures.
2. Expose the interior of the left atrium and left ventricle in the following manner.
   - Using a scalpel or the sharp end of a pair of scissors, make a small incision in the lateral wall of the left atrium.
   - Insert the blunt end of the scissors into the incision and cut along the lateral margin of the left atrium.
   - Continue to cut along the lateral margin of the left ventricle to the apex.
3. Identify the interatrial septum that separates the left and right atria. The shallow, oval depression along the right atrial side of the septum is the fossa ovalis. Inside the atrial chambers, identify the pectinate muscles. Locate the atrioventricular orifices that lead into the ventricles. These openings mark the locations of the atrioventricular valves, which will be examined later.
4. Identify the interventricular septum that divides the two ventricles. Notice how the inferior portion is thick and muscular, and the superior portion is thin and membranous.
5. Beginning at the apex, cut through the interventricular septum with a scalpel or scissors. Continue cutting through the interatrial septum until you have completed a coronal section of the heart (Figure 20.7c).
6. In the coronal section, identify the following structures (Figure 20.7c).
   - The atrioventricular (AV) valves are located between the atria and ventricles. The tricuspid valve, with three cusps, is on the right side and the bicuspid valve, with two cusps, is on the left. For each valve, observe that the cusps are connected to the papillary muscles by the chordae tendinae.
   - The trabeculae carneae are muscular elevations along the walls of both ventricles. Notice that they are found predominately in the inferior portions of these chambers.
   - The superior portions of the ventricles are narrow, smooth-walled corridors that lead to the great arteries. In the right ventricle, the conus arteriosus leads to the pulmonary trunk. In the left ventricle, the aortic vestibule leads to the aorta.
7. From the severed free margins of the aorta and pulmonary trunk, cut along the walls of these blood vessels toward the ventricles until you reach the semilunar valves. Observe that each valve is composed of three crescent-shaped cusps.
8. Along the wall of the aorta, just superior to the aortic semilunar valve, use a blunt probe to find the openings to the right and left coronary arteries.

The Heart Wall

1. Observe the left ventricular wall and identify the three layers of the heart wall (Figure 20.7c).
   - The inner endocardium is a thin serous membrane. It appears as a smooth, shiny surface lining the internal walls of the heart chambers.
   - The middle myocardium is the thickest layer. It is composed primarily of cardiac muscle fibers.
   - The outer epicardium, which is the visceral pericardium, is also a serous membrane.
2. Examine the relative thickness of the walls surrounding the heart chambers (Figure 20.7c). Observe that the atrial walls are much thinner than the ventricular walls.
3. Compare the thickness of the two ventricular walls and observe that the left ventricular wall is thicker than the right ventricular wall.

QUESTIONS TO CONSIDER

1. On a model of the human heart, observe the thickness of the heart wall in the various chambers. In terms of relative thickness, how do the heart walls in the chambers of the human heart compare with the sheep heart?

2. Provide an explanation to account for the fact that the heart wall is thicker in the left ventricle than in the right ventricle.

Exercise 20 Review Sheet

Gross Anatomy of the Heart

Name ______________________

Lab Section ______________________

Date ______________________

Questions 1–4: Define the following terms.

1. Mediastinum

____

____

2. Atria versus ventricles

____

____

3. Apex of the heart

____

____

4. Base of the heart

____

____

5. Describe the circulatory pathways and the primary functions of the pulmonary and systemic circulations.

____

____

6. Describe the three tissue layers that comprise the heart wall.

____

____

7. Describe the organization of the pericardium.

____

____

Questions 8–11: Each heart valve is located at the junction of an atrium and ventricle, or a ventricle and great artery. Use this concept to describe the location of the heart valves.

8. Bicuspid valve

____

____

9. Tricuspid valve

____

____

10. Pulmonary semilunar valve

____

____

11. Aortic semilunar valve

____

____

12. Describe the function of the chordae tendinae and papillary muscles.

____

____

Questions 13–16: Complete the following table.

Artery	Vessel from which Artery Branches	Groove in which Artery Travels	Regions Supplied by the Artery
13.	Ascending aorta		Right side of the heart
14. Anterior interventricular artery			
15.	Left coronary artery		Left ventricle and left atrium
16.		Posterior interventricular groove	

Questions 17–20: Match the coronary artery in column A with the cardiac vein in column B that travels with it. The answers in column B may be used more than once or not at all.

A	B
17. Marginal artery ________	a. Great cardiac vein
18. Anterior interventricular artery ________	b. Middle cardiac vein
19. Circumflex artery ________	c. Small cardiac vein
20. Posterior interventricular artery ________	d. Coronary sinus

Questions 21–31: In the following diagram, identify the structures by labeling with the color that is indicated.

21. Right atrium = **yellow**
22. Left ventricle = **gray**
23. Aorta = **red**
24. Left atrium = **green**
25. Pulmonary trunk = **blue**
26. Superior vena cava = **purple**
27. Right ventricle = **orange**
28. Inferior vena cava = **pink**
29. Coronary sinus = **blue**
30. Pulmonary arteries = **brown**
31. Pulmonary veins = **black**

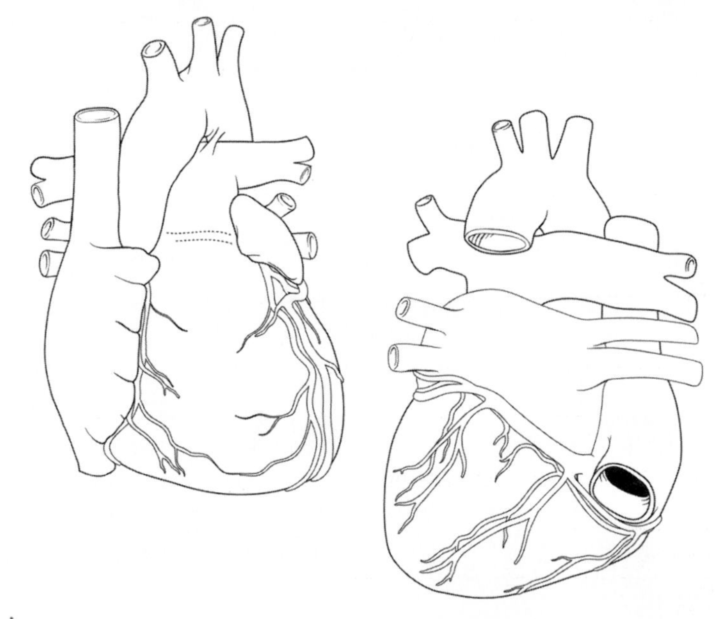

Questions 32–41: In the following diagram, identify the labeled structures. Select your answers from the following list.

32. ______________________
33. ______________________
34. ______________________
35. ______________________
36. ______________________
37. ______________________
38. ______________________
39. ______________________
40. ______________________
41. ______________________

a. Apex of the heart
b. Cusp of tricuspid valve
c. Inferior vena cava
d. Chordae tendinae
e. Middle cardiac vein
f. Trabeculae carneae
g. Papillary muscle
h. Interventricular septum
i. Great cardiac vein
j. Cusp of bicuspid valve
k. Ascending aorta
l. Coronary sinus

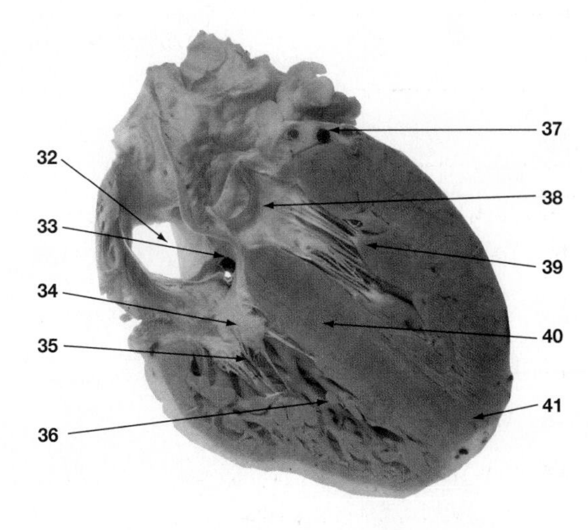

Exercise 21

Anatomy of Blood Vessels

Laboratory Objectives

On completion of the activities in this exercise, you will be able to:

- Identify the three tissue layers in the wall of a blood vessel.
- Describe the difference between an artery and a vein when viewed in cross section.
- Compare the structure and function of the various types of blood vessels.
- Compare the pathways and functions of the pulmonary and systemic circuits.
- Identify the major arteries and veins in the pulmonary circulation.
- Identify the major arteries and veins in the systemic circulation.
- Define collateral circulation and describe its functional significance.

Materials

- Compound light microscopes
- Prepared microscope slides
  - Arteries, veins, and capillaries
  - Liver sinusoids
- Anatomical models that illustrate the major blood vessels of the blood circulatory system
  - Heart
  - Torso or whole body
  - Head and neck
  - Upper extremity
  - Lower extremity
  - Brain

Blood vessels form an extensive network that delivers blood to all the cells and tissues in the body. **Arteries** are blood vessels that transport blood away from the heart. The largest arteries include the great arteries (pulmonary trunk and aorta) that are connected to the heart ventricles and their primary branches. They give rise to several generations of medium-sized arteries, small arteries, and arterioles. The arterioles deliver blood to capillary beds, where gas and nutrient exchange occurs. **Veins** are blood vessels that transport blood toward the heart. **Venules** are the smallest veins that directly receive the blood that flows out of capillary beds. Venules converge to form small veins, and small veins in turn give rise to larger and larger veins. Ultimately, blood returns to the right atrium via the two great veins, the superior and inferior vena cavae, and the left atrium via the pulmonary veins.

Microscopic Structure of Blood Vessels

Blood vessel walls have three distinct tissue layers (Figure 21.1). The innermost layer is called the **tunica intima.** It is a simple squamous epithelium, referred to as the **endothelium**, which lines the **lumen** (internal space) of the blood vessel, a basement membrane, and a thin layer of loose connective tissue. The middle layer is known as the **tunica media.** It is usually the thickest layer and is composed of smooth muscle and elastic fibers. The outermost layer, the **tunica externa (adventitia)**, is a connective tissue layer with numerous elastic and collagen fibers. The adventitial layer is typically continuous with the connective tissue of adjacent structures.

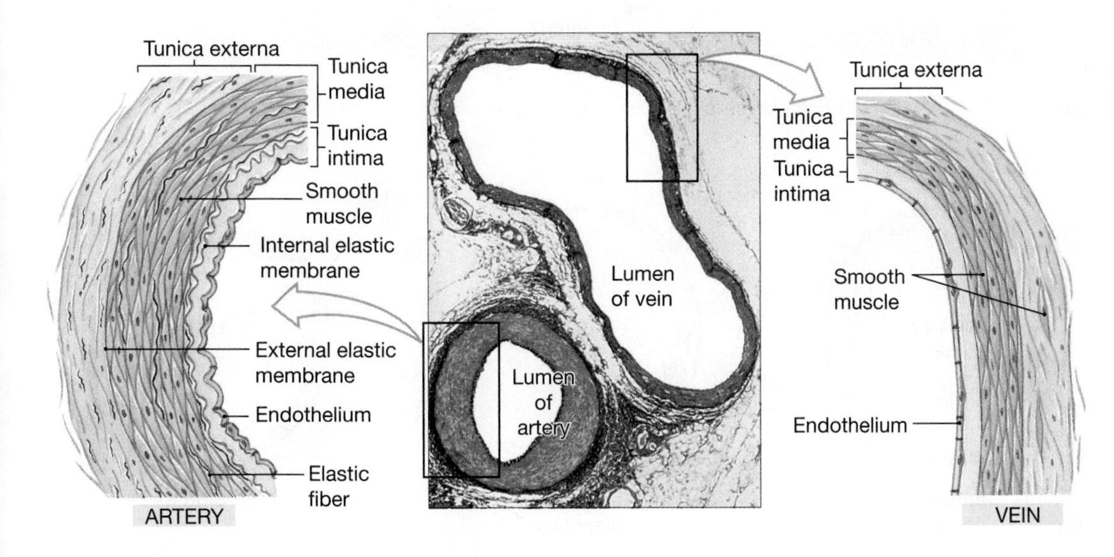

Figure 21.1 Comparative structure of an artery and a vein. The walls of arteries and veins contain three distinct tissue layers: tunica intima, tunica media, and tunica externa. An artery has a more developed wall (particularly the tunica media) and a rounder, smaller lumen than its corresponding vein.

The arterial system contains three basic types of arteries (Figure 21.2a).

- **Elastic arteries** include the great arteries (aorta and pulmonary trunk) and some of their primary branches. Compared to other arteries, they have the largest luminal diameters but their walls are relatively thin. Elastic fibers are abundant in all three tunics, but particularly in the tunica media.
- **Muscular arteries** deliver blood to specific body regions or organs. For example, the brachial artery supplies the arm and the renal artery supplies the kidney. Proportionately, muscular arteries have the thickest walls of all blood vessels. The tunica media contains mostly smooth muscle and relatively few (compared to elastic arteries) elastic fibers.
- **Arterioles** are very small arteries that deliver blood to capillary beds. The walls of arterioles are very thin but contain all three tunics. The tunica media consists mostly of smooth muscle with very few elastic fibers. As arterioles get closer and closer to capillary beds, their walls become progressively thinner. The smallest arterioles contain only an endothelium and a single layer of smooth muscle fibers.

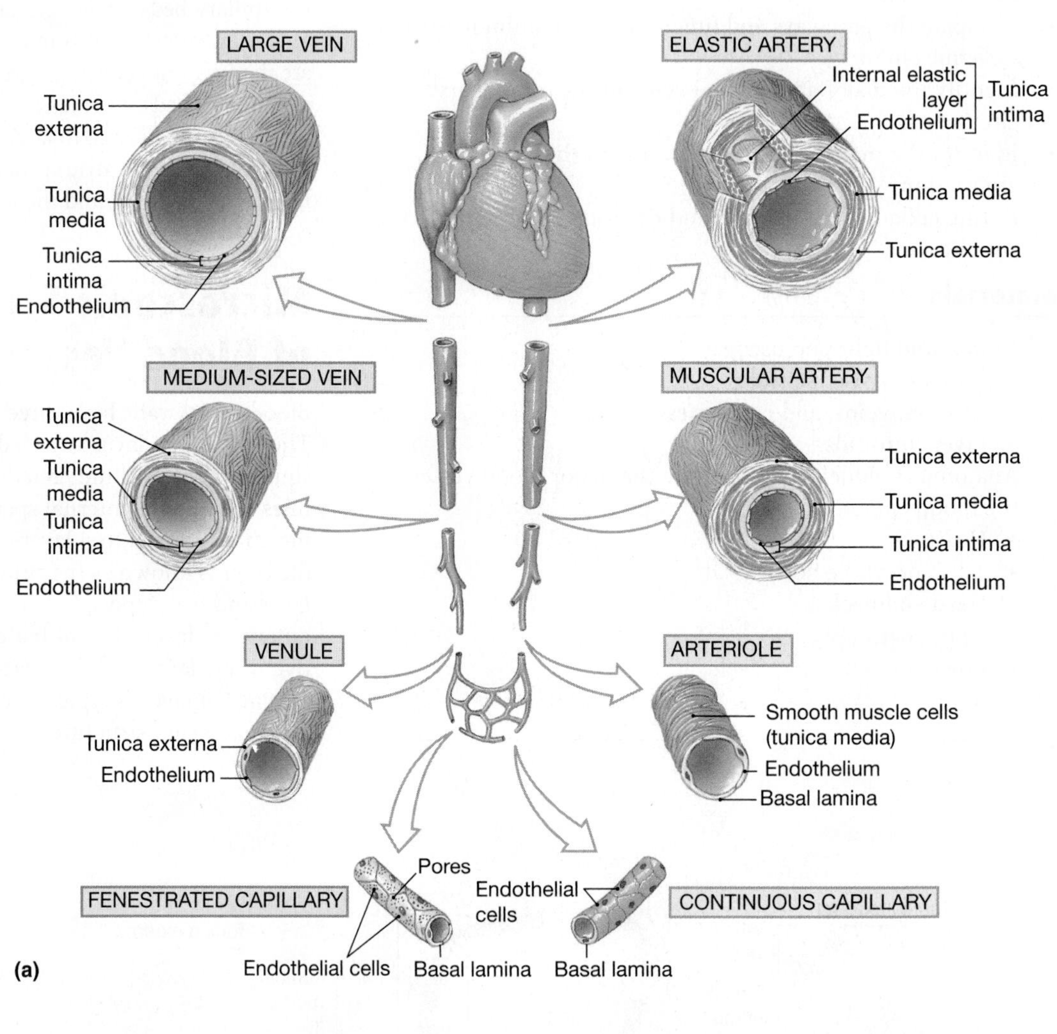

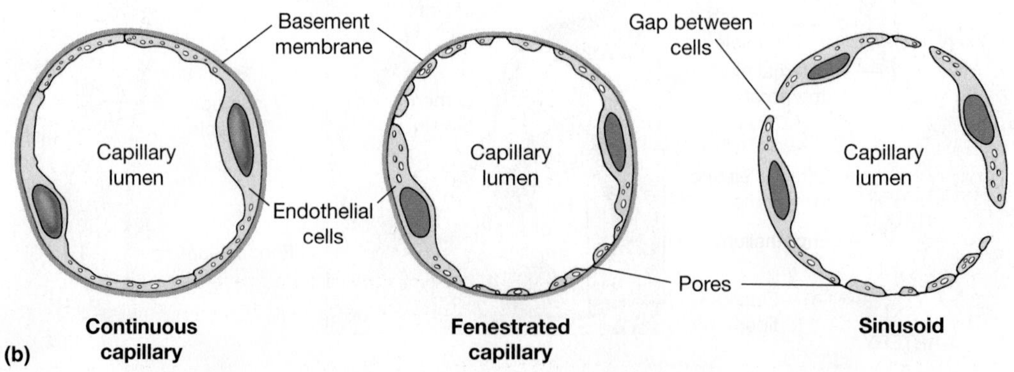

Figure 21.2 Blood vessel types. **a)** Overview of the various types of arteries and veins; **b)** cross sections of the three capillary types.

CLINICAL CORRELATION

Elastic fibers and smooth muscle significantly influence the function of the various types of arteries. For example, the abundance of elastic fibers in the pulmonary trunk and aorta (elastic arteries) allows the walls of these vessels to act like rubber bands as the heart pumps blood. When the ventricles contract and eject blood into the great arteries, the walls of these vessels can expand to accommodate the dramatic increase in pressure. When the ventricles relax, the pressure declines and the arterial walls recoil to their original condition.

The activity of smooth muscle in the tunica media of muscular arteries and arterioles is influenced by sympathetic nerve fibers and various hormones. When smooth muscle fibers contract, the vessel diameter decreases. This is known as **vasoconstriction.** When the muscle relaxes, the diameter increases. This is called **vasodilation.** In muscular arteries, vasodilation and vasoconstriction help to regulate the volume of blood flowing to a particular structure. In arterioles, changes in vessel diameter are directly linked to changes in blood pressure. When arterioles dilate, the resistance to blood flow declines and less pressure is required to move blood forward. When arterioles constrict, resistance increases, and more pressure is required to force blood through the vessels. Thus, changes in the resistance to blood flow in arterioles, referred to as **peripheral resistance,** can have a noticeable effect on an individual's blood pressure.

Veins possess the same three tissue layers as arteries, but they differ in their relative thickness (Figure 21.2a). For example, consider a corresponding artery and vein, which travel together and serve the same structure or body area. The three tissue layers are typically more developed in the artery, but the lumen is usually larger in the vein (Figures 21.1 and 21.2a). Many veins have valves that prevent the backflow of blood. The valves are continuations of the tunica intima and are similar to the cardiac semilunar valves, both structurally and functionally. Valves are most abundant in veins of the extremities, in which blood must flow back to the heart against the force of gravity.

Capillaries, the smallest of all the blood vessels (Figure 21.2), connect the arterial and venous circulatory networks. Capillaries provide a blood supply to all the tissues in the body and are sites where nutrients and wastes are exchanged between the blood and body cells. To accomplish this task, the walls of capillaries are very thin, consisting only of an endothelium and a basement membrane in the tunica intima; the tunica media and tunica adventitia are absent. Some capillaries are more permeable than others, and thus allow larger amounts of substances to be exchanged. Based on differences in permeability, three categories of capillaries can be identified (Figure 21.2b).

- **Continuous capillaries** are the least permeable because the endothelial lining is uninterrupted and the cells are held together by tight junctions. Continuous capillaries are abundant in skin, skeletal muscles, and the brain.
- **Fenestrated capillaries** are more permeable than the continuous variety because their endothelial cells contain pores (fenestrations) that are covered by a very thin layer of cytoplasm. They are commonly found in the digestive organs and endocrine glands.
- **Sinusoids** are the most permeable capillaries. Their endothelial lining is highly irregular and loosely arranged, with many pores and spaces between cells. These highly specialized blood vessels are found in the liver and spleen.

ACTIVITY 21.1 Examining the Microscopic Structure of Blood Vessels

Arteries

1. Obtain a prepared slide that shows an elastic artery in cross section. The slide will likely have a corresponding large vein, which you will examine more closely later.
2. Scan the slide under low magnification until you can locate the circular profile of an elastic artery. The lumen may contain some pink-staining red blood cells. Identify the three tissue layers in the arterial wall (Figure 21.3a): the

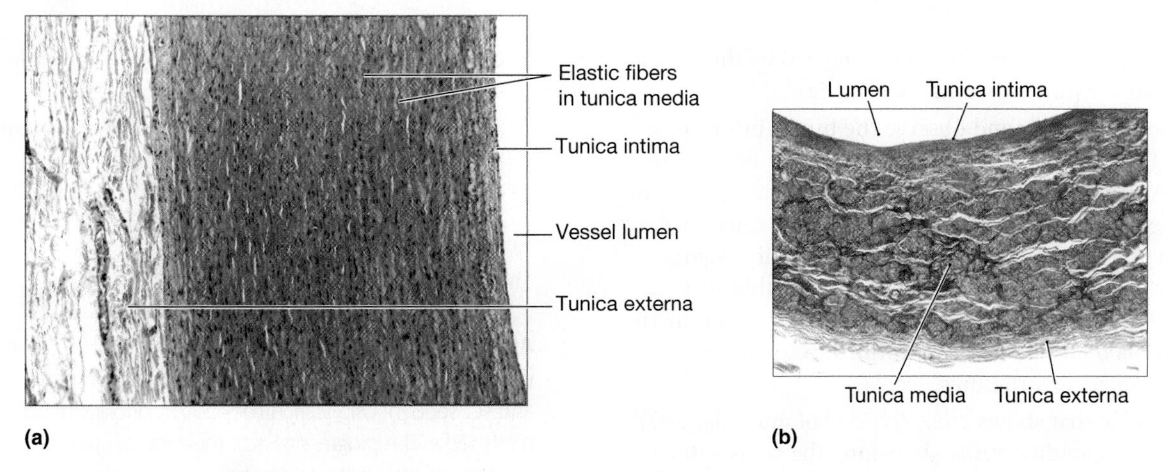

Figure 21.3 Light microscopic structure of the aorta and vena cava. a) The aorta is an example of an elastic artery (LM × 100). Notice the numerous elastic fibers in the tunica media. **b)** The wall of the superior vena cava (LM × 40).

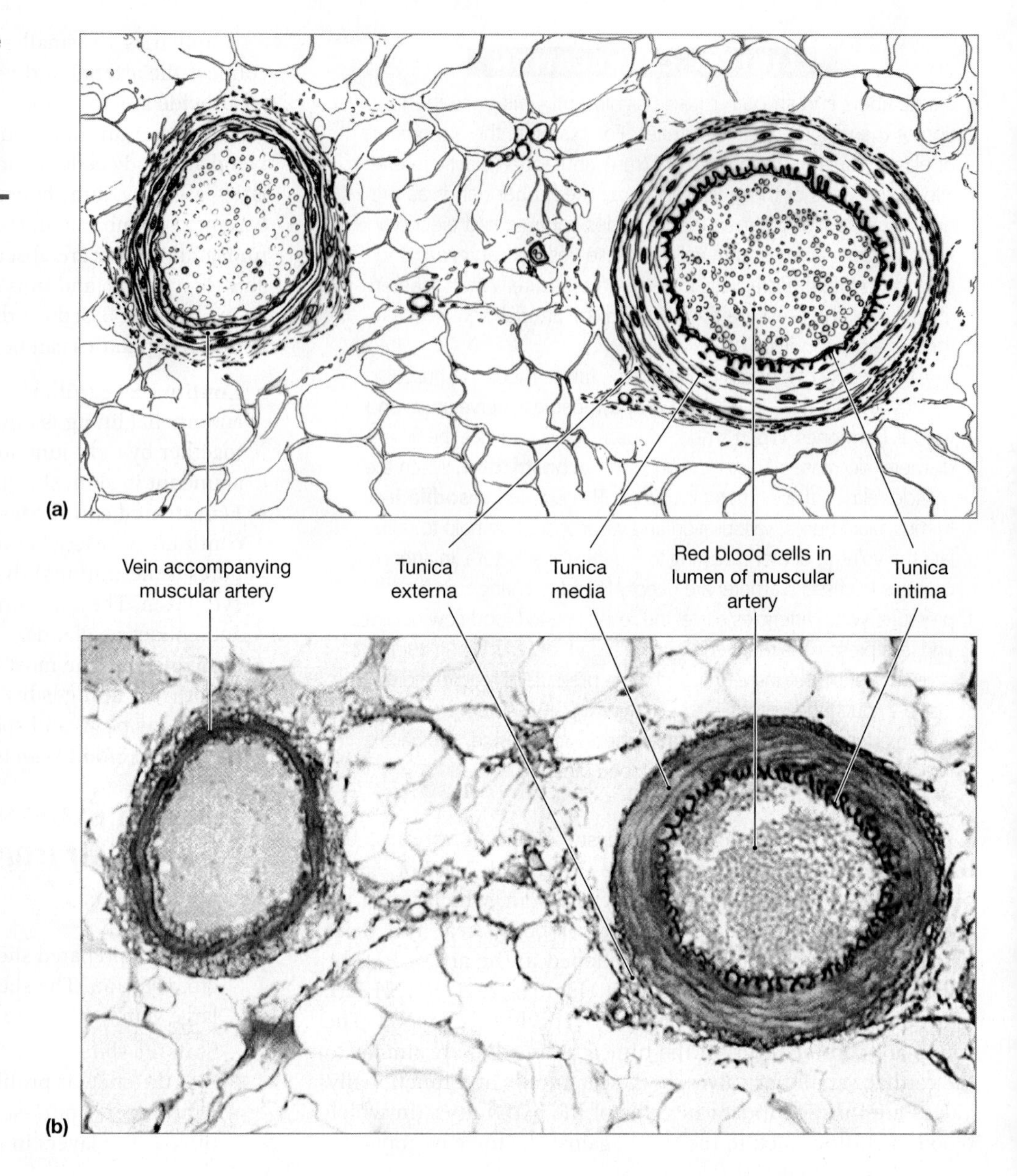

Figure 21.4 Light microscopic structure of a muscular artery and accompanying vein. a) Illustration demonstrating the comparative structure of an artery and a vein; **b)** corresponding light micrograph (LM × 100).

innermost tunica intima (interna), followed by the tunica media, and the tunica externa (adventitia).

3. Switch to high power and observe the tunica intima more closely. This layer may appear as a dark wavy line due to the contraction of elastic fibers. Notice that the epithelium (endothelium) in the tunica intima is simple squamous. Move the slide to the tunica media and identify elastic fibers, which are the dominant structures in this area (Figure 21.3a). Notice that the connective tissue in the tunica externa blends in with the connective tissue surrounding adjacent structures.
4. Obtain a slide that shows cross sections of muscular arteries and corresponding veins. As before, the veins will be studied later.
5. Under low power, identify the three tissue layers in the artery and compare their structure with the layers in the elastic artery. In particular, notice the multiple layers of smooth muscle fibers in the tunica media (Figure 21.4).
6. Switch to high power. Scan the slide slowly and attempt to find examples of arterioles. In cross section, these vessels will have small diameters and the tunica media will have one to five layers of smooth muscle (Figure 21.5).

CLINICAL CORRELATION

Atherosclerosis is the accumulation of fatty deposits in the tunica intima of arteries. This condition is usually accompanied by damage and subsequent calcification (deposition of calcium salts) of the tunica media. The progressive accumulation of lipids diminishes the diameter of the lumen and reduces blood flow. Atherosclerosis of the coronary arteries can lead to **coronary artery disease;** if damage occurs to arteries supplying the brain, it can lead to a **stroke.**

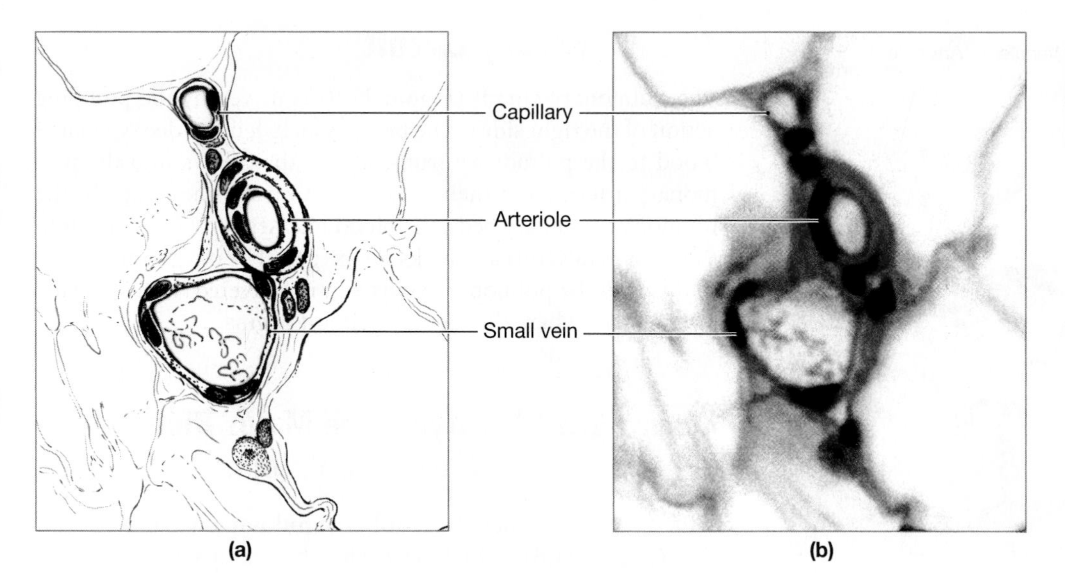

Figure 21.5 Light microscopic structure of an arteriole and a small vein. Notice the more organized structure in the arteriole. **a)** Illustration of an arteriole, a small vein and a capillary; **b)** corresponding light micrograph (LM × 800).

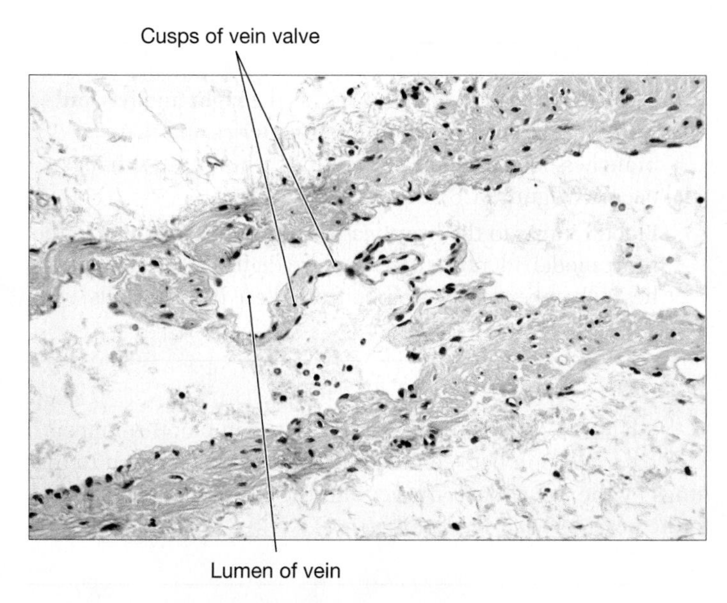

Figure 21.6 Valve in the lumen of a vein. The cusps of the valve are extensions of the tunica intima.

Veins

1. Using the same slides as before, identify cross sections of various types of veins (Figures 21.3b, 21.4 and 21.5). In standard slide preparations, the lumen of a vein is usually collapsed.
2. Under low power, identify the three tissue layers and compare their structure with the layers in the corresponding artery in the slide.
3. The tunica intima may be difficult to identify at low power. However, if you switch to high power, you may be able to identify the blue-purple nuclei of the endothelium.
4. In some slide sections, a valve may be identified extending across the lumen of the vein (Figure 21.6).

CLINICAL CORRELATION

In the lower extremities and other regions where venous return proceeds against gravity, the veins have numerous valves to prevent backflow. If the valves do not work properly, blood tends to flow in the reverse direction and becomes stagnant in some locations. The slowdown in venous return creates areas of blood pooling and vessel dilation. This condition, referred to as **varicose veins,** can be caused by a number of factors, including genetic inheritance, pregnancy, and growth of abdominal tumors.

Capillaries

1. Obtain a slide that illustrates capillaries along with accompanying small arterioles and venules. These small blood vessels are often referred to as the **microcirculation.**
2. Examine the slide under low power. Scan the slide and look for regions of loosely arranged connective tissue surrounded by adipose tissue.
3. Switch to high power and attempt to locate capillaries and other components of the microcirculation in cross or longitudinal section (Figure 21.7). The lumina of some vessels may be filled with pink-staining red blood cells. The capillary walls are very thin because only the tunica intima is present. Identify the blue-purple nuclei of the endothelial cells.

QUESTION TO CONSIDER How would you describe the structure of a vein in relation to its corresponding artery? Can you think of a functional significance for this difference in structure?

____

____

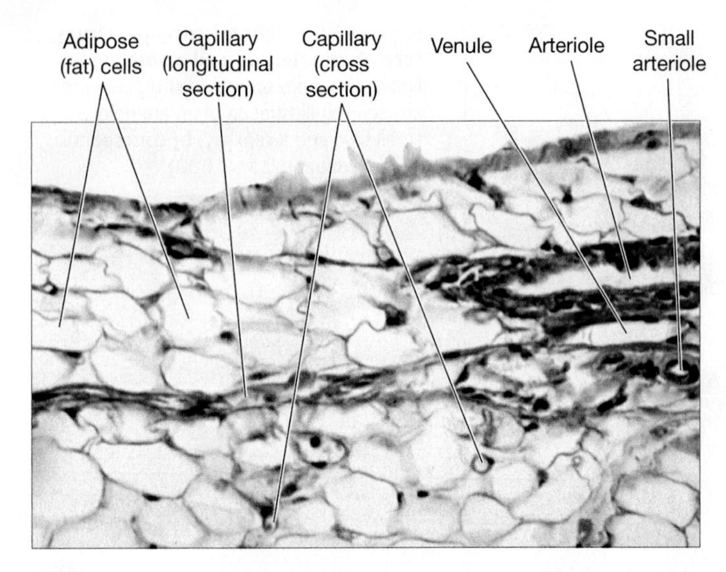

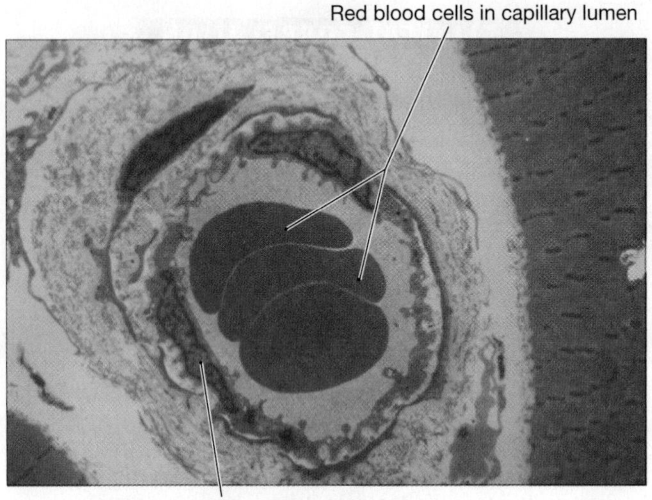

Figure 21.7 Structure of a capillary. a) Light micrograph showing capillaries in longitudinal and cross sections (LM × 250); **b)** electron micrograph of a capillary in cross section (LM × 2000).

Identifying the Major Arteries and Veins in the Human Body

In the activities that follow, you will be using a variety of anatomical models that are available in the laboratory to identify the major arteries and veins in the blood circulatory system. Use Figures 21.8 through 21.20 to guide you through these activities.

The presentation of blood vessels varies considerably among anatomical models. Be aware that some arteries and veins described in the following activities may not be visible on the models in your laboratory.

The Pulmonary Circuit

The pulmonary circuit (Figure 21.8) is driven by the pumping action of the right side of the heart, which delivers deoxygenated blood to the pulmonary trunk. Blood then flows into the pulmonary arteries and their tributaries to reach the lungs. In the pulmonary capillary beds, the blood releases carbon dioxide into the air sacs (alveoli) and picks up oxygen. The oxygenated blood drains into the pulmonary veins and is delivered to the left side of the heart, where the systemic circuit begins.

ACTIVITY 21.2 Identifying the Major Blood Vessels in the Pulmonary Circuit

1. On the heart model, identify the **pulmonary trunk** (Figure 21.8). Which heart chamber pumps blood into this blood vessel?

   ____

2. The pulmonary trunk gives rise to the **right** and **left pulmonary arteries.** The pulmonary arteries and their branches deliver blood to the lungs where gas exchange occurs (Figure 21.8).
3. Blood returns to the heart via the **pulmonary veins**. On the heart model, identify these vessels (Figure 21.8). Which heart chamber receives blood from these blood vessels?

   ____

QUESTION TO CONSIDER What is the major difference in blood composition between the blood that is transported to the lungs by the pulmonary arteries and the blood that is returned to the heart by the pulmonary veins? ______________________________

____

____

The Systemic Circuit

The systemic circuit is regulated by the pumping action of the left side of the heart, which delivers oxygenated blood to the aorta. Blood is distributed to all the body tissues by way of the aorta and its branches. In the systemic capillary beds, oxygen and nutrients are delivered to cells and tissues, and carbon dioxide and other metabolic wastes are released. The deoxygenated blood is then returned to the right side of the heart by way of the systemic venous system. The major blood vessels in the systemic circuit are illustrated in Figures 21.9 to 21.18.

Systemic Arteries

The principal artery in the systemic circuit is the aorta. Along its course, the aorta gives off numerous branches that deliver oxy-

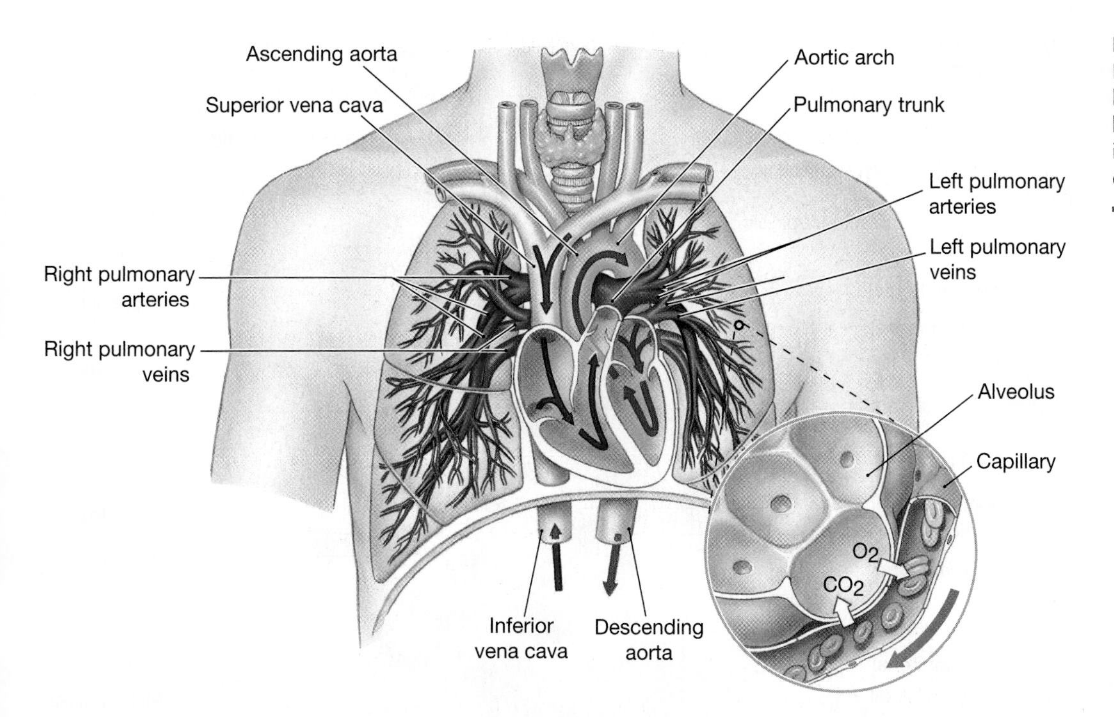

Figure 21.8 The pulmonary circuit. Pulmonary arteries deliver blood to the lungs, where oxygen is gained and carbon dioxide is released (magnified view in circle). Pulmonary veins transport the oxygenated blood back to the heart.

genated blood to specific body regions and internal organs. Most systemic arteries travel deeply and are well protected by surrounding structures.

WHAT'S IN A WORD Names of the systemic arteries offer clues that will help you during the learning process. For example, some arteries are named according to the body region through which they travel. Examples include the axillary, brachial, common iliac, and popliteal arteries. Others, such as the ovarian, renal, bronchial, and esophageal arteries, describe the organ that is supplied. Additional arteries, for example, the occipital, ulnar, radial, and femoral arteries, are named according to an adjacent bone. Use these helpful hints to your advantage as you study the arterial pathways.

ACTIVITY 21.3 Identifying the Major Arteries of the Systemic Circulation

The Aorta and Its Branches

1. On the heart model, identify the **ascending aorta** (Figure 21.8). Which heart chamber pumps blood into this blood vessel?

____

2. Near its origin, the ascending aorta gives rise to the **left** and **right coronary arteries.** The coronary arteries and their branches supply blood to the heart. Return to Exercise 20 to review the major arteries of the coronary circulation.

3. The **arch of the aorta** (Figures 21.8 and 21.9a) is the superior, curved portion of the aorta. It begins posterior to the manubrium and anterior to the trachea and ends just to the left of the T4 vertebra. Along its course it gives rise to three branches in the following order (Figure 21.9a).
   - The **brachiocephalic artery** travels a short distance and then bifurcates (splits into two branches) to form the **right common carotid artery**, which supplies the head and neck on the right side, and the **right subclavian artery**, which supplies the neck and upper extremity on the right side.
   - The **left common carotid artery** supplies the head and neck on the left side.
   - The **left subclavian artery** supplies the neck and upper extremity on the left side.

4. On a whole body or torso model, identify the **thoracic division** of the **descending aorta (thoracic aorta)** as it travels along the posterior wall of the thoracic cavity (Figure 21.9a). Notice the relationship of the thoracic aorta with the vertebral column. It begins, as a continuation of the arch, to the left of the T4 vertebra. As it descends through the thorax, it gradually shifts to the right. As it passes through the **aortic opening** in the **diaphragm**, the thoracic aorta is anterior to the T12 vertebra. Identify the series of paired arteries that branch off the thoracic aorta and travel within the intercostal spaces. These vessels are the **posterior intercostal arteries.** They supply the intercostal muscles and ribs. There are several other branches of the thoracic aorta that are not shown on most anatomical models. These arterial branches are illustrated in Figures 21.9a and b.

Figure 21.9 The aorta and its branches. a) Illustration of the aorta and branches with all thoracic and most abdominal organs removed; *(continues)*

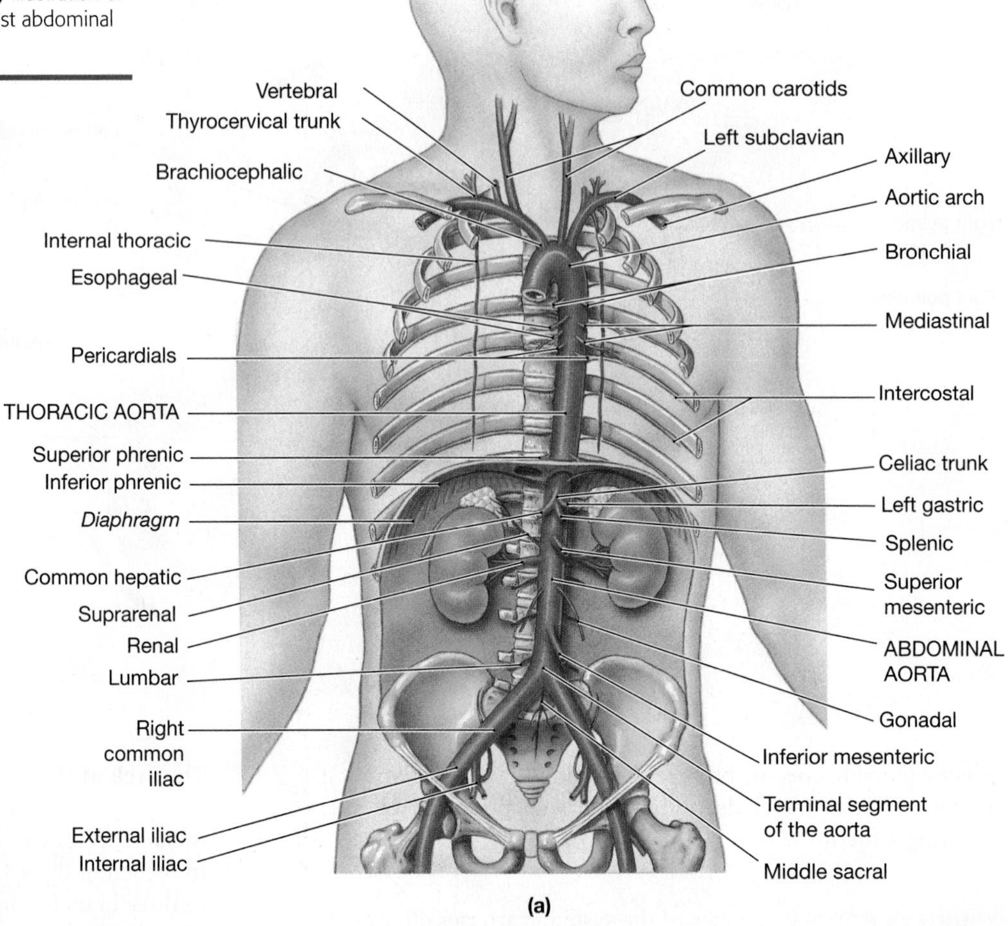

5. Identify the **abdominal division** of the descending aorta **(abdominal aorta)** as it passes along the posterior wall of the abdominal cavity (Figure 21.9a). At the T12 vertebra, it begins by passing through the aortic opening of the diaphragm. As the abdominal aorta descends through the abdominal cavity it lies anterior to the vertebral column. It ends at the body of the L4 vertebra, where it gives off its two terminal branches, the **left** and **right common iliac arteries**.
6. Along the anterior surface of the abdominal aorta, identify the three unpaired arteries that supply the digestive organs (Figure 21.9a).
   - The first of these branches is the **celiac artery (trunk)**. It arises from the aorta just inferior to the diaphragm. It is a short vessel that quickly gives rise to several smaller branches.
   - The **superior mesenteric artery** originates about 2.5 cm inferior to the celiac trunk.
   - The **inferior mesenteric artery** is the smallest and most inferior of the three branches. It originates about 3 to 4 cm superior to the termination of the abdominal aorta.

   The specific structures supplied by these arteries are listed in Figure 21.9b.
7. Identify the four paired arterial branches of the abdominal aorta (Figure 21.9a).
   - The **inferior phrenic arteries** are the first branches of the abdominal aorta, arising just superior to the celiac trunk.
   - The **suprarenal arteries** originate near the origin of the superior mesenteric artery.
   - The **renal arteries** come off the aorta just inferior to the suprarenal and superior mesenteric arteries.
   - The **gonadal (ovarian** or **testicular) arteries** branch off the aorta between the superior and inferior mesenteric arteries.
8. Along the posterolateral surface of the abdominal aorta, identify the four pairs of **lumbar arteries.** If you observe carefully on the torso model, you will notice that the lumbar arteries are a continuation of the series of posterior intercostal arteries arising from the thoracic aorta.

   The specific structures supplied by the paired branches of the abdominal aorta are identified in Figure 21.9b.
9. A small single branch, the **median sacral artery**, originates on the posterior surface of the abdominal aorta, just superior to its termination (Figure 21.9a). Identify this artery, if it appears on your model, as it descends from the terminal bifurcation of the aorta. The median sacral artery supplies the sacrum and coccyx.

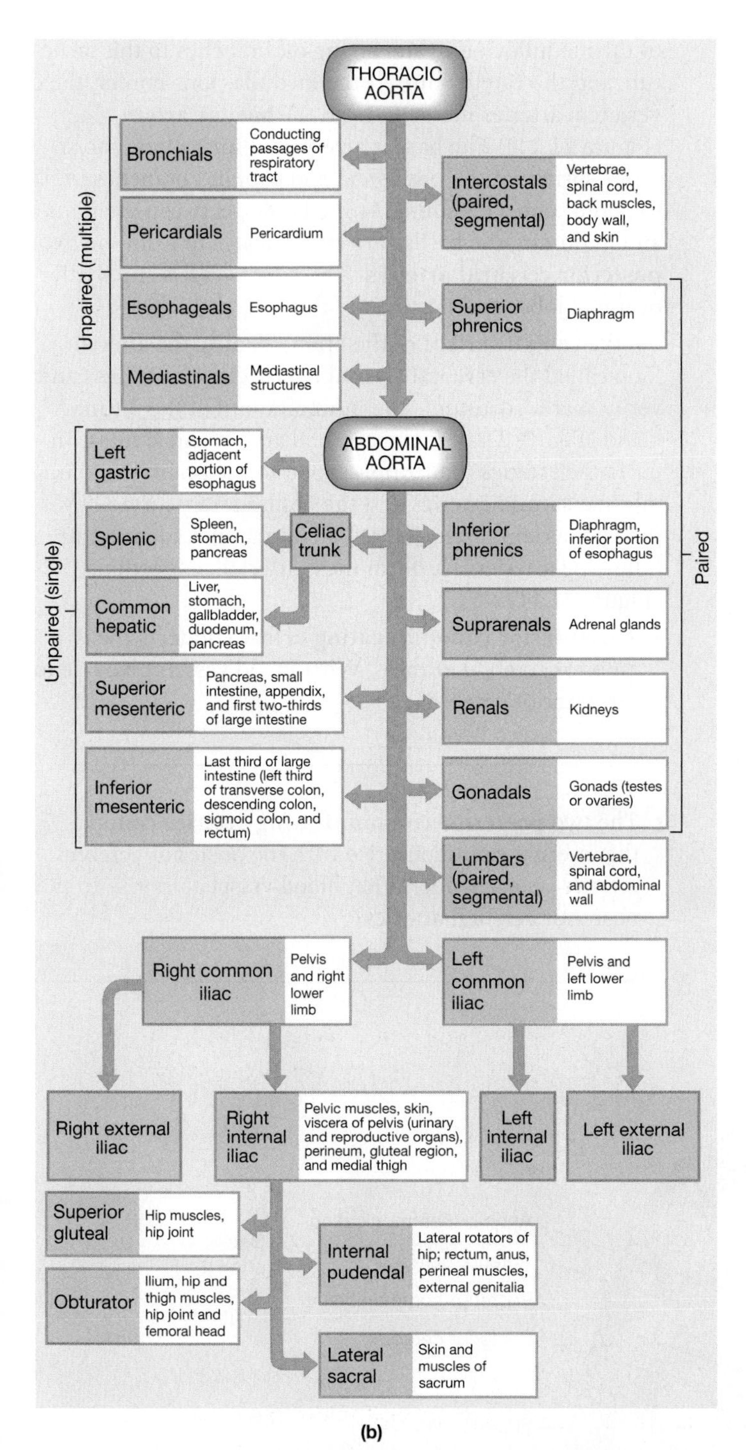

Figure 21.9 The aorta and its branches. *(Continued)* b) flow chart listing the structures supplied by the various aortic branches.

Blood Supply to the Head and Neck

1. On a model, identify the left and right common carotid arteries. Verify that the right common carotid artery is a branch of the brachiocephalic artery (Figures 21.9a and 21.10), while the left common carotid artery branches directly off the arch of the aorta (Figure 21.9a). The common carotid arteries and their branches distribute the main blood supply to structures in the head and neck.

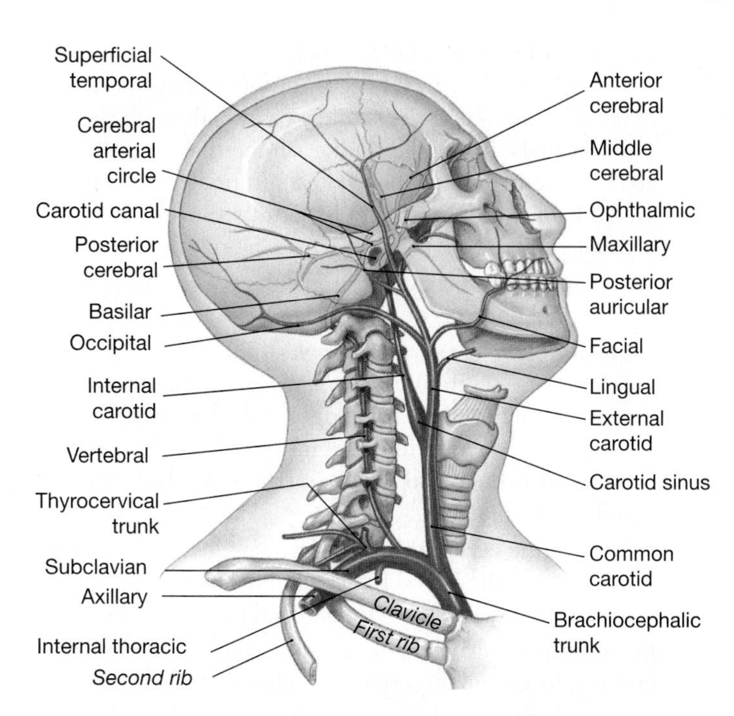

Figure 21.10 Arteries that supply the head and neck. The blood supply to this region originates from the carotid arteries and branches of the subclavian artery.

2. Identify the common carotid arteries as they travel superiorly and laterally through the neck. At the superior border of the thyroid cartilage, each common carotid artery bifurcates and gives rise to the **external carotid artery** and the **internal carotid artery** (Figure 21.10). At the region of bifurcation, there is a dilated sac called the **carotid sinus**. The wall of the sinus contains baroreceptors that monitor and control blood pressure. Also in this region is the **carotid body**. It contains chemoreceptors that regulate the respiratory rate by monitoring carbon dioxide and oxygen concentrations in the blood.
3. The external carotid arteries give rise to several smaller arteries that supply the face, scalp, and neck. If a model that demonstrates them is available in your lab, attempt to identify the following branches of the external carotid artery (Figure 21.10), as follows:
   - The **lingual artery** supplies the tongue, soft palate, epiglottis, lingual tonsils, and sublingual salivary gland.
   - The **facial artery** supplies skin and muscles of the face.
   - The **occipital artery** supplies the posterior scalp.
   - The **posterior auricular artery** supplies the scalp superior and posterior to the ear, portions of the middle and inner ear, the auricle, parotid salivary gland, and some muscles in the neck.
   - The **maxillary artery** supplies the upper and lower jaws, muscles of mastication, teeth, nasal cavity, and inner surfaces of the parietal and temporal bones and adjacent dura mater.
   - The **superficial temporal artery** supplies most of the scalp and the parotid salivary gland.

4. The internal carotid arteries supply the orbits and most of the cerebrum. They travel deeply in the superior portion of the neck and enter the skull by passing through the carotid canals. Inside the skull, each internal carotid artery gives rise to three branches. If an appropriate model is available in your lab, attempt to identify the following branches (Figure 21.10).
   - The **ophthalmic arteries** supply the eyes, walls of the orbit, and portions of the nose and forehead.
   - The **anterior cerebral arteries** supply the medial sides of the cerebral hemispheres. On a sagittal section of the brain, identify this artery and its many branches. The anterior cerebral arteries contribute to the **cerebral arterial circle (circle of Willis)**, which is described later.
   - The **middle cerebral arteries** can be identified as they travel along the lateral fissures of the brain. They supply the lateral sides of the temporal and parietal lobes.
5. Identify the subclavian arteries once again. These vessels give rise to several important branches that supply structures in the head and neck. The first of these branches are the **vertebral arteries**, which travel superiorly through the neck by passing through the series of transverse foramina on each side of the cervical vertebrae (Figure 21.10). On a model of the vertebral column, identify the two vertebral arteries. As the arteries ascend toward the brain, they give off branches to the spinal cord and vertebrae. They enter the skull by passing through the foramen magnum. On a model of the brain, locate the vertebral arteries as they travel along the anterolateral aspects of the medulla oblongata, giving off branches to this structure and the cerebellum. At the medulla-pons border, the vertebral arteries merge to form the **basilar artery** (Figure 21.11). The basilar artery continues along the anterior surface of the brainstem and supplies branches to the pons and cerebellum. At the border between the pons and midbrain, the basilar artery bifurcates to form the two **posterior cerebral arteries.** These two vessels supply the occipital lobes and portions of the temporal lobes.
6. On the ventral surface of the brain, identify the arterial loop called the cerebral arterial circle (circle of Willis) and verify that it surrounds the infundibulum of the pituitary gland (Figure 21.11). The cerebral arterial circle interconnects the arteries that supply blood to the brain: the two internal carotid arteries and the basilar artery formed by the two vertebral arteries. On a brain model, identify the following arteries that form these arterial connections (Figure 21.11).
   - The **anterior communicating artery** connects the two anterior cerebral arteries. What blood vessels give rise to the anterior cerebral arteries?

     ____

   - The two **posterior communicating arteries** connect the internal carotid arteries with the posterior cerebral arteries on each side. What blood vessel gives rise to the posterior cerebral arteries?

     ____

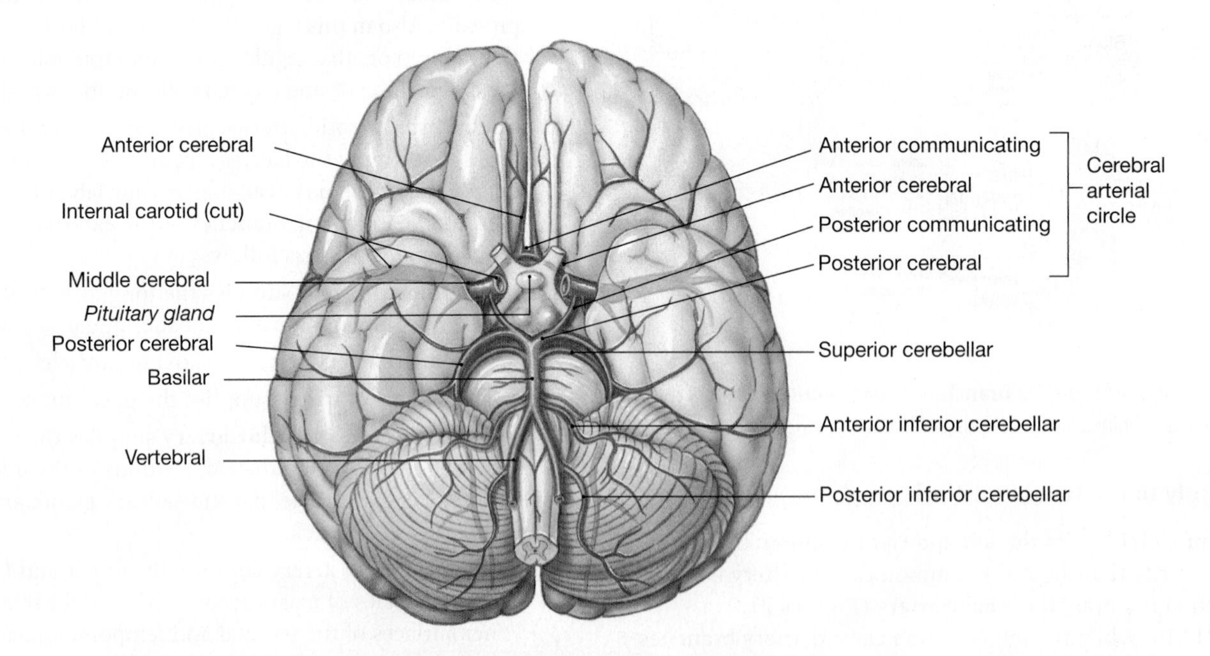

Figure 21.11 Arteries that supply the brain. The cerebral arterial circle, which surrounds the pituitary gland, interconnects the internal carotid arteries and the basilar artery.

7. Just lateral to the point where the vertebral artery originates, identify the following arterial branches of the subclavian artery. (Depending on the models that are available in your lab, these branches may not be demonstrated.)
   - The **thyrocervical trunk** (Figures 21.10 and 21.12) sends branches to the thyroid gland, larynx, and some shoulder muscles.
   - The **costocervical trunk** supplies deep back muscles in the neck and the first two intercostal spaces.

Blood Supply to the Anterior Thoracic Wall and Upper Extremities

1. Remove the anterior body wall from a torso model. On its posterior surface, identify the artery that travels just lateral to the sternum. This is the **internal thoracic (mammary) artery.** Notice that this artery is a branch of the subclavian artery (Figure 21.12). Each internal thoracic artery gives rise to a series of **anterior intercostal arteries.** Identify

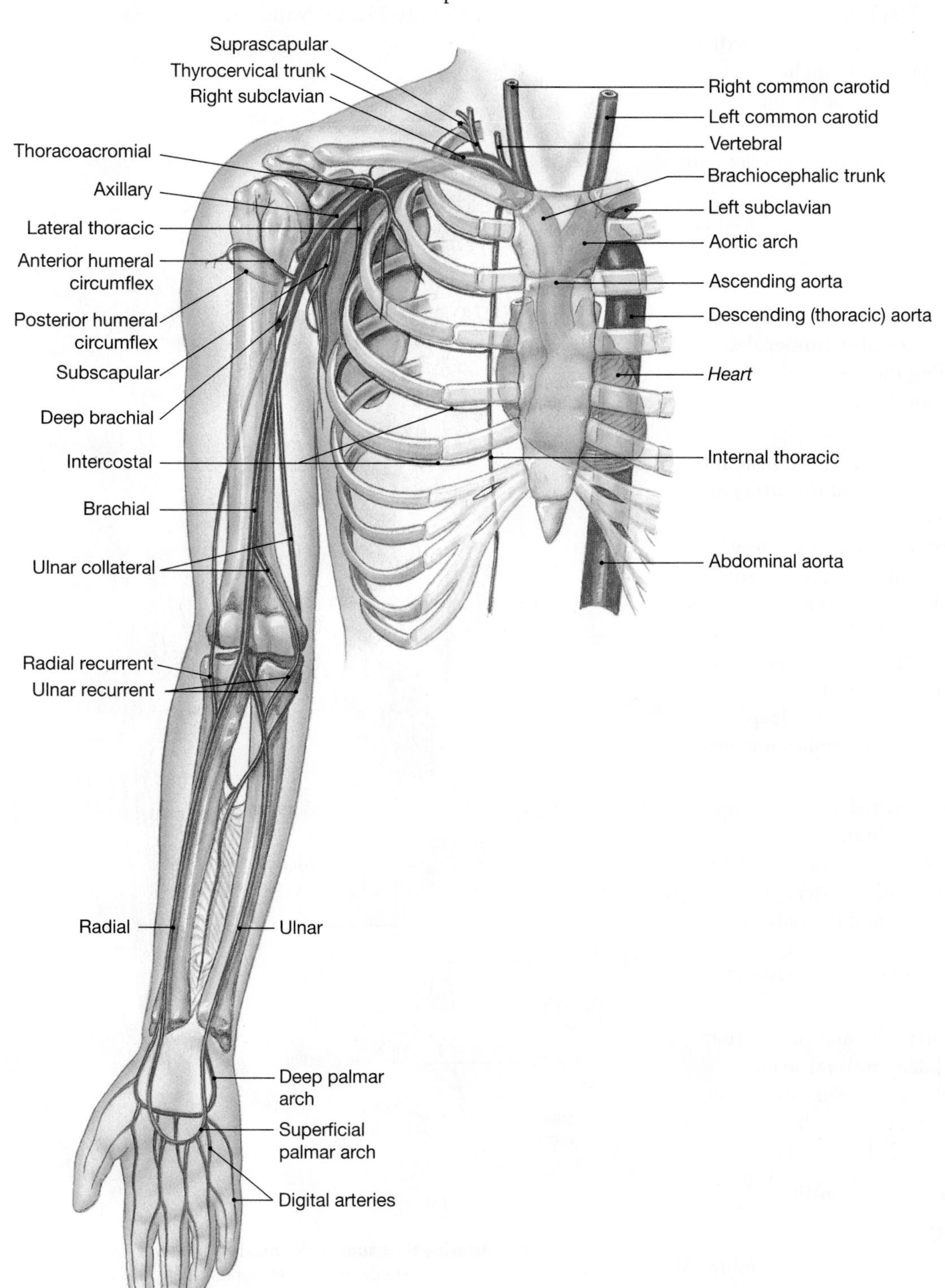

Figure 21.12 Arteries that supply the anterior thoracic wall and upper extremity. Arterial branches to these regions are derived from the subclavian artery.

these blood vessels as they travel through the intercostal spaces. They supply the intercostal muscles and ribs, the mammary glands, pectoral muscles, and skin on the anterior thoracic wall.

2. Each upper extremity is supplied by blood vessels that arise from the subclavian artery. On a torso model, identify the subclavian artery on one side of the body. Notice that this artery passes between the clavicle and the first rib. At the lateral border of the first rib, the **axillary artery** begins as a direct continuation of the subclavian artery (Figure 21.12). As it travels through the axilla, the axillary artery gives off the following important branches (Figure 21.12). These branches may not be visible on the models in your laboratory.
   - The **thoracoacromial artery** supplies the shoulder and pectoral region.
   - The **lateral thoracic artery** supplies the lateral wall of the thorax.
   - The **subscapular artery** supplies the scapula and the posterior wall of the thorax.
   - The **anterior** and **posterior circumflex humeral arteries** form an arterial loop around the humerus, immediately distal to the shoulder joint. It supplies the shoulder and deltoid muscle.
3. On a model of the upper extremity, identify the **brachial artery** (Figure 21.12). Locate the origin of the artery at the inferior border of the teres major. Verify that the brachial artery is a continuation of the axillary artery. It begins by traveling along the medial aspect of the anterior muscle compartment of the arm. As it descends, it curves laterally and passes through the cubital fossa, the region anterior to the elbow joint. The brachial artery gives off several branches that supply the muscles and other structures of the arm. One important branch is the **deep brachial artery** (Figure 21.12), which supplies the posterior region of the arm.
4. Near the head of the radius, the brachial artery divides into its two terminal branches, the **ulnar** and **radial arteries** (Figure 21.12). Identify these blood vessels on the upper extremity model. Note that the ulnar artery travels along the medial side of the forearm and the radial artery passes along the lateral side. Branches of the ulnar and radial arteries supply the muscles and other structures in the forearm.
5. The ulnar and radial arteries enter the hand, where they give rise to the **superficial** and **deep palmar arches** (Figure 21.12). Identify the palmar arches on the upper extremity model. The arches and their branches supply the palm of the hand and the digits.

Blood Supply to the Pelvis and Lower Extremities

1. On a model of the torso, identify the termination of the abdominal aorta anterior to the body of the L4 vertebra. At this position, the aorta gives rise to its two terminal branches, the left and right common iliac arteries (Figure 21.9a). On one side of the torso model, trace the common iliac artery to its termination (at the level of the lumbosacral joint) where it bifurcates to form the **internal** and **external iliac arteries** (Figures 21.9a and 21.13a). The internal iliacs give rise to several branches that supply the pelvic wall, urinary bladder, external genitalia, and gluteal muscles. In females, branches of the internal iliac also supply the uterus and vagina.
2. On a torso or lower extremity model, identify the external iliac artery (Figure 21.13). Notice that it travels inferolat-

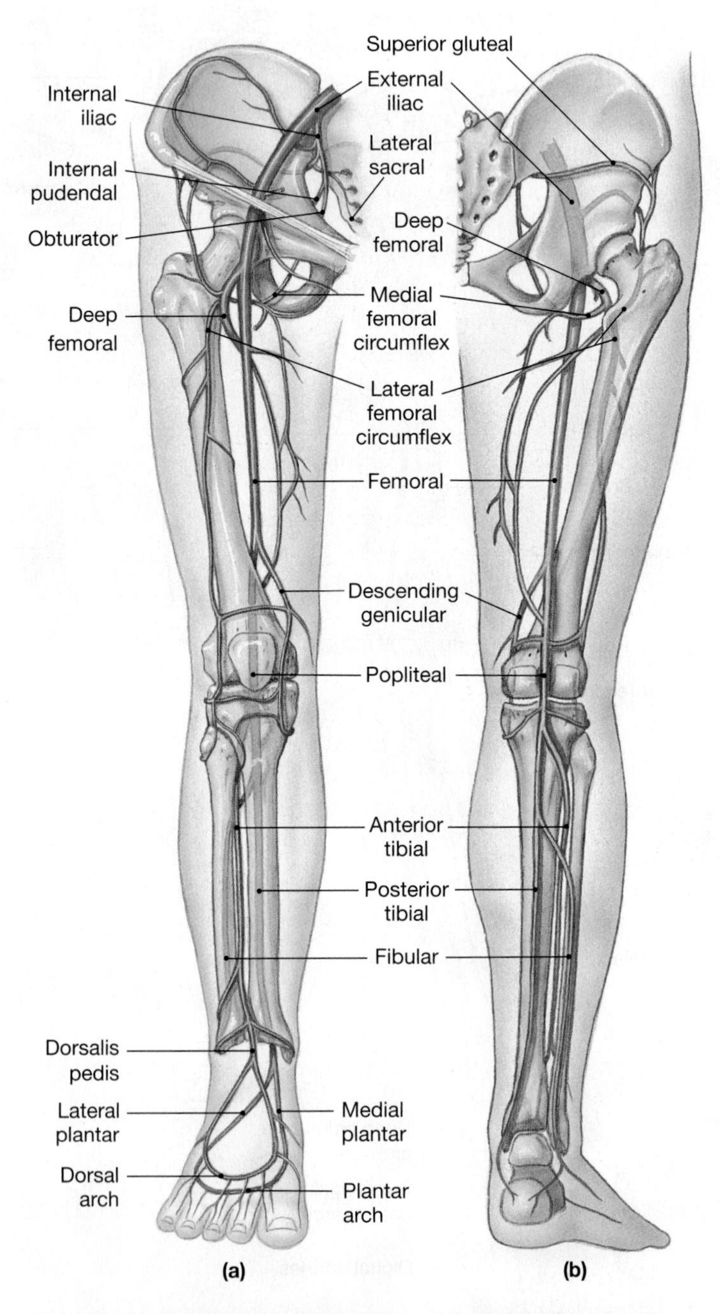

Figure 21.13 Arteries that supply the pelvis and lower extremity. Branches to these regions are derived from the internal and external iliac arteries. **a)** Anterior view; **b)** posterior view.

erally through the pelvic cavity, along the surface of the iliopsoas muscle. As it travels into the anterior thigh, the external iliac artery becomes the **femoral artery.** The femoral artery gives off numerous branches that supply the structures of the thigh, as well as the hip and knee joints. Following are the more important branches.

- The **medial** and **lateral femoral circumflex arteries** supply the head and neck of the femur and the hamstring muscles.
- The **profunda (deep) femoral artery** gives off branches to the posterior thigh and hip joint.

3. On the lower extremity model, identify the **popliteal fossa,** the shallow depression posterior to the knee joint. Identify the **popliteal artery** passing through the fossa (Figure 21.13b). Notice that the popliteal artery is a direct continuation of the femoral artery.
4. Trace the popliteal artery to its termination, where it gives rise to the **anterior** and **posterior tibial arteries** (Figure 21.13). Identify these arteries on the model.
   - The anterior tibial artery travels through the anterior compartment of the leg and supplies the muscles and other structures in this region. On the dorsum of the foot the anterior tibial artery becomes the **dorsalis pedis artery** (Figure 21.13), which forms the dorsal arch of the foot. The dorsal arch gives off digital branches to the toes.
   - The posterior tibial artery travels through the posterior compartment of the leg and supplies all the structures in this region. It gives rise to the **peroneal (fibular) artery** that supplies the lateral compartment of the leg. On the lower extremity model, verify that the peroneal artery is a branch of the posterior tibial artery. Along the medial side of the calcaneus, the posterior tibial artery bifurcates to form its two terminal branches, the **lateral** and **medial plantar arteries** (Figure 21.13a). The larger lateral plantar artery gives rise to the plantar arch, which gives off digital arteries to the toes.

QUESTIONS TO CONSIDER 1. What is the functional difference between pulmonary arteries and bronchial arteries?

____

____

2. Explain why all the major arteries in the systemic circulation have a deep position in the body.

____

____

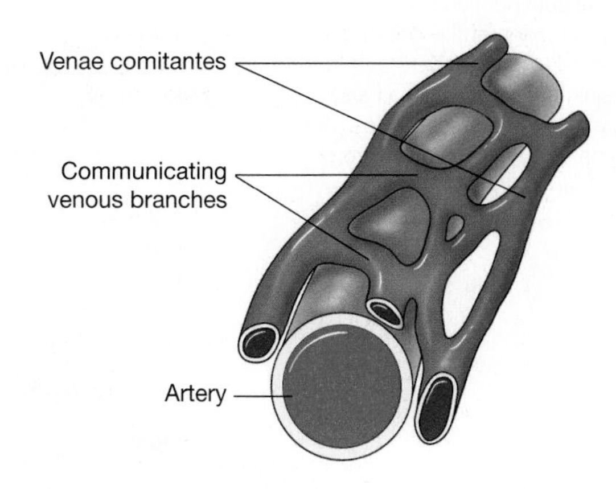

Figure 21.14 Venae comitantes. Several veins linked by communicating branches surround a single artery.

Systemic Veins

Veins in the systemic circulation are in one of two sets. The superficial veins travel in the superficial fascia (the hypodermis of the skin) and are quite variable in their pathways. The deep veins are typically surrounded by the same connective tissue sheath as their accompanying arteries. Most veins that accompany the arteries in the trunk occur as single blood vessels; however, veins that travel with arteries in the extremities are arranged as two or more vessels connected by several small communicating branches. These groups of veins, arranged around a common artery, are called **venae comitantes**, or **companion veins** (Figure 21.14).

ACTIVITY 21.4 Identifying the Major Veins of the Systemic Circulation

The Superior Vena Cava

On a torso model, identify the **superior vena cava.** Verify that this large blood vessel empties into the right atrium. The superior vena cava receives blood from the head and neck, upper extremities, and thorax. It is formed by the union of the **right** and **left brachiocephalic veins** (Figure 21.15). Identify these two blood vessels on the torso model. Notice that the left brachiocephalic vein is longer than its counterpart on the right.

Venous Drainage of the Head and Neck

1. On a torso model, identify the **internal jugular veins** as they descend through the neck. Notice that they run parallel to the common carotid arteries. At the base of the neck, on each side, the internal jugular vein and **subclavian vein** merge to form the brachiocephalic veins (Figures 21.15 and 21.16). The internal jugular veins receive blood from the network of dural sinuses that drain blood from the brain. They also drain blood from the face via the facial veins (Figure 21.16).
2. Locate the external jugular veins on a torso model. These veins drain blood from the face, scalp, and neck. They are formed by the union of the maxillary and temporal veins (Figure 21.16) and descend through the neck, superficial to the sternocleidomastoid muscles. They usually drain

Figure 21.15 Veins that drain the abdomen, thorax, and upper extremity. The superior vena cava is formed by the union of the two brachiocephalic veins, which receive blood from smaller veins that drain the head and neck, thorax, and upper extremities. The inferior vena cava receives blood from veins that drain abdominal structures.

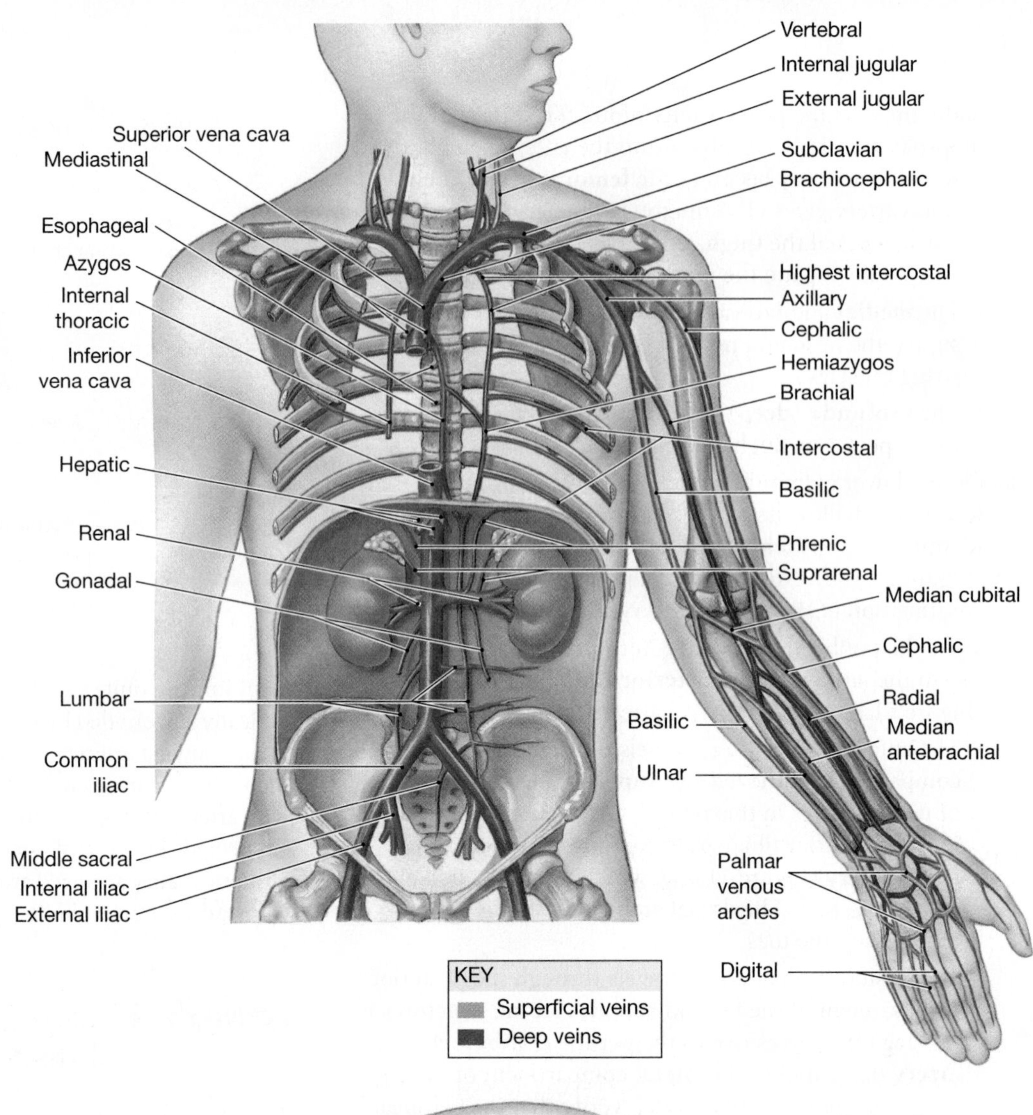

Figure 21.16 Veins that drain the head and neck. The internal jugular veins drain most blood from the brain. The external jugular veins drain blood from the scalp, face, and neck.

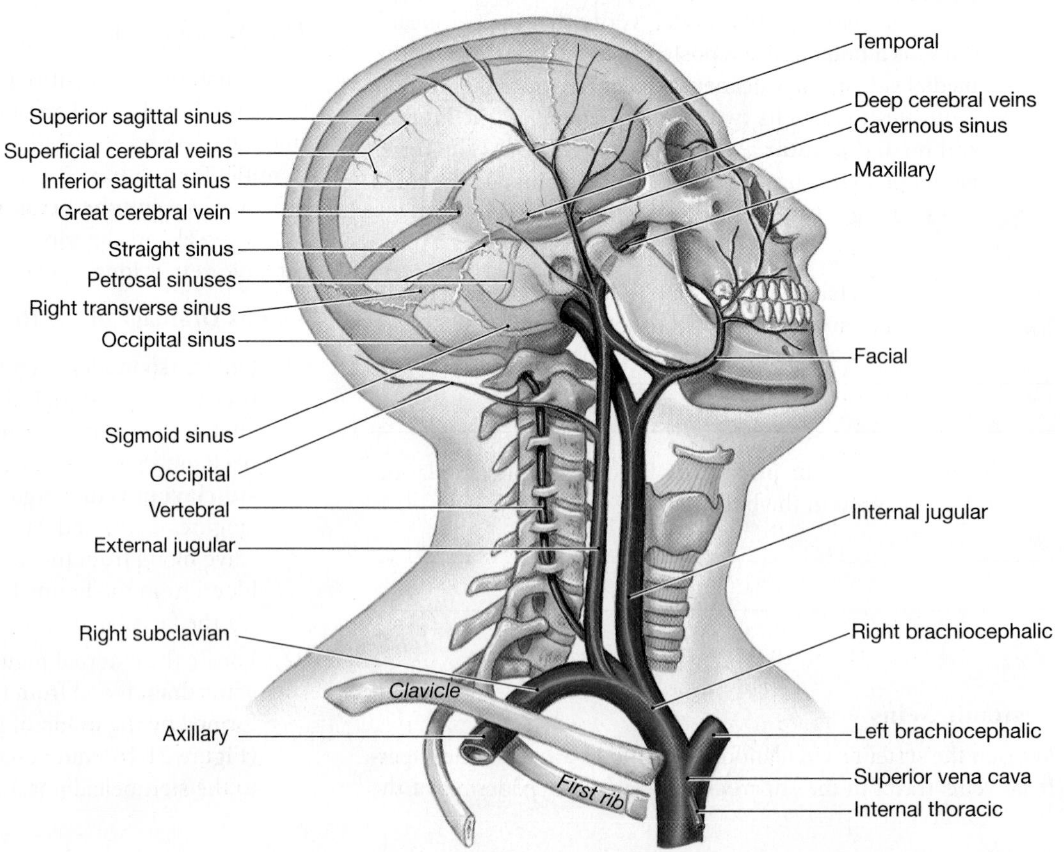

into the subclavian veins (Figure 21.16), posterior to the clavicles. The external jugular veins are relatively large superficial veins that can be easily seen and palpated in many individuals. Attempt to palpate one of these veins on yourself or your lab partner.

3. On a model of the vertebral column, identify the vertebral arteries, described earlier, as they travel through the bony canals formed by the transverse foramina of cervical vertebrae. **Vertebral veins**, which are usually not demonstrated on models, travel alongside these arteries (Figure 21.16). These veins drain blood from the brain, posterior skull bones, and cervical vertebrae and empty into the brachiocephalic veins.

Venous Drainage of the Upper Extremities

1. On a torso or upper extremity model, identify the **axillary vein** (Figure 21.15). Notice that this blood vessel travels through the axilla, medial to the axillary artery. Observe the axillary vein as it travels toward the base of the neck, and passes between the first rib and the clavicle. At this position, the axillary vein becomes the subclavian vein (Figure 21.16).
2. If present, identify the **brachial**, **radial**, and **ulnar veins** (Figure 21.15) traveling alongside their corresponding arteries. If they are not shown on your model, you can determine their approximate position by identifying the brachial, radial, and ulnar arteries.
   - In the arm, the brachial vein forms a venae comitantes around the brachial artery. As it travels toward the axilla, it merges with the **basilic vein** (a major superficial vein that is discussed below) to form the axillary vein.
   - In the forearm, the radial and ulnar veins form venae comitantes around their corresponding arteries. These veins arise from venous arches in the hand and travel superiorly through the forearm. Anterior to the elbow, the radial and ulnar veins merge to form the brachial vein.
3. The distribution of superficial veins in the upper extremity is quite variable, with numerous connections between them. In addition, they give off numerous short branches that freely communicate with the deep veins. Identify the major superficial veins on a torso or upper extremity model.
   - The **cephalic vein** originates from an extensive venous network in the hand and travels along the lateral aspect of the forearm and arm. Locate its terminal portion traveling along the **deltopectoral groove**, a shallow depression between the deltoid and pectoralis major muscles. The vein ends by piercing the deep fascia and draining into the axillary vein (Figure 21.15).
   - The **basilic vein** travels along the medial side of the forearm and arm. Identify this blood vessel where it joins the brachial vein to form the axillary vein (Figure 21.15).
   - The **median cubital vein** travels obliquely through the cubital fossa, the shallow depression anterior to the elbow joint. It begins as a branch of the cephalic vein, and travels superomedially to join the basilic vein (Figure 21.15). Make a tight fist and try to identify the median cubital vein traveling superficially through your own cubital fossa.

CLINICAL CORRELATION

The superficial veins of the upper extremity are easily accessible, and therefore, routinely used for drawing blood or injecting solutions (**venipuncture**). For example, when you go to a doctor's office and a blood sample is taken, your median cubital vein is often the blood vessel used. If you have just finished running a marathon on a hot day and are suffering from dehydration, you will be taken to the medical tent for treatment. In such a situation, the cephalic vein, basilic vein, or the venous network in the dorsum of the hand might be used to introduce fluids and electrolytes over a long period of time. This is referred to as **intravenous feeding.**

Venous Drainage of the Thorax

1. The **azygos system of veins** is a highly variable system of blood vessels that drains blood from most thoracic structures. It includes the **azygos vein**, **hemiazygos vein**, and their tributaries (Figure 21.15). Identify the azygos and hemiazygos veins on a torso model. (Note: Some anatomical models do not clearly demonstrate these vessels.) Observe how the azygos vein ascends along the anterior surface of the vertebral column, just to the right of the midline, and empties into the superior vena cava. The hemiazygos vein ascends along the left side of the vertebral column. Its distribution is highly variable, but it is usually connected to the azygos vein by one or more communicating veins. More superiorly, the hemiazygos vein communicates with the highest intercostal vein, which empties into the left brachiocephalic vein.
2. Identify the **posterior intercostal veins** as they travel through the intercostal spaces. Verify that most of these veins drain into the azygos vein on the right side and the hemiazygos vein on the left side.
3. Earlier in this exercise, you observed the internal thoracic and anterior intercostal arteries. Once again, identify these vessels on the posterior surface of the anterior body wall. There are corresponding veins that travel with these arteries. **Anterior intercostal veins** drain blood from the intercostal spaces and empty into the **internal thoracic veins.** The internal thoracic veins travel along each side of the sternum and empty into the brachiocephalic veins (Figure 21.15).

Venous Drainage of the Abdomen

1. On a model of the torso or abdominal cavity, identify the **inferior vena cava** as it ascends to the right of the aorta along the posterior abdominal wall (Figure 21.15). Observe that it is formed by the union of the two **common iliac**

veins at the level of the L5 vertebra. It exits the abdominal cavity by passing through the diaphragm at the level of the T10 vertebra and empties into the right atrium.

2. Most of the veins emptying into the inferior vena cava travel with their corresponding arteries (Figure 21.15). Identify these veins on a model. (Note: In some models, the corresponding arteries, but not the veins, will be demonstrated.)
   - Several pairs of **lumbar veins** empty into the lateral aspect of the inferior vena cava in the lumbar region. These veins drain the posterior abdominal wall.
   - The **gonadal veins** drain the ovaries (**ovarian veins**) in the female, and the testes (**testicular veins**) in the male. Observe these veins traveling obliquely along the posterior abdominal wall. The right gonadal vein empties into the inferior vena cava; however, the left gonadal vein often drains into the left renal vein.
   - The **renal veins** drain blood from the kidneys. Observe that both of these relatively large vessels empty directly into the inferior vena cava.
   - The **suprarenal veins** drain blood from the adrenal glands. Notice that the right vein empties directly into the inferior vena cava. The left vein usually drains into the left renal vein.
   - Observe that the **hepatic veins** are relatively large vessels and empty directly into the inferior vena cava. They drain blood from the liver sinusoids of the hepatic portal system, which is described in step three.
   - The **inferior phrenic veins** are relatively small vessels that drain the inferior surface of the diaphragm. Both empty directly into the inferior vena cava.
3. A **portal system** is a modified portion of the systemic circulation, in which blood passes through an extra capillary bed before entering the veins that return blood to the heart. In the **hepatic portal system** (Figure 21.17a), blood is drained from the capillaries of the digestive organs and enters a set of veins that lead into the **hepatic portal vein.** The hepatic portal vein directs blood into the liver sinusoids, which serve as the second capillary bed (Figure 21.17b). As the blood percolates through the sinusoids, liver cells metabolize nutrients and detoxify poisons that are present in the blood. Ultimately, blood drains into the hepatic veins and the inferior vena cava for its return to the heart. On a torso model, identify the following components of the hepatic portal system. If the models in your laboratory do not adequately demonstrate these structures, refer to Figure 21.17a.
   - The **inferior mesenteric vein** drains blood from the second half of the large intestine.
   - The **superior mesenteric vein** receives blood from the small intestine and the first half of the large intestine.
   - The **splenic vein** drains blood from the spleen and portions of the stomach and pancreas. It also receives blood from the inferior mesenteric vein.
   - The **hepatic portal vein** is formed by the union of the superior mesenteric and splenic veins. It transports blood to the sinusoids in the liver.
   - **Gastric veins**, which drain blood from the stomach, empty into the hepatic portal vein.
   - As described earlier, **liver sinusoids** are drained by the hepatic veins, which lead directly to the inferior vena cava. Obtain a microscope slide of the liver and focus the slide with the low-power objective lens in position on your microscope. Switch to the high-power objective and identify the columns of **liver cells (hepatocytes).** The clear channels traveling between the rows of cells are liver sinusoids (Figure 21.17b).

Venous Drainage of the Lower Extremity

1. On a torso model, identify the two common iliac veins that form the inferior vena cava (Figure 21.15). Blood returning to the heart from the pelvis and lower extremity flows into these vessels before draining into the inferior vena cava.
2. The common iliac veins are formed by the union of the **internal** and **external iliac veins** (Figures 21.15 and 21.18). Identify these blood vessels on a torso model. The internal iliac veins drain blood from structures along the pelvic wall, the reproductive organs, and urinary bladder in the pelvic cavity. The external iliac veins drain blood from the lower extremities.
3. The major deep veins of the lower extremity include the **femoral vein** in the thigh, the **popliteal vein** in the popliteal fossa, and the **anterior tibial**, **posterior tibial**, and **peroneal (fibular) veins** in the leg (Figure 21.18).
   - The anatomical models in your laboratory may not clearly illustrate the popliteal vein and the deep veins in the leg. If this is the case, identify the corresponding arteries on a model and realize that the veins form venae comitantes, and have a distribution pattern similar to their corresponding arteries. If they are present, identify them traveling with the arteries.
   - On a model of the lower extremity, identify the femoral vein as it travels through the anterior thigh, medial to the femoral artery. Trace the vein as it ascends through the thigh and passes into the pelvis to become the external iliac vein. The femoral vein receives a number of branches that drain blood from structures in the thigh.
4. The **great saphenous vein**, which is the longest vein in the body, is the main superificial vein in the lower extremity. It originates on the medial side of the foot, passes anterior to the medial malleolus, and ascends along the medial aspect of the leg and thigh (Figure 21.18).
   - Sit comfortably on a laboratory stool so that your feet are hanging freely. Remove your shoe and sock from one foot. Do not allow your bare foot to contact the

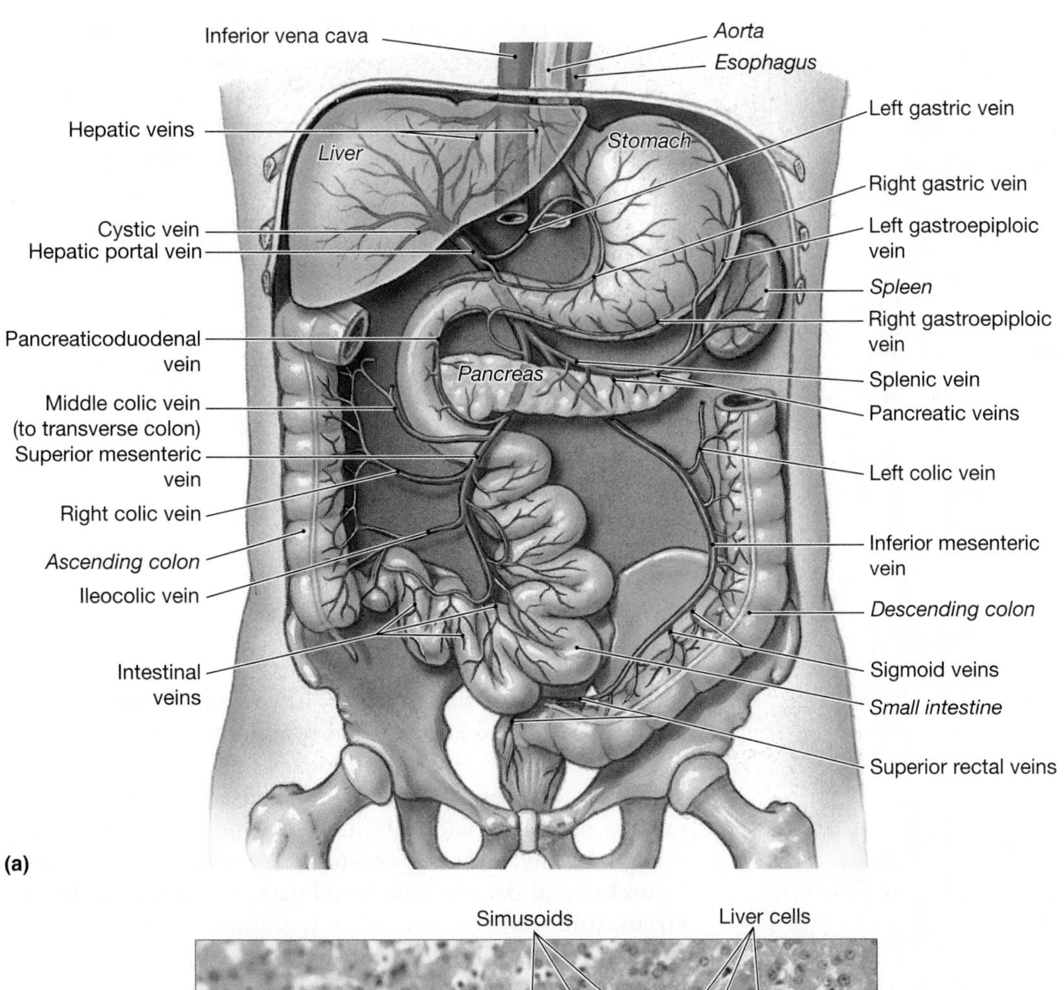

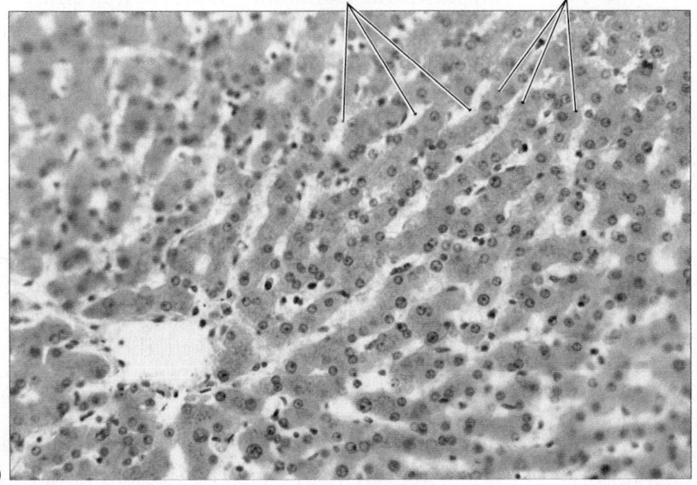

Figure 21.17 The hepatic portal system. a) Veins that drain abdominal organs and deliver blood to the liver; **b)** light micrograph of the liver showing sinusoids traveling between rows of liver cells (LM × 200).

laboratory floor. Attempt to locate the great saphenous vein near its origin, anterior to the medial malleolus. You should be able to feel the vein as it travels subcutaneously at this location.

- At its termination in the anterior thigh, the great saphenous vein pierces the deep fascia at the **saphenous opening** and empties into the femoral vein (Figure 21.18). On a torso model, locate the saphenous opening near the superior border of the anterior thigh. Identify the great saphenous vein where it joins the femoral vein.

CLINICAL CORRELATION

Since the great saphenous vein is quite long and easily accessible, it is often used for vein graft surgical procedures to bypass obstructions in blood vessels. For example, in a coronary bypass operation, a section of the great saphenous vein is attached before and after the obstruction in a damaged coronary artery, thus establishing a circulatory pathway that travels around the obstruction. Removal of the great saphenous vein does not produce significant problems with blood drainage because numerous additional veins will take over its function.

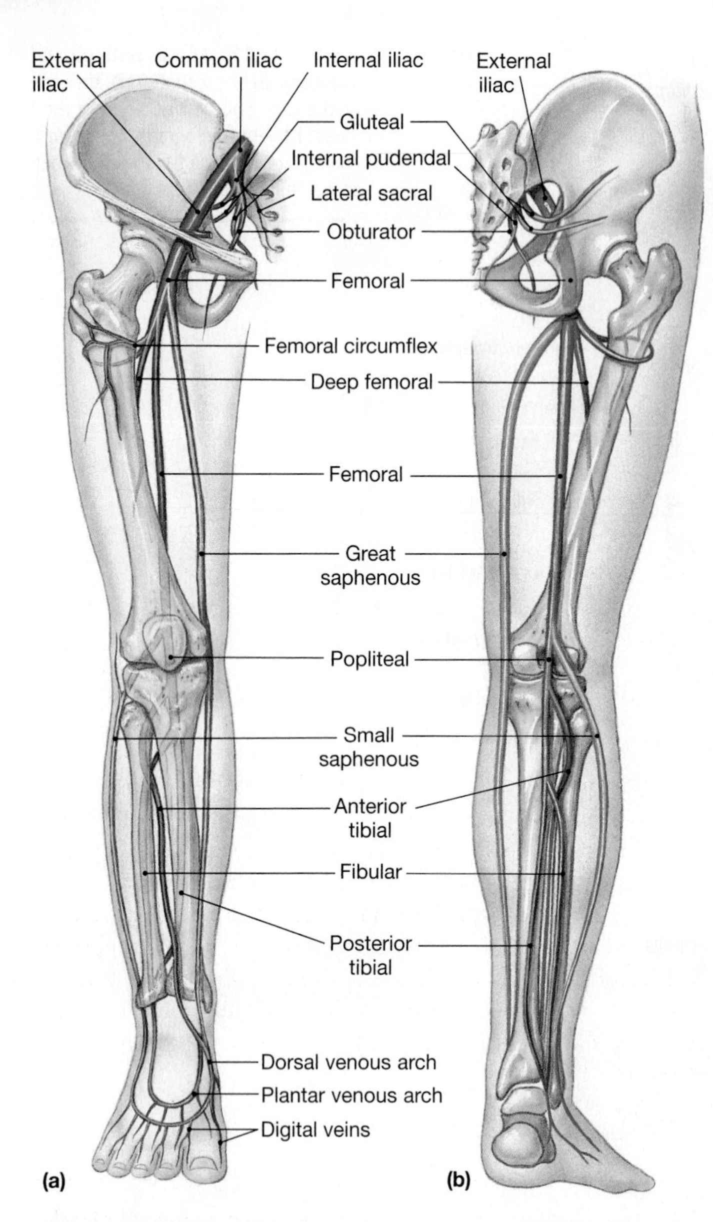

Figure 21.18 Veins that drain the lower extremities. a) Anterior view; **b)** posterior view.

QUESTIONS TO CONSIDER

1. The superior vena cava is formed by the merging of the left and right brachiocephalic veins. Explain why the left brachiocephalic vein is longer than the right brachiocephalic vein.

____

____

2. Can you explain why a pregnant woman or an individual with an abdominal tumor has a greater risk of developing varicose veins?

____

____

3. In addition to draining blood from the thoracic wall, the azygos system of veins also receives venous blood from veins that drain a number of internal thoracic structures. Two examples are the pericardial veins, which drain blood from the pericardial wall, and the bronchial veins, which drain blood from the lungs.

- What is the functional difference between pericardial veins and cardiac veins? ____________________________

  ____

  ____

- What is the functional difference between bronchial veins and pulmonary veins? ____________________________

  ____

  ____

Collateral Circulation

It is common for the branches of two or more arteries with different origins to form natural communications. These vascular unions are called **anastomoses** (singular = **anastomosis**). Anastomoses are important because they provide body regions or organs with alternate supplies of blood. Arterial branches that form an anastomosis are called **collateral branches** and the alternate blood pathways form a **collateral circulation.** Collateral circulation is clinically significant because if normal blood flow in an artery is blocked due to injury, disease, or surgery, alternate pathways can be utilized so that the blood supply to the affected structure is maintained (Figure 21.19).

Most structures in the body receive blood from collateral circulations. Arterial anastomoses are common in abdominal organs, the heart, and around joints, where active movement (e.g., flexion of the elbow or knee) can occlude the main channel. However, some structures are supplied by only one artery that does not form anastomoses with other vessels. If these so-called **end arteries** are blocked, the blood supply to the affected organ will be interrupted and tissue necrosis (death) will result (Figure 21.19). An example of an end artery is the central artery of the retina, a branch of the ophthalmic artery. The central artery represents the only source of blood to the retina. Occlusion of this vessel can lead to permanent blindness in the affected eye.

A simplified collateral circulation is schematically illustrated in Figure 21.19. In Figure 21.19a, arteries 1 and 2 lack collateral branches. Thus, structures X and Y are supplied by end arteries. In Figure 21.19b, both arteries give off a collateral branch. The collateral branches merge to form an anastomosis. As a result, a collateral circulation between arteries 1 and 2 is established. In Figure 21.19c, artery 1 is blocked and cannot deliver blood to structure X. Blood from artery 2 will reach structure X via the collateral circulation.

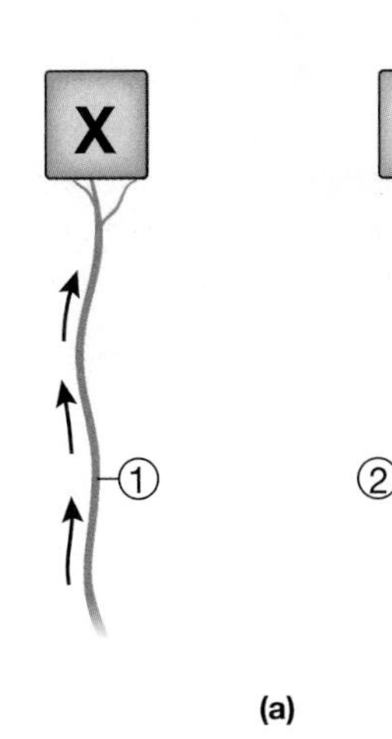

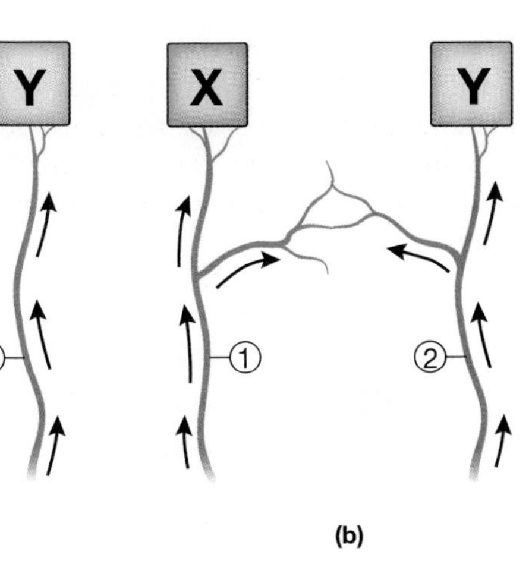

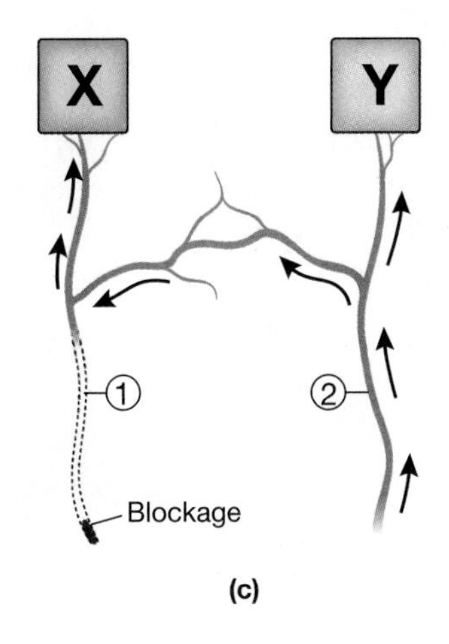

Figure 21.19 Collateral circulation. **a)** Arteries 1 and 2 lack collateral branches and represent the only blood supply (end arteries) to structures X and Y. **b)** Collateral branches from arteries 1 and 2 form an anastomosis and establish a collateral circulation. **c)** If artery 1 is blocked, structure X will receive blood from artery 2 and the collateral circulation.

ACTIVITY 21.5 Identifying Collateral Pathways in the Systemic Circulation

1. On a model of the brain, identify the anastomoses that form the cerebral arterial circle (Figure 21.11). This arterial loop forms important collateral blood pathways for the brain and lowers the risk of interrupting blood flow to cranial regions.
   - The anterior cerebral arteries and the anterior communicating artery connect the left and right internal carotid arteries.
   - The basilar artery is connected to the two internal carotid arteries by posterior communicating arteries.
2. On a model of the upper extremity, identify the anastomoses that form a collateral circulation around the elbow joint (Figure 21.12).
   - The **ulnar collateral arteries**, branching off the brachial artery, communicate with the **ulnar recurrent arteries**, originating from the ulnar artery.
   - The **deep brachial artery**, a branch off the brachial artery, forms an anastomosis with the **radial recurrent artery**, coming from the radial artery.

QUESTIONS TO CONSIDER

1. The primary blood flow to the posterior cerebral arteries is by way of the vertebral and basilar arteries (Figure 21.11). Describe the collateral circulation to these arteries if, due to atherosclerosis, blood flow along the basilar artery is reduced. ______________________

____

____

2. If you are sitting at your desk, studying for an important anatomy and physiology exam, you might support your head with your forearm and hand as you review your lecture notes and required textbook readings. In this position, your forearm is flexed for an extended period of time, and the brachial artery is partially occluded. Consequently, the main blood supply to the forearm and hands is reduced. Refer to Figure 22.12 and describe the collateral arterial pathways around the elbow that will deliver an adequate amount of blood to structures in the forearm and hand.

____

____

3. In the previous activity, you observed two examples of collateral circulation. Using the models in the laboratory, identify and describe two other examples of collateral circulation.

____

____

Exercise 21 Review Sheet

Anatomy of Blood Vessels

Name ______________________

Lab Section ______________________

Date ______________________

1. Describe and compare the two main circulatory pathways: the pulmonary circulation and the systemic circulation.

____

____

2. Describe the basic structure of a blood vessel. How does the anatomy of a corresponding artery and vein differ?

____

____

3. Describe the main structural difference between elastic and muscular arteries. What is the functional significance of this structural change?

____

____

4. Compare the structure of the three main types of capillaries.

____

____

Questions 5–14: Identify the labeled arteries in the following figure.

5. ______________________
6. ______________________
7. ______________________
8. ______________________
9. ______________________
10. ______________________
11. ______________________
12. ______________________
13. ______________________
14. ______________________

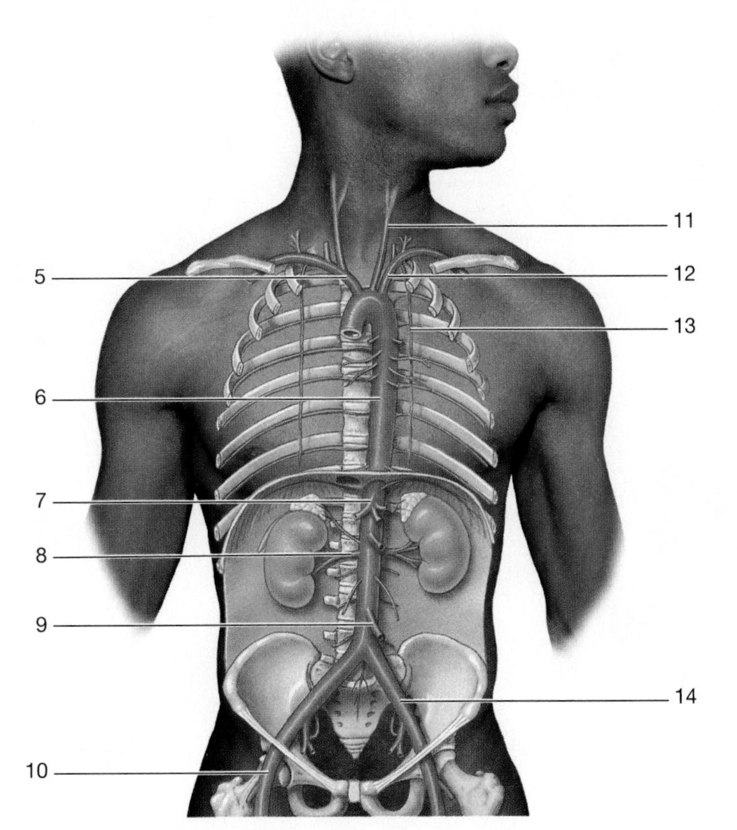

Questions 15–25: Identify the labeled blood vessels in the following figure.

15. ______________________________
16. ______________________________
17. ______________________________
18. ______________________________
19. ______________________________
20. ______________________________
21. ______________________________
22. ______________________________
23. ______________________________
24. ______________________________
25. ______________________________

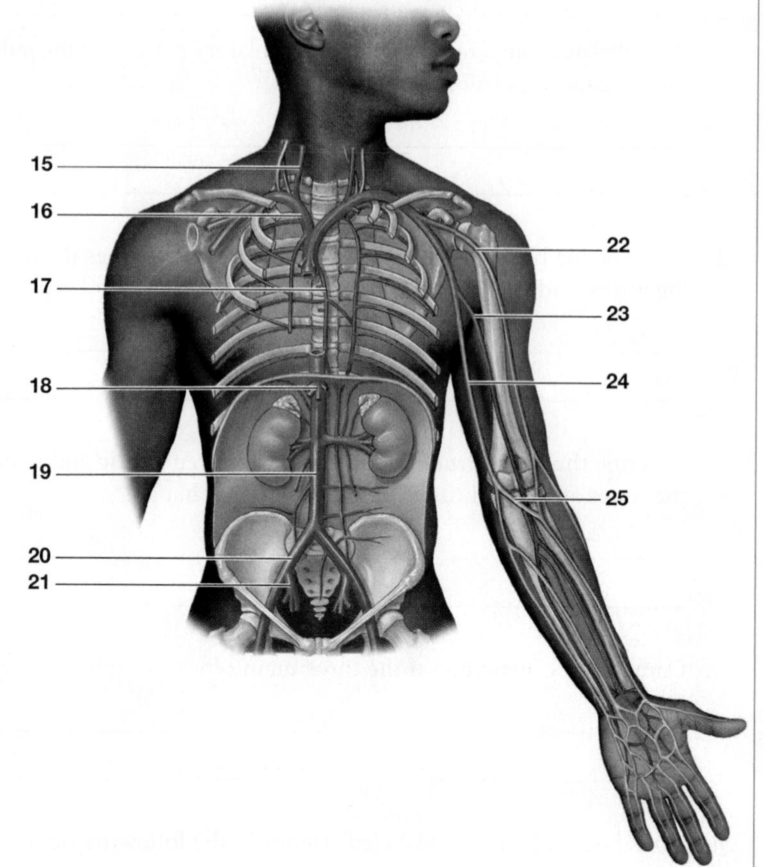

Exercise 22

Cardiovascular Physiology

Laboratory Objectives

On completion of the activities in this exercise, you will be able to:

- Locate the auscultation areas for the heart.
- Use a stethoscope to listen to the heart sounds at the auscultation areas.
- Measure blood pressure at rest, during exercise, and during the recovery period after exercise.
- Calculate the pulse pressure and mean arterial pressure (MAP).
- Measure the pulse rate at rest, during exercise, and during the recovery period after exercise.
- Describe the function of the cardiac conducting system.
- Measure and evaluate the electrical activity of the heart (ECG).

Materials

- Stethoscopes
- Alcohol swabs
- Sphygmomanometers
- Stopwatch or clock with a second hand
- Stationary cycle
- Biopac Student Lab system

The **cardiac cycle** refers to the series of events that occurs during one heartbeat. During one cycle, the two atria will contract at the same time. As the atria relax, the two ventricles will contract simultaneously. A period of contraction in a heart chamber is called **systole** (**atrial systole, ventricular systole**), and a period of relaxation is called **diastole** (**atrial diastole, ventricular diastole**). In clinical use, these terms typically refer to events in the ventricles, because they are the larger and more powerful chambers that pump blood into the great arteries.

The events of the cardiac cycle are illustrated and described in Figure 22.1. During the cycle, changes in blood pressure inside the chambers and great arteries cause the heart valves to open and close. These events regulate the flow of blood through the heart and into the systemic and pulmonary circuits. During this laboratory exercise you will investigate some of the physiological events—heart valve function, pulse, blood pressure—that characterize the cardiac cycle. You will also record measurements of the electrical activity of the heart (an electrocardiogram or ECG) and evaluate the results.

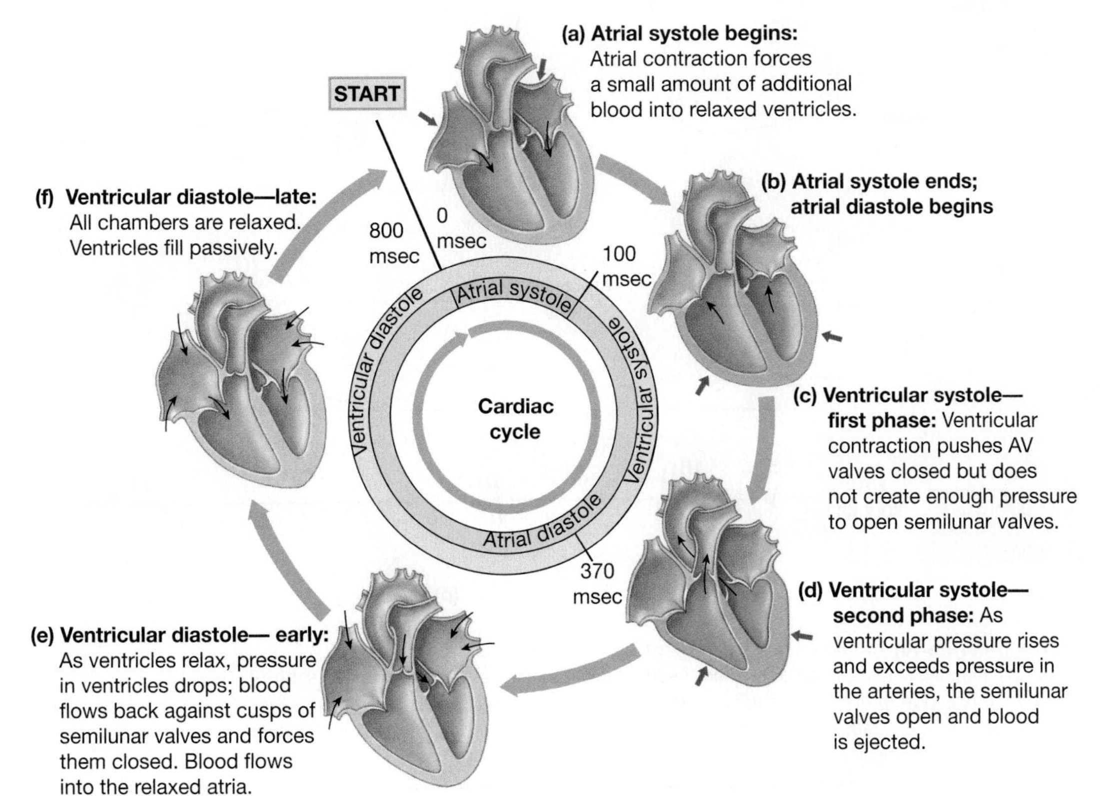

START
(a) Atrial systole begins: Atrial contraction forces a small amount of additional blood into relaxed ventricles.
(b) Atrial systole ends; atrial diastole begins
(c) Ventricular systole—first phase: Ventricular contraction pushes AV valves closed but does not create enough pressure to open semilunar valves.
(d) Ventricular systole—second phase: As ventricular pressure rises and exceeds pressure in the arteries, the semilunar valves open and blood is ejected.
(e) Ventricular diastole— early: As ventricles relax, pressure in ventricles drops; blood flows back against cusps of semilunar valves and forces them closed. Blood flows into the relaxed atria.
(f) Ventricular diastole—late: All chambers are relaxed. Ventricles fill passively.
0 msec
100 msec
370 msec
800 msec
Atrial systole
Ventricular systole
Atrial diastole
Ventricular diastole
Cardiac cycle

Figure 22.1 The cardiac cycle. The illustrated steps describe the events that occur during one heartbeat. During each cycle, the atria contract together and the ventricles contract together.

Heart Sounds

During the cardiac cycle, when blood passes from the atria to the ventricles and from the ventricles to the great vessels, the heart valves open and close. The closing of the values produces two distinctive heart sounds. The sounds can be heard, with the aid of a stethoscope, as **"lub-dup"** vibrations. The cardiac cycle begins when a small volume of blood is pumped into each ventricle during atrial systole (Figure 22.1a). The first "lub" sound is caused by vibrations that follow the closing of the atrioventricular (AV) valves (Figure 22.2b). This occurs at the end of atrial systole (Figure 22.1b). This point in the cardiac cycle also marks the beginning of ventricular systole (Figure 22.1c), known as **isovolumetric contraction**, when all heart valves are closed. As the pressure in the ventricles increases, the semilunar valves open and blood is pumped into the great vessels. This is the second phase of ventricular systole (Figure 22.1d), known as **ventricular ejection.** The closing of the semilunar valves occurs at the beginning of ventricular diastole (Figure 22.1e). Vibrations, generated by the closing of these valves, creates the second "dup" sound (Figure 22.2b). At this time, pressure in the ventricles is decreasing and falls below the pressure in the great arteries. At the end of the cycle, ventricular pressure falls below the atrial pressure. As a result, the AV valves open and passive filling of the ventricles begins. A new cycle begins with the initiation of atrial systole.

CLINICAL CORRELATION

Incomplete closure of the AV valves can cause regurgitation or backflow of blood into the atria. This can cause an abnormal gurgling sound known as a **heart murmur.** On the left side of the heart, incomplete closure of the biscupid (mitral) valve is called a **mitral valve prolapse.** Minor prolapses are fairly common and most people live with them and do not experience adverse effects. However, a major prolapse, possibly caused by rupturing of the chordae tendinae or severe damage to the cusps, can have serious if not life-threatening consequences.

ACTIVITY 22.1 Listening for Heart Sounds

1. The best locations to hear heart sounds are the **auscultation areas for the heart** (Figure 22.2a) on the anterior thoracic wall. These areas are named after the heart valve that can best be heard. Locate the following auscultation areas on yourself or your lab partner.
   - The **biscupid area** is located in the left fifth intercostal space, where the apex of the heart is located. To find this region, locate the inferior end of the sternum by finger palpation. From this point, move your finger approximately 7 cm (2.75 in) to the left, where you

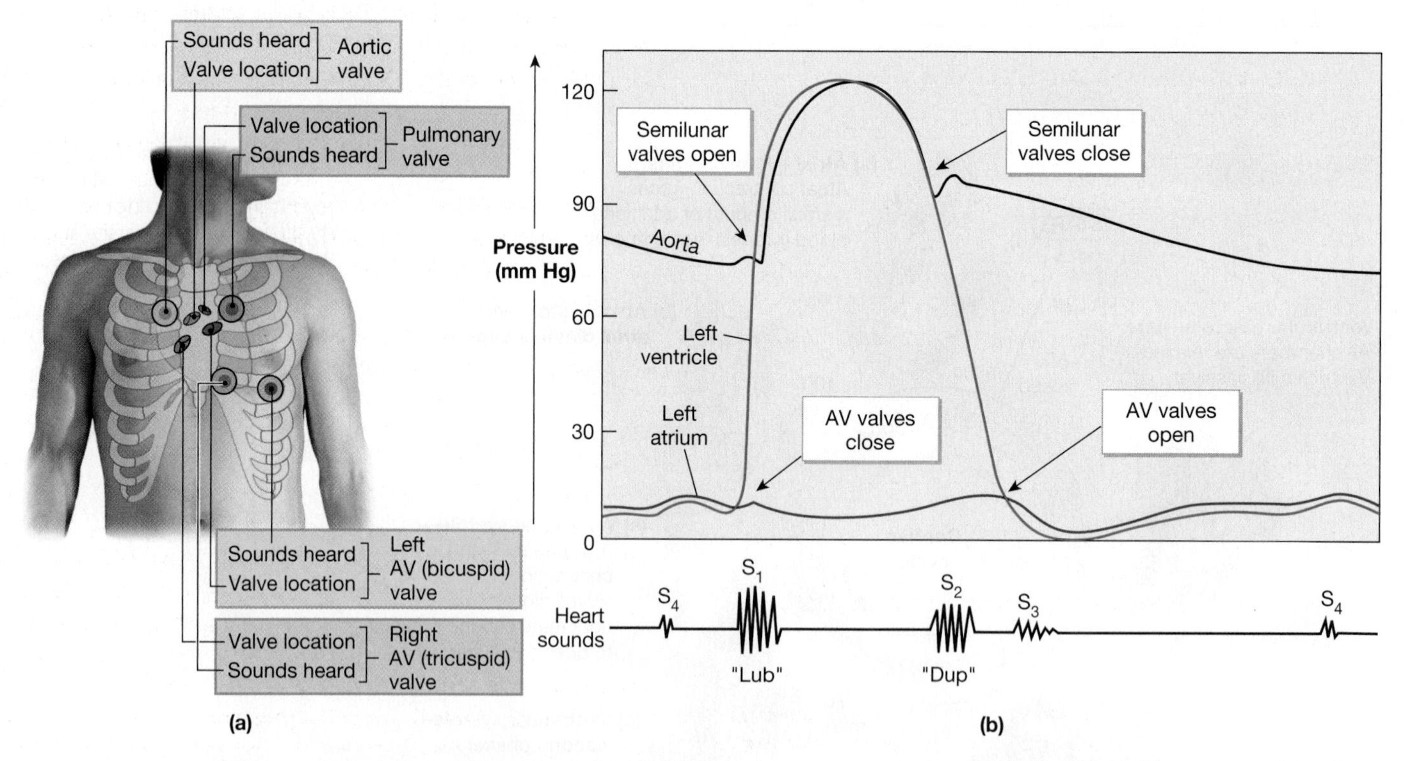

Figure 22.2 Auscultation areas for the heart. a) Diagram that illustrates the locations of the heart valves (oval areas) and the auscultation areas (circular areas), where heart sounds can best be detected with a stethoscope. **b)** Graph showing the relationship of the heart sounds with events in the cardiac cycle. The two primary heart sounds, S_1 and S_2 ("lub-dup"), are caused by the closing of heart valves. Two minor sounds, S_3 and S_4, are not related to valve function and are difficult to hear.

can feel the fifth intercostal space between the fifth and sixth ribs.

- To locate the **tricuspid area**, palpate the inferior end of the sternum and place your stethoscope just to the left of that point.
- The **aortic semilunar area** is located in the second intercostal space, just to the right of the sternum. The second intercostal space can be located by first palpating the superior margin of the manubrium. From this margin, move your finger inferiorly until you can feel the junction between the manubrium and the body of the sternum. If you move your finger laterally to the right of the sternum, you can feel the costal cartilage of the second rib. The second intercostal space is just inferior to this rib. Place your stethoscope in this space, just lateral to the sternum.
- The **pulmonary semilunar area** is located in the second intercostal space, just to the left of the sternum. Follow the same procedure for locating the aortic semilunar area, but this time, move to the left side of the sternum.

WHAT'S IN A WORD The term *auscultation* is derived from the Latin word *ausculto,* which means "to listen." Auscultation is an important diagnostic tool used by doctors and other health care providers. It involves listening to the sounds made by various organs in the thoracic or abdominal cavities, such as the closing of heart valves.

2. Obtain a stethoscope and sterilize the earpieces with an alcohol swab.
3. Place the stethoscope on the bicuspid area and listen for the heart sounds. If the background noise is too high and you are experiencing difficulty in detecting the sounds, move to a quieter area in the laboratory or to another room. Can you hear both heart sounds ("lub-dub") when the stethoscope is placed over the biscuspid area? Can you hear one sound better than the other?

____

____

4. Listen to the heart sounds at the other auscultation areas. How do the sounds compare at each area? Answer the same two questions that were asked for the biscuspid area in the previous step.

	Can you hear both heart sounds?	Can you hear one sound better than the other?
• Tricuspid area	________	________
• Aortic semilunar area	________	________
• Pulmonary semilunar area	________	________

QUESTION TO CONSIDER During a physical examination, why does the doctor listen to the heart sounds at all four auscultation areas?

____

____

Blood Pressure

Blood pressure is the force exerted by blood on the walls of blood vessels. It is a function of the pumping action of the heart and the resistance to flow as blood moves through the blood vessels. Blood flows throughout the circulatory pathways due to the existence of a pressure gradient that allows blood to move from areas of high pressure to areas of low pressure. Blood flow in arteries and veins begins with ventricular ejection into the great arteries and ends with venous return to the atria. In the systemic circulation, blood pressure is highest in the aorta due to the force of ejection by the left ventricle. Pressure gradually declines throughout the circulatory pathway, and is close to zero when blood enters the right atrium.

In large elastic arteries, the blood pressure fluctuates between a maximum and minimum value, which correspond to the cardiac cycle. For example, during ventricular systole, blood is ejected into the aorta from the left ventricle. The force of ejection causes the elastic walls of the aorta to stretch, and the pressure inside the aorta reaches a peak. This maximum pressure is called the **systolic pressure.** During ventricular diastole, the aortic semilunar valve closes, and the elastic fibers in the wall of the aorta recoil to force blood forward. At this time, the aortic pressure declines to a minimum level, referred to as the **diastolic pressure.** Thus, blood pressure in the aorta is not smooth or constant, but pulsatile in nature. This characteristic is also true for other elastic arteries, but it diminishes in the smaller arteries and arterioles as the number of elastic fibers in the vessel walls diminishes. Blood, flowing through capillaries and veins, travels under relatively low pressure with little or no fluctuation.

Blood pressure is measured in units called millimeters of mercury (mm Hg). If the pressure in a blood vessel is 95 mm Hg, it means that the force exerted by the blood will cause a column of mercury to rise 95 millimeters. When blood pressure is measured, it is the **arterial blood pressure** in the systemic circulation that is recorded. Usually, the **brachial artery** is used to measure arterial blood pressure (Figure 22.3), because it is at the same level as the heart, so the effects of gravity are negligible. Thus, blood pressure measurements taken from the brachial artery are fairly close to the blood pressure in the aorta.

Since the pressure in arteries is pulsatile, both systolic and diastolic pressures are measured. If a person's blood pressure is 120/80, it means that the systolic pressure is 120 mm Hg and the diastolic pressure is 80 mm Hg. The systolic pressure represents the force exerted by the left ventricle when it pumps blood into the aorta. The diastolic pressure measures the resistance to blood flow in the arteries.

The difference between the systolic and diastolic pressures is the **pulse pressure**. Thus, a person with a blood pressure of 120/80 will have a pulse pressure of 40 mm Hg:

120 mm Hg	−	80 mm Hg	=	40 mm Hg
(systolic pressure)		**(diastolic pressure)**		**(pulse pressure)**

The average pressure that drives blood through the systemic circulation is the **mean arterial pressure (MAP)**. MAP is crucial for maintaining a steady blood flow from the heart to the capillaries. It is equal to the diastolic pressure plus one third of the pulse pressure and is represented by the following equation:

MAP = diastolic pressure + 1/3 pulse pressure

For an individual with a blood pressure of 120/80, the MAP would be:

$$\text{MAP} = 80\ \text{mm Hg} + 1/3\ (40\ \text{mm Hg}) = 93.3\ \text{mm Hg}$$

CLINICAL CORRELATION

Temporary elevations in blood pressure are normal when the body is adapting to changing conditions that occur during physical exercise, fever, emotional upset, or fear. On the other hand, chronic or persistent elevations in blood pressure, called **hypertension,** can cause serious disease. Hypertension is usually defined as a sustained arterial blood pressure of 140/90 or higher. Many people do not realize that they have high blood pressure because the disease can progress without symptoms for 10 to 20 years. During this time, the heart can be weakened and effects of arteriosclerosis can progress rapidly, leading to stroke, kidney failure, or blindness. As a general rule, the higher the blood pressure, the greater the chance of serious cardiovascular damage. Of the two blood pressure readings, elevated diastolic pressure is considered to be more critical because it suggests that damage, due to arteriosclerosis, has occurred to the arterial walls. However, more recent evidence has suggested that systolic pressure might be a more reliable indicator of cardiovascular disease than was once believed.

ACTIVITY 22.2 Measuring Blood Pressure

1. Obtain a **sphygmomanometer** and stethoscope. A sphygmomanometer is an instrument used to measure blood pressure. It consists of a rubber cuff with two attached rubber tubes. At the end of one tube is a compressible hand bulb; at the end of the other tube is a dial that records pressure in millimeters of mercury (Figure 22.3).
2. Have your lab partner sit quietly in a chair. Wrap the cuff around his or her arm and position the stethoscope over the brachial artery, distal to the cuff (Figure 22.3c).
3. Squeeze the hand bulb to inflate the cuff until the cuff pressure is greater than the pressure in the brachial artery. At this point, the walls of the brachial artery are compressed and blood is prevented from passing to the forearm (Figure 22.3b). For most people, full compression of the artery can be accomplished by inflating the cuff to a pressure that is greater than 120 mm Hg. You can verify

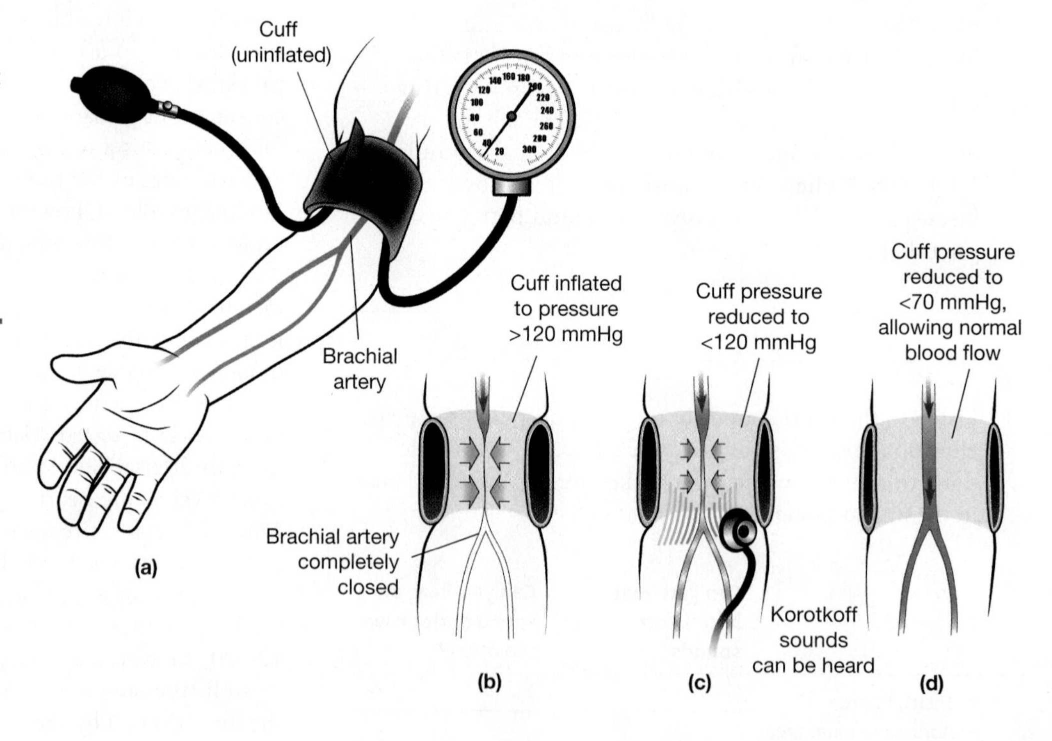

Figure 22.3 Procedure for measuring blood pressure. The cuff of the sphygmomanometer is placed around the arm so that blood pressure at the brachial artery can be measured. **a)** The position of the brachial artery; **b)** the blood pressure cuff in position with full compression of the brachial artery; **c)** the brachial artery partially open to allow some blood to pass through; **d)** additional opening of the brachial artery to allow blood to flow normally.

this by listening to the brachial artery as you squeeze the cuff. When you hear no sound, compression is complete.

4. Slowly deflate the cuff while listening to the brachial artery with the stethoscope. The cuff can be deflated by turning the knob at the base of the hand bulb. Try to deflate the cuff at a rate of 2 to 3 mm Hg per second.
5. As the cuff pressure slowly decreases, it will eventually become less than the arterial pressure. At this point, the artery is partially open and blood will begin to pass through (Figure 22.3c). Because the blood flow is turbulent at this time, thumping sounds will be heard. These sounds, known as **Korotkoff's sounds**, correspond to the systolic blood pressure. Make a mental note of the pressure reading at the time that you begin to hear Korotkoff's sounds.
6. Continue to slowly deflate the cuff. As you do so, the constriction in the artery is reduced and blood flow becomes less turbulent. As a consequence, the thumping sounds will become faint and eventually disappear. The pressure at the time when the thumping sounds stop corresponds to the diastolic pressure (Figure 22.3d). Make a mental note of the pressure reading at this time.
7. Record your partner's blood pressure in Table 22.1. Calculate your partners' pulse pressure and mean arterial pressure and record these values in Table 22.1.
8. With your lab partner's assistance, repeat steps 2 through 6 to determine your own blood pressure. Record this value in Table 22.1. Calculate your pulse pressure and mean arterial pressure and record these values in Table 22.1.

QUESTION TO CONSIDER The pulse pressure is a good diagnostic tool for predicting the condition of the arteries. For example, **arteriosclerosis** causes a decrease in the elasticity of arterial walls and an increase in the resistance to blood flow. Under these conditions, how would you expect the pulse pressure to change so that normal blood flow is maintained? Explain.

____

____

ACTIVITY 22.3 Examining the Effect of Exercise on Blood Pressure

1. Using the instructions from Activity 22.2, measure your partner's blood pressure after he or she sits quietly for 2 to 3 minutes. Record these results in Table 22.2.

   Form a Hypothesis **Before you begin, predict what effect, if any, a period of exercise will have on blood pressure, pulse pressure, and mean arterial pressure (MAP).**

   a. Effect on blood pressure _______________

   b. Effect on pulse pressure _______________

   c. Effect on MAP _______________

2. Have your partner exercise for 10 to 15 minutes on a stationary bicycle. If a bicycle is not available, another form of exercise, such as running in place, can be substituted. Keep the sphygmomanometer and stethoscope in position during the exercise period.

   Warning: *Anyone who is physically or medically unable to perform physical exercise should not participate in this portion of the laboratory activity.*

3. At 3-minute intervals during the exercise period, and immediately after exercise ends (15 minutes), measure your lab partner's blood pressure. Calculate the pulse pressure and MAP. Record the results in Table 22.2.
4. Measure your lab partner's blood pressure 1 minute after the exercise period. Take additional measurements at 3-minute intervals until the return to resting blood pressure. For each measurement, calculate the pulse pressure and MAP. Record the results in Table 22.2.

Assess the Outcome **Did your actual results agree with the prediction you made earlier? Explain.**

____

____

Table 22.1 Blood Pressure Readings

Subject	Blood pressure (mm Hg)	Pulse pressure (mm Hg)	Mean arterial pressure (mm Hg)
1. Population average *	120/80	40	93.3
2.			
3.			

* Keep in mind that these readings are only average values. You should expect variation in the population.

Table 22.2 Effect of Exercise on Blood Pressure

Time	Blood pressure (mm Hg)	Pulse pressure (mm Hg)	Mean arterial pressure (mm Hg)
1. Preexercise:			
At rest			
2. Exercise period:			
3 minutes			
6 minutes			
9 minutes			
12 minutes			
15 minutes			
3. Postexercise:			
1 minute			
4 minutes			
7 minutes			
10 minutes			
13 minutes			

Pulse Rate

In the systemic circulation, waves of pressure are initiated when the left ventricle pumps blood into the aorta. The pulse pressure generates these pressure waves. The waves then travel along other elastic arteries whose walls expand and recoil at a frequency that corresponds to the heartbeat. The rhythmic expansion and recoil of the arteries is known as the **pulse.** Pulses can be felt at various locations, such as the radial artery in the wrist and the femoral artery in the thigh. They tend to diminish in smaller arteries and are absent in capillaries and veins.

ACTIVITY 22.4 Measuring the Pulse Rate

Using your index and middle fingers, apply light pressure to the pulse points at the following locations.

1. **Temporal artery pulse** on the side of the head in the temporal area
2. **Carotid artery pulse** in the neck, about 1 cm inferomedial to the angle of the jaw
3. **Radial artery pulse** on the anterior surface of the lateral wrist
4. **Popliteal artery pulse** in the popliteal fossa, posterior to the knee

QUESTIONS TO CONSIDER

1. While measuring the pulse, you were instructed to apply light pressure to the various pulse points. Exerting too much pressure will stimulate the vagus nerve. How will this affect your pulse measurement?

____

____

2. The average resting pulse is between 70 and 80 beats per minute, but there is considerable variation in the population. For example, it is not unusual for well-trained athletes to have pulse rates as low as 40 to 50 beats per minute. (The pulse rates of some world-class marathon runners are 35 to 40 beats per minute!) Why do athletes tend to have slower pulses?

____

____

ACTIVITY 22.5 Examining the Effect of Exercise on Pulse Rate

1. Have your lab partner sit quietly for about 3 minutes. After this period, take his or her resting pulse. The most accurate measurement would be to count the number of beats for 1 full minute; however, a resting pulse can usually be measured accurately by counting for 15 seconds and multiplying by 4. Record your results in Table 22.3.
2. Have your partner exercise for 10 to 15 minutes on a stationary bicycle. If a bicycle is not available, substitute another form of exercise (running in place or running up and down stairs).

Table 22.3 Measuring Pulse Rates

	Subject	
	Yourself	**Lab partner**
Resting pulse:		
Pulse immediately after exercise:		
Pulse rate at various times after exercise: 2 minutes:		
4 minutes:		
6 minutes:		
8 minutes:		
10 minutes:		
12 minutes:		
14 minutes:		
16 minutes:		

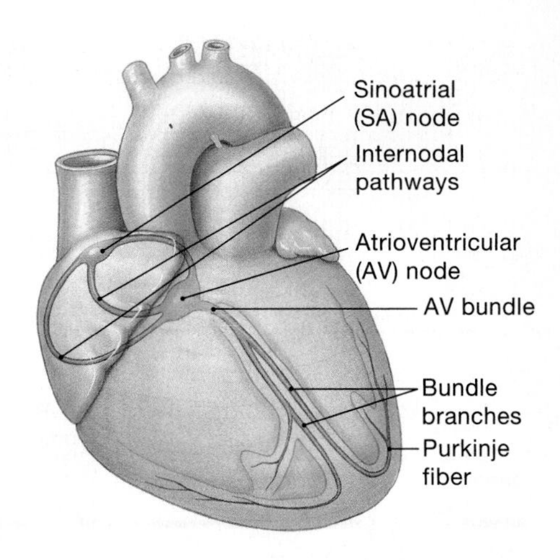

Figure 22.4 The cardiac conduction system. Action potentials are spontaneously generated at the sinoatrial (SA) node. The other components of the conduction system spread depolarizing electric impulses along the atrial and ventricular walls, and thus, regulate the heart's pumping action.

Warning: *Anyone who is physically or medically unable to perform physical exercise should not participate in this portion of the laboratory activity.*

3. Immediately after the exercise period, measure your lab partner's pulse by counting beats for 10 seconds and multiplying by 6. Record the results in Table 22.3.
4. Measure your partner's pulse every 2 minutes after exercise until she or he returns to the resting pulse. Record the results in Table 22.3. How long does it take for the resting pulse to return?

___

5. Repeat steps 1 through 4 for yourself and record the results in Table 22.3.

QUESTIONS TO CONSIDER 1. After a period of exercise, counting the pulse for 10 seconds and multiplying by 6 is more accurate than counting for a full minute. Explain why.

___

___

2. The length of time it takes for your pulse to recover to a resting level is an indication of your physical fitness and cardiovascular efficiency. As a general rule, the shorter the recovery time, the better your fitness. How do you compare with your lab partner and with other members of the class?

___

___

Electrocardiography

The heart contains a network of specialized cardiac muscle cells, known as the **cardiac conduction system** (Figure 22.4), which is able to generate and conduct action potentials without neural or hormonal stimulation. The **sinoatrial (SA) node** is located in the posterior wall of the right atrium, adjacent to the opening for the superior vena cava. Cells in the SA node serve as the heart's pacemaker by spontaneously generating electrical impulses. These depolarizing impulses spread across the walls of the atria, resulting in atrial contraction (systole). As the impulse reaches the **atrioventricular (AV) node** in the floor of the right atrium, the atrial muscle cells repolarize and relax. From the AV node, impulses spread along the **AV bundle**, left and right **bundle branches**, and **Purkinje fibers**, resulting in ventricular contraction. In this manner, these waves of electrical impulses cause the mechanical pumping activity of the heart. Faint traces of these electric impulses spread through the rest of the body and create changes in the electrical potential of the skin. These electrical changes can be detected by electrodes on the skin and recorded in a procedure known as electrocardiography. From the resulting **electrocardiogram** (ECG) it is possible to examine the heart's mechanical activity. If a medical condition causes disruptions in the electrical activity of the heart, these may be reflected in the ECG and useful for diagnosis.

The normal ECG tracing is a flat baseline interrupted by a series of waves (Figure 22.5). In a single cardiac cycle, the **P wave** indicates the depolarization of the atria just prior to the beginning of atrial contraction or systole. The **QRS complex** (QRS interval) represents the depolarization of the ventricles, which precedes ventricular systole. The **T wave** results from

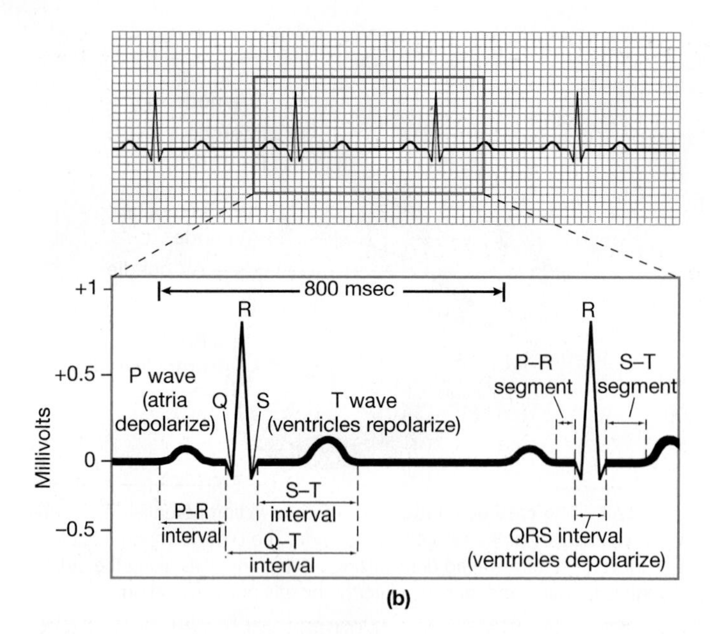

Figure 22.5 Components of the ECG. Intervals, such as the P-R interval, are measured from the beginning of the first wave to the beginning of the second wave. Segments, such as the S-T segment, are measured from the end of the first wave to the beginning of the second wave.

ventricular repolarization, which occurs before ventricular relaxation or diastole. The wave associated with repolarization of the atria is hidden by the much larger QRS complex.

In the following activity, you will record the ECG of your lab partner under varying conditions, correlate electrical and mechanical events of the heart, and observe the changes in heart rate associated with body position and breathing pattern.

CLINICAL CORRELATION

Variations in the size and length of the various waves and wave segments in an ECG are useful in detecting abnormalities in the heart. For example, a higher than normal P wave may indicate an enlarged atrium, and an increase in the height of the R wave (part of the QRS complex) suggests that the ventricles are enlarged. The position of the ST segment (Figure 22.5) can identify the previous occurrence of a heart attack (myocardial infarction). If the segment is above its normal horizontal position, a heart attack likely occurred.

ACTIVITY 22.6 Using the Biopac Student Lab System to Measure and Evaluate the Electrical Activity of the Heart

BIOPAC Systems, Inc.

Setup and Calibration

1. Instruct your lab partner to remove all jewelry, watches, and other metal objects. Ask your partner to lie down and relax.

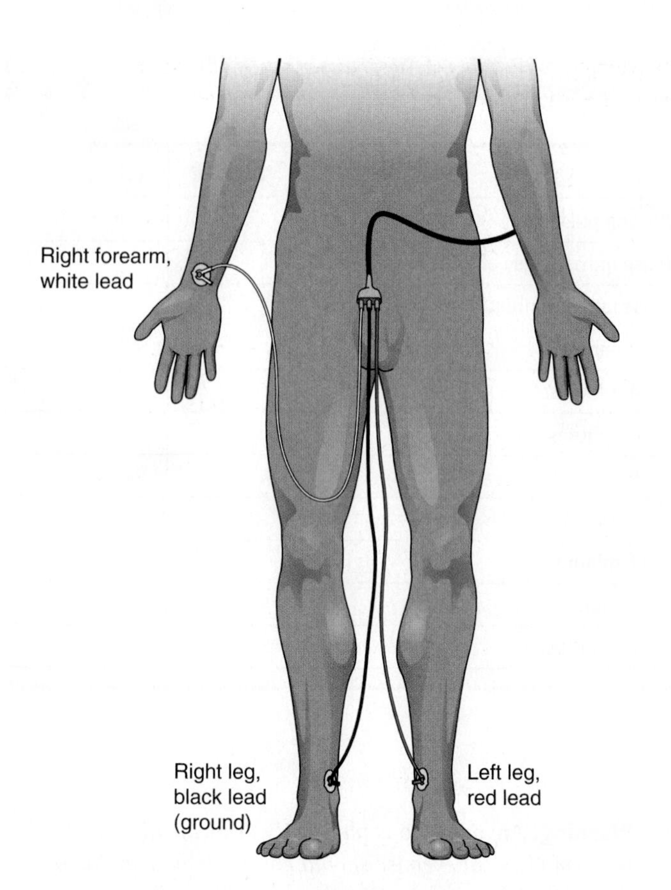

Figure 22.6 Proper electrode placement and connection for Lead II ECG.

2. Attach electrode leads (SS2L) to Channel 2 of the Acquisition Unit and then turn the unit on.
3. Use Figure 22.6 as a guide to attach three electrodes to your lab partner. Place one electrode on the medial aspect of each leg just above the ankle and a third electrode on the anterior wrist of the right forearm. Attach the white electrode lead to the right forearm, the red lead to the left ankle, and the black lead to the right ankle.
4. Start the Biopac Student Lab System, choose Lesson 5 (**L05-ECG-1**), and click **OK.** When prompted, enter a unique filename and click **OK.**
5. Click on the **Calibrate** button and wait while the computer adjusts for optimal recording. The calibration procedure will automatically stop after 8 seconds.
6. If there is a small ECG waveform with a relatively flat baseline, then calibration was successful. If not, click on the **Redo Calibration** button, check all connections, and repeat step 5.

Recording Data

In this activity, you will record the ECG from your lab partner under four conditions: lying down, after sitting up, while breathing deeply, and after exercise. In each case, it is important that the actual recording be done while your partner is physically still. Electrical activity from muscular movements, like that recorded in an EMG, can corrupt the ECG signal.

1. With your lab partner lying comfortably, click on the **Record** button.
2. Wait 20 seconds and then click on the **Suspend** button. If the data do not appear correct or have significant baseline drift, then click on the **Redo** button and repeat this recording segment.
3. Instruct your lab partner to quickly sit upright. Immediately click on the **Resume** button as soon as he or she is sitting fully upright. A marker labeled **After sitting up** will be automatically inserted into the recording.
4. Record for 20 seconds and then click on the **Suspend** button. If the data do not appear correct, then click on the **Redo** button and repeat this recording segment.
5. With your partner remaining seated, click on the **Resume** button. A marker labeled **Deep breathing** will be automatically inserted into the recording.
6. Record for 20 seconds and then have your lab partner take five long, slow, deep breaths. You should insert markers at the beginning of each inhale and at the subsequent exhale. Markers are inserted by pressing the **Esc** key (Mac) or **F9** key (PC).
7. Click on the **Suspend** button and inspect the data. Deep breathing may cause some baseline drift. As long as it is not excessive, you do not need to redo the recording. If the data do not appear correct, then click on the **Redo** button, check the electrodes and connections, and repeat this recording segment.
8. Have your lab partner perform an exercise, such as push-ups, jumping jacks, or running in place to raise the heart rate. The exercise period should last for about 3 minutes. Be careful that the electrode leads are not pulled loose and are not obstructing your partner's movements. If necessary, you may remove the leads, but do not remove the electrodes. If you do remove the leads, you must reconnect them properly immediately after the exercise is complete.

   Warning: *Anyone who is physically or medically unable to perform physical exercise should not participate in this portion of the laboratory activity.*

9. Once your lab partner is seated, immediately click on the **Resume** button to capture the ECG while he or she is recovering from exercise.
10. Record for 60 seconds and click on the **Suspend** button. Inspect the data. Some baseline drift is normal. As long as it is not excessive, click the **Done** button. If the data do not appear correct, click the **Redo** button, check the electrodes and connections, and repeat this segment.

Data Analysis

1. Enter the **Review Saved Data** mode and select the appropriate data file for Lesson 5, which ends in-LO5.

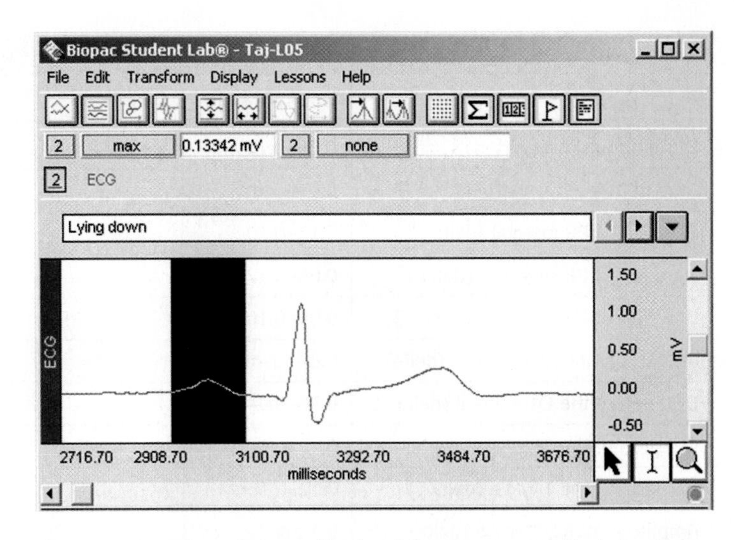

Figure 22.7 Measuring amplitude of the P wave. Amplitude is measured by computing the difference in values between the last and first points of the selected area.

2. Set three channel measurement boxes as follows:

   Ch 2 delta T

   Ch 2 delta

   Ch 2 BPM

3. Measure the duration and amplitude of various components of an ECG. The duration is calculated using the **delta T** measurement. This measurement computes the difference in time between the ending and beginning points of the area selected by the I-beam cursor tool. The amplitude (mV) of an ECG component is calculated using the **delta** measurement.
   a. When you use the I-beam cursor tool to select the peak and adjacent baseline of an ECG component, the measurement computes the difference in values between the last and first points of the selected area (Figure 22.7).
   b. In order to accurately select the proper regions of the ECG, you will need to use the magnifying glass tool to zoom in on the desired heart cycles. Then, if necessary, from the **Display** menu, use **Autoscale waveforms** to scale and position the ECG for optimal analysis.
   c. Refer to Figure 22.5 and use the I-beam cursor tool to select the appropriate regions from three different complete heart cycles in Segment 1 (at rest, lying down). Calculate the mean and record the results in Table 22.4.

QUESTION TO CONSIDER Do the wave durations and amplitudes fall within normal ranges? If any are abnormal, what mechanical or electrical events are occurring in the heart at this time?

Note: *Accurately interpreting ECGs requires significant training and practice. A trained health care professional is best able to*

Table 22.4 Components of the ECG at Rest (Segment 1)

Component	Normal range	Cycle 1	Cycle 2	Cycle 3	Mean
Duration of P wave (delta T)	0.06–0.12 s				
Amplitude of P wave (delta)	0.1–0.3 mV				
Duration of PR interval (delta T)	0.12–0.20 s				
Duration of PR segment (delta T)	0.06–0.12 s				
Duration of QRS complex (delta T)	0.06–0.10 s				
Amplitude of QRS complex (delta)	0.8–1.2 mV				
Duration of the QT interval (delta T)	0.36–0.44 s				
Duration of the ST segment (delta T)	0.12 s				
Duration of the T wave (delta T)	0.12–0.16 s				
Amplitude of the T wave (delta)	0.3 mV				

determine what abnormalities are due to normal variation, experimental noise, and medical conditions. Do not be alarmed if your ECG is different from those illustrated or from the normal values in the table.

4. Measure the duration (delta T) of the cardiac cycle. This measurement reflects the amount of time between heartbeats. The computer can automatically convert this measurement to beats per minute (BPM) by dividing the delta T value by 60 seconds. **BPM** will be displayed in the appropriate channel measurement box.
   a. In Segment 1, use the I-beam cursor tool to select a single cardiac cycle from R wave peak to R wave peak (Figure 22.8). Record the duration (delta T) and heart rate (BPM) in Table 22.5. Repeat for two additional cycles in Segment 1 and calculate the mean.
   b. Repeat step (a) for each of the remaining three recording segments and record your results in Table 22.5. In Segment 3, **Deep Breathing**, select three cycles that occurred during inspirations and then repeat on three cycles that occurred during expirations.

QUESTIONS TO CONSIDER 1. How does the heart rate (BPM) change during each of the four experimental conditions? Describe the physiological mechanisms causing these changes.

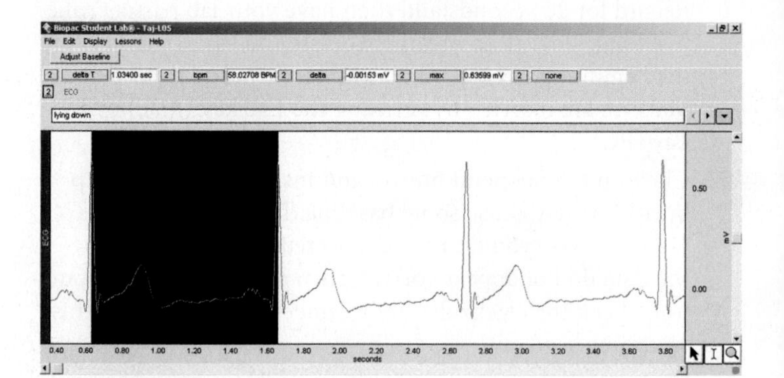

Figure 22.8 Measuring beats per minute. Heart rate is measured by computing the change in time between the last and first points of the selected area.

2. How does the duration (delta T) of the cardiac cycle change during inspiration and expiration? What causes this change?

5. Measure changes in the duration of ventricular systole and diastole that occur during exercise. The **QT interval** is defined as the period from the Q wave to the end of the T wave. It corresponds to ventricular systole. Ventricular diastole, then, is measured from the end of the T wave to the subsequent R wave.
   a. Transfer the data you recorded in Table 22.4 for the QT interval into the appropriate cells of Table 22.6 for Segment 1.

Table 22.5 Changes in Heart Rate and Duration of Cardiac Cycle

Segment	Measurement	Cardiac cycle			Mean	Range
		1	2	3		
1—Resting, lying down	Delta T					
	BPM					
2—Sitting up	Delta T					
	BPM					
3—Seated, inspiration	Delta T					
	BPM					
3—Seated, expiration	Delta T					
	BPM					
4—After exercise	Delta T					
	BPM					

Table 22.6 Changes in Duration (delta T) of Ventricular Systole and Diastole

Segment	Measurement	Cardiac cycle			Mean
		1	2	3	
1—Resting, lying down	QT interval (ventricular systole)				
	End of T to subsequent R (ventricular diastole)				
4—After exercise	QT interval (ventricular systole)				
	End of T to subsequent R (ventricular diastole)				

b. Scroll to the ECG waveform from Segment 1. Using the I-beam cursor tool, select the region corresponding to ventricular diastole (from the end of T to next R) and record the delta T measurement. Repeat for two additional cycles.

c. Scroll to Segment 4, **After Exercise**, and measure the duration of ventricular diastole and systole from three cardiac cycles. Record the results in Table 22.6 and calculate the means.

QUESTION TO CONSIDER How does the duration (delta T) of ventricular systole and diastole change from resting to after exercise?

____

____

____

Exercise 22 Review Sheet

Cardiovascular Physiology

Name ____________________

Lab Section ____________________

Date ____________________

1. Describe the physical basis for the first ("lub") and second ("dup") heart sounds.

2. Why is it important for the walls of large arteries to have an abundant supply of elastic fibers?

3. What is meant by systolic pressure and diastolic pressure?

4. What is the pulse pressure? How is this value used as a diagnostic tool?

5. Calculate the mean arterial pressure of an individual with a blood pressure of 115/70.

Questions 6–11: Define the following terms:

6. Cardiac cycle

7. Hypertension

8. Pulse

9. Auscultation areas for the heart

10. Sphygmomanometer

___

___

11. Korotkoff's sounds

___

___

12. Identify the components of the cardiac conduction system and describe their function.

___

___

13. What is an electrocardiogram (ECG)? Correlate the various wave patterns on a normal ECG with events that occur during the cardiac cycle.

___

___

Exercise 23

The Lymphatic System

Laboratory Objectives

On completion of the activities in this exercise, you will be able to:

- List the main functions of the lymphatic system.
- Explain the functional relationship between the lymphatic system and the cardiovascular system.
- Identify the locations of lymphatic organs and nodules in the human body.
- Describe the anatomical relations of lymphatic organs and nodules to adjacent structures.
- Describe the microscopic structure of important lymphatic structures.
- Describe the structure of an antibody molecule.
- Use immunodiffusion (ID) to observe an antigen–antibody reaction.

Materials

- Anatomical models
  - Midsaggital section of the head and neck
  - Human torso
- Compound light microscopes
- Prepared microscope slides
  - Thymus
  - Spleen
  - Lymph node
  - Tonsils
  - Small intestine (ileum)
  - Lymphatic vessels
- Petri dishes divided into three compartments
- Pipettes with rubber bulbs
- Micropipettes with rubber bulbs
- Beakers
- Graduated cylinders
- Hot plates
- Wax marking pencils
- 1% agar, containing 8% sodium chloride (NaCl)
- Animal sera containing natural concentrations of albumins
  - Bovine (cow) serum
  - Horse serum
  - Swine (pig) serum
- Goat antisera containing antibodies for albumins
  - Goat anti-bovine albumin
  - Goat anti-horse albumin
  - Goat anti-swine albumin
- Physiological saline

The lymphatic system consists of an extensive network of **lymphatic vessels** and various **lymphatic organs** and **lymphatic nodules** distributed throughout the body. Functionally, this system is closely connected to the cardiovascular system in the following ways.

- Lymphatic capillaries provide drainage for accumulating fluid that leaks from blood capillaries into the surrounding tissue spaces. This fluid, known as **lymph**, is similar in composition to blood plasma, and contains many **lymphocytes** and smaller numbers of other white blood cells such as neutrophils and monocytes. Lymph is transported along a system of lymphatic vessels and is ultimately returned to the bloodstream.
- After a meal, dietary lipids and lipid-soluble vitamins are absorbed by the small intestine and packaged within vesicles called chylomicrons. Lymphatic vessels transport the chylomicrons, and the fats they contain, to the blood circulation.
- Lymphocytes are found in both the lymphatic and cardiovascular systems. These cells initiate specific immune responses that provide protection for the body against disease and infection.

WHAT'S IN A WORD The word *lymph* is derived from the Latin word *lympha,* which means "clear spring water." Typically, lymph is a clear fluid, although in the small intestine, it has a milky appearance due to the high fat content.

Gross Anatomy of the Lymphatic System

An overview of the lymphatic system is illustrated in Figure 23.1. The system includes the following components.

- Lymphatic organs are organized areas of lymphoid tissue surrounded by a connective tissue capsule. These structures include the **lymph nodes**, **spleen**, and **thymus**.
- Lymphatic nodules are concentrated regions of lymphoid tissue that lack a connective tissue capsule. They are found in the walls of the respiratory, digestive, and urinary tracts. Examples include the **tonsils** in the wall of the pharynx, **aggregated lymphoid nodules (Peyer's patches)** in the wall of the small intestine, and the nodules located in the **appendix**.
- The **lymphatic circulation**, which transports lymph, includes **lymphatic capillaries**, **lymphatic collecting vessels**, **lymphatic trunks**, and **lymphatic ducts**.

ACTIVITY 23.1 Examining the Gross Anatomical Structure of the Lymphatic System

Lymphatic Organs

1. Obtain a torso model and remove the anterior body wall to expose the organs in the thoracic and abdominopelvic cavities.
2. Identify the following lymphatic organs.

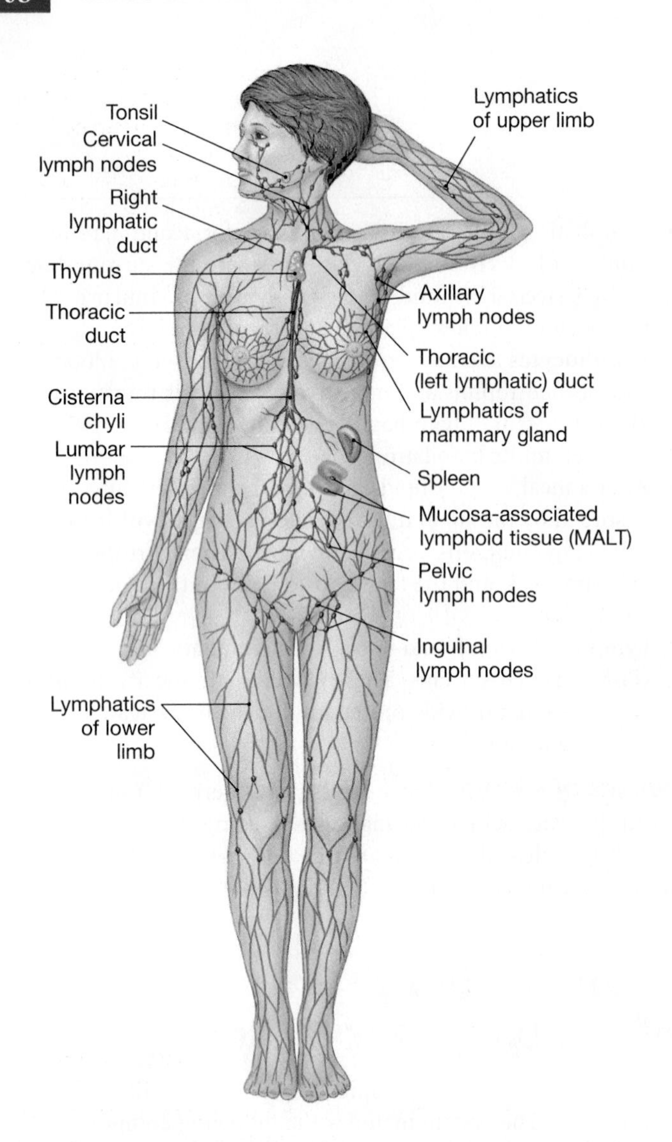

Figure 23.1 Overview of the lymphatic system. Components of the lymphatic system include lymphoid organs, lymphoid nodules, and lymphatic vessels.

- The thymus gland is a bilobed organ located in the mediastinum. If present on the models in your lab, observe how it covers the superior portion of the heart and the great vessels. Also notice that it extends superiorly into the base of the neck, where it blankets the inferior end of the trachea (Figure 23.2a). As you learned earlier when you studied the endocrine system (Exercise 18), the thymus produces a group of hormones called **thymosins** (thymic hormones). Lymphoid stem cells from bone marrow migrate to the thymus and, under the influence of these hormones, become mature **T lymphocytes (T cells)**. From the thymus, mature cells are transported by the bloodstream to the spleen, lymph nodes, and other lymphoid structures, thus establishing T-cell populations throughout the body. These populations are vital for coordinating the body's immune response against diseases and infections. The thymus is most active during childhood and adolescence. Consequently, it is relatively large in newborns and young children. Shortly after puberty, however, the organ becomes less active and begins to atrophy. As the thymus progressively degenerates, glandular tissue is gradually replaced by fat and fibrous connective tissue.

WHAT'S IN A WORD The term *thymus* is the Latin word for *sweetbread*. If you go to a restaurant and order sweetbread from the menu, you will probably be eating the thymus gland.

CLINICAL CORRELATION

The primary target for the **human immunodeficiency virus (HIV),** the virus that causes AIDS, is a type of T lymphocyte called a **T-helper cell.** These cells are needed to stimulate the actions of all other immune functions. As HIV disease progresses to AIDS, the T-helper cell population declines and the body's immune system is suppressed. As a result, an individual's ability to fight off infections (opportunistic infections) is compromised. AIDS-related deaths occur when individuals succumb to infections that a suppressed immune system can no longer control.

- Locate the spleen in the upper left quadrant of the abdominal cavity. It is posterolateral to the stomach and is positioned between ribs 9 and 11 along the posterior body wall (Figures 23.1 and 23.3a). Remove the stomach to expose the anterior or visceral surface of the spleen. Along the visceral surface, locate the **hilus**, an area where blood vessels, lymphatic vessels, and nerves enter and exit (Figure 23.3b). With the stomach removed, notice that the spleen is just lateral to the left kidney and the tail of the pancreas (Figure 23.3a).

  The spleen performs the following important functions that define its close relationship to the blood.

  - It acts as a "blood filter." Macrophages and monocytes in the spleen will attack and eliminate bacteria, viruses, or other infectious agents that are present in the blood. In addition, lymphocytes are activated to begin a specific immune response.
  - Splenic macrophages break down old red blood cells and platelets, and the components are ultimately recycled.
  - It produces lymphocytes in the adult and red blood cells during fetal development.
  - It serves as a blood reservoir by storing red blood cells and platelets that can be added to the circulation when needed.

- On a torso model, identify lymph nodes in the axillary or inguinal regions. Notice how they are positioned along the course of lymphatic collecting vessels, like beads on a string (Figures 23.1 and 23.4a). Lymph nodes are ovoid structures that are typically less than 2.5 cm in length. They are found deep and superficially throughout the body, but are aggregated into clusters in specific body regions such as the axilla (axillary lymph nodes) and groin (inguinal lymph nodes). Lymph is constantly passing through lymph nodes. As the fluid percolates through the network of lymph sinuses, macrophages attack any foreign microorganisms that might be present. In addition,

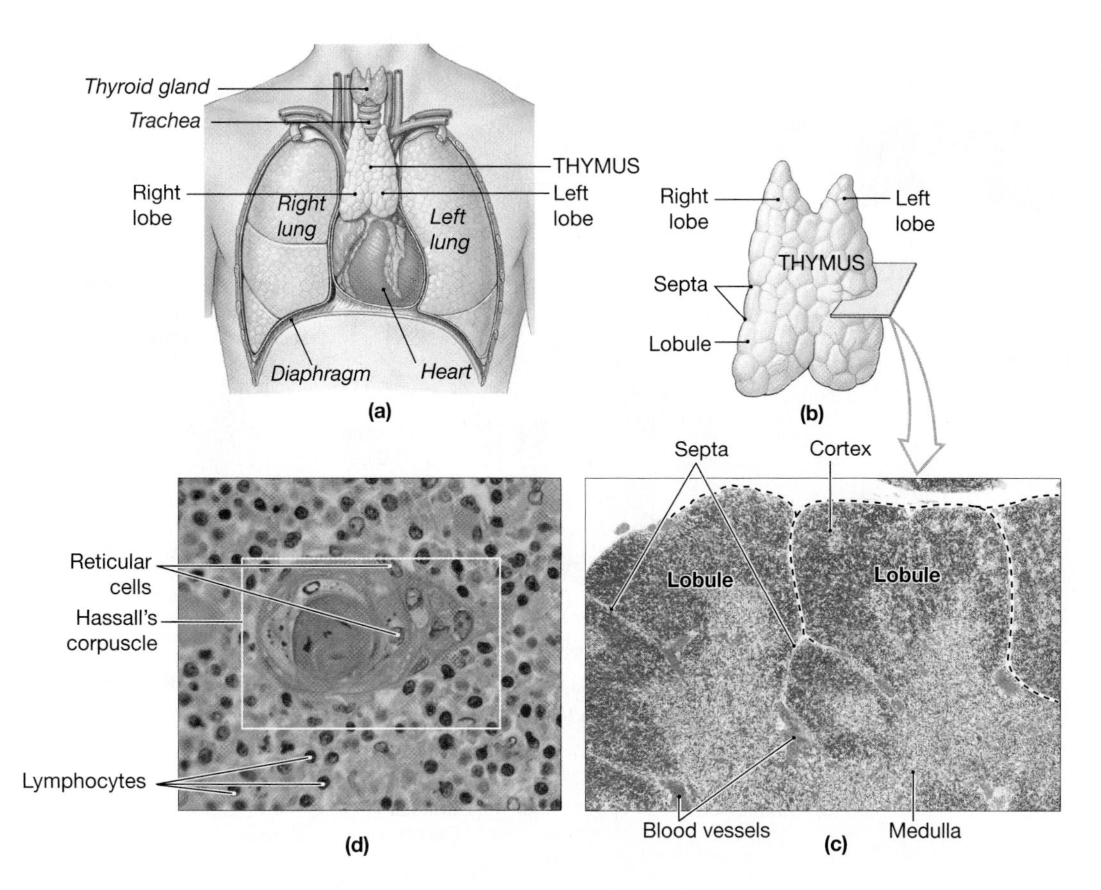

Figure 23.2 Anatomy of the thymus. a) Position of the thymus in relation to the heart and trachea; **b)** the thymus, in isolation, illustrating its bilobed appearance; **c)** low-power light micrograph illustrating the microscopic structure of the thymus (LM × 40); **d)** high-power light micrograph showing a Hassall's body surrounded by lymphocytes (LM × 400).

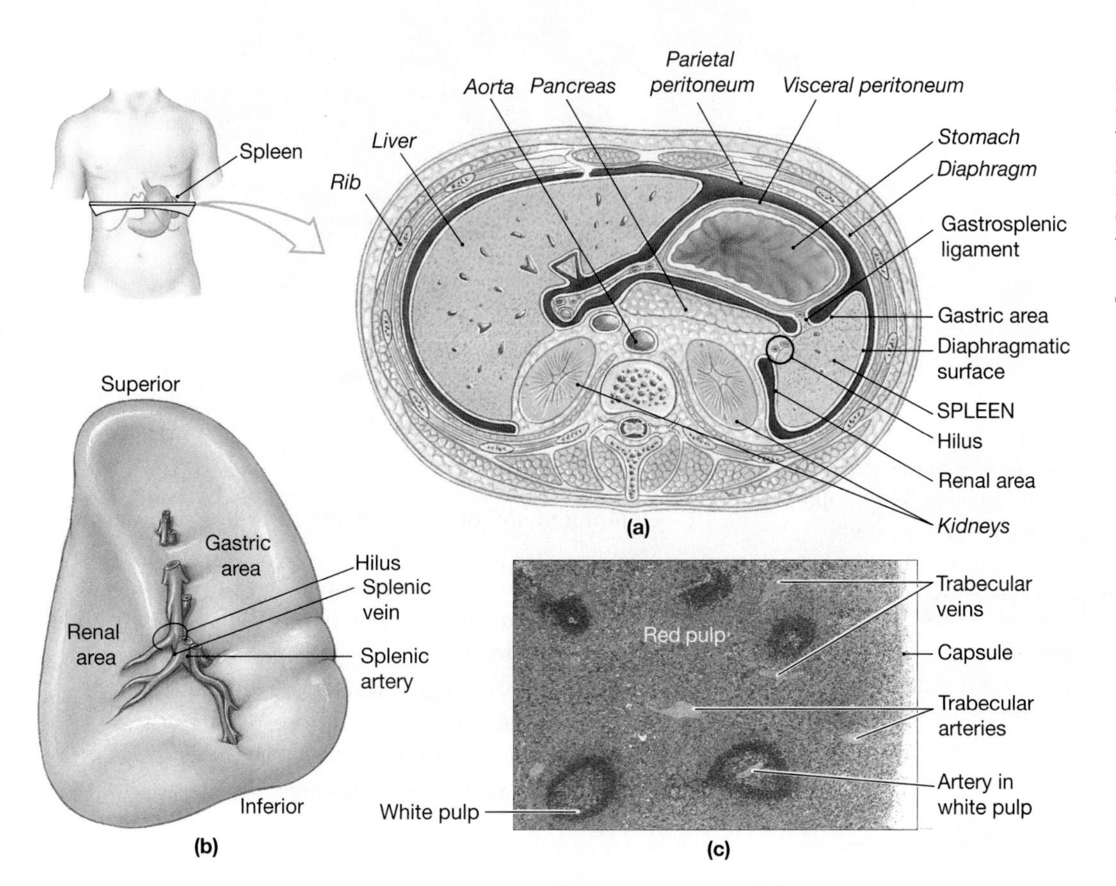

Figure 23.3 Anatomy of the spleen. a) Cross section of the abdominal cavity showing the position of the spleen in relation to adjacent organs; **b)** visceral surface of the spleen showing the location of the hilus; **c)** light micrograph of the spleen illustrating areas of white and red pulp (LM × 40).

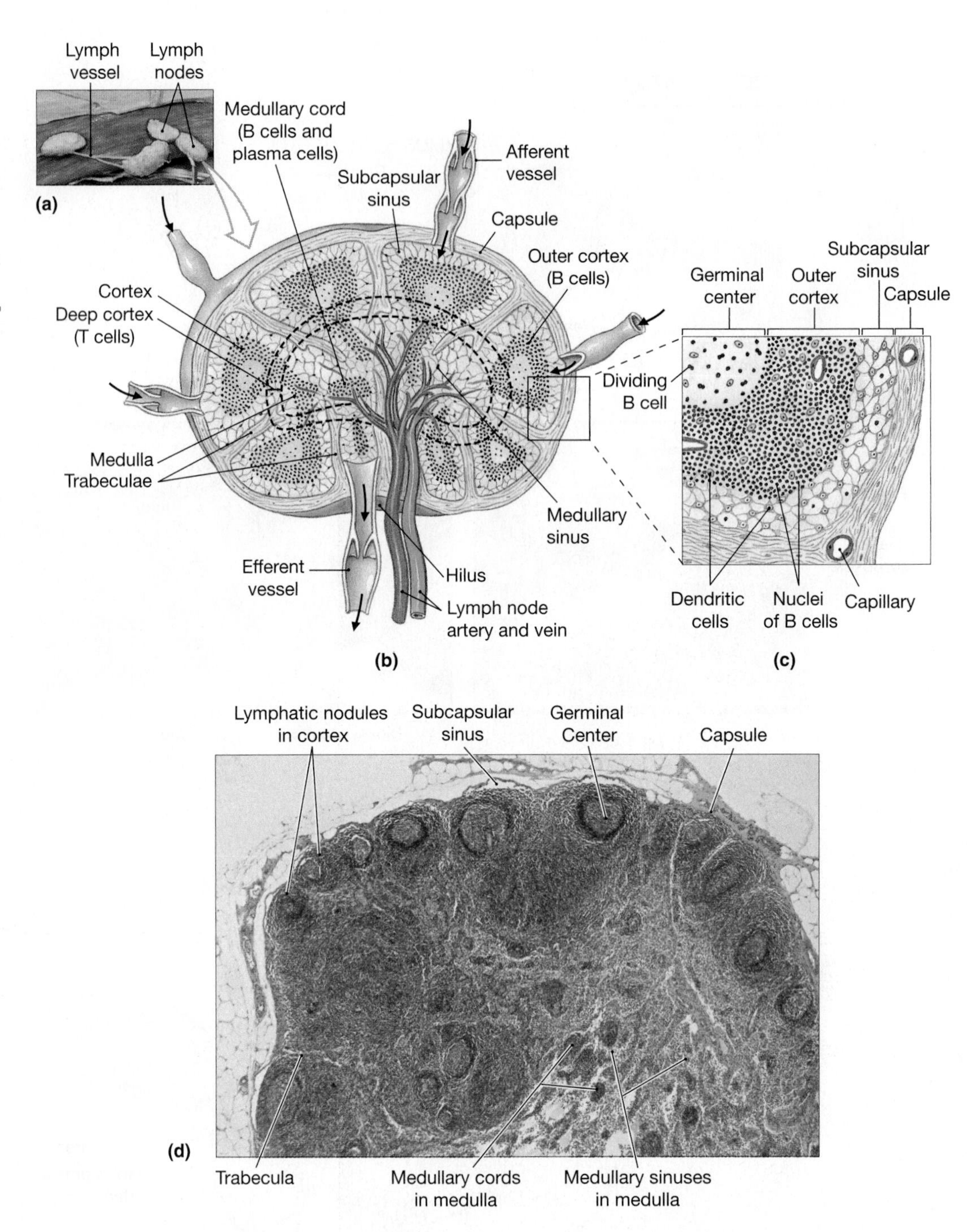

Figure 23.4 Structure of a lymph node. a) Lymph nodes positioned along the course of lymphatic vessels. **b)** Illustration of the microscopic structure of a lymph node. **c)** Illustration of the area enclosed by the box in (b). A closer view of the cortex with part of a germinal center is shown. **d)** Low-power light micrograph of a lymph node illustrating structures in the cortex and medulla (LM × 40).

T lymphocytes are activated to initiate an immune response. Thus, lymph nodes serve as efficient lymph filters, much like the spleen serves as a blood filter.

CLINICAL CORRELATION

Lymphadenopathy refers to an enlargement or swelling of lymph nodes. It can be caused by a number of diseases, and in most cases, the swelling declines when the illness ends. However, for people who are HIV positive, lymph nodes can remain swollen for several months even if there are no other indications or symptoms of a disease.

Lymphatic Nodules

1. Obtain a model of the head and neck in a midsagittal section.
2. Identify the tonsils, which form a discontinuous ring of lymphoid tissue near the junction of the oral cavity and pharynx (Figure 23.5a).
   - The single **pharyngeal tonsil (adenoid)** is located along the posterior wall of the most superior region of the pharynx, known as the nasopharynx.
   - The two **lingual tonsils** are situated at the base of the tongue.

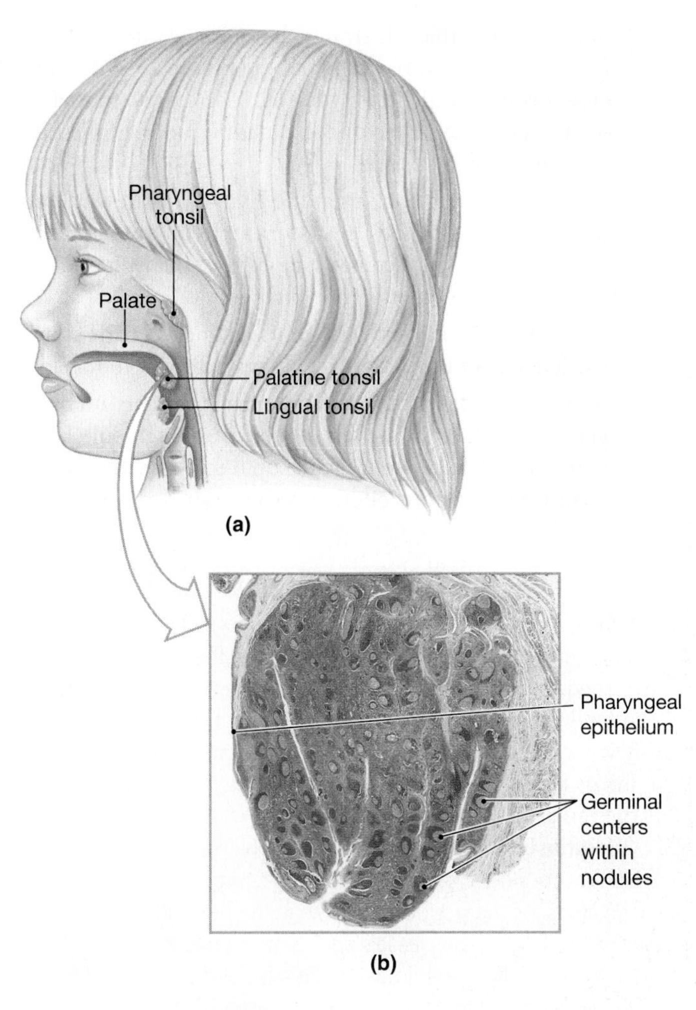

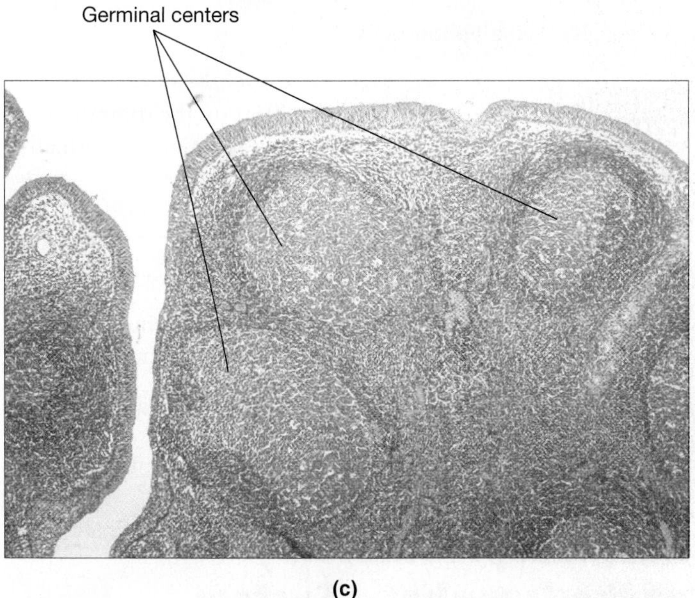

Figure 23.5 Location and structure of the tonsils. a) Illustration showing the positions of the tonsils in the pharynx; **b)** low-power light micrograph of a tonsil (LM × 40); **c)** high-power light micrograph illustrating germinal centers in a tonsil (LM × 100).

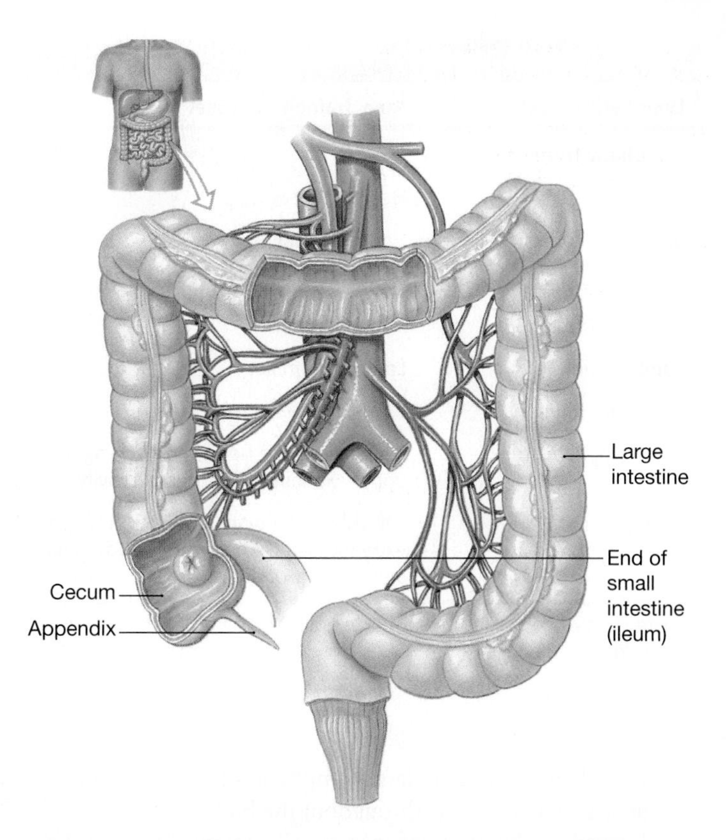

Figure 23.6 Location of the appendix. The appendix extends from the wall of the cecum, near the junction of the small and large intestines.

- The two **palatine tonsils** are found along the lateral walls of the pharynx near the posterior end of the palate.
- Observe how the tonsils are strategically positioned to fight off infections that develop in the upper respiratory and digestive tracts.

3. On a torso model, locate the junction between the small and large intestines in the lower right quadrant of the abdominopelvic cavity (Figure 23.6).
4. The first part of the large intestine is called the cecum. Identify the wormlike appendage, known as the appendix, extending from the wall of the cecum (Figure 23.6). The appendix contains aggregated lymphoid nodules, similar to the tonsils.
5. Other locations of aggregated lymphoid nodules include the walls of the small and large intestines and the bronchial passageways in the lungs.

Lymphatic Circulation

1. On a torso model, identify networks of lymphatic collecting vessels in the axillary or inguinal regions. As you observed earlier, lymph nodes are distributed along the course of these vessels. Lymphatic collecting vessels receive lymph from lymphatic capillaries, which drain excess fluid released from blood capillaries.

Table 23.1 Major Lymphatic Trunks and Ducts

Lymphatic vessel	Area drained by vessel
Lymphatic trunks*	
Jugular trunks	Head and neck
Subclavian trunks	Upper extremities
Bronchomediastinal trunks	Thoracic cavity
Intestinal trunk	Abdominal cavity
Lumbar trunks	Lower extremities
Lymphatic ducts	
Right lymphatic trunk	Right side of the head and neck, right upper extremity, right thoracic cavity
Thoracic trunk	Left side of the head and neck, left upper extremity, left thoracic cavity, abdominal and pelvic cavities, lower extremities

* There is only one intestinal trunk. All the other trunks occur in pairs—one on each side of the body.

2. Lymphatic trunks are large lymphatic vessels, positioned at strategic locations throughout the body. They are formed by the union of several collecting vessels and drain the lymph from a specific region of the body. Review the regions that are drained by each trunk (Table 23.1). Identify three structures from which the following lymphatic trunks will receive lymph.
   - Intestinal trunk

     ____
   - Jugular trunk

     ____
   - Lumbar trunk

     ____
   - Bronchomediastinal trunk

     ____
   - Subclavian trunk

     ____

3. The lymphatic trunks unite to form one of two **lymphatic ducts**, which drain lymph into the venous circulation. On each side of a torso model, find the location where the subclavian and internal jugular veins join to form the brachiocephalic vein (Figure 23.7b). On the right side, the **right lymphatic duct** empties into the bloodstream near this venous junction (Figure 23.7b). This duct drains lymph from the upper right quadrant of the body (Figure 23.7a; Table 23.1). The much larger **thoracic duct** has a similar drainage pattern on the left side (Figure 23.7b). It drains lymph from all other body regions (Figure 23.7a; Table 23.1).

4. Obtain a model that illustrates the thoracic duct and trace its path along the vertebral column (Figure 23.7b). If such a model is not available in your lab, use a torso model and trace the position of the duct by identifying the following key anatomical structures.
   - The thoracic duct begins at the level of the L2 vertebra, as a dilated sac called the **cisterna chyli**. Notice that the cisterna chyli receives lymph from the right and left lumbar trunks and the intestinal trunk.
   - From the cisterna chyli, the thoracic duct ascends along the left side of the vertebral column, adjacent to the aorta.
   - At about the level of the clavicle, the thoracic duct receives lymph from the left jugular trunk, left subclavian trunk, and left bronchomediastinal trunk. It then empties into the venous circulation as described earlier.

CLINICAL CORRELATION

If there is a blockage of normal lymph drainage, it can cause swelling due to the accumulation of fluids in the tissue spaces. This condition, referred to as **edema,** can be caused in a variety of ways. If an individual wears tight clothing or sleeps in a certain position, edema will be temporary. Surgical procedures that result in the destruction of lymphatic vessels or the formation of scar tissue can cause a more chronic edema.

Filiariasis is an infection of the lymphatic system caused by a parasitic roundworm. Roundworm larvae are transmitted by mosquitoes. As the population grows in the host, adult roundworms block lymphatic vessels and lymph nodes, causing edema, especially in the limbs. Progressive and extreme edema of the limbs and external genitalia, as a result of this parasitic infection, is called **elephantiasis.**

QUESTIONS TO CONSIDER

1. What would be the effect on the immune system if the thymus was not functioning normally during childhood?

____

____

2. Why is it important to have an abundance of lymphatic nodules located along the respiratory and digestive tracts?

____

____

Microscopic Anatomy of the Lymphatic System

Lymphoid tissue is a specialized form of loose connective tissue that consists of a dense network of **reticular fibers** and a variety of cell types that are largely involved in protecting the body from infection and disease. Lymphocytes are the principal cells found in lymphoid tissue. T lymphocytes are responsible for the **cell-mediated immune response**. They directly attack and destroy

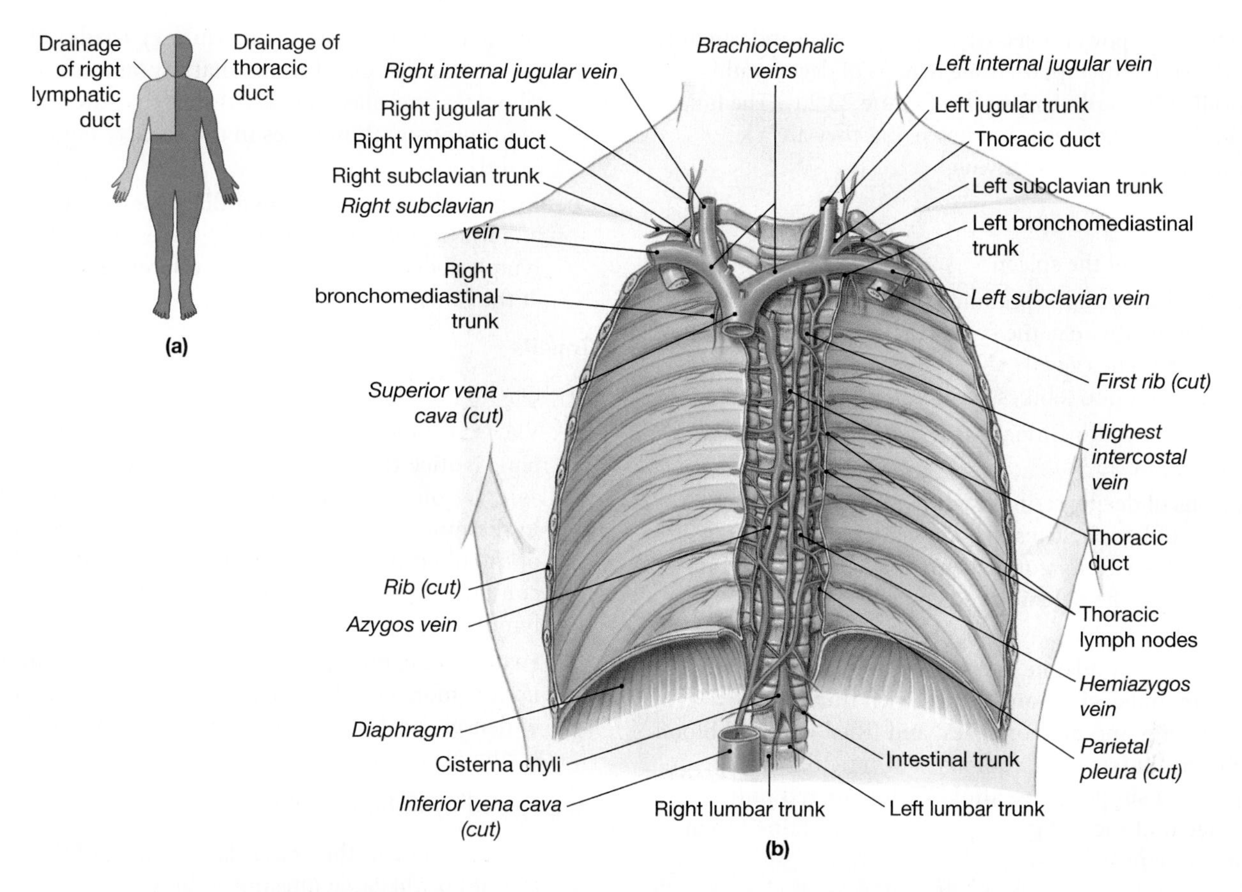

Figure 23.7 Lymph drainage into the venous circulation. a) Illustration of the lymphatic drainage areas for the two lymphatic ducts. The right lymphatic duct drains lymph from the upper right quadrant of the body. The much larger thoracic duct drains lymph from the upper left quadrant and all regions inferior to the diaphragm. **b)** Diagram of the posterior thoracic wall showing the drainage of lymph into the bloodstream. On the right side of the body, the right lymphatic duct empties into the venous circulation near the junction of the subclavian and internal jugular veins. The thoracic duct follows a similar pattern on the left side.

foreign microorganisms by phagocytosis or by releasing chemicals. One type of T lymphocyte, the **T-helper cell**, activates all immune activity. **B lymphocytes** initiate the **humoral immune response** by differentiating into **plasma cells**, which produce **antibodies**. As they circulate through the blood and lymph, antibodies bind to and destroy foreign **antigens**, such as disease-causing microorganisms, certain foods and drugs, or transplanted organs. Other cells found in lymphoid tissue include **macrophages** that attack foreign cells by phagocytosis and activate T lymphocytes, **reticular cells** that produce the reticular fibers, and **dendritic cells** that also play a role in T-lymphocyte activation.

CLINICAL CORRELATION

Even when the tissue match between donor and recipient is very close, a transplanted organ is recognized as foreign tissue and it will induce an immune response. To reduce the risk of tissue rejection, patients who receive an organ transplant must take drugs that suppress the immune system. The danger of taking these immunosuppressant drugs is that it increases an individual's vulnerability to disease and infection.

ACTIVITY 23.2 Examining the Microscopic Structure of Lymphatic Structures

Thymus

1. Obtain a slide of the thymus.
2. View the slide with the scanning or low-power objective lens. Notice that the thymus is covered by a connective tissue capsule. The capsule gives rise to connective tissue septa that subdivide each lobe (Figure 23.2b) of the thymus into lobules (Figure 23.2c). Each lobule consists of a dark-staining outer **cortex** with a dense population of lymphocytes, and a light-staining inner **medulla** with far fewer lymphocytes (Figure 23.2c).
3. Use the high-power objective lens to examine the thymus more closely, and identify the dense aggregations of T lymphocytes. In a standard staining preparation, these cells will have round, deeply stained nuclei, surrounded by a lightly stained cytoplasm. Other cells, called **thymocytes**, produce the thymic hormones. They will be difficult to find.

4. Keep the high-power objective lens in position and carefully scan the slide. Attempt to locate regions of degenerating cells called **Hassall's corpuscles** (Figure 23.2d). The function of these structures is unknown, but they are a key identifying feature of the thymus.

Spleen

1. Obtain a slide of the spleen.
2. View the slide with the scanning or low-power objective lens. Like the thymus, the spleen is surrounded by a thin connective tissue capsule with trabeculae (septa) that divide the organ into lobules.
3. Identify the two functional tissue components of the spleen (Figure 23.3c).
   - Regions of deeply staining **white pulp** are compact masses of lymphocytes that are scattered throughout the spleen. They form around **central arteries**, which are branches of the **splenic artery**.
   - Most of the spleen consists of **red pulp**, which completely surrounds the areas of white pulp. The red pulp includes **sinusoids**, lined with numerous monocytes and macrophages, and filled with red blood cells. **Splenic cords**, composed of reticular tissue, provide a supportive scaffolding for the red pulp. Notice that the red pulp is more lightly stained than the white pulp.
4. Move the slide so that an area of white pulp is in the center of the field of view. Switch to the high-power objective lens to examine the region more closely. Identify the dense aggregations of lymphocytes. As in the thymus, these cells can be detected by their deeply stained nuclei.
5. With the low-power objective lens in place, move the slide to an area of red pulp. Switch back to high power and observe the vast network of sinusoids. Most of the important functions of the spleen occur in the red pulp as blood flows through the sinusoids.

Lymph Node

1. Obtain a slide of a lymph node.
2. View the slide with the scanning or low-power objective lens. Locate the connective tissue capsule that surrounds the entire node and the trabeculae that extend inward and form partitions (Figures 23.4b and d). Just beneath the capsule, identify a **subcapsular sinus. Afferent lymphatic vessels** deliver lymph to a node by draining into these sinuses (Figure 23.4b).
3. Like the thymus, lymph nodes contain two distinct regions: the outer cortex and the inner medulla (Figures 23.4b and d). Identify these two regions.
4. In the cortex, identify the lymphatic nodules. Each nodule is an aggregation of lymphocytes. In the center are the lighter staining **germinal centers**, where new lymphocytes are produced (Figures 23.4b, c, and d).
5. In the medulla locate the **medullary cords**. These structures are columns of tissue that contain masses of lymphocytes and plasma cells. Notice that the cords are continuous with nodules in the cortex (Figures 23.4b and d).
6. Identify the network of **medullary sinuses** (Figures 23.4b and d). Lymph traveling through these sinuses exits a lymph node by flowing into an **efferent lymphatic vessel** at the hilus (Figure 23.4b).

Tonsils

1. Obtain a slide of a tonsil.
2. View the slide with the scanning or low-power objective lens. Notice that the tonsil lacks a connective tissue capsule. It consists of a mass of lymphoid tissue embedded in the mucous membrane and covered by the epithelium of the pharyngeal wall (Figure 23.5b). Notice that it contains numerous lymphatic nodules, similar to a lymph node.
3. Switch to the high-power objective lens to examine the nodule more closely. Notice that it contains an outer ring of densely packed lymphocytes and a centrally located germinal center (Figure 23.5c).

Aggregated Lymphoid Nodules

1. Obtain a slide of the ileum (last segment of the small intestine) or the large intestine (colon).
2. View the slide with the scanning or low-power objective lens. Scan the slide and locate large masses of deeply stained lymphocytes. These regions are called aggregated lymphoid nodules (Figure 23.8). In the ileum, these structures are called Peyer's patches.

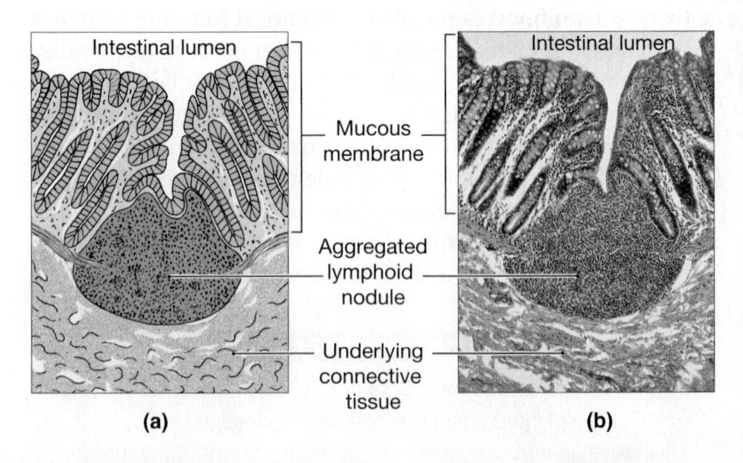

Figure 23.8 Microscopic structure of aggregated lymphoid nodules. **a)** Illustration of a lymphoid nodule in the intestinal wall; **b)** corresponding light micrograph (LM × 20).

3. Aggregated lymphoid tissues are structurally similar to tonsils. Notice that they lack a connective tissue capsule and that many display a pale-staining germinal center. Aggregated lymphatic nodules are also located in the appendix and the bronchial tubes.

Lymphatic Collecting Vessels

1. Obtain a slide of a lymphatic collecting vessel with valves.
2. View the slide with the scanning or low-power objective lens. Lymphatic collecting vessels often travel with arteries and veins (Figure 23.9a). They are structurally similar to veins, but their walls are thinner and contain a smaller amount of smooth muscle.
3. Locate the position of a valve in the lymphatic vessel. The two leaflets or cusps of the valve are more deeply stained than the vessel wall and resemble the letter V. Lymph travels in one direction. Actions of the valves prevent lymph from flowing back toward lymphatic capillaries.
4. Examine the valve more closely with the high-power objective lens. Notice that the valve leaflets are continuous with the wall of the lymphatic vessel.

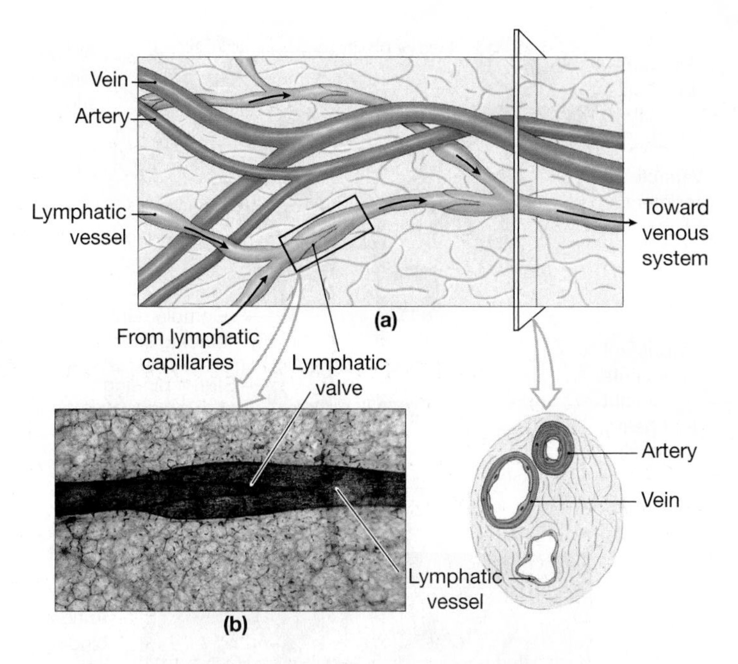

Figure 23.9 Lymphatic collecting vessels. a) Illustration showing relationship of lymphatic collecting vessels with small arteries and veins. The accompanying cross section compares the relative thickness of the walls of an artery, vein, and lymphatic vessel. **b)** Light micrograph of a lymphatic collecting vessel, illustrating the position of a valve (LM × 40).

QUESTIONS TO CONSIDER 1. Based on your microscopic observations in the previous activity, identify structural similarities and differences in the various lymphatic structures. Focus your attention on the arrangement and structure of the lymphoid tissue in each structure.

Similarities:

____

____

Differences:

____

____

2. From your microscopic observations of the spleen, you know that the white pulp does not appear white. Why do you think these splenic regions are called "white" pulp?

____

____

3. In the previous activity, you observed lymphatic valves, which prevent the backflow of lymph. Why is it important that lymph not flow back toward lymphatic capillaries?

____

____

The Antigen-Antibody Reaction

Antibodies are protein molecules known as **immunoglobulins (Ig)** that act to eliminate foreign tissues, cells, or microbes, known as antigens. The typical antibody is a Y-shaped molecule that consists of two large polypeptides, the **heavy chains**, and two small polypeptides, the **light chains** (Figures 23.10a and b). The heavy and light chains contribute to the formation of one **constant segment** and two **variable segments.** The constant segment comprises the base and the first part of each arm on the antibody molecule (Figure 23.10a). The structure of these regions is used as one criterion to categorize antibodies into five distinct classes: IgG, IgE, IgD, IgM, and IgA. Within each class, the amino acid sequences in the constant segments are identical.

The variable segments are located at the ends of each arm on the antibody molecule, and contain the **antigen binding sites** (Figure 23.10a). The changeable structure of these regions is the principal reason for the specificity of the **antigen-antibody reaction.** During an antigen-antibody reaction, antigen binding sites on the antibody bind to **antigenic determinant sites** on the antigen, forming an **antigen-antibody complex** (Figure 23.10c). For each type of antigen that gains entrance into the body, a unique antibody is produced to fight it off. The high degree of specificity is due to the three-dimensional molecular structure at the binding sites that requires a precise fit between antigen and antibody.

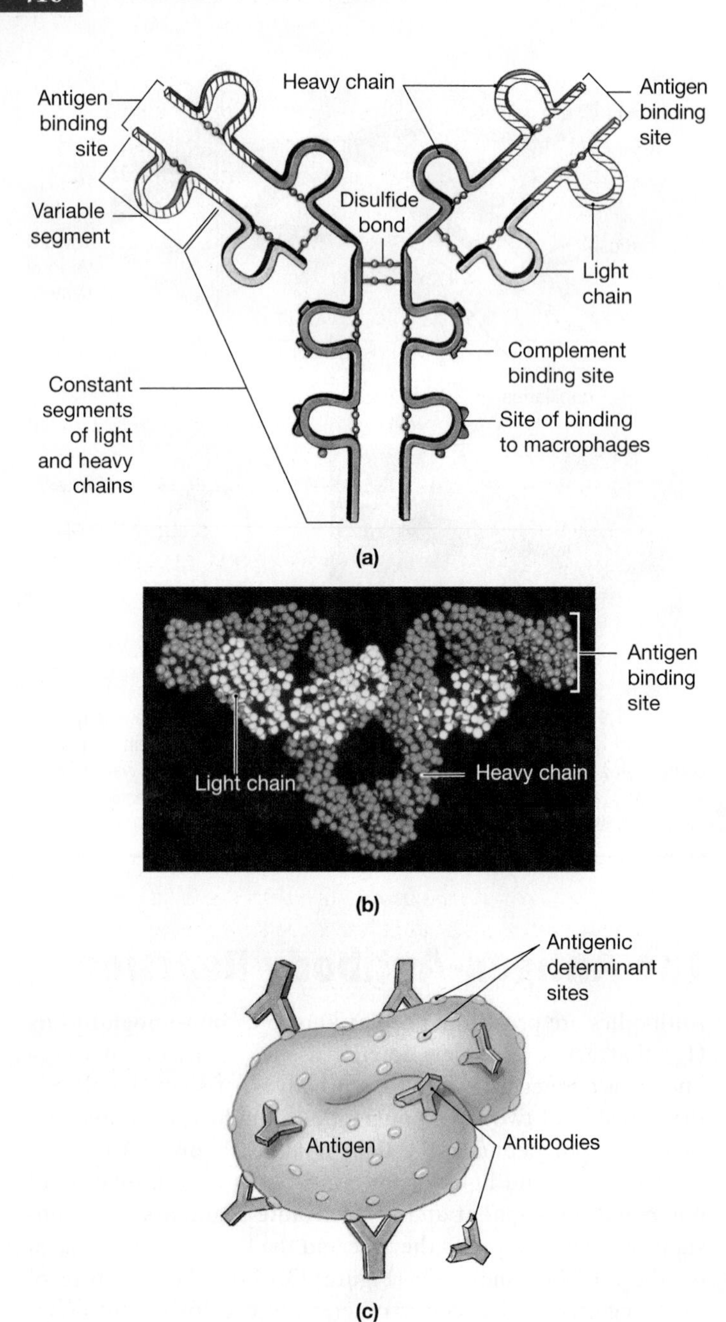

Figure 23.10 Structure of an antibody. a) Diagram showing the Y-shaped configuration of an antibody; **b)** computer-generated image of an antibody molecule; **c)** diagram showing an antigen-antibody complex. Notice that several antibodies can bind to a single antigen.

In the following activity you will use **immunodiffusion** (ID) or the **Ouchterlony technique** to detect the presence of antigens. Antigens and antibodies will be allowed to diffuse toward each other through a saline agar gel in petri dishes. A **line of identity**, or **precipitin line**, will form if a specific antigen-antibody reaction occurs.

CLINICAL CORRELATION

Typically, the immune system attacks cells and tissues that are foreign to the body. However, if immune function is not operating normally, an **autoimmune response** can be triggered during which antibodies against the body's own tissues are produced. Examples of **autoimmune diseases** include Type I diabetes in which antibodies attack the insulin-producing beta cells in the pancreatic islets, rheumatoid arthritis in which connective tissues around joints are destroyed, and multiple sclerosis in which the myelin sheath is targeted.

ACTIVITY 23.3 Observing the Antigen-Antibody Reaction Using Immunodiffusion

Preparing the Petri Dishes

1. Prepare an 8% sodium chloride (NaCl) solution in distilled water.
2. Use the NaCl solution to make a 1% agar mixture.
3. Place the agar mixture in a gently boiling water bath and stir occasionally until the agar has completely dissolved.
4. Obtain a petri dish that is divided into three compartments. Alternatively, use a petri dish with a single compartment and mark off three sections with a wax marking pencil as illustrated in Figure 23.11. Label the three compartments with the letters A, B, and C.
5. Pour 6 ml of the warm agar mixture equally into the three petri dish compartments.

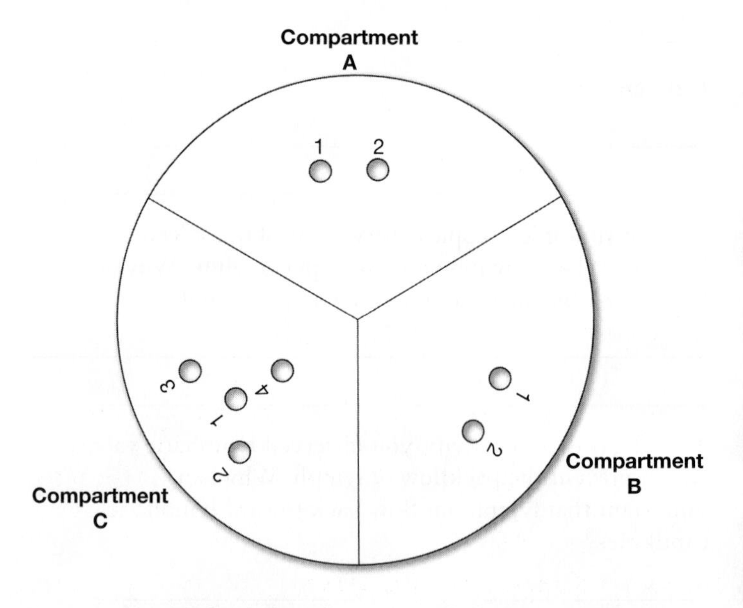

Figure 23.11 Configuration of sample wells for immunodiffusion. Compartments A and B will be used for control tests. Compartment C will be used to determine the identity of an unknown antigen.

6. When the agar has solidified, you will cut out sample wells, 4 mm in diameter, in the agar. Each small circle in Figure 23.11 marks a location where a sample well will be positioned. Place the petri dish over the figure to guide you during this process. Alternatively, you can place a copy of Figure 23.11 on your lab bench and use it as your template.
7. To make the sample wells, use the following microsuction technique.
   a. Squeeze the rubber bulb of a glass pipette or medicine dropper and lower it in a vertical position until the tip is just touching the surface of the agar.
   b. Simultaneously, release the bulb and plunge the pipette into the agar to the bottom of the petri dish. Keep the pipette in a vertical position at all times. As you perform this action, a small amount of agar will be suctioned into the pipette.
   c. Carefully lift the pipette, vertically, out of the newly formed sample well. As you do this, be careful not to damage the sides of the well or to lift the agar off the bottom of the petri dish. If this occurs, there is a risk that your sample will leak into adjacent wells.
8. The petri dish can be prepared in advance and refrigerated in an airtight container. To store for longer periods, place damp paper towels along the bottom of the container. The moisture will extend the storage time.

Preparing the Animal Sera

1. Obtain bovine (cow), horse, and swine (pig) sera, each containing natural concentrations of albumins.
2. Dilute each serum to a 20% solution in physiological saline.
3. The sera can be prepared in advance and refrigerated in covered flasks or bottles until they are used.

Filling the Sample Wells

1. Fine-tipped pipettes or micropipettes should be used when filling the wells. Be sure to use a different pipette for each solution. Fill each well with equal amounts of solution, but avoid overfilling.
2. Compartment A in the petri dish will serve as a **positive control**. In this test, you will expect positive results by observing a precipitin line. Fill the wells according to the following list.
   - Well 1: 20% horse serum albumin
   - Well 2: goat anti-horse albumin
3. Compartment B will serve as a **negative control**. In this test, you will expect negative results. A precipitin line will not appear. Fill the wells according to the following list.
   - Well 1: 20% horse serum albumin
   - Well 2: goat anti-bovine albumin
4. Compartment C will serve as an **unknown test**. In this test, you will determine the type of antigen (albumin) in an unknown serum based on the presence or absence of a precipitin line when it reacts with various antisera. Your instructor will give you an unknown serum to test. Fill the wells according to the following list.
   - Well 1: 20% unknown serum albumin
   - Well 2: goat anti-bovine albumin
   - Well 3: goat anti-horse albumin
   - Well 4: goat anti-swine albumin
5. When all the wells are filled, place the cover over the petri dish and store in a moist, airtight container at room temperature. Allow incubation to proceed for 16 to 24 hours.
6. After the incubation period, observe the results. Precipitin lines will be easier to see if you place your petri dish over a light source. Record your results in Table 23.2.

QUESTIONS TO CONSIDER

1. Why does a precipitin line form (or not form) between two wells?

____

____

____

2. Explain why compartment A in the petri dish is a positive control test, and compartment B is a negative control test.

____

____

3. What antigen was present in your unknown serum? How did your results allow you to determine the answer?

____

____

Table 23.2 Results of Immunodiffusion to Detect the Presence of Antigens

Test	Presence of a precipitin line (+/−)	Comments
	Between Sample Wells	
A. Positive control	1 & 2: ______	
B. Negative control	1 & 2: ______	
C. Unknown test	1 & 2: ______	
	1 & 3: ______	
	1 & 4: ______	

Exercise 23 Review Sheet

The Lymphatic System

Name ______________________________

Lab Section ______________________________

Date ______________________________

1. What is the main structural difference between a lymphatic organ and a lymphatic nodule?

____

____

2. Explain the meaning of the following statement: *The spleen acts as a blood filter and lymph nodes act as lymph filters.*

____

____

3. Explain why normal lymphatic circulation is important.

____

____

4. Explain why the thymus is both an endocrine and a lymphatic organ.

____

____

5. Briefly explain the difference between the cell-mediated immune response and the humoral immune response.

____

____

Questions 6–10: Define the following terms.

6. Aggregated lymphoid nodules

____

____

7. Cisterna chyli

____

____

8. White pulp versus red pulp

____

____

9. Afferent lymphatic vessel versus efferent lymphatic vessel

10. Germinal center

11. Describe the basic structure of an antibody molecule.

12. What is an antigen-antibody complex? Explain why the reaction between an antigen and antibody is very specific.

Exercise 24

Anatomy of the Respiratory System

Laboratory Objectives

On completion of the activities in this exercise, you will be able to:

- Identify the gross and microscopic anatomy of the upper and lower respiratory tracts.
- Understand specializations of the respiratory tract at both the gross and microscopic levels.
- Trace the path of air from where it enters the nasal cavity to where gas exchange occurs within the alveoli.

Materials

- Human skull
- Anatomical models
  - Torso
  - Head and neck
  - Larynx and trachea
  - Bronchial tree
  - Right and left lungs
- Prepared microscope slides
  - Any structure with a respiratory epithelium
  - Trachea
  - Inflated lung

The structures comprising the respiratory system can be categorized into two divisions. The **upper respiratory tract** includes the nose, nasal cavity, paranasal sinuses, and pharynx. The **lower respiratory tract** includes the larynx, trachea, and bronchial tree within the lungs.

The Nose and Nasal Cavity

The **nose** is the only externally visible component of the respiratory system. It is covered by skin and supported by cartilage and bone. The **nostrils (external nares)** provide an entrance for air to pass into the **nasal cavity**. These passageways are protected by internal hairs that filter out dust particles and other coarse particulate matter from the incoming air. The nasal cavity is the open space posterior to the nose. It is divided into right and left halves by a bony and cartilaginous partition called the **nasal septum**. The roof of the nasal cavity is formed by the cribriform plate of the ethmoid bone and a portion of the sphenoid bone. The bony roof separates the nasal cavity from the cranial cavity. The **palate**, formed by an anterior bony portion **(hard palate)** and a posterior soft tissue portion **(soft palate)**, separates the nasal cavity from the oral cavity. Posteriorly, the nasal cavity is continuous with the nasopharynx via the **internal nares**.

ACTIVITY 24.1 Examining the Gross Anatomy of the Nose and Nasal Cavity

1. On a skull, identify the bones that comprise the bony portion of the nose. Superiorly, a portion of the **frontal bone** and the two **nasal bones** form the nasal bridge, and processes of the **maxillae** make up the lateral walls. Several small plates of hyaline cartilage, not seen on a skull, complete the nose inferior to the bones (Figure 24.1a).
2. On yourself, palpate the bony and cartilaginous portions of your nose.
3. On a midsagittal model of the head and neck, identify the nose and nasal cavity (Figure 24.1c).
4. Verify that the nostrils (external nares) provide a passageway for air to enter the nasal cavity.
5. Posteriorly, identify the area where the nasal cavity leads directly to the pharynx through openings called the internal nares, or **posterior nasal apertures.**
6. Identify the bony part of the nasal septum, which divides the nasal cavity into right and left halves. Verify that it is formed by the **vomer** and the **perpendicular plate of the ethmoid.** The nasal septum is completed by cartilage that is not found on the skull. Due to the sectioning, the septum cannot be seen on the head model, but it can be identified in Figure 24.1b.
7. On a head model and on a skull, identify the superior border or **roof of the nasal cavity.** Verify that the cribriform plate separates the nasal cavity from the cranial cavity.
8. On the skull, identify the foramina that traverse the cribriform plate. What nerve passes through these openings?

   ____

9. On the head model, identify the palate and examine how it forms the floor of the nasal cavity and separates it from the oral cavity.
10. Identify the hard palate on both the head model and the skull. What two bones form this structure?

    ____

11. On the head model, identify the soft palate. Its posterior tip is the **uvula**. During swallowing, the soft palate elevates and the uvula closes off the internal nares, preventing food or fluids from entering the nasal cavity.
12. Ask your lab partner to open his or her mouth and say, "Ahhhhh." Look inside the mouth to easily identify the uvula.
13. On both the skull and the head model, identify the **superior**, **middle**, and **inferior conchae** along the lateral wall

Figure 24.1 Structure of the upper respiratory tract. a) Nasal cartilages in the nose; **b)** frontal section of the head showing the relationship between the nasal cavity and the paranasal sinuses; **c)** midsagittal section of the nasal cavity and pharynx.

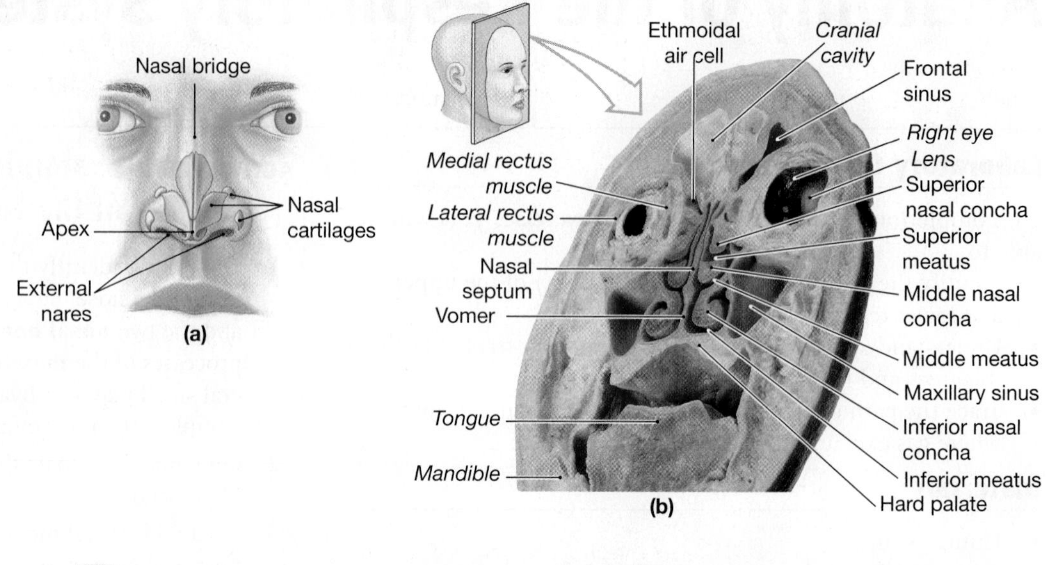

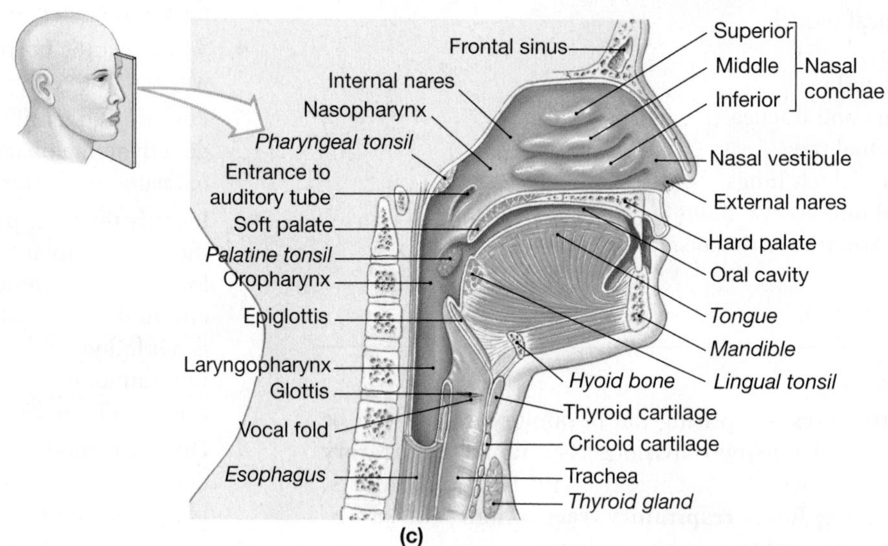

of the nasal cavity. The inferior concha is a separate bone. To what bone do the superior and middle conchae belong?

____

14. Locate the space inferior to each concha. The space is called a **meatus.** Identify the **superior**, **middle**, and **inferior meatuses** on both the skull and the model.
15. Notice that the configuration of the conchae and meatuses forms irregular twists and turns along the wall of the nasal cavity (Figures 24.1b and c). This arrangement maximizes the surface area of mucous membrane that is exposed to inspired air. When air passes over this asymmetrical surface, turbulence reduces flow and increases the ability of the mucous membrane to trap particulate matter.
16. On the head and neck model, identify the **paranasal sinuses** in the frontal bone **(frontal sinus)** and the sphenoid bone **(sphenoidal sinus).** Other paranasal sinuses are located in the maxillary and ethmoid bones but may not be visible on your model. All the sinuses are continuous with the nasal cavity and are lined by a mucous membrane. Mucous secretions originating within the sinuses drain into the nasal cavities.

CLINICAL CORRELATION

If drainage from the paranasal sinuses is blocked due to infection-induced inflammation (**sinusitis**), mucus will accumulate in the sinuses. The resulting increase in sinus pressure can cause severe sinus headaches.

The sinuses act as resonance chambers that affect the quality of the voice. Thus, inflammation of these chambers also changes the nature of an individual's voice, which explains the "nasal" quality of the voice when a person has a head cold.

QUESTION TO CONSIDER The nasal septum separates the nasal cavity into right and left sides. Ideally it will run along the midline and divide the cavity into equal parts. It is common for the nasal septum to be slightly off center. A **deviated septum**, however, is a condition in which the nasal septum is shifted significantly away from the midline. Use your anatomical knowledge of the nasal cavity to explain how this abnormality may cause difficulty in breathing. ____________________

____

____

The Pharynx

The **pharynx**, or throat, is the cavity posterior to the nasal and oral cavities. It serves as a passageway for air to the larynx and food to the esophagus. The pharynx is divided into three distinct regions: the **nasopharynx**, **oropharynx**, and **laryngopharynx**. Aggregates of lymphatic nodules called **tonsils** form a discontinuous ring of lymphoid tissue along the wall of the pharynx near the junction with the oral cavity (Figure 24.1c).

ACTIVITY 24.2 Examining the Gross Anatomy of the Pharynx

1. On the head model, identify the three pharyngeal regions (Figure 24.1c).
   - The nasopharynx stretches from the internal nares to the posterior end of the soft palate.
   - The oropharynx extends from the soft palate to the superior margin of the epiglottis. Describe its position relative to the oral cavity.

     ____

   - The laryngopharynx extends from the superior margin of the epiglottis to the openings for the larynx and esophagus.
2. Observe the positions of the three pharyngeal regions in relation to the nasal cavity, oral cavity, larynx, and esophagus. Verify that the nasopharynx serves as a passageway for air only, and thus, is solely a respiratory structure. Notice that the oropharynx and laryngopharynx serve dual functions, conveying both air for respiration and food for digestion.
3. Identify the three sets of tonsils along the wall of the pharynx (Figure 24.1c).
   - The single **pharyngeal tonsil (adenoid)** is located posteriorly in the nasopharynx. Locate the opening of the **auditory tube** near this tonsil. This tube connects the nasopharynx with what structure?

     ____

   - The two **palatine tonsils** are located laterally in the oropharynx, near the posterior aspect of the soft palate.
   - The two **lingual tonsils** are located at the base of the tongue.

CLINICAL CORRELATION

The tonsils contain dense populations of lymphocytes and germinal centers, where new lymphocytes are produced. Their primary function is to fight off infections in the upper respiratory and digestive tracts. Serious inflammation and swelling of the tonsils (**tonsillitis**) may require their removal. The tonsils most often removed during a **tonsillectomy** are the palatine tonsils; however, the pharyngeal tonsil may also be removed if it is enlarged enough to block the passageway between the pharynx and nasal cavity.

Review the locations of the tonsils and explain how tonsillitis might affect breathing.

____

____

The Larynx

The **larynx**, or voice box (Figure 24.1c), is approximately 5 cm (2 in) long, extending from the C4 to C6 vertebrae. The wall of the larynx is composed of several pieces of cartilage that are connected by ligaments and muscles. It provides a passageway for air to enter and exit the trachea during pulmonary ventilation, and contains the **vocal cords**, which are used for sound production.

ACTIVITY 24.3 Examining the Gross Anatomy of the Larynx

1. On a head and neck model, locate the larynx. Verify that it is directly continuous with the laryngopharynx, superiorly, and the trachea, inferiorly.
2. Obtain an enlarged model of the larynx and observe its anterior surface (Figure 24.2a).
3. Identify the large **thyroid cartilage**, which forms an anterior protective wall for the larynx (Figure 24.2a). It is composed of two cartilage plates that fuse anteriorly to form the **laryngeal prominence**, or **Adam's apple.** Notice that the **hyoid bone** is positioned superior to the thyroid cartilage and attached to it by connective tissue membranes.
4. Gently palpate the laryngeal prominence on yourself. Due to the influence of male sex hormones, the laryngeal prominence is larger in males than in females. Verify this

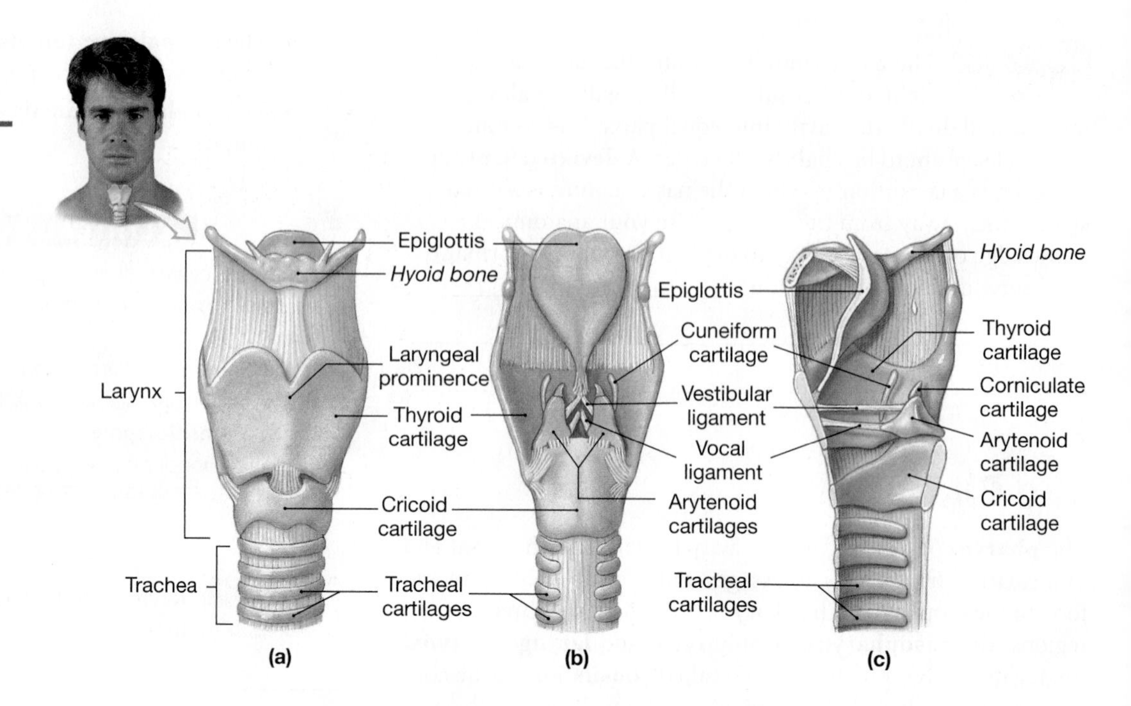

Figure 24.2 Structure of the larynx. a) Anterior view; **b)** posterior view; **c)** midsagittal section.

difference by observing the laryngeal prominence on both genders in your class.

5. Identify the ringlike **cricoid cartilage**, just inferior to the thyroid cartilage (Figure 24.2a). This cartilage forms the inferior border of the larynx. It is connected by connective tissue to the thyroid cartilage, superiorly, and the first tracheal cartilage, inferiorly.

WHAT'S IN A WORD The word *cricoid* originates from the Greek word *krikos,* which means "a ring." The cricoid cartilage forms a complete ring, unlike the thyroid cartilage, which is open posteriorly.

6. Observe the larynx from its posterior aspect (Figure 24.2b). Notice that the cricoid cartilage forms a complete ring, but the thyroid cartilage does not. From this view, identify the **epiglottis**, a cartilaginous shield attached to the superior aspect of the thyroid cartilage and extending to the base of the tongue.
7. The model illustrates the position of the epiglottis during normal breathing. It projects superiorly and the entrance to the larynx remains open to allow air to pass into the trachea and bronchial tubes. Pull the epiglottis inferiorly to verify that it blocks the entrance to the larynx during swallowing. When you swallow, the larynx is elevated and the epiglottis bends posteriorly to cover the laryngeal opening.

CLINICAL CORRELATION

During normal swallowing, the elevation of the larynx and the bending of the epiglottis channel food and fluids into the esophagus rather than the trachea. Occasionally, the swallowing reflex will not operate properly and a small amount of food may enter the larynx and trachea. This is commonly referred to as "going down the wrong pipe."

8. Three pairs of small cartilages are located along the posterior and lateral walls of the larynx (Figures 24.2b and c).
   - Identify the paired pyramid-shaped **arytenoid cartilages** that rest on the superior margin of the cricoid cartilage. Along with the thyroid cartilage, they provide attachment sites for the vocal cords (described below).
   - The cone-shaped **corniculate cartilages** are connected to the superior tips of the arytenoid cartilages. They serve as attachments for muscles that regulate the tension of the vocal cords during speech.
   - The cylinder-shaped **cuneiform cartilages** are embedded in the mucous membrane that lines the larynx. During sound production, they function to stiffen the soft tissue along the lateral laryngeal wall. These cartilages cannot be seen on the model.
9. Inside the larynx, locate the two vocal cords. Each vocal cord includes a band of elastic tissue, the **vocal ligament** (Figure 24.2c), which is covered by a fold of epithelium, the **vocal fold.** The vocal cords extend from the thyroid cartilage, anteriorly, to the arytenoid cartilages, posteriorly. The opening between the vocal cords is called the **glottis.** Sound is produced when air travels through the glottis and causes the vocal cords to vibrate. Laryngeal muscles that move the thyroid and arytenoid cartilages control the tension in the vocal cords. When the cords are taut, high-pitched sounds are produced; when they are relaxed, low-pitched sounds are produced. Sound amplification is determined by the amount of vibration of the cords. Loud sounds, such as yelling, are produced when the cords are separated so that the glottis is wide open and a relatively large volume of air is allowed to pass through. Soft sounds, such as whispering, are produced when the vocal cords are brought close together so that the glottis is reduced to a narrow slit and a smaller volume of air passes over the cords.

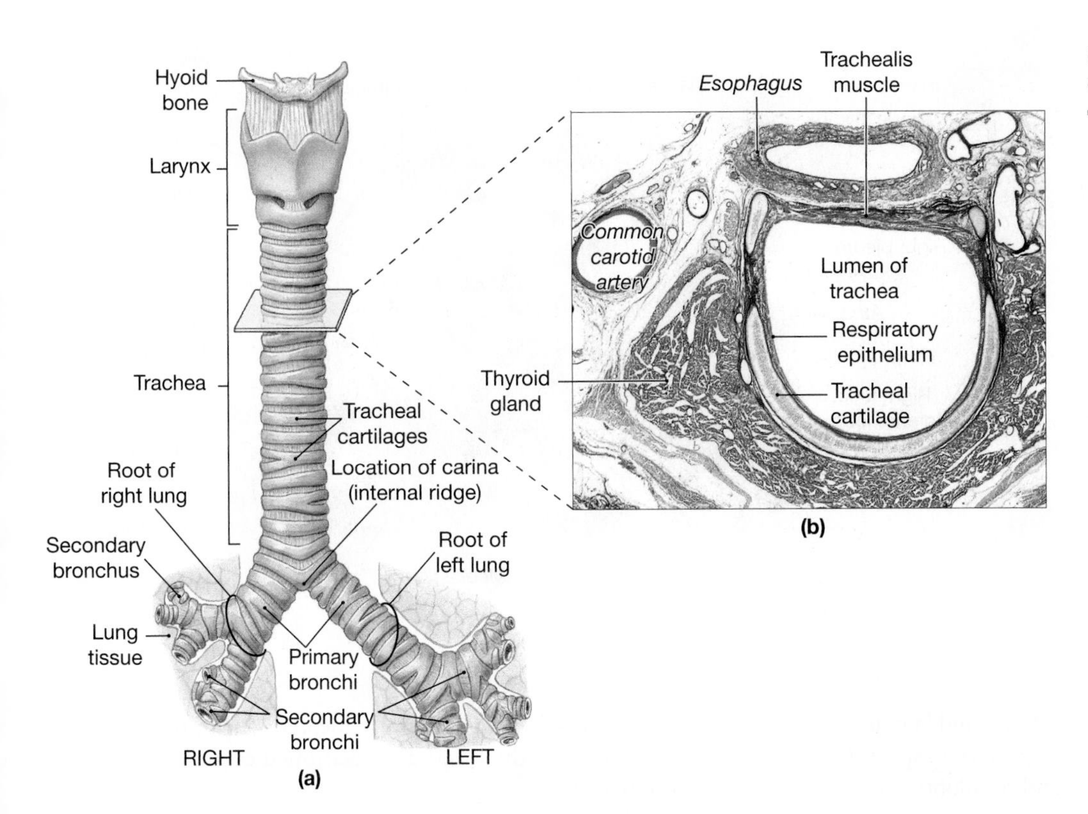

Figure 24.3 Structure of the trachea. **a)** Anterior view; **b)** cross section.

10. On the larynx model, rotate the two arytenoid cartilages. What happens to the glottis when you perform this action?

____

CLINICAL CORRELATION

In both genders, children's vocal cords are relatively narrow and short, so the vibrations are rapid and produce high-pitched sounds. At puberty, when testosterone levels increase in boys, the vocal cords elongate and thicken. As a result, cord vibrations become slower and produce lower pitched, deeper sounds. As a general rule, due to the influence of testosterone, adult males have thicker and longer vocal cords than females. This anatomical difference is the primary reason for gender differences in voice.

QUESTION TO CONSIDER Explain how the rotation of the arytenoid cartilages affects sound production. _______________

____

____

The Trachea

The **trachea**, commonly referred to as the windpipe (Figure 24.3), is a tubular structure that delivers air to the bronchial tree. It is approximately 12 cm (4.5 in) long and 2.5 cm (1 in) in diameter. It begins in the neck as a continuation of the larynx and travels into the mediastinum, anterior to the esophagus. At its inferior end, the trachea bifurcates (divides into two branches) to form the right and left primary bronchi (Figure 24.3a).

ACTIVITY 24.4 Examining the Gross Anatomy of the Trachea

1. On a model of the torso or thoracic cavity, identify the trachea (Figure 24.3). Observe that it begins at the level of the C6 vertebra, where it is directly continuous with the larynx and ends at the level of T5, where it bifurcates.
2. Identify the approximately 20 C-shaped cartilage rings. The soft tissue that connects the cartilage rings contains connective tissue and smooth muscle.
3. Identify the last cartilage, the **carina**, which separates the entrances to the left and right primary bronchi at the tracheal bifurcation (Figure 24.3a). The mucous membrane that covers the carina contains many sensory receptors that strongly respond to any dust particles that reach this far into the respiratory tract. If these receptors are stimulated, they can initiate a strong cough reflex to prevent foreign debris from entering the bronchial tree.

WHAT'S IN A WORD *Carina* is a Latin word which means "the keel of a boat." The carina at the inferior end of the trachea is a cartilaginous ridge that resembles a keel.

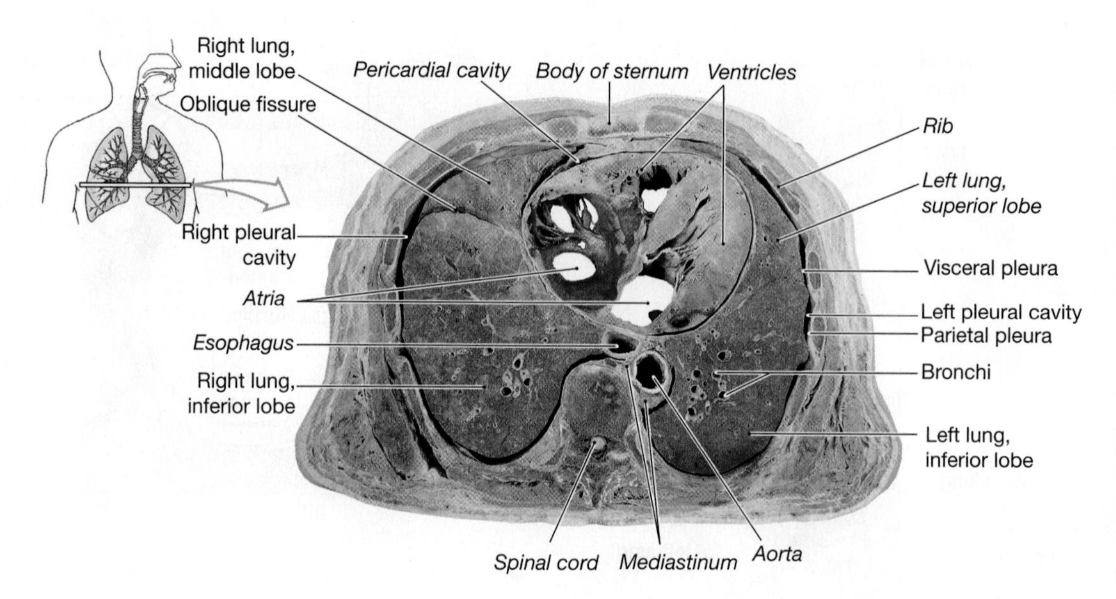

Figure 24.4 Transverse section of the thoracic cavity. The heart is located in the centrally located mediastinum. The lungs are located on each side of the heart.

4. Verify that the cartilage covers the anterior and lateral aspects of the tracheal wall, but not the posterior aspect. Along the posterior surface, identify the band of smooth muscle known as the **trachealis muscle**. Notice that this muscle connects the open ends of the tracheal rings (Figure 24.3b).

QUESTION TO CONSIDER Why do you think cartilage is missing along the posterior tracheal wall? (*Hint:* Examine the anatomical relations of the trachea on models in the laboratory and in Figure 24.3b.) ____________________

____

____

The Lungs

The **lungs** are pyramid or cone-shaped organs located in the left and right sides of the thoracic cavity. They are separated by the **mediastinum**, which is the central space in the thorax where the heart is found. The lungs are lined by a double serous membrane known as the **pleural membrane**. The pleural cavity is the narrow, fluid-filled space between the two membranes. The **apex** of each lung forms a domed surface that extends superiorly into the neck, about 2.4 cm superior to the first rib. The **base** forms a concave surface that rests on the superior surface of the diaphragm.

ACTIVITY 24.5 Examining the Gross Anatomy of the Lungs

1. On a model of the thoracic cavity, observe the position of the lungs in relation to the heart (Figures 24.4 and 24.5a). The heart is centrally located within the mediastinum, and the lungs are positioned on either side of the heart.
2. Observe that each lung has a superior apex (Figure 24.5) that extends into the root of the neck, and an inferior base that rests on the diaphragm.
3. Identify the lobes and the fissures of each lung (Figure 24.5b). The left lung is divided into two lobes (superior and inferior) that are separated by an **oblique fissure**. The right lung has three lobes (superior, inferior, and middle) that are separated by an oblique fissure and a **horizontal fissure**.
4. For each lung, identify the following surfaces.
   - The **costal surface** travels along the surface of the rib cage. It forms the anterior, lateral, and posterior aspects of the lung.
   - The inferior **diaphragmatic surface** rests on the diaphragm.
   - The medial **mediastinal surface** borders the central mediastinum. On the left lung, the mediastinal surface forms the **cardiac impression** to accommodate the heart. The anterior border of the cardiac impression is called the **cardiac notch** (Figure 24.5b).
5. The **hilus of the lung** is located on the mediastinal surface (Figure 24.5c). This region serves as an area of entry or exit for structures that comprise the **root of the lung**, such as the bronchi, nerves, blood vessels, and lymphatics. Identify the **main bronchus** and the **pulmonary artery** and **vein** where they enter the hilus of each lung. Recall that the pulmonary artery contains deoxygenated blood entering the lungs to drop off carbon dioxide, while the pulmonary veins contain freshly oxygenated blood heading back to the heart to be distributed throughout the body.

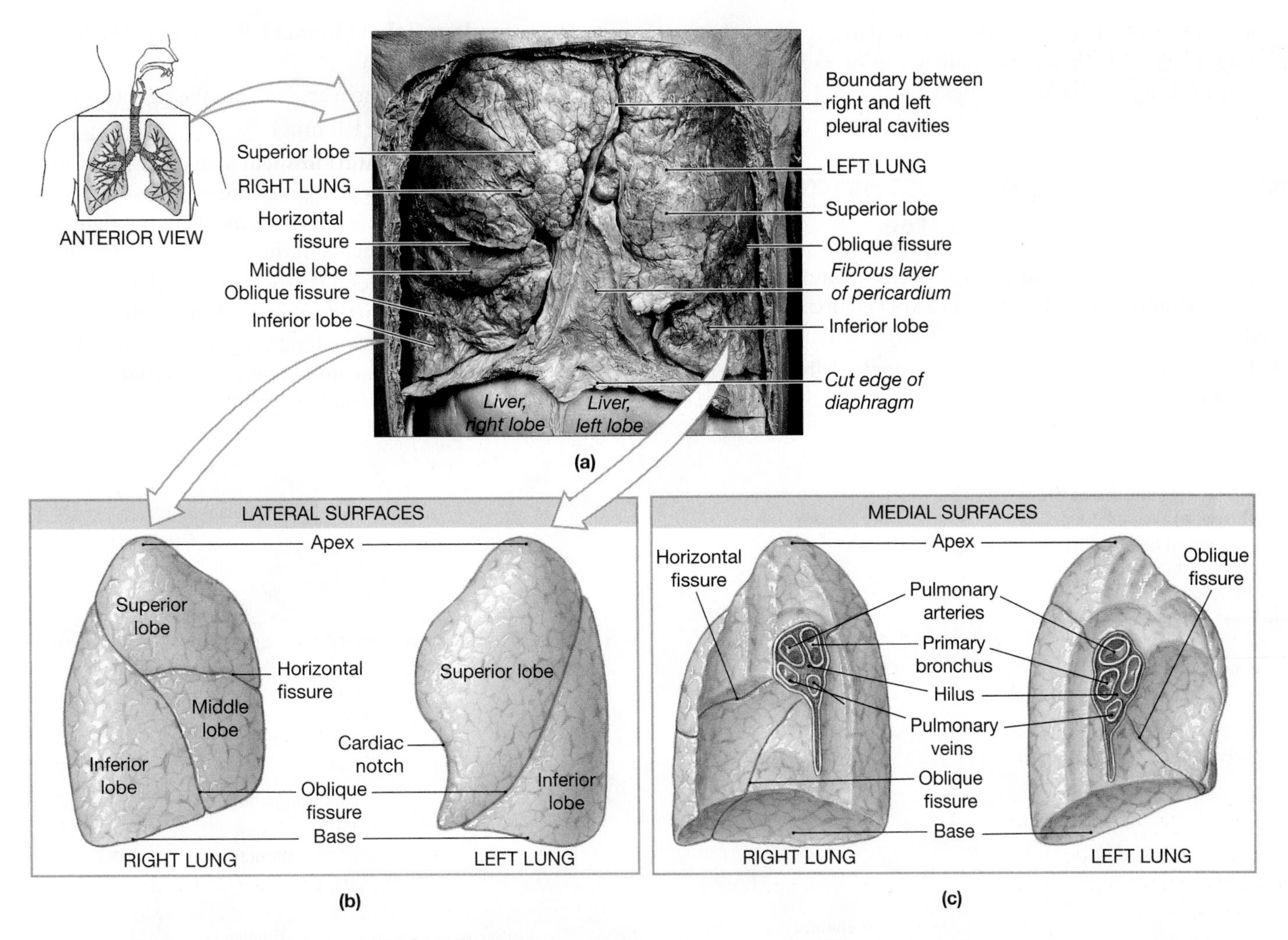

Figure 24.5 Structure of the lungs. a) Superficial cadaver dissection of the thoracic cavity, anterior view; **b)** diagrams of the lateral surfaces of the lungs; **c)** diagrams of the medial surfaces of the lungs.

6. On a torso model, observe the lungs in their anatomical position within the thoracic cavity (Figure 24.5a). Identify, on the model, the surfaces that are covered by the pleural membrane and the orientation of its components.
   - The **visceral pleura** covers the outside surfaces of the lungs.
   - The **parietal pleura** lines the ribs and intercostal spaces, mediastinum, and diaphragm.
   - The parietal and visceral pleura are continuous at the root of the lung.
   - The **pleural cavity** is the narrow space between the two pleural membranes.

QUESTION TO CONSIDER The visceral and parietal pleural membranes produce a fluid that fills the pleural cavity. The fluid provides lubrication that reduces friction and allows the two membranes to slide easily along each other during breathing. **Pleurisy** is an inflammatory condition that causes abnormal adhesions to form between the pleural membranes. Explain why this condition can disrupt normal breathing. ____________

____

____

The Bronchial Tree

The **bronchial tree** refers to the treelike branching of the airways in the lungs. It begins at the tracheal bifurcation (split), with the right and left **primary bronchi** (singular = **bronchus**) and ends within the substance of the lung with the **alveoli** (singular = **alveolus**). The bronchial tree contains two functional categories of airways. The **conducting airways** are responsible for distributing air to a particular region of a lung. They deliver air to the **respiratory airways**, where the exchange of carbon dioxide and oxygen occurs between the lungs and pulmonary capillaries.

WHAT'S IN A WORD The term *alveolus* is derived from the Latin word *alveus,* which means "hollow sac or cavity." Each alveolus in the lung is a hollow air-filled sac.

ACTIVITY 24.6 Examining the Gross Anatomy of the Bronchial Tree

1. On a model of the torso or thoracic cavity, identify the first three generations of conducting airways (Figures 24.3a and 24.6a).
   - The left and right primary (main) bronchi deliver air to their respective lungs.
   - The **secondary (lobar) bronchi** deliver air to the lobes of the lungs.

   How many lobar bronchi are found in the right lung? ___

   How many are in the left lung? ___
   - The **tertiary (segmental) bronchi** deliver air to **bronchopulmonary segments**, which are subdivisions within each lobe. There are 10 segments in the right lung and 8 to 10 segments in the left lung.
2. From each tertiary bronchus there are approximately 20 generations of conducting airways. With each order of branching, the airway diameter gradually decreases. Airways with a diameter of 1 mm or less are referred to as **bronchioles.** Each bronchiole supplies air to a small re-

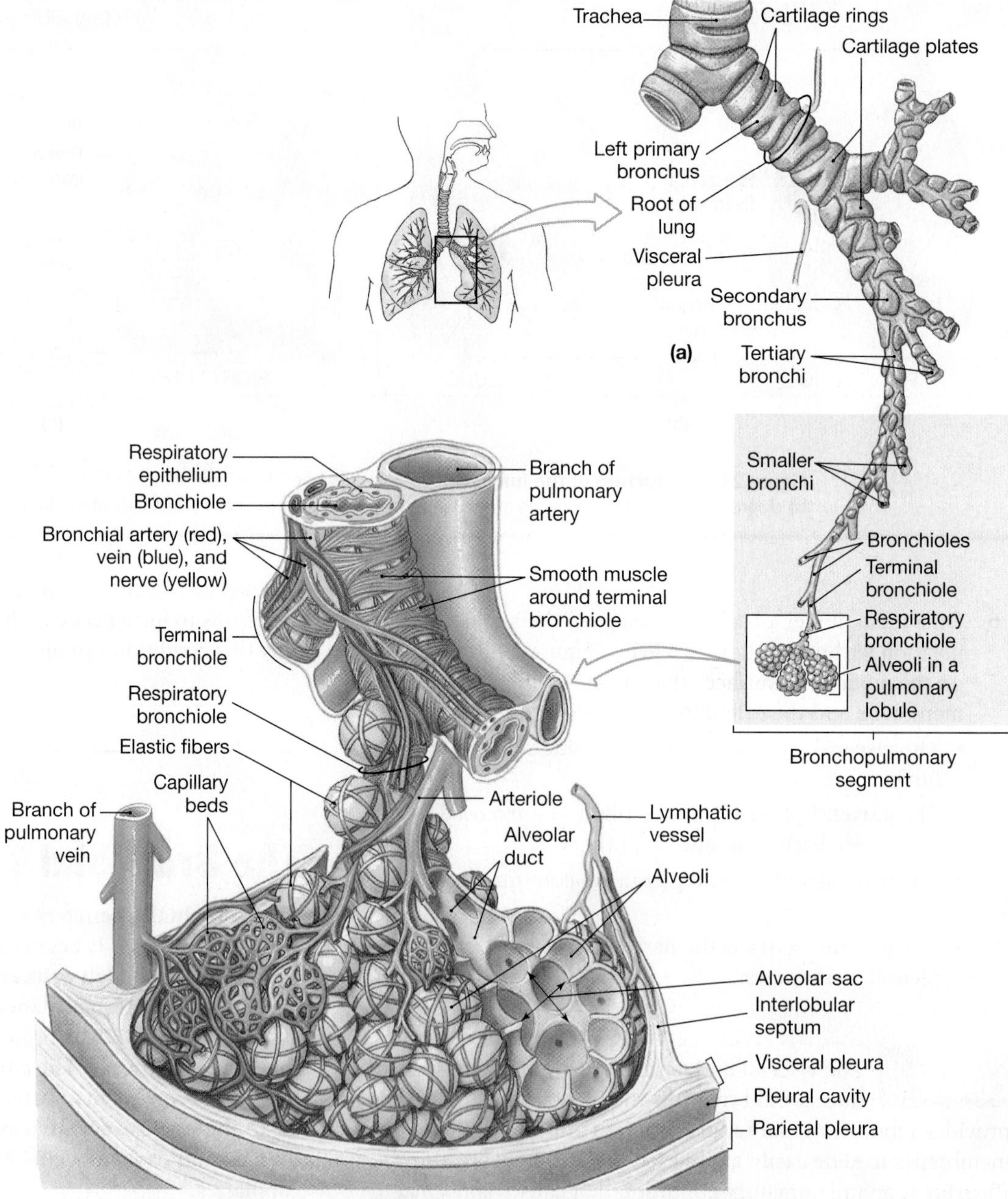

Figure 24.6 Components of the bronchial tree. a) A generalized branching pattern of the conducting airways in the left lung; **b)** components of the respiratory airways.

gion known as a **lung lobule** and gives rise to between 50 and 80 **terminal bronchioles**. The terminal bronchioles are less than 0.5 mm in diameter and mark the termination of the conducting airways (Figure 24.6). On a torso or thoracic cavity model, remove the anterior portion of one lung to produce a frontal section. Observe the cross sections of the small bronchi and bronchioles of the conducting airways. Notice that the conducting airways comprise a relatively small proportion of lung volume and surface. The vast majority of the lung parenchyma (functional components of the lung) consists of the thin-walled respiratory airways. On the model (and in fresh lung tissue) these regions will have a spongelike appearance.

QUESTION TO CONSIDER In the lung, why do you think the respiratory airways predominate over the conducting airways?

____

____

Microscopic Anatomy of the Respiratory System

Most of the respiratory passageways leading to the lungs are lined by a mucous membrane with mucous-secreting cells and ciliated cells. In addition to conducting air, these airways filter out most particulate matter before it can reach the lungs. The small respiratory airways (the alveoli) have thin walls to maximize the exchange of carbon dioxide and oxygen with pulmonary capillaries.

Respiratory Mucosa

The respiratory mucosa is the mucous membrane that lines the conducting portions of the respiratory tract. In several structures, the mucosa contains the highly specialized **respiratory epithelium**, which is a pseudostratified, ciliated columnar epithelium with mucous-secreting goblet cells. Deep to the epithelium is a layer of connective tissue which contains mucus and serous glands. Mucous secretions entrap dust and other foreign particles, while serous secretions contain **lysozyme**, an enzyme that can kill potentially harmful bacteria that enter the respiratory tract. The cilia act to push the mucus to the pharynx where it can be swallowed. Any foreign particles in the mucus are then digested by stomach secretions.

The nasal cavity contains two functionally distinct mucous membranes. In addition to the respiratory mucosa which lines most of the nasal cavity wall, there is the **olfactory mucosa** (See Exercise 17). Recall that it is located in the most superior region of the nasal cavity and it contains sensory neurons for the sense of smell (olfaction). The predominant respiratory mucosa contains a vast supply of capillaries. On cold days, heat from these capillaries radiates into the nasal cavity and quickly warms incoming air to body temperature. This function has practical significance for people who work or exercise outside during cold winter weather.

ACTIVITY 24.7 Examining the Microscopic Anatomy of the Respiratory Mucosa

1. Obtain a slide of any one of the following structures: nasal cavity, nasopharynx, larynx, or trachea. All of these respiratory structures have a mucous membrane with the respiratory epithelium.
2. Observe the tissue with the low-power objective lens. Move the slide so that the mucous membrane and underlying submucosa are centered.
3. Switch to the high-power objective lens and examine these two layers more carefully. In the mucous membrane, identify the **pseudostratified columnar epithelium.** Observe the placement of the cell nuclei. Notice that they are located at various levels, giving the appearance of a stratified epithelium (Figure 24.7). In fact, all the cells rest on the basement membrane and most extend to the lumen.
4. Identify the numerous **goblet (mucous) cells** (Figure 24.7). These are the cells with goblet-shaped regions near the lumen that are filled with mucous granules (stained purple or blue in some sections, clear in other sections).
5. Observe that many of the epithelial cells contain **cilia** along their luminal surfaces. Carefully adjust the fine focus to see these structures more clearly. You may need to reduce the light as well.
6. Attempt to identify the **lamina propria**, a thin layer of loose connective tissue deep to the epithelium. This layer may be difficult to isolate because the boundary between it and the deeper submucosa is hard to identify. The lamina propria and the epithelium comprise the respiratory mucosa.
7. Identify the **submucosa**, which is the prominent layer of connective tissue deep to the mucosa. Scan the slide to locate mucous and serous glands in this layer. Note also the many blood vessels traveling through the submucosa. Most of these are capillaries and some may be filled with red blood cells. Attempt to identify the endothelial cells that form the capillary walls.

QUESTIONS TO CONSIDER 1. The mucous membrane covering the nasopharynx contains a respiratory epithelium. However, the epithelium lining the wall of the oropharynx and laryngopharynx is stratified squamous. In terms of function, explain the significance of this change in the epithelium? (*Hint:* Review the gross anatomy of the upper respiratory tract and compare the function of the three pharyngeal regions.)

____

____

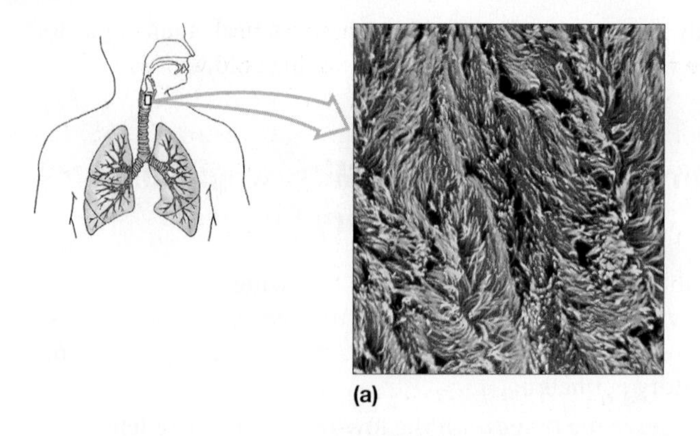

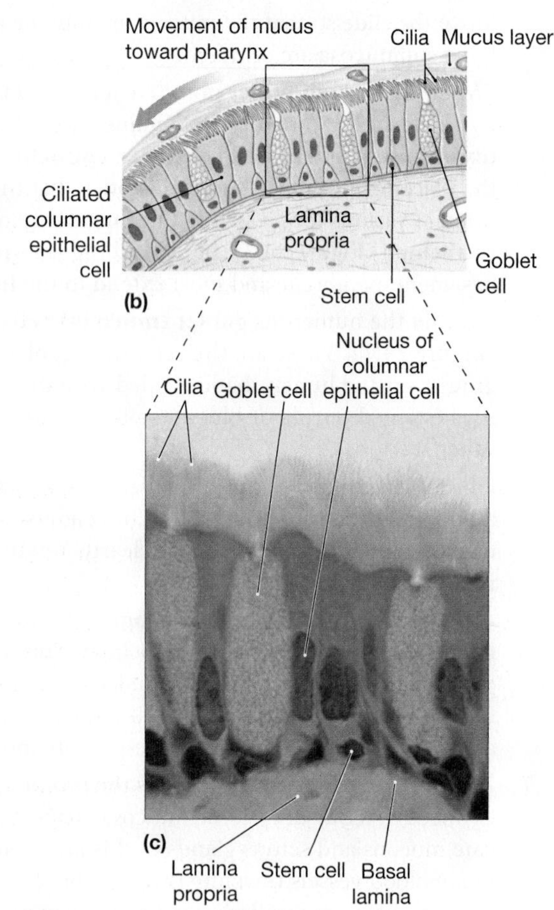

Figure 24.7 Microscopic structure of the respiratory mucosa. a) Surface view of ciliated cells in the respiratory epithelium (LM × 1600); **b)** diagram of the respiratory mucosa; **c)** light micrograph of the respiratory epithelium (LM × 1000).

2. Chemicals in cigarette smoke can destroy cilia, increase mucous secretions, and cause swelling along the trachea and bronchial tree. Explain how these changes can affect the function of the respiratory mucosa. ______________________

____

____

Figure 24.8 Microscopic structure of the trachea. The three tissue layers in the tracheal wall are the respiratory mucosa, submucosa, and adventitia. The C-shaped tracheal rings are embedded in the adventitia (LM × 100).

The Trachea

The tracheal wall has three tissue layers. The mucous membrane contains a respiratory epithelium. The submucosa, deep to the mucous membrane, contains mucous glands and blood vessels. The cartilage plates that keep the trachea open at all times are located in the adventitia.

ACTIVITY 24.8 Examining the Microscopic Anatomy of the Trachea

1. Obtain a slide of the trachea. You may have previously examined the respiratory mucosa of the trachea. Now you will examine its complete structure.
2. Scan the slide with the low-power objective lens. You are observing a small section of the tracheal wall. Notice that the trachea consists of several tissue layers (Figure 24.8).
   - The respiratory mucosa contains the pseudostratified, ciliated columnar epithelium (Figures 24.7 and 24.8) and a thin lamina propria. At this magnification, the epithelium can be clearly identified, but the lamina propria may be difficult to observe.
   - The submucosa is the connective tissue layer deep to the mucosa. It contains numerous blood vessels that supply the tracheal wall. Also observe the **tracheal glands** (mucous glands). These structures are lined by stratified cuboidal epithelium.
   - Beneath the submucosa, locate the large area of **cartilage** that is part of the **adventitia**. The cartilage that you are observing is a portion of one C-shaped cartilage ring. Observe the **lacunae** distributed throughout the purple-staining cartilaginous matrix. A **chondrocyte** is located within each lacuna.

- Identify the connective tissue layer that is present just deep to the cartilage. This layer is also part of the adventitia. It is continuous with the connective tissue of surrounding structures that are adjacent to the trachea.

QUESTION TO CONSIDER Why do you think it is essential to have cartilage plates in the tracheal wall. (*Hint:* Consider the primary function of the trachea.)

____

____

CLINICAL CORRELATION

Obstruction of the trachea due to swollen tissues, excessive glandular secretions, or a foreign object can be a dangerous condition. If the obstruction prevents breathing, a tracheostomy must be performed. During this procedure, an incision is made in the anterior wall of the trachea and a tube is inserted to allow air to enter the lungs. If a foreign object, such as a piece of food, obstructs the trachea, the Heimlich maneuver may successfully dislodge the item. In this procedure, compression of the abdomen elevates the diaphragm. If the force is sufficient, the pressure generated will remove the object.

The Lungs

As you observe the various conducting and respiratory airways, be aware of the following structural changes that occur as bronchial tubes become smaller and smaller.

- *A gradual decline in the amount of cartilage.* As described earlier, the trachea contains a series of C-shaped cartilaginous rings. In the main bronchi, the amount of cartilage is reduced and takes the form of irregularly shaped plates that surround the airways. With each successive airway generation, the amount of cartilage declines. At the level of the bronchioles, cartilage is completely absent.
- *The presence of smooth muscle.* As the amount of cartilage decreases in the conducting airways, smooth muscle and connective tissue fill the spaces between the plates. Thus, the relative amount of smooth muscle increases. In the bronchioles, smooth muscle forms a continuous layer around the circumference of the airways. In the respiratory airways, there is a gradual decline in smooth muscle and it is absent in the alveoli.

  Autonomic nerves supplying the muscle layer around bronchioles can influence the volume of air reaching a particular region of the lung by constricting (**bronchoconstriction**) or dilating (**bronchodilation**) the airway lumen. In this sense, control of air flow by bronchioles is similar to the control of blood flow by small, muscular arteries and arterioles.
- *A transition in the type of epithelium.* The epithelium in the trachea and primary bronchi is pseudostratified, ciliated columnar. With successive airway generations, the height of the epithelial cells gradually declines. The epithelium changes to simple columnar in the smaller bronchi, simple cuboidal in the bronchioles, and simple squamous in the alveolar passages. Furthermore, there is a gradual decline in the numbers of ciliated and goblet cells until they are completely absent in the bronchioles and successive generations.

ACTIVITY 24.9 Examining the Microscopic Anatomy of the Lungs

1. Obtain a slide of an inflated lung. Scan the slide with the low-power objective lens. Notice that most of the lung's volume is occupied by **alveolar ducts**, **alveolar sacs**, and **alveoli**, the thin-walled **respiratory airways** where gas exchange between blood and air occurs (Figures 24.6b and 24.9). The arrangement and distribution of the alveolar airways give the lung a spongelike appearance. The lungs contain an estimated 30 million alveoli that provide a total surface area of 70 to 80 m^2 (750 to 850 ft^2) for gas exchange.
2. Identify examples of bronchi and bronchioles. These airways are branches of the conducting airways that distribute air to the alveoli.
   - Large bronchi may not be present in some microscope sections. The walls of these airways contain the typical respiratory epithelium, varying numbers of cartilage plates, and a layer of smooth muscle (Figure 24.9a).
   - Small bronchi will have columnar epithelium with a reduced number of ciliated and goblet cells. The cartilage plates will be smaller in size and fewer in number but smooth muscle will be a more prominent feature (Figure 24.9b).
   - Bronchioles are lined by a cuboidal or low columnar epithelium with no ciliated and globlet cells. These airways are surrounded by a thin but prominent layer of smooth muscle, and cartilage plates are absent. The presence or absence of cartilage plates is a good way to distinguish between bronchi and bronchioles (Figure 24.9c).
3. Attempt to locate an area where a terminal bronchiole gives rise to a respiratory bronchiole. The junction between these two airways marks the termination of the conducting airways and the beginning of the respiratory airways. Look for an area where the airway epithelium abruptly changes. Terminal bronchioles have a simple cuboidal epithelium. Respiratory bronchioles have stretches of cuboidal epithelium that are frequently interrupted by openings leading to alveoli (Figure 24.9d). The presence of these alveoli is the reason why the respiratory bronchioles are categorized as respiratory airways. You may also observe the respiratory bronchioles giving rise to the thin-walled alveolar ducts (Figure 24.9d).

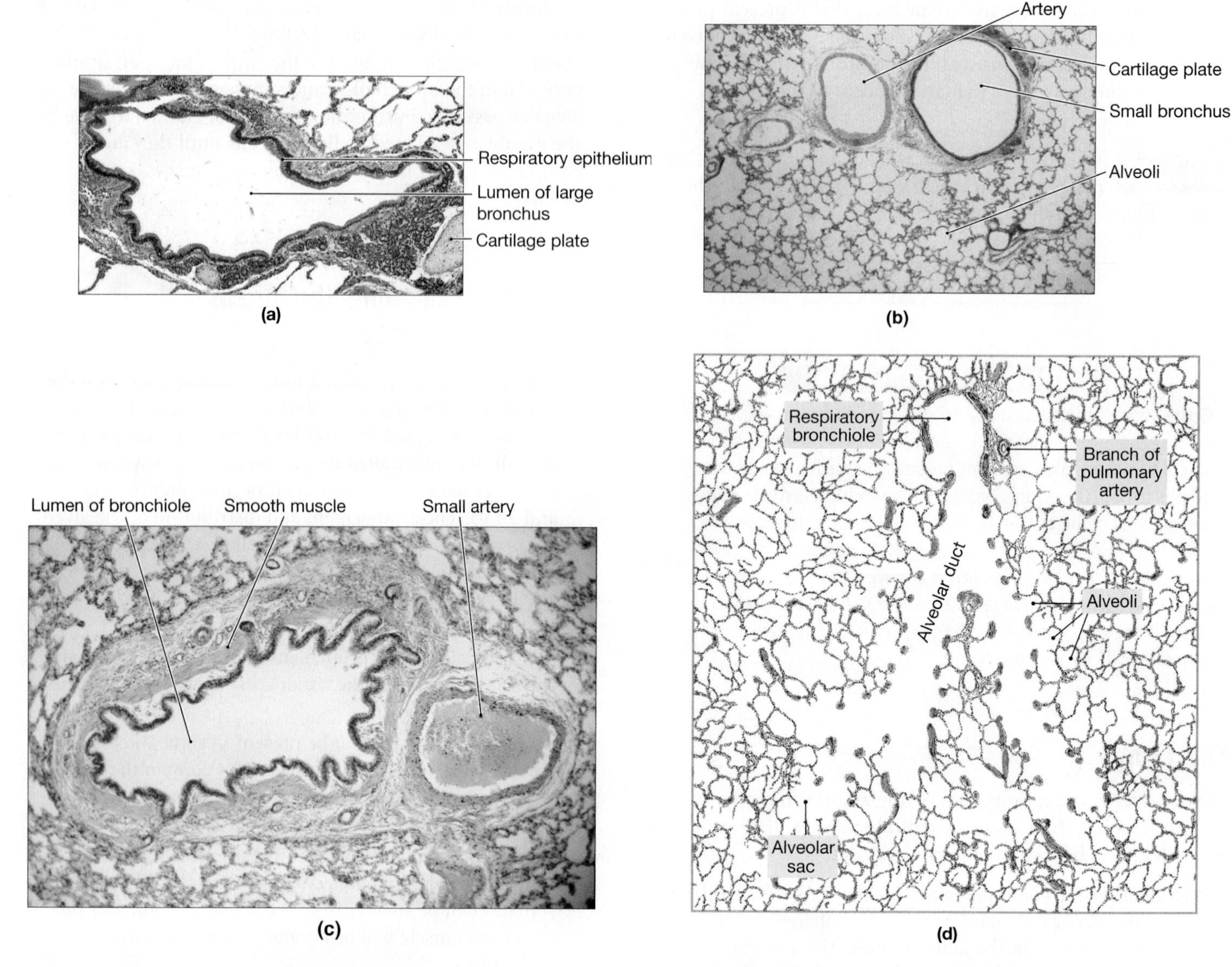

Figure 24.9 Light microscopic structure of the lung. a) Large bronchus (LM × 100); **b)** small bronchus and artery (LM × 40); **c)** bronchiole and small artery (LM × 100); **d)** respiratory airways (LM × 40).

4. Identify an alveolar duct and notice that it gives rise to a number of alveolar sacs. Each alveolar sac contains several alveoli (Figures 24.6b and 24.9d). To help you visualize the organization of the respiratory airways, consider the arrangement of grapes on their stems. An alveolar duct is analogous to a stem leading to a bunch of grapes. An alveolar sac is similar to a bunch of grapes, and each alveolus is an individual grape.
5. Using the high-power objective lens, examine the structure of the alveoli more closely. Notice that the epithelium lining the alveoli has a single layer of squamous cells. These cells are called **type I alveolar cells.** Note the close association between these cells and the **pulmonary capillaries** (Figure 24.10 on page 433). Type I alveolar cells are specially designed to facilitate rapid diffusion of carbon dioxide and oxygen between the blood and the lungs. The **respiratory membrane** is the barrier through which oxygen and carbon dioxide diffuse. It contains the following layers (Figure 24.10b).
   - Squamous type I alveolar cells
   - Basement membrane beneath the alveolar cells
   - Basement membrane beneath the capillary endothelial cells
   - Squamous endothelial cells in pulmonary capillaries

   Although there are several layers, the thickness of the respiratory membrane is only 0.5 μm (1/16 the diameter of a red blood cell).
6. With the high-power objective lens, carefully scan the epithelia of several alveoli and attempt to identify cells that are cuboidal rather than squamous. The cuboidal cells, known as **type II alveolar cells** (Figure 24.10a), are scat-

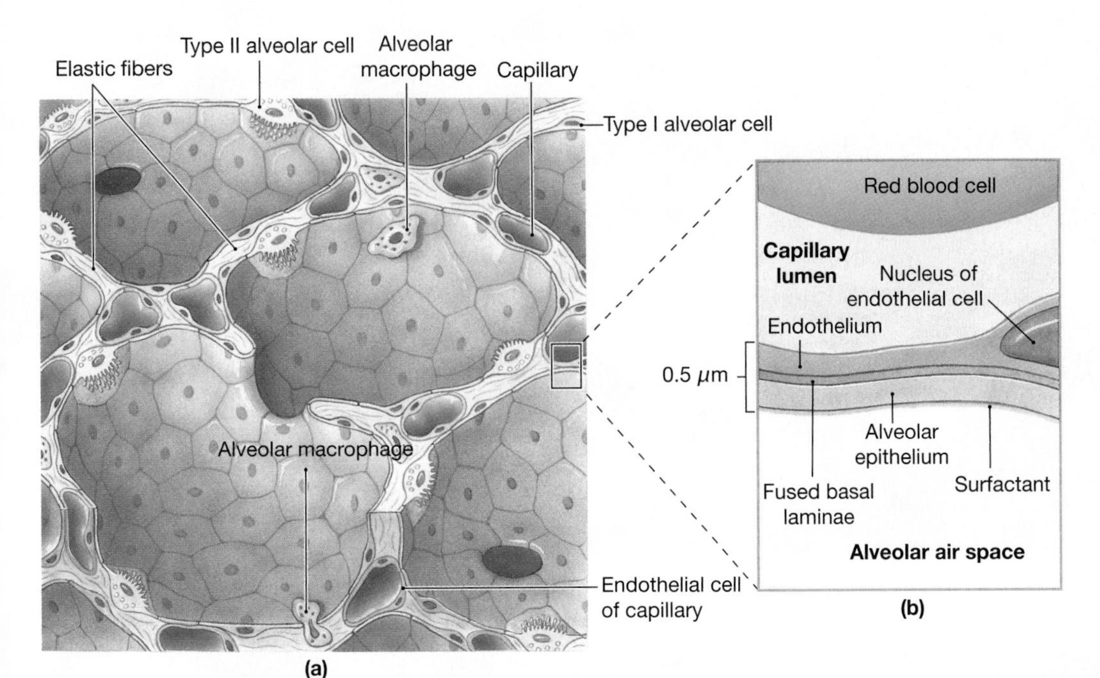

Figure 24.10 Organization and structure of the alveoli and the respiratory membrane. a) diagram illustrating the structure of alveoli; **b)** structure of the respiratory membrane.

tered among the more prevalent type I cells. Type II cells secrete a phospholipid substance called **surfactant**, which lowers the surface tension along the surface of alveoli and prevents the air sacs from collapsing.

7. In some alveoli you might identify cells within the alveolar spaces. These cells are **alveolar macrophages** (Figure 24.10a). They act as scavengers to remove dust particles or infectious microorganisms that were not filtered out by the mucous secretions in the larger airways.
8. Search for arterioles and venules. At first glance, these blood vessels are easily confused with bronchioles, but closer examination at low and high magnifications reveals the anatomical differences between these structures. A blood vessel has a simple squamous epithelium lining the lumen (endothelial cells in the tunica intima), while a bronchiole has a simple cuboidal epithelium. In addition, the smooth muscle in the tunica media of the blood vessel will be more prominent than the relatively thin layer of smooth muscle around a bronchiole (Figures 24.9b and c).

QUESTIONS TO CONSIDER

1. Nicotine in cigarette smoke promotes bronchoconstriction of small bronchioles. Explain how this can affect breathing.

____

____

2. Emphysema is a condition that is characterized by damage to alveolar walls. It can develop as a result of long-term exposure to air pollutants or by cigarette smoking. Explain how emphysema will affect gas exchange between the lungs and pulmonary capillaries.

____

____

Exercise 24 Review Sheet

Anatomy of the Respiratory System

Name ____________________

Lab Section ____________________

Date ____________________

1. Identify the structures that comprise the upper respiratory tract and the lower respiratory tract.

____

____

Questions 2–7: Match the term in column A with the correct description in column B.

A	B
2. Hard palate ____________	a. Space in the nasal cavity
3. Nasal septum ____________	b. Formed by portions of the maxillae and palatine bones
4. Internal nares ____________	c. Divides the nasal cavity into right and left parts
5. Meatus ____________	d. Opening at the posterior end of the nasal cavity
6. Concha ____________	e. Separates the nasal cavity from the cranial cavity
7. Cribriform plate ____________	f. Bony projection attached to the lateral wall of the nasal cavity

8. Describe the structure of the respiratory mucous membrane (respiratory mucosa).

____

____

9. Describe the anatomic positions of the three regions of the pharynx.

____

____

10. Describe the basic structure of the trachea.

____

____

11. What is the functional difference between the conducting airways and the respiratory airways?

____

____

12. Describe the basic branching pattern of the bronchial tree.

____

____

Questions 13–15: Describe the following changes that occur in the airway walls as the bronchial tubes become progressively smaller in diameter.

13. Presence of cartilage:

14. Presence of smooth muscle:

15. Airway epithelium:

16. Describe the structure and function of the respiratory membrane.

Questions 17–19: Describe the functions of the following cells.

17. Type I alveolar cells

18. Type II alveolar cells

19. Alveolar macrophages

20. Compare and contrast the gross anatomical structure of the left and right lungs.

Questions 21–31: Identify the labeled structures.

21. ______________________________

22. ______________________________

23. ______________________________

24. ______________________________

25. ______________________________

26. ______________________________

27. ______________________________

28. ______________________________

29. ______________________________

30. ______________________________

31. ______________________________

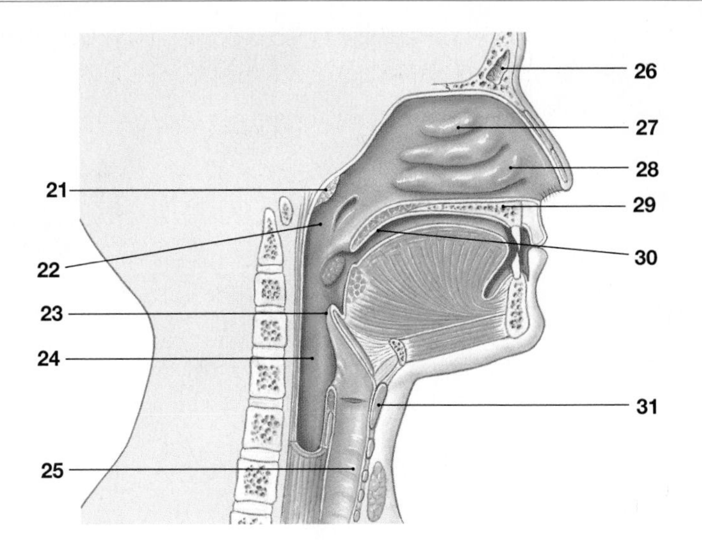

Questions 32–40: In the following diagram, identify the structures by labeling with the color that is indicated. After completing the coloring exercise, label all structures that are *not* colored.

32. Trachea = **blue**
33. Left primary bronchus = **yellow**
34. Thyroid cartilage = **green**
35. Right superior lobe = **red**
36. Left lobar bronchi = **purple**
37. Left inferior lobe = **black**
38. Right segmental bronchi = **brown**
39. Right middle lobe = **light blue**
40. Cricoid cartilage = **orange**

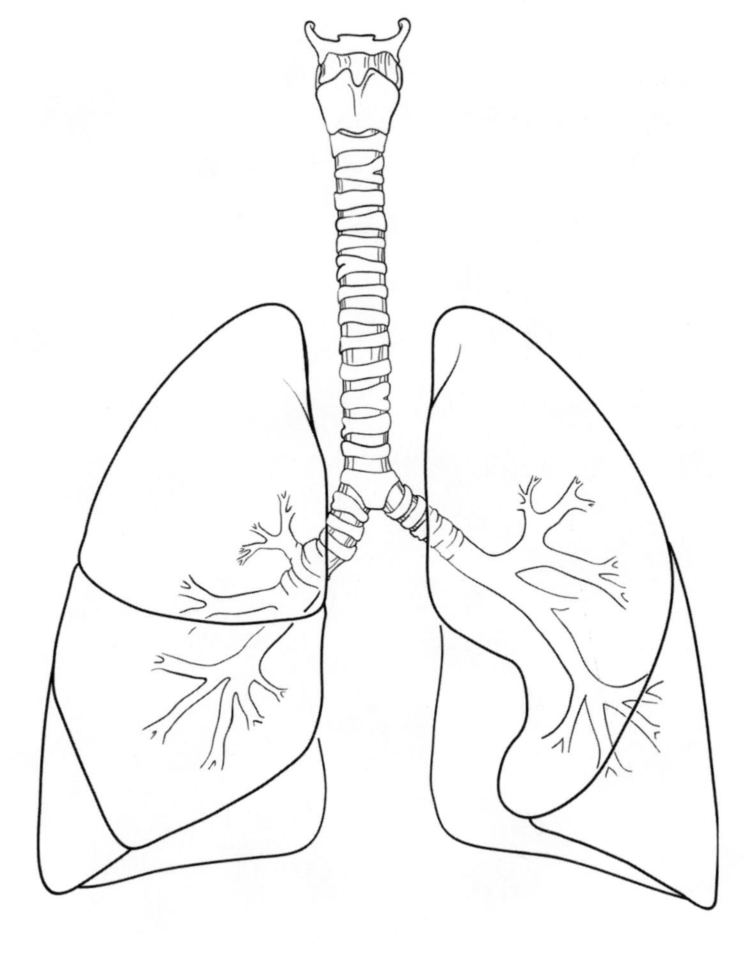

Exercise 25

Respiratory Physiology

Laboratory Objectives

On completion of the activities in this exercise, you will be able to:

- Describe the activities required for respiration.
- Identify and discuss the functions of the muscles involved in pulmonary ventilation and explain the role of elastic recoil.
- Explain the importance of Boyle's law, Dalton's law, and pressure gradients during respiration.
- Differentiate between the various respiratory volumes and capacities.
- Measure or calculate the various respiratory volumes and capacities.
- Compare predicted and actual vital capacities and explain differences between the two.
- Measure the depth of ventilation and the breathing rate during various patterns of breathing.
- Evaluate changes in temperature of inspired and expired air.

Materials

- Torso model
- Balloons
- Handheld spirometer (or another model with instructions)
- Disposable mouthpieces
- Biopac Student Lab system

Respiration involves a series of activities that are the result of an intimate functional relationship between the respiratory and cardiovascular systems. These activities include:

- **Pulmonary ventilation (breathing)**, which is the movement of air into and out of the lungs
- **External respiration**, which is the exchange of oxygen and carbon dioxide between the lungs and the blood
- **Transport of gases**, which is how oxygen and carbon dioxide are carried by the blood between the lungs and all tissues
- **Internal respiration**, which is the exchange of oxygen and carbon dioxide between the blood and the body cells

The respiratory process operates according to the principles of **Dalton's law**, which states that the total pressure of a gas mixture is equal to the **partial pressures** of all the individual gases in the compound. For example, the atmosphere contains approximately 78% nitrogen, 21% oxygen, and 0.04% carbon dioxide. (The remainder contains other gases.) If normal atmospheric pressure at sea level is 760 mm of mercury (mm Hg), then the partial pressures of each component would be as follows:

Gas	Partial pressure
Nitrogen	0.78×760 mm Hg = 592.8 mm Hg
Oxygen	0.21×760 mm Hg = 159.6 mm Hg
Carbon dioxide	0.0004×760 mm Hg = 0.3 mm Hg

During respiration, oxygen and carbon dioxide will move along a pressure gradient from a region of high pressure to a region of low pressure. In all cases, the two gases will be moving in opposite directions.

WHAT'S IN A WORD In 1643, Italian physicist Evangelista Torricelli discovered that atmospheric pressure causes a fluid to rise in a tube inverted over the same liquid. This led to the development of the mercury barometer, an instrument that measures air pressure in **millimeters of mercury (mm Hg)**. As mentioned, normal atmospheric pressure at sea level is 760 mm Hg, which means that mercury will rise 760 mm in an inverted tube. Sometimes the term **torr** (named after Torricelli) is used for pressure measurements. Fortunately, 1 torr = 1 mm Hg, so the numbers do not change even if the unit of measure does.

External respiration refers to the exchange of gases that occurs inside the lungs, between the alveoli and the blood capillaries in the pulmonary circulation. The partial pressure of oxygen (P_{O_2}) in the alveoli is greater than the P_{O_2} in the blood of pulmonary capillaries. Consequently, oxygen will move along its pressure gradient, from the alveoli to the capillaries, and the blood becomes oxygenated. At the same time, the partial pressure of carbon dioxide (P_{CO_2}) is greater in the blood than in the alveoli. Thus, carbon dioxide will move in the opposite direction, from the capillaries into the airways (Figure 25.1a). Carbon dioxide is a waste product of cellular metabolism and is excreted during expiration.

Internal respiration is the exchange of gases between blood in the systemic capillaries and individual cells in the body. During this process, the oxygenated blood in the capillaries delivers oxygen to the cells, while the cells release carbon dioxide that enters the capillaries. Once again, the two gases are moving in opposite directions and along their respective pressure gradients (Figure 25.1b).

Respiration would not occur without the efficient transport of oxygen and carbon dioxide between the lungs and individual cells. Almost all oxygen (98%) is transported by binding to hemoglobin in red blood cells. The remainder simply dissolves in the blood plasma. On the other hand, most carbon dioxide (68% to 78%) diffuses into red blood cells where it is converted to carbonic acid (H_2CO_3). Carbonic acid quickly dissociates into

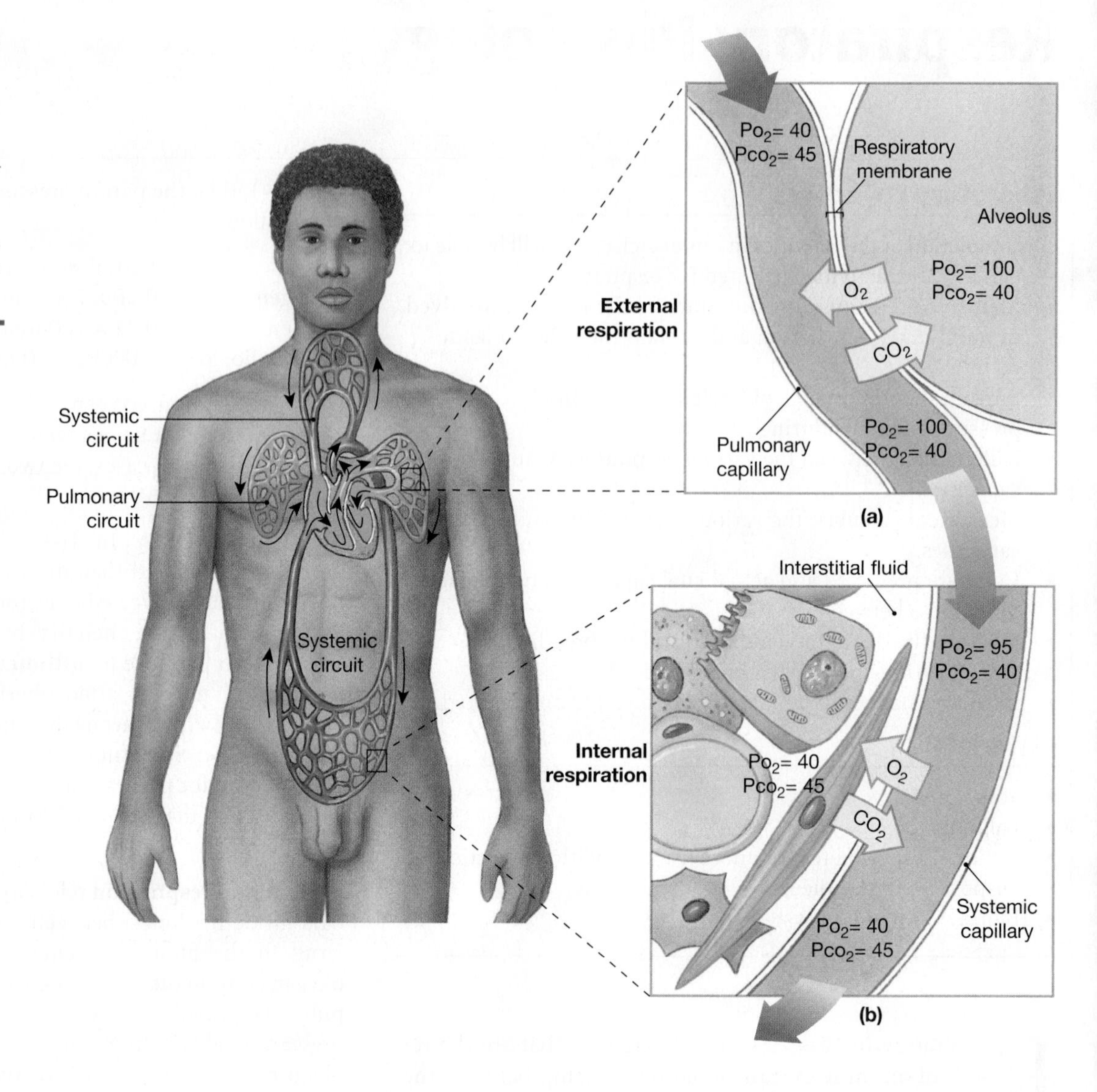

Figure 25.1 External and internal respiration. Oxygen and carbon dioxide diffuse across membranes by moving along pressure gradients. **a)** During external respiration, oxygen spreads into pulmonary capillaries and carbon dioxide moves into alveoli. **b)** During internal respiration, oxygen migrates into cells and carbon dioxide travels into systemic capillaries.

bicarbonate ions (HCO_3^-) and hydrogen ions (H^+). The following chemical equation illustrates this relationship.

$$CO_2 + H_2O \rightleftharpoons H_2CO_3 \rightleftharpoons HCO_3^- + H^+$$

Smaller amounts of carbon dioxide bind to hemoglobin (15% to 25%) or dissolve in the plasma (7%).

In the previous chemical equation, notice that the arrows point in two directions, indicating that the reaction is reversible. This means that carbon dioxide and water can combine to form carbonic acid and, in turn, bicarbonate and hydrogen ions (left to right), or bicarbonate and hydrogen ions can combine to produce carbonic acid and, in turn, carbon dioxide and water (right to left). Although the reaction can proceed either way at any given time, one direction will dominate over the other, depending on the local conditions. For example, during internal respiration, carbon dioxide from cells is entering the systemic capillaries. As carbon dioxide diffuses into the blood, it combines with water to produce bicarbonate and hydrogen ions. Under these conditions, the reaction tends to proceed from left to right.

Notice the relationship between blood concentrations of carbon dioxide and hydrogen ions. During internal respiration, how will the pH of the blood change?

____

During external respiration, carbon dioxide is transferred from the blood in the pulmonary capillaries to the air in the alveoli. In this case, the chemical reaction tends to proceed from right to left.

Explain why the chemical reaction will shift to the left.

____

____

How will the pH of the blood change in this situation?

____

____

Pulmonary Ventilation

Pulmonary ventilation is governed by pressure gradients that form between the air in the atmosphere (atmospheric pressure) and the air in the airways. According to **Boyle's law**, the pressure of a gas is inversely proportional to its volume. Therefore, if the volume of the thoracic cavity increases, the air pressure in the airways decreases below atmospheric pressure (Figure 25.2b). The resulting pressure gradient that forms between the atmosphere (area of high pressure) and the airways (area of low pressure) forces air to move into the lungs **(inspiration)**. Conversely, if thoracic volume decreases (Figure 25.2c), the pressure gradient is reversed and air is forced out of the lungs **(expiration)**.

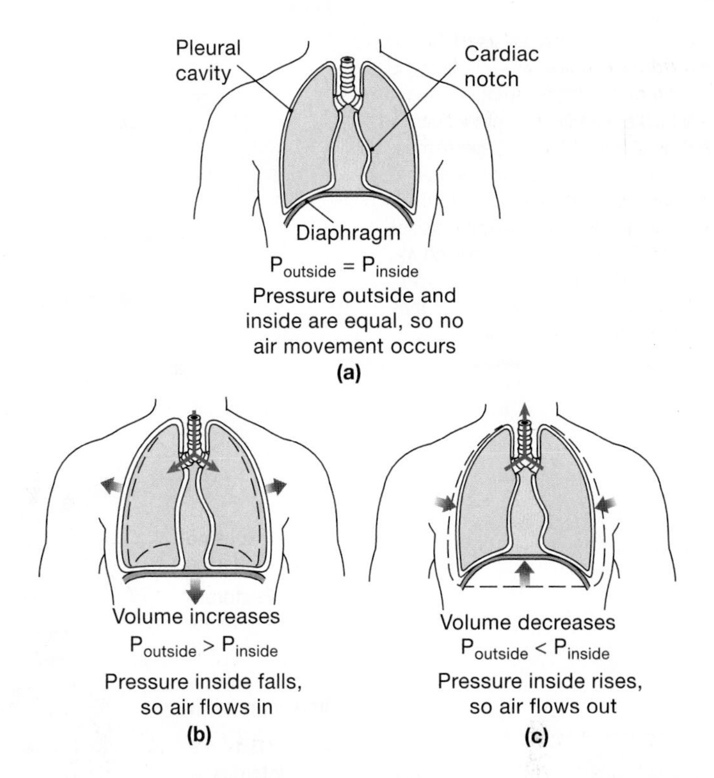

Figure 25.2 Pulmonary ventilation. a) At rest, the outside pressure (atmospheric pressure) equals the inside pressure (pressure within the thoracic cavity). **b)** During inspiration, thoracic volume increases and the pressure within the thoracic cavity falls below atmospheric pressure. The resulting pressure gradient forces air to move into the lungs. **c)** During expiration, the thoracic volume decreases and internal thoracic pressure rises above atmospheric pressure. As a result, air is forced out of the lungs.

CLINICAL CORRELATION

Once our lungs have been expanded with the very first breath we take at birth, they will always contain some air. For this reason, fresh, healthy lung tissue will float when placed in water. Lungs are not filled with air until that first breath at birth, so lung tissue from an unborn infant will not float. This has significant medical and legal significance when an infant is found dead. If the lung tissue does not float, the infant was likely stillborn. If it does float, the child was born alive.

ACTIVITY 25.1 Examining Pulmonary Ventilation

1. Obtain a torso model.
2. Locate the attachment of the ribs to the sternum. Notice that the ribs angle inferiorly and then curve posteriorly and superiorly as they approach the thoracic vertebrae. During inspiration, the ribs are elevated and the sternum moves outward (Figure 25.3a). During expiration, the ribs are depressed and the sternum moves inward.
3. Locate the following muscles involved in **normal** or **quiet inspiration**.
   - The **diaphragm** is the primary inspiratory muscle in the body. It forms a dome-shaped muscular partition between the thoracic and abdominal cavities (Figure 25.3b). Its convex surface forms the floor to the thoracic cavity. When the diaphragm contracts during inspiration, it tightens into a flat sheet that pushes the abdominal organs inferiorly. This action will increase the vertical (superior–inferior) dimensions of the thoracic cavity.
   - The **external intercostal muscles** are the outermost muscular layers in the intercostal spaces (Figures 25.3b and c). Their contractions act to pull the ribs and sternum upward and outward, increasing the transverse and anterior–posterior dimensions of the thoracic cavity.

   During normal or quiet inspiration, the diaphragm and external intercostal muscles act to increase thoracic volume, which in turn decreases the air pressure inside the lungs. Air flows into the lungs along the pressure gradient that has formed (Figures 25.2b and 25.3c).
4. Locate the following muscles involved in **forced inspiration** (Figure 25.3).
   - The **scalenes** are located in the neck. They extend from cervical vertebrae to the first and second ribs. (The scalenes may not be visible on your model.)
   - The **sternocleidomastoid** is a superficial neck muscle. From the mastoid process on the temporal bone, it travels inferiorly and medially through the neck and attaches to the clavicle and manubrium of the sternum.
   - The **pectoralis major** is the large superficial muscle on the anterior chest wall. It extends from the sternum and clavicle medially to the greater tubercle of the humerus.
   - The **pectoralis minor** lies deep to the pectoralis major. From its attachment to ribs 3 through 5, it passes superiorly and medially to the coracoid process of the scapula.
   - The **serratus anterior** is attached to the first eight or nine ribs. From there, the muscle passes posteriorly along the rib cage and connects to the medial border of the scapula.

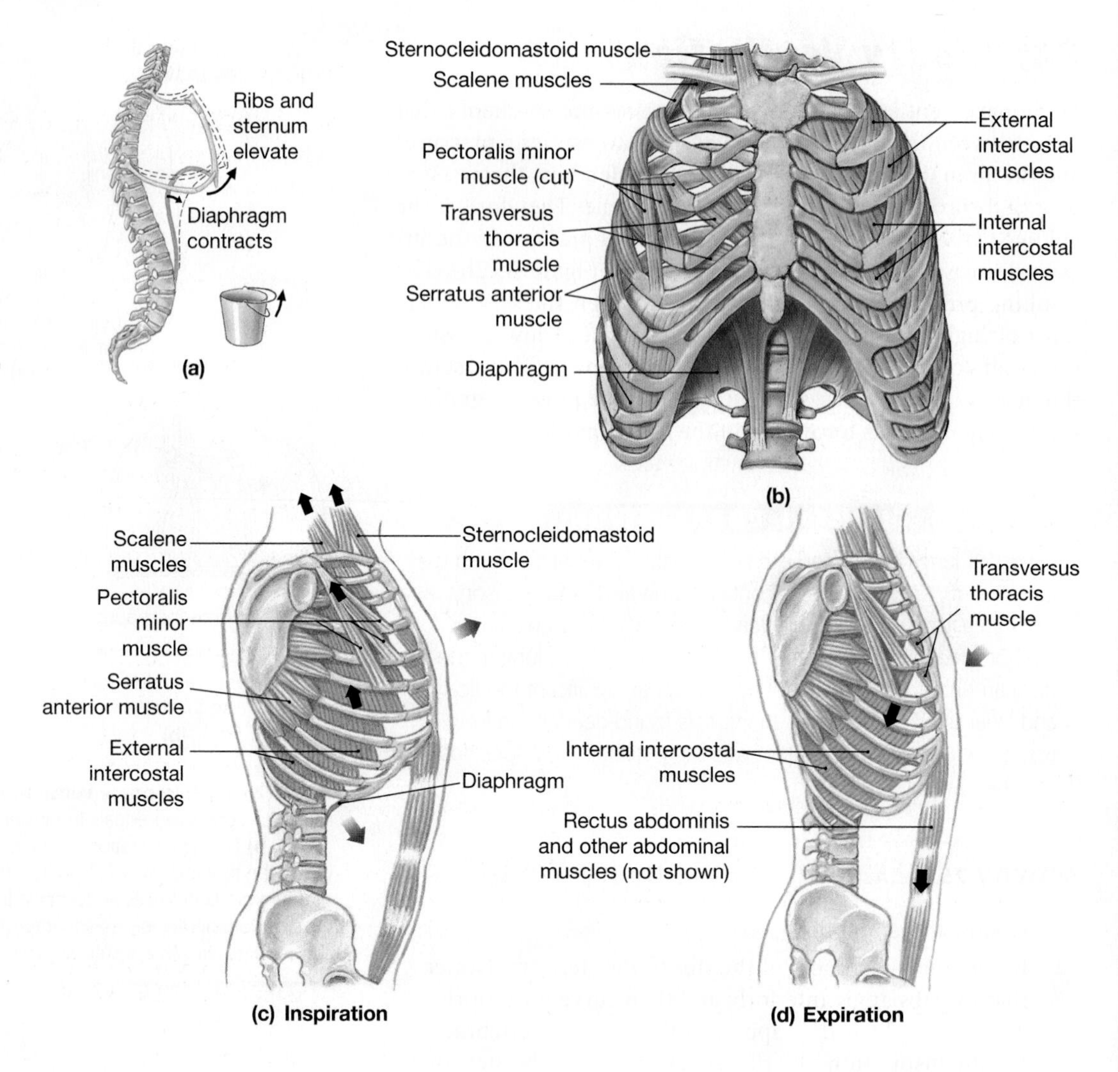

Figure 25.3 Muscles used for respiration. a) During inspiration, the elevation of the ribs resembles the lifting of a bucket handle. **b)** Anterior view of the thoracic wall at rest. Several muscles that are active during respiration are illustrated. **c)** Lateral view of the right thoracic wall, illustrating muscles that act during normal and forced inspiration. **d)** Lateral view of the right thoracic wall, illustrating muscles that act during forced expiration.

During forced inspiration, these muscles create additional increases in thoracic volume by pulling the ribs superiorly and laterally. The resulting increase in the pressure gradient allows a greater volume of air to enter the lungs.

5. Unlike normal inspiration, which occurs when certain muscles are actively contracted, **normal expiration** is a passive process that is initiated by relaxing the muscles of inspiration and through the **elastic recoil** of the lungs. During inspiration, elastic fibers in the walls of bronchial tubes are stretched as the lungs are expanded. When the muscles of inspiration relax, the elastic fibers recoil and the diaphragm, ribs, and sternum return to their original positions. As a result, thoracic volume decreases, and the thoracic pressure increases above atmospheric pressure. Air moves out of the lungs, following the pressure gradient. Elastic recoil in the lungs can be demonstrated by performing the following simple procedure.
   a. Take a balloon and blow it up. Compress the opening of the balloon with your fingers to prevent air from leaking out. The air-filled balloon represents your expanded lungs at the end of an inspiration.
   b. Slowly release air from the balloon (do not release the balloon from your hand). As air is released, the balloon will recoil to its original position, just as the lungs do when the muscles of inspiration relax.
6. Locate the following muscles involved in **forced expiration** (Figure 25.3).
   - The **internal intercostals** are the second layer of muscles in the intercostal spaces. They are deep to the external intercostal muscles.
   - The abdominal muscles—**rectus abdominus, external oblique, internal oblique**, and **transversus abdominis**—form a strong muscular wall that protects the abdominal viscera.

   During forced expiration, the internal intercostal muscles depress the ribs and indirectly cause the sternum to move downward (inferiorly) and inward. These movements decrease the transverse and anterior–posterior dimensions of the thoracic cavity. Contractions of the abdominal muscles push the abdominal organs up against the diaphragm, causing a reduction in the superior–inferior dimensions of the thoracic cavity. The rapid reduction in thoracic volume leads to an additional increase in thoracic pressure and forces more air out of the lungs.

7. Observe the breathing movements of your lab partner while he or she performs a forced inspiration. Focus on changes in the dimensions of the thorax. Observe the changes that occur in thoracic dimensions as your lab partner performs a forced expiration.
8. Gently palpate the sternocleidomastoid muscle on one side of your neck. Feel the contraction of the muscle as you perform a forced inspiration. Feel the muscle relax as you exhale.
9. Palpate your rectus abdominis muscle. Perform a forced inspiration followed by a forced expiration. Feel the contraction of the muscle during the forced expiration.

QUESTION TO CONSIDER The pectoralis minor is one of several respiratory muscles that is used during forced inspiration. In Exercise 11, you learned that the pectoralis minor also depresses and inferiorly rotates the scapula. The muscle has attachments to the coracoid process of the scapula and to ribs 3 through 5. Return the Exercise 11 and review the definitions of the origin and insertion of a muscle. List the attachment of the pectoralis minor that serves as the origin, and as the insertion when the muscle performs the following:

a. Acts on the scapula:

____

____

b. Acts as a respiratory muscle:

____

____

c. Provide a rationale for your answer.

____

____

Respiratory Volumes

Spirometry is a diagnostic technique used to measure respiratory volumes. The instrument used to measure these volumes is called a **spirometer**. Today, spirometry is usually accomplished through the use of sophisticated computerized airflow transducers. A more traditional device, known as a wet spirometer, consisted of a mouthpiece connected with tubing to an air-filled bell inverted in a container of water. When a subject exhaled into the mouthpiece, the bell would rise up in the water. When the subject inhaled, the air in the bell would return to the subject's lungs, causing the bell to sink in the water. A pen attached to the bell would record the bell's movements as a series of waves, called a **spirogram**. This wavelike recording is a graphic representation of an individual's respiratory volumes (Figure 25.4).

WHAT'S IN A WORD The term *spirometer* is derived from the Latin word *spiro*, which means "to breathe," and the Greek word *metron*, which means "measure." The health of a person's respiratory system can be assessed by using a spirometer to measure the volume of air exchanged during inspiration (inhalation) and expiration (exhalation).

The health of a person's respiratory system can be assessed by measuring four **primary lung volumes** (Figure 25.4).

- **Tidal volume (TV)**
- **Expiratory reserve volume (ERV)**
- **Inspiratory reserve volume (IRV)**
- **Residual volume (RV)**

The sum of two or more of these volumes is known as a **pulmonary capacity**. The five pulmonary capacities are as follows (Figure 25.4):

- **Vital capacity (VC)**
- **Expiratory capacity (EC)**
- **Inspiratory capacity (IC)**
- **Functional residual capacity (FRC)**
- **Total lung capacity (TLC)**

CLINICAL CORRELATION

Primary lung volumes and capacities depend on many factors, including the age, height, and sex of the individual. For this reason, measurements that are within 80% of the predicted values are considered normal. However, more significant reductions in these capacities may indicate a serious pulmonary condition. For example, the diagnosis of **pulmonary fibrosis** includes the finding of a significant decrease in inspiratory and expiratory reserve volumes. Fibrosis is the abnormal buildup of fibrous connective tissue along the airways, and can be caused by chronic inhalation of irritants such as coal dust, silicon, or asbestos.

Not all respiratory diseases are associated with reductions in pulmonary volumes. In some cases, the volume of air flowing through the respiratory system is normal, but the air is not flowing quickly enough to meet the body's needs. The restricted airways present in a **chronic obstructive pulmonary disorder (COPD),** such as **asthma,** reduce the rate that air can flow in and out of the lungs per minute. While an asthmatic's vital capacity may be normal, or near normal, the excessive mucous accumulation and narrow airways slow the time it takes for complete exhalation of that vital capacity. Therefore, diagnosis of COPD requires measuring the rate of airflow (ventilation) in addition to measuring the pulmonary volumes.

In the following activity, you will use a simple handheld spirometer to measure or calculate the primary lung volumes and pulmonary capacities (Figure 25.4). Consult your instructor for specific instructions if other types of spirometers are available in your laboratory.

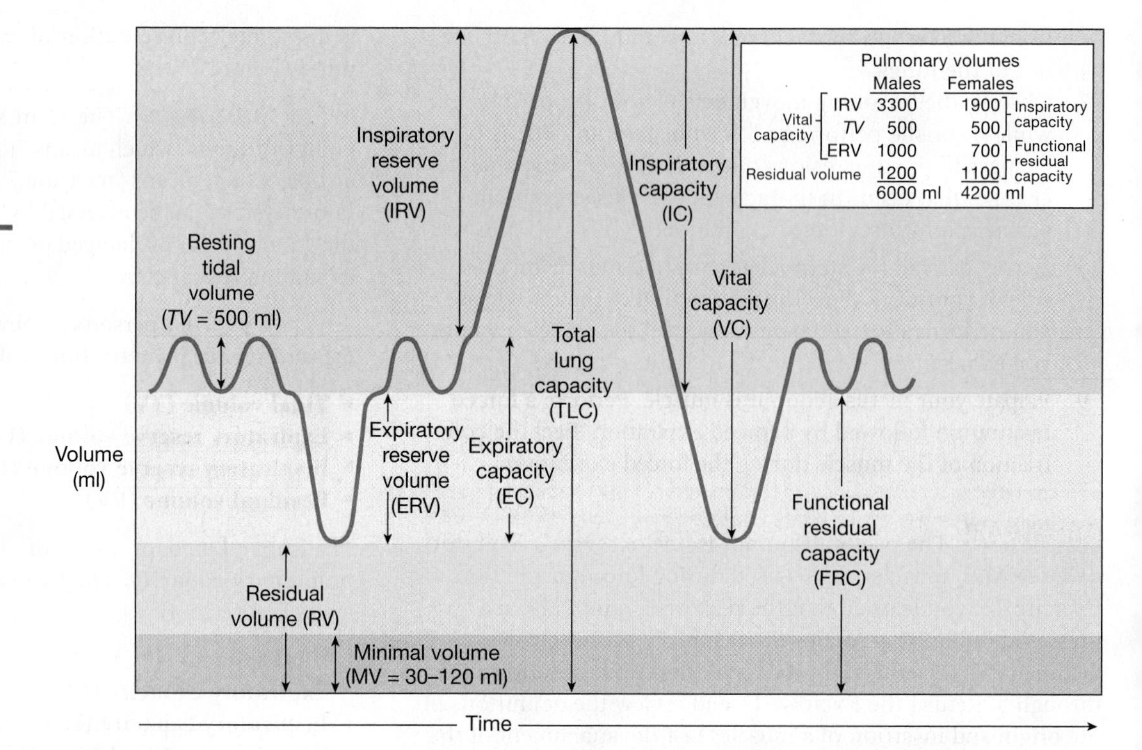

Figure 25.4 Spirogram illustrating respiratory volumes and capacities. On the graph, the curve moves up during an inspiration and down during an expiration. Average values of lung volumes for men and women are listed in the upper right.

ACTIVITY 25.2 Measuring Respiratory Volumes

1. Obtain a handheld spirometer and perform the respiratory volume tests described herein.
2. Each person should attach a disposable mouthpiece before using the spirometer.
3. Before each volume measurement, set the spirometer dial to the zero mark by turning the silver knob at the top of the cylinder.

Expiratory Capacity

The expiratory capacity (EC) is equal to the sum of the tidal volume (TV) and the expiratory reserve volume (ERV). Tidal volume is the amount of air that moves in (or out) of the lungs during normal inspiration (or expiration). Expiratory reserve volume is the maximum volume of air that can be forcefully exhaled after a normal exhalation (after exhaling the tidal volume). The expiratory capacity is, therefore, the maximum volume of air that can be forcefully exhaled after a normal inhalation (after inhaling the tidal volume).

1. Do not begin until you are breathing normally and relaxed.
2. Inhale normally then, blow into the mouthpiece of the spirometer, and forcefully exhale the maximum volume of air that you can release from your lungs.
3. Repeat this procedure two more times and calculate the average value. Record your results in Table 25.1.

Expiratory Reserve Volume

1. Do not begin until you are breathing normally and relaxed.

Table 25.1 Spirometric Measurements of Respiratory Volumes

	Measurements (ml)			
Volume Test	**Trial 1**	**Trial 2**	**Trial 3**	**Average score**
1. Expiratory capacity (EC)				
2. Expiratory reserve volume (ERV)				
3. Tidal volume (TV)	XX	XX	XX	
4. Vital capacity (VC)				

2. At the end of a normal exhalation, forcefully exhale into the spirometer as much air as you can release from your lungs.
3. Repeat this procedure two more times and calculate the average value of your expiratory reserve volume (ERV). Record your results in Table 25.1.

Tidal Volume

1. Calculate your tidal volume (TV) by subtracting your average ERV from your average EC.
2. Can you explain why this calculation will determine your tidal volume? (Hint: Refer to the spirogram in Figure 25.4.)

____

____

3. Record the results in Table 25.1.

Vital Capacity

Vital capacity (VC) is the maximum amount of air that a person can exhale (or inhale). It is equal to the sum of the tidal volume, the expiratory reserve volume, and the inspiratory reserve volume, as shown.

$$VC = TV + ERV + IRV$$

1. Do not begin until you are breathing normally and relaxed.
2. At the end of a normal exhalation, forcefully inhale until the maximum volume of air enters the lungs.
3. Immediately after the forced inhalation, forcefully exhale into the spirometer as much air as possible.
4. Repeat this procedure two more times and calculate the average value. Record your results in Table 25.1.

QUESTION TO CONSIDER Explain why a person suffering from emphysema or pneumonia will most likely have a reduced vital capacity. To answer this question, you must first understand how emphysema and pneumonia affect the lungs. To learn about these diseases, consult your textbook or another reliable source.

____

____

ACTIVITY 25.3 Calculating Respiratory Volumes and Capacities

Several respiratory volumes are not easy to measure, but can be calculated mathematically. In this activity we will examine these volumes and calculations.

Inspiratory Reserve Volume

Inspiratory reserve volume (IRV) is the volume of air that can be forcefully inhaled after a normal inhalation (after inhaling the tidal volume). Inspiratory reserve volume cannot be determined directly with the spirometer. However, it can be calculated mathematically by using the results of two volume tests that you performed earlier. The inspiratory reserve volume is equal to the vital capacity minus the expiratory capacity. The following equation summarizes this calculation.

$$IRV = VC - EC$$

1. Refer to the spirogram in Figure 25.4 to verify that this calculation is correct.
2. Take the average values that you calculated for vital capacity and expiratory capacity to compute your inspiratory reserve volume.
3. Record your results in Table 25.2.

Table 25.2 Mathematical Calculations of Respiratory Volumes

Volume test	Calculation	Measurement (ml)
1. Inspiratory reserve volume (IRV)	IRV = VC − EC	
2. Inspiratory capacity (IC)	IC = TV + IRV	
3. Minute respiratory volume (MRV)	MRV = TV × respiratory rate	
4. Functional residual capacity (FRC)	FRC = ERV + RV	
5. Total lung capacity (TLC)	TLC = VC + RV	

Inspiratory Capacity

Inspiratory capacity (IC) is the maximum volume of air that can be inhaled after a normal (tidal) exhalation. IC can also be determined indirectly by mathematical calculation. It is equal to the tidal volume plus the inspiratory reserve volume. The following equation summarizes this calculation.

$$IC = TV + IRV$$

1. Once again, refer to the spirogram in Figure 25.4 to verify that this calculation is correct.
2. Use the data from the appropriate previous tests to calculate your inspiratory capacity. Record the result in Table 25.2.

Minute Respiratory Volume

Minute respiratory volume (MRV) is the volume of air that moves in (or out) of the lungs per minute during normal inspiration (or expiration).

1. Sit quietly and breathe normally.
2. When you are relaxed, start reading or drawing to distract your attention from your breathing. Have a lab partner count the number of normal respiratory cycles you complete in 1 minute (one cycle = one normal inspiration and one normal expiration). This value is your respiratory rate. Record the result here.

   Respiratory rate: _____________ breaths/minute

3. Calculate your minute respiratory volume by multiplying your average tidal volume (see Table 25.1) by your respiratory rate.

$$MRV = TV \times \text{respiratory rate}$$

4. Record your result in Table 25.2.

CLINICAL CORRELATION

Tidal volume can be obtained with the spirometer, but the results may not be accurate. It is difficult to breathe normally when you are consciously aware of your breathing. To confirm this, try breathing normally while counting your breaths per minute. It is not easy to do this. Therefore, during a physical examination, your respiratory rate is usually taken at the same time that your pulse is being taken, when you are not aware of it.

For the last two volume tests, you must know the value of your residual volume (RV), which is the amount of air that remains in the lungs after the most forceful expiration. Since there is no practical way to measure this volume, you can assume that, on average, the residual volume is 1100 ml (1.1 L) for females and 1200 ml (1.2 L) for males.

Functional Residual Capacity

Functional residual capacity (FRC) is the volume of air remaining in the lungs after a normal expiration. The functional residual capacity can be calculated by adding the expiratory reserve volume and the residual volume. The following equation summarizes this calculation.

$$FRC = ERV + RV$$

1. Refer to the spirogram in Figure 25.4 to verify that this calculation is correct.
2. Use your average expiratory reserve volume (see Table 25.1) to determine your functional residual capacity.
3. Record your result in Table 25.2.

Total Lung Capacity

Total lung capacity (TLC) is the maximum amount of air that the lungs can hold after a maximum forced inspiration. The total lung capacity can be calculated by adding the vital capacity and the residual volume. The following equation summarizes this calculation.

$$TLC = VC + RV$$

1. Refer to the spirogram in Figure 25.4 to verify that this calculation is correct.
2. Use your average vital capacity (see Table 25.1) to determine your total lung capacity. Record your result in Table 25.2.

QUESTION TO CONSIDER During the previous activity, you were able to calculate inspiratory reserve volume (IRV) and inspiratory capacity (IC) mathematically. Explain why these volumes could not be determined directly with the spirometer.

____

____

Prediction of Vital Capacity

Vital capacity is a reliable diagnostic indicator of pulmonary function. A person's vital capacity should be at least 80% of the predicted vital capacity. Healthy individuals who exercise regularly will usually have vital capacities that are greater than their predicted values. On the other hand, individuals who smoke or have a pulmonary disease, such as emphysema, have reduced vital capacities.

Vital capacity varies, depending on an individual's height, age, and gender. As a general rule, vital capacity increases with height and decreases with age. However, women usually have smaller vital capacities than men of comparable height and age. The following equations can be used to predict any individual's vital capacity.

- Female: $VC = 0.041H - 0.018A - 2.69$
- Male: $VC = 0.052H - 0.022A - 3.60$

VC = vital capacity in liters (L)
H = height in centimeters (cm)
A = age in years (yrs)

For example, the predicted vital capacity for a woman who is 66 in (5 ft, 6 in or 167 cm) tall and 28 years old can be calculated as follows:

$$\begin{aligned} VC &= 0.041H - 0.018A - 2.69 \\ &= 0.041(167) - 0.018(28) - 2.69 \\ &= 3.66 \text{ L } (3660 \text{ ml}) \end{aligned}$$

Suppose the actual vital capacity of the woman in this example is 3.5 L. What percent of the predicted vital capacity is her actual value? We can answer this question by dividing the actual value by the predicted value, and multiplying by 100:

$$\text{Actual value / predicted value} \times 100$$
$$(3.50 \div 3.66) \times 100 = 95.6\%$$

Based on this result, we can predict that this woman has excellent pulmonary function. She probably does not smoke, has no serious pulmonary diseases, and leads a healthy lifestyle.

Let's try a second example: a man 71 in tall (5 ft, 11 in or 180 cm), 49 years old, who has an actual vital capacity of 4.42 L (4420 ml). What is the man's predicted vital capacity, and what percent of the predicted vital capacity is his actual value?

1. *Predicted vital capacity*

$$\begin{aligned} VC &= 0.052H - 0.022A - 3.60 \\ &= 0.052(180) - 0.022(49) - 3.60 \\ &= 4.68 \text{ L } (4680 \text{ ml}) \end{aligned}$$

2. *Percent of predicted vital capacity*
   a. Predicted vital capacity = 4.68 L
   b. Actual vital capacity = 4.42 L
   c. Percent of predicted vital capacity: $(4.42 \div 4.68) \times 100 = 94.4\%$

The man's vital capacity is close to 95% of his predicted value. Presumably, this can be attributed to the fact that he is a nonsmoker and leads a healthy lifestyle that includes daily exercise.

ACTIVITY 25.4 Predicting and Measuring Vital Capacity

Form a Hypothesis **Before you begin, speculate on the relationship between your actual and predicted vital capacities by selecting one of the following choices.**

- **My actual vital capacity will be less than 80% of my predicted vital capacity.**
- **My actual vital capacity will be between 80% and 100% of my predicted vital capacity.**
- **My actual vital capacity will be greater than 100% of my predicted vital capacity.**

____

____

1. Use the procedure just described to predict the vital capacities for you and your partner.
2. Compare your actual vital capacity (see Actvity 25.2; recorded in Table 25.1) as a percentage of the predicted value.
3. Record your results in Table 25.3.

Assess the Outcome **Did your results agree or disagree with your hypothesis? If your actual values differ from your predicted values, what might be some of the reasons for the variation?**

____

____

QUESTION TO CONSIDER Measurements of vital capacity that are consistently below 80% of the predicted value on repeated tests suggest the presence of a restrictive lung disease. Are there any individuals in your laboratory whose lung volumes are of concern?

____

____

Depth and Rate of Ventilation

A normal respiratory rhythm during quiet breathing is called **eupnea.** The average resting respiratory rate is 12 to 14 breaths per minute (bpm). This rate is controlled by respiratory centers in the medulla oblongata. These centers primarily respond to changes in the pH of the blood and cerebrospinal fluid. If a person is insufficiently ventilating, carbon dioxide levels in the blood rise, causing blood pH to fall. This condition, called **hypoventilation**, is caused by breathing either too shallowly or too slowly, or a combination of both. In either case, the body is producing carbon dioxide faster than it is eliminating it through ventilation. The respiratory control centers respond to the elevated carbon dioxide and low pH by increasing ventilation.

In **hyperventilation**, a person is overventilating and eliminating carbon dioxide faster than it is being produced. Breathing more deeply than necessary is one way a person can hyperventilate. When deep breathing is accompanied by an elevated rate, even more carbon dioxide is eliminated from the body. With low levels of carbon dioxide, the pH of the body's fluids increases. However, since the respiratory control center is primarily sensitive to low pH, there is a reduced stimulus to breathe and the center depresses the respiration rate. After significant voluntary hyperventilation, there may be a temporary cessation of breathing called **apnea vera.** Over time, the reduced respiration rate will cause carbon dioxide levels to rise, returning the fluid pH, and resulting respiration rate, to normal.

WHAT'S IN A WORD *Eupnea* refers to free, easy respiration while at rest. The term is derived from the Greek word *eupnoea,* which means "good breath." *Apnea* refers to the absence of breathing. It originates from the Greek word *apnoia,* which means "want of breath."

In the following activity, you will record the depth and rate of your lab partner's ventilation by using a pneumograph transducer. This transducer monitors the physical expansion of your chest upon inspiration and converts the mechanical signal to an electrical signal of increasing voltage. Expiration relaxes tension on the transducer, resulting in a signal of decreasing voltage.

You will also record the temperature of the inspired and expired air passing through the nostrils of your lab partner. During inspiration, relatively cool ambient air is breathed in. Once inside the body, the air is warmed by contact with the airways of the lungs. Upon expiration, this warmer air will be detected by the temperature probe as it passes out of the body.

Table 25.3 Comparison of Predicted and Actual Vital Capacities

Subject	Predicted vital capacity (ml)	Actual vital capacity (ml)	Actual vital capacity as a % of the predicted vital capacity
Yourself			
Your partner			

ACTIVITY 25.5 Using the Biopac Student Lab System to Measure Depth of Ventilation and Breathing Rate

BIOPAC Systems, Inc.

Setup and Calibration

1. Attach a respiratory transducer (SS5L) to Channel 1 and a temperature transducer (SS6L) to Channel 2 of the Biopac Acquisition Unit and then turn the unit on.
2. Using Figure 25.5 as a guide, attach the respiratory transducer around your lab partner's chest between the armpits and the nipples. Be sure that slight tension remains on the transducer when your lab partner exhales completely.
3. Locate the probe end of the temperature transducer and make a small loop in the cable about 2 in from the tip. Tape this loop to your lab partner's face just below the right nostril (Figure 25.6). Be sure the tip of the transducer probe does not touch the skin. In this position, the probe will measure the temperature of the air passing in and out of the nostril.
4. Start the Biopac Student Lab System, choose Lesson 8 (**L08-Resp-1**), and click **OK**. When prompted, enter a unique filename and click **OK**.
5. Click on the **Calibrate** button, wait 2 seconds and instruct your partner to take one deep breath, exhale, and then breathe normally until the calibration procedure automatically stops after 8 seconds.
6. If there are gradual vertical deflections in the two waveforms, then calibration was successful. If not, click on the **Redo Calibration** button, check all connections, and repeat step 5.

Recording Data

You will record airflow and chest expansion from your lab partner under four conditions: normal breathing, hyperventilation and recovery, hypoventilation and recovery, and coughing and reading aloud. If your lab partner becomes dizzy at any time during this activity, immediately stop the experiment.

1. With your lab partner sitting comfortably and breathing normally, click on the **Record** button.
2. Wait 15 seconds and then click on the **Suspend** button. If the data do not appear correct, or both channels have only flat lines, then click on the **Redo** button, check the transducers, and repeat this recording segment.
3. Click on the **Resume** button. A marker labeled **Hyperventilation and recovery** will automatically be inserted. Immediately instruct your lab partner to breathe rapidly and deeply through the mouth for a maximum of 30 seconds, then resume breathing through the nose. Record for 30 to 60 seconds and then click on the **Suspend** button. If the data do not appear correct, then click on the **Redo** button and repeat this recording segment. Once your lab partner has reestablished a normal breathing pattern, you are ready to proceed to the next step.
4. Click on the **Resume** button. A marker labeled **Hypoventilation and recovery** will be automatically inserted into the recording. Immediately instruct your partner to breathe slowly and shallowly through the mouth for a maximum of 30 seconds and then return to breathing normally through the nose. Continue recording for 30 to 60 seconds and then click on the **Suspend** button. If the data do not appear correct, then click on the **Redo** button and repeat this recording segment. Once your partner has reestablished a normal breathing pattern, you are ready to proceed to the next step.
5. Click on the **Resume** button. A marker labeled **Cough, then read aloud** will be automatically inserted into the

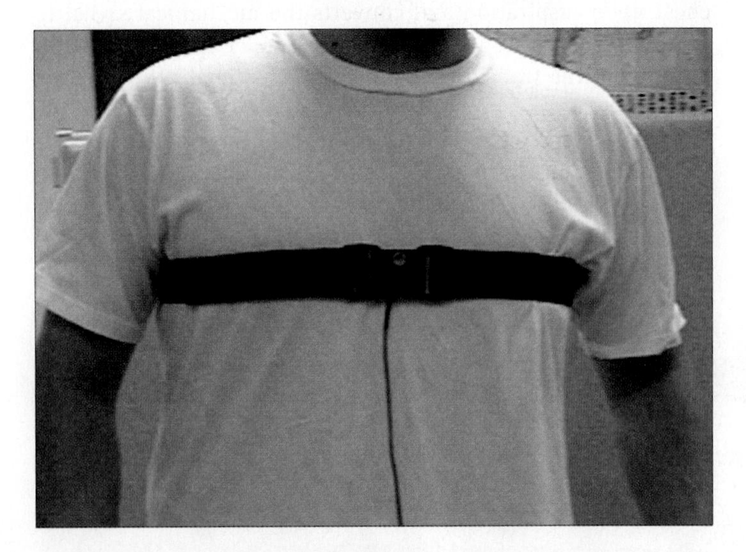

Figure 25.5 Placement of the BIOPAC respiratory transducer. The respiratory transducer should be placed around the chest, below the armpits and above the nipples.

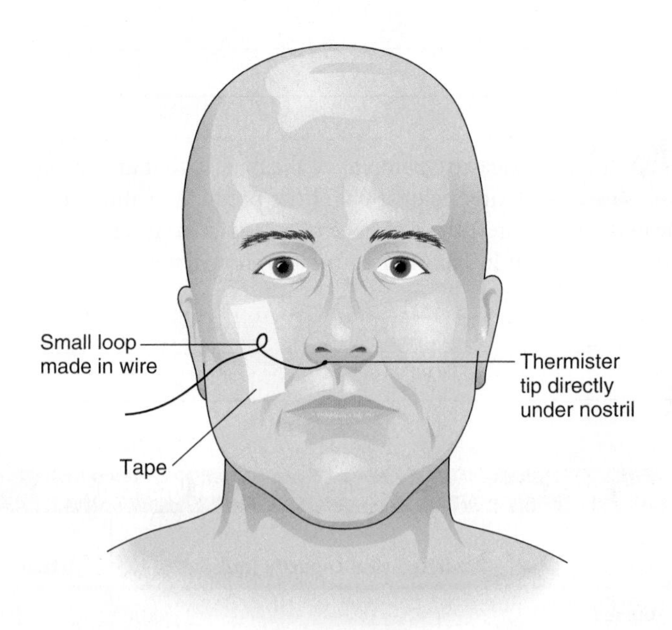

Figure 25.6 Placement of the BIOPAC temperature transducer. The temperature transducer should be positioned with the probe tip below the nostril but not touching the skin.

recording. Have your lab partner cough once and then begin reading aloud for 15 seconds.

6. Click on the **Suspend** button. If the data do not appear correct, then click on the **Redo** button and repeat this recording segment. Otherwise, click on the **Done** button.

Analyzing the Data

1. Enter the **Review Saved Data** mode and select the appropriate data file for Biopac Lesson 8, which ends in -LO8.
2. Inspect the recording. The data should appear similar to Figure 25.7 with Channel 2 displaying airflow waveforms and Channel 40 displaying respiration waveforms.
3. Set three channel measurement boxes as follows:
   - Ch 40 p-p
   - Ch 40 BPM
   - Ch 2 p-p
4. In the following analysis you will measure the depth of ventilation and the breathing rate during normal breathing (Segment 1), the recovery period after hyperventilation (Segment 2), and the recovery period after hypoventilation (Segment 3). The depth of each breath is obtained by using the **p-p** measurement with Ch 40. When you use the I-beam cursor tool to select a single breathing cycle, from the middle of one expiration to the middle of the next expiration, the measurement computes the difference in amplitude between the maximum and minimum peaks within the selected area. The breathing rate (breaths per minute) is calculated using the **BPM** measurement. When you use the I-beam cursor tool to select a single breathing cycle, from the beginning of inspiration to the end of expiration, the measurement computes the duration and divides this into 60 seconds per minute. Figure 25.8 illustrates the selection of a complete breathing cycle for measuring depth and breathing rate (BPM). To accurately select these regions of the waveforms, you will need to use the magnifying glass tool to zoom in on the desired breath cycles. Then, if necessary, from the **Display** menu, use **Autoscale waveforms** to scale and position the cycle for optimal analysis.
   a. Zoom in on a series of three single breath cycles from Segment 1. Using the I-beam cursor tool and the preceding instructions, measure the ventilation depth and breathing rate of three individual cycles. Record in Table 25.4 and calculate the mean values.
   b. Locate the first marker labeled **Hyperventilation and recovery.** Note the long, flat region in the Airflow (Ch 2) waveform. This flat region corresponds to the period when your lab partner was hyperventilating

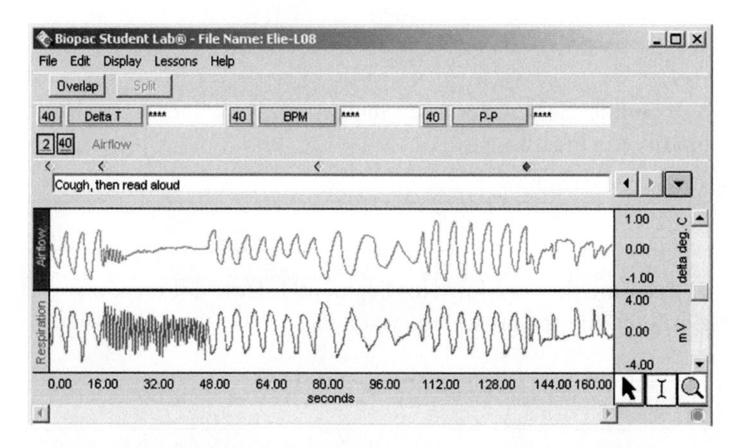

Figure 25.7 Sample BIOPAC data for depth of ventilation and breathing rate. Your data file should appear similar to the window illustrated.

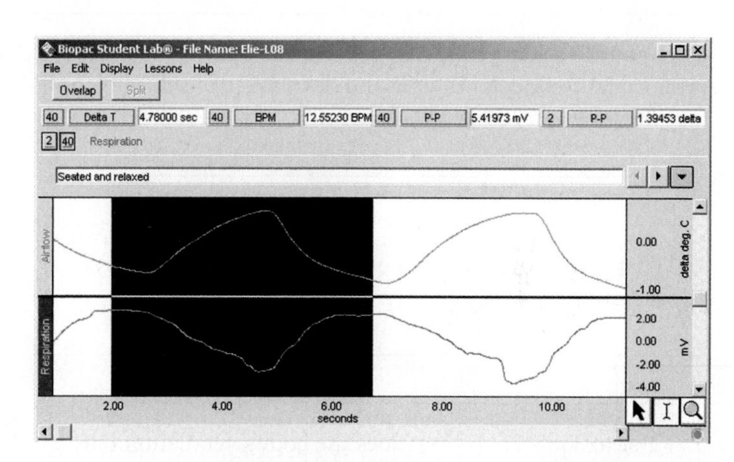

Figure 25.8 Selection of a single breath cycle for measuring depth of ventilation and breathing rate. The I-beam cursor is used to select a complete breath cycle. The **p-p** measurement of Ch 40 provides the ventilation depth of a cycle. From the same selected area, the **BPM** measurement of Ch 40 provides the breaths per minute.

Table 25.4 Comparison of Ventilation Rates after Various Breathing Patterns

	Normal breathing (segment 1)		After hyperventilation (segment 2)		After hypoventilation (segment 3)	
	p-p Ch 40	**BPM Ch 40**	**p-p Ch 40**	**BPM Ch 40**	**p-p Ch 40**	**BPM Ch 40**
Cycle 1						
Cycle 2						
Cycle 3						
Mean						

through the mouth. The recovery period, when your partner again breathes through the nose, begins when the Airflow waveform resumes its cycles.

c. Select three individual cycles within the Respiration (Ch 40) waveform during the first 15 seconds of the recovery period after hyperventilating. Measure the depth and breathing rate for each cycle. Record in Table 25.4 and calculate the mean values.

d. Repeat step c for three individual breath cycles from the first 15 seconds of the recovery period after hypoventilating (Segment 3). Record in Table 25.4 and calculate the mean values.

QUESTIONS TO CONSIDER 1. What physiologic changes occur in the body during hyperventilation?

____

____

2. Refer to Table 25.4. How does the body's ventilation rate and depth respond to a brief period of hyperventilation? Is this as expected? Explain.

____

____

3. Examine the data from Segment 2. Did a period of apnea vera occur? Define apnea vera and describe the feedback loop that causes it.

____

____

4. What physiologic changes occur in the body during hypoventilation?

____

____

5. Refer to Table 25.4. How does the body's ventilation rate and depth respond to a brief period of hypoventilation? Is this as expected? Explain.

____

____

5. Locate a single breathing cycle in the Respiration (Ch 40) waveform of Segment 1. Using the I-beam cursor tool, click on the lowest point of the cycle representing the beginning of inspiration. Note the corresponding point in the Airflow (Ch 2) waveform in the panel above. As inspiration progresses, the Respiration waveform deflects upwards toward a peak. At the same time, in which direction does the corresponding Airflow waveform proceed?

____

6. During expiration, the Respiration waveform deflects downwards toward a valley. At the same time, in which direction does the corresponding Airflow waveform proceed?

____

7. In the following analysis you will examine the changes in temperature of inspired and expired air. The change in temperature is obtained by using the **p-p** measurement with Airflow (Ch 2). When you use the I-beam cursor tool to select a single breathing cycle, the measurement computes the difference in amplitude between the maximum and minimum peaks within the selected area. Using the Airflow (Ch 2) waveform, select three individual respiratory cycles from Segments 1 through 3 and measure the depth (p-p) of each cycle. Record the results in Table 25.5 and calculate the mean values.

QUESTIONS TO CONSIDER 1. What is the relationship between airflow temperature and inspiration/expiration? Why does temperature vary in this manner with the stage of the respiratory cycle?

____

____

2. Describe the relationship between change in air temperature and breathing pattern. _______________

____

____

3. Examine the waveform for Segment 4. How does a cough modify the breathing cycle?

____

____

4. Examine the waveform for Segment 4. How does reading aloud modify the breathing cycle?

____

____

Table 25.5 Relative Change in Temperature after Various Breathing Patterns

Breathing pattern	Cycle 1 (p-p ch 2)	Cycle 2 (p-p ch 2)	Cycle 3 (p-p ch 2)	Mean
Normal breathing (Segment 1)				
After hyperventilating (Segment 2 recovery)				
After hypoventilating (Segment 3 recovery)				

Exercise 25 Review Sheet

Respiratory Physiology

Name ______________________

Lab Section ______________________

Date ______________________

1. Explain what is meant by the partial pressure of a gas and how it applies to respiration.

2. What is meant by external and internal respiration? How does gas exchange differ in each case?

3. Explain how pulmonary ventilation operates according to the principles of Boyle's law.

4. Explain how the actions of the muscles of inspiration create a pressure gradient that allows air to enter the lungs.

5. Normal or quiet expiration is caused by elastic recoil of the lungs. What is meant by elastic recoil? Why is elastic recoil referred to as a passive process?

6. How do carbon dioxide and pH levels in the blood change during hypoventilation and hyperventilation?

7. What is apnea vera? Why does apnea vera occur after a period of hyperventilation?

Exercise 26

Anatomy of the Digestive System

Laboratory Objectives

On completion of the activities in this exercise, you will be able to:

- Identify the various main and accessory organs of the digestive system and explain their roles in the digestive process.
- Recognize the microscopic features of various portions of the digestive system.
- Understand the structure of the wall of the alimentary canal, including specializations found in certain regions.
- Describe the peritoneum and its relationships within the abdominal cavity.
- Trace the path of food through the alimentary canal and explain what digestive activities occur at each location.

Materials

- Torso model
- Clear plastic bag
- Head and neck model, midsagittal section
- Human skull
- Tooth model
- Sagittal section of the male and female pelvis
- Prepared microscope slides
  - Tongue
  - Submandibular gland
  - Esophagus, middle region
  - Esophagus—stomach
  - Stomach, fundic
  - Stomach—duodenum
  - Duodenum, jejunum, ileum
  - Liver
  - Liver, injected
  - Pancreas, islet cells
  - Colon or large intestine

The digestive system includes the organs of the **alimentary canal (gastrointestinal tract)** and various **accessory digestive organs**. The organs of the alimentary canal include the **mouth**, **pharynx**, **esophagus**, **stomach**, **small intestine**, and **large intestine**. The accessory digestive organs include the **teeth**, **tongue**, **salivary glands**, **liver**, **gallbladder**, and **pancreas** (Figure 26.1).

Digestion is the process by which foods are broken down into simpler forms so that nutrients can be delivered to all areas in the body. To perform this function, the digestive system carries out a variety of activities.

- **Ingestion** brings food in as we eat our daily meals.
- **Chewing** chops the food into smaller pieces. This activity makes the food easier to move along the alimentary canal and increases the accessible surface area upon which digestive enzymes can act.
- **Muscular actions** include **peristalsis, mixing movements**, and **sphincter contractions.** Peristalsis pushes the food lengthwise through the alimentary canal in a wavelike fashion. Mixing movements mix the food with digestive enzymes. This activity especially occurs in the stomach (churning) and small intestine (segmentation). At certain locations along the alimentary canal, particularly at the junctions between two organs, the circular muscle layer thickens to form **sphincters.** Sphincter muscles control the amount of material that passes from one organ to the next and also prevent backflow, much like the valves in veins prevent the backflow of blood.
- **Enzymatic breakdown of food (chemical digestion)** is the metabolic conversion of large, complex nutrient molecules into simpler molecules that are small enough to be absorbed from the small intestine into the circulation. The activity relies on the **secretion** of digestive enzymes and begins in the mouth, where salivary amylase splits polysaccharides into disaccharides. However, most enzymatic breakdown occurs in the stomach and small intestine. Most **absorption** occurs in the small intestine.
- Any undigested material passes into the large intestine to form the bulk of the **feces.** Additional water and some vitamins are absorbed by the epithelium of the large intestine before the feces is eliminated during **defecation**, which also allows **excretion** of some wastes.

The Peritoneum

The **peritoneum** (Figure 26.2) is the serous membrane associated with the abdominopelvic cavity and organs. The **parietal peritoneum** covers the walls of the abdominopelvic cavity and is continuous with the **visceral peritoneum** that covers the outside of most abdominal organs. The narrow potential space between the peritoneal layers is the **peritoneal cavity.** It is filled with a watery fluid that is produced by the peritoneal

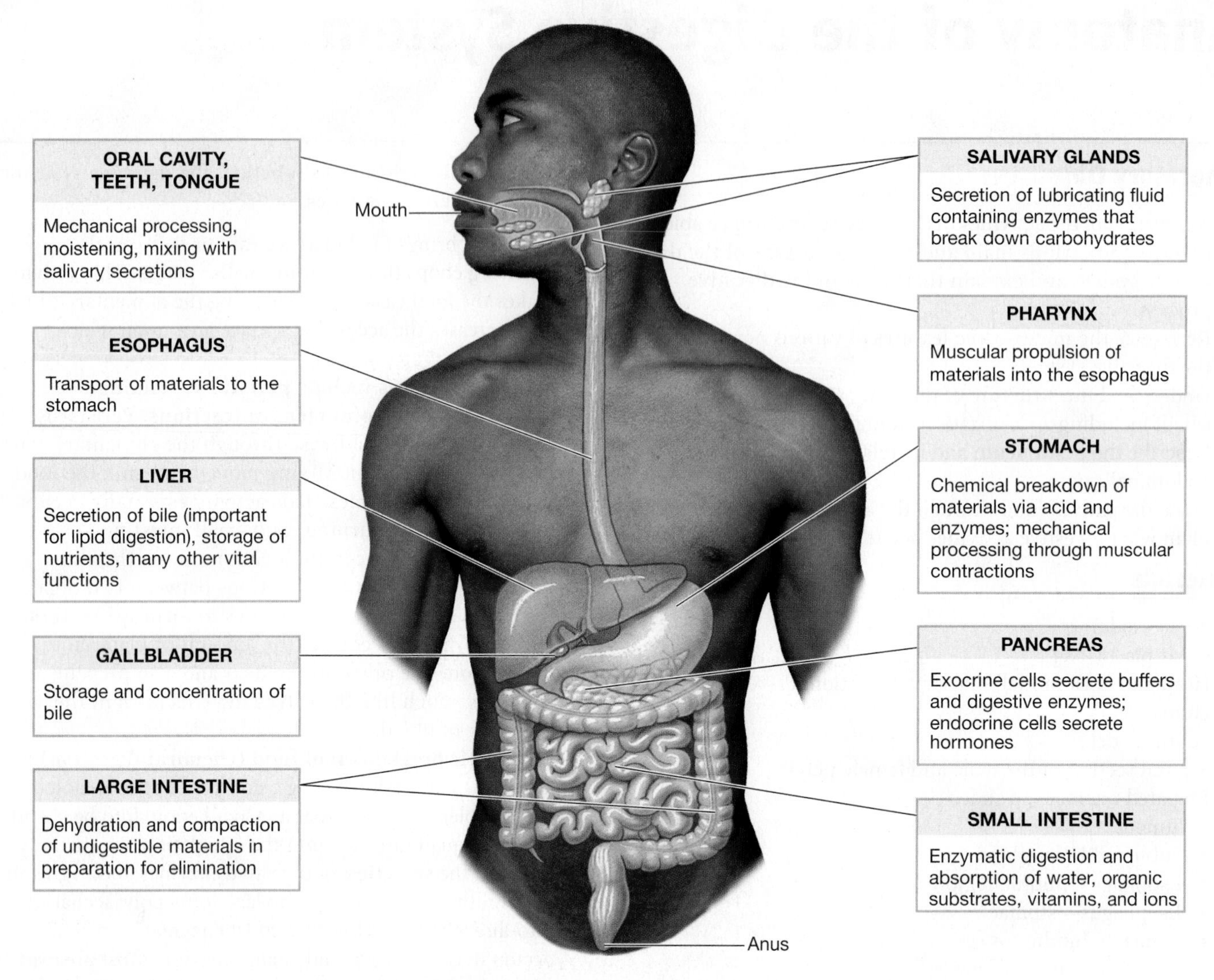

Figure 26.1 Overview of the digestive system. Organs of the alimentary canal and accessory structures are illustrated.

membranes. The fluid lubricates the membrane surfaces so that the organs can move along each other and the body wall during digestive movements.

Mesenteries are double layers of parietal peritoneum that attach some digestive organs to the body wall and provide a passageway for blood vessels, lymphatic vessels, and nerves. Mesenteries can be classified as dorsal (posterior) or ventral (anterior), depending on their connections to the body wall during fetal development (Figure 26.2a). In the adult, the **dorsal mesenteries** include the **mesentery proper**, the **mesocolons**, and the **greater omentum**. The ventral mesenteries include the **lesser omentum** and the **falciform ligament** (Figures 26.2b, c, and d).

Some organs, such as the first part of the small intestine, some regions of the large intestine, and most of the pancreas, are not supported by mesenteries. Instead, they are positioned along the posterior body wall, behind the peritoneum, and are said to be **retroperitoneal** (Figures 26.2c, and d).

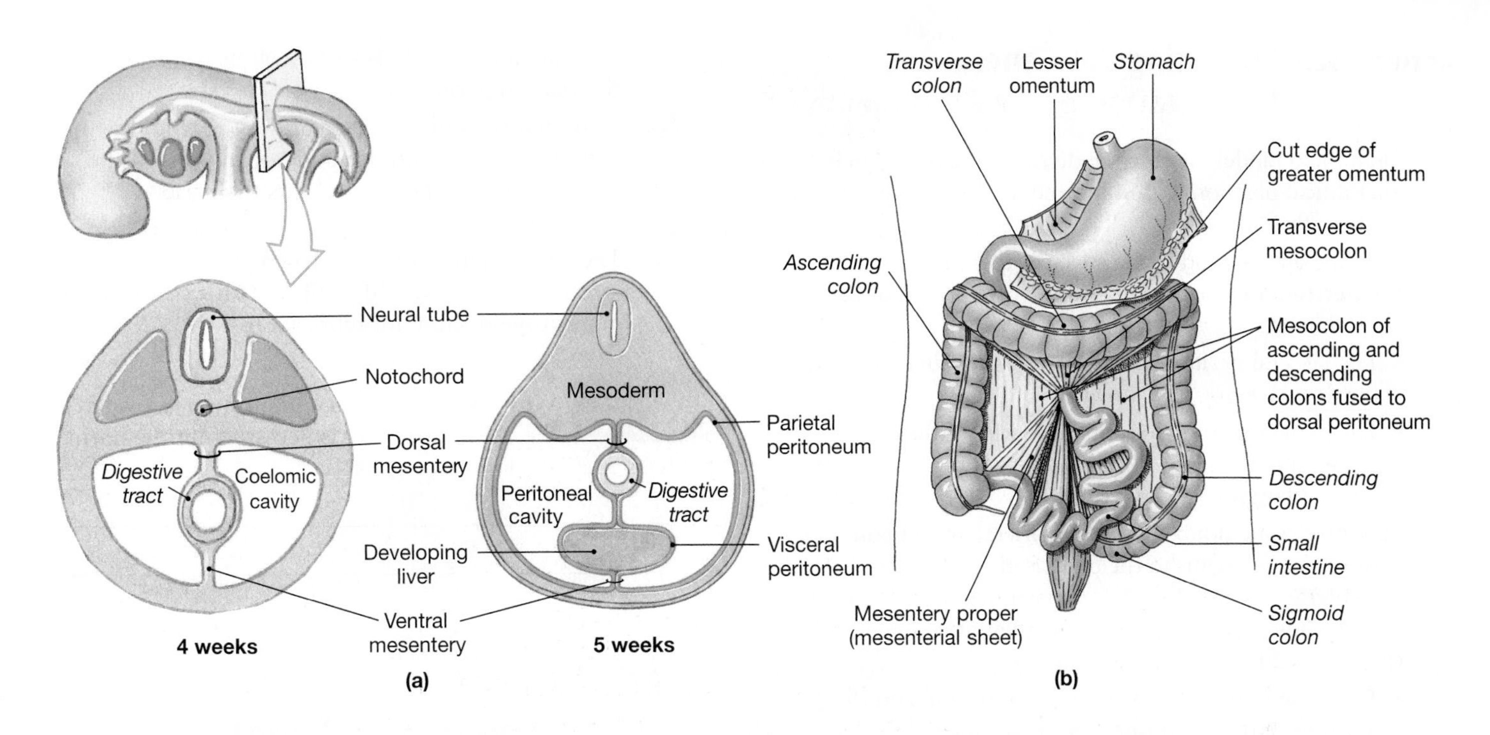

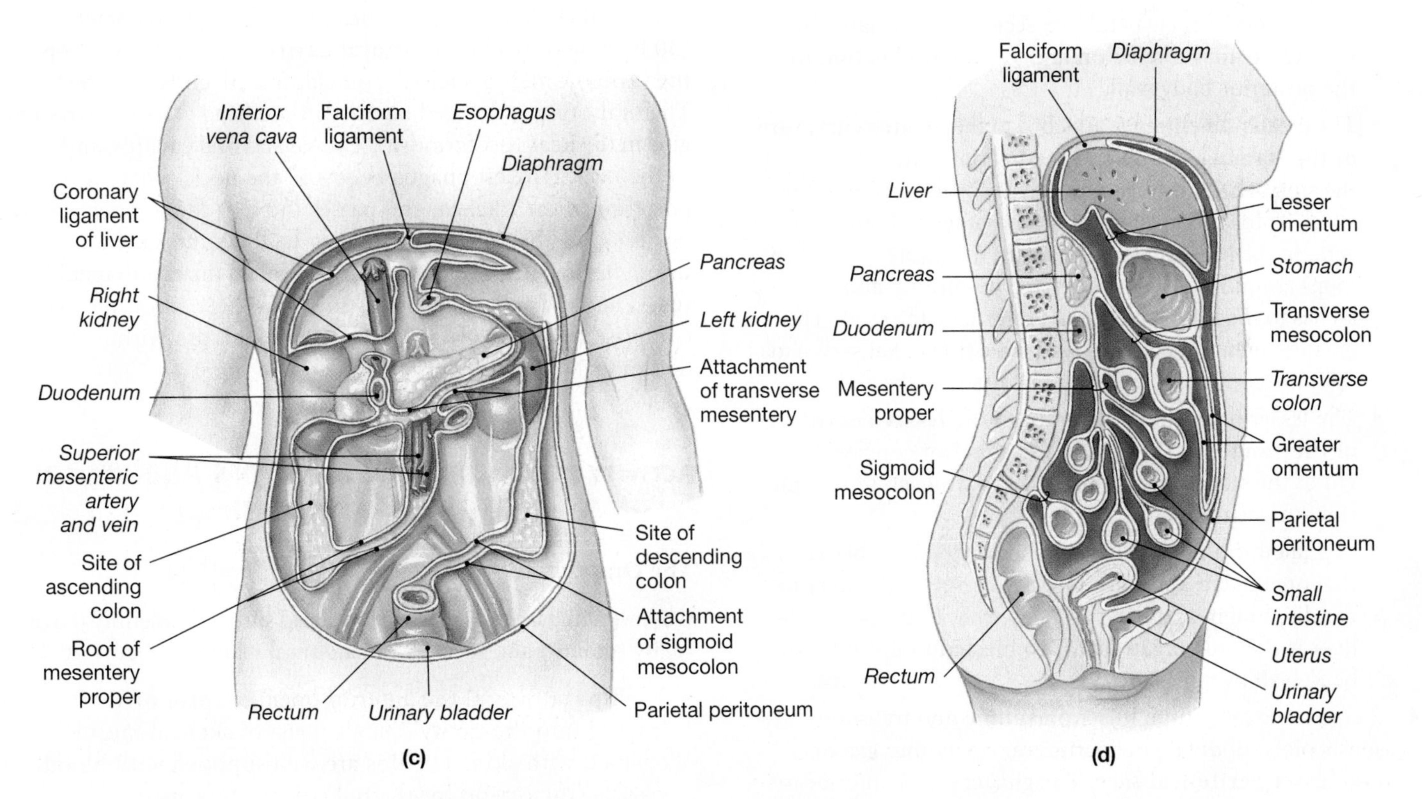

Figure 26.2 Organization of the mesenteries. a) In the embryo, the developing digestive tract is supported by ventral and dorsal mesenteries. In the adult, all ventral mesenteries, except the lesser omentum and falciform ligament, are lost. **b)** Anterior view of the abdominopelvic cavity, showing the arrangement of the mesenteries in the adult. **c)** Anterior view of the abdominopelvic cavity with most organs removed to illustrate the attachments of the mesenteries. **d)** Lateral view of the abdominopelvic cavity illustrating the arrangement of the mesenteries.

ACTIVITY 26.1 Examining the General Organization of the Peritoneum

1. On a torso model observe the digestive organs in their anatomical position within the abdominopelvic cavity (Figure 26.1).
2. Identify, on the model, the surfaces that are covered by the **peritoneal membrane** and the orientation of its components.
   - The visceral peritoneum covers the outside surfaces of the digestive organs.
   - The parietal peritoneum lines the wall of the abdominopelvic cavity, including the inferior surface of the diaphragm.
3. Mesenteries are double layers of parietal peritoneum that connect many organs to the body wall (Figure 26.2). The mesenteries are usually not demonstrated on anatomical models; however, you can identify their attachment sites (Figure 26.2d).
   - The mesentery proper suspends the second and third portions of the small intestine (**jejunum** and **ileum**) to the posterior body wall.
   - The mesocolons suspend two sections of the large intestine, the **transverse colon** and the **sigmoid colon**, from the posterior body wall.
   - The greater omentum is attached to the **greater curvature of the stomach** (the lateral and inferior convex margin of the stomach). It travels inferiorly to form a membranous covering for the small intestine and other abdominal viscera. In the lower abdominal cavity, the greater omentum loops around, runs superiorly and posteriorly, and attaches to the transverse colon of the large intestine. The greater omentum contains lymphatic tissue that serves to fight infections and repair injured tissue in the abdomen.
   - The lesser omentum is attached to the **lesser curvature of the stomach** (the medial and superior concave margin of the stomach). It connects the stomach to the inferior surface of the liver.
   - The falciform ligament is attached to the liver between that organ's left and right lobes. It connects the liver to the diaphragm and anterior body wall. In the adult, the liver is the only organ that is connected to the anterior body wall by a mesentery.
4. The greater omentum, lesser omentum, and transverse mesocolon subdivide the peritoneal cavity into **greater** and **lesser peritoneal sacs**. The greater sac is anterior to the greater omentum, and the lesser sac is posterior to the lesser omentum (Figure 26.2d). To illustrate what a peritoneal sac is, perform the following procedure.
   a. Obtain a clear plastic bag. Fold the top so the bag is closed, but with some air inside. This represents a peritoneal sac.
   b. On the torso model, position the bag in the abdominal cavity so that its inner surface is in contact with the abdominal organs. This surface of the bag represents the visceral peritoneum.
   c. Note that the outer surface of the bag would be in contact with the anterior and lateral parts of the abdominal wall. This represents the parietal peritoneum.
   d. The air-filled interior of the bag represents the peritoneal cavity and would be filled with a small amount of **peritoneal fluid**, for lubrication.

QUESTION TO CONSIDER Why do you think that lubrication, provided by the peritoneal fluid, is necessary for the normal function of the abdominal organs?

____

____

Gross Anatomy of the Alimentary Canal

The alimentary canal is a muscular tube approximately 9 m (30 ft) long. It begins at the **oral cavity**, ends at the **anal opening (anus)**, and is open to the outside at both ends (Figure 26.1). The oral cavity is located entirely in the head. The pharynx begins in the head, posterior to the nasal and oral cavities, and ends in the neck. The esophagus begins in the neck, where it travels posterior to the trachea, and passes through the thoracic cavity, posterior to the heart. It connects to the stomach by passing through an opening in the diaphragm called the **esophageal hiatus**. The stomach, small intestine, and large intestine, which comprise the bulk of the alimentary canal, are within the abdominopelvic cavity.

ACTIVITY 26.2 Examining the Gross Anatomical Structure of the Alimentary Canal

The Oral Cavity

Observe a midsagittal section of the head and neck. Identify the oral cavity (mouth) and review its anatomical relations (Figure 26.3).

- The **lips** surround the anterior opening **(oral orifice)** leading into the cavity. They consist of skeletal muscle covered with skin. The lips are well supplied with blood vessels (the reason for the red color) and sensory receptors.
- Posteriorly, the **fauces** is the opening leading to the oropharynx.
- The **cheeks** form the lateral walls of the oral cavity. Deep to the skin, they contain subcutaneous fat and skeletal muscles (muscles of facial expression and mastication).
- The **tongue** occupies the floor of the oral cavity. It is connected in the midline to the floor of the mouth by a

membranous fold called the **lingual frenulum**. The root of the tongue is anchored to the hyoid bone and covered with lymphatic tissue that forms the **lingual tonsils**.

- The **palate** forms the roof of the oral cavity. The bony, anterior portion is the **hard palate**, composed of portions of the maxillary and palatine bones. The soft tissue posterior portion is the **soft palate**, which is a muscular arch that extends posteriorly and inferiorly. The cone-shaped projection at the posterior tip of the soft palate is the **uvula**. During swallowing, muscular action draws the soft palate and uvula posteriorly and superiorly. This action prevents food from entering the nasal cavity while eating.

The Pharynx

1. Earlier, when the anatomy of the respiratory system was studied (Exercise 24), you examined the structure of the pharynx and its associated structures. Since the pharynx has both a respiratory and digestive function, it will be beneficial to review the anatomy once again.
2. Observe a midsagittal section of the head and neck. Identify the pharynx, which is a muscular tube located posterior to the oral cavity (Figure 26.3b).
3. Identify the three regions of the pharynx (Figure 26.3b).
   - The **nasopharynx** extends from the posterior margin of the nasal cavity to the tip of the uvula.
   - The **oropharynx** extends from the tip of the uvula to the superior margin of the **epiglottis**.
   - The **laryngopharynx** extends from the superior margin of the epiglottis to the bifurcation that gives rise to the larynx anteriorly and the esophagus posteriorly.
4. Locate the **opening to the auditory (Eustachian) tube.** In what region of the pharynx is this located?

____

5. On the model, locate positions of the tonsils (Figure 26.3b).
   - The two **palatine tonsils** are located in the posterior region of each cheek, near the border with the oropharynx.
   - The single **pharyngeal tonsil (adenoid)** is located on the posterior wall of the pharynx, superior to the soft palate.
   - The two **lingual tonsils** can be found at the base of the tongue.
6. Inspect the oral cavity of your lab partner and attempt to identify the palatine tonsils (if they have not been removed). They appear as elevated mounds of soft tissue on the walls of the cheeks in the posterior region of the oral cavity (Figure 26.3a).

The Esophagus

1. On a midsagittal section of the head and neck, locate the inferior border of the laryngopharynx. Identify the split that gives rise to the larynx and trachea, anteriorly, and the esophagus, posteriorly. The esophagus begins in the neck at the level of the C6 vertebra.
2. On a torso model, identify the esophagus and trace its path from the neck into the thoracic cavity (Figure 26.1). The esophagus is a straight muscular tube, about 25 cm (10 in) long that is flattened when not transporting food. Notice that as it descends through the mediastinum, it passes posterior to the trachea and heart, and anterior to the vertebral column.
3. Identify the esophageal hiatus, which is the opening in the diaphragm that allows the esophagus to enter the abdominopelvic cavity. This opening is located at the level of the T10 vertebra. After passing through the diaphragm, notice that the esophagus descends for a very short distance before it connects directly to the stomach at the **cardiac orifice**.

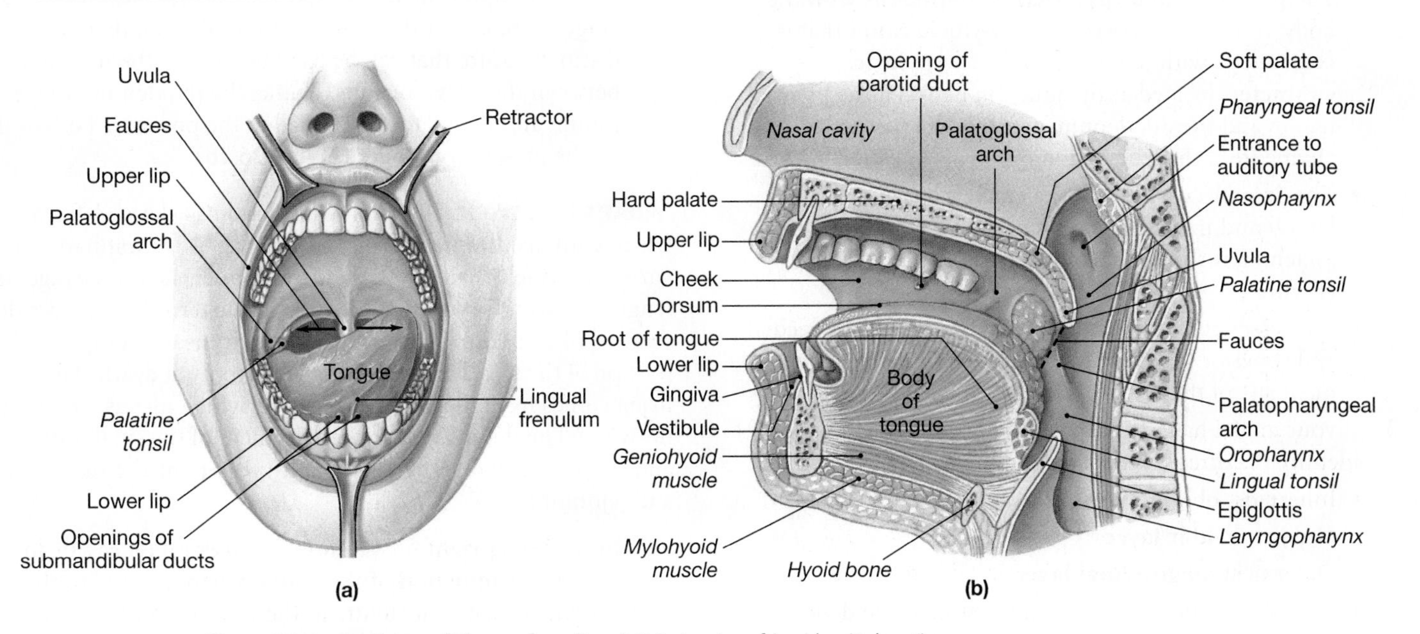

Figure 26.3 Anatomy of the oral cavity. a) Anterior view; **b)** midsagittal section.

CLINICAL CORRELATION

At the junction between the esophagus and stomach, the smooth muscle functions like a sphincter. Even though a true sphincter is absent, it is called the **cardiac** or **lower esophageal sphincter.**

When food passes down the esophagus, the sphincter relaxes to allow food to enter the stomach. If the esophagus is empty, however, the sphincter remains closed to prevent the backflow of gastric contents. Sometimes, after a heavy meal, regurgitation of acidic gastric material does occur. This causes a burning sensation in the chest called **heartburn.** If this condition becomes chronic, it is called **gastroesophageal reflux disorder,** or **GERD.** With time, the damage done by GERD can lead to esophageal cancer.

The Stomach

1. On a torso model, locate the stomach, a J-shaped pouch that is almost entirely located within the upper left quadrant of the abdominal cavity, just inferior to the diaphragm (Figure 26.1). Inside the stomach, food is mixed with gastric juices to form a pastelike material called **chyme**. Gastric digestive enzymes initiate the digestion of proteins before the chyme is transported to the small intestine.
2. Remove the stomach from the torso model and identify the following components (Figure 26.4).
   - The **cardiac region (cardia)** is a small area near the junction with the esophagus.
   - The **fundic region (fundus)** is the domed-shaped area lateral to the cardiac region.
   - The **body** is the main portion of the stomach, and is located inferior to the cardiac and fundic regions.
   - The **pyloric region (pylorus)** is continuous with the body. It narrows to become the **pyloric canal** that is continuous with the duodenum. The **pyloric sphincter**, located at the junction between the pyloric region and the duodenum, controls the passage of chyme from the stomach to the small intestine.
   - The greater curvature is the convex curve along the lateral and inferior margins of the stomach. It is the attachment site for the greater omentum (Figure 26.4a).
   - The lesser curvature is the concave curve on the medial and superior margins of the stomach. It is the attachment site of the lesser omentum.
3. If your model has a cutaway view of the stomach wall, identify the three layers of smooth muscle (Figure 26.4b).
   - Innermost **oblique muscle layer**
   - Middle **circular layer**
   - Outermost **longitudinal layer**
4. If possible on your model, open the stomach and view the numerous folds along the internal surface. These are the **rugae** (Figure 26.4b). After eating a heavy meal, the stomach wall is distended. What do you think happens to the rugae as stomach volume increases?

   ____

WHAT'S IN A WORD In Latin, *ruga* (plural = *rugae*) means "a wrinkle." The rugae form wrinkles or creases along the inner wall of the stomach.

Small Intestine

1. In the torso model, identify the small intestine. Notice that this organ is a highly folded tubular structure that occupies much of the abdominal cavity (Figure 26.1). It is about 6 m (20 ft) long and 2.5 cm (1 in) in diameter (Figure 26.5). The enzymatic breakdown of food is completed in this organ. The nutrient endproducts are absorbed into the blood and lymphatic circulations. Undigested material is eventually transported to the large intestine.
2. Most of the small intestine is suspended from the posterior body wall by the mesentery proper. Remove the bulk of the small intestine from the abdominal cavity and identify the mesentery that supports it.
3. Identify the **duodenum**, which is the first portion of the small intestine (Figure 26.5). Notice that it is continuous with the pylorus of the stomach and forms a C-shaped tube as it travels around the head of the pancreas (Figure 26.6), anterior to the right kidney and the first three lumbar vertebrae. Realize that, with the exception of the initial portion connected to the stomach, the duodenum is retroperitoneal.
4. Identify the second and third portions of the small intestine. The second region, the jejunum, occupies the superior portion of the intestinal folds. The third region, the ileum, occupies the inferior portion of the intestinal folds (Figure 26.5a). At the gross anatomical level, there is no distinct feature that can be used to identify the junction between these two regions. Unlike the duodenum, the jejunum and ileum are suspended to the posterior body wall by the mesentery proper (Figure 26.2d).

WHAT'S IN A WORD The term *duodenum* is derived from the Latin word *duodeni,* which means "twelve." The first part of the small intestine is called the duodenum because, on average, its length is about the width of 12 fingers. The term *jejunum* is a derivative of the Latin word *jejunus,* which means "empty." This portion of the small intestine is usually empty at death. The term *ileum* comes from the Greek word *eileo,* which means "to roll up or twist." The ileum is the longest segment of the small intestine and, therefore, has more twists and turns than the duodenum and jejunum.

5. In the lower right quadrant of the abdominal cavity, locate the termination of the ileum, where it joins to the **cecum**, a saclike structure at the beginning of the large intestine. This junction, known as the **ileocecal junction**, contains a muscular sphincter called the

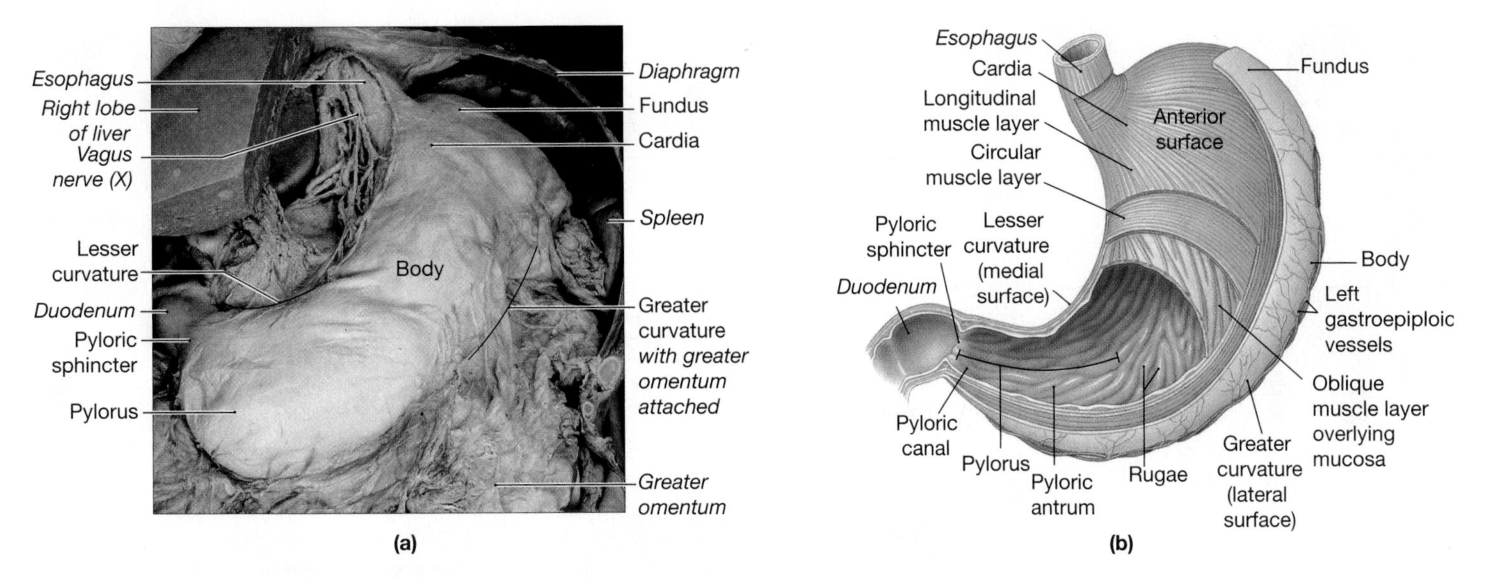

Figure 26.4 Anatomy of the stomach. a) Cadaver dissection of the stomach with the greater omentum attached to the greater curvature; **b)** illustration of the muscle layers in the stomach wall and the internal folds known as rugae.

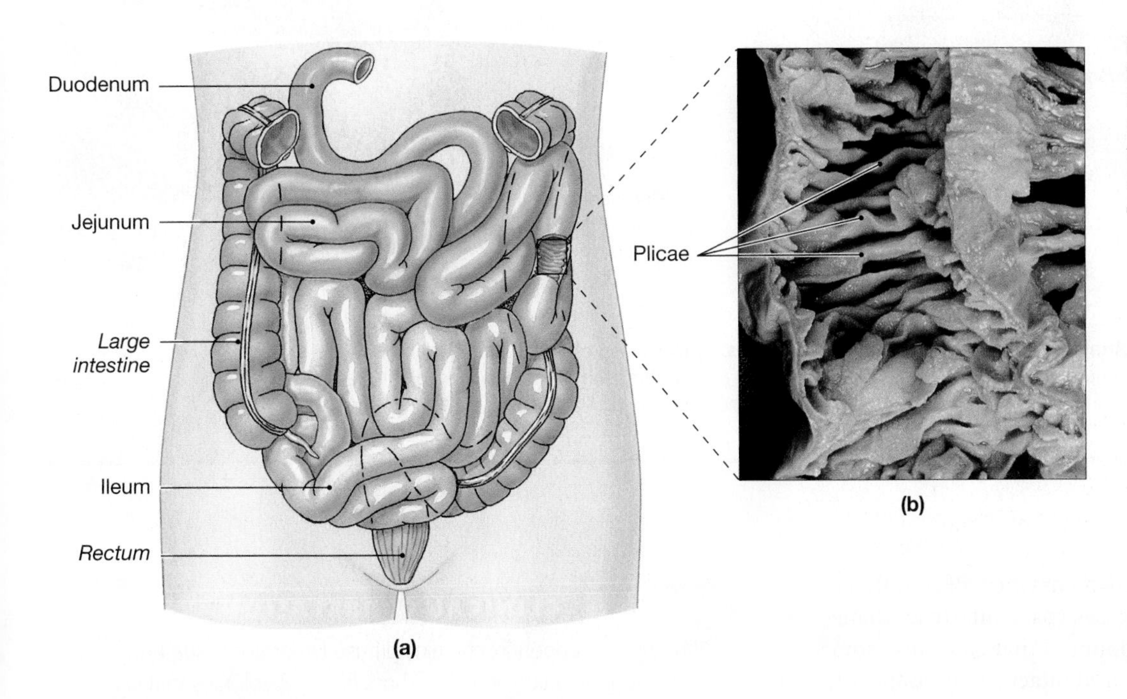

Figure 26.5 Anatomy of the small intestine. a) Illustration of the three regions of the small intestine; **b)** internal view of the jejunum, illustrating the transverse folds known as plicae (plicae circulares).

ileocecal valve that controls the entrance of material into the large intestine (Figure 26.7).

6. If possible on your model, open the small intestine and view the transverse folds along the inner wall. These structures, known as **plicae** or **plicae circulares** (Figure 26.5b), cause a slow, spiraling movement of chyme through the small intestine, thus allowing more time for digestive enzymes to act and for nutrients to be absorbed.

WHAT'S IN A WORD In Latin, *plica* (plural = plicae) means "a fold." The plicae form numerous circular folds along the inner wall of the small intestine.

Large Intestine

1. On a torso model, identify the large intestine (Figures 26.1 and 26.7a). Notice that it is larger in diameter (6.5 cm or 2.5 in) but shorter in length (1.5 m or 5 ft) than the small intestine. Undigested material and anything not absorbed that enters the large intestine is known as feces. Although there is no digestive activity in the large intestine, many electrolytes and most of the water (about 90%) are absorbed from the feces before defecation. The large intestine also provides a suitable environment for beneficial bacteria that synthesize certain vitamins and break down cellulose (fiber), a material

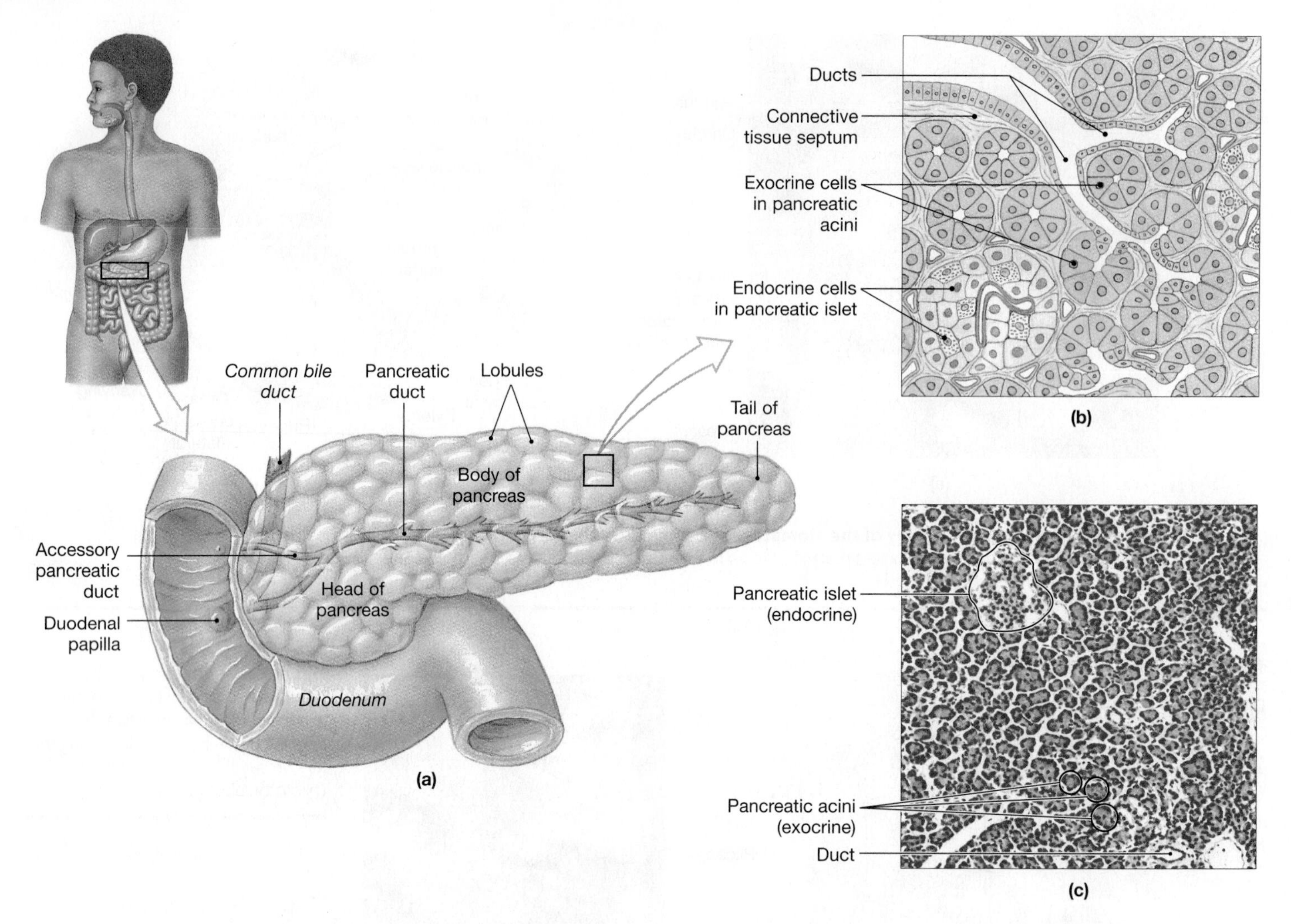

Figure 26.6 Anatomy of the duodenum and pancreas. a) Illustration of the anatomic relationship between the duodenum and pancreas. The duodenum curves around the head of the pancreas. **b)** Diagram; and **c)** light micrograph illustrating the microscopic structure of the pancreas (LM × 100). Notice that the pancreatic islet cells stain lighter than the pancreatic acinar cells.

that is not digested by our own enzymes. Bacterial metabolism produces various gases (particularly methane and sulfur-containing compounds) that can cause some physical discomfort for you and olfactory discomfort for others who might be nearby!

2. In the lower right quadrant of the abdominal cavity, locate the ileocecal junction, where the ileum of the small intestine joins the cecum of the large intestine. If your model can be opened at this junction, open it to view the ileocecal valve, the sphincter that regulates entry into the large intestine (Figures 26.7a and b).
3. The cecum is a pouchlike structure that marks the beginning of the large intestine. Notice that the **vermiform (worm-shaped) appendix** is attached to the posteromedial wall of the cecum (Figure 26.7a).

CLINICAL CORRELATION

Although the appendix contains diffuse lymphatic tissue and has an immune function, it also harbors bacterial colonies that can lead to inflammation **(appendicitis).** If inflammation is severe, removal of the appendix **(appendectomy)** is necessary, hopefully before the structure ruptures and disperses toxic substances into the peritoneal cavity **(peritonitis).**

4. From the cecum, identify the other components of the large intestine (Figure 26.7a), as follows:
   - The **ascending colon** is connected to the cecum. It travels superiorly on the right side of the abdominal cavity, against the posterior body wall. Just inferior to the liver and anterior to the right kidney, the ascending colon

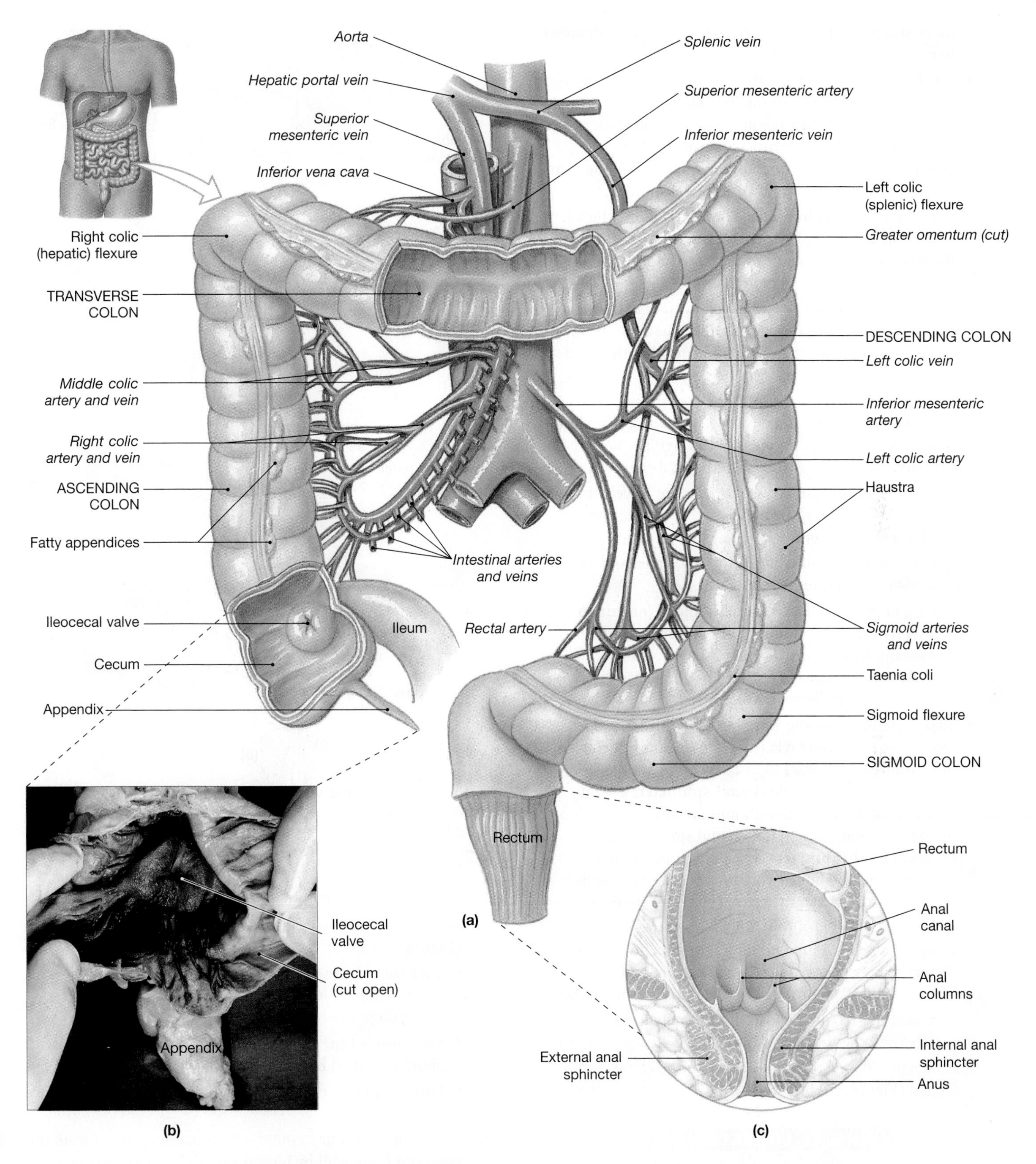

Figure 26.7 Anatomy of the large intestine. a) Illustration of the regions of the large intestine; **b)** cadaver dissection of the ileocecal valve and the appendix; **c)** diagram of the rectum and anus, showing the positions of the internal and external sphincters.

turns sharply to the left **(right colic** or **hepatic flexure)** to become the transverse colon.

- The **transverse colon** passes to the left across the abdominal cavity, just inferior to the liver and stomach. Just anterior to the spleen, the colon makes a second sharp turn **(left colic** or **splenic flexure)** to form the descending colon. (Note the greater omentum draped off of the stomach and transverse colon, if your model shows it.)
- The **descending colon** travels inferiorly along the left side of the abdominal cavity. At the brim of the pelvis, the descending colon curves to the right at the **sigmoid flexure** and joins the sigmoid colon.
- The **sigmoid colon** is an S-shaped portion of the large intestine that begins close to the left iliac crest. It travels medially and joins the rectum at the level of the third sacral vertebra.
- The **rectum** travels anterior to the sacrum and follows the bone's curvature. Approximately 5 cm (2 in) inferior to the coccyx, the rectum passes through the muscle on the pelvic floor and gives rise to the anal canal.
- The **anal canal** begins where the rectum travels through the muscular floor of the pelvic cavity and ends at the **anal opening (anus)**.

5. If possible on your model, open the anal canal and identify the following structures (Figure 26.7c).
   - The inner wall of the anal canal is folded into six to eight longitudinal columns (**anal columns**) and contains many small veins. The abrasive forces of bowel movements can damage these veins and cause discomfort and bleeding **(hemorrhoids)**.
   - Two muscular sphincters are located at the anal opening. These sphincters relax to open the anus during defecation and contract to keep it closed at other times. One sphincter, the **internal anal sphincter**, is composed of involuntary smooth muscle and is controlled by the defecation reflex; the second sphincter, the **external anal sphincter**, is composed of voluntary skeletal muscle. Thus, if you are not close to a bathroom, you can use your external sphincter to delay defecation until an appropriate time.
6. Obtain models that illustrate midsagittal sections of the male and female pelves. Identify the rectum and note its position relative to reproductive structures in the pelvis.
   - In males, the rectum is posterior to the two seminal vesicles and the prostate gland (Figure 26.8a).
   - In females, the rectum is posterior to the cervix and vagina (Figure 26.8b).

CLINICAL CORRELATION

The fact that the rectum is positioned next to several pelvic organs is clinically significant. A doctor can examine the conditions of these organs by inserting his or her finger into the rectum and palpating along its anterior wall. This procedure is known as a **rectal exam.**

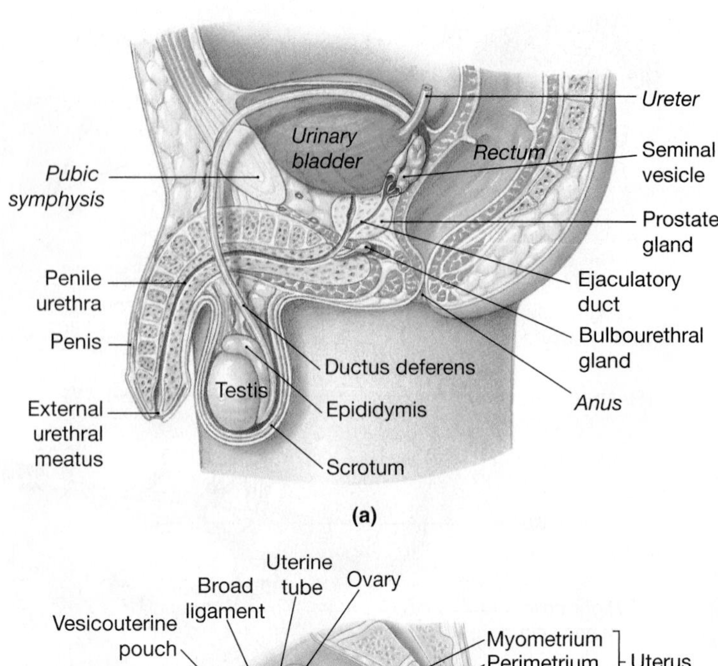

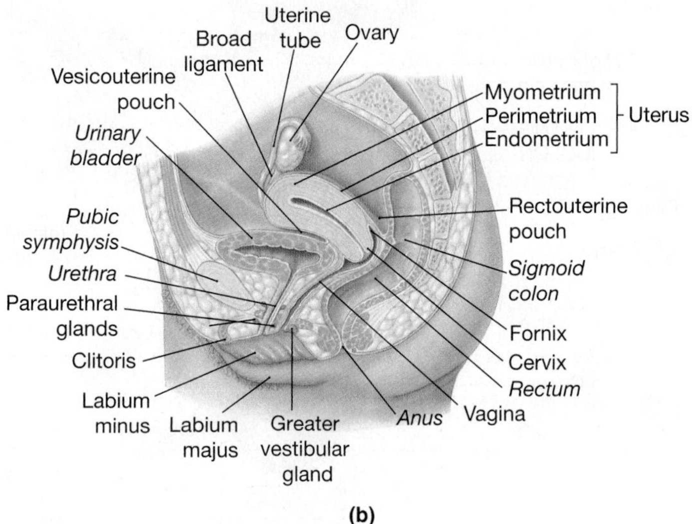

Figure 26.8 Midsagittal section of the pelvic cavity. The rectum travels posterior to male and female reproductive organs. **a)** Male pelvis; **b)** female pelvis.

7. Identify the three specializations of the large intestine (Figure 26.7a).
   - **Taeniae coli** are the three distinct longitudinal bands of smooth muscle found on the wall of all the large intestine, except the rectum and anal canal.
   - The muscle tone of the taeniae coli creates a series of pouches, called **haustra** (singular = haustrum).
   - **Fatty appendices** are fat-filled tabs suspended from the wall.
8. The transverse and sigmoid colons are suspended from the posterior body wall by mesenteries known as mesocolons. On the torso model, identify the mesocolon for the transverse colon (Figure 26.2d). All the other parts of the large intestine are retroperitoneal except for the anal canal which lies completely outside the abdominopelvic cavity.

QUESTIONS TO CONSIDER 1. Review the positions of the nasopharynx, oropharynx, and laryngopharynx (Figure 26.3b). Which regions serve a dual function for respiration and digestion? Provide an anatomical explanation for your answer.

___

___

2. What are the functions of three layers of smooth muscle in the stomach wall during digestion? ___

___

___

3. Explain why the duodenum has a more fixed position in the abdominal cavity than the other two segments of the small intestine, the jejunum and ileum. ___

___

___

4. When unfolded and stretched end to end, the small intestine is about 6 m (20 ft) and the large intestine is 1.5 m (5 ft) in length. Compare the functions of these structures and suggest a reason why the small intestine is so much longer than the large intenstine.___

___

___

5. Although many types of bacteria cause serious diseases, they are not all harmful to humans. In terms of digestion, explain why this statement is true. ___

___

___

Gross Anatomy of Accessory Digestive Structures Associated with the Oral Cavity

The accessory structures linked to the oral cavity include the teeth, tongue, and salivary glands. The tongue forms a large part of the floor of the oral cavity. It consists mostly of skeletal muscle and is covered by a mucous membrane. The tongue is responsible for maneuvering food while chewing and to shape it into a round mass (a bolus) that is easy to swallow.

The teeth are embedded in the sockets formed by the alveolar processes of the mandible and maxillary bones. They are essential for grinding and chewing food before it is swallowed.

Numerous small salivary glands are scattered throughout the mucosa of the tongue, palate, and cheeks. These glands secrete saliva continuously to keep the lining of the mouth moist. In addition, there are three pairs of major salivary glands, as follows:

- Parotid glands
- Submandibular glands
- Sublingual glands

Salivary glands produce two types of secretions.

- Watery serous secretions contain the digestive enzyme, **salivary amylase**, which breaks starch down to disaccharides.
- Thick, mucous secretions bind food particles together and act as a lubricant during swallowing.

ACTIVITY 26.3 Examining the Gross Anatomy of the Teeth, Salivary Glands, and Tongue

The Tongue

1. Obtain a model of a midsagittal section of the head and neck.
2. Identify the tongue and notice that it forms the floor of the oral cavity. The tongue is composed of skeletal muscle. These **intrinsic tongue muscles** are important for changing the shape of the tongue while speaking and swallowing.
3. On the model identify the soft palate and **mandible.** On a skeleton locate the **hyoid bone**, just inferior to the mandible, and the **styloid process on the temporal bone** on the skull. These structures serve as the origins of **extrinsic tongue muscles.** From their origins, these muscles will insert into the connective tissue of the tongue. They act to move the tongue in various directions while chewing food and vocalizing (e.g., speaking and whistling).

Teeth

1. On a skull, observe how the teeth are embedded in the sockets of the mandibular and maxillary alveolar processes. Each quadrant contains 8 teeth (a total of 32 teeth) in the following order from anterior to posterior (Figure 26.9b): central incisor, lateral incisor, cuspid (canine), first bicuspid (premolar), second bicuspid, first molar, second molar, third molar (**wisdom tooth**). Wisdom teeth often become impacted and must be surgically removed. Some individuals do not have wisdom teeth, or may have fewer than the usual four.
2. On the skull, identify an incisor, a canine, and a molar tooth. Compare their sizes and shapes. Each type of tooth has a different shape and provides a special function while chewing. The incisors are chisel shaped and bite off pieces of food, canines are cone shaped to grasp and tear food, and premolars and molars have flattened surfaces to grind food (Figure 26.9b).
3. The alveolar processes of the mandible and maxillae are covered by a mucous membrane called the **gingiva (gums).** Each region of a tooth is identified according to its relationship to the gingival margin (gum line). On a model of a tooth, identify the **crown**, **root**, and **neck** (Figure 26.10). The crown is the visible portion of a tooth, above the gum line. The root is the portion that is below the gum line. It is

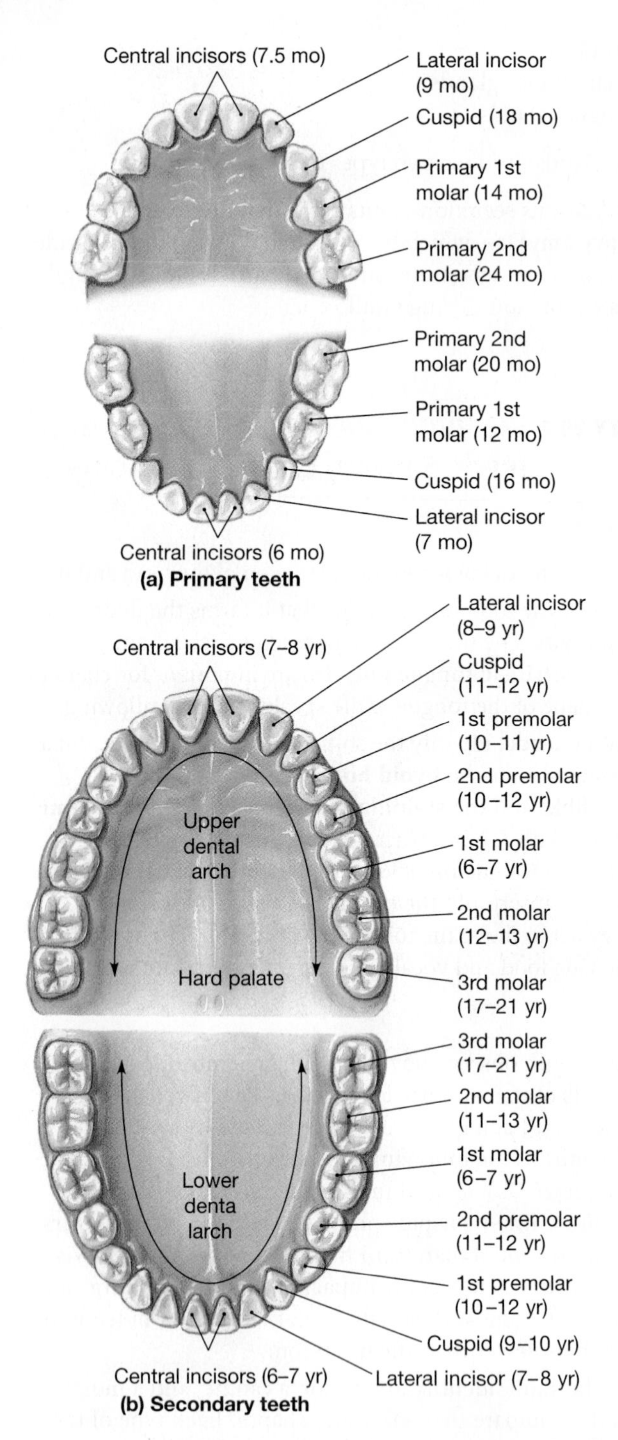

Figure 26.9 The arrangement of teeth in a child and an adult. a) The primary teeth with the average age of appearance, in months, shown in parentheses; **b)** the secondary teeth with the average age of appearance, in years, shown in parentheses.

anchored to the bone by a periodontal ligament and is normally not seen. In some older individuals who have progressive gum disease, the gingival margin recedes and the roots of some teeth become visible. The neck is the region at the gum line where the crown and root meet.

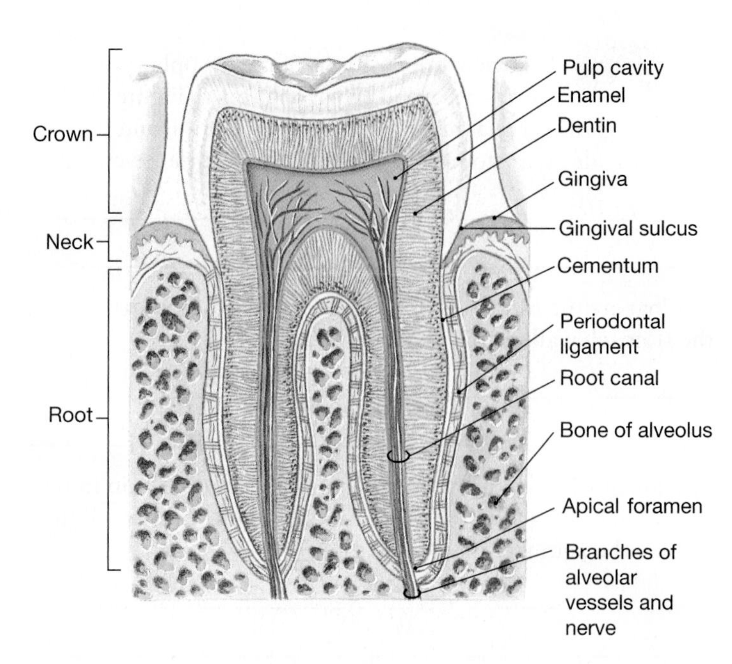

Figure 26.10 The structure of a tooth. The root of the tooth is anchored firmly to bone by a periodontal ligament.

What type of joint is formed by the articulation of the root of a tooth and its bony socket?

____

4. Notice on a model of the tooth that the bulk of each tooth consists of an acellular, bonelike material called **dentin**. The dentin covers the **central (pulp) cavity** in the crown and the **root canal** in the root. In the crown, the dentin layer is covered by a very hard, calcified layer called **enamel**. In the root, **cementum** forms the outer layer.
5. On the tooth model, locate the central cavity and root canal. The central cavity contains blood vessels, nerves, and connective tissue (pulp). Blood vessels and nerves reach the central cavity by way of the root canal, which extends through the interior of the root (Figure 26.10).

CLINICAL CORRELATION

Enamel is the hardest substance in the body, but it tends to wear away with age and is not replaced when it is damaged. The progressive decay of enamel and underlying dentin by decalcification forms **dental caries,** or tooth decay. Tooth decay is caused by the metabolic acids released by bacteria, which feed on sugars and other carbohydrates in the food that we eat. If a bacterial infection spreads into the root canal, all the pulp must be removed and the root canal must be sterilized and filled. This dental procedure is referred to as **root canal therapy** or, simply, a **root canal.**

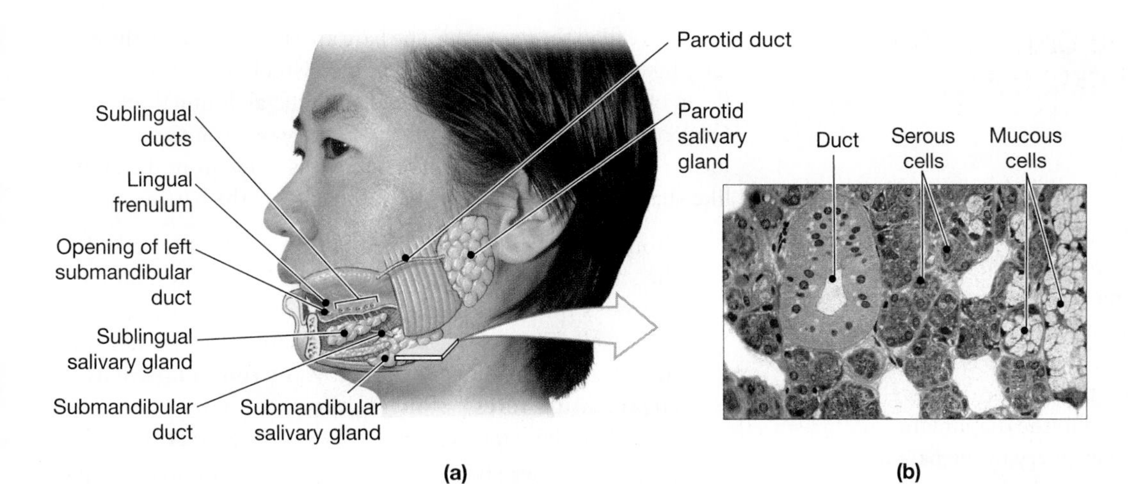

Figure 26.11 The salivary glands. a) Diagram showing the locations of the three major salivary glands on the left side of the head; **b)** light micrograph of the submandibular gland, which produces both serous and mucous secretions (LM × 300).

6. If a model is available, observe the arrangement of the **primary** or **deciduous teeth** in the skull of a young child (Figure 26.9a). The deciduous teeth appear at intervals between 6 months and 2 years of age. Ten deciduous teeth emerge from each jaw (a total of 20). Years later, the roots of the deciduous teeth are reabsorbed and the teeth are pushed out by the emerging **secondary** or **permanent teeth**, which first appear at age 6, but are not completely in place until 17 to 21 years of age.

Salivary Glands

1. On the torso model of the head and neck, locate the **parotid glands** (Figure 26.11a). These largest salivary glands are located anterior to each ear, between the skin of the cheek and the masseter muscle. They secrete mostly serous fluid with a high concentration of salivary amylase.
2. Locate the **parotid duct** that transports saliva from the parotid gland to the oral cavity. It passes through the buccinator muscle and enters the mouth just opposite the upper second molar on each side of the jaw (Figure 26.11a).
3. The **submandibular glands** are located on the floor of the mouth along the medial surfaces of the mandible (Figure 26.11a). The ducts to the submandibular glands enter the oral cavity inferior to the tongue, near the frenulum. They secrete a mixture of serous and mucous secretions.
4. The **sublingual glands** are the smallest of the major salivary glands and are located on the floor of the mouth near the tongue (Figure 26.11a). They are predominately mucus-secreting glands and transport saliva to the oral cavity via several small ducts.

QUESTIONS TO CONSIDER

1. When the enamel and dentin on the crown decay, the area of decay must be cleaned and filled with an inert substance. Some individuals are more prone to developing dental caries (cavities) than others. Inspect the oral cavities of other students in the laboratory and note the relative number of fillings in each individual. How much variation exists in your class?

____

____

2. Using models or Figure 26.9, compare the arrangement of primary teeth in a child and secondary teeth in an adult. What teeth are present in an adult that are missing in a child?

____

____

3. A disease called *mumps* is caused by a virus that infects and inflames the parotid salivary glands. Review the location of the parotid glands and explain why a person who has mumps will probably experience pain while eating a meal.

____

____

Gross Anatomy of Accessory Organs Associated with the Small Intestine

The small intestine has the main task of completing the chemical digestion of our food and absorbing the nutrients. Once these tasks are completed, the "leftovers" move on into the large intestine. The accessory organs that assist the small intestine with its vital functions include the liver, gallbladder, and pancreas.

ACTIVITY 26.4 Examining the Gross Anatomy of the Liver, Gallbladder, and Pancreas

Liver and Gallbladder

1. On a torso model, identify the liver just inferior to the diaphragm (Figure 26.1). The liver is the largest internal organ in the body. Notice that most of the organ is positioned within the upper right quadrant of the abdominal cavity and is protected by the rib cage. The liver is covered by a visceral peritoneum except on the superior surface, known as the **bare area**, which is fused to the diaphragm (Figure 26.12c). Around the periphery of the bare area, the visceral peritoneum folds onto the diaphragm to form the **coronary ligaments** (Figures 26.12b and c).

WHAT'S IN A WORD The superior surface of the liver is against the diaphragm and is devoid of peritoneum. For this reason, this region is called the *bare area*. The bare area is encircled by the coronary ligaments, which mark the point at which the peritoneum reflects off the surface of the liver to line the abdominal wall. The term *coronary* is derived from the Latin word *coronaries,* which means "crown." These ligaments take this name from the crown-like shape they form as they mark the edge of the bare area.

2. From an anterior view, identify the large **right lobe** and the smaller **left lobe**. Realize that the **falciform ligament** separates these lobes and connects the liver to the anterior body wall (Figures 26.12a and b). Within the inferior free margin of the falciform ligament is the **round ligament (ligamentum teres)**, which is a remnant of the umbilical vein from the fetal circulatory system (Figure 26.12b).
3. Remove the liver from its anatomical position to examine its posteroinferior surface. Identify the two smaller lobes: the more posterior **caudate lobe** and the more anterior **quadrate lobe** (Figure 26.12c). Notice that the **inferior vena cava** travels along the right side of the caudate lobe,

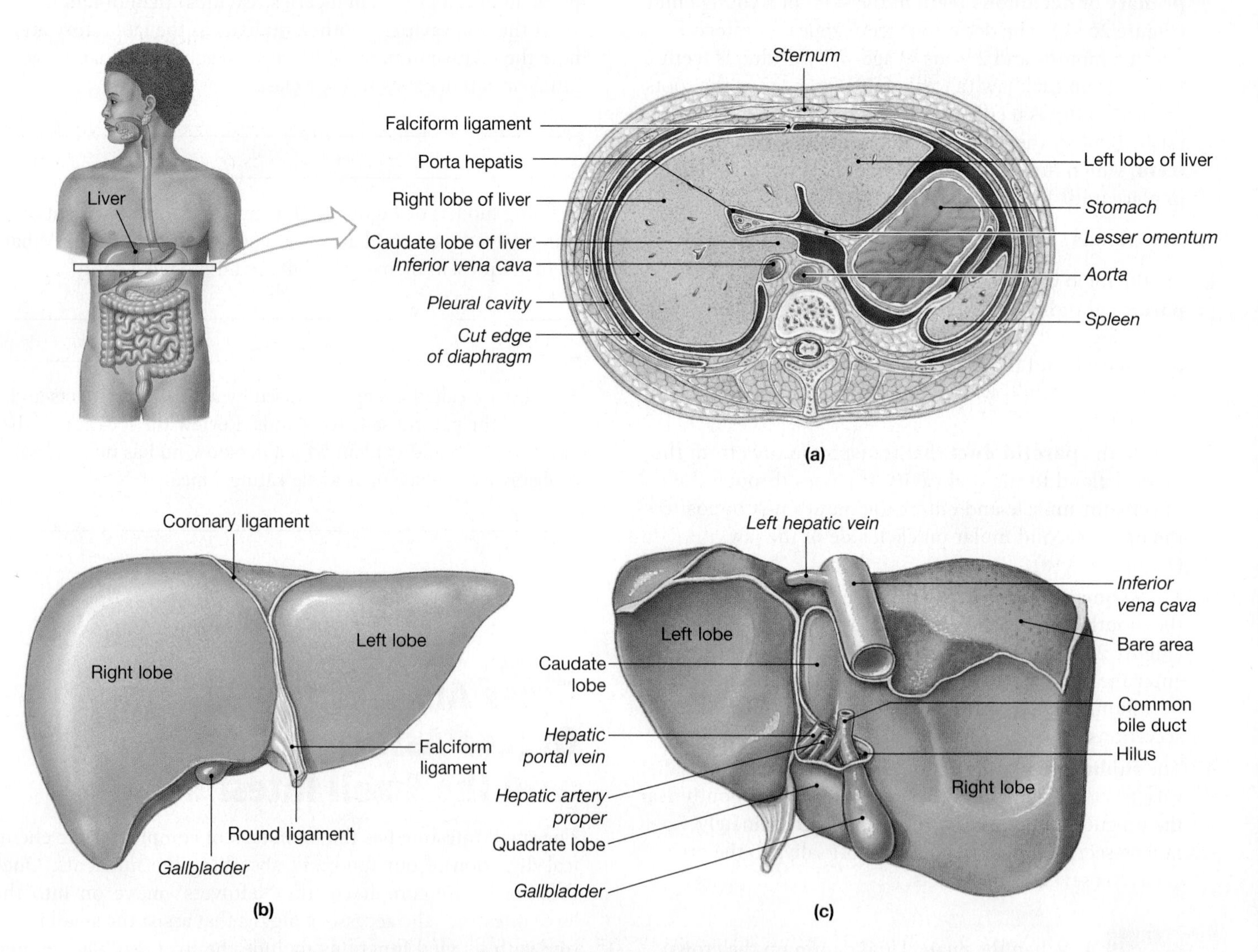

Figure 26.12 The gross anatomy of the liver. a) Transverse section illustrating the anatomic relationship of the liver to other abdominal structures; **b)** anterior view of the liver; **c)** posteroinferior view of the liver.

and the gallbladder is located in a depression on the right side of the quadrate lobe (Figure 26.12c).

4. Along the inferior surface, identify the **porta hepatis.** Notice all four liver lobes meet here (Figures 26.12a and c). The hepatic artery, portal vein, lymphatics, nerves, and hepatic ducts enter or exit the liver at the porta hepatis.
5. Also along the inferior surface of the liver, identify the **right hepatic duct** originating from the right lobe and the **left hepatic duct** from the left lobe (Figure 26.13a). Notice that the right and left hepatic ducts merge to form the **common hepatic duct. Bile**, a derivative of cholesterol, is produced by liver cells (**hepatocytes**) and exits the liver by traveling along the various hepatic ducts. Bile is used to emulsify fat globules during the early stages of fat digestion.
6. Locate the dark green gallbladder at the anteroinferior edge of the liver, slightly right of the midline (Figures 26.12c and 26.13a). Identify the **cystic duct** that originates from the gallbladder and follow it to where it unites with the common hepatic duct to form the **common bile duct**. The gallbladder provides a temporary storage site for bile. While in storage, the bile becomes concentrated as water is absorbed by the gallbladder epithelium. After a heavy meal, particularly a meal high in fat, enteroendocrine cells in the small intestine release **cholecystokinin**. This hormone promotes muscular contractions in the gallbladder, to move bile into the ducts to the duodenum.

CLINICAL CORRELATION

If bile is excessively concentrated (too much water is absorbed) or if the bile contains too much cholesterol, the cholesterol can crystallize to form **gallstones.** These structures can block the passage of bile in the ducts and cause excessive pain. Gallstones can be dissolved with drugs or pulverized with ultrasound or lasers. If they are surgically removed, the gallbladder is sometimes removed as well.

7. Return the liver to its anatomical position in the torso model and trace the common bile duct as it passes inferiorly toward the concave surface of the duodenum. Identify the location where it merges with the pancreatic duct to form the **duodenal (hepatopancreatic) ampulla.** Notice that the ampulla empties directly into the duodenum. At this location, the passage of bile (and pancreatic juice) is controlled by the **hepatopancreatic sphincter** (Figure 26.13b).

Pancreas

1. On the torso model, remove the stomach to reveal the pancreas. (Figure 26.1). The pancreas is about 12.5 cm

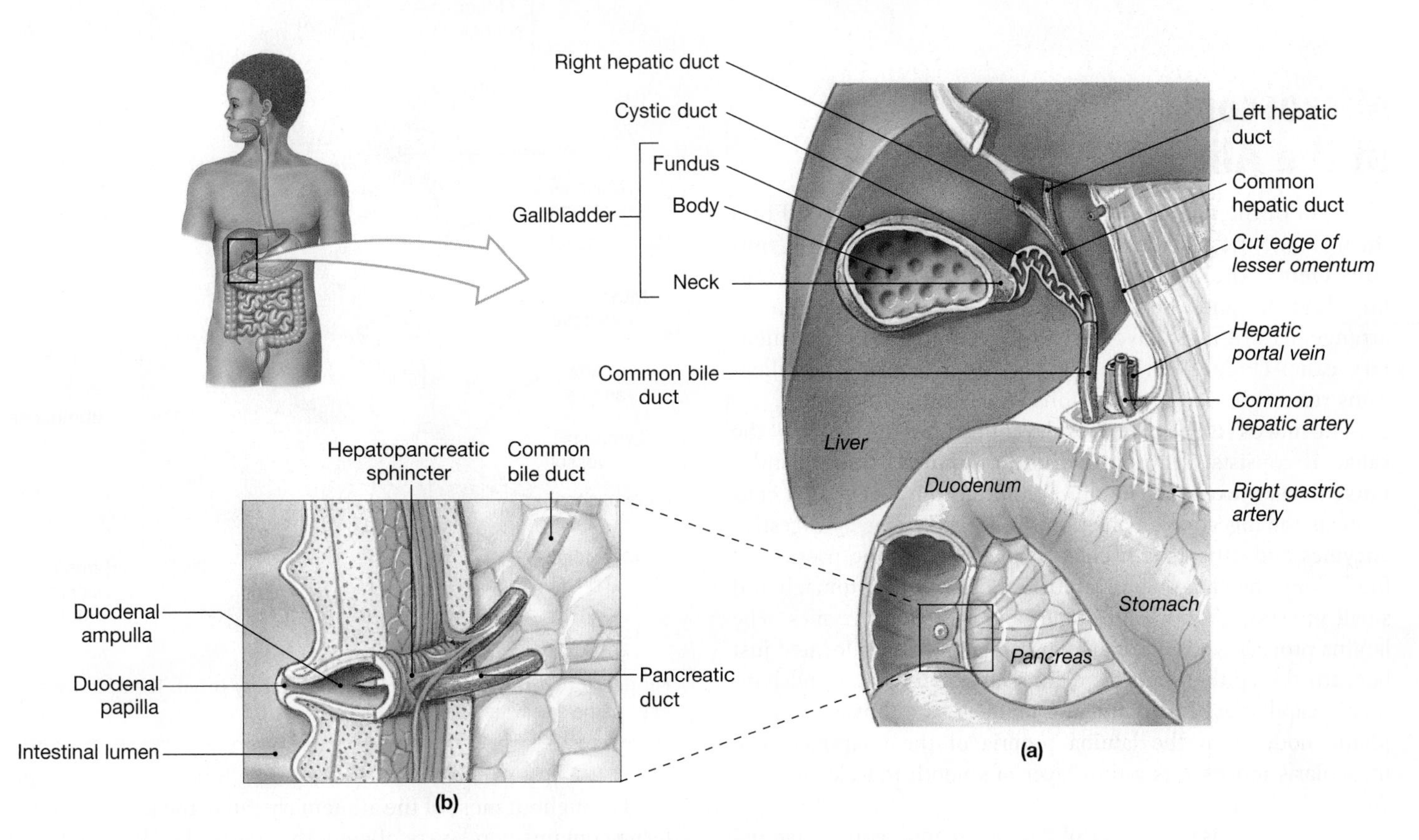

Figure 26.13 The gallbladder and bile duct system. a) Position of the gallbladder and the bile duct system that transports bile to the duodenum; **b)** closer view of the duodenal ampulla emptying into the duodenum.

(5 in) long and is positioned horizontally along the posterior abdominal wall.

2. From right to left, identify the three sections of the pancreas (Figure 26.6a).
   - The **head** is the broad right portion of the pancreas. It is surrounded by the C-shaped duodenum.
   - The **body** is the middle portion of the pancreas.
   - The elongated **tail** extends to the left side of the abdominal cavity toward the spleen.

   Except for a small portion of the head, the entire pancreas is retroperitoneal.
3. Notice that the **pancreatic duct** travels along the entire length of the pancreas (Figure 26.6a). As described earlier, the pancreatic duct merges with the common bile duct to form the duodenal (hepatopancreatic) ampulla. Identify the ampulla as it empties into the duodenum along its concave surface.

QUESTION TO CONSIDER If the gallbladder is surgically removed, what function will no longer be possible? How will the transport of bile be altered? ______________________

____

____

Microscopic Structure of the Alimentary Canal

The wall of the alimentary canal has four tissue layers. From the inside to the outside these layers include the **mucosa (mucous membrane)**, **submucosa**, **muscularis externa (muscular layer)**, and **serosa (serous layer)**. The general arrangement of these layers is similar throughout the alimentary canal (Figure 26.14), but certain structural specializations reflect the unique functions of each digestive organ.

The mucosa is the innermost layer that lines the lumen of the canal. It consists of an **epithelium**, a **lamina propria**, and a **muscularis mucosae**. The mucus produced by the goblet cells protects the digestive organs from possible damage by digestive enzymes and other harsh chemicals, and eases the passage of food along the canal. Some epithelial cells in the stomach and small intestine produce hormones and digestive enzymes. The lamina propria is a layer of areolar connective tissue located just beneath the epithelium. It contains numerous blood and lymphatic capillaries, and lymphatic nodules (the tonsils are lymphatic nodules in the lamina propria of the pharynx). The muscularis mucosae is a thin layer of smooth muscle that underlies the lamina propria.

The submucosa is a layer of dense, irregular connective tissue situated just below the mucosa. It contains a vast supply of blood and lymphatic vessels and nerves that supply the surrounding tissues.

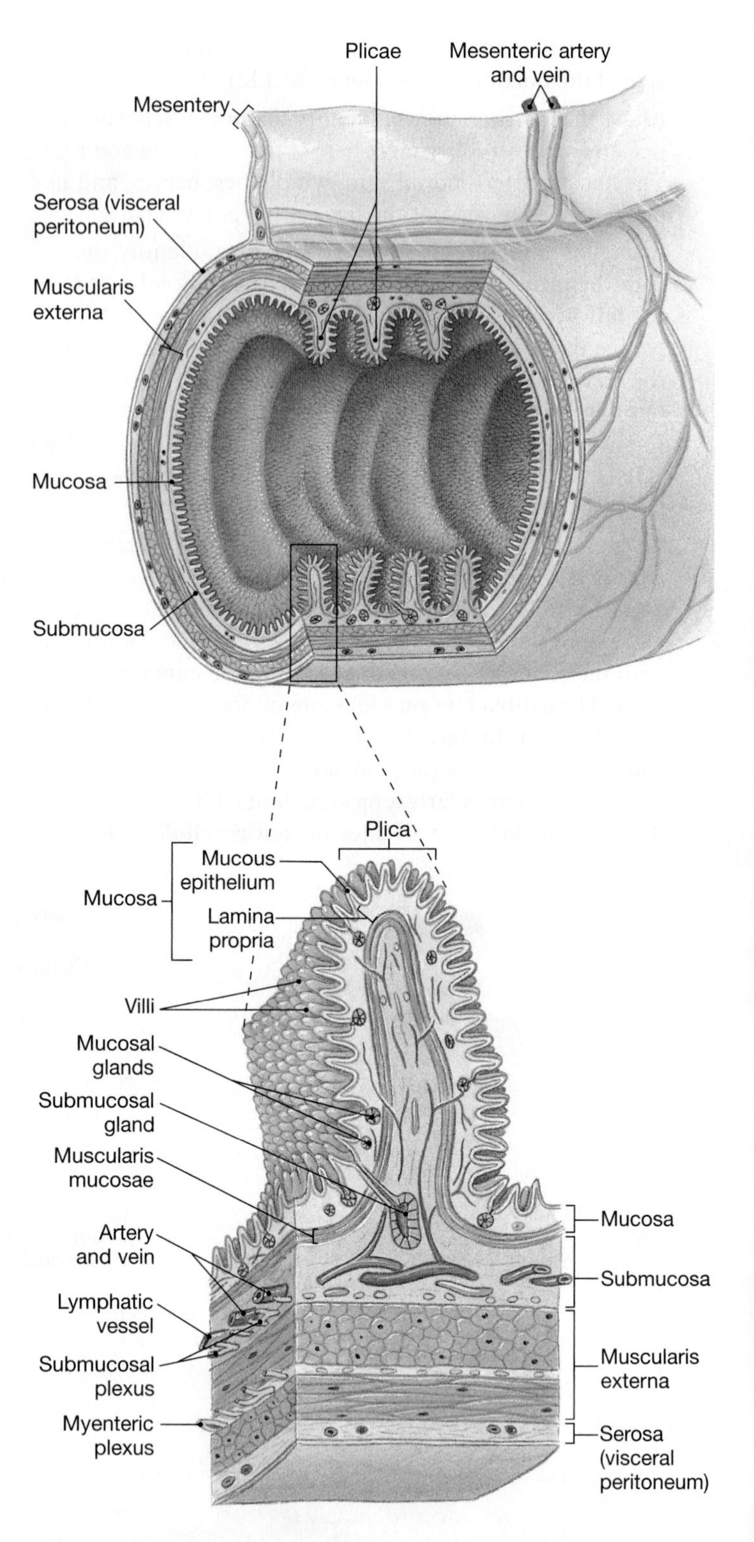

Figure 26.14 The structure of the alimentary canal. The special features of the small intestine are illustrated.

Throughout most of the alimentary canal, the muscularis externa contains two layers of smooth muscle. The inner circular layer controls the diameter of the digestive tube, while the outer longitudinal layer controls the length of the tube. The actions of these muscle layers are responsible for mixing movements and

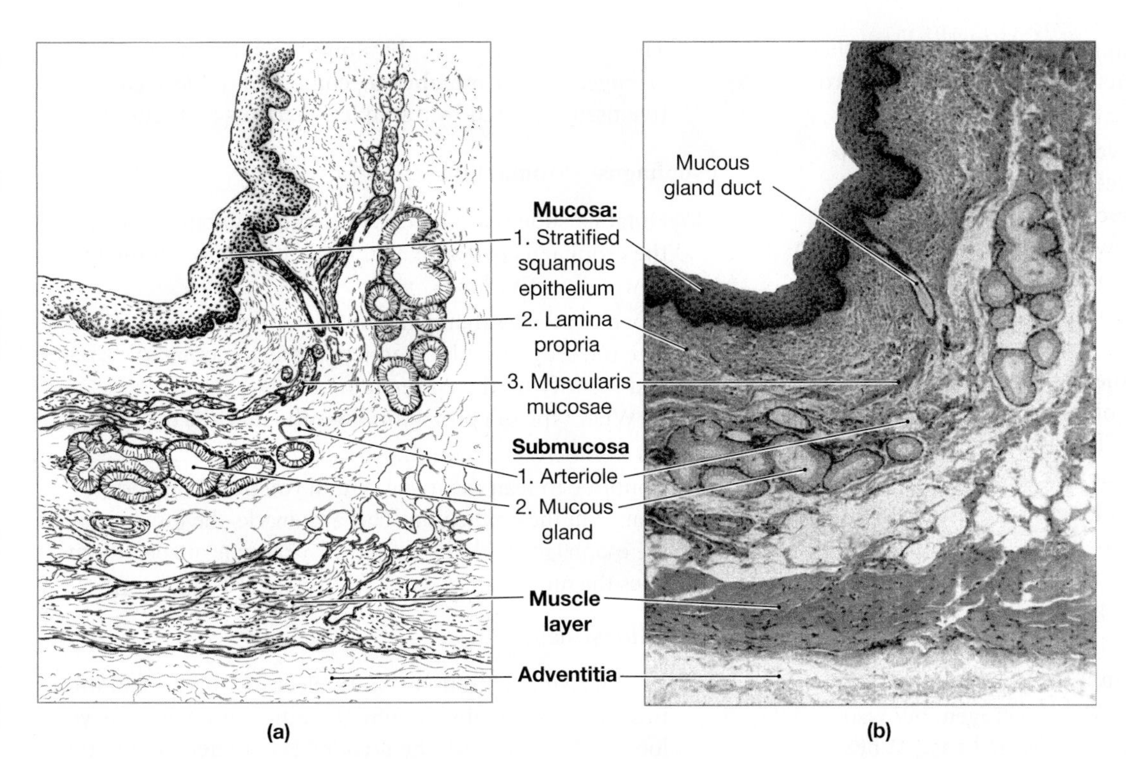

Figure 26.15 The microscopic structure of the esophagus. **a)** Diagram of the esophageal wall; **b)** corresponding light micrograph (LM × 80).

peristalsis. At certain locations along the alimentary canal, particularly at the junctions between two organs, the circular muscle layer thickens to form **sphincters**. Sphincter muscles control the amount of material that passes from one organ to the next and also prevent backflow, much like the valves in veins prevent the backflow of blood.

The outermost **serosa** consists of a thin layer of loose connective tissue that is covered by a simple squamous epithelium. In areas where the peritoneum exists, the serosa and the visceral peritoneum are the same structure. In the walls of the pharynx and esophagus, the serosa is replaced by an **adventitial layer**. Similar to the tunica adventitia in blood vessels, the adventitial layer is composed of fibrous connective tissue that is continuous with adjacent structures. In the abdominal cavity, retroperitoneal structures have a serosa on surfaces facing the cavity, and an adventitial layer on surfaces embedded in the body wall.

In the next activity, you will examine the layers of the wall as you explore the various parts of the alimentary canal.

ACTIVITY 26.5 Examining the Microscopic Structure of the Alimentary Canal

The Esophagus

1. Obtain a slide with a longitudinal section of the middle region of the esophagus.
2. Scan the slide with the low-power objective lens. Identify the mucosa along the top of the section. Note that the epithelium that lines the lumen is stratified squamous. The other subdivisions of the mucosa, the lamina propria and muscularis mucosae, are also apparent just below the epithelium (Figure 26.15).
3. Identify the submucosa, which lies deep to the muscularis mucosae. The submucosa is a connective tissue layer that contains mucus-secreting **esophageal glands** that secrete mucus onto the luminal surface to facilitate food transport. Attempt to locate these glands, although they may not be visible on all slides. Blood vessels and nerves are also found in the submucosa. Use the high-power objective lens to identify small blood vessels such as arterioles and capillaries (Figure 26.15).
4. Just below the submucosa, locate the muscular layer. It consists of two layers of muscle: an inner circular layer and an outer longitudinal layer (Figure 26.15). Along the middle portion of the esophagus, the muscle layer contains a combination of smooth and skeletal muscle. However, at the superior end, the muscle layer is entirely skeletal muscle; at the inferior end, only smooth muscle is present.
5. The esophagus does not have a serous layer like other organs of the alimentary canal that are located in the abdominal cavity. Instead, the outermost layer is the adventitia. It is composed of loose connective tissue that blends in with the connective tissue of adjacent structures.

Stomach

1. Obtain a slide of the fundic or pyloric region of the stomach.
2. Observe the mucosa and submucosa of the stomach (Figure 26.16) with the scanning or low-power objective. On some slides these two layers might have a folding or rolling pattern. The folds that you are observing are the rugae that you identified earlier at the gross anatomical level.
3. Observe a region of mucosa that is not folded. As you scan along the luminal surface, note that numerous invaginations extend deep into the mucosa. These are the **gastric pits**,

which are lined by light-staining, simple columnar epithelial cells. Most of these cells secrete mucus onto the surface of the mucosa.

4. Note that the gastric pits give rise to cell clusters deep within the mucosa. These cells comprise the **gastric glands** (Figure 26.16). These glands produce **gastric juice**, a mixture of mucus, hydrochloric acid, digestive enzymes, and digestive hormones.
5. Switch to the high-power objective lens and observe the cells of the gastric pits and gastric glands more closely. Verify that the cells lining the pits are columnar. As you move to the deeper glands, observe that the cells are either low columnar or cuboidal. The gastric gland cells include the following types (Figure 26.16b).
   - **Goblet cells (mucous neck cells)** produce large amounts of mucus that protect the stomach wall from being digested by its own enzymes.
   - **Parietal cells** produce hydrochloric acid (HCl), used for protein digestion, and intrinsic factor, which is needed for the absorption of vitamin B_{12}.
   - **Chief cells** primarily produce pepsinogen, but also manufacture small amounts of gastric lipase. When pepsinogen comes in contact with HCl in the gastric lumen, it is converted to pepsin, which digests proteins. Gastric lipase digests lipids.
   - **Enteroendocrine cells** produce a variety of digestive hormones and secrete their products into capillaries traveling in the lamina propria. The hormones produced by these cells are as follows:
     - **Gastrin**, which stimulates gastric gland activity
     - **Histamine**, which stimulates HCl secretion by parietal cells
     - **Serotonin**, which stimulates contraction of smooth muscle in the stomach
     - **Somatostatin**, which inhibits digestive activity
6. Identify the lamina propria between the cells of the gastric glands and along the base of the mucosa. The connective tissue fibers usually stain red or dark pink. Examine this connective tissue under low power, then switch to high power to obtain a more detailed observation.
7. At the base of the mucosa, try to identify a thin band of pink-staining smooth muscle. This is the muscularis mucosae. This layer is somewhat difficult to identify but, with care and patience, you will be successful.
8. Under low power, once again observe the submucosa, a layer of loose connective tissue and fat. This layer usually stains pink and has a frothy appearance.
9. Just outside the submucosa is the muscle layer of the stomach (Figure 26.16). As you observed earlier, the stomach wall contains three, rather than two, muscle layers. In addition to the **circular** and **longitudinal layers** that are typical of the alimentary canal, is an innermost **oblique layer**. The oblique layer is hard to identify on the slides, but, try to identify the circular and longitudinal layers.
10. The outermost layer of the stomach wall is the loosely arranged **serous membrane**, composed of loose connective tissue covered by a simple squamous epithelium.

Esophagus—Stomach Junction

1. Obtain a slide that has a section of the junction between the esophagus and stomach. View the slide with the low-power objective lens (Figure 26.17).
2. Move to an area that contains the esophagus. You can locate the esophagus by scanning along the mucosal surface and identifying the epithelium.

   What type of epithelium lines the mucous membrane of the esophagus? ______________________________
3. Continue to scan the mucosa until you notice an abrupt change in the epithelium. This area marks the junction of the esophagus with the stomach. What type of epithelium lines the mucous membrane of the stomach? ______________

Small Intestine

1. Obtain a slide with tissue sections of the three small intestinal regions: the duodenum, jejunum, and ileum. As you look at the slide with the unaided eye, understand that the top section is the ileum, the middle section is the jejunum, and the bottom section is duodenum. However, when you view the slide with the microscope, remember that the images are inverted, so the top section in the microscope field will be the duodenum and the bottom section will be the ileum.
2. View the middle section on the slide under low magnification. You are observing a longitudinal section of the jejunum. Observe the fingerlike projections along the top of the section. These structures, called **intestinal villi**, are a distinguishing feature of the mucosa (Figure 26.18). They are most prominent in the jejunum (Figure 26.19b), where the absorption of nutrients is most active. In some regions of your tissue section, the villi may be difficult to identify due to unusual planes of section. Thus, on the same slide, the villi will appear in longitudinal section as well as cross and oblique sections. It will be easier to study the villi as they appear in a longitudinal section (Figures 26.18c and d).
3. Deep to the mucosa, observe the submucosa and muscle layer. The muscle layer contains both inner circular and outer longitudinal fibers (Figure 26.18a). The serosa may not be present on your slide.
4. Switch to the high-power objective lens to observe the intestinal villi, in longitudinal section, more closely. Note that the villi are covered by a simple columnar epithelium of mostly **absorptive cells**, interspersed with mucus-secreting **goblet cells** and hormone-producing **enteroendocrine cells** (Figures 28.18d and e). The epithelial cells are usually stained pink with dark blue nuclei at the bases. The goblet cells are usually light staining (whitish or gray), with ovoid apical (upper) regions filled with mucus (Figure 28.18e), but on some slides they could be a deep blue or red. Unless

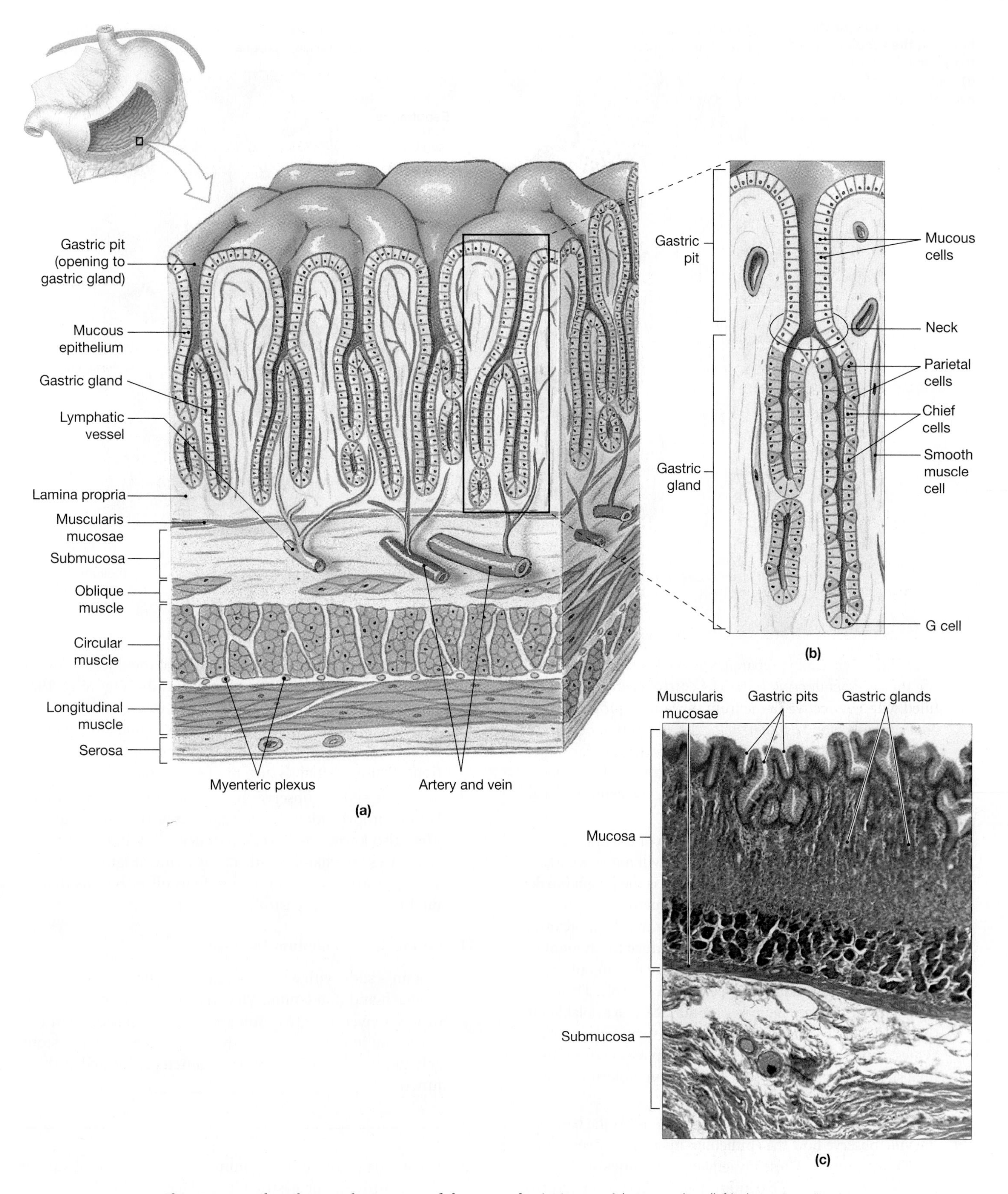

Figure 26.16 The microscopic structure of the stomach. a) Diagram of the stomach wall; **b)** closer view of a gastric pit and gastric gland; **c)** light micrograph of the stomach wall, showing a similar view as in (a) (LM × 100).

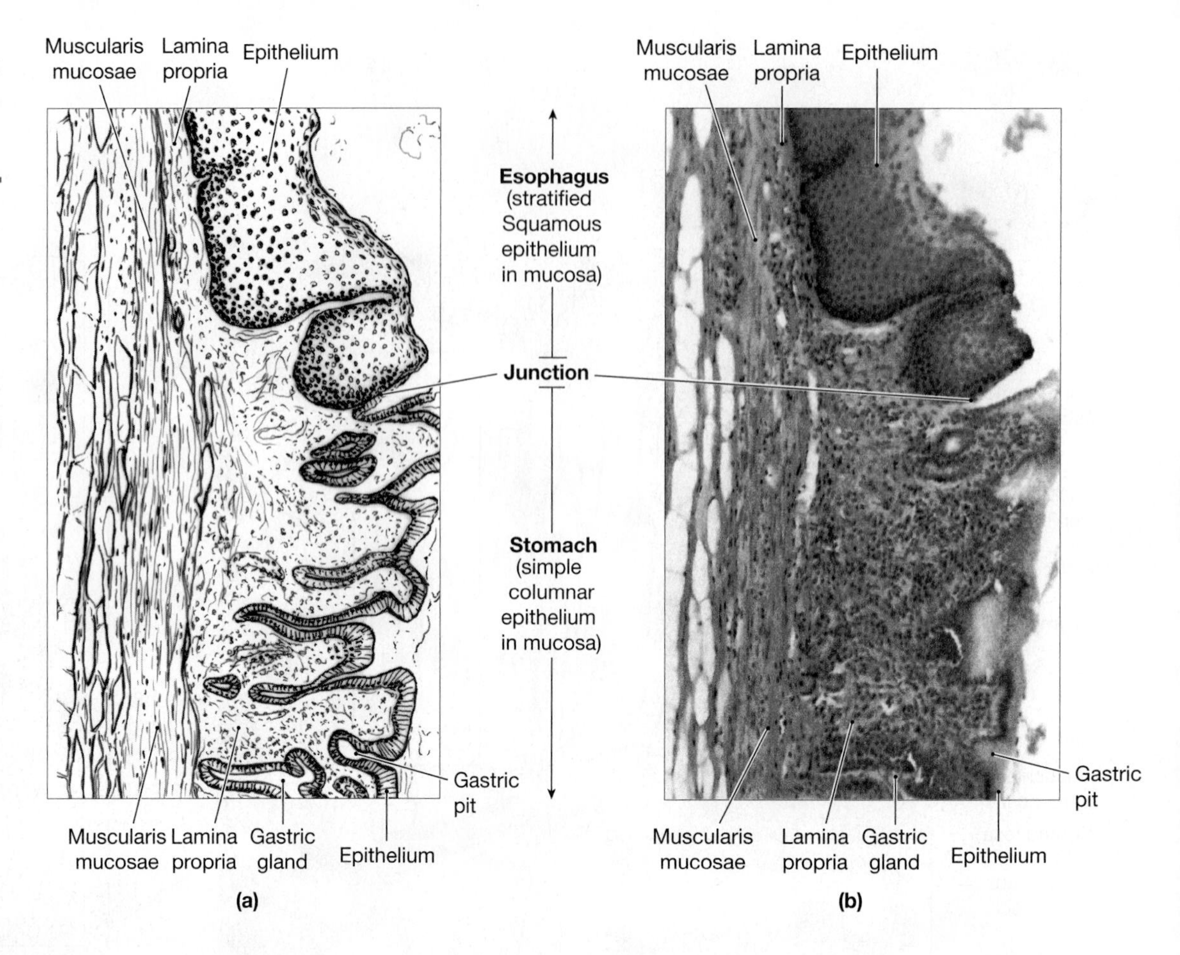

Figure 26.17 The junction between the esophagus and the stomach. a) Illustration; **b)** corresponding light micrograph (LM × 100).

your slide is specially prepared, you will not be able to identify the enteroendocrine cells. The core of each villus is filled with connective tissue from the lamina propria. This area contains blood capillaries for the absorption of amino acids and monosaccharides, and a lymphatic capillary known as a **central lacteal** for the absorption of fatty acids and other lipids (Figures 26.18d and e). The central lacteal may be difficult to identify on your slide.

5. Observe the simple columnar epithelium of the mucosa under high power. Look carefully and you will notice a dark fringe along the surface of the cells. This is the **brush border** (Figure 26.18e), which is composed of microscopic projections called microvilli. (The brush border may be more obvious if you adjust the iris diaphragm to reduce the amount of light being transmitted.) Individual microvilli can only be seen with an electron microscope (Figure 26.18f). Their presence significantly increases the surface area available for the absorption of nutrients. They also possess digestive enzymes along their membranes. These enzymes perform the final steps of digestion, producing nutrient molecules that are small enough to be absorbed.
6. Switch back to the low-power objective lens. At the base of the villi, observe how the epithelium invaginates deep into the lamina propria. These invaginations are **intestinal glands** or **crypts of Lieberkuhn** (Figures 26.18c and 26.19), which secrete mucus and numerous digestive enzymes.
7. Move the slide to observe the section of the duodenum. In the submucosa, identify **Brunner's glands** (Figure 26.19a). These mucous glands are found only in the duodenum. Their alkaline secretions neutralize the acidic chyme that enters the small intestine from the stomach. In some sections, Brunner's glands may not be present.
8. Move the slide to observe the section of the ileum. In the lamina propria, identify the **aggregated lymphatic nodules**, also known as **Peyer's patches** (dark blue). These structures are characteristic of the ileum (Figure 26.19c). They are strategically located to fight off infections that might occur in the gastrointestinal tract.

The Stomach—Duodenum Junction

1. Obtain a slide with a section of the junction between the stomach and duodenum. View the slide with the scanning or low-power objective. You are observing a section of tissue at the junction between the pyloric region of the stomach and the duodenum. What sphincter is located at this junction? ____________________

____
2. Locate the stomach by scanning along the mucosal surface and identifying the gastric pits and gastric glands (Figure 26.16c).

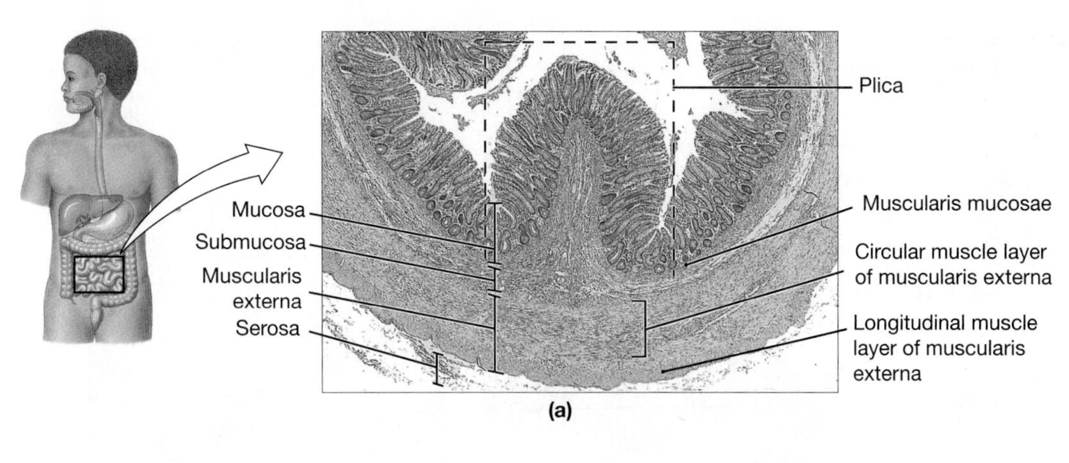

Figure 26.18 The microscopic structure of the small intestine. a) Low-power light micrograph of the intestinal wall illustrating the various tissue layers common to the alimentary canal and the formation of a plica (LM × 40); **b)** diagram of the intestinal wall showing a plica and many villi; **c)** diagram of the intestinal wall showing the organization of intestinal villi in the mucosa and the arrangement of the submucosa, muscularis externa, and serosa; **d)** diagram of a single intestinal villus; **e)** light micrograph of the region enclosed by the box in (d). Note the presence of the brush border which marks the presence of microvilli along the surface of absorptive cells (LM × 600); **f)** electron micrograph illustrating the microvilli along the surface of an absorptive cell (LM × 20,000).

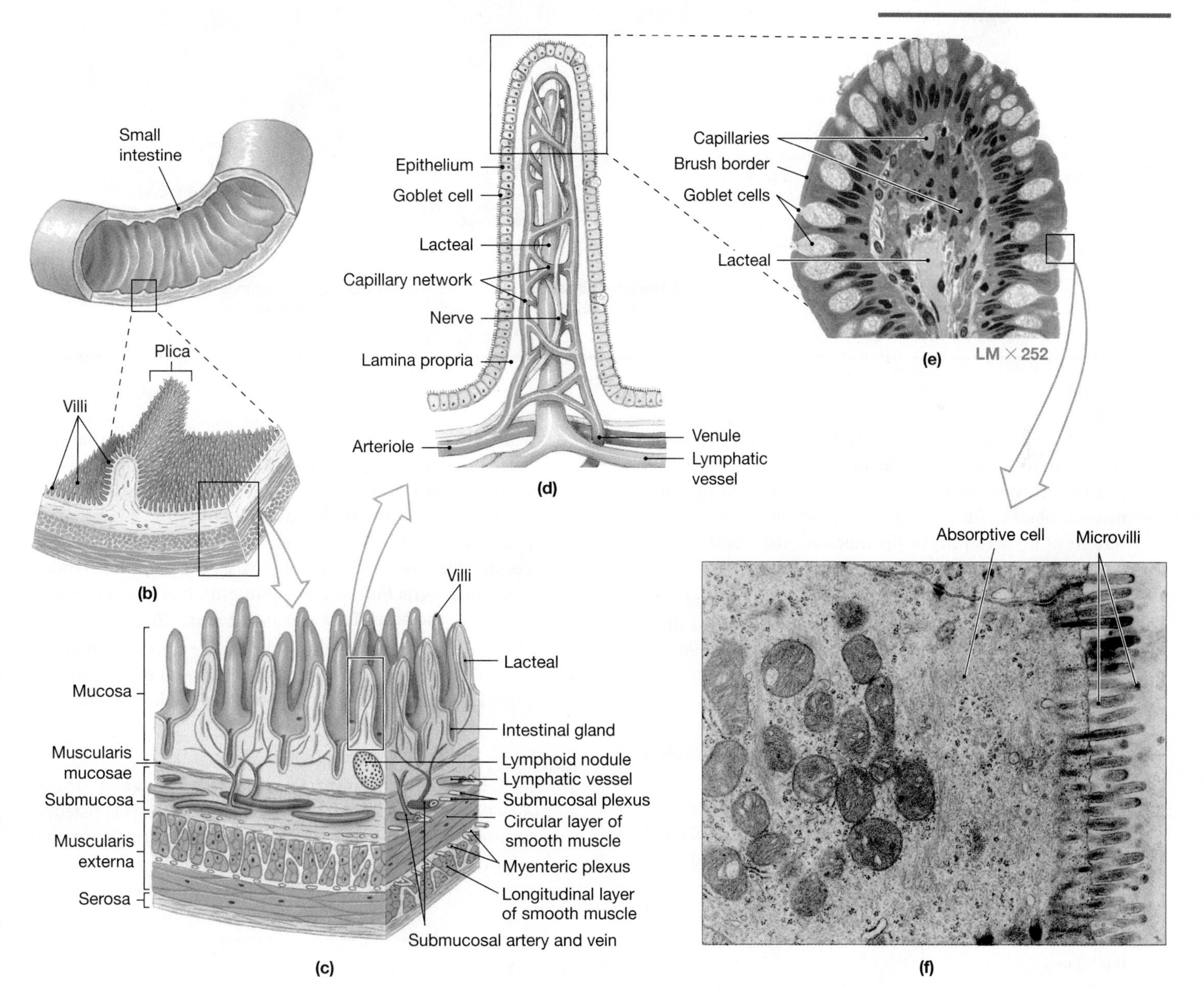

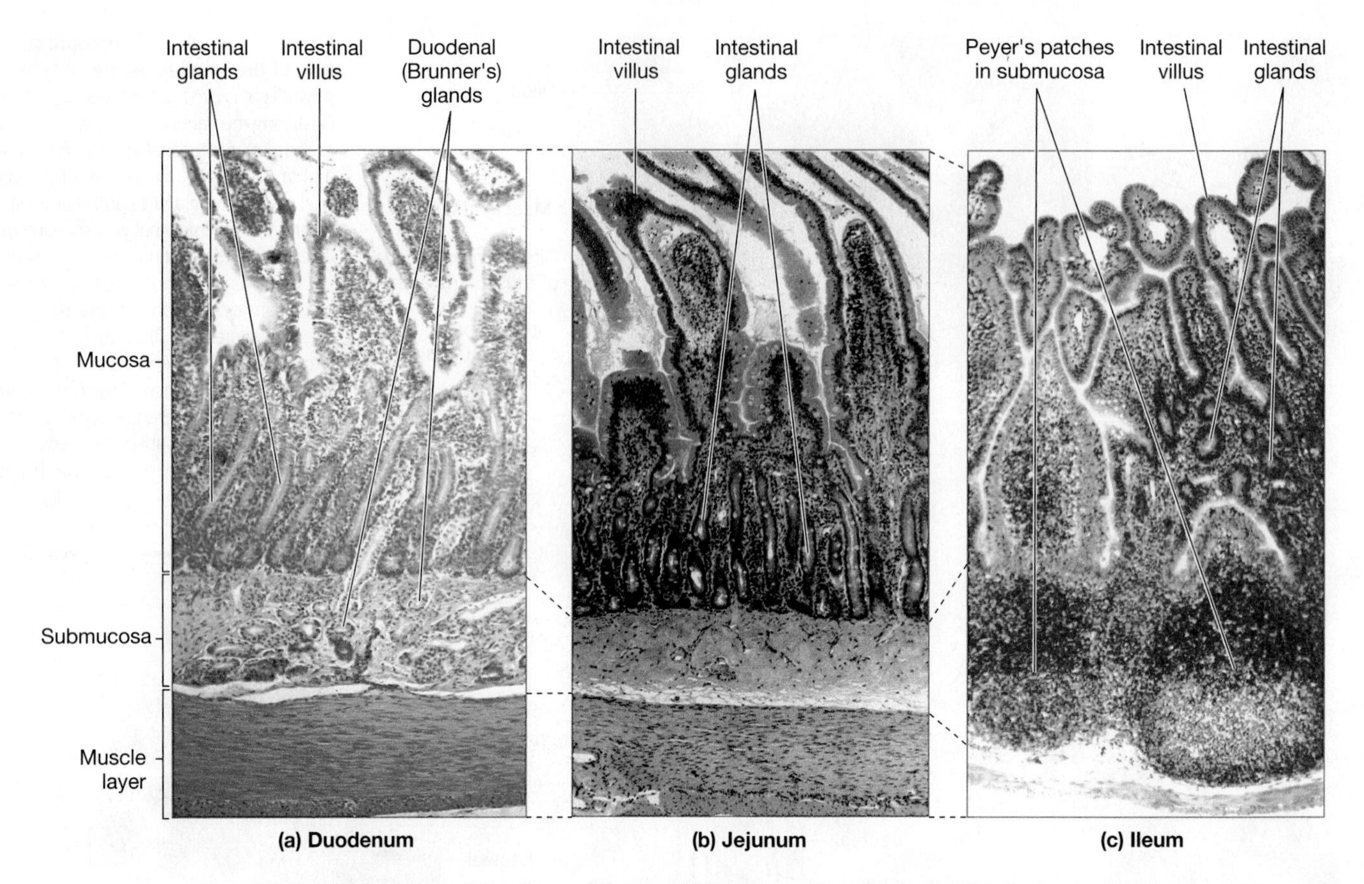

Figure 26.19 Regional specializations of the small intestine. a) The duodenum with Brunner's glands in the submucosa; **b)** the jejunum with prominent intestinal villi; **c)** the ileum with aggregated lymphoid nodules (Peyer's patches) in the lamina propria.

3. Continue to scan the mucosa until you notice an abrupt change in the structure of the mucosa. This area marks the junction of the stomach with the duodenum. In the duodenum, observe the intestinal villi and intestinal glands (crypts of Lieberkuhn) in the mucosa. Also observe the goblet cells in the epithelium (Figure 26.19a).
4. Identify the submucosa and muscle layers for both the stomach and the duodenum. The submucosa of the duodenum is filled with Brunner's glands (Figure 26.19a). These glands are absent in the submucosa of the stomach.

Large Intestine

1. Obtain a slide with a section of the large intestine (colon). View the slide with the low-power objective lens.
2. Observe the mucosa and verify that the epithelium is simple columnar with many goblet cells. The goblet cells are usually unstained and can be clearly identified (Figure 26.20). The mucus produced by the goblet cells protects the intestinal wall against abrasion, holds fecal matter together, and neutralizes the acids released by bacteria.
3. In the mucosa, notice that both the intestinal villi and the brush border (microvilli) along the cell surfaces are absent. The lack of these structures reflects the fact that the large intestine has little or no digestive function. Scan along the surface and identify the numerous invaginations, which lead to **intestinal glands.** Identify the lamina propria and muscularis mucosae. The lamina propria is the connective tissue that fills the spaces between the glands. The muscularis mucosae is a thin pink band of smooth muscle at the base of the mucosa (Figure 26.20).
4. Deep to the mucosa, identify the lightly stained submucosa. Deep to this layer are the circular and longitudinal muscle layers (dark pink) followed by the loose connective tissue in the serosa.

QUESTION TO CONSIDER Identify the special structural features in the mucous membranes of the digestive organs listed, and comment on the significance of these structures in relation to function.

Organ	Special Structural Features	Significance
Stomach	____________	____________
Small intestine	____________	____________
Large intestine	____________	____________

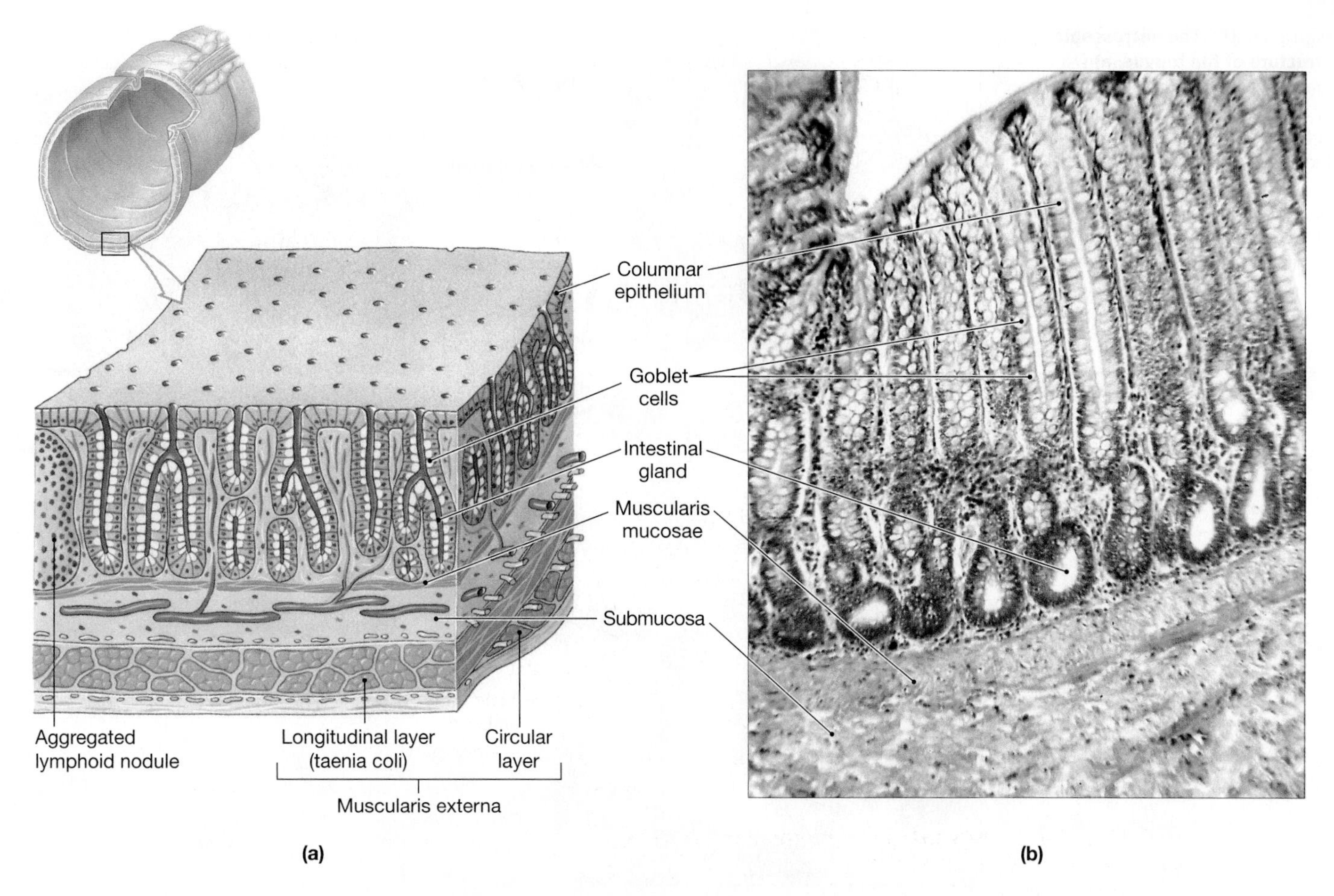

Figure 26.20 The microscopic structure of the large intestine. a) Diagram of the large intestinal wall; **b)** light micrograph of the mucosa and submucosa of the large intestine (LM × 100).

ACTIVITY 26.6 Examining the Microscopic Structure of Accessory Digestive Organs

The Tongue

1. Obtain a microscope slide of the tongue.
2. Scan the slide with the low-power objective lens and observe the mucosa of the tongue. It is lined by a stratified squamous epithelium. Just below the epithelium is a layer of loose connective tissue that contains numerous blood vessels, nerve fibers, and mucous glands (Figures 26.21a and b).
3. Note that the mucosa of the tongue contains **papillae.** These projections give the surface of the tongue an uneven or roughened appearance. The papillae provide friction while handling food, and contain the **taste buds.**
4. Taste buds are located along the surface of the papillae, embedded within the epithelium. As you learned, in Exercise 17, these ovoid structures contain sensory receptors that allow us to taste the food we eat. In the slide you are observing, taste buds are likely difficult to identify. If you have a slide that is specially prepared to view taste buds, they will appear lighter than the surrounding epithelium (Figure 26.21c).
5. The majority of the tongue is composed of skeletal muscle fibers. These intrinsic muscles of the tongue can be observed below the mucous membrane. Note that the muscle fibers are traveling in many directions, and that bundles of muscle fibers are separated by connective tissue. As discussed earlier, the intrinsic muscles can change the *shape* of the tongue. The extrinsic muscles of the tongue (not seen on this slide) connect the tongue to other structures such as the palate or hyoid bone. These muscles can change the *position* of the tongue.
6. Observe an area of tongue muscle that is cut in a longitudinal section. Switch to high power and identify the striations that characterize skeletal muscle.

   What is the anatomical basis for the striations?

   ____

   ____

Salivary Glands

1. Obtain a microscope slide of the **submandibular salivary gland.**
2. Observe the gland with the low-power objective lens (Figure 26.11b). The glandular cells are arranged in small

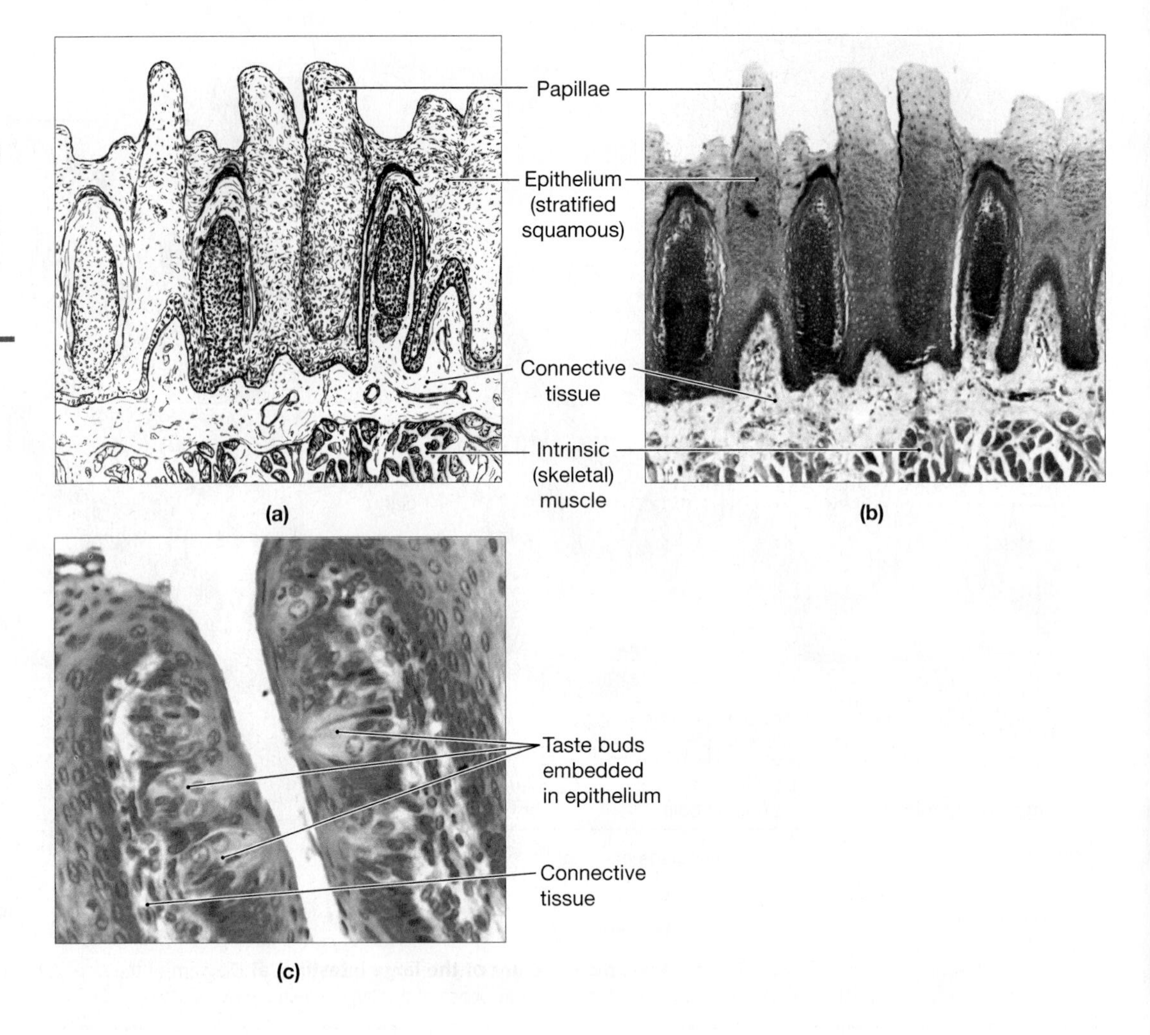

Figure 26.21 The microscopic structure of the tongue. a) Diagram; **b)** corresponding light micrograph of the tongue illustrating the stratified squamous epithelium along the surface and the deeper skeletal muscle layer (LM × 100); **c)** high-power light micrograph of taste buds embedded in the epithelium of a papilla on the surface of the tongue (LM × 200).

clusters called **acini** (singular = acinus) with a small lumen in the center. The lumen will often be obscured due to the plane of section.

3. You might be able to distinguish between serous-secreting cells and mucus-secreting cells. Depending on the stain, the serous cells will have a deep blue or red-staining cytoplasm and centrally located, blue-purple nuclei. The mucous cells usually have a pale, frothy-looking cytoplasm with blue-purple nuclei pushed to the cell's base (Figure 26.11b).
4. Identify some of the salivary gland ducts that are also apparent on the slide. A typical duct has a relatively large lumen that is lined by a simple cuboidal epithelium (Figure 26.11b).

The Liver

1. Obtain a slide of the liver and view it with the scanning or low power objective lens in position.
2. The functional unit of the liver is the **liver lobule**. Each lobule is a hexagonal-shaped column of tissue with a **central vein** running down the middle (Figures 26.22a and d). On your slide identify a liver lobule and the central vein.
3. Identify the **liver cells (hepatocytes)** and **liver sinusoids**. Hepatocytes are arranged in rows that radiate from the central vein, like the spokes of a bicycle wheel. The sinusoids can be seen meandering between the rows of hepatocytes (Figures 26.22a and b).
4. A **portal area (hepatic triad)** is located at each corner along the periphery of a liver lobule. Locate a portal area and identify the three structures within it (Figures 26.22b and c).
   - A **branch of the portal vein** (a part of the hepatic portal system) that transports venous blood from the gastrointestinal tract to the liver. This blood is rich in nutrients that were absorbed by the small intestine.
   - A **branch of the hepatic artery** that brings oxygenated arterial blood to the liver.
   - A **bile duct** that transports bile to the duodenum or gallbladder.

Blood from the two blood vessels in the triad percolates through the sinusoids and drains into the central vein. All the central veins drain into the hepatic vein, which leads into the inferior vena cava. As blood flows through the sinusoids, bacteria, old blood cells, and other debris are filtered out by phagocytic cells known as **Kupffer cells** (Figure 26.22b).

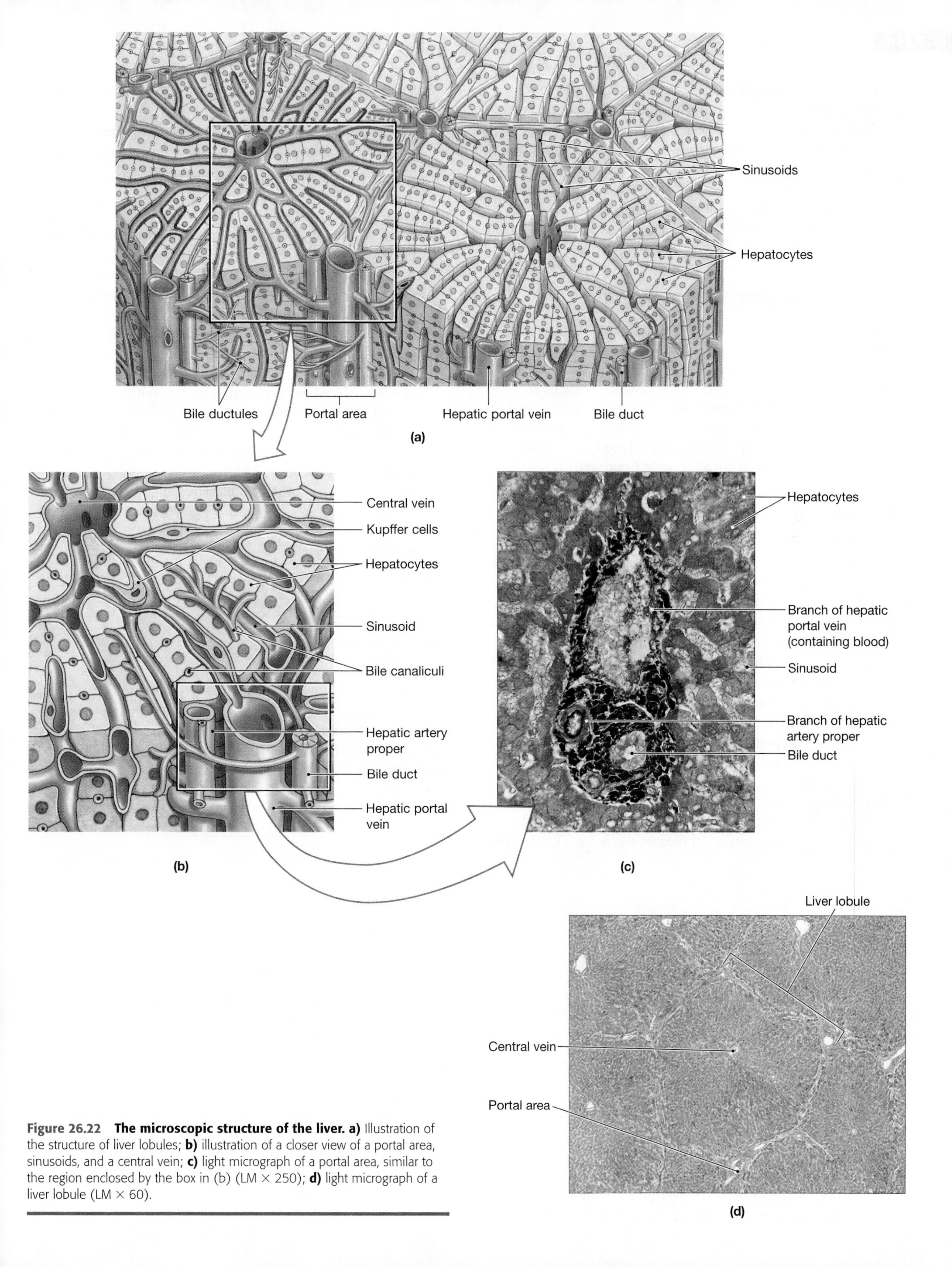

Figure 26.22 The microscopic structure of the liver. a) Illustration of the structure of liver lobules; **b)** illustration of a closer view of a portal area, sinusoids, and a central vein; **c)** light micrograph of a portal area, similar to the region enclosed by the box in (b) (LM × 250); **d)** light micrograph of a liver lobule (LM × 60).

5. Obtain a slide that is specially prepared to illustrate the transport of bile through the liver. Examine the slide with the low-power objective lens. Bile, produced by the hepatocytes, is secreted into tiny ducts called **bile canaliculi** and flows into the bile ducts located in the portal areas (Figure 26.22b). The black-staining material marks the pathways of the canaliculi traveling to the bile ducts in the triads. The solid black, circular structures in the triads are bile ducts.

CLINICAL CORRELATION

In addition to producing bile, hepatocytes have many other vital functions. They detoxify poisons and break down biologically active molecules such as hormones and drugs. They can reduce blood sugar levels by converting glucose to glycogen for storage or to noncarbohydrate molecules such as triglycerides. Conversely, hepatocytes can increase blood sugar by breaking down glycogen and converting fats and amino acids to glucose. These versatile cells are also responsible for the production of most blood proteins.

The Pancreas

1. Obtain a slide of the pancreas and view it with the low-power objective lens. Recall that you examined the microscopic structure of the pancreas in Exercise 18. Review its structure once again.
2. Identify the **pancreatic acini**. Each acinus contains a cluster of cuboidal cells **(pancreatic acinar cells)**, arranged around a central lumen (Figures 26.6b and c). The acini represent the exocrine portion of the pancreas. The acinar cells produce **pancreatic juice** that is transported along a network of ducts leading to the **pancreatic duct**. You identified the pancreatic duct earlier. Recall that it extends the entire length of the pancreas and delivers pancreatic juice to the duodenum. The pancreatic juice is a watery secretion that contains a variety of enzymes for digesting proteins, carbohydrates, fats, and nucleic acids, as well as a substantial concentration of bicarbonate ions. The alkaline bicarbonate ions neutralize the acidity of chyme coming into the duodenum from the stomach.
3. Observe that the darkly stained acinar cells comprise the vast majority of the pancreas. Notice that groups of pancreatic acini are separated by connective tissue (Figure 26.6b).
4. Scattered among the pancreatic acini, identify the islands of lighter staining cells. These are the **pancreatic islets** or **islets of Langerhans** (Figures 26.6b and c). These cells represent the endocrine portion of the pancreas. They produce the hormones **insulin**, **glucagon**, **somatostatin**, and **pancreatic polypeptide.** Insulin and glucagon regulate blood glucose levels. Specifically, insulin promotes cellular uptake of glucose from the blood and stimulates the formation of glycogen. Glucagon, on the other hand, promotes the breakdown of glycogen, the synthesis of glucose from noncarbohydrate sources (fats and proteins), and the release of glucose into the blood (primarily by liver cells). Somatostatin acts as an important regulating hormone by inhibiting the secretion of both insulin and glucagons, and pancreatic polypeptide inhibits muscular contractions in the gallbladder.

QUESTIONS TO CONSIDER

1. Explain why the liver sinusoids contain a mixture of arterial and venous blood. ____________

____

____

2. Cystic fibrosis is a genetic disease that causes ducts in several organs to be blocked by an excessive production of thick mucus. Although most cystic fibrosis deaths are due to lung damage, the disease also clogs bile ducts in the liver and pancreatic ducts in the pancreas. Discuss how the disease's effects on the liver and pancreas will impact digestive activity. _______

____

____

Exercise 26 Review Sheet

Anatomy of the Digestive System

Name ____________________

Lab Section ____________________

Date ____________________

1. Review the organs of the digestive system and identify those that are components of the alimentary canal and those that are accessory structures for digestion.

2. Identify and describe the various activities that occur during the digestion of food.

3. Describe the basic organization of the peritoneum. What are the functions of the mesenteries and the omenta? What is a retroperitoneal organ?

4. Describe the basic arrangement of tissue layers in the wall of the alimentary canal.

Questions 5–17: For the following structures, identify the region or organ of the alimentary canal in which they are found. Some structures are found in more than one organ.

5. Lingual tonsils ____________________
6. Gingiva ____________________
7. Goblet cells ____________________
8. Vermiform appendix ____________________
9. Peyer's patches ____________________
10. Gastric glands ____________________
11. Uvula ____________________
12. Brush border ____________________
13. Taeniae coli ____________________
14. Taste buds ____________________
15. Parietal cells ____________________
16. Brunner's glands ____________________
17. Rugae ____________________

18. What are sphincter muscles? From what layer of the alimentary canal are they derived? What important functions do they perform?

Questions 19–23: Describe the anatomical and/or physiological basis for the following conditions.

19. Heartburn

20. Hemorrhoids

____

____

21. Dental caries

____

____

22. Appendicitis

____

____

23. Gallstones

____

____

Questions 24–33: Match the organ in column A with the appropriate description in column B.

A

24. Oral cavity ____________________
25. Esophagus ____________________
26. Salivary gland ____________________
27. Sigmoid colon ____________________
28. Pharynx ____________________
29. Pancreas ____________________
30. Liver ____________________
31. Ascending colon ____________________
32. Stomach ____________________
33. Ileum ____________________

B

a. Peyer's patches are located in the submucosa of this structure.
b. The cells in this organ produce bile.
c. The skeletal muscle in the walls of these two structures functions involuntarily during the swallowing reflex.
d. The cardiac sphincter is located at the junction of the stomach and this organ.
e. Brunner's glands are located in the submucosa of this structure.
f. This structure is the S-shaped portion of the large intestine.
g. Islets of Langerhans are found in this organ.
h. The parietal cells in this organ produce hydrochloric acid.
i. This structure produces amylase that breaks down starch to disaccharides in the oral cavity.
j. The right colic (hepatic) flexure represents the junction between this structure and the transverse colon.
k. The pharynx is located posteriorly to this structure.

34. Briefly describe how the secretions of the three major pairs of salivary glands differ.

____

____

35. Describe the structure of a liver lobule.

____

____

36. Describe the system of ducts that transports bile.

____

____

37. Briefly discuss the functions of the pancreatic acinar cells and the islets of Langerhans.

____

____

Questions 38–50: In the following diagram, identify the structures by labeling with the color that is indicated.

38. Transverse colon = **green**
39. Liver = **blue**
40. Jejunum = **yellow**
41. Cecum = **orange**
42. Descending colon = **red**
43. Stomach = **brown**
44. Ascending colon = **purple**
45. Sigmoid colon = **light blue**
46. Gallbladder = **black**
47. Ileum = **light brown**
48. Rectum = **light green**
49. Veriform appendix = **pink**
50. Spleen = **gray**

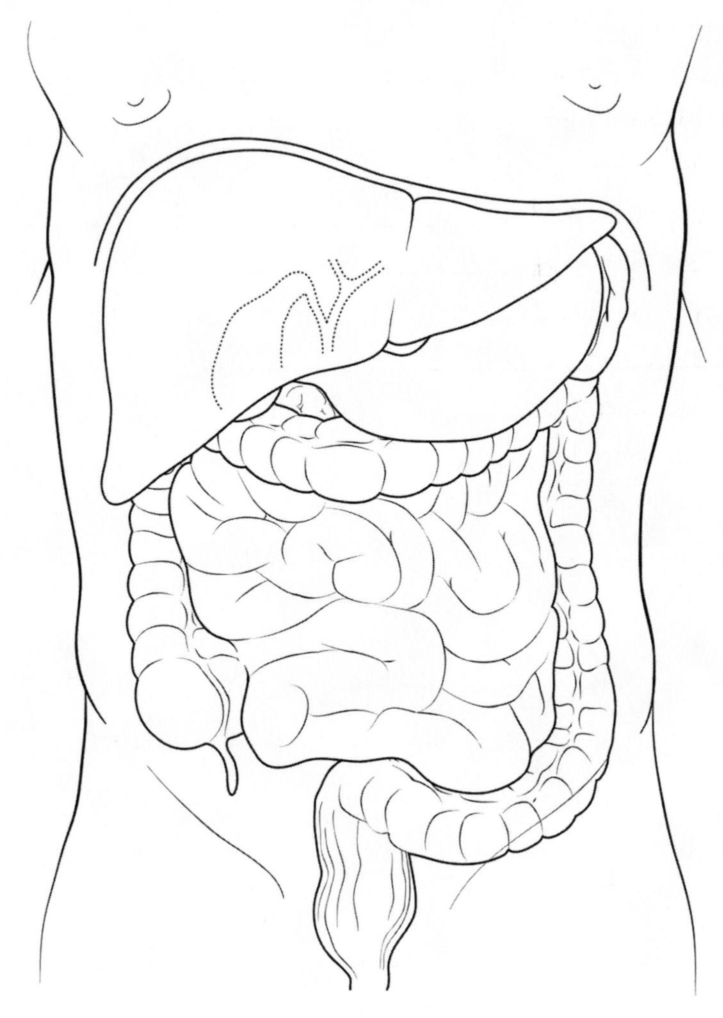

Exercise 27

Actions of a Digestive Enzyme

Laboratory Objectives

On completion of the activities in this exercise, you will be able to:

- Explain the roles of enzymes in metabolic processes.
- Discuss the specific digestion of starch by the enzyme amylase.
- Correctly perform the IKI and Benedict's tests to assess the presence of starch and disaccharides.
- Test a solution to determine the amount of starch digestion that has occurred.
- Explain the role of pH in starch digestion.

Materials

- Test tubes
- Test tube rack
- Laboratory marking pencil
- Hot water bath (hot plate and beaker with water)
- Test tube holder/tongs
- Small cups
- Gauze for filtering
- 1% solutions of the following:
  - Potato starch (pH 4)
  - Potato starch (pH 7)
  - Potato starch (pH 10)
  - Maltose
- Benedict's solution
- 1% iodine, 2% potassium iodide (IKI) solution
- Distilled water
- Cola
- Milk
- Orange juice
- Amylase solution (to be made by the students)

Enzymes are proteins that act as biological **catalysts.** Catalysts are substances that increase the rate at which chemical reactions occur. In this role, enzymes are essential for regulating all the biochemical reactions in the cells of our bodies.

Enzymes are precise in their actions. A particular enzyme will act only upon a specific reactant substance, known as the **substrate**, so it is responsible for catalyzing one type of reaction. The substrate binds to a particular region on the enzyme, called the **active site**, and forms the **enzyme-substrate complex.** The active site has a specific three-dimensional shape that is designed for binding only to the substrate. Once the enzyme-substrate complex has formed, the enzyme acts to accelerate the formation of a specific product. Subsequently, the newly formed product detaches from the active site (Figure 27.1). The enzyme is not changed, chemically, during the overall reaction and is recycled for repeated use. The reaction between an enzyme and a substrate can be summarized by the following sequence of events.

substrate + enzyme → enzyme-substrate complex → product + enzyme

↑ enzyme is recycled for repeated use ↓

Before a substrate can bind to an enzyme, smaller molecules or ions known as **cofactors** must first bind to the active site or some other region of the enzyme. When cofactors attach to the enzyme, the shape of the active site is altered so that the substrate can attach properly. Without cofactors, the enzyme-substrate complex will not form and the enzyme is nonfunctional. Examples of cofactors include various ions, such as Ca^{+2} and Mg^{+2}, and vitamins.

The following factors have an effect on enzyme activity.

- pH
- Temperature
- Concentration of the enzyme
- Concentration of the substrate
- Length of reaction

In this laboratory exercise, you will study the activity of **amylase**, an enzyme found in saliva and pancreatic juice, to learn how pH affects amylase activity.

Amylase initiates the process of carbohydrate digestion in the mouth. In this experiment, the carbohydrate that will be digested by amylase is **starch** (the substrate). Starch is a **polysaccharide** that contains repeating units of **glucose.**

When starch is digested by amylase, it is broken down by **hydrolysis** into smaller units called **dextrins.** As hydrolysis continues, the dextrins are broken down to **maltose**, which is a **disaccharide.** During its final digestion, maltose is broken down to glucose, which is a **monosaccharide.** This reaction is carried out by another enzyme in the small intestine.

WHAT'S IN A WORD Amylase is the enzyme involved in chemical breakdown of starch. The two forms of starch are **amylopectin** and **amylose.** Notice that the three words *amylase, amylopectin,* and *amylose* begin with the prefix *-amyl,* which is derived from the Greek word for starch *(amylon).* The *-ase* suffix on a name indicates that the substance is an enzyme. Thus, *amylase* means "starch enzyme." ■

You will be able to observe the activity of amylase during starch digestion by using an indicator solution that contains 1% iodine and 2% potassium iodide (**IKI solution**). This solution is amber in color. When IKI solution is added to starch, the mixture takes on a color that ranges from blue-black to purple. If amylase is added, the digestion of starch will begin. As the digestion progresses, the mixture will undergo a series of color

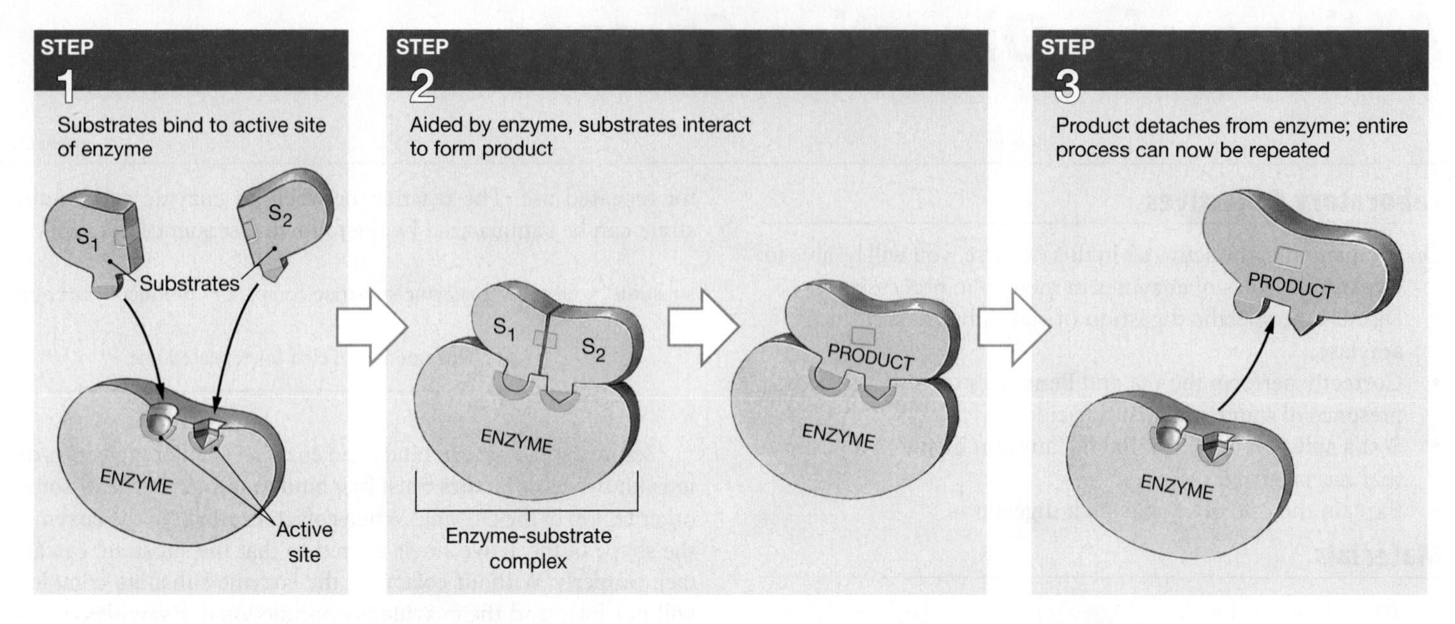

Figure 27.1 The function of an enzyme. Each enzyme catalyzes a specific biochemical reaction. The substrate molecule binds to a specific active site on the surface of the enzyme. After the product is formed, the enzyme can be used again.

changes, from blue-black to red and finally to colorless. The red color is characteristic of the reaction of the IKI solution with dextrins, and it indicates that the starch is beginning to break down. (Recall that dextrins are polysaccharides that are smaller in size than starch molecules.) Further digestion of the dextrins to maltose is indicated when the achromatic (colorless) point is reached. The reaction occurs in the following sequence.

Digestion of starch with amylase:	starch (large polysaccharide)	→	dextrins (small polysaccharide)	→	maltose (disaccharide)
Color of solution with IKI:	blue-black or purple	→	red	→	colorless

Control Tests for Starch and Reducing Sugars

Before investigating the digestion of starch to maltose by the digestive enzyme salivary amylase, it is important that you be able to identify the presence of starch and maltose in a solution. To do this, you will prepare control solutions and observe specific color changes that indicate the presence of these substances. These control tests will provide known standards against which comparisons can be made. Before you proceed, be sure you understand what a control test is and why it is important.

When conducting these trials, it is important to remember the following:

- The test tubes containing your control samples should be clearly labeled. They should be kept in your test tube rack so you can compare them to your later results.
- IKI and Benedict's solutions are used for different tests, so these two solutions should never be added to the same test tube.
- Use separate and clean test tubes for each trial.

ACTIVITY 27.1 Conducting the Control Tests

Positive Control for Starch

1. Use a marking pencil to label a test tube "Starch (+)."
2. Place 1 ml of starch solution (pH 7) in the test tube. The solution will be colorless.
3. Add 4 drops of IKI solution to the test tube. The presence of a blue-black color when IKI is added indicates the presence of starch.
4. Note the color of your sample and record the results of your observations in Table 27.1.

Negative Control for Starch

1. Label a test tube "Starch (−)."
2. Place 1 ml of distilled water in the test tube.
3. Add 4 drops of IKI solution to the test tube. The solution will be faintly yellow or amber, the color of the IKI solution. Note this color.
4. The lack of any additional color change indicates the absence of starch.
5. Record the results of your observations in Table 27.1.

Table 27.1 Control Tests for Starch and Reducing Sugars (Disaccharides)

Test	Results
1. Starch tests (IKI solution)	
a. Positive control	______
b. Negative control	______
2. Disaccharide tests (Benedict's solution)	
a. Positive control	______
b. Negative control	______

Positive Control for Disaccharides (Reducing Sugars)—Benedict's Test

1. Label a test tube "Disaccharides (+)." Be sure the label will be well above the water line once the test tube is placed in the hot water bath, or it may be soaked or steamed off.
2. Place 1 ml of 1% maltose in the test tube.
3. Add 10 drops of Benedict's solution to the test tube. The solution will be blue, the color of the Benedict's solution.
4. Mix well by swirling the test tube, and then place it in a hot water bath heated to just below boiling.
5. Keep the test tube in the water bath for a maximum of 5 minutes or until a color change occurs.
6. A color change from blue to another color, ranging from bright yellow to deep red, indicates the presence of disaccharides. The greater the color change, the more disaccharides that are present.
7. Record the results of your observations in Table 27.1.
8. Remove your test tube from the hot water bath using a test tube holder. **(Caution: The tube may be very hot.)** Place it in your test tube rack for later comparisons.

Negative Control for Disaccharides (Reducing Sugars)

1. Label a test tube "Disaccharides, (−)." Be sure the label will be well above the water line once the test tube is placed in the hot water bath, or it may be soaked or steamed off.
2. Place 1 ml of distilled water in the test tube.
3. Add 10 drops of Benedict's solution to the test tube. The solution will be blue in color.
4. Mix well by swirling the test tube, and then place it in a hot water bath.
5. Keep the test tube in the water bath for a maximum of 5 minutes.
6. Observe the presence or absence of a color change from the original blue color. The lack of a color change indicates the absence of disaccharides.
7. Record the results of your observations in Table 27.1.

QUESTIONS TO CONSIDER

1. Why is it necessary to conduct a positive control test?

______

______

2. Why is it necessary to conduct a negative control test?

______

______

Starch and Disaccharides in Other Solutions

In the next activity, you will conduct the IKI and Benedict's tests on various beverages. By observing specific color changes you will be able to determine the presence or absence of starch or disaccharide sugars in these solutions.

ACTIVITY 27.2 Testing for the Presence of Starch and Disaccharides in Various Beverages

1. Obtain six clean test tubes. Label each one with the name of one of the sample solutions, so you will have two sets of three test tubes, labeled for each of the following:
   - Cola
   - Milk
   - Orange juice
2. Place 1 ml of cola in each of the two test tubes labeled "Cola."
3. Perform the IKI test on one of these two tubes and record your observations in Table 27.2.
4. Perform the Benedict's test for the other test tube that contains cola, remembering to heat it for no more than 5 minutes. Record your results in Table 27.2.
5. Place 1 ml of milk into each of the test tubes labeled "milk," then perform the IKI and Benedict's tests on each of these two test tubes. Record your results in Table 27.2.
6. Repeat this procedure for the two remaining test tubes, adding 1 ml of orange juice to each, and performing the IKI and Benedict's tests on these tubes. Again, record your results in Table 27.2.
7. The specific color changes reveal the presence or absence of starch or disaccharide sugars in these solutions. Analyze the results of your observations in Table 27.2 to see which solutions contain which types of carbohydrates.

Table 27.2 Starch (IKI) and Disaccharide (Benedict's) Tests for Other Solutions

Solution	Test	Results
1. Cola	a. Starch test	
	b. Disaccharide test	
2. Milk	a. Starch test	
	b. Disaccharide test	
3. Orange juice	a. Starch test	
	b. Disaccharide test	

QUESTION TO CONSIDER Did the colors of the beverages that you tested in the previous activity have any effect on your ability to determine the presence or absence of starch or disaccharides? Explain. ______

______

______

Table 27.3 Effect of pH on Amylase Activity

pH of starch solution	Time for solution to become colorless	Comments
pH 4		
pH 7		
pH 10		

The Effect of pH on Amylase Activity

When amylase is added to a starch solution, it begins the chemical digestion of the starch. As a consequence, the concentration of starch decreases and the concentration of dissacharide increases. The process can be summarized by the following chemical reaction:

$$\underset{\text{(polysaccharide)}}{\text{STARCH}} \xrightarrow{\textbf{In the presence of amylase}} \underset{\text{(disaccharide)}}{\text{MALTOSE}}$$

If this reaction proceeds in the presence of IKI solution, the color of the solution will change from blue-black to red to a colorless solution. **The time it takes for the solution to become colorless can be used as an indicator of the enzyme's effectiveness.** Thus, the enzyme would be considered very effective if the transition to a colorless solution occurred in a relatively short time. In this experiment, you will test the effect of pH on enzyme activity.

ACTIVITY 27.3 Testing the Effect of pH on Amylase Activity

Amylase Activity at Varying pH

1. Begin by making your own amylase solution. Recall that amylase is present in saliva. Collect 10 ml of your own saliva is a small cup.
2. Add an equal volume of distilled water and mix well.
3. Filter the mixture through a gauze filtering system. Lacking any special filtration system, you can simply cover your cup with gauze, then pour the solution from your cup into a second cup. The resulting **filtrate** is the amylase solution you will use for this activity. It contains your own amylase, the digestive enzyme that you will be testing.
4. Mark three test tubes as follows:
   - pH 4
   - pH 7
   - pH 10
5. Add 4 drops of IKI solution to each test tube.
6. Add 0.5 ml of starch solution at pH 4 to the test tube labeled accordingly. Repeat with the pH 7 starch solution and the pH 10 solution.
7. Add 2 ml of your amylase solution to each tube. Mix well by swirling.
8. Place the tubes in a 37° C water bath, to speed the process so you can observe it easily within your lab period.
9. Observe the color changes and record in Table 27.3 the time it takes for the solutions in the tubes to become colorless.

Testing for the Presence of Starch after Amylase Digestion

1. Label three new test tubes "pH 4," "pH 7," and "pH 10."
2. Transfer 1 ml of each solution from the previous activity to the designated clean test tube, keeping them separate.
3. Perform the IKI test for each tube, adding 4 drops of IKI solution to each.
4. Observe and record the results of this test in Table 27.4.

   Is there any starch remaining in the test solutions?

   ______

Table 27.4 Starch (IKI) and Disaccharide (Benedict's) Tests for Test Solutions after Amylase Digestion

pH of test solution	Starch (IKI) test (+/−)	Disaccharide (Benedict's) test (+/−)	Comments
pH 4			
pH 7			
pH 10			

Testing for the Presence of Disaccharides after Amylase Digestion

1. Label three new test tubes with the three pH values, as in the last trial.
2. Transfer 1 ml of each solution from the first part of this activity to the appropriately labeled clean test tube and perform the Benedict's test for each: Add 10 drops of Benedict's solution to each test tube, mix by swirling, then heat for no more than 5 minutes in a hot water bath.
3. Observe and record the results of this test in Table 27.4.

   Are dissaccharides (maltose) present in the test solutions?

   ____

QUESTIONS TO CONSIDER

1. Based on your results, at what pH is the enzyme amylase most effective? At what pH is it least effective?

____

____

2. What effect did pH have on starch digestion by salivary amylase?

____

____

Writing a Laboratory Report

Prepare a laboratory report in which you carefully and thoughtfully examine the results of your "research." Your report should include the following:

- Introduction and rationale for the experiment
- Description of the materials and methods
- Recording and explanation of the results
- Discussion that interprets and critically analyzes the data

For more detailed instructions on how to write your laboratory report, refer to the section "Writing a Laboratory Report" in the Appendix.

CLINICAL CORRELATION

Salivary amylase is, of course, secreted in the mouth, which has a normal pH value around 7. Starch digestion begins immediately. Once swallowed, the food enters the stomach, which contains hydrochloric acid. The hydrochloric acid is needed to activate the protein-digesting enzyme pepsinogen, but it also dramatically lowers the pH of the stomach contents to around 2.5. This large change in pH denatures the amylase so it can no longer function, and thus halts starch digestion. Consequently, the pancreas also secretes an amylase, which enters the small intestine where the pH is again elevated. This amylase can then complete whatever starch digestion was left undone by the salivary amylase.

Exercise 27 Review Sheet

Actions of a Digestive Enzyme

Name ______________________

Lab Section ______________________

Date ______________________

1. Why are enzymes required for the normal functioning of biochemical reactions in cells?

____

____

2. What is the active site? What role does the active site play in defining the specificity of an enzyme's action?

____

____

3. What is a cofactor? Explain why cofactors are essential for enzymes to function properly.

____

____

4. Briefly describe the steps in the digestion of starch by amylase

____

____

Exercise 28

Anatomy of the Urinary System

Laboratory Objectives

On completion of the activities in this exercise, you will be able to:

- Describe the gross and microscopic anatomy of the urinary system.
- Trace the path of materials through the urinary system, from when they enter the kidneys to when they leave the body.
- Describe the functions of the various components of the urinary system.
- Explain the positions of the two types of nephrons.
- Detail the blood supply of the kidneys.

Materials

- Compound light microscope
- Prepared microscope slides
  - Human kidney
  - Kidney, injected to demonstrate renal blood supply
  - Ureter
  - Urinary bladder
  - Urethra
- Torso model
- Dissectable kidney model, coronal section
- Model of a nephron
- Model of the female pelvis
- Model of the male pelvis
- Injected sheep or pig kidney for dissection
- Dissection tools (scalpel, probe, forceps, dissecting pan)

The **kidneys** (Figure 28.1), which are the principal organs of the urinary system, perform several essential regulatory functions. As they produce urine, the kidneys remove metabolic wastes and toxins from the blood and conserve glucose, water, and many essential electrolytes. These activities are critical for maintaining osmotic and pH balance in the blood plasma. The kidneys also have an endocrine function. They secrete erythropoietin, which promotes red blood cell production in the bone marrow, and renin, which plays a role in maintaining normal blood pressure.

The **ureters**, **urinary bladder**, and **urethra** are accessory organs of the urinary system (Figure 28.1). Collectively, these structures transport, store, and excrete urine.

Anatomy of the Urinary System

The two retroperitoneal kidneys are located on each side of the vertebral column on the posterior abdominal wall. They extend, approximately, from the T12 vertebra to the L3 vertebra. The right kidney rests 1.5 to 2 cm (0.5 to 0.7 in) lower than the left kidney due to the large area taken by the liver (Figure 28.1a). Each kidney is approximately 12 cm (5 in) long, 6 cm (2.5 in) wide, and 3 cm (1.25 in) thick. Its lateral surface is convex and its medial surface is concave. A central region along the medial surface, called the **hilus**, serves as a point of entry and exit for the renal artery and vein, ureter, nerves, and lymphatic vessels (Figures 28.1 and 28.2).

Each kidney is surrounded by three connective tissue coverings (Figure 28.1b). The **renal fascia** is the outer fibrous connective tissue covering that anchors the kidney to the peritoneum and abdominal wall. The middle **adipose capsule** is a fatty layer that provides shock-absorbing protection for the kidney. The inner **renal capsule** is a strong fibrous layer that tightly adheres to the kidney surface. It provides a protective barrier that prevents infections from spreading to renal tissue.

In a coronal section of the kidney, three regions can be identified (Figure 28.2). The outer **renal cortex** and the middle **renal medulla** contain the various urine-producing tubules that comprise the **nephrons**. The inner region of the kidney is a cavity known as the **renal sinus**. It serves to collect urine that is produced in the nephrons, and to transport this fluid to the ureter.

The ureters are retroperitoneal, tubular organs that are approximately 25 cm (10 in) long. They begin at the renal pelvis (Figure 28.2) and travel inferiorly along the posterior abdominal wall. They enter the pelvic cavity and terminate at the posterior aspect of the urinary bladder (Figure 28.1a). The primary function of the ureters is to transport urine from the kidneys to the urinary bladder.

The urinary bladder is located within the pelvic cavity, posterior to the pubic symphysis and anterior to the rectum (Figures 28.3a and b). It serves as a temporary storage sac for urine, and its shape is determined by the volume of urine it contains. An empty bladder has a pyramidal shape, with the base forming its posterior surface and the apex oriented anteroinferiorly. When the bladder is full, it becomes ovoid and the superior surface bulges into the abdominal cavity.

The urethra is a muscular tube that conveys urine from the bladder to the outside of the body (Figure 28.3). At the junction with the bladder, a thick band of smooth muscle surrounds the urethra to form the **internal urethral sphincter** (Figure 28.3c). This sphincter is involuntary and keeps the urethra closed to prevent urine from leaking between periods of micturition (urination). A second sphincter, the **external urethral sphincter**, consists of skeletal muscle. In females, this voluntary sphincter is located near the **external urethral orifice** (Figure 28.3b); in males, it is found just below the prostate gland (Figures 28.3a and c).

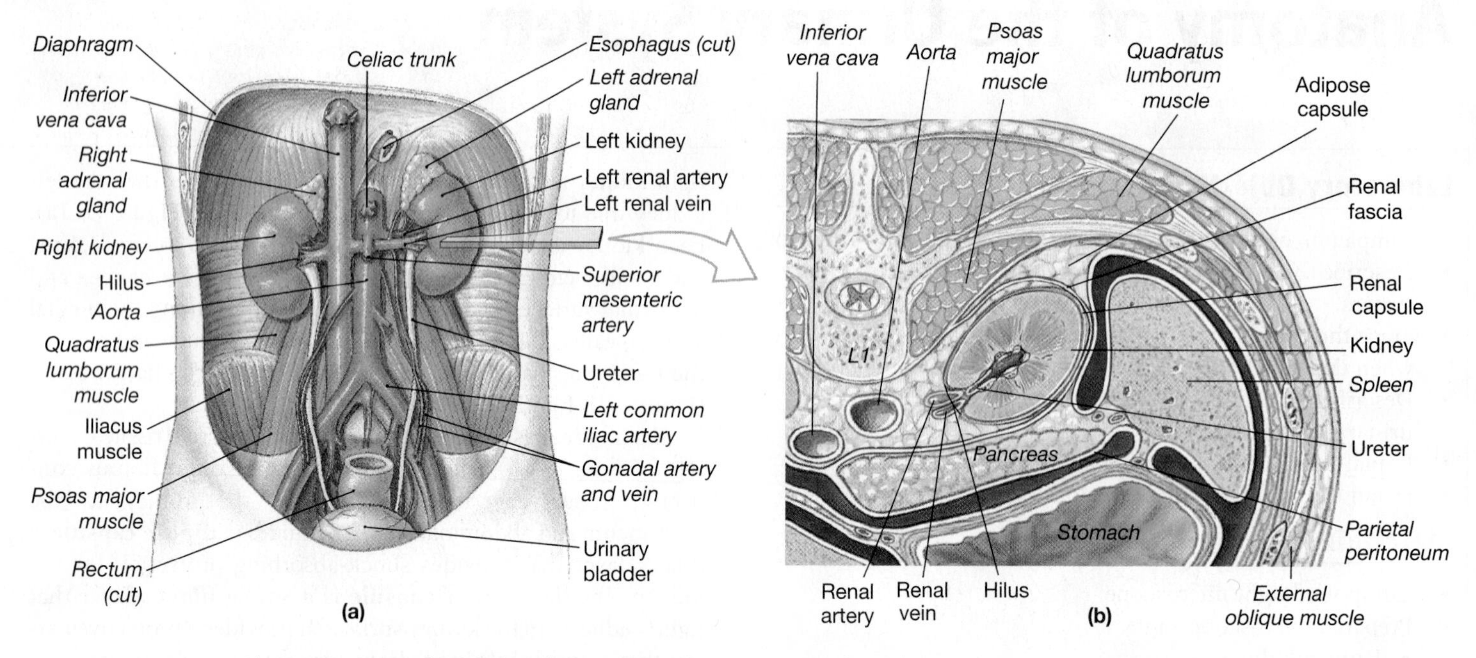

Figure 28.1 Overview of the urinary system. a) Anterior view in the male illustrating anatomical relationships of the kidneys and ureters with posterior abdominal muscles, and major abdominal blood vessels; **b)** transverse section of the left posterior abdominal wall, showing the position of the kidney in relation to other abdominal organs.

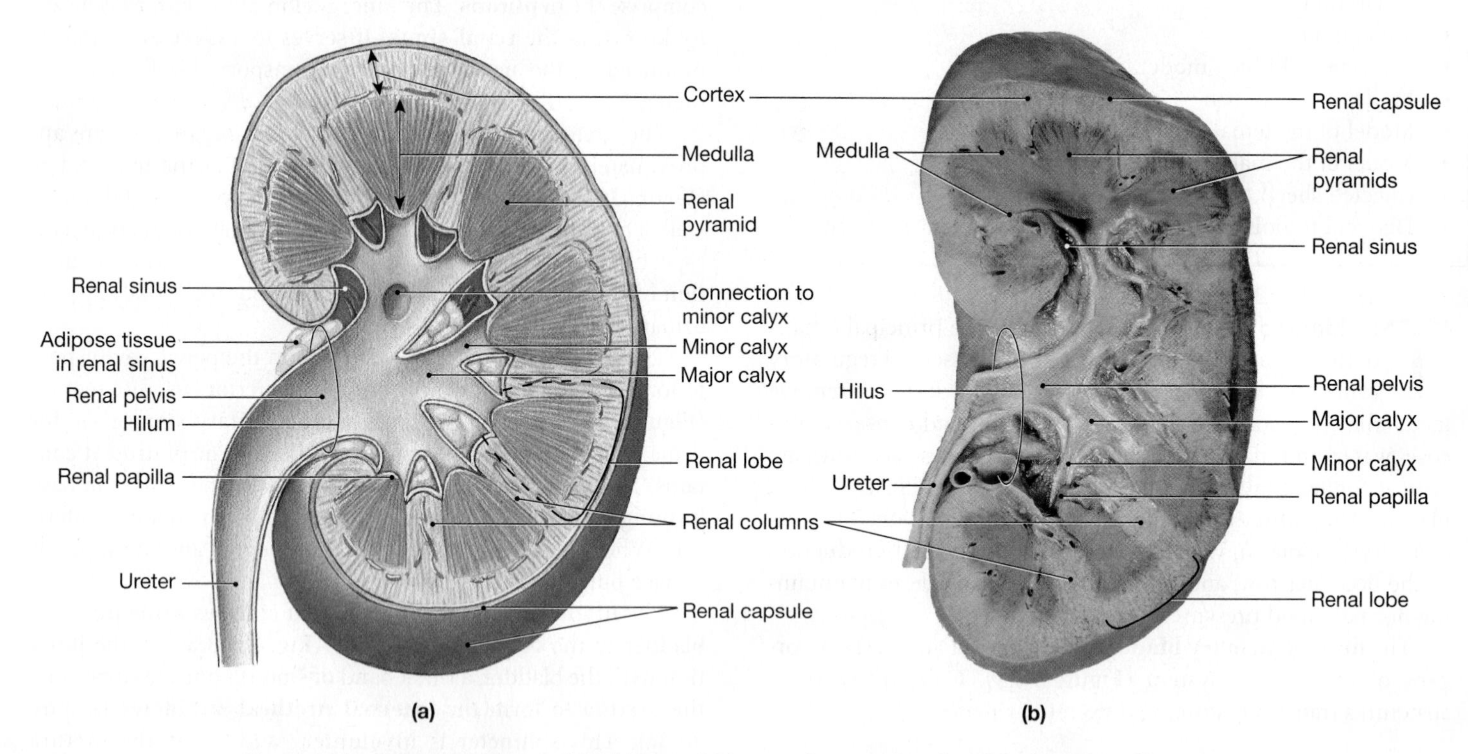

Figure 28.2 Internal structure of the kidney. a) Diagram of a coronal section of the left kidney; **b)** corresponding cadaver dissection.

ACTIVITY 28.1 Examining Gross Anatomy of the Urinary System

1. On the torso model, identify the two kidneys along the posterior abdominal wall (Figure 28.1). Note their positions relative to the vertebrae. Also observe that the position of the liver causes the right kidney to be more inferior than the left kidney.
2. Observe the lateral convex surface and medial concave surface, and identify the **hilus** along the center of the con-

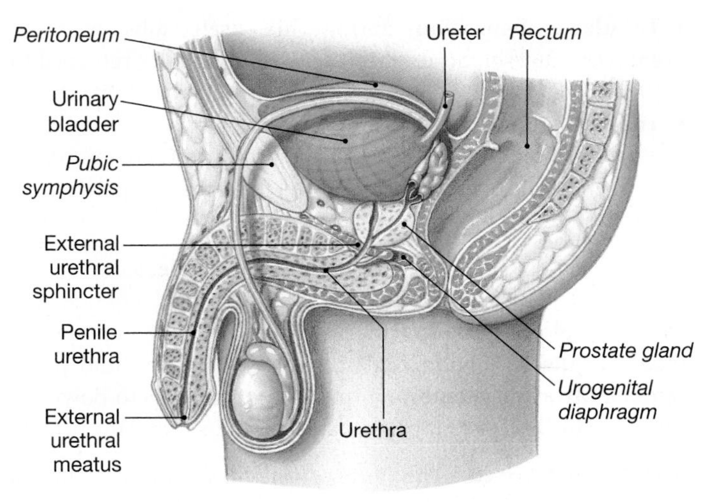

(a) Male

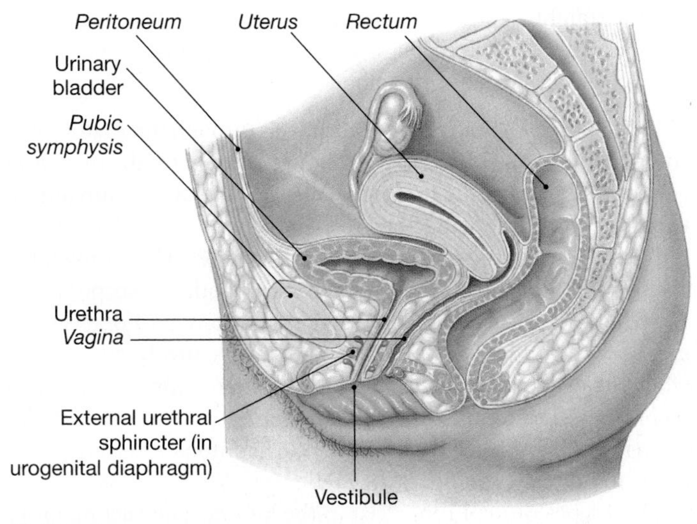

(b) Female

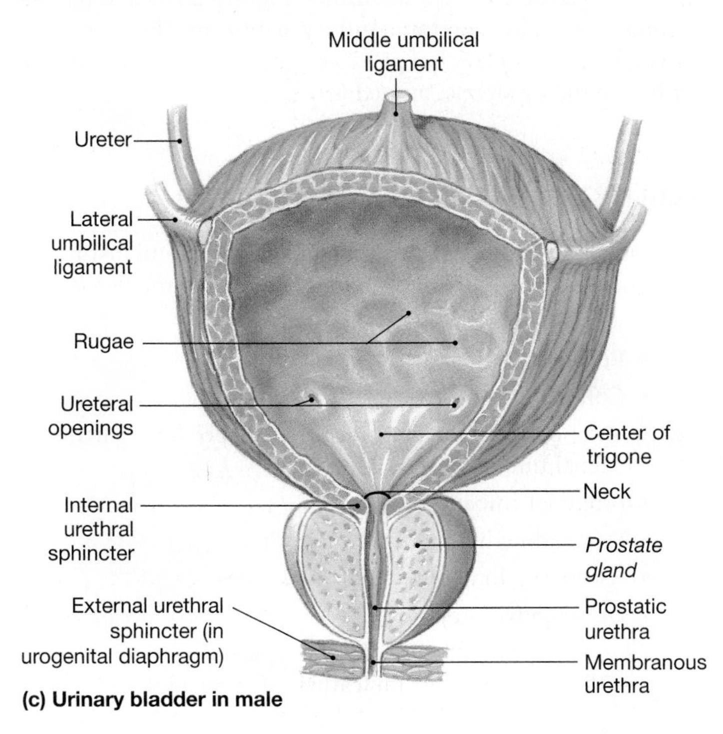

(c) Urinary bladder in male

Figure 28.3 Position of the urinary bladder. Midsagittal sections of **a)** the male pelvic cavity, and **b)** the female pelvic cavity, illustrating the position of the urinary bladder in relation to other pelvic structures. **c)** Internal view of the male urinary bladder, illustrating the rugae and the trigone.

cave surface (Figures 28.1a and 28.2). At the hilus identify the ureter, renal artery, and renal vein entering or exiting the kidney.

3. Identify the **adrenal glands** that rest on the superior surface of each kidney.
4. Examine a coronal section of a kidney and identify the three internal regions (Figure 28.2): renal cortex, renal medulla, and renal sinus.
5. Locate the outer renal cortex and note its appearance.
6. Locate the middle portion, the renal medulla, and identify the cone-shaped **renal pyramids**. They are separated by inward extensions of cortical tissue known as **renal columns**. Note that the base of each renal pyramid forms a border with the cortex, while the apex, known as a **renal papilla**, projects inwardly.
7. Identify the inner region of the kidney, which is called the renal sinus. It consists of the **calyces** (singular = **calyx**) and the **renal pelvis** (Figure 28.2).
8. Locate a **minor calyx**. Urine formed within a renal pyramid drains through the renal papilla to enter the minor calyx.
9. Follow the minor calyx inward until it joins at least one more minor calyx. The larger area marked by this junction is a **major calyx**.
10. Locate the renal pelvis, a funnel-shaped cavity continuous with the ureters. The renal pelvis collects urine from three to six major calyces and transmits it to the ureter.
11. On the torso model, identify the two ureters where they exit the kidneys. Observe that these tubes travel inferiorly along the posterior abdominal wall and enter the pelvic cavity (Figure 28.1a).
12. In the pelvic cavity, locate the termination of the ureters at the posterior aspect of the urinary bladder (Figures 28.3a and c).
13. On models of the female and male pelvic cavities, observe the relative position of the urinary bladder to other abdominopelvic organs. In both sexes, the urinary bladder is posterior to the **pubic symphysis** and anterior to the **rectum** (Figures 28.3a and b). In females, it is anterior to the **uterus** and the **vagina** (Figure 28.3b). In males, it is superior to the **prostate gland** (Figure 28.3a).
14. Obtain a model with an open bladder. Along the internal surface of the contracted bladder, identify the numerous **rugae** (Figure 28.3c), which are folds of the mucous membrane. They are present when the bladder is empty. When the bladder fills with urine, the wall stretches and the rugae disappear.

15. Identify the **trigone**, which is a triangular region marked by the connections of the two ureters and the urethra to the bladder wall (Figure 28.3c). This region lacks rugae and remains smooth at all times. It is noteworthy because bladder infections seem to occur most often in this area. The reason for this is not known.
16. Identify the urethra where it emerges from the inferior aspect of the urinary bladder. Notice that in females, the urethra is relatively short (approximately 3.5 cm or 1.5 in) and travels anterior to the vagina. The external urethral orifice is located anterior to the **vaginal orifice** and posterior to the **clitoris** (Figure 28.3b). Observe that the male urethra is relatively long (approximately 20 cm or 8 in) and has the following three sections (Figures 28.3a and c).
    - The **prostatic urethra** passes through the prostate gland.
    - The **membranous urethra** passes through a band of muscle called the **urogenital diaphragm**.
    - The **penile urethra** travels through the shaft of the penis.

CLINICAL CORRELATION

For two reasons, females are more susceptible to urinary tract infections than males. First, the female urethra is much shorter and inflammatory infections of this structure (**urethritis**) can easily spread to the bladder (**cystitis**) and possibly to the kidneys (**pyelitis** or **pyelonephritis**). Second, the external urethral orifice is very close to the anal opening. Fecal bacteria can be transferred to the urethra particularly if an individual wipes with toilet paper in a posterior to anterior direction after defecation.

QUESTION TO CONSIDER On a torso model, identify the two kidneys and describe their anatomical relations to other abdominal organs.

Right kidney:

____

____

Left kidney:

____

____

The Nephron

The functional units of the kidney are the nephrons (Figure 28.4). They consist of various types of tubules that closely interact with capillaries during urine formation. To produce urine, the nephrons perform three basic activities.

- **Glomerular filtration**, during which water, glucose, amino acids, and nitrogen-containing wastes are passed from blood to the nephrons along a pressure gradient
- **Tubular reabsorption**, during which useful substances such as water, glucose, amino acids, and ions are returned to the blood
- **Tubular secretion**, during which unwanted substances such as excess potassium and hydrogen ions, drugs, creatinine, and metabolic acids are transported from the capillaries into the tubules

Each nephron begins with **Bowman's capsule**, a cup-shaped structure that surrounds a tuft of capillaries called a **glomerulus** (plural = **glomeruli**). Together, the two structures comprise a **renal corpuscle** (Figure 28.4). In each corpuscle, fluid passes across a **filtration membrane** from the glomerulus to Bowman's capsule. The fluid entering Bowman's capsule is called the **filtrate.**

WHAT'S IN A WORD The term *glomerulus* comes from the Latin term meaning "ball" or "globe." Indeed, the glomerulus is like a ball or tuft of capillaries. This arrangement, rather than a single straight capillary, dramatically increases the amount of capillary surface area through which filtration can occur, allowing the kidneys to filter the blood much more rapidly.

After Bowman's capsule, filtrate travels through three additional components of a nephron in the following sequence: the **proximal convoluted tubule**, the **loop of Henle**, and the **distal convoluted tubule** (Figure 28.4). Together, these structures form a long, twisting passageway known as the **renal tubule.** The convoluted tubules and the loop of Henle are capable of both reabsorption and secretion. However, there are regional specializations. For example, most reabsorption occurs in the proximal convoluted tubules, while the distal convoluted tubules have a more important role in secretion. Other tubules, the **collecting ducts**, are straight tubular passageways that receive urine from the nephrons and transport it to the minor calyces (Figure 28.4a).

Two types of nephrons exist in the kidney. The vast majority (80% to 85%) are **cortical nephrons** (Figures 28.4a and b). The second type, called **juxtamedullary nephrons** (Figures 28.4a and c), is less abundant but plays a critical role in conserving water by producing a concentrated urine.

ACTIVITY 28.2 Examining the Nephron

1. Obtain a model or illustration of a nephron and distinguish between the following structures (Figure 28.4).
    - Renal corpuscle
    - Renal tubule
    - Collecting duct
2. In their correct sequence, identify the specific segments of the renal tubule (Figure 28.4a).
    - Proximal convoluted tubule
    - Descending limb of the loop of Henle
    - Ascending limb of the loop of Henle
    - Distal convoluted tubule
3. Closely examine the pattern of a typical nephron and notice the following general features (Figure 28.4).

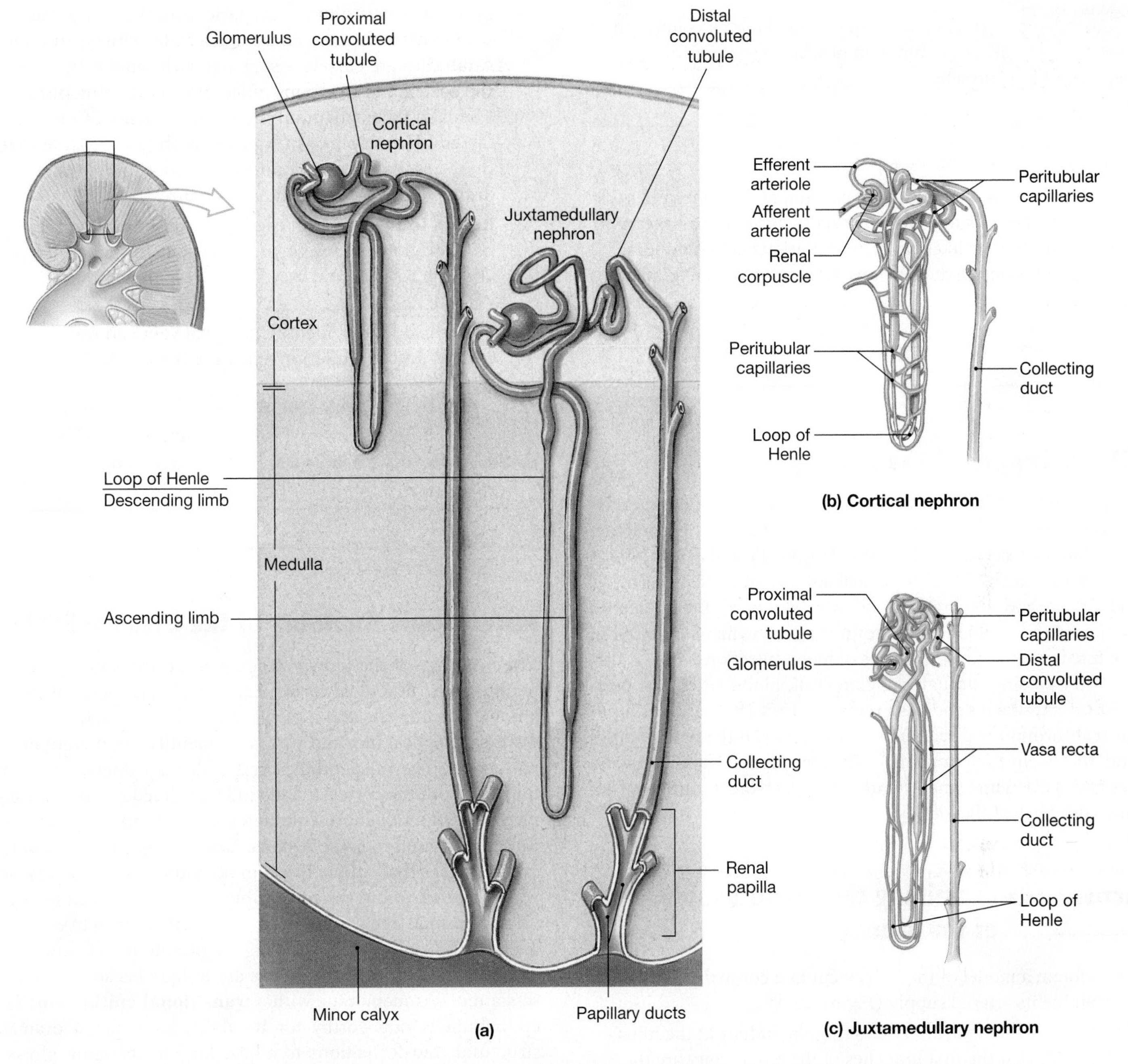

Figure 28.4 The kidney nephrons. a) Positions of cortical and juxtamedullary nephrons in relation to the renal cortex and medulla; **b)** blood supply to a cortical nephron; **c)** blood supply to a juxtamedullary nephron.

- All renal corpuscles are located in the cortex.
- The cortex also contains the proximal and distal convoluted tubules.
- The loops of Henle begin and end in the cortex but descend and ascend in the medulla.
- The distal convoluted tubules drain urine directly into collecting ducts.

4. Identify a collecting duct and notice that it receives urine from several nephrons. Trace its path as it begins in the cortex and descends through a renal pyramid in the medulla. At the renal papilla, the collecting duct empties into a minor calyx.
5. Observe that the kidney has two types of nephrons (Figures 28.4b and c).
   - In cortical nephrons, renal corpuscles originate in the outer two thirds of the cortex and the loops of Henle do not travel deeply into the medulla.
   - In juxtamedullary nephrons, renal corpuscles originate in the inner one third of the cortex and the loops of Henle extend deep within the medulla.

QUESTIONS TO CONSIDER 1. Inflammation of the renal cortex is likely to affect the function of which segments of the nephrons? Explain why. ______________________

____

____

2. **Kidney stones** are deposits of calcium and magnesium salts or uric acid crystals that can obstruct the urinary passageways. Which condition is likely to be more serious: a kidney stone blocking a collecting duct, or one blocking a ureter? Explain why.

____

____

Blood Supply of the Kidney

Under normal conditions, about 25% of total cardiac output travels through the kidneys at a rate of about 1.2 liters per minute. Each kidney receives its blood supply from a **renal artery**, which is a direct branch of the abdominal aorta (Figure 28.1a). The renal artery and its branches deliver blood to the glomeruli (Figure 28.5a), which are surrounded by Bowman's capsules in the renal cortex and are the sites of blood filtration.

After passing through the glomeruli, blood enters the **peritubular capillaries** and **vasa recta** (Figures 28.4b and c). Tubular reabsorption and secretion occurs between the renal tubules and these capillary networks. The peritubular capillaries and vasa recta drain into small venules that mark the beginning of venous drainage of the kidney.

ACTIVITY 28.3 Examining the Blood Supply of the Kidney

1. Obtain a model of the kidney cut in a coronal plane to examine its arterial supply (Figure 28.5).
2. Identify the renal artery entering the kidney at the hilus. Notice that the first branches of the renal artery are the **segmental arteries.**
3. The segmental arteries give rise to a number of **interlobar arteries** that travel toward the cortex through the renal columns.
4. At the junction between the cortex and medulla, observe how interlobar arteries branch to form the **arcuate arteries.** The arcuate arteries provide a good landmark for identifying the **corticomedullary boundary.**
5. The arcuate arteries give off several **interlobular arteries**, which can be seen ascending into the cortex.
6. Obtain a model or illustration of a nephron. Identify the **afferent arteriole** leading to a glomerulus, and the **efferent arteriole** which exits the glomerulus (Figure 28.4).
7. Locate the peritubular capillaries and the vasa recta, which arise from the efferent arteriole. The peritubular capillaries are closely associated with renal tubules in the cortex and the loops of Henle of cortical nephrons. The vasa recta surround the loops of Henle of juxtamedullary nephrons (Figures 28.4b and c). Note that these capillaries empty into the venous system that drains blood from the kidney (Figure 28.5c).
8. Return to the model of the kidney and examine the venous drainage of the organ. The veins in the kidney parallel the arteries and are assigned identical names (Figure 28.5).
9. On a torso model, identify the renal vein and notice that it empties into the inferior vena cava (Figure 28.1a).

QUESTION TO CONSIDER Explain why it is important for the kidneys to receive a large daily volume of blood.

____

____

Microscopic Anatomy of the Urinary System

The histology of the kidney is centered on the structure of the nephrons. As described earlier, each renal corpuscle consists of a Bowman's capsule and a glomerulus. These structures have a distinctive appearance and are easy to identify throughout the renal cortex. The renal tubules and collecting ducts, which are found in both the cortex and medulla, are lined mostly by a simple cuboidal epithelium, although squamous and columnar epithelia are found in some regions. You will also observe arteries and arterioles that supply blood to the kidney and capillary networks that surround the renal tubules. The capillaries are critical for normal renal function because of their active role in reabsorption and secretion during the production of urine.

The accessory urinary organs are unique because they possess a mucous membrane with a **transitional epithelium**. This epithelium is noteworthy for its ability to undergo dramatic structural transformations to adjust for the changing physical conditions of the organs where it is located.

ACTIVITY 28.4 Examining the Microscopic Anatomy of the Urinary System

The Kidney

1. Obtain a slide of the kidney.
2. Scan the slide under low magnification until your field of view is within the renal cortex. The cortex can be easily identified because it contains numerous renal corpuscles, each consisting of a glomerulus surrounded by Bowman's capsule (Figure 28.6a). Each glomerulus will appear as a

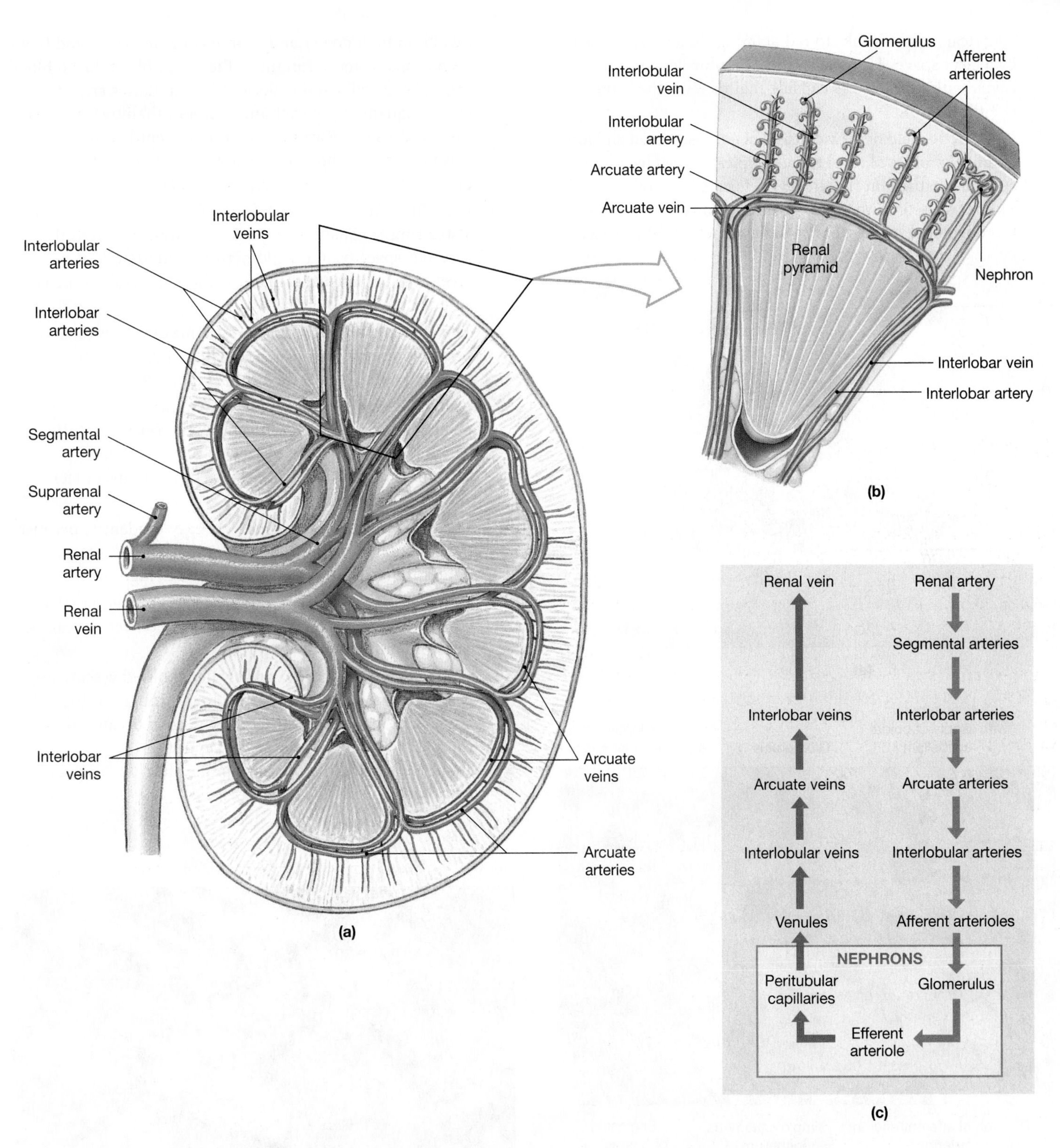

Figure 28.5 Blood supply to the kidney. a) Coronal section of the kidney, illustrating the major arteries and veins; **b)** a closer view of the blood supply to a renal pyramid and the adjacent cortex; **c)** summary chart of the blood flow through the kidney.

spherical mass of pink- to red-staining tissue surrounded by a clear space. This space, the **capsular space**, is the cavity within Bowman's capsule that receives the filtrate (Figure 28.6b).

3. Much of the field of view consists of cross-sectional and longitudinal profiles of renal tubules typically lined by a simple cuboidal epithelium (Figure 28.6). Locate these tubules. Most of these are either proximal convoluted tubules or distal convoluted tubules, but collecting ducts are also present.

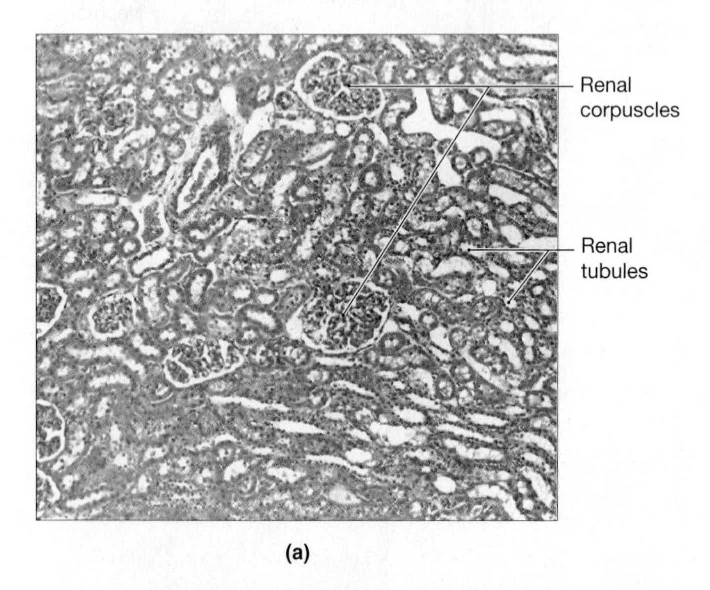

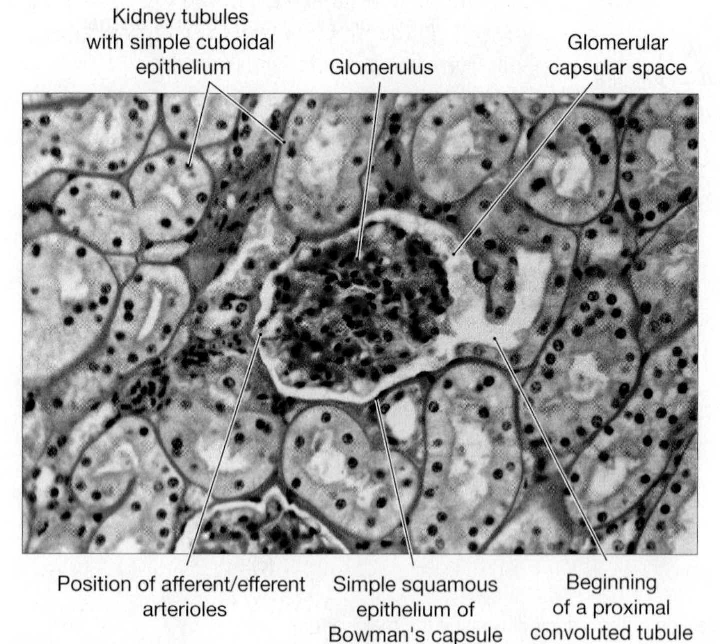

Figure 28.6 The renal cortex. a) Low-power light micrograph of the renal cortex. Renal corpuscles are distributed throughout the field of view (LM × 100). **b)** High-power light micrograph of a renal corpuscle, which includes a glomerulus enclosed by a Bowman's capsule. Kidney tubules surround the corpuscle (LM × 300).

4. Switch to high power and examine a glomerulus and Bowman's capsule more carefully (Figure 28.6b). Like all blood vessels, the walls of the glomerular capillaries are lined by simple squamous epithelium (the endothelium). You cannot clearly see the arrangement of this epithelium with this preparation; however, many of the nuclei in the glomerulus belong to the endothelial cells.
5. Examine a Bowman's capsule. The epithelium lining it is also a simple squamous type. Look carefully across the capsular space from the glomerulus and identify the single layer of squamous cells that line the outer wall of the capsule (Figure 28.6b).
6. The afferent and efferent arterioles may be identified on some glomeruli; however, you will be unable to distinguish one vessel from the other (Figure 28.6b).
7. Switch back to low power and move the slide to a region where glomeruli are absent. This is the **renal medulla** (Figure 28.7). The field of view is filled with cross-sectional and longitudinal profiles of the loops of Henle and collecting ducts.
8. The collecting ducts are relatively easy to identify because the cuboidal epithelium that lines these tubules stains lighter than the other cells on the slide (Figure 28.7).
9. Obtain a slide labeled, "kidney, injected." This slide has been specially prepared so that renal blood vessels can be easily identified.
10. Scan the slide under low power. Renal blood vessels appear as bright red channels traveling throughout the field of view. Kidney tubules are barely visible, because they stain very faintly (Figure 28.8). Do not be concerned about identifying the tubules on this slide. Focus your attention on the blood vessels.

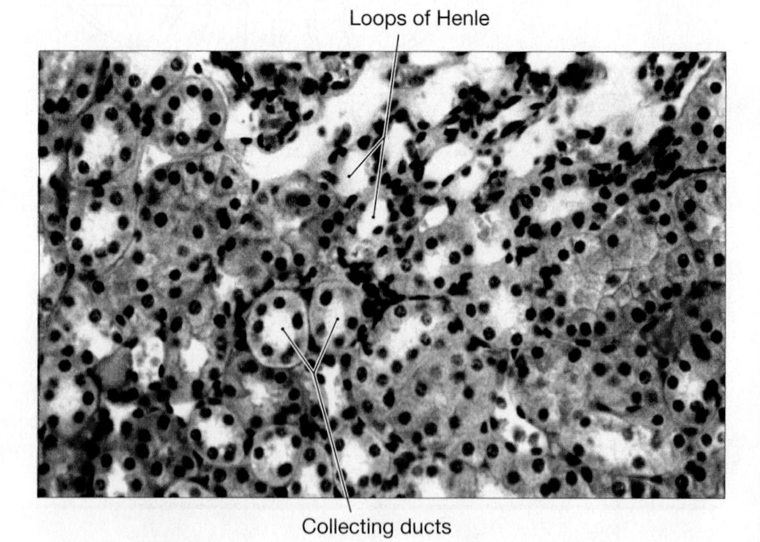

Figure 28.7 The renal medulla. Light micrograph of renal tubules in the medulla of the kidney. Notice the lack of renal corpuscles in this region of the kidney (LM × 300).

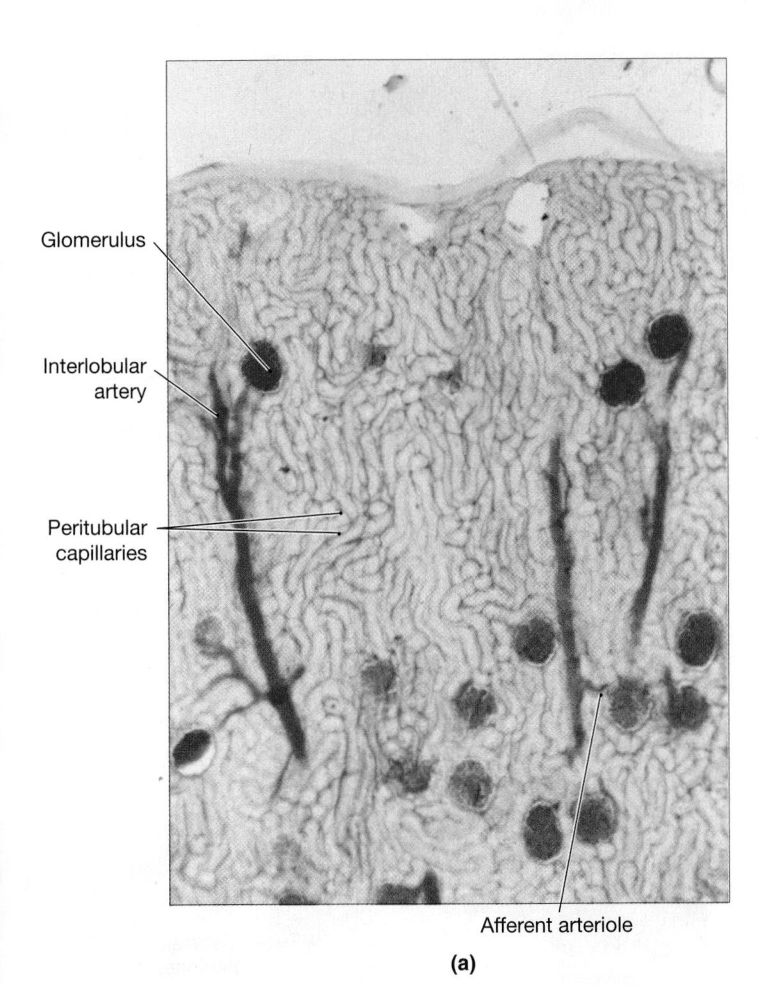

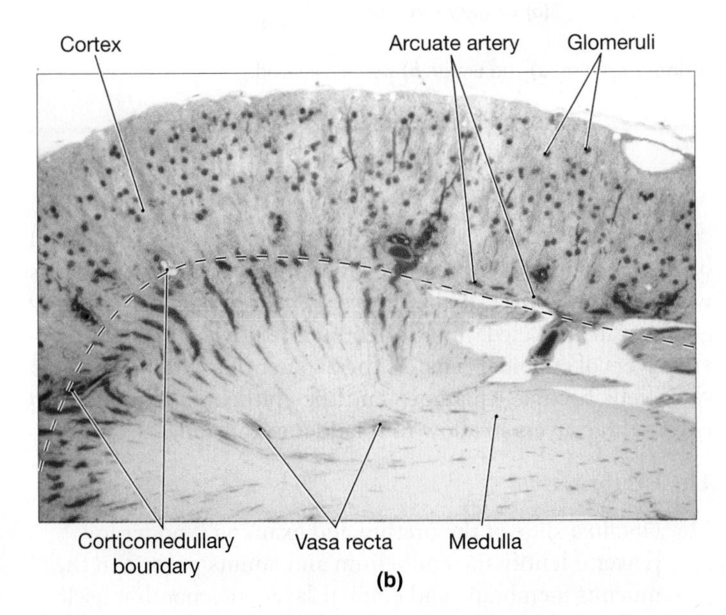

Figure 28.8 Renal blood vessels. a) Light micrograph of the blood vessels in the renal cortex (LM × 60); **b)** light micrograph of blood vessels at the border between the renal cortex and renal medulla (LM × 10).

11. Locate the cortex by looking for glomeruli, which will now appear as bright red spherical structures (Figure 28.8).
12. Identify the afferent and efferent arterioles. Depending on the plane of section, you may see both vessels, only one vessel, or no vessels at all. In most cases, it is not possible to distinguish these vessels from each other, but if you scan the slide carefully, you may see the afferent arteriole branching off an interlobular artery and traveling to a glomerulus (Figure 28.8a).
13. Other blood vessels in the cortex are the interlobular arteries and peritubular capillaries. These blood vessels can be seen throughout a typical field of view of the cortex. The interlobular arteries have a larger diameter and are easily distinguishable from the narrower and more abundant peritubular capillaries (Figure 28.8a).
14. Move the slide to a region where the glomeruli abruptly stop occurring. This is the border between the cortex and medulla. Scan the slide and attempt to find blood vessels traveling along this border. These are the arcuate arteries (Figure 28.8b). These vessels may or may not be visible, depending on the plane of section of the tissue you are viewing.
15. Move the slide deeper into the medulla. The blood vessels that you see here are the vasa recta, which are intimately associated with the loops of Henle (Figure 28.8b).

The Ureters

1. Obtain a slide of the ureter and examine it under low power. Identify the three tissue layers in the wall of the ureter (Figure 28.9a): the **mucosa (mucous membrane)**, the **muscularis (smooth muscle layer)**, and the **adventitia (connective tissue layer)**.
2. Switch to high power and confirm the following (Figure 28.9a).
   - The inner mucous membrane contains a transitional epithelium and an underlying lamina propria.
   - The middle muscle layer has an inner longitudinal and an outer circular layer of smooth muscle.
   - The outer adventitia is composed of fibrous connective tissue.

The Urinary Bladder

1. Obtain a slide of the bladder. This slide has two sections of tissue. Both are longitudinal sections of the urinary bladder wall. View both sections under low power and notice that one is much thicker than the other. The thicker section illustrates the contracted wall of an empty bladder (Figure 28.9b); the thinner section is the distended wall of a full bladder.
2. Examine the section of the contracted bladder wall. Identify the mucous membrane, which consists of transitional epithelium and a layer of connective tissue called the lamina propria. Observe that the epithelium is relatively thick, with several layers of cuboidal cells. Also notice that the mucous membrane is arranged into a number of tissue folds known as rugae (Figure 28.9b).

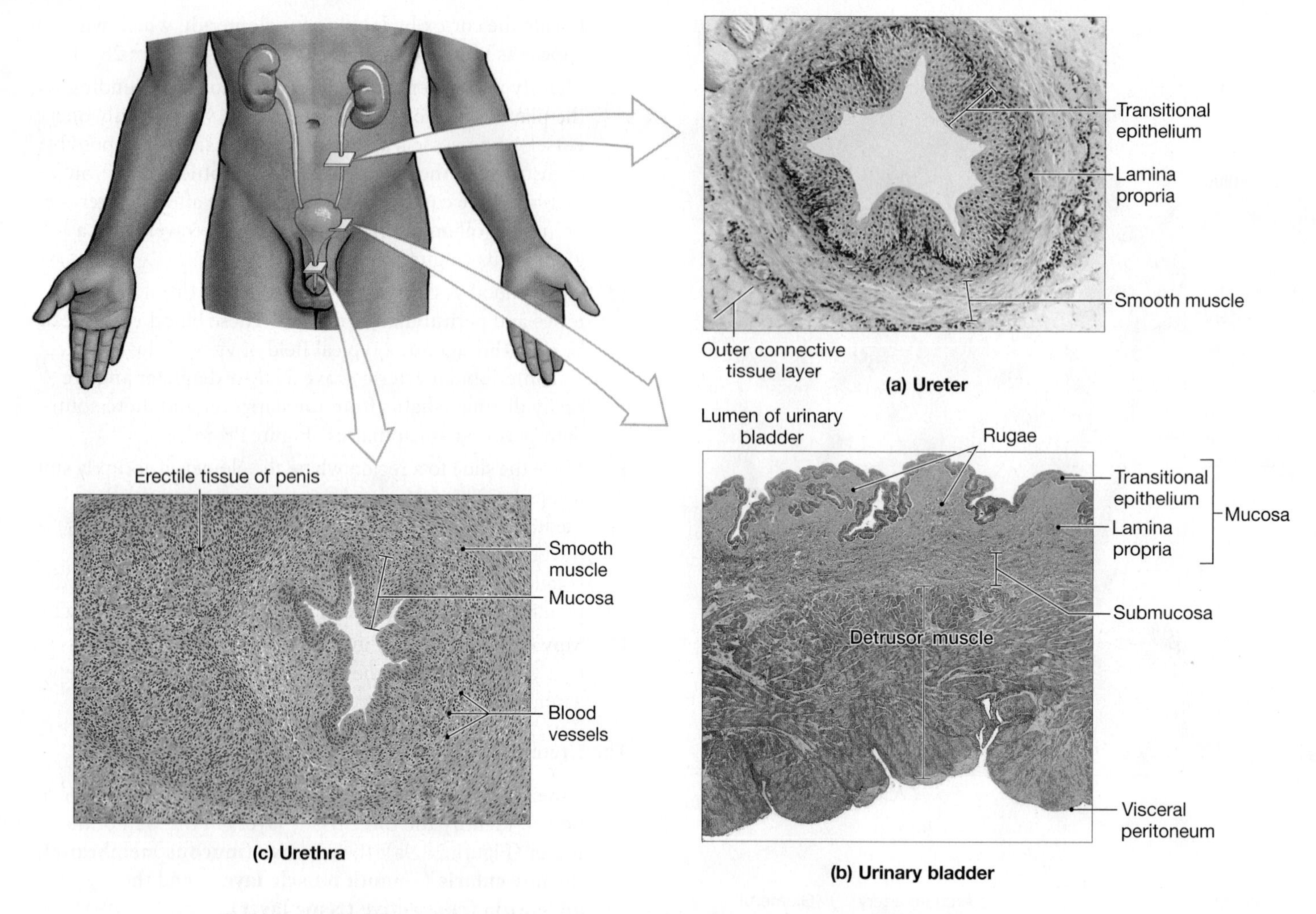

Figure 28.9 Histology of the accessory organs of the urinary system. a) The ureter; **b)** the urinary bladder in the contracted state; **c)** the urethra.

3. Just below the mucous membrane, identify a second layer of connective tissue, called the submucosa, and a region of smooth muscle, known as the muscularis. The muscularis consists of three layers of intersecting smooth muscle fibers, collectively referred to as the **detrusor muscle** (Figure 28.9b).
4. Identify the **serosa** (visceral peritoneum) that covers the outer wall of the bladder (Figure 28.9b). This is a very thin serous membrane, which may not be present on your slide.
5. Examine the section of the distended bladder. Notice that the bladder wall here lacks rugae.
6. Examine the epithelium on this section. It is strikingly different on this section, consisting of only two or three layers of squamous cells. The transitional epithelium that you are viewing on this slide and in the previous slide of the ureter is unique to the urinary tract.
7. Identify the other layers in the bladder wall described previously.

WHAT'S IN A WORD The term *transitional* implies some kind of change. The cells in transitional epithelium alter their shapes with changing physical conditions. When the bladder is empty and its wall is relaxed, the epithelium is relatively thick with several layers of cuboidal cells. As the bladder fills with urine, these same cells become squamous and the epithelium is reduced to two to three layers to allow for bladder expansion.

The Urethra

1. Obtain a slide of the urethra and examine it under low power. Identify the epithelium and lamina propria in the mucous membrane and circular layers of smooth muscle (Figure 28.9c).
2. Switch to high power and examine the mucous membrane more closely. The epithelium lining the mucous membrane will vary depending on the region of the urethra that you are viewing.
   - Near the junction with the urinary bladder, the epithelium is transitional.

- There is a gradual change to pseudostratified columnar toward the middle section of the urethra. This is the primary epithelial type.
- Near the urethra orifice, the epithelium changes to stratified squamous.

Based on the epithelial type, what region of the urethra are you viewing?

___

QUESTIONS TO CONSIDER 1. When examining a microscopic structure of the kidney, the presence or absence of renal corpuscles is a reliable criterion for determining whether you are viewing the renal cortex or the renal medulla. Explain why.

___

___

2. Based on your microscopic observations in the previous activity, identify structural similarities and differences in the structure of the ureters, urinary bladder, and urethra. Focus your attention on the organization of the mucous membrane and muscle layers.

Similarities:

___

___

Differences:

___

___

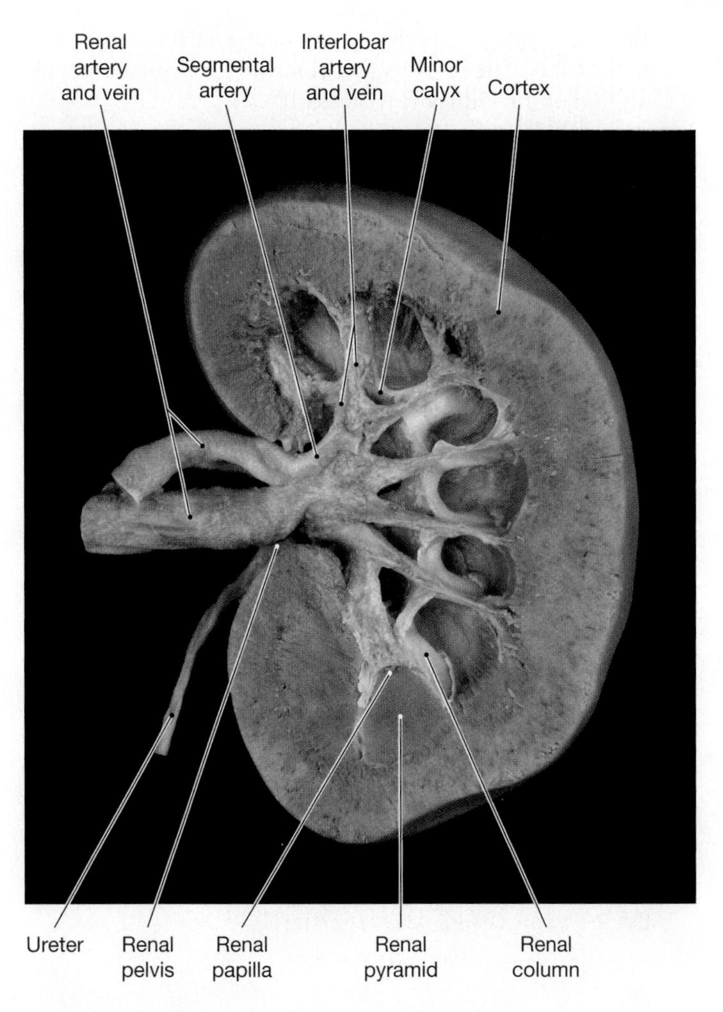

Figure 28.10 Dissection of the sheep kidney. The kidney has been cut along a frontal plane to expose internal structures. The arteries have been injected with red latex and the veins with blue latex.

Kidney Dissection

In the following activity, you will dissect an intact sheep or pig kidney. Pig kidneys are much larger and flatter than sheep kidneys. However, both types are quite similar, anatomically, to the human kidney and are excellent models for studying renal structure. As you dissect, have models, illustrations, or photographs of the human kidney readily available so that you can make structural comparisons.

ACTIVITY 28.5 Dissecting a Kidney

1. Obtain a preserved kidney from the laboratory instructor. The kidneys may be double-injected specimens, meaning the blood vessels have been injected with colored latex for easier identification (red = arteries; blue = veins).
2. Before making any incisions, observe the external structures of the kidney. Similar to the human kidney, the sheep (or pig) kidney has a lateral convex surface and a medial concave surface. The hilus is located along the medial surface. Identify the ureter, renal artery, and renal vein at the hilus (Figure 28.10). From superior to inferior, these structures are positioned in the following order.
   - Renal artery
   - Renal vein
   - Ureter
3. Of the three connective tissue capsules that cover the kidney, only the innermost renal capsule is intact. It appears as a thin, shiny membrane that adheres tightly to the surface of the kidney. To observe this capsule more closely, lift a portion of the membrane from the surface with a forceps. The other two coverings, the renal fascia and adipose capsule, have been removed, but remnants may remain near the hilus.
4. With a scalpel, make an incision that will divide the kidney into anterior and posterior halves. Begin at the superior margin and cut along the lateral convex surface, toward the medial concave surface. Try to cut as close to the midline as possible. You can cut completely through

the organ or, if you prefer, keep the two halves connected at the hilus. The resulting incision will give you a coronal section of the kidney (Figure 28.10).

5. Identify the following internal structures (Figure 28.10).
   - The outer layer is the renal cortex, which has a granular texture and light color.
   - The middle layer is the renal medulla. Observe the alternating renal pyramids and renal columns. The pyramids have a darker color than the cortex and have a striped appearance due to the longitudinal arrangement of kidney tubules. The apex of each pyramid (renal papilla) projects inward and the base borders the cortex. The renal columns, located between the pyramids, are extensions of and have the same appearance as the cortex.
   - The inner region is the renal sinus, into which urine drains before flowing into the ureter. The sinus consists of minor calyces, major calyces, and the renal pelvis. Identify several minor calyces, located adjacent to the renal papillae. Locate the major calyces as they are formed by the convergence of several minor calyces. The renal sinus is formed by the merging of the major calyces. Locate this funnel-shaped structure as it joins the ureter.
6. Attempt to identify some of the blood vessels that serve the kidney (Figure 28.10). Carefully dissect within a renal column and locate an interlobar artery and vein. Along the corticomedullary border, identify examples of the arcuate blood vessels. Finally, try to identify the fragile interlobular arteries and veins as they travel through the cortex.

QUESTION TO CONSIDER Compare the anatomy of the human kidney with the sheep or pig kidney and identify structural similarities and differences.

Similarities:

____

____

Differences:

____

____

Exercise 28 Review Sheet
Anatomy of the Urinary System

Name ______________________

Lab Section ______________________

Date ______________________

Questions 1–9: Match the structure in column A with the appropriate description in column B.

A

1. Renal artery ______________
2. Bowman's capsule ______________
3. Renal pyramid ______________
4. Afferent arteriole ______________
5. Minor calyx ______________
6. Renal column ______________
7. Distal convoluted tubule ______________
8. Arcuate artery ______________
9. Renal cortex ______________

B

a. This structure is a branch of an interlobular artery and delivers blood to a glomerulus.

b. An interlobar artery travels in this region of the kidney.

c. This structure is the first portion of a nephron. It receives the blood filtrate.

d. This blood vessel travels along the border between the renal medulla and cortex.

e. This portion of the renal sinus receives urine directly from collecting ducts at a renal papilla.

f. This blood vessel is a direct branch of the aorta, and delivers blood to the kidney.

g. This cone-shaped structure, located in the renal medulla, contains loops of Henle and collecting ducts.

h. Renal corpuscles are found in this region of the kidney.

i. This portion of a nephron drains urine directly into a collecting duct.

10. Briefly describe the three functions that occur in the nephron during the process of urine production.

____

____

11. Compare the two types of nephrons that are found in the kidney: cortical nephrons and juxtamedullary nephrons.

____

____

12. Describe the pathway of blood through the kidney.

____

____

13. Why is transitional epithelium unique to other epithelial types? Why is it important that this type of epithelium be located in the wall of the urinary bladder?

____

____

14. Compare the anatomical relationships of the urinary bladder in the male and female.

____

____

15. Explain why females are more vulnerable to urinary tract infections than males.

____

____

Questions 16–23: Identify the labeled structures in the following diagram.

16. ______________________

17. ______________________

18. ______________________

19. ______________________

20. ______________________

21. ______________________

22. ______________________

23. ______________________

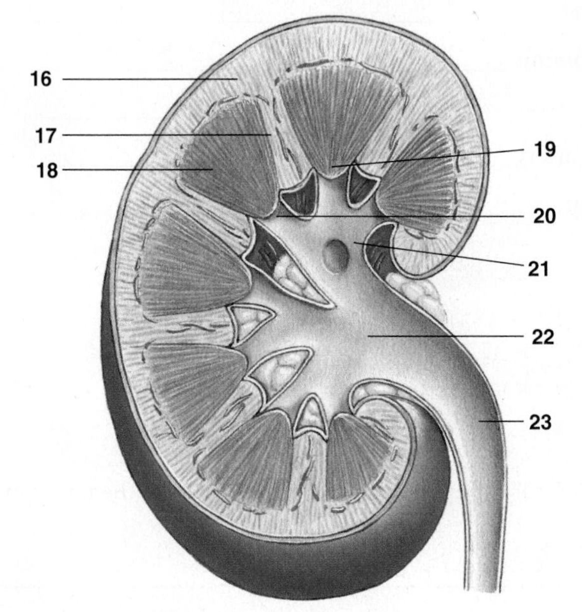

Questions 24–33: In the following diagram, identify the structures by labeling with the color that is indicated.

24. Descending limb of the loop of Henle = **green**

25. Distal convoluted tubule = **blue**

26. Glomerulus = **red**

27. Efferent arteriole = **orange**

28. Proximal convoluted tubule = **brown**

29. Ascending limb of the loop of Henle = **purple**

30. Afferent arteriole = **black**

31. Bowman's capsule = **pink**

32. Interlobular artery = **tan**

33. Collecting duct = **yellow**

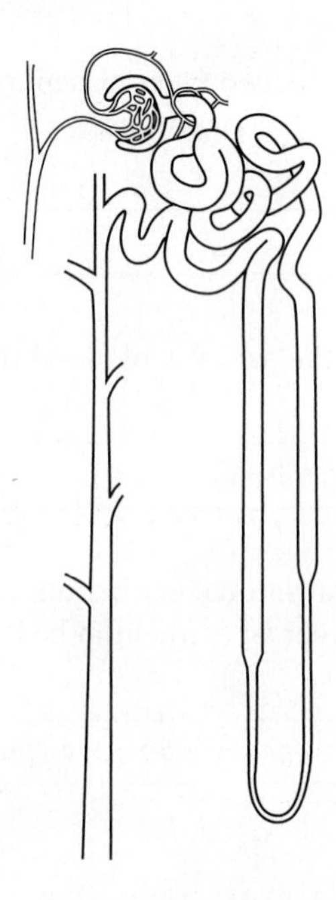

Exercise 29

Urinary Physiology

Laboratory Objectives

On completion of the activities in this exercise, you will be able to:

- Explain how the kidney produces urine.
- List the normal chemical components of urine.
- Collect a urine sample to perform a urinalysis test.
- Handle a urine sample by using safety procedures to prevent the spread of infectious microorganisms.
- Analyze and describe the physical characteristics of urine.
- Identify abnormal chemical characteristics of urine.
- Measure the specific gravity of urine.
- Identify microscopic sediments in urine.

Materials

- Disposable gloves
- Safety eyewear
- Face masks
- Autoclave bags
- Autoclave (if available)
- 70% alcohol solution
- 10% bleach solution
- Student urine samples, collected in 500 ml collection cups
- Individual or combination urinalysis dipsticks
- pH paper (if pH test is not included on the dipsticks)
- Urinometer
- Test tube rack
- Centrifuge and centrifuge tubes
- Pasteur pipettes
- Wax marking pencils
- Compound light microscope
- Microscope slides and coverslips
- 10% methylene blue solution

The production and elimination of urine is carried out by the kidneys and the accessory urinary structures (ureters, urinary bladder, and urethra). Each kidney contains over 1 million **nephrons** (Figure 29.1), the microscopic renal tubules in which urine is produced. The production of urine requires the following three processes.

- During **filtration**, which occurs in the **renal corpuscles**, blood pressure forces water and small dissolved molecules to move from the **glomerular capillaries** to the lumen of **Bowman's capsule (capsular space)**. The resulting fluid, called the **filtrate**, has the same concentration of small dissolved substances as blood, but normally lacks blood cells and large plasma proteins. It contains mostly water along with excess ions (mostly sodium and potassium), glucose, amino acids, and nitrogenous (nitrogen-containing) metabolic waste products.
- During **reabsorption**, almost all the water (99%) and other useful substances, such as glucose, amino acids, and various ions, return to the blood by passing from the renal tubules to the **peritubular capillaries** or **vasa recta**. Reabsorption occurs throughout the nephrons, but mostly occurs in the **proximal convoluted tubules** and **loops of Henle**.
- During **secretion**, unwanted substances such as metabolic wastes, drugs, and excess ions (hydrogen and potassium) are removed from the blood and enter the renal tubules. This process is just the opposite of reabsorption, and supplements glomerular filtration. Most secretory activity occurs in the **distal convoluted tubules** and **collecting ducts**.

Normally, urine contains 95% water and a wide variety of dissolved solutes including various ions (sodium, potassium, sulfate, calcium, magnesium, bicarbonate), small amounts of amino acids, lipids, carbohydrates, and nitrogenous metabolic wastes such as urea, uric acid, and creatinine.

CLINICAL CORRELATION

The two kidneys, combined, filter blood at an average rate of 125 ml/min or 180 l/day. This process provides the **glomerular filtration rate (GFR).** Fortunately, the kidneys do not excrete this much water (43 gallons) in the urine since 99% of the filtrate's fluid volume is reabsorbed into the blood. Typically, the kidneys produce only 0.5 to 2 l of urine each day.

The production of urine and its elimination is the basis for the following essential kidney functions.

- Maintaining normal blood pressure
- Keeping fluid concentrations of electrolytes such as sodium, potassium, and calcium within normal metabolic ranges
- Closely monitoring fluid osmolarity (number of dissolved ions and molecules per liter of solution) to protect cells from taking in or losing too much water
- Keeping the pH of body fluids within a narrow physiological range (7.35 to 7.45), which is optimal for the function of enzymes and other biological molecules
- Eliminating waste products from cellular metabolism as well as drugs and toxic substances

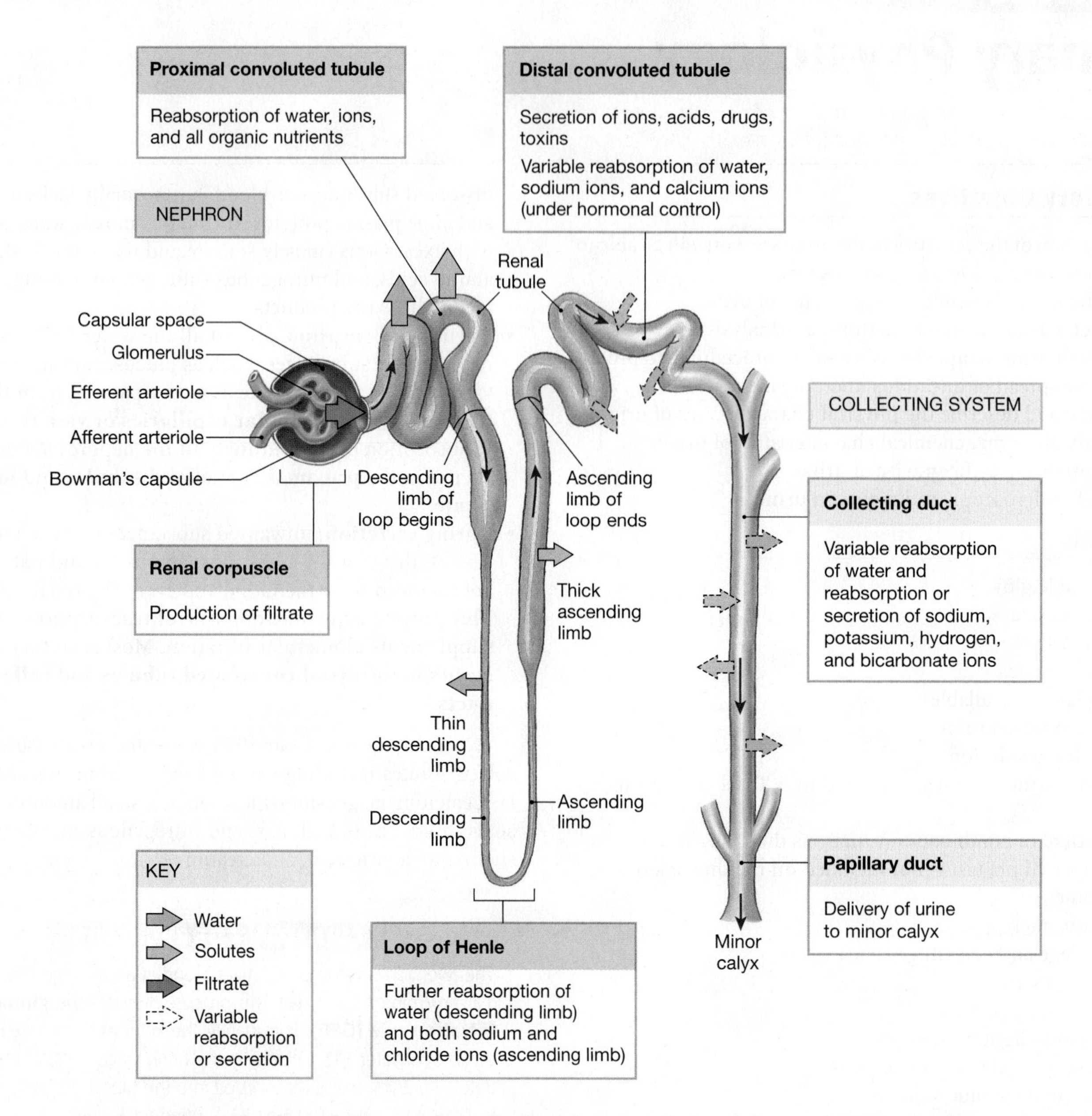

Figure 29.1 Urine formation in a nephron. The filtrate is produced in the renal corpuscle. Most reabsorption occurs in the proximal convoluted tubule and loop of Henle. Most secretion occurs in the distal convoluted tubule.

Universal Precautions for Handling Biospecimens

In the following activities, you will be examining the physical and chemical characteristics of your own urine sample (**urinalysis**). Since urine is a body fluid that may contain infectious organisms, the following safety procedures must be followed while handling your sample. These **universal precautions** should be used when handling or potentially being exposed to any human body fluids. [*Note: Although you cannot contract any illness from your own specimen, you should always assume there is a risk of coming in contact with someone else's specimen and use utmost caution by adhering to these rules.*]

- Always assume that the urine with which you are working is infected with a disease-causing organism. This attitude will put you in the right frame of mind to work with extreme caution.
- Work with your own urine *only*. Under no circumstances should you collect or conduct experiments with the urine of another individual.
- Always wear gloves, safety eyewear, and a mask when working with urine. Never allow urine to come in contact

with unprotected skin. Protective gear should be worn throughout the activities and during cleanup.

- If any part of your skin is accidentally contaminated with urine, disinfect the area immediately with a 70% alcohol solution for at least 30 seconds, followed by a 1-minute soap scrub, then rinsing.
- Any urine that spills onto your work area should immediately be disinfected with a 10% bleach solution or a commercially prepared disinfectant. The contaminated area should remain covered with the bleach/disinfectant for at least 30 minutes before wiping it off.
- All reusable glassware and other instruments should be disinfected in a 10% bleach solution for 30 minutes and then washed in hot, soapy water. If available, autoclaving this equipment before washing, rather than disinfecting with bleach, is recommended.
- Work areas should be cleaned with a 10% bleach solution or a commercially prepared disinfectant before and after any laboratory activity during which urine is used.
- Gloves, masks, and any toweling used for cleanup should be deposited in an appropriate biohazard container, and should never be placed into the regular garbage.
- As an alternative, simulated urine samples are commercially available. These samples will allow you to conduct urinalysis experiments without the risk of being exposed to human body fluids.

Table 29.1 Physical and Chemical Analysis of a Urine Sample

	Characteristic	Results: Normal	Results: Your sample
Physical analysis	Color	Colorless to pale yellow	
	Transparency/turbidity	Clear	
	Odor	Slightly aromatic	
	Specific gravity	1.003–1.030	
Chemical analysis	pH	4.5–8	
	Nitrites	Negative*	
	Bilirubin	Negative	
	Urobilinogen	Positive	
	Leukocytes	Negative	
	Erythrocytes/blood	Negative	
	Protein	Negative	
	Glucose	Negative	
	Ketone bodies	Negative	

*A negative result does not necessarily indicate a complete absence of the substance, but that its concentration is too low to be detected.

Physical Characteristics of Urine

The color of urine ranges from colorless to deep yellow. The yellow color is due to the presence of the pigment **urochrome**, which is produced when hemoglobin is broken down. The color deepens as the concentration of dissolved solutes increases. Thus, the color of urine can be used as a qualitative measure of solute concentration. Normal urine will be clear, but various types of suspended particles, such as microbes, dead cells, and crystals, can cause **turbidity** (cloudiness).

Freshly voided urine usually has an aromatic odor. If left to stand, chemical breakdown by bacteria can produce a strong odor of ammonia. Drugs, vitamins, and certain foods may also affect the odor.

ACTIVITY 29.1 Examining the Physical Characteristics of Urine

1. Obtain a sterile 500 ml collection cup for your urine sample.
2. Void the first 2 to 3 ml of urine before collecting your sample. This will flush the urethra of bacteria and other foreign materials and reduce the risk of contamination.
3. Observe the **color** of your urine sample. Atypical colors are possible. For example, a brown or red color indicates the presence of blood. Other colors, such as blue, orange, or green, may be caused by drugs, vitamins, or certain foods. Record your results, both the actual color and the lightness and darkness of your sample, in Table 29.1.
4. Observe the **transparency** of your sample. Bacteria, mucus, crystals of calcium salts and cholesterol, epithelial cells, and cell casts (hardened cell fragments) are sources of turbidity. Record your results in Table 29.1.
5. Hold your urine sample about 30 cm (12 in) from your nose. To identify an **odor** from your sample, wave your hand over the opening of the collection cup in the direction of your nose. If you cannot detect an odor, slowly move the cup closer to your nose, but not closer than 20 cm (8 in). Record your results in Table 29.1 by describing the odor (if any) that you can detect.

QUESTION TO CONSIDER Explain why cloudiness (turbidity) in a urine sample could indicate the presence of a urinary tract infection.

____

____

Specific Gravity of Urine

A reliable qualitative method to examine the solute concentration of urine is measuring its **specific gravity**, which is a weight comparison between a given volume of urine and an equal volume of distilled water. Distilled water has a specific gravity of 1.000. The normal range of the specific gravity for urine is 1.003 to 1.030, depending on the concentration of solutes: the more concentrated a urine sample, the greater its specific gravity.

ACTIVITY 29.2 Measuring the Specific Gravity of Urine

1. Obtain a **urinometer** to measure the specific gravity of your urine sample. A urinometer consists of two parts: a glass container to hold your urine sample and a flotation device called a **hydrometer**, which is used to measure specific gravity (Figure 29.2).
2. Before you begin, gently swirl your urine sample so that any substances that settled to the bottom are suspended in the fluid.
3. Add a portion of your urine sample to the glass container to at least two thirds full.
4. Gently place the hydrometer into the urine with the long stem directed upward (Figure 29.2). The hydrometer must not be resting on the bottom of the container. If it is not freely floating, add more urine to the glass container.
5. Determine the specific gravity of your sample by reading the position of the lower margin of the meniscus (curved line on the surface of a fluid) on the hydrometer stem's calibrated scale.
6. Record your results in Table 29.1.

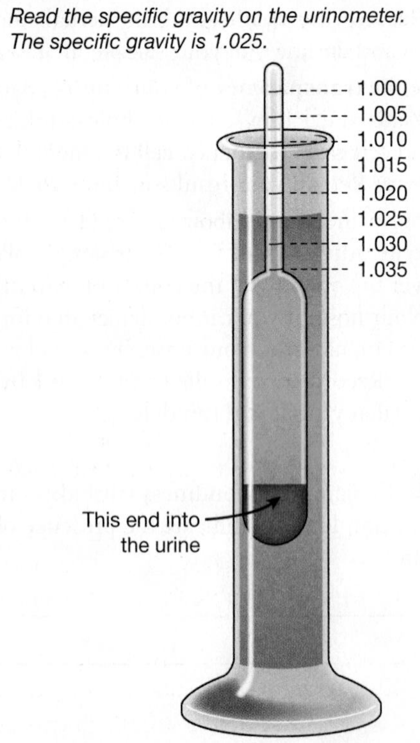

Figure 29.2 Measuring the specific gravity of urine. The hydrometer is placed into the urine sample with the long stem directed upward. The specific gravity is determined by reading the position of the meniscus on the hydrometer scale.

QUESTION TO CONSIDER When measuring the specific gravity of urine, or any other fluid, why is distilled water always used as a standard reference for comparison?

____

____

Chemical Analysis of Urine

As described earlier, urine normally contains a wide range of dissolved ions, small molecules, and nitrogenous wastes. However, a number of other chemical substances may be present abnormally, as a result of disease or infection. In the following activity, you will use individual or combination dipsticks to test for the presence of abnormal chemical constituents in your urine sample. The individual dipsticks have a single test pad to detect the presence of one specific chemical component; the combination dipsticks have several test pads to identify several components simultaneously. Depending on the type of dipsticks that are available in your laboratory, you will be able to test for the following chemical constituents.

- *Hydrogen ion concentration (pH).* Urine has a pH that ranges from 4.5 to 8.0 (average value of 6.0), but diet and certain diseases can alter this measurement. For example, a high-protein diet will tend to lower the pH of urine. On the other hand, vegetarian diets or urinary tract infections may cause the pH to increase.
- *Nitrites.* The presence of nitrites is a reliable indicator of a bacterial infection of the urinary tract, particularly the bladder. Urine is normally sterile. If bacteria are present, they convert nitrates, a normal constituent of urine, to nitrites.
- *Bilirubin.* This pigment, which is found in bile, is produced in the liver as a result of the breakdown and recycling of old red blood cells. A small amount of bilirubin in urine is normal. An excessive amount (**bilirubinuria**) could indicate a problem with normal liver function (blockage of a bile duct, hepatitis, cirrhosis).
- *Urobilinogen.* In the intestines, bilirubin is broken down to urobilinogen, which gives feces its brown color. Some urobilinogen is absorbed into the bloodstream, transported to the liver, and eventually, excreted in the urine. A small quantity of urobilinogen in urine is normal, but an excessive amount (**urobilinogenuria**) could indicate liver disease such as infectious hepatitis, cirrhosis, or jaundice.
- *Leukocytes.* The presence of leukocytes and pus (a fluid containing leukocytes and the debris of dead cells) in the urine is called **pyuria**. It is an indication of inflammation caused by a urinary tract infection.

- *Erythrocytes/blood.* The presence of red blood cells in the urine is called **hematuria.** It usually indicates inflammation or infection of the urinary tract which results in bleeding. If enough blood is present, the urine will have a reddish or brown color.
- *Protein (albumin).* Blood proteins are usually too large to be filtered and are rarely found in urine. Albumin is the most abundant blood protein. Its presence in urine (**albuminuria**) suggests that a problem with glomerular filtration exists. Albuminuria can be temporarily caused by physical exercise, high-protein meals, or pregnancy. A more permanent condition is brought about by high blood pressure, ingestion of poisons, or kidney infections.
- *Glucose.* Normally, glucose is absent from urine or is present in trace amounts. **Glucosuria**, the presence of glucose in the urine, is caused by a sharp rise in blood glucose levels (high blood sugar). When this occurs, the amount of glucose that is filtered across the glomerular membrane exceeds the amount that is reabsorbed back into the blood. The glucose that remains in the nephron tubules is secreted in the urine. Glucosuria can occur temporarily after eating a high-carbohydrate meal or during a period of stress. Chronic glucosuria is a common symptom of diabetes mellitus.
- *Ketone bodies.* These products of fat metabolism are normally present in very small amounts in urine. An abnormally high concentration of ketone bodies, known as **ketonuria**, will occur when the body is forced to rely on its fat reserves for energy due to starvation, eating low-carbohydrate foods, or diabetes mellitus.

WHAT'S IN A WORD You may have noticed that, in many cases, the name given to an abnormally high amount of a substance in urine ends with the suffix *-uria.* This term is a derivation of *ouron,* which is the Greek word for urine. Thus, *glucosuria* means "glucose in urine," and *ketonuria* means "ketones in urine."

ACTIVITY 29.3 Examining the Chemical Characteristics of Urine

1. Obtain individual or combination dipsticks to analyze the chemical components of your urine sample.
2. Before you begin your analysis, read the instructions carefully. In general, the test pads on the dipstick react with the urine, resulting in a color change. You will analyze the results by comparing the color of the test pads with a color chart. Depending on the test, results must be taken immediately or after a brief period of time, but never exceeding 2 minutes. Carefully review the times required for each test before proceeding.
3. Gently swirl your sample to suspend any sediment that settled to the bottom of the collection cup.
4. Immerse a test strip into the urine. If you are using a multiple test strip, be sure that all the test pads are completely submerged in the urine.
5. Remove the test strip and place it on a paper towel with the test pads facing up.
6. After the appropriate time has elapsed, read the results of each test by comparing the color on the test pad with the color chart that is available.
7. Record your results in Table 29.1.

QUESTION TO CONSIDER Diabetes mellitus is a disease in which the pancreatic islets do not produce adequate amounts of insulin, or insulin receptors on cell membranes are missing or not functioning properly. In either case, cells are unable to take up glucose from the blood. Explain why diabetes mellitus can cause glucosuria and ketonuria.

____

____

Urine Sediments

If a urine sample is centrifuged, the sediment can be viewed microscopically to examine the solid components that are present. Although some solids are normally found in urine, this procedure is a useful diagnostic tool for revealing abnormal sediments linked to various kidney diseases and infections.

In the following activity, you will view a microscope preparation of sediments from your urine sample. The sediments that you will observe can be organized into three major groups: **cells**, **casts**, and **crystals** (Figure 29.3).

Small numbers of epithelial cells that are shed from various regions of the urinary tract are normally found in urine sediments. These include transitional epithelium from the urinary bladder and ureters, and squamous epithelium from the urethra. Large numbers of white blood cells (pus) and any amount of red blood cells are abnormal and usually indicate some type of disease or infection.

Casts are hardened organic materials that usually form when cells clump together. They typically form when urine is acidic or contains unusually high levels of proteins or salts. Thus, their presence usually indicates an abnormal or pathological condition. Cellular casts can be formed from epithelial cells, white blood cells, and red blood cells. Granular and waxy casts represent various stages in the breakdown of cellular casts. Hyaline casts form when proteins secreted from nephron tubules clump together.

Crystals are small, geometric particles of calcium salts, uric acid, cholesterol, and phosphates. Small amounts of crystals are normal, but various urinary tract infections can increase their presence.

A small number of bacteria or other types of microorganisms (Figure 29.3) may also be present if your urine is contaminated as it passes along the urethra. A large presence of microbes may indicate an infection.

Figure 29.3 Urine sediments. The three main sediment types are cells, casts, and crystals.

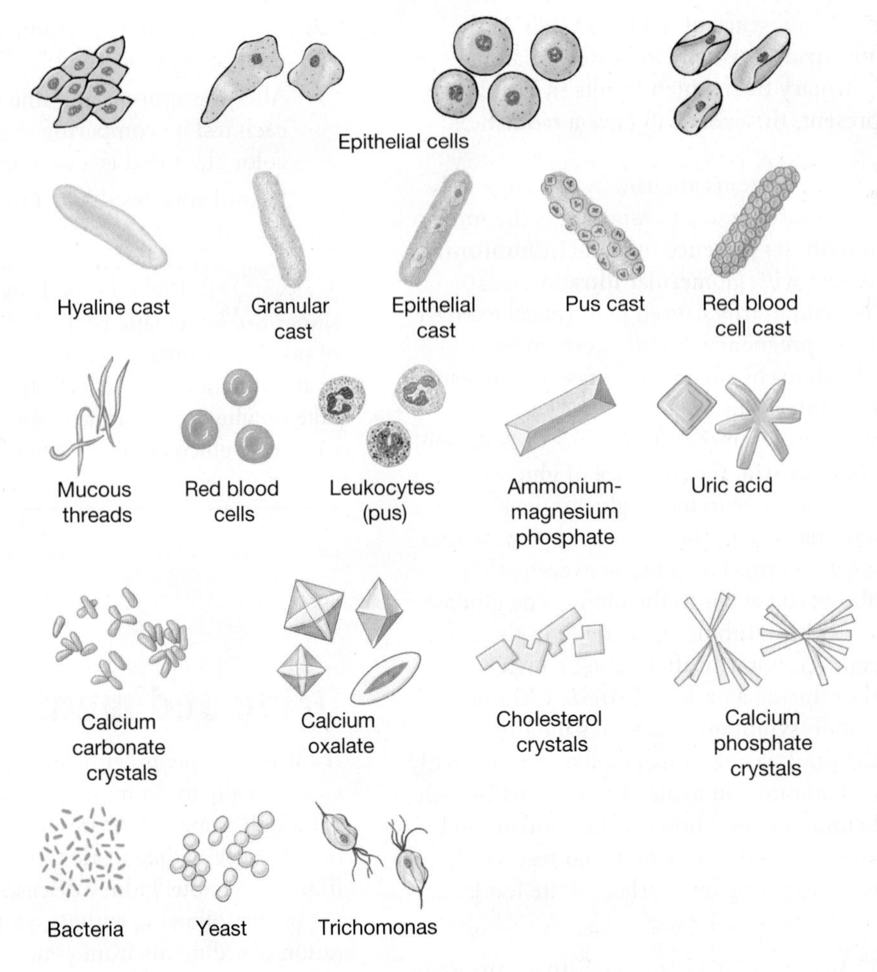

ACTIVITY 29.4 Examining a Urine Sample Microscopically

1. Gently swirl your urine sample to suspend any solids that settled.
2. Pour a small volume of urine into a centrifuge tube until it is about two thirds full.
3. Add an equal volume of distilled water to a second centrifuge tube.
4. Place both tubes in centrifuge slots that are opposite each other.
5. Centrifuge the urine sample for about 10 minutes.
6. When centrifugation is complete, carefully pour off the supernatant (liquid portion) in the urine sample and use a Pasteur pipette to obtain a small amount of the sediment.
7. Transfer the sediment to a glass microscope slide, add a drop of 10% methylene blue solution, and place a coverslip over the preparation.
8. View your preparation with the low-power objective lens. Reduce the amount of light by decreasing the aperture of the iris diaphragm. Attempt to identify cells, casts, and crystals (Figure 29.3). To observe crystals more clearly, switch to the high-power objective lens and increase the amount of light.
9. In the space provided, draw some of the sediments that you observe and make comparisons with the examples in Figure 29.3.

QUESTIONS TO CONSIDER **Kidney stones** are composed of calcium and magnesium salts, or uric acid crystals. If large enough, they can create a blockage of the urinary tract and cause severe pain during urination.

1. If a person has a kidney stone, you would expect to find an increased amount of what sediment type in the urine sample? Explain.

____

____

2. A urinary blockage in a ureter by a kidney stone can significantly decrease the glomerular filtration rate in the affected kidney. Explain how this could occur. ______________

____

____

Exercise 29 Review Sheet

Urinary Physiology

Name ______________________

Lab Section ______________________

Date ______________________

1. Summarize the important functions of the kidney.

2. Explain why urine contains substances that are filtered and secreted by the nephrons, but not substances that are reabsorbed.

3. Why is it important to follow strict safety procedures when handling a urine sample?

4. Define the following terms.

   a. Urinalysis

   b. Specific gravity

   c. Ketonuria

5. Describe the different types of sediments that can be found in a urine sample.

Exercise 30

The Male Reproductive System

Laboratory Objectives

On completion of the activities in this exercise, you will be able to:

- Describe general characteristics and functions common to both the male and female reproductive systems.
- Describe the overall anatomy and physiology of the male reproductive system.
- Explain the process of spermatogenesis.
- Trace the path of sperm cells from their formation through delivery into the female tract.
- Explain the descent of the testes and the clinical relevance of this process.
- Discuss the roles of the male accessory reproductive structures.
- Describe the composition of semen.
- Discuss the anatomy of the external male genitalia.
- Explain the events that occur during the male sexual response.

Materials

- Prepared microscope slides
  - Testis
  - Epididymis
  - Vas deferens
  - Prostate gland
  - Penis
  - Sperm smear
- Anatomical model of a sperm cell
- Midsagittal section model of the male reproductive system

The reproductive system is unique to all other organ systems in the body because it is not necessary for the survival of the individual, but its activities are absolutely required for sustaining the human species. Unlike other organs that are functional throughout life, the reproductive organs (also referred to as the **genitalia**) are inactive until **puberty**. At this time, which normally occurs between 11 and 15 years of age, the reproductive organs respond to increased levels of **sex hormones** (**androgens** in males, **estrogens** in females) by growing rapidly and becoming functionally mature structures.

Both the male and female reproductive systems develop from similar embryonic tissue. In fact, during the first few weeks of development, male embryos are indistinguishable from female embryos. As development proceeds, however, two distinct and special organ systems form. Despite the obvious structural differences between the sexes, the adult reproductive systems share some functional similarities.

- Both systems produce mature sex cells or **gametes**: the **spermatozoon** (plural = spermatozoa) or **sperm cell** in males and the **ovum** (plural = ova) or **egg cell** in females.
- Both systems can store, nourish, and transport gametes so that a sperm cell and an egg cell can fuse together to form a **zygote** through a process called **fertilization**. The formation of the zygote, also called **conception**, marks the beginning of a 9-month period of development that produces a new individual.
- Organs in both systems produce sex hormones and, thus, act as endocrine glands. These hormones are essential for the normal development and function of the reproductive organs.

Both male and female reproductive systems consist of **primary sex organs**, or **gonads**, and **accessory sex organs**. The gonads are the **testes** in the male and the **ovaries** in the female, which produce both the gametes and sex hormones. The accessory sex organs include various internal glands and ducts as well as the **external genitalia** that are found outside the pelvic cavity. These structures nourish, support, and transport the gametes. In addition, the female accessory organs support the development of the fetus during pregnancy and facilitate the birth process.

Gross Anatomy of the Male Reproductive System

The testes and accessory sex organs of the male reproductive system are illustrated in Figures 30.1 and 30.2. As a system, these organs function to produce and store sperm cells, to transport sperm cells along with supporting fluids to the female reproductive tract during sexual intercourse, and to produce male sex hormones.

The **testes** are ovoid structures, about 5 cm long and 3 cm wide, located within the **scrotal sac (scrotum)**. During fetal development, they develop from a retroperitoneal position along the posterior body wall, close to the kidneys. As the fetus grows, the testes slowly move to a more inferior position in the abdominal cavity. During the seventh month of development, they make their final descent into the scrotal sac by traveling through the **inguinal canals**, passageways that connect the abdominal cavity with the scrotum. This so-called **descent of the testes** is usually complete near the time of birth.

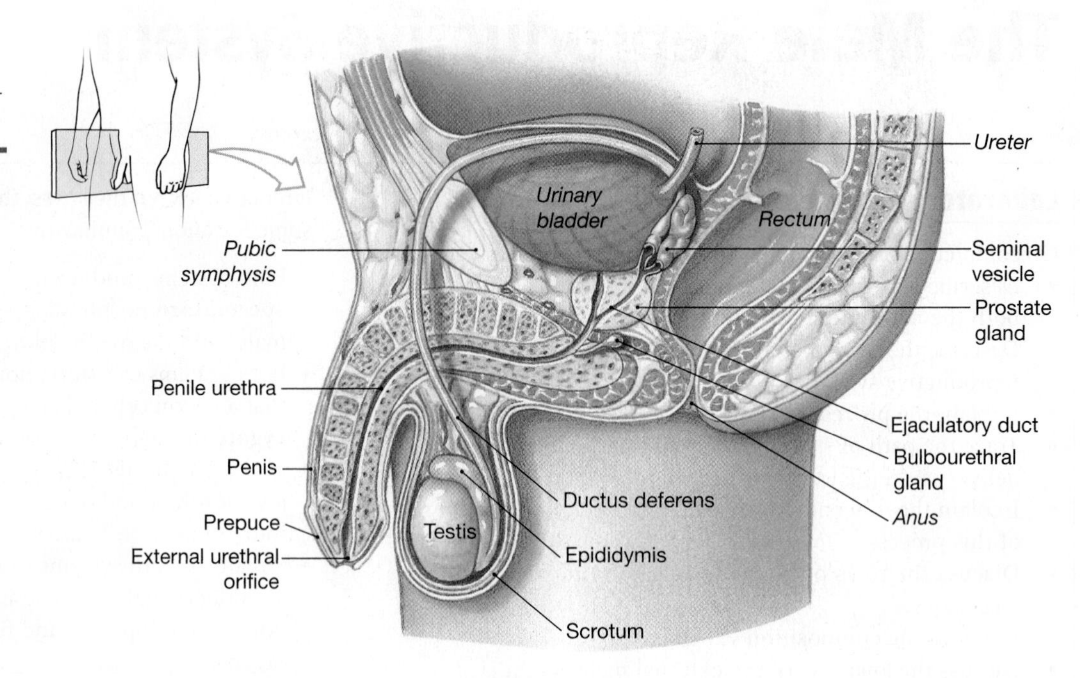

Figure 30.1 The male reproductive system, midsagittal view. Note the anatomical relations of reproductive structures with other pelvic organs.

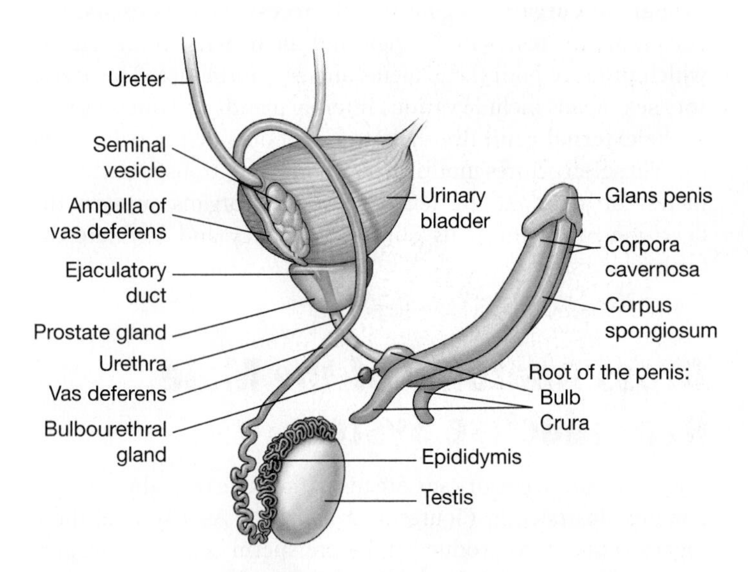

Figure 30.2 The male urogenital system, dissected view. Note the pathway of the accessory ducts and the components of the penis.

CLINICAL CORRELATION

In some cases, particularly in premature births, one or both testes do not complete the migration into the scrotum and remain in the abdominal cavity or inguinal canals. These so-called **cryptorchid testes** will usually descend into the testes within a few weeks after birth. If normal descent does not occur, the testes can be surgically moved into the scrotum. If the testes remain in the abdominal cavity, the individual will become sterile because the internal body temperature is 1° to 3° C too high for normal sperm development. In addition, males with cryptorchid testes have a higher risk of developing testicular cancer.

As the testes move into the scrotum, associated structures such as the vas deferens, blood vessels, nerves, and lymphatics travel with them. These other structures are enclosed by connective tissue and muscle to form the **spermatic cords**.

Inside the scrotum, each testis is enclosed by the **tunica vaginalis**, a continuation of the peritoneum that lines the abdominopelvic cavity. Deep to the tunica vaginalis, a fibrous capsule, the **tunica albuginea**, covers each testis. The tunica albuginea gives rise to connective tissue septa (partitions) that divide the testis into approximately 250 lobules. Each lobule contains three or four highly coiled **seminiferous tubules** (Figure 30.3). Along the posterior aspect of each testis, the seminiferous tubules converge to form a tubular network called the **rete testis**. The rete testis transports sperm to about 10 to 12 relatively straight **efferent ductules** that lead directly into the **epididymis** (Figure 30.4a).

The **accessory glands** include the **seminal vesicles**, **prostate gland**, and **bulbourethral (Cowper's) gland**. They produce substances that provide nourishment and support for the sperm, particularly while they are traveling through the female reproductive tract after sexual intercourse.

The **accessory ducts** include the **epididymides** (singular = epididymis), **vasa deferentia** (singular = **vas deferens**), and **seminal vesicles.** During or just prior to ejaculation, sperm cells and glandular secretions, which comprise the **semen**, are released into the accessory ducts by muscular contractions and transported to the outside.

The **external genitalia** include those structures that are found outside the body cavity. They include the **scrotal sac** and the **penis**. Although the testes, epididymides, and portions of the vasa deferentia are also external structures, they are not considered to be external genitalia because they are derived from embryonic tissue found inside the abdominal cavity.

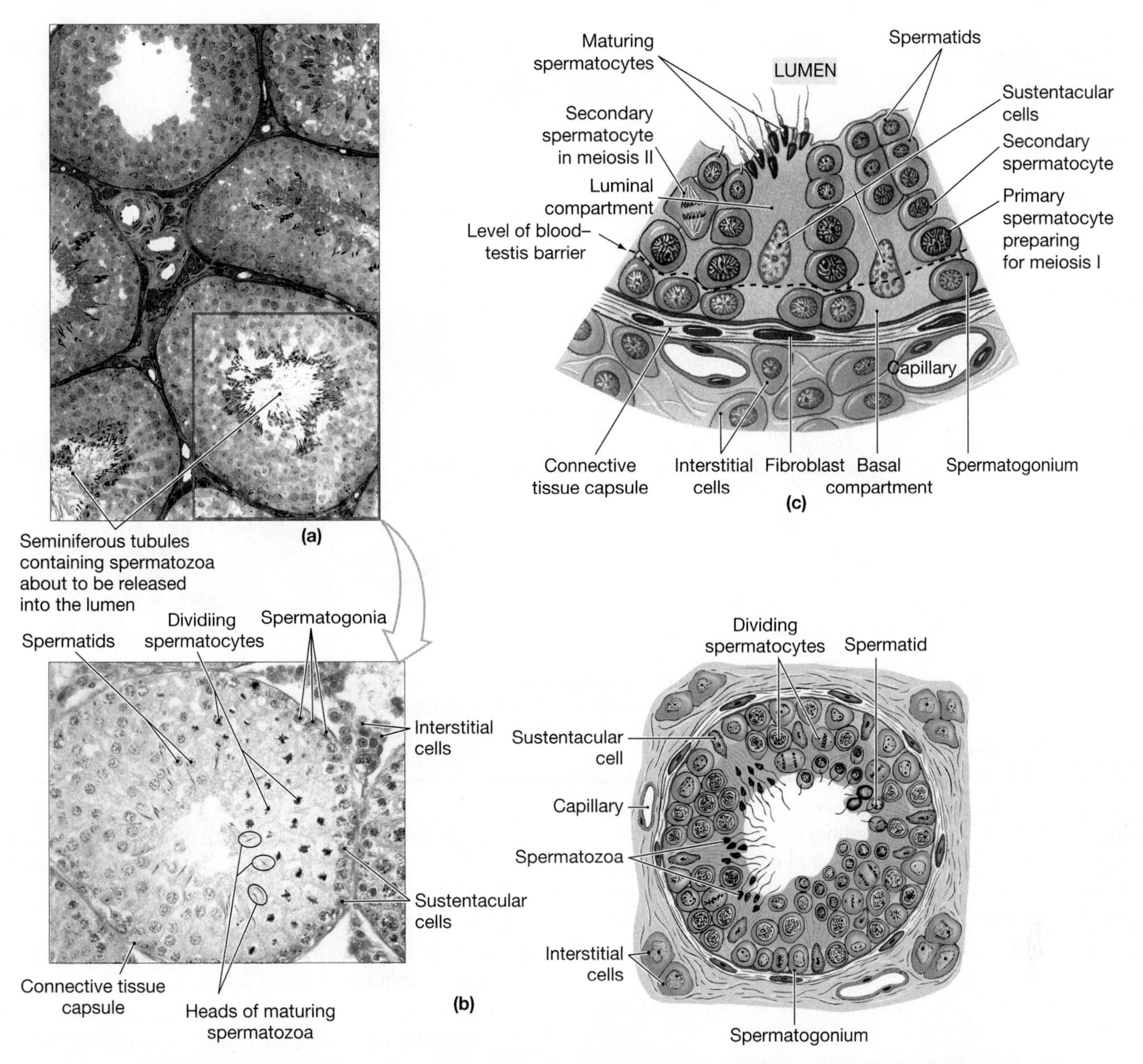

Figure 30.3 Microscopic structure of the testes. a) Low-power light micrograph illustrating seminiferous tubules (LM × 100); **b)** high-power light micrograph and corresponding illustration showing developing sperm cells and sustentacular cells in the wall of a seminiferous tubule, and interstitial cells outside the tubule (LM × 300); **c)** close-up view of the seminiferous tubule showing the relationship between developing sperm cells and sustentacular cells.

ACTIVITY 30.1 Examining the Gross Anatomy of the Male Reproductive System

1. Obtain a model of a midsagittal section of the male reproductive system (Figure 30.1).
2. Within the scotal sac, identify the ovoid testes. If possible on your model, observe a sectional view of the testes and identify the lobules that contain the seminiferous tubules, where **spermatogenesis** (sperm cell production) occurs (Figure 30.4a).
3. Identify the epididymis (Figure 30.1). This single, highly convoluted, tubular structure begins at the superior margin of the testes. Observe how the epididymis descends along the posterior surface of the testes (Figures 30.1 and 30.2). While in the epididymis, sperm cells mature and become motile. During ejaculation, sperm are propelled into the vas deferens by smooth muscular contractions along the epididymal wall. Sperm can be stored in the epididymis for several months, after which they are taken up and digested by the epithelial cells lining the lumen.
4. Locate the inferior margin of the testes. At this location, the epididymis gives rise to the vas deferens (ductus deferens). Realize that this long muscular tube transports sperm to the urethra, which begins within the prostate gland.
5. From its beginning, follow the path of the vas deferens to the urethra (Figures 30.1 and 30.2). As it passes superiorly into the abdominal cavity, the vas deferens travels with associated blood vessels and nerves within the **spermatic**

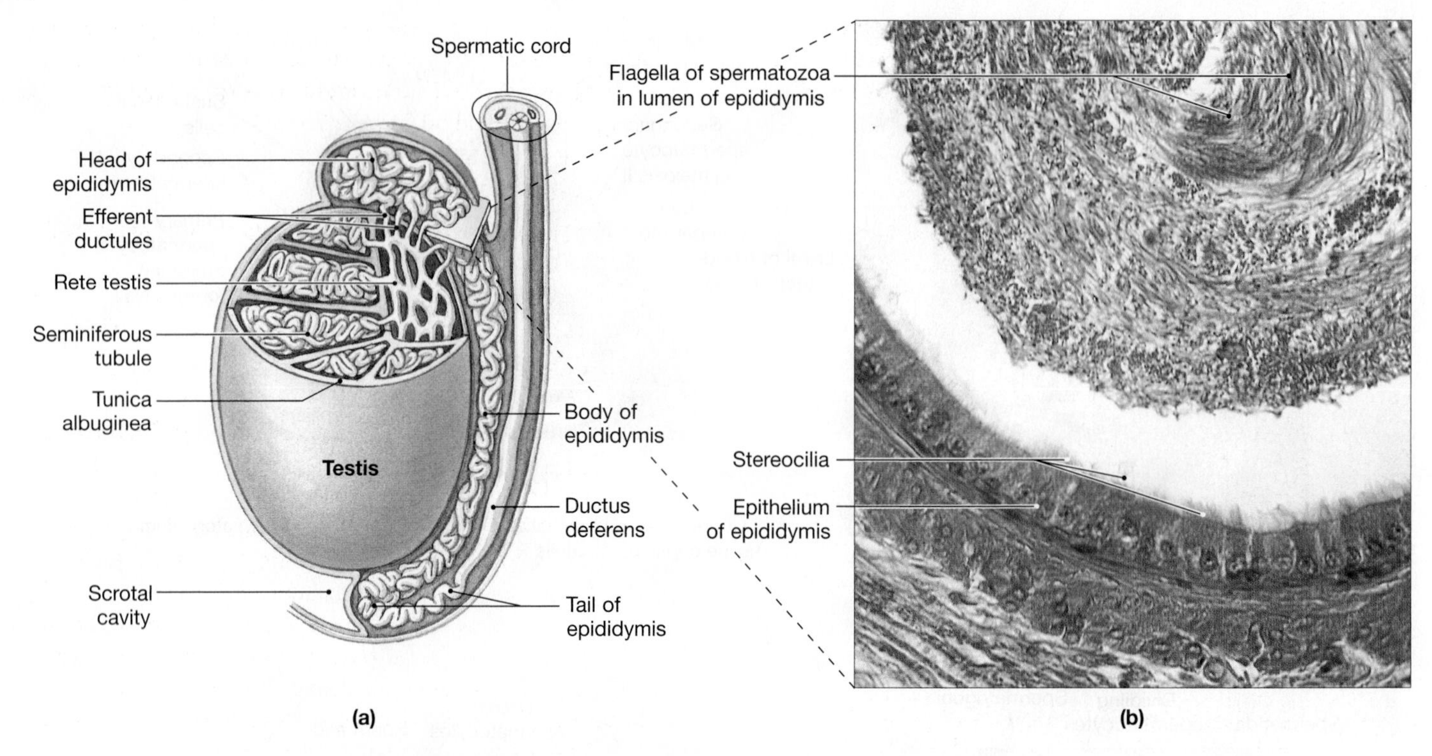

Figure 30.4 Microscopic structure of the epididymis. a) Illustration of the epididymis, testes, and the initial portion of the ductus deferens; **b)** light micrograph of the epididymis (LM × 400). The epididymis is a highly convoluted tubule lined by a pseudostratified epithelium with stereocilia projecting into the lumen. A large mass of maturing sperm cells fills the lumen.

cord. It travels through the inguinal canal to enter the abdomen. Next, it passes posteriorly and medially into the pelvic cavity and travels along the lateral aspect of the urinary bladder. After looping over the ureter, it turns inferomedially along the posterior surface of the bladder and unites with the duct of the seminal vesicle to form the ejaculatory duct.

CLINICAL CORRELATION

The presence of the spermatic cords creates areas of weakness along the abdominal wall where the inguinal canals are located. As a result, loops of the small intestine may protrude into the canal and scrotum, resulting in an **inguinal hernia.** Inguinal hernias are rare in females because spermatic cords are absent.

6. The **ejaculatory duct** travels into the prostate gland where it empties into the urethra (Figure 30.1).
7. The urethra is a relatively long tube that conveys both urine and sperm to the outside. Identify its three regions (Figure 30.1), as follows:
   - The **prostatic urethra** begins at the internal urethral orifice in the bladder wall and passes through the prostate gland.
   - The **membranous urethra** passes through the muscular urogenital diaphragm along the anterior floor of the pelvic cavity.
   - The **penile urethra (spongy urethra)** passes through the corpus spongiosum in the shaft of the penis. It ends at the external urethral orifice at the tip of the penis (Figure 30.1).
8. Identify the three major glands of the male reproductive system (Figures 30.1 and 30.2).
   - The paired seminal vesicles are convoluted saclike structures located along the posterior surface of the bladder. As described earlier, the ducts from these glands unite with the vasa deferentia to form the ejaculatory ducts. The secretions of the seminal vesicles include fructose as a source of energy for sperm, prostaglandins that promote muscular contractions of female accessory ducts and the uterus, and substances that promote sperm motility and protect sperm cells from the female's immune system.
   - The prostate gland is a chestnut-shaped structure located just inferior to the urinary bladder. The ejaculatory ducts and the first segment of the urethra (prostatic urethra) pass through the gland. The prostate produces a milky secretion with an alkaline (basic) pH, which protects sperm cells from their own acidic waste products and neutralizes the pH of vaginal secretions. Other substances produced by the prostate enhance sperm motility.
   - The bulbourethral (Cowper's) glands are pea-size structures located inferior to the prostate gland and within the urogenital diaphragm (Figure 30.1). Upon sexual

arousal, the glands' mucous secretions are released into the penile urethra. The mucus neutralizes any urine that remains in the urethra and lubricates the glans penis in preparation for sexual intercourse.

CLINICAL CORRELATION

Diseases of the prostate gland are common, particularly among men over age 50. **Benign prostatic hyperplasia (BPH)** is a noncancerous tumor that causes enlargement of the prostate and constriction of the urethra. Individuals with this condition can experience difficulty with micturition (urination), retention of urine in the bladder, urinary tract infections, and development of kidney stones. Since the prostate is located anterior to the rectum, a digital rectal exam can be done to determine if the prostate is enlarged. Treatment for BPH ranges from drug therapies that reduce the tumor to surgical removal of the prostate.

Prostate cancer, a far more serious and potentially deadly disease, also causes enlargement of the prostate and constriction of the urethra. This type of cancer is usually without symptoms during the early stages. If left undetected, secondary tumors (metastases) can form in nearby lymph nodes, pelvic bones, and lumbar vertebrae. Early detection is now possible by performing a blood test for elevated levels of a prostate enzyme known as prostate-specific antigen (PSA). Treatment for prostate cancer includes surgical removal of the gland or radiation therapy, in which small radioactive pellets are placed into the gland and specifically attack the cancerous tumor. The prognosis is very encouraging if treatment occurs before a secondary tumor forms. If a metastasis is detected, treatments are not as effective. However, prostate cancer grows so slowly that if it occurs after the age of 70, men usually die of natural causes or other diseases related to aging rather than the cancer itself. For this reason, sometimes the best treatment for prostate cancer may be no treatment at all.

WHAT'S IN A WORD Although their name is rather lengthy, the tiny *bulbourethral glands* are aptly named. The term describes their positions—they are located between the *bulb* of the penis and the *urethra.*

9. Identify the scrotal sac (scrotum), and notice that it is suspended inferiorly from the floor of the pelvis, posterior to the penis, and anterior to the anus. Internally, the scrotum is divided into two chambers by a connective tissue partition. Within each chamber are a testis, an epididymis, and the initial portion of a spermatic cord (Figure 30.1). The wall of the scrotum consists of skin and superficial fascia. In addition, two layers of muscle are present. In the dermis, there is a thin layer of smooth muscle known as the **dartos muscle.** Contractions of this muscle create wrinkling in the skin that covers the scrotum. Deep to the dermis is a thicker layer of skeletal muscle called the **cremaster muscle.** If air or body temperature increases, the cremaster muscle relaxes and the testes are drawn farther from the body's core. Conversely, a decline in temperature causes the muscle to contract and brings the testes closer to the body.

10. The penis conveys urine to the outside during micturition (urination) and sperm into the female reproductive tract during sexual intercourse. Identify its three regions (Figures 30.1 and 30.2), as follows:
    - The **root of the penis** is attached to the pelvic bone and urogenital diaphragm.
    - The **shaft (body of the penis)** is the elongated cylindrical portion.
    - The **glans penis** is the expanded tip of the shaft.
11. Identify the three columns of **erectile tissue** inside the penis. They include two dorsal columns known as the **corpora cavernosa**, and a single ventral column called the **corpus spongiosum**, which transmits the penile urethra (Figures 30.1 and 30.2). In the shaft, the corpora cavernosa form two cylindrical structures that travel parallel to each other. At the root of the penis, they diverge to form the **crura of the penis** (Figure 30.2) and attach to the pelvis. The corpus spongiosum also has a cylindrical shape as it passes through the shaft. At the root of the penis, it enlarges as the **bulb of the penis**, which attaches to the urogenital diaphragm. Distally, the corpus spongiosum expands to form the glans penis, which bears the **external urethral orifice** (Figures 30.1 and 30.2).
12. The skin covering the penis is relatively thin and moves freely over the surface. Identify the fold of skin called the **prepuce (foreskin)** that covers the glans penis (Figure 30.1). The inner surface of the prepuce contains glands that secrete a waxy material called **smegma.**

CLINICAL CORRELATION

Smegma, secreted from the prepuce, provides a favorable environment for bacteria. Consequently, bacterial infections are common in this area. To reduce the risk of infections, the prepuce can be surgically removed, a procedure known as **circumcision.** Removal of the prepuce is commonly performed in newborn babies in the United States. The procedure is controversial, however, because opponents believe that it is medically unnecessary and causes undue pain to the baby.

QUESTIONS TO CONSIDER 1. During a **vasectomy**, incisions are made on each side of the scrotum and the two vasa deferentia are cut and tied off within the spermatic cord. A man's testes will still function normally, but he will be infertile. Explain why. What other structures could potentially be damaged during a vasectomy? ______________________________

____

____

2. If the prostate gland is surgically removed, the functions of what other structures could be affected by the procedure?

____

____

Microscopic Anatomy of the Male Reproductive System

In the following activity, you will examine the microscopic structure of various organs in the male reproductive system. As you examine these structures, be aware that they possess unique anatomical features that reflect their characteristic functions.

ACTIVITY 30.2 Examining the Microscopic Anatomy of Male Reproductive Organs

The Testes

1. Obtain a slide of a human or mammalian testis.
2. Scan the slide under low magnification. You are viewing a cross section of the testis, the primary sex organ in the male (Figure 30.3).
3. Scan along the edge of the section and observe the fibrous connective tissue covering called the tunica albuginea. Connective tissue partitions derived from the tunica albuginea divide the testis into **lobules.**
4. Within each lobule of the testes are three or four seminiferous tubules. As you scan the slide under low power, the lobules can be observed throughout the field of view. Each tubule is surrounded by connective tissue and contains several layers of cells surrounding a central lumen. Because of the plane of section, the lumen may not be evident in some tubule profiles (Figure 30.3a).
5. Observe a seminiferous tubule under high power (Figures 30.3b and c). Most of the cells in the walls of the tubules are developing sperm cells, also known as **spermatogenic cells.** These cells are referred to by several names, according to their stage of development. At the base of the tubules are the **spermatogonia.** These undifferentiated cells are continuously dividing by **mitosis.** (Recall from Exercise 3 that cells formed by mitosis have the exact genetic composition of the original cell.) Some spermatogonia will differentiate to become **primary spermatocytes** and begin **meiosis,** a process of sexual cell division that forms **secondary spermatocytes, spermatids,** and finally, **spermatozoa (sperm cells).** Cells formed by meiosis have half the amount of DNA as the original cell; they are not exact genetic copies.
6. As the sperm cells form (spermatogenesis), they move from the base to the lumen of the seminiferous tubules. Observe these various cells on the slide and notice the change in their appearance as you move from the base of the tubule wall to the lumen. Realize that the spermatogonia are located along the base and the spermatozoa are in the lumen (Figures 30.3b and c).
7. In addition to the spermatogenic cells, **sustentacular (Sertoli) cells** are found in the seminiferous tubules (Figures 30.3b and c), extending from the base to the lumen of the tubules and completely surrounding the spermatogenic cells. They nourish the spermatogenic cells and regulate spermatogenesis. On the slides that you are observing, it is difficult to identify sustentacular cells, but you should realize that they are present.
8. Scan the slide under high power and observe areas of connective tissue between seminiferous tubules. These interstitial areas contain the **interstitial (Leydig) cells** (Figures 30.3b and c) that produce and secrete male sex hormones, or androgens. The primary hormone produced is **testosterone.**

WHAT'S IN A WORD The term *androgen* is derived from the Greek words *andros,* which means "male," and *genos,* which means "birth or give rise to." Androgens are hormones that are responsible for (give rise to) maleness. Both men and women make androgens, but these hormones can also be converted into the female hormones called estrogens.

The Epididymis

1. Obtain a slide of the epididymis and scan the slide with the low-power objective lens.
2. Observe the numerous sectional profiles of tubules in the epididymis (Figure 30.4b). You are actually viewing one continuous tubule that is highly convoluted.
3. Switch to the high-power objective lens. Notice that the epithelium lining the lumen is pseudostratified columnar. Identify the elongated microvilli, known as **stereocilia** (despite the name, these structures are not cilia) extending into the lumen from the surfaces of the epithelial cells (Figure 30.4b). The stereocilia absorb testicular fluid and provide nutrients to the sperm cells.
4. In the lumina of some tubule profiles, you will see masses of sperm cells. Observe the sperm masses under higher magnification (Figure 30.4). The sperm that enter the epididymis from the testes are nonmotile. While they move along the tubule of the epididymis, spermatozoa become mature and gain the ability to swim.

Vas Deferens

1. Obtain a slide of the vas deferens (ductus deferens) in cross section.
2. Scan the slide with the low-power objective lens and identify the following tissue layers (Figure 30.5).
   - The mucous membrane lines the lumen. It consists of the epithelium and underlying connective tissue, the lamina propria.
   - The relatively thick smooth muscle layer consists of three layers. The inner and outer layers are longitudinal and the middle layer is circular. The muscle layer contracts rhythmically to propel sperm forward during ejaculation.
   - The adventitia is a layer of fibrous and loose connective tissue that covers the smooth muscle layer. It blends in with the connective tissue of adjacent structures.

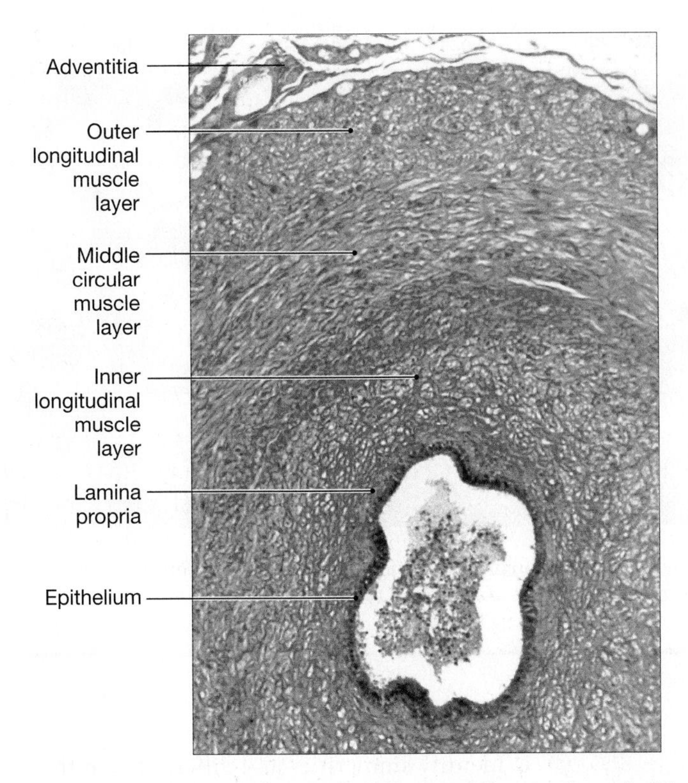

Figure 30.5 Microscopic structure of the vas deferens. The wall of the vas deferens contains three layers of smooth muscle (LM × 100).

3. Switch to the high-power objective lens and view the epithelium in the mucous membrane more closely. Similar to the epididymis, the vas deferens possesses a pseudostratified columnar epithelium with stereocilia projecting from the free surface of the cells.

Prostate Gland

1. Obtain a slide of the prostate gland.
2. Scan the slide with the low-power objective lens. Notice that the prostate contains numerous irregularly shaped glands, separated by areas of smooth muscle and connective tissue (Figure 30.6).
3. View a glandular region with the high-power objective lens. Notice that the epithelium in the glands is simple columnar or pseudostratified columnar.
4. Switch back to the low-power objective lens and try to locate the prostatic urethra (this structure may be absent on your slide). Notice that the epithelium lining the urethra is transitional.

   What other structures have a transitional epithelium?

   ____

   ____

Penis

1. Obtain a slide of the penis, cut in cross section.
2. Scan the slide with the low-power objective lens and observe the general organization of the penis (Figure 30.7).

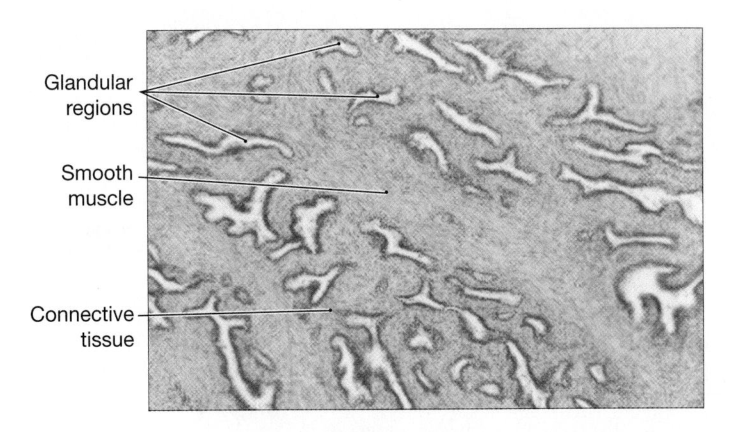

Figure 30.6 Microscopic structure of the prostate gland. The glands in the prostate are irregularly shaped and separated by regions that contain connective tissue and smooth muscle (LM × 100).

3. Identify the three columns of erectile tissue: the paired dorsal **corpora cavernosa** and the single ventral **corpus spongiosum** (Figure 30.7).
4. Identify the tunica albuginea, a tough fibrous connective sheath that surrounds the corpora cavernosa and the corpus spongiosum.
5. Notice that the penile urethra passes through the corpus spongiosum (Figure 30.7).
6. Switch to the high-power objective lens and examine the erectile tissue in a corpus cavernosum more closely. Erectile tissue contains a vast network of venous sinusoids and arteries within a meshwork of connective tissue and smooth muscle. Upon sexual arousal, parasympathetic nerves relax the smooth muscle surrounding the erectile tissue and the arteries supplying the penis. As the arteries dilate, blood flow to the penis increases, the venous sinusoids become engorged with blood, and the penis swells, elongates, and becomes erect. The erection is maintained because the swollen columns of erectile tissue compress the veins that would normally drain blood from the penis. At the height of sexual arousal **(orgasm)**, sympathetic nerves promote the expulsion of semen to the outside **(ejaculation).** Under sympathetic control, smooth muscle along the accessory ducts and glands contracts to force semen into the penile urethra. Then, skeletal muscles at the base of the penis contract to force semen through the penile urethra and to the outside. Shortly after ejaculation, sympathetic nerves cause constriction of the penile arteries and contraction of the smooth muscle surrounding the erectile tissue columns. As a result, blood flow to the penis is reduced, the erectile tissue is gradually drained, and the penis becomes flaccid.

Sperm

1. Obtain a slide of a human (or mammalian) sperm smear.
2. Carefully focus the slide under low magnification. When the spermatozoa (sperm cells) can be identified, switch to high magnification.

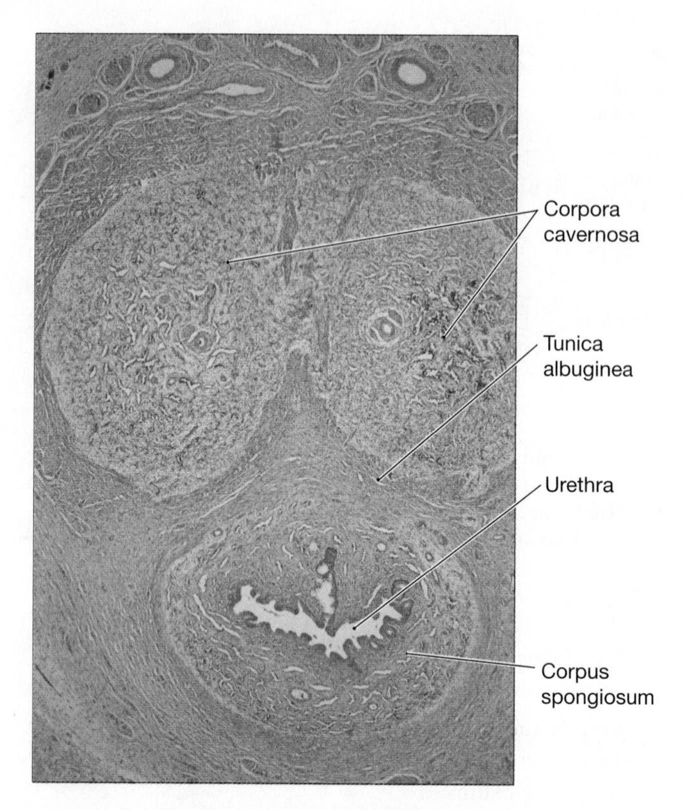

Figure 30.7 Microscopic structure of the penis. Erectile tissue in the penis is found in three columns of tissue: the paired dorsal corpora cavernosa and one ventral corpus spongiosum, through which the penile urethra passes (LM × 40).

3. Under high magnification, attempt to identify the three parts of a spermatozoon: the head, neck (body), and tail (Figure 30.8).
4. Obtain and observe a model that illustrates the structure of a mature spermatozoon. Identify the three parts of a sperm cell (Figure 30.8).
   - The head contains the nucleus with the male genetic material (DNA). The **acrosome** is located at the tip of the head. It contains enzymes that allow a sperm cell to penetrate the egg during fertilization. You may not be able to identify the acrosome on the microscope slides you are viewing.
   - The neck, or body, contains mitochondria that produce the ATP needed for sperm motility.
   - The tail is a flagellum that propels the sperm forward.

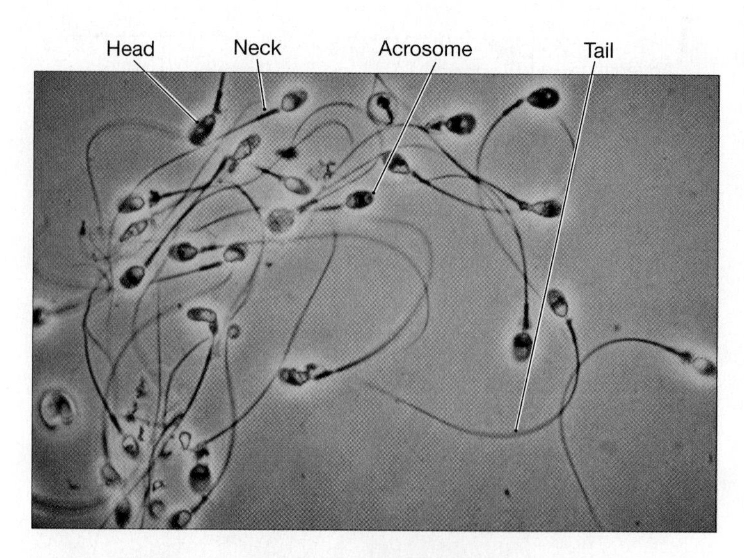

Figure 30.8 Structure of human sperm cells. A sperm cell is the only human cell to possess a flagellum (the tail portion). Notice the lighter staining acrosome covering the tip of the sperm head.

QUESTIONS TO CONSIDER

1. Identify similarities and differences in the microscopic structure of the epididymis and vas deferens.

Similarities:

____

____

Differences:

____

____

2. In the penis, the tunica albuginea surrounding the corpus spongiosum is less dense and more elastic compared to the same structure surrounding the corpora cavernosa. As a result, the corpus spongiosum is less turgid (swollen) upon erection than the corpora cavernosa. What is the functional significance of this difference? (Hint: Consider the position of the penile urethra.)

____

____

Exercise 30 Review Sheet

The Male Reproductive System

Name ______________________________

Lab Section ______________________________

Date ______________________________

1. Describe the general functions that can be attributed to the reproductive systems in both sexes.

____

____

2. From a functional perspective, why is it necessary for the testes to descend into the scrotum?

____

____

3. What is an inguinal hernia? Why are inguinal hernias more common in males than in females?

____

____

Questions 4–8: Match the structure in column A with the correct description in column B.

A	B
4. Testes ______________	a. The vas deferens and the duct from the seminal vesicle merge to form this structure.
5. Vas deferens ______________	b. A portion of the urethra passes through this structure.
6. Epididymis ______________	c. Spermatozoa become mature and motile while in this structure.
7. Prostate gland ______________	d. The corpus spongiosum is found in this structure.
8. Ejaculatory duct ______________	e. Interstitial (Leydig) cells, found in this structure, produce testosterone.
	f. A portion of this structure is located within the spermatic cord.

9. What is the clinical significance of prostate-specific antigen (PSA)?

____

____

10. The secretions from the male accessory glands have an alkaline (basic) pH. What would be the consequences if the pH of these fluids became more acidic?

____

____

11. What is the cremaster muscle? Describe how the actions of this muscle protect the testes.

____

____

12. Describe the arrangement of the erectile tissue in the penis. How does the penis become erect? If the blood flow along the penile arteries were reduced, how would this affect a man's ability to have an erection?

____

____

13. Review Figure 30.1 and explain why a digital rectal exam is a good procedure for diagnosing an enlarged prostate gland.

____

____

Questions 14–24: In the following diagram, identify the structures by labeling with the color that is indicated.

14. Prostate = **green**
15. Epididymis = **blue**
16. Corpus cavernosum = **red**
17. Ejaculatory duct = **brown**
18. Testes = **yellow**
19. Seminal vesicle = **orange**
20. Corpus spongiosum = **light blue**
21. Bulbourethral gland = **tan**
22. Vas deferens = **pink**
23. Scrotum = **black**
24. Urinary bladder = **purple**

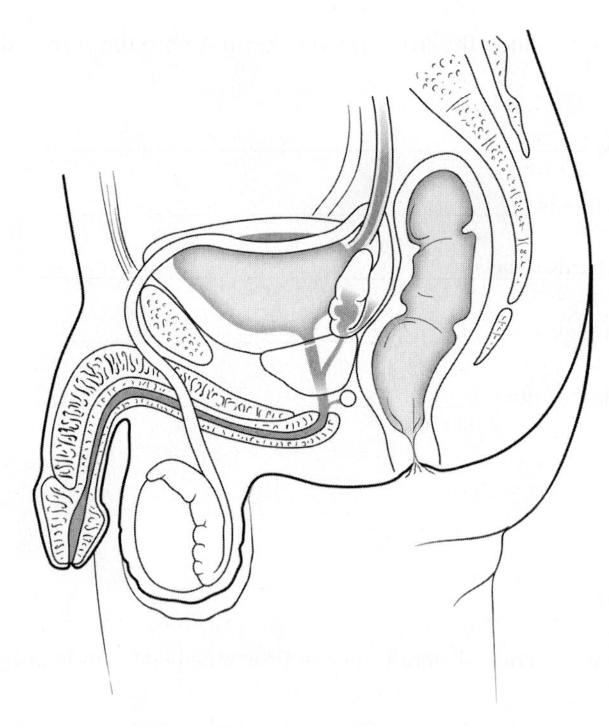

Exercise 31

The Female Reproductive System

Laboratory Objectives

On completion of the activities in this exercise, you will be able to:

- Describe the overall anatomy and physiology of the female reproductive system.
- Explain the processes of oogenesis and ovulation.
- Trace the path of egg cells from their formation through ovulation and fertilization.
- Discuss the structure and functions of the uterus.
- Correlate events in the ovarian and uterine cycles.
- Describe the functions of the internal female reproductive structures.
- Discuss the anatomy of the external female genitalia.
- Explain the events that occur during the female sexual response.
- Correlate male and female homologous reproductive structures.

Materials

- Prepared microscope slides
  - Ovary
  - Uterine tube, cross section
  - Uterus, proliferative
  - Uterus, secretory
  - Uterus, menstrual
  - Vagina, cross section
- Anatomical models
  - Ovary, showing follicular development
  - Female pelvis, midsagittal section
  - Female external genitalia
  - Female torso
- Whole skeleton or bone model of the bony pelvis
- Pieces of string, approximately 30 cm in length
- Masking or cellophane tape

Unlike males, who are able to produce sperm cells throughout their reproductive lives, females produce a finite number of egg cells. During early fetal development, germ cells (early embryonic cells) migrate into the ovaries and differentiate into cells called **oogonia**. The oogonia divide by mitosis for the next few months and some differentiate into **primary oocytes** (immature egg cells). By the fifth month of development, about 7 million primary oocytes have formed, but most of them degenerate over the next 2 months. Those that remain are surrounded by a single layer of squamous epithelial (follicular) cells. Each primary oocyte with its associated follicular cells is called a **primordial follicle**. Degeneration of primary oocytes continues for the remainder of fetal development and during childhood. At birth, about 1 million primordial follicles remain, and approximately 400,000 are left at puberty. During a woman's reproductive life, only 400 to 500 follicles reach maturity and release their eggs at ovulation.

The **ovarian cycle** refers to the cyclic monthly changes that occur in the ovary during a woman's reproductive life. Each month about 6 to 12 primordial follicles are induced by follicle-stimulating hormone (FSH) from the anterior pituitary to grow and mature. This is known as the **follicular phase** of the cycle. Of those follicles that begin maturation, usually only one reaches full maturity (a **tertiary** or **Graafian follicle**) and discharges its egg into the oviduct. The release of the egg from the ovary, known as **ovulation**, is influenced by luteinizing hormone (LH) from the anterior pituitary. It typically occurs at about the midpoint of the cycle, which is roughly 14 days. After ovulation, the remains of the follicle are transformed into a structure called the **corpus luteum** to begin the **luteal phase**. The corpus luteum produces hormones, mainly progesterone, that maintain the integrity of the uterine wall in preparation for implantation by an early embryo. If fertilization does not occur, the corpus luteum degenerates to scar tissue within 2 weeks. The various stages of follicular development during the ovarian cycle will be examined in more detail during Activity 31.3.

Gross Anatomy of the Female Reproductive System

The ovaries and the accessory sex organs of the female reproductive system are illustrated in Figures 31.1, 31.2, and 31.3a. As a system, these organs function to produce and maintain egg cells, transport egg cells to the site of fertilization, provide a favorable environment for the developing fetus, facilitate the birth process, and produce female sex hormones.

The **ovaries** are solid, ovoid structures, about 2 cm in length and 1 cm in width. Like the testes, they develop from embryonic tissue along the posterior abdominal wall, near the kidneys, and descend to their adult position along the lateral wall of the pelvis. Unlike the testes, the ovaries do not leave the pelvic cavity.

The accessory organs include the **uterine (fallopian) tubes** or **oviducts**, the **uterus**, and the **vagina** (Figures 31.1, 31.2, and 31.3a). The fallopian tubes are muscular tubes, approximately 13 cm in length. Their primary function is to convey the ovulated egg into the uterine cavity.

The uterus is located in the anterior portion of the pelvic cavity, superior to the vagina and bent anteriorly to lie directly behind the urinary bladder (Figure 31.1). In a nonpregnant adult female, the uterus is shaped like an inverted pear and is approximately 7 cm in length and 5 cm in width along its widest plane. Its primary functions are to receive the embryo that results from fertilization, nourish and support the embryo/fetus

Figure 31.1 The female reproductive system, midsagittal view. Note the anatomical relations of reproductive structures with other pelvic organs.

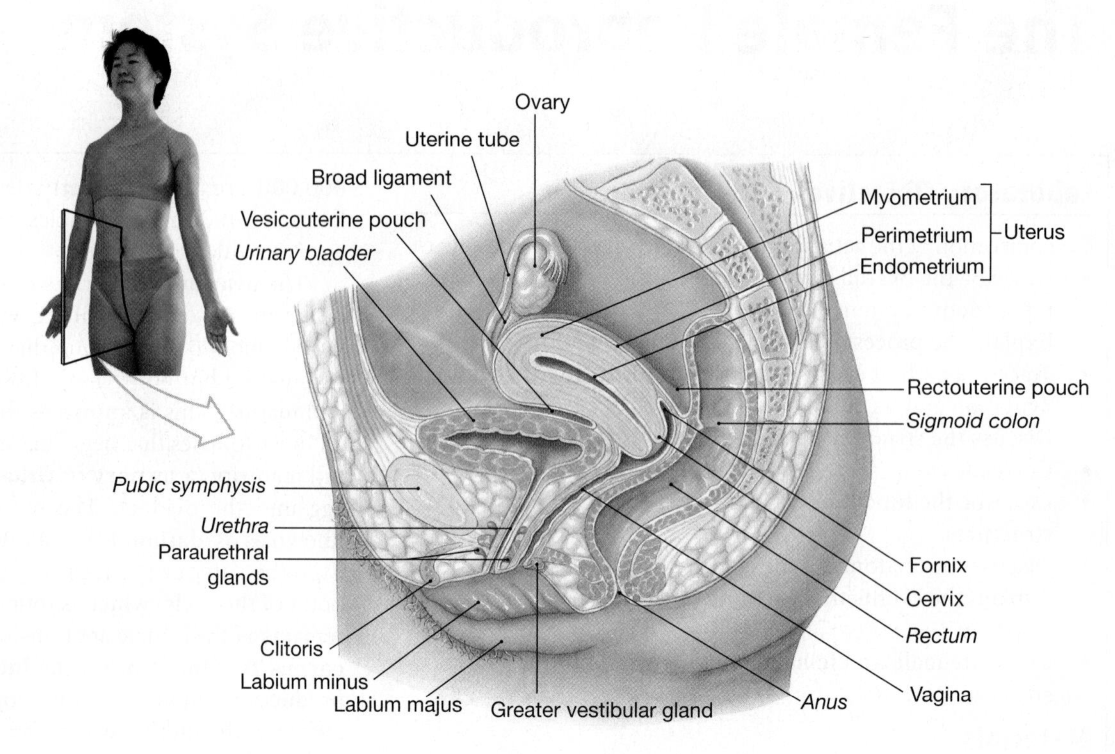

Figure 31.2 The ovaries, uterine (fallopian) tubes, and uterus. **a)** Posterior view with supporting ligaments; **b)** sectional view of an ovary and uterine tube, cut in the plane indicated in (a).

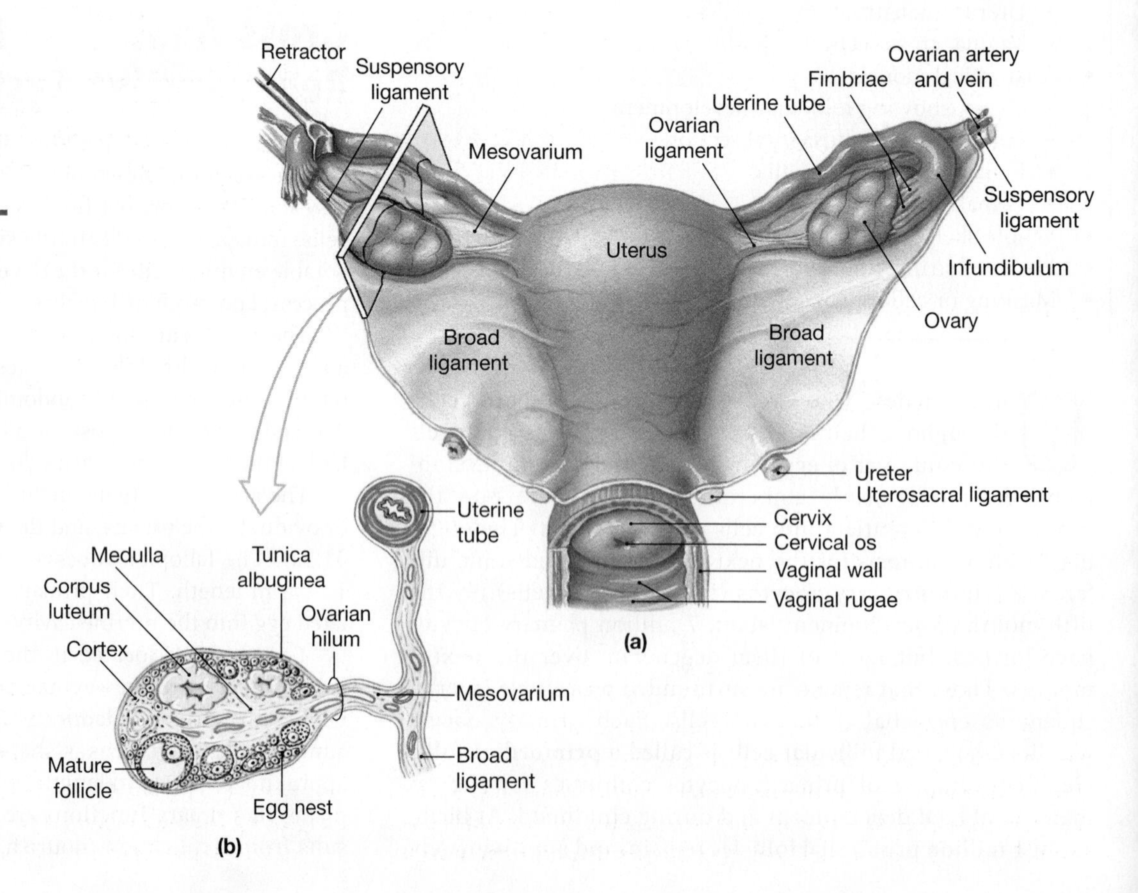

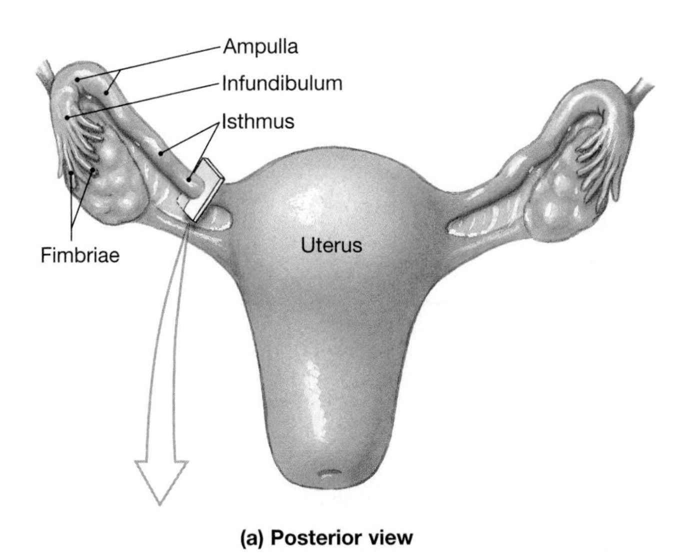

(a) Posterior view

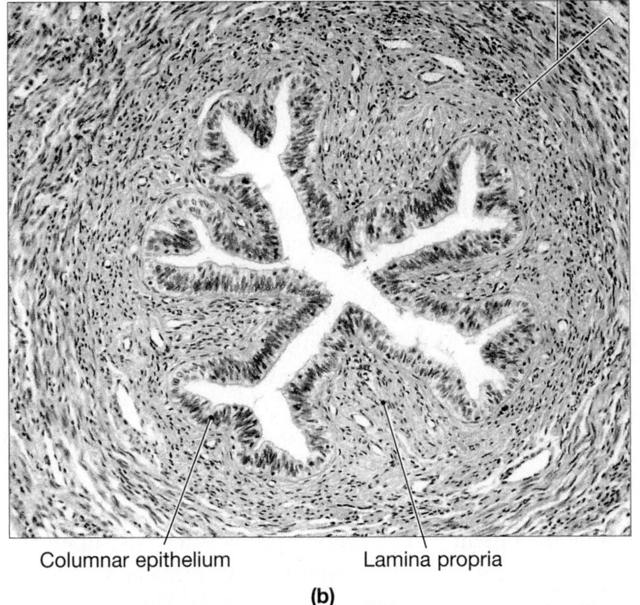

(b)

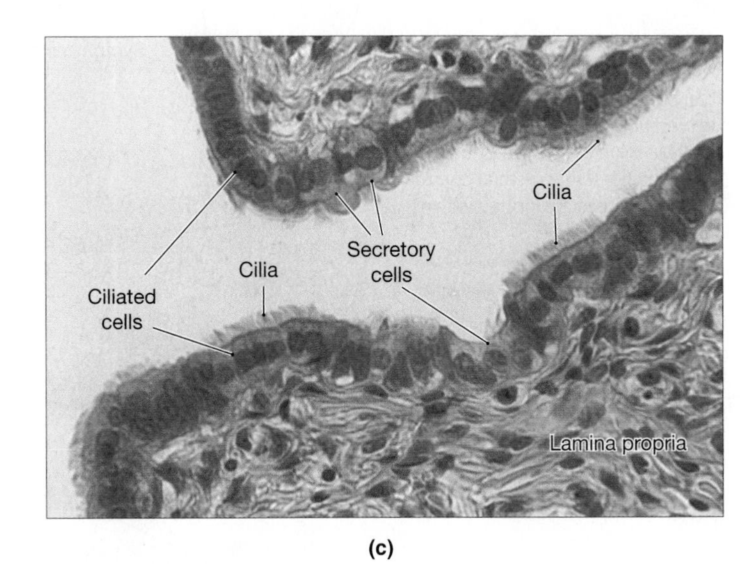

(c)

Figure 31.3 Structure of the uterine tube. a) Posterior view of the uterine tube, ovaries, and uterus. The three regions of the uterine tube are identified. **b)** Low-power light micrograph of the isthmus in cross section (LM × 100); **c)** high-power light micrograph of the mucous membrane (LM × 800).

during its development, and facilitate the birth, by muscular contractions, of the full-term fetus.

CLINICAL CORRELATION

The anterior bend, known as anteflexion, is the normal position of the uterus. In some women, however, the uterus bends posteriorly toward the sacrum. This is referred to as retroflexion of the uterus. A posterior bend of the uterus has no negative clinical consequences. In fact, a retroflexed uterus typically becomes anteflexed during pregnancy.

The vagina is a muscular tube between 7.5 cm and 9 cm in length, extending from the uterus to the vaginal orifice, which opens to the outside. It passes posterior to the urinary bladder and urethra and anterior to the rectum (Figures 31.1 and 31.2). The vagina serves as a passageway that allows the flow of menstrual fluids to the outside, receives the penis and temporarily holds sperm cells during vaginal intercourse, and serves as the birth canal for the fetus during birth.

The external genitalia (the **vulva**) include the **clitoris**, **vestibular glands** and **bulbs**, two pairs of skin folds (**labia majora** and **labia minora**), and the **vestibule**. These structures are located within a triangular region known as the **urogenital triangle** (Figure 31.4).

The **mammary glands** are modified sweat glands located in the subcutaneous tissue on the anterior thoracic wall. Embryologically, these glands can develop anywhere along the **milk** or **mammary line**, a thickened ridge of embryonic tissue that extends from the axilla to the groin. Usually, only a small segment of the milk line along the midthoracic region persists. It is from this tissue that the mammary glands form. They develop in both males and females, but are only functional in females during periods of breastfeeding. Occasionally, other portions of the milk line will remain active and give rise to accessory nipples (**polythelia**) or complete mammary glands (**polymastia**).

ACTIVITY 31.1 Examining the Gross Anatomy of the Female Reproductive System

1. Obtain a model that illustrates a midsagittal section of the female reproductive system (Figure 31.1).
2. Identify the uterus in the pelvic cavity and observe its relationship to the urinary bladder.
3. Identify the main portion of the uterus, called the **body** or **corpus.** Implantation of the early embryo typically occurs along the wall of this region. The **fundus** is the superior dome-shaped portion of the body. Inferiorly, observe how

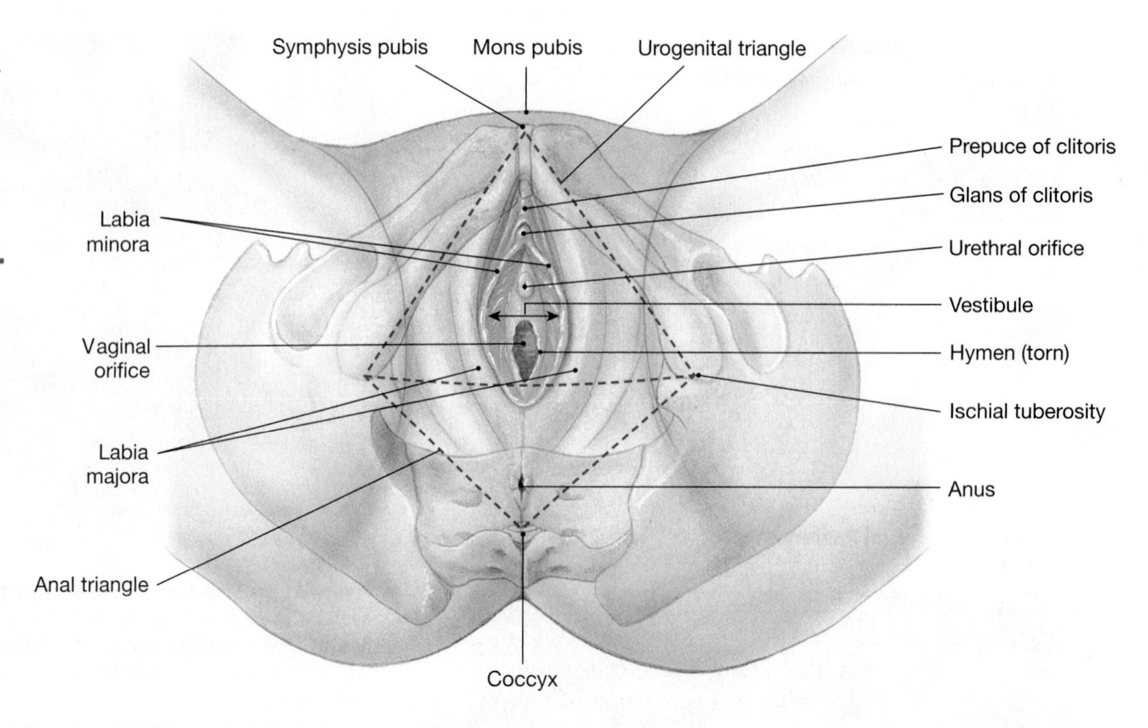

Figure 31.4 The female external genitalia. The vaginal and urethral openings are located within the vestibule, which is enclosed by the labia minora. The clitoris is located anterior to the urethral opening. All structures are located within the urogenital triangle.

the body tapers to form the **isthmus**. At the inferior end of the uterus, the isthmus gives rise to the **cervix**. Note that the cervix is adjacent to the posterior wall of the urinary bladder and projects into the superior end of the **vagina** (Figures 31.1 and 31.2).

WHAT'S IN A WORD In Latin, the word *cervix* means "neck," and in anatomy it refers to any necklike structure. In the uterus, the cervix is the inferior neck that leads to the vagina.

CLINICAL CORRELATION

Cervical cancer usually develops in the epithelium near the cervical orifice leading into the vagina. The majority of cases result from previous infection by the human papillomavirus (HPV), the virus that causes genital warts. Early detection is possible by doing a **cervical** or **Pap smear.** Cervical epithelial cells are scraped off the surface and a biopsy is performed to detect any cancerous cells. The prognosis is very good if the tumor is confined to the cervix, but worsens significantly if secondary tumors spread to the pelvic area. Consequently, it is highly recommended that women have an annual Pap test.

4. Identify the ovaries on either side of the uterus (Figures 31.1 and 31.2). Notice that each ovary is located along the lateral wall of the pelvis in the region between the internal and external iliac arteries.
5. Locate the uterine tubes (oviducts, fallopian tubes) where they connect to the uterine wall. Each uterine tube consists of the following regions (Figure 31.3a).
   - The **isthmus** is the narrow medial segment that attaches to the superolateral wall of the uterus.
   - The **ampulla**, the middle region, is wider than the isthmus. It is the longest segment of the fallopian tube, comprising at least half its length.
   - The **infundibulum** is the funnel-shaped lateral portion of the tube. At the free margin of the infundibulum, fingerlike projections called **fimbriae** hang freely over the ovary. During ovulation, the actions of the fimbriae draw the ovulated egg into the fallopian tube to begin its journey to the uterus.
6. Identify the vagina (vaginal canal). Once again, notice how the opening of the cervix projects into its superior end. Identify the recess (**fornix**) that forms between the vaginal and cervical walls (Figure 31.1). Verify that the vagina passes posterior to the urethra and anterior to the rectum (Figure 31.1). It opens to the outside at the **vaginal orifice**, which is partially closed by the **hymen**, a thin connective tissue membrane. The hymen usually ruptures and causes bleeding after the first sexual intercourse or other activity that exerts excess pressure on the membrane (e.g., inserting a tampon or physical exercise).

WHAT'S IN A WORD In Latin, the word *vagina* means "sheath." The vagina serves as a sheath or covering that surrounds the erect penis during sexual intercourse. It also forms a sheath around the full-term fetus as it passes from the uterus to the outside during childbirth.

7. Obtain a model of the female external genitalia (the vulva) and identify the following structures (Figure 31.4).
   - The **mons pubis** is a mound of fat and skin that covers the symphysis pubis. After puberty, the skin is covered with pubic hair.

- The **labia majora** are two fatty skin folds that are covered with pubic hair. They project posteriorly from the mons pubis.
- The **labia minora** are two smaller skin folds that are located within the borders of the labia majora. These structures lack hair and have a pinkish color due to their rich vascular supply. Posteriorly, the two labia minora merge to form the **fourchette**.

WHAT'S IN A WORD The word *labium* (plural-*labia*) means "lip" in Latin. Anatomists use the word to describe any lip-shaped structure. The two pairs of skin folds that are parts of the female external genitalia are good examples. Thus, the larger pair is called the *labia majora* and the smaller pair the *labia minora*.

- The vestibule is the space enclosed by the labia minora. Along the floor of the vestibule locate the anterior **urethral orifice** and the posterior vaginal orifice.
- The **greater vestibular glands** (Figure 31.1) are located deep to the posterior portion of the labia minora. The secretions from these glands moisten the area around the vaginal orifice to facilitate the insertion of the penis during sexual intercourse.
- The vestibular bulbs are elongated masses of erectile tissue located on either side of the vestibule and extend from the vaginal opening to the clitoris. The vestibular bulbs contain erectile tissue that is structurally equivalent to the corpus spongiosum in the penis. During sexual arousal, the vestibular bulbs become engorged with blood.

  As the corpus spongiosum develops, it begins as a single column of erectile tissue in both sexes. In males, it remains as a single column and encloses the penile urethra. In females, it splits along the midline as the vestibular bulbs and vaginal canal form. As a result, the urethra is not surrounded by erectile tissue.
- The clitoris is located anterior to the vestibule, adjacent to the area where the labia minora meet. Similar to the penis, it is a cylindrical structure composed mostly of erectile tissue. The body of the clitoris contains two cylindrical masses of erectile tissue called the corpora cavernosa which diverge at the base to form the **crura**. The **glans clitoris**, located at the distal end of the body, is formed by the fusion of erectile tissue from the two vestibular bulbs. Similar to the glans penis, the glans clitoris is covered by a fold of skin, the **prepuce**.

  During sexual arousal in the female, the erectile tissues in the corpus cavernosa (crura and clitoris) and in the corpus spongiosum (vestibular bulbs) become engorged with blood. This results from parasympathetic stimulation, which dilates the arteriole walls, allowing more blood to enter these tissues. The increased blood flow and tissue engorgement shut the venous pathways, trap blood in the tissues, and cause the clitoris to become erect. At the conclusion of orgasm, sympathetic nerves constrict the arteriole walls, blood flow is reduced, and veins drain blood from the erectile tissues.

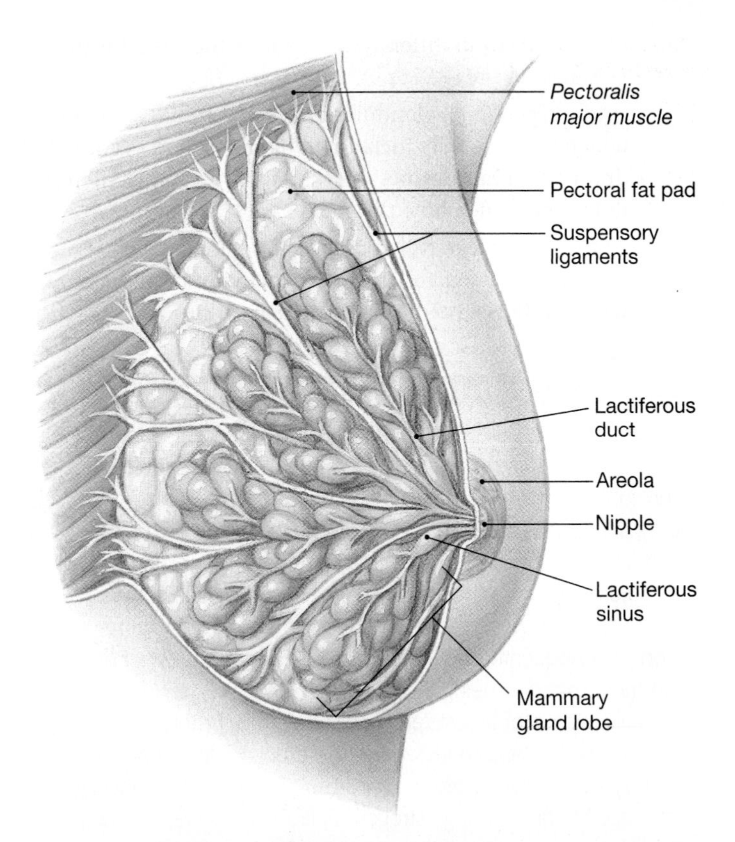

Figure 31.5 Structure of the mammary gland. The gland is divided into several lobes, each containing lobules of secretory cells that empty into a lactiferous duct. The lactiferous ducts form sinuses that converge at the nipple.

8. The external genitalia are located in the urogenital triangle. On a skeleton or bone model of the bony pelvis, connect the two ischial tuberosities with a piece of string. With two additional pieces of string, connect each ischial tuberosity to the pubic symphysis. The area enclosed by the string is the urogenital triangle (Figure 31.4).
9. On a torso model, identify the mammary glands (Figure 31.5). Notice that each gland overlies the pectoralis major muscle. The circular base of the gland extends from the second to sixth ribs and from the sternum to the axilla. It is separated from the pectoralis major by deep fascia and a more superficial layer of loose connective tissue. Each gland is supported by **suspensory ligaments** that run from the skin to the deep fascia. Identify the following structures of a mammary gland (Figure 31.5).
   - A circular area of pigmented skin called the **areola** is located at the apex of each breast. The **nipple**, projecting from the center of the areola, is the structure from which the baby sucks milk.
   - Internally, the mammary gland has several (15 to 20) lobes of glandular tissue divided by connective tissue and fat. The amount of glandular material is fairly constant among nonpregnant females. However, the amount

of fat deposition differs greatly and is the basis for the variation in breast size.

- Within each lobe, glandular cells secrete milk into ducts, which progressively increase in diameter, and eventually drain into a **lactiferous duct**. Identify several lactiferous ducts and notice how they converge at the nipple.
- At the nipple, carefully observe how each lactiferous duct forms an expanded **lactiferous sinus** (Figure 31.5), where milk is stored prior to breastfeeding.

CLINICAL CORRELATION

Breast cancer is a common female cancer, affecting about one in eight women in the United States. The most common type occurs in the epithelium that lines the ducts in the lobules. Potentially, it can spread quickly if cancerous cells dislodge from the primary tumor and enter lymph vessels to the axillary and *parasternal* (next to the sternum) lymph nodes. Free communication exists between these lymph nodes and other body regions. Consequently, metastases can spread to a variety of areas, the most common being the vertebrae, brain, liver, and lungs.

Risk factors for breast cancer include family history and various factors that tend to increase one's exposure to estrogens such as early puberty, late menopause, having a first pregnancy after age 35, or receiving estrogen replacement therapy. Recent evidence has indicated that exposure to environmental toxins that mimic estrogen may also increase one's risk.

The most common treatment for early-stage breast cancer is a lumpectomy, in which the cancerous tumor and a small amount of adjacent tissue are removed. This surgery is followed by radiation therapy and chemotherapy.

For more advanced tumors, various types of surgery in which the breast is removed (**mastectomy**) are performed. In a **simple mastectomy,** only the breast is removed. A **modified radical mastectomy** involves removing the breast along with all or some axillary lymph nodes, to see if the cancer cells have spread. The most drastic procedure is a **radical mastectomy** in which the breast, underlying deep fascia, pectoral muscles, and all the axillary lymph nodes are taken out. It is rarely performed today.

Early detection is the primary weapon against breast cancer because the survival rate declines dramatically if the cancer has spread to the axillary lymph nodes. Therefore, women over the age of 40 are advised to have an annual mammogram (breast X-ray), and to conduct self-examinations on a monthly basis.

10. Various supporting mesenteries and ligaments connect the female reproductive organs to the body wall (Figure 31.2). On a model of the female reproductive system, locate the **broad ligament**, which is a fold of peritoneum that supports the ovaries, fallopian tubes, and uterus. It contains several subdivisions.
    - The **mesometrium**, the largest portion of the broad ligament, attaches the uterus to the pelvic wall.
    - The **mesovarium** attaches each ovary to the mesometrium.
    - The **mesosalpinx** attaches each fallopian tube to the mesometrium.

11. Identify three other supporting ligaments (Figure 31.2).
    - The **suspensory ligament of the ovary** is a lateral continuation of the broad ligament. It attaches the superior aspect of each ovary to the pelvic wall.
    - The **ovarian ligament** is a cordlike thickening within the broad ligament that attaches the ovary to the uterus.
    - The **round ligament** is a thickened band of tissue within the broad ligament. It extends from the superolateral wall of the uterus to the anterior pelvic wall, then passes through the inguinal canal to anchor in the tissue of the mons pubis.
    - The **uterosacral ligament** is a band of tissue within the broad ligament that attaches the cervix to the sacrum.

CLINICAL CORRELATION

A major cause of sterility in women is **pelvic inflammatory disease (PID),** an infection that typically first develops in the vagina, but usually spreads to the uterus, fallopian tubes, and ovaries. In severe cases, the infection can spread to other organs in the abdominopelvic cavity. The bacteria responsible for the sexually transmitted diseases gonorrhea and chlamydia often cause the infection. Symptoms include fever, lower abdominal pain, elevated white blood cell counts, and abnormal vaginal discharge. Treatment includes the administration of antibiotics, but abdominal pain may persist for a prolonged period. Infertility may result if scarring of the fallopian tubes blocks the passageway leading to the uterus.

QUESTIONS TO CONSIDER

1. Examine Figures 31.1 and 31.2 and explain how a bacterial infection that develops in the vagina can spread to the abdominopelvic cavity. How can such an infection (e.g., pelvic inflammatory disease) cause infertility?

___

___

2. In the previous activity, you identified female reproductive structures in the urogenital triangle. What male reproductive structures would you find in this region?

___

___

Homologous Structures in the Reproductive Systems

Sex hormones (androgens or estrogens) that are present in the fetal circulation influence the development and differentiation of the reproductive systems. The genital duct systems in each sex develop from different embryonic tissues. However, the primary sex organs and the external genitalia have similar origins and differentiate into male or female structures later in development. Structures that are derived from the same embryonic tissue but are not necessarily similar in function are called **homologous structures**.

Table 31.1 Homologous Structures in the Male and Female Reproductive Systems

Male structure	Function of male structure	Homologous female structure	Function of female structure
Testes		Ovaries	
Scrotum		Labia majora	
Penile urethra		Labia minora and vestibule	
Corpora cavernosa		Body of clitoris	
Corpus spongiosum		Vestibular bulbs; glans clitoris	
Bulbourethral glands		Greater vestibular glands	

ACTIVITY 31.2 Identifying Homologous Reproductive Structures

1. Obtain models of the male and female reproductive systems. Place them side by side so that the anatomy of each system can be easily compared.
2. Table 31.1 lists the homologous structures in the male and female reproductive systems. Identify the pairs of homologous structures listed in the table.
3. Record the function of each structure in the space provided in Table 31.1.

QUESTION TO CONSIDER Explain why knowledge of embryological origins will give you a better understanding of similarities and differences in the structure and function of male and female reproductive organs.

____

____

Microscopic Anatomy of the Female Reproductive System

In the following activity, you will examine the microscopic structure of various organs in the female reproductive system. As you examine these structures, be aware that the histology of female reproductive structures is generally more complex than the male, presumably because they perform more intricate functions.

ACTIVITY 31.3 Examining the Microscopic Structure of Female Reproductive Organs

The Ovaries

1. Obtain a slide of a mammalian ovary.
2. Scan the slide under low magnification. You are examining a sectional view of the ovary, the primary sex organ in the female (Figure 31.6).
3. Identify the two regions of the ovary. The outer **ovarian cortex** contains the developing egg cells (oocytes) that are enclosed by multicellular capsules called follicles. The inner **ovarian medulla** is a core of loose connective tissue that contains blood vessels, nerves, and lymphatics that supply the ovary (Figures 31.2b and 31.6a).
4. Switch to high power and observe the outer surface of the ovary. Note that this surface is covered by a single layer of cuboidal epithelial cells called the **germinal epithelium.** Just below this cell layer is a connective tissue covering called the **tunica albuginea**.
5. Switch back to low power and scan the slide through the cortex. Identify the various stages of follicular development (ovarian cycle) as described here (Figure 31.6b). It is unlikely that all the stages will appear on one slide, so viewing several slides will be necessary.
   - *Primordial follicle.* These are produced during fetal development and located in the outer portion of the cortex, just below the tunica albuginea. Each primordial follicle contains an oocyte surrounded by a single layer of squamous **follicle (follicular) cells.**
   - *Primary follicle.* In the transition to a primary follicle, the follicular cells divide to form multiple layers and change shape to become cuboidal or columnar. These cells are now referred to as **granulosa cells.** The granulosa cells and the **thecal cells**, which form a layer around the follicle, begin to produce female sex hormones known as **estrogens.** A protective layer of glycoprotein, called the **zona pellucida**, forms around the egg, separating it from the granulosa cells. The zona pellucida is instrumental in preventing more than one sperm cell from entering the egg during fertilization.
   - *Secondary follicle.* A secondary follicle forms when fluid-filled spaces appear between the granulosa cells. These spaces eventually merge to form a single fluid-filled chamber known as the **antrum.**
   - *Graafian (tertiary) follicle.* In a Graafian follicle, granulosa cells surround the large antrum. The egg is located at one end of the antrum. It is surrounded by a ring of granulosa cells called the **corona radiata**, and is connected to the wall of the follicle by an additional stalk of granulosa cells. Graafian follicles form a bulge on the

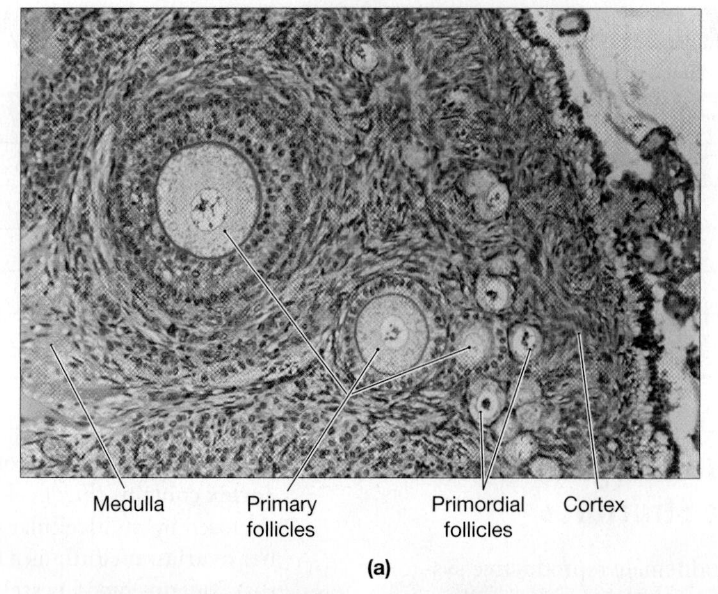

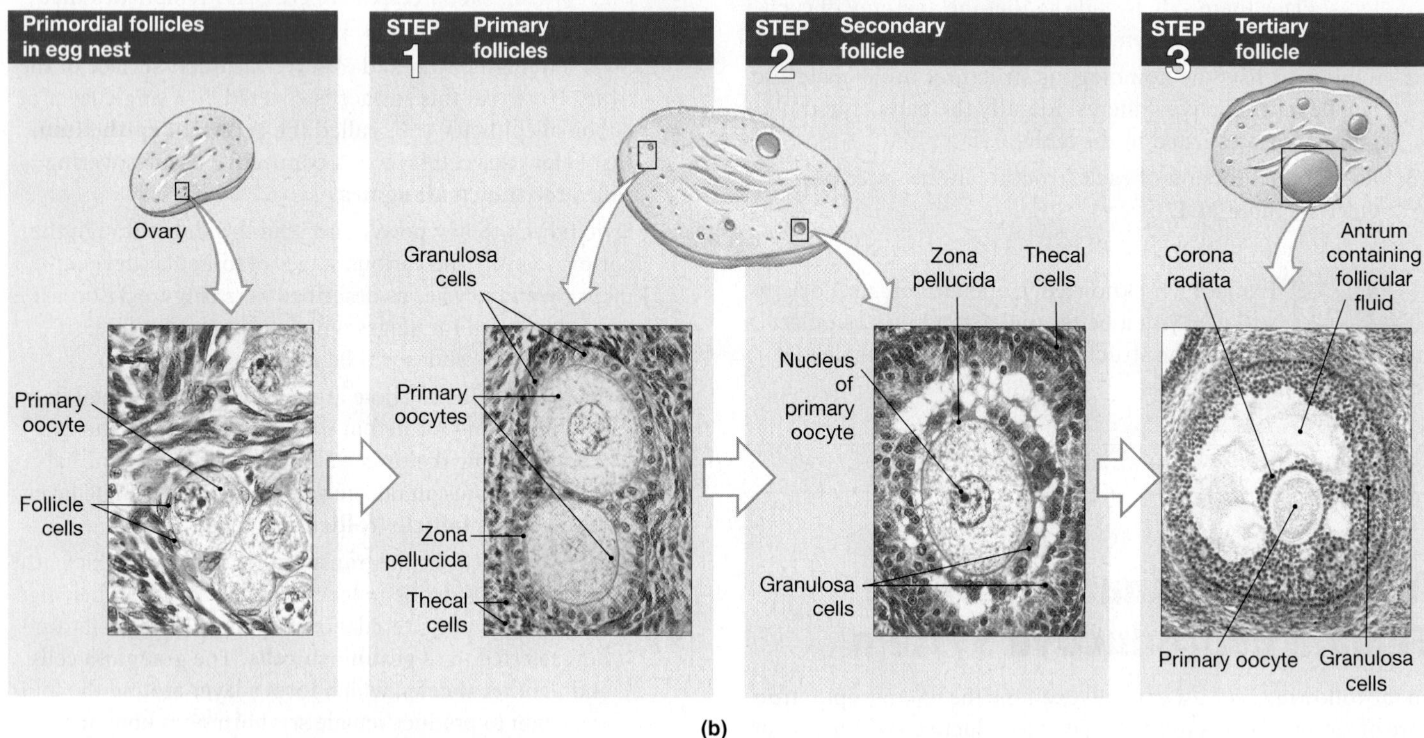

Figure 31.6 Microscopic structure of the ovary. a) Light micrograph of the ovary showing the outer cortex with developing follicles and the inner medulla with numerous blood vessels; **b)** different stages of follicular development during the ovarian cycle.

surface of the ovary. At the midpoint of the menstrual cycle, a Graafian follicle releases its egg during ovulation (Figure 31.7c).

- *Corpus luteum.* After ovulation, the remains of the follicle are transformed into a structure called the corpus luteum. If a pregnancy occurs, the corpus luteum produces **progesterone** to maintain the wall of the uterus during the early period of development. The corpus luteum may not be present on your slides.

6. If available, examine a model of the ovarian cycle. Identify the various development stages of the ovarian follicles that you previously observed with the microscope.

The Uterine Tubes

1. Obtain a slide of the uterine tube, in cross section.
2. Scan the slide with the low-power objective lens and identify the irregularly shaped lumen. Identify the following tissue layers (Figure 31.3b).

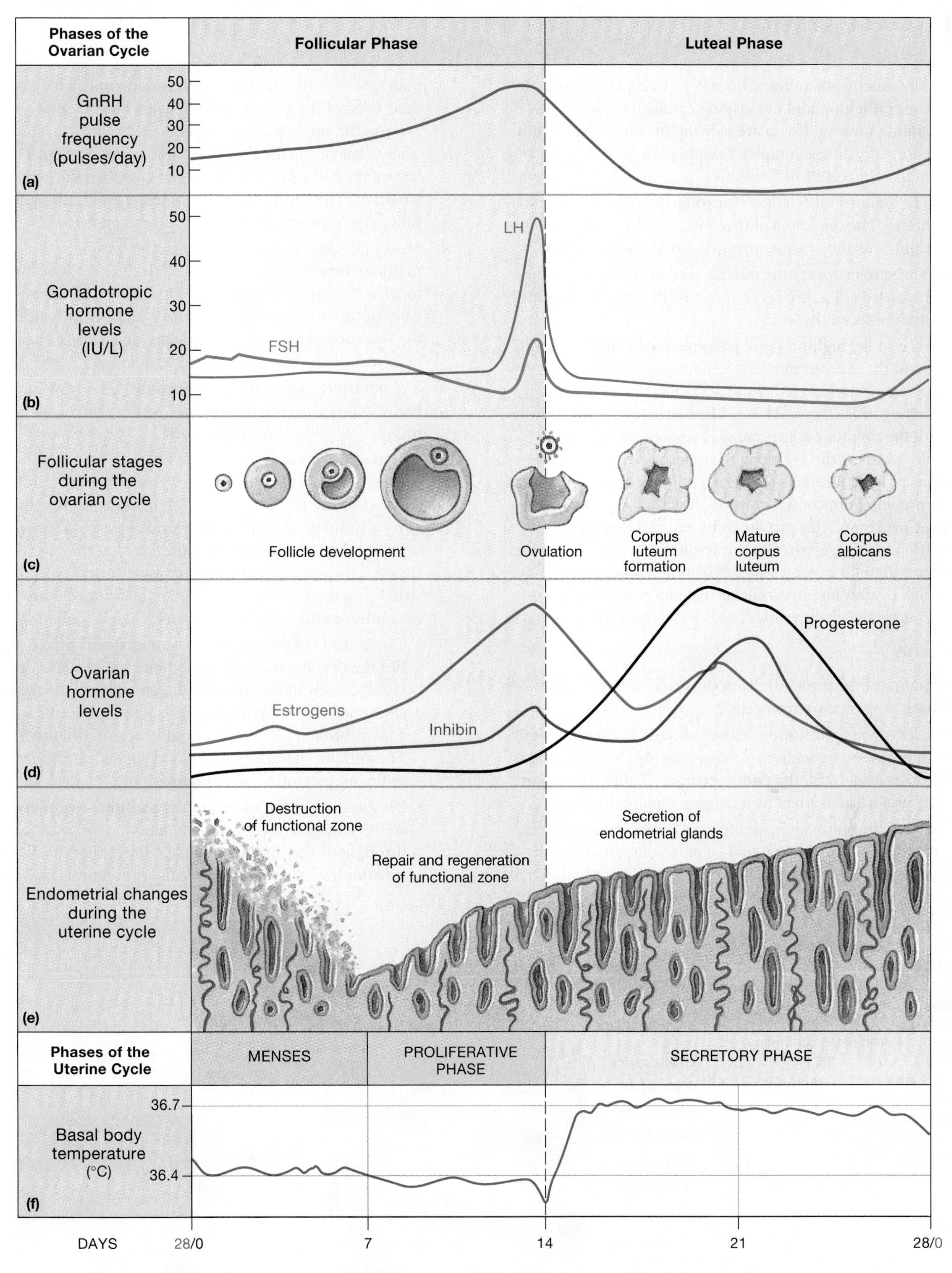

Figure 31.7 Major events of the ovarian and uterine cycles. The cycles are regulated by a number of hormones. The exact timing of these cycles varies greatly between individuals, so the time scale that is shown is approximate.

- The mucous membrane lines the lumen. It consists of the epithelium and underlying connective tissue, the lamina propria. If you are viewing the ampulla, the mucosa will show a complex folding pattern. The pattern is much simpler in the isthmus.
- The smooth muscle layer surrounds the mucous membrane. The thickness of this layer becomes progressively thicker as the uterine tube gets closer to the uterus.
- The **serosa** covers the outside wall of the uterine tube. It consists of a thin layer of connective tissue and simple squamous epithelium.

3. Switch to the high-power objective lens and view the epithelium in the mucous membrane more closely. Notice that it is a simple columnar epithelium with ciliated and mucus-secreting cells (Figure 31.3c). The beating of the cilia, along with the rhythmic contractions of smooth muscle in the wall, transport the egg along the tube. The mucous secretions provide nutrients for the egg and any sperm that might be present. Fertilization must occur within 12 to 24 hours after ovulation. This process most often takes place in the ampulla. The egg, whether it is fertilized or not, reaches the uterus about 3 to 4 days after ovulation. Fertilized eggs, which by now are multicellular structures, will implant in the uterine wall; unfertilized eggs will degenerate.

The Uterus

1. Obtain a slide of the uterus in the proliferative or secretory phase of the menstrual cycle.
2. Under low magnification, move the slide to the upper portion of the tissue section. You are viewing the inner layer of the uterine wall, the **endometrium** (Figure 31.8). Note that its surface is lined by a columnar epithelium.
3. Beneath the epithelium, identify the lamina propria, a thick layer of connective tissue that contains numerous uterine glands and blood vessels (Figure 31.8). Attempt to differentiate between the superficial **functional zone**, which contains most of the glands, and the deeper **basal zone**, which borders the smooth muscle layer or myometrium. The functional zone forms the majority of the endometrium. This layer goes through cyclic changes and is shed during menstruation. The basal zone remains intact and forms a new functional layer when menstruation is completed.
4. Move the slide to the next layer in the uterine wall, the **myometrium**, and identify its interlacing fibers of smooth muscle (Figure 31.8). The myometrium is the thickest layer of the uterine wall. The strong labor contractions of the myometrial smooth muscle provide much of the force required for the birth of the full-term fetus.
5. The outermost layer, the **perimetrium**, is a serous membrane and is a part of the visceral peritoneum. This layer may not be present on your slides.
6. The **uterine** or **menstrual cycle** refers to the cyclic changes that occur in the endometrium over a period of approximately 28 days (Figure 31.7e). The events that occur are influenced by hormones and correspond to specific stages of the ovarian cycle (Figure 31.7c). Observe the slides of the uterus in the proliferative, secretory, and menstrual phases of the cycle. Note the following changes in the endometrium during each phase.
   - On the slide of the uterus in the **menstrual phase**, notice the degeneration of the endometrial glands and blood vessels in the functional zone. During the menstrual phase (approximately days 1 to 5), the functional zone is broken down and bleeding occurs (Figure 31.7e). The endometrial tissue and blood passes out the vagina as the **menstrual flow** or **menses**.
   - On the slide of the uterus in the **proliferative phase**, identify the uterine glands and blood vessels scattered throughout the functional zone. Notice that the glands are simple in structure with little or no branching. Dur-

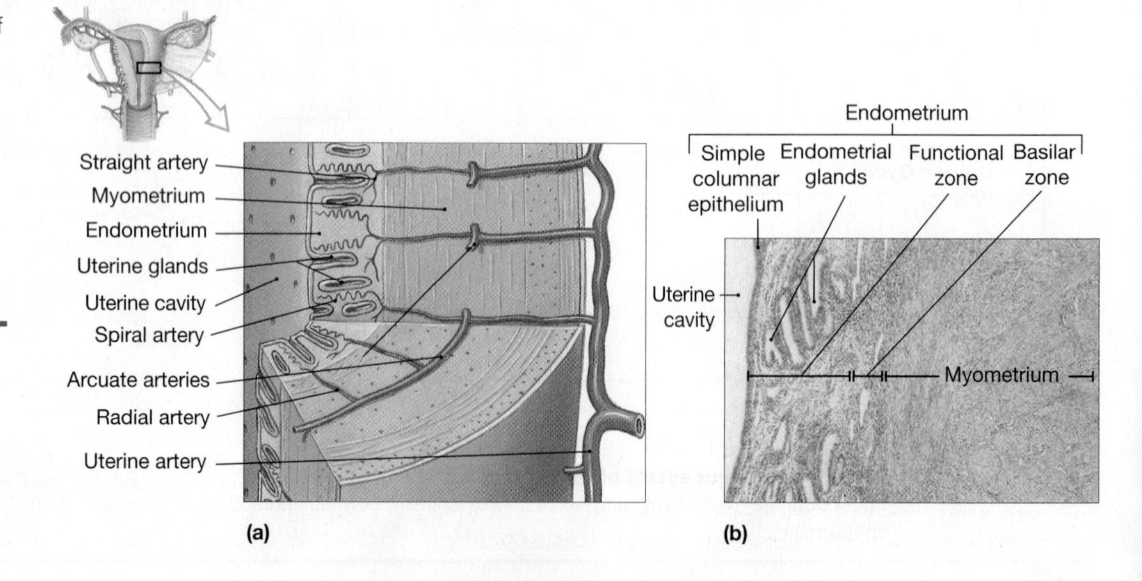

Figure 31.8 Microscopic structure of the uterus. a) Illustration of the uterine wall. The uterine artery supplies branches to the myometrium and endometrium. **b)** Light micrograph demonstrating the general structure of the uterine wall. The functional zone of the endometrium undergoes structural changes during the uterine cycle, whereas the basilar zone remains relatively intact (LM × 40).

ing the proliferative phase (approximately days 6 to 14), the basal zone gives rise to a new functional zone. During this process, new glands form, the endometrium thickens, and blood vessels infiltrate the new tissue (Figure 31.7e). Estrogens that are produced by cells in the ovarian follicles induce these rebuilding activities.

- On the slide of the uterus in the **secretory phase**, once again identify uterine glands and blood vessels in the functional zone. Notice that the glands are now enlarged and have numerous branches. The secretory phase (Figure 31.7e) is the final stage of the uterine cycle (approximately days 15 to 28). The events during this period are designed to prepare the endometrium for implantation. Endometrial glands are larger, more complex, and begin to secrete glycoproteins and other nutrients. The arteries provide a rich resource of maternal blood to support the early embryo. These activities are supported by progesterone that is produced by the corpus luteum. If implantation occurs, the endometrium remains in the secretory phase for the entire pregnancy. Otherwise, progesterone production eventually declines when the corpus luteum degenerates, and the endometrium breaks down to begin another menstrual phase (Figure 31.7).

CLINICAL CORRELATION

Sometimes regions of endometrial tissue can appear in areas outside the uterus, such as on the ovaries or along the peritoneum that covers the pelvic cavity. This condition is called **endometriosis.** These extra endometrial areas go through the same cyclic changes as the normal endometrium. Thus, blood and misplaced endometrial tissue can accumulate in the pelvic cavity, and blood-filled cysts called endometriomas may form on the ovaries or throughout the pelvis, causing severe pain during menstruation. In addition, the immune system may recognize that this tissue should not be where it is and attack these lesions, leading to scar tissue and adhesions as well. Treatments for endometriosis include hormonal therapy to reduce menstruation and laser or traditional surgery to remove endometrial tissue. The success of treatment varies with the severity of the disease. This disorder remains poorly understood.

The Vagina

1. Obtain a slide of the vagina, in cross section.
2. Scan the slide with the low-power objective lens and identify the lumen.
3. Identify the following tissue layers in the vaginal wall surrounding the lumen (Figure 31.9).
   - The mucous membrane consists of a stratified squamous epithelium and underlying lamina propria, which contains the numerous blood vessels.
   - The smooth muscle layer is deep to the lamina propria. Notice that bundles of smooth muscle are arranged in both circular and longitudinal fashion. Fibrous connective tissue is interspersed between the muscle bundles.
   - The adventitia covers the outside wall of the vagina. It consists of fibrous connective tissue that blends in with connective tissue from neighboring structures.

CLINICAL CORRELATION

The vaginal wall can become irritated and inflamed (**vaginitis**) by various types of fungal, bacterial, and viral infections. These infections not only create pain and discomfort, but also cause infertility by reducing the survival rate of sperm.

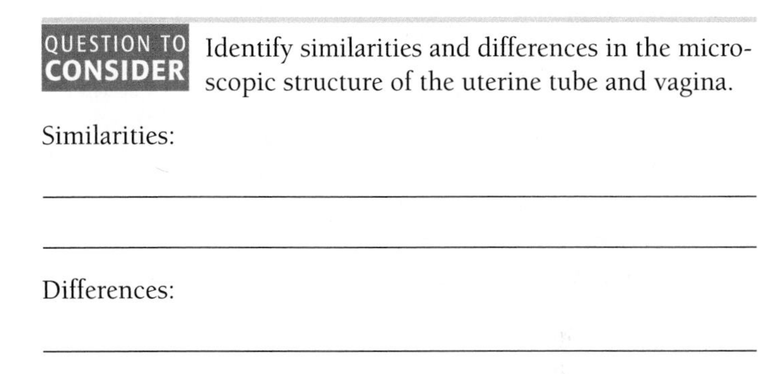

QUESTION TO CONSIDER Identify similarities and differences in the microscopic structure of the uterine tube and vagina.

Similarities:

____

____

Differences:

____

____

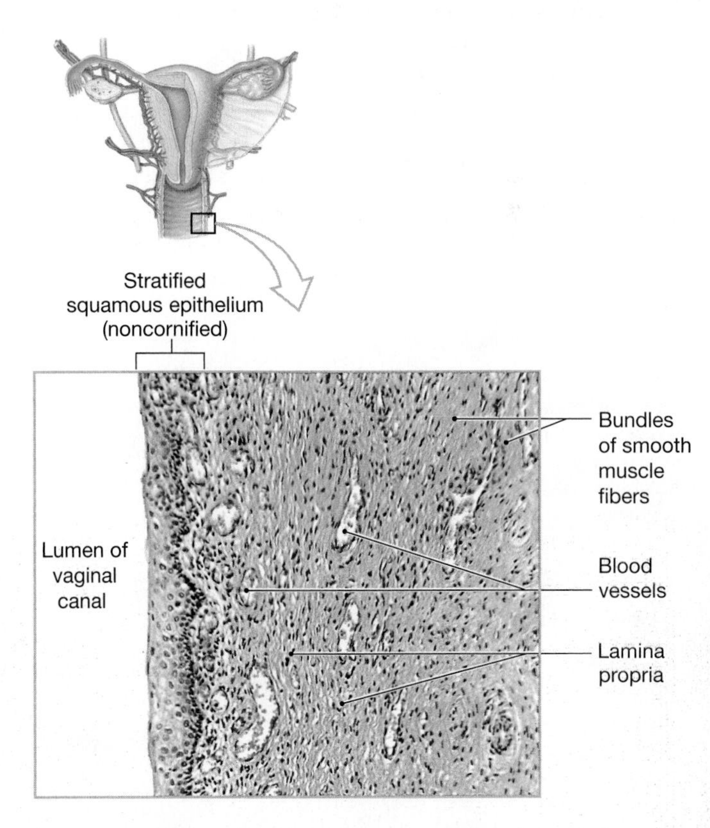

Figure 31.9 Microscopic structure of the vagina. The vaginal canal is lined by a stratified squamous epithelium. The muscle layer contains a meshwork of smooth muscle fibers and fibrous connective tissue (LM × 25).

Exercise 31 Review Sheet

The Female Reproductive System

Name ______________________________

Lab Section ______________________________

Date ______________________________

1. An ectopic pregnancy occurs when the early embryo implants at a location outside the uterus. A common site for an ectopic pregnancy is in the fallopian tubes. Examine Figure 31.3 and explain why an ectopic pregnancy that occurs in a fallopian tube will not be successful.

____

____

2. The surfaces of the ovaries in a young girl are smooth. However, several years after puberty they become rough and uneven, with numerous regions of scar tissue. Provide an explanation for this transformation.

____

____

3. Review Figures 31.2a and 31.3a. Discuss the significance of the fimbriae during ovulation of the egg. What could happen to the egg if the fimbriae did not function properly?

____

____

4. Review the events of the ovarian and uterine cycles (Figure 31.7). What would happen if production of FSH by the anterior pituitary was significantly reduced? What would occur if fertilization occurred, but the corpus luteum did not produce progesterone?

____

____

5. Review the supporting membranes and ligaments of the female reproductive system (Figure 31.2a). Predict what might happen if these structures were damaged by infection or injury.

____

____

Questions 6–10: Match the structure in column A with the correct description in column B.

A	B
6. Fallopian tube ______	a. A Pap smear is a test for detecting cancer of this structure.
7. Clitoris ______	b. Fertilization usually occurs in the ampulla of this structure.
8. Cervix ______	c. This structure encloses the vestibule.
9. Ovary ______	d. Like the penis, this structure has corpora cavernosa.
10. Vagina ______	e. This structure passes posterior to the urinary bladder and urethra, and anterior to the rectum.
	f. During fetal development, primordial follicles are produced by this structure.

11. What is meant by a homologous structure? Provide examples of homologous structures in the reproductive systems.

______

______

Questions 12–22: In the following diagram, identify the labeled structures.

12. ______
13. ______
14. ______
15. ______
16. ______
17. ______
18. ______
19. ______
20. ______
21. ______
22. ______

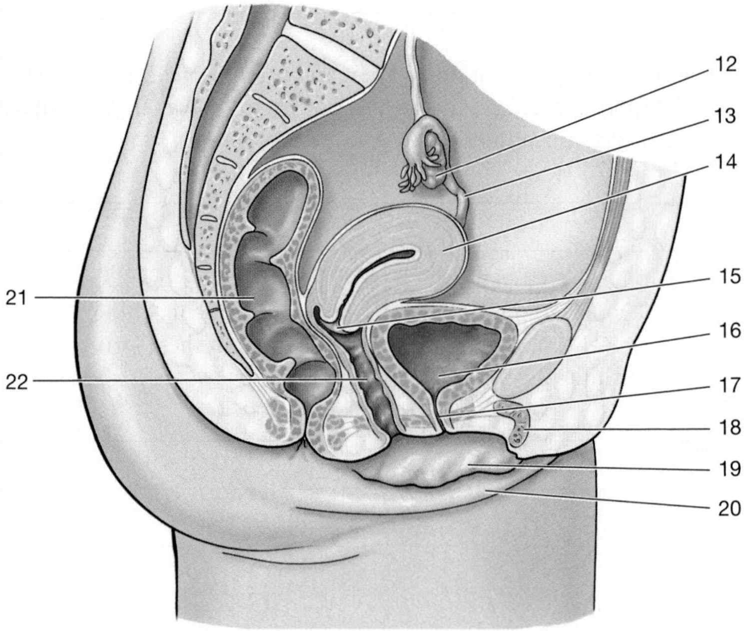

Notes

Notes

Notes

Notes

Notes

Notes

Appendix A

Writing a Laboratory Report

Introduction

In order to conduct good scientific laboratory work you must meticulously record the process that you follow. This is often done with a written laboratory report. Your instructor may require a specific format. However, the following guidelines may be used if no specific instructions are provided.

When writing a laboratory report, the primary goals are:

- to clearly articulate the rationale for performing the experiment and the procedure that was used
- to provide an accurate account of the data
- to critically analyze the data and provide a reasonable interpretation of the results

Your report will be a chronicle of the important events and outcomes of your work. It will indicate how well you carried out the experiment, if you understood the fundamental concepts under examination, if you can clearly explain the results, and how logically you can derive and present your conclusions. While great eloquence in style is not important, clarity and thoroughness are. In addition, you should be vigilant in using correct grammar, spelling, and sentence structure.

Format of a Laboratory Report

Introduction This portion of the report should clearly indicate what questions you are trying to answer, and state specifically the purpose of your experiment. If you do several experiments in one lab, the purpose of each should be given. Copying what is stated in the laboratory exercise is not acceptable; present the introduction in your own words.

Materials and Methods In this section you should describe the procedures used to carry out your experiments and the equipment and supplies needed to successfully complete the work. You should describe your methods clearly and with enough detail so that another person could perform the experiment based on what you have written. Use your own words and if you can, improve on the way the author of your laboratory manual explains techniques! **No results should be included in this section.**

Results In this section you should include all tables, graphs, diagrams, and any other data that illustrate what happened during the experiment. In addition, you should supplement your data with descriptive observations. Any drawings should accurately demonstrate what you saw; they should not be copied from another source. Do *not* interpret your data in this section (see Discussion). All information (results) presented in this section should be interpreted *objectively*, not *subjectively*.

Discussion This section includes an interpretation (or analysis) of your results and a discussion of what these results mean. If something did not turn out the way you expected, it should be addressed in this section. If you think your experiment could have been designed differently for better results, discuss the improvements you would employ. You can also compare your results with others in the laboratory or with similar results published in books or research articles. Information obtained from outside sources should be properly referenced.

A Note on Plagiarism

Plagiarism refers to the copying of statements or ideas of others (whether exact words or paraphrased) without acknowledging where you obtained them. It is a very serious crime in academia. Therefore, you must learn how to properly acknowledge information as described above. If you use the exact words of an author you must make that clear by using quotation marks and citing the source. For the most part, however, you should not directly quote others. Putting an idea into your own words will indicate how well you understand a concept.

Plagiarism may also refer to the copying of papers or lab reports. This is an equally serious academic crime and will not be tolerated. Lab partners should **never** copy each other's laboratory reports. You may do an experiment together, but all data should be accumulated, analyzed, and presented individually. The laboratory report should clearly reflect your own attempt to make sense out of what you did in the laboratory. Of course, students may consult with each other when they are trying to analyze data. Judgment and experience will tell you how to (and how often to) acknowledge a source of information.

Appendix B

Weights and Measures

Table 1 The U.S. System of Measurement

Physical property	Unit	Relationship to other U.S. units	Relationship to household units
Length	inch. (in.)	1 in. = 0.083 ft	
	foot (ft)	1 ft = 12 in. = 0.33 yd	
	yard (yd)	1 yd = 36 in. = 3 ft	
	mile (mi)	1 mi = 5,280 ft = 1,760 yd	
Volume	fluidram (fl dr)	1 fl dr = 0.125 fl oz	
	fluid ounce (fl oz)	1 fl oz = 8 fl dr = 0.0625 pt	= 6 teaspoons (tsp) = 2 tablespoons (tbsp)
	pint (pt)	1 pt = 128 fl dr = 16 fl oz = 0.5 qt	= 32 tbsp = 2 cups (c)
	quart (qt)	1 qt = 256 fl dr = 32 fl oz = 2 pt = 0.25 gal	= 4 c
	gallon (gal)	1 gal = 128 fl oz = 8 pt = 4 qt	
Mass	grain (gr)	1 gr = 0.002 oz	
	dram (dr)	1 dr = 27.3 gr = 0.063 oz	
	ounce (oz)	1 oz = 437.5 gr = 16 dr	
	pound (lb)	1 lb = 7000 gr = 256 dr = 16 oz	
	ton (t)	1 t = 2000 lb	

Table 2 The Metric System of Measurement

Physical property	Unit	Relationship to standard metric units	Conversion to U.S. units	
Length	nanometer (nm)	1 nm = 0.000000001 m (10^{-9})	= 3.94 × 10^{-8} in.	25,400,000 nm = 1 in.
	micrometer (μm)	1 μm = 0.000001 m (10^{-6})	= 3.94 × 10^{-5} in.	25,400 mm = 1 in.
	millimeter (mm)	1 mm = 0.001 m (10^{-3})	= 0.0394 in.	25.4 mm = 1 in.
	centimeter (cm)	1 cm = 0.01 m (10^{-2})	= 0.394 in.	2.54 cm = 1 in.
	decimeter (dm)	1 dm = 0.1 m (10^{-1})	= 3.94 in.	0.25 dm = 1 in.
	meter (m)	standard unit of length	= 39.4 in. = 3.28 ft = 1.093 yd	0.0254 m = 1 in. 0.3048 m = 1 ft 0.914 m = 1 yd
	kilometer (km)	1 km = 1000 m	= 3280 ft = 1093 yd = 0.62 mi	1.609 km = 1 mi
Volume	microliter (μl)	1 μl = 0.000001 l (10^{-6}) = 1 cubic millimeter (mm^3)		
	milliliter (ml)	1 ml = 0.001 l (10^{-3}) = 1 cubic centimeter (cm^3 or cc)	= 0.0338 fl oz	5 ml = 1 tsp 15 ml = 1 tbsp 30 ml = 1 fl oz
	centiliter (cl)	1 cl = 0.01 l (10^{-2})	= 0.338 fl oz	2.95 cl = 1 fl oz
	deciliter (dl)	1 dl = 0.1 l (10^{-1})	= 3.38 fl oz	0.295 dl = 1 fl oz
	liter (l)	standard unit of volume	= 33.8 fl oz = 2.11 pt = 1.06 qt	0.0295 l = 1 fl oz 0.473 l = 1 pt 0.946 l = 1 qt
Mass	picogram (pg)	1 pg = 0.000000000001 g (10^{-12})		
	nanogram (ng)	1 ng = 0.000000001 g (10^{-9})	= 0.000000015 gr	66,666,666 mg = 1 gr
	microgram (μg)	1 μg = 0.000001 g (10^{-6})	= 0.000015 gr	66,666 mg = 1 gr
	milligram (mg)	1 mg = 0.001 g (10^{-3})	= 0.015 gr	66.7 mg = 1 gr
	centigram (cg)	1 cg = 0.01 g (10^{-2})	= 0.15 gr	6.67 cg = 1 gr
	decigram (dg)	1 dg = 0.1 g (10^{-1})	= 1.5 gr	0.667 dg = 1 gr
	gram (g)	standard unit of mass	= 0.035 oz = 0.0022 lb	28.4 g = 1 oz 454 g = 1 lb
	dekagram (dag)	1 dag = 10 g		
	hectogram (hg)	1 hg = 100 g		
	kilogram (kg)	1 kg = 1000 g	= 2.2 lb	0.454 kg = 1 lb
	metric ton (kt)	1 mt = 1000 kg	= 1.1 t = 2205 lb	0.907 kt = 1 t

Temperature	Centigrade	Fahrenheit
Freezing point of pure water	0°	32°
Normal body temperature	36.8°	98.6°
Boiling point of pure water	100°	212°
Conversion	°C → °F: °F = (1.8 × °C) + 32	°F → °C: °C = (°F − 32) × 0.56

Photo Credits

Frontmatter Marlena Yannetti

Exercise 1 **1.06a**, **1.06b,c** Custom Medical Stock Photo, Inc.

Exercise 2 **2.01** Olympus America, Inc.

Exercise 3 **3.03b** Dr. Birgit H. Satir, **3.04b** Fawcett/Hirokawa/Heuser/Science Source/Photo Researchers, **3.05a** Fawcett/de Harven/Kalnins/Photo Researchers, **3.06a** Biophoto Associates/Photo Researchers, **3.06b**, **3.08**, **3.08**, **3.08**, **3.08** Ed Reschke/Peter Arnold, Inc., **3.08** James Solliday/Biological Photo Service

Exercise 5 **5.01b**, **5.02a**, **5.02c**, **5.03b**, **5.04b** Robert B. Tallitsch, Ph.D., Augustana College, Rock Island, IL, **5.05b** John D. Cunningham/Visuals Unlimited, **5.06b** Robert B. Tallitsch, Ph.D., Augustana College, Rock Island, IL, **5.07a**, **5.07b** Frederic H. Martini, **5.08b** Project Masters, Inc./The Bergman Collection, **5.09b** Ward's Natural Science Establishment, Inc., **5.10b** Frederic H. Martini, **5.11b** Ward's Natural Science Establishment, Inc., **5.12b** John D. Cunningham /Visuals Unlimited, **5.13b** Frederic H. Martini, **5.14b** Bruce Iverson/Visuals Unlimited, **5.15b** Ed Reschke/Peter Arnold, Inc., **5.15d** Robert Brons/Biological Photo Service, **5.15f** Science Source/Photo Researchers, **5.16b** Frederic H. Martini, **5.17b** Michael Ross/Photo Researchers

Exercise 6 **6.02a** John D. Cunningham/Visuals Unlimited, **6.03b**, **6.03c** Frederic H. Martini, **6.04** Dr. Richard Kessel and Dr. Randy H. Kardon/Visuals Unlimited, **6.05**, **6.06a**, **6.07** Robert B. Tallitsch, Ph.D., Augustana College, Rock Island, IL, **6.08b**, **6.08c** Frederic H. Martini

Exercise 7 **7.02a left**, **7.02a right**, **7.04a**, **7.08**, **7.11a**, **7.11b**, **7.11c**, **7.12a**, **7.12b**, **7.12c**, **7.13a**, **7.13b**, **7.13c**, **7.14a**, **7.14b**, **7.14c**, **7.15a**, **7.15b** Ralph T. Hutchings

Exercise 8 **8.02b**, **8.02c**, **8.03a**, **8.03b**, **8.03c**, **8.04a**, **8.04b**, **8.05a**, **8.05b**, **8.06a**, **8.06b**, **8.07a**, **8.07b**, **8.10a**, **8.10b**, **8.11a**, **8.11b**, **8.12a**, **8.12b**, **8.13** Ralph T. Hutchings

Exercise 9 **9.07** Ralph T. Hutchings, **9.08a**, **9.08b** Project Masters, Inc./The Bergman Collection, **9.11a**, **9.11b**, **9.13c**, **9.13d** Ralph T. Hutchings

Exercise 10 **10.01a**, **10.01b**, **10.02a**, **10.02b**, **10.03a**, **10.03b** Robert B. Tallitsch, Ph.D., Augustana College, Rock Island, IL, **10.04d** Photo Researchers, **10.06** Kent Wood/Photo Researchers

Exercise 11 **11.02c**, **11.10**, **11.15** Ralph T. Hutchings

Exercise 13 **13.02b** David Scott/Phototake NYC, **13.03a** Biophoto Associates/ Photo Researchers, **13.03b** Photo Researchers, **13.06** Stephen Whitefield, Dalhousie University, **13.07a** David G. King, Southern Illinois University School of Medicine. Used with permission., **13.07b** Robert B. Tallitsch, Ph.D., Augustana College, Rock Island, IL, **13.08** Biophoto Associates/Photo Researchers, **13.09a**, **13.09b**, **13.10b**, **13.10c**, **13.10d**, **13.10e** Robert B. Tallitsch, Ph.D., Augustana College, Rock Island, IL, **13.11b** Dr. Richard Kessel and Dr. Randy H. Kardon/Visuals Unlimited

Exercise 14 **14.01**, **14.02** Ralph T. Hutchings, **14.03b** Michael J. Timmons, **14.03c** Pat Lynch/Photo Researchers, **14.03**, **14.04**, **14.05a** Ralph T. Hutchings, **14.08b**, **14.08c** Shawn Miller & Mark Nielsen/Pearson Education/Benjamin Cummings Publishing Company, **14.08d** Shawn Miller & Mark Nielsen/Pearson Education/Benjamin Cummings Publishing Company

Exercise 15 **15.01b** Ralph T. Hutchings

Exercise 17 **17.01c** David G. King, Southern Illinois University School of Medicine. Used with permission., **17.05c** G. W. Willis/Terraphotographics/Visuals Unlimited, **17.05c** Pearson Education/PH College (photomicrograph), **17.07c** Custom Medical Stock Photo, Inc., **17.07a** Ed Reschke/Peter Arnold, Inc., **17.09a** Ralph T. Hutchings, **17.10** Shawn Miller & Mark Nielsen/Pearson Education/Benjamin Cummings Publishing Company, **17.15 left**, **17.15 right** Diane Schiumo/ Fundamental Photographs, NYC, **17.19b** Michael J. Timmons, **17.20b** Ward's Natural Science Establishment, Inc.

Exercise 18 **18.03b** Manfred Kage/Peter Arnold, Inc., **18.04c**, **18.05b**, **18.05c**, **18.07c** Frederic H. Martini, **18.08c**, **18.08c**, **18.08c**, **18.08c** David G. King, Southern Illinois University School of Medicine. Used with permission., **18.09c** Robert B. Tallitsch, Ph.D., Augustana College, Rock Island, IL, **18.10b** Don W. Fawcett, **18.10c** Ward's Natural Science Establishment, Inc.

Exercise 19 **19.01** Michael Ross/Photo Researchers, **19.02a** David Scharf/Peter Arnold, Inc., **19.03a**, **19.03b**, **19.03c**, **19.03d**, **19.03e** Michael Ross/Photo Researchers, **19.06a**, **19.06b**, **19.06c**, **19.06d** Pearson Education/Benjamin Cummings Publishing Company

Exercise 20 **20.02a**, **20.02**, **20.05c** Ralph T. Hutchings, **20.07a**, **20.07b**, **20.07c** Shawn Miller & Mark Nielsen/Pearson Education/Benjamin Cummings Publishing Company

Exercise 21 **21.01** Biophoto Associates/Photo Researchers, **21.03b** Dr. John D. Cunningham/Visuals Unlimited, **21.03a**, **21.04b**, **21.05b** Robert B. Tallitsch, Ph.D., Augustana College, Rock Island, IL, **21.06** Dr. Gladden Willis/Visuals Unlimited, **21.07b** Robert B. Tallitsch, Ph.D., Augustana College, Rock Island, IL, **21.07a** Robert Caughey/Visuals Unlimited, **21.18b** David G. King, Southern Illinois University School of Medicine. Used with permission.

Exercise 23 **23.02c**, **23.02d**, **23.03c** Frederic H. Martini, **23.04d** Ed Reschke/Peter Arnold, Inc., **23.05b** Biophoto Associates/ Photo Researchers, **23.05c** Dr. Fred Hossler/Visuals Unlimited, **23.08** David M. Phillips/ Visuals Unlimited, **23.09b** Frederic H. Martini, **23.10b** Leonard Lessin/ Peter Arnold, Inc.

Exercise 24 **24.01b** Ralph T. Hutchings, **24.03b** John D. Cunningham/Visuals Unlimited, **24.04**, **24.05a** Ralph T. Hutchings, **24.07a** Frederic H. Martini, **24.07c** Photo Researchers, **24.08** Robert B. Tallitsch, Ph.D., Augustana College, Rock Island, IL, **24.09b** David G. King, Southern Illinois University School of Medicine. Used with permission., **24.09c** Ed Reschke/Peter Arnold, Inc., **24.09a**, **24.10d** Robert B. Tallitsch, Ph.D., Augustana College, Rock Island, IL

Exercise 26 **26.04b**, **26.05** Ralph T. Hutchings, **26.06c** Frederic H. Martini, **26.07b** Ralph T. Hutchings, **26.11b** Frederic H. Martini, **26.15b**, **26.16c**, **26.17b** Robert B. Tallitsch, Ph.D., Augustana College, Rock Island, IL, **26.18f** From the collection of the author, **26.18e** M.I. Walker/Photo Researchers, **26.18a** Phototake NYC, **26.19a**, **26.19b**, **26.19c** Robert B. Tallitsch, Ph.D., Augustana College, Rock Island, IL, **26.21b** Ward's Natural Science Establishment, Inc., **26.22b**, **26.22c** Robert B. Tallitsch, Ph.D., Augustana College, Rock Island, IL, **26.23c** Michael J. Timmons, **26.23d** Stephen Whitefield, Dalhousie University

Exercise 28 **28.02b** Ralph T. Hutchings, **28.06a**, **28.06b**, **28.07**, **28.08a**, **28.08b** David G. King, Southern Illinois University School of Medicine. Used with permission., **28.09a** Ward's Natural Science Establishment, Inc., **28.09b** Frederic H. Martini, **28.09c** G. W. Willis, MD/ Visuals Unlimited, **28.10** Shawn Miller & Mark Nielsen/Pearson Education/ Benjamin Cummings Publishing Company

Exercise 30 30.03a Don W. Fawcett, M.D., Harvard Medical School, 30.03b Ward's Natural Science Establishment, Inc., 30.04b Frederic H. Martini, 30.05 Robert B. Tallitsch, Ph.D., Augustana College, Rock Island, IL, 30.06 David G. King, Southern Illinois University School of Medicine. Used with permission., 30.07 Dr. Frederick Skvara/Visuals Unlimited, 30.08 John Walsh/Photo Researchers

Exercise 31 31.04c David G. King, Southern Illinois University School of Medicine. Used with permission., 31.04b, 31.07b1, 31.07b2, 31.07b3, 31.07b4 Frederic H. Martini, 31.07a Science Pictures Ltd/Photo Researchers, 31.08b Ward's Natural Science Establishment, Inc., 31.09 Michael J. Timmons

Index

Note: Page numbers followed by *f* or *t* indicate figure or table entry.